Large Air Cooled Engines Service Manual

(Third Edition)

Only engines of 15 cubic inch displacement and over are contained in this manual.

CONTENTS

GENERAL

ENGINE SHOP MANUALS

Published by **TECHNICAL PUBLICATIONS DIV.**
INTERTEC PUBLISHING CORP.

P.O. Box 12901, Overland Park, Kansas 66212

Technical Publications, a division of Intertec Publishing Corporation, wishes to acknowledge the assistance and contribution of technical information and illustrations by the following manufacturers:

BRIGGS & STRATTON
CLINTON
CRAFTSMAN
KOHLER
ONAN
TECUMSEH
TELEDYNE WISCONSIN

Cover photo courtesy of Kohler Co.

This service manual provides specifications in both the U.S. Customary and Metric (SI) system of measurements. The first specification is given in the measuring system used during manufacture, while the second specification (given in parenthesis) is the converted measurement. For instance, a specification of "0.011 inch (0.279 mm)" would indicate that the equipment was manufactured using the U.S. system of measurement and the metric equivalent of 0.011 inch is 0.279 mm.

All instructions and diagrams have been checked for accuracy and ease of application, however, success and safety in working with tools depend to a great extent upon individual accuracy, skill and caution. For this reason the publishers are not able to guarantee the result of any procedure contained herein. Nor can they assume responsibility for any damage to property or injury to persons occasioned from the procedures. Persons engaging in the procedures do so entirely at their own risk.

FUNDAMENTALS SECTION

ENGINE FUNDAMENTALS

OPERATING PRINCIPLES

The one, two or four cylinder engines used to power riding lawn mowers, garden tractors, pumps, generators, welders, mixers, windrowers, hay balers and many other items of power equipment in use today are basically similar. All are technically known as "Internal Combustion Reciprocating Engines."

The source of power is heat formed by the burning of a combustible mixture of petroleum products and air. In a reciprocating engine, this burning takes place in a closed cylinder containing a piston. Expansion resulting from the heat of combustion applies pressure on the piston to turn a shaft by means of a crank and connecting rod.

The fuel-air mixture may be ignited by means of an electric spark (Otto Cycle Engine) or by heat formed from compression of air in the engine cylinder (Diesel Cycle Engine). The complete series of events which must take place in order for the engine to run occurs in two revolutions of the crankshaft (four strokes of the piston in cylinder) and is referred to as a "Four-Stroke Cycle Engine."

OTTO CYCLE. In a spark ignited engine, a series of five events is required in order for the engine to provide power. This series of events is called the "Cycle" (or "Work Cycle") and is repeated in each cylinder of the engine as long as work is being done. This series of events which comprise the "Cycle" is as follows:

1. The mixture of fuel and air is pushed into the cylinder by atmospheric pressure when the pressure within the engine cylinder is reduced by the piston moving downward in the cylinder.
2. The mixture of fuel and air is compressed by the piston moving upward in the cylinder.
3. The compressed fuel-air mixture is ignited by a timed electric spark.
4. The burning fuel-air mixture expands, forcing the piston downward in the cylinder thus converting the chemical energy generated by combustion into mechanical power.
5. The gaseous products formed by the burned fuel-air mixture are exhausted from the cylinder so that a new "Cycle" can begin.

The above described five events which comprise the work cycle of an engine are commonly referred to as (1), INTAKE; (2), COMPRESSION; (3), IGNITION; (4), EXPANSION (POWER); and (5), EXHAUST.

DIESEL CYCLE. The Diesel Cycle differs from the Otto Cycle in that air alone is drawn into the cylinder during the intake period. The air is heated from being compressed by the piston moving upward in the cylinder, then a finely atomized charge of fuel is injected into the cylinder where it mixes with the air and is ignited by the heat of the compressed air. In order to create sufficient heat to ignite the injected fuel, an engine operating on the Diesel Cycle must compress the air to a much greater degree than an engine operating on the Otto Cycle where the fuel-air mixture is ignited by an electric spark. The power and exhaust events of the Diesel Cycle are similar to the power and exhaust events of the Otto Cycle.

FOUR-STROKE CYCLE. In a four-stroke cycle engine operating on the Otto Cycle (spark ignition), the five events of the cycle take place in four strokes of the piston, or in two revolutions of the engine crankshaft. Thus, a power stroke occurs only on alternate downward strokes of the piston.

In view "A" of Fig. 1-1, the piston is on the first downward stroke of the cycle. The mechanically operated intake valve has opened the intake port and, as the downward movement of the piston has reduced the air pressure in the cylinder to below atmospheric pressure, air is forced through the carburetor, where fuel is mixed with the air, and into the cylinder through the open intake port. The intake valve remains open and the fuel-air mixture continues to flow into the cylinder until the piston reaches the bottom of its downward stroke. As the piston starts on its first upward stroke, the mechanically operated intake valve closes and since the exhaust valve is closed, the fuel-air mixture is compressed as in view "B".

Just before the piston reaches the top of its first upward stroke, a spark at the spark plug electrodes ignites the compressed fuel-air mixture. As the engine crankshaft turns past top center, the burning fuel-air mixture expands rapidly and forces the piston downward on its power stroke as shown in view "C". As the piston reaches the bottom of the power stroke, the mechanically operated exhaust valve starts to open and as the

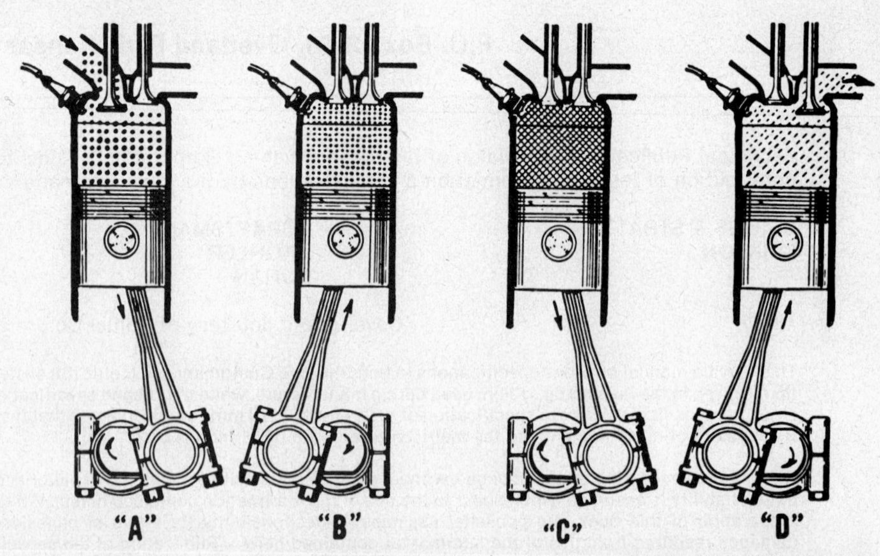

Fig. 1-1 – Schematic diagram of four-stroke cycle engine operating on the Otto (spark ignition) cycle. In view "A", piston is on first downward (intake) stroke and atmospheric pressure is forcing fuel-air mixture from carburetor into cylinder through the open intake valve. In view "B", both valves are closed and piston is on its first upward stroke compressing the fuel-air mixture in cylinder. In view "C", spark across electrodes of spark plug has ignited fuel-air mixture and heat of combustion rapidly expands the burning gaseous mixture forcing the piston on its second downward (expansion or power) stroke. In view "D", exhaust valve is open and piston on its second upward (exhaust) stroke forces the burned mixture from cylinder. A new cycle then starts as in view "A".

pressure of the burned fuel-air mixture is higher than atmospheric pressure, it starts to flow out the open exhaust port. As the engine crankshaft turns past bottom center, the exhaust valve is almost completely open and remains open during the upward stroke of the piston as shown in view "D". Upward movement of the piston pushes the remaining burned fuel-air mixture out of the exhaust port. Just before the piston reaches the top of its second upward or exhaust stroke, the intake valve opens and the exhaust valve closes. The cycle is completed as the crankshaft turns past top center and a new cycle begins as the piston starts downward as shown in view "A".

In a four-stroke cycle engine operating on the Diesel Cycle, the sequence of events of the cycle is similar to that described for operation on the Otto Cycle, but with the following exceptions: On the intake stroke, air only is taken into the cylinder. On the compression stroke, the air is highly compressed which raises the temperature of the air. Just before the piston reaches the top dead center, fuel is injected into the cylinder and is ignited by the heated, compressed air. The remainder of the cycle is similar to that of the Otto Cycle.

CARBURETOR FUNDAMENTALS

OPERATING PRINCIPLES

Function of the carburetor on a spark-ignition engine is to atomize the fuel and mix the atomized fuel in proper proportions with air flowing to the engine intake port or intake manifold. Carburetors used on engines that are to be operated at constant speeds and under even loads are of simple design since they only have to mix fuel and air in a relatively constant ratio. On engines operating at varying speeds and loads, the carburetors must be more complex because different fuel-air mixtures are required to meet the varying demands of the engine.

FUEL-AIR MIXTURE RATIO REQUIREMENTS. To meet the demands of an engine being operated at varying speeds and loads, the carburetor must mix fuel and air at different mixture ratios. Fuel-air mixture ratios required for different operating conditions are approximately as follows.

	Fuel	Air
Starting, cold weather	1 lb.	7 lbs.
Accelerating	1 lb.	9 lbs.
Idling (no load)	1 lb.	11 lbs.
Part open throttle	1 lb.	15 lbs.
Full load, open throttle	1 lb.	13 lbs.

BASIC DESIGN. Carburetor design is based on the venturi principle which simply means that a gas or liquid flowing through a necked-down section (venturi) in a passage undergoes an increase in velocity (speed) and a decrease in pressure as compared to the velocity and pressure in full size sections of the passage. The principle is illustrated in Fig. 2-1, which shows air passing through a carburetor venturi. The figures given for air speeds and vacuum are approximate for a typical wide-open throttle operating condition. Due to low pressure (high vacuum) in the venturi, fuel is forced out through the fuel nozzle by the atmospheric pressure (0 vacuum) on the fuel; as fuel is emitted from the nozzle, it is atomized by the high velocity air flow and mixes with the air.

In Fig. 2-2, the carburetor choke plate and throttle plate are shown in relation to the venturi. Downward pointing arrows indicate air flow through the carburetor.

At cranking speeds, air flows through the carburetor venturi at a slow speed; thus, the pressure in the venturi does not usually decrease to the extent that atmospheric pressure on the fuel will force fuel from the nozzle. If the choke plate is closed as shown by dotted line in Fig. 2-2, air cannot enter into the carburetor and pressure in the carburetor decreases greatly as the engine is turned at cranking speed. Fuel can then flow from the fuel nozzle. In manufacturing the carburetor choke plate or disc, a small hole or notch is cut in the plate so that some air can flow through the plate when it is in closed position to provide air for the starting fuel-air mixture. In some instances, after starting a cold engine, it is advantageous to leave the choke plate in a partly closed position as the restriction of air flow will decrease the air pressure in carburetor venturi, thus causing more fuel to flow from the nozzle, resulting in a richer fuel-air mixture. The choke plate or disc should be in fully open position for normal engine operation.

If, after the engine has been started, the throttle plate is in the wide-open position as shown by the solid line in Fig. 2-2, the engine can obtain enough fuel and air to run at dangerously high speeds. Thus, the throttle plate or disc must be partly closed as shown by the dotted lines to control engine speed. At no load, the engine requires very little air and fuel to run at its rated speed and the throttle must be moved on toward the closed position as shown by the dash lines. As more load is placed on the engine, more fuel and air are required for the engine to operate at its rated speed and the throttle must be moved closer to the wide open position as shown by the solid line. When the engine

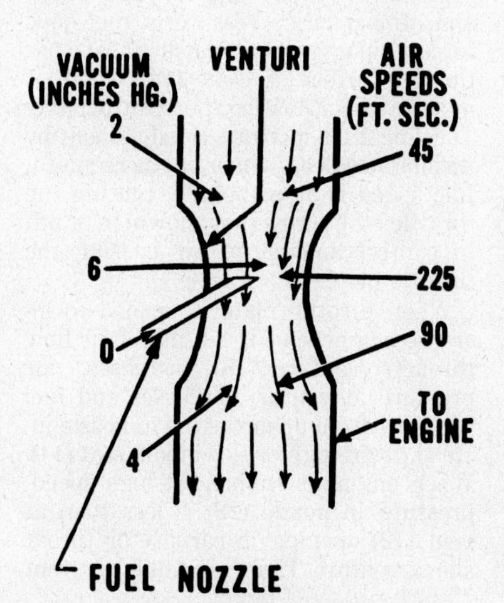

Fig. 2-1 — Drawing illustrating the venturi principle upon which carburetor design is based. Figures at left are inches of mercury vacuum and those at right are air speeds in feet per second that are typical of conditions found in a carburetor operating at wide open throttle. Zero vacuum in fuel nozzle corresponds to atmospheric pressure.

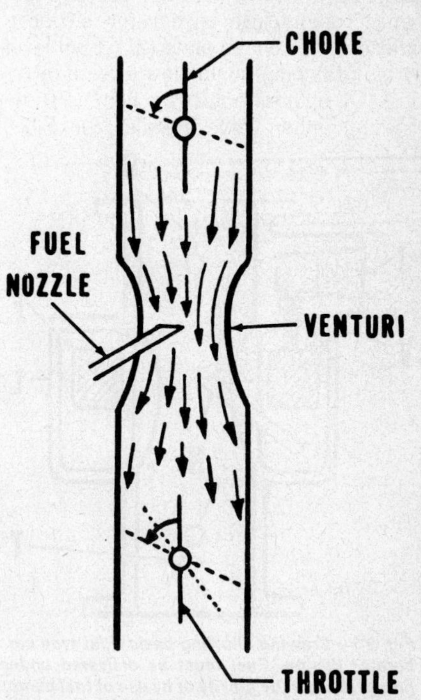

Fig. 2-2 — Drawing showing basic carburetor design. Text explains operation of the choke and throttle valves. In some carburetors, a primer pump may be used instead of the choke valve to provide fuel for the starting fuel-air mixture.

is required to develop maximum power or speed the throttle must be in the wide open position.

Although some carburetors may be as simple as the basic design just described, most engines require more complex design features to provide variable fuel-air mixture ratios for different operating conditions. These design features will be described in the following paragraph.

CARBURETOR TYPE

Carburetors are of the float type and are either of the downdraft or side draft design. The following paragraph describes the features and operating principles of the float type carburetor.

FLOAT TYPE CARBURETOR. The principle of float type carburetor operation is illustrated in Fig. 2–3. Fuel is delivered at inlet (I) by gravity with fuel tank placed above carburetor, or by a fuel lift pump when tank is located below carburetor inlet. Fuel flows into the open inlet valve (V) until fuel level (L) in bowl lifts float against fuel valve needle and closes the valve. As fuel is emitted from the nozzle (N) when engine is running, fuel level will drop, lowering the float and allowing valve to open so that fuel will enter the carburetor to meet the requirements of the engine.

In Fig. 2–4, a cut-away view of a well known make of float type carburetor is shown. Atmospheric pressure is maintained in fuel bowl through passage (20) which opens into carburetor air horn ahead of the choke plate (21). Fuel level is maintained at just below level of opening (O) in nozzle (22) by float (19) actuating inlet valve needle (8). Float height can be adjusted by bending float tang (5).

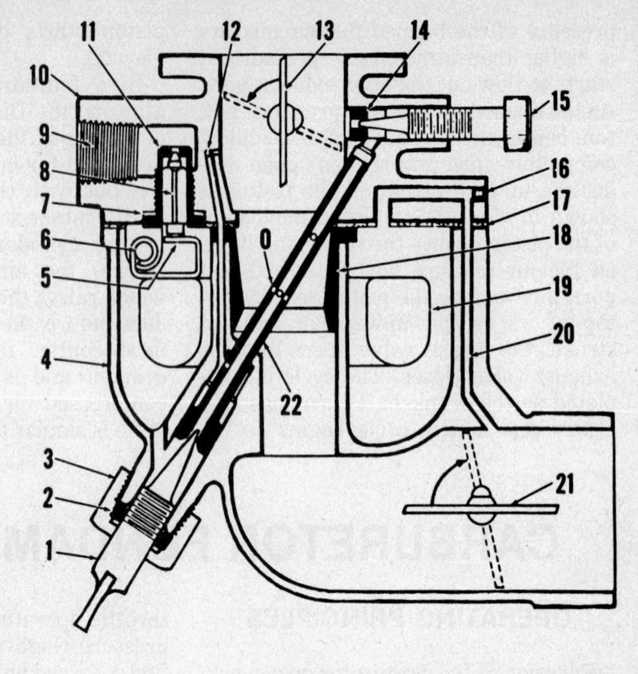

Fig. 2-4 — Cross-sectional drawing of float type carburetor used on some engines.

O. Orifice
1. Main fuel needle
2. Packing
3. Packing nut
4. Carburetor bowl
5. Float tang
6. Float hinge pin
7. Gasket
8. Inlet valve
9. Fuel inlet
10. Carburetor body
11. Inlet valve seat
12. Vent
13. Throttle plate
14. Idle orifice
15. Idle fuel needle
16. Plug
17. Gasket
18. Venturi
19. Float
20. Fuel bowl vent
21. Choke
22. Fuel nozzle

When starting a cold engine, it is necessary to close the choke plate (21) as shown by dotted lines so as to lower the air pressure in carburetor venturi (18) as engine is cranked. Then, fuel will flow up through nozzle (22) and will be emitted from openings (O) in nozzle. When an engine is hot, it will start on a leaner fuel-air mixture than when cold and may start without the choke plate being closed.

When engine is running at slow idle speed (throttle plate nearly closed as indicated by dotted lines in Fig. 2–4), air pressure above the throttle plate is low and atmospheric pressure in fuel bowl forces fuel up through the nozzle and out through orifice in seat (14) where it mixes with air passing the throttle plate. The idle fuel mixture is adjustable by turning needle (15) in or out as required. Idle speed is adjustable by turning the throttle stop screw (not shown) in or out to control amount of air passing the throttle plate.

When throttle plate is opened to increase engine speed, velocity of air flow through venturi (18) increases, air pressure at venturi decreases and fuel will flow from openings (O) in nozzle instead of through orifice in idle seat (14). When engine is running at high speed, pressure in nozzle (22) is less than at vent (12) opening in carburetor throat above venturi. Thus, air will enter vent and travel down the vent into the nozzle and mix with the fuel in the nozzle. This is referred to as air bleeding and is illustrated in Fig. 2–5.

Many different designs of float type carburetors will be found when servicing the different makes and models of engines. Reference should be made to the engine repair section of this manual for adjustment and overhaul specifications. Refer to carburetor servicing paragraphs in fundamentals section for service hints.

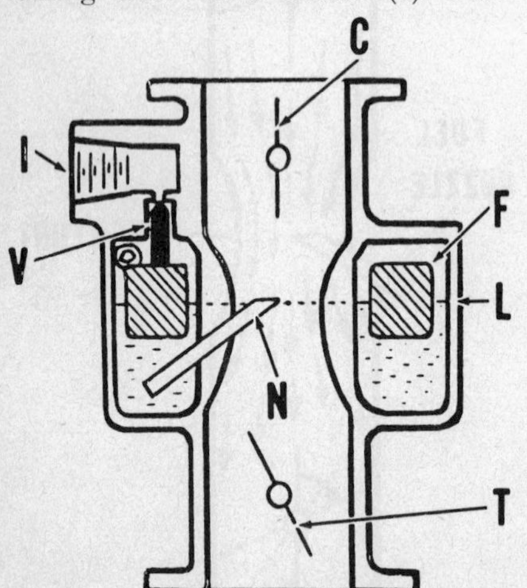

Fig. 2-3 — Drawing showing basic float type carburetor design. Fuel must be delivered under pressure either by gravity or by use of fuel pump, to the carburetor fuel inlet (I). Fuel level (L) operates float (F) to open and close inlet valve (V) to control amount of fuel entering carburetor. Also shown are the fuel nozzle (N), throttle (T) and choke (C).

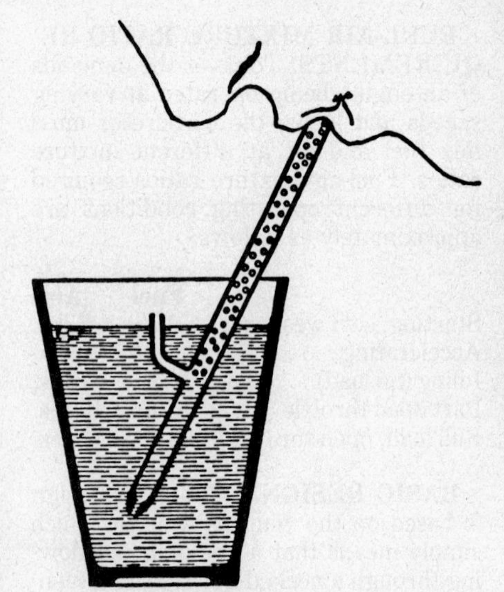

Fig. 2-5 — Illustration of air bleed principle explained in text.

IGNITION SYSTEM FUNDAMENTALS

The ignition system provides a properly timed surge of extremely high voltage electrical energy which flows across the spark plug electrode gap to create the ignition spark. Engines may be equipped with either a magneto or battery ignition system. A magneto ignition system generates electrical energy, intensifies (transforms) this electrical energy to the extremely high voltage required and delivers this electrical energy at the proper time for the ignition spark. In a battery ignition system, a storage battery is used as a source of electrical energy and the system transforms the relatively low electrical voltage from the battery into the high voltage required and delivers the high voltage at proper time for the ignition spark. Thus, the function of the two systems is somewhat similar except for the basic source of electrical energy. The fundamental operating principles of ignition systems are explained in the following paragraphs.

MAGNETISM AND ELECTRICITY

The fundamental principles upon which ignition systems are designed are presented in this section. As the study of magnetism and electricity is an entire scientific field, it is beyond the scope of this manual to fully explore these subjects. However, the following information will impart a working knowledge of basic principles which should be of value in servicing engines.

MAGNETISM. The effects of magnetism can be shown easily while the theory of magnetism is too complex to be presented here. The effects of magnetism were discovered many years ago when fragments of iron ore were found to attract each other and also attract other pieces of iron. Further, it was found that when suspended in air, one end of the iron ore fragment would always point in the directin of the North Star. The end of iron ore fragment pointing north was called the "north pole" and the opposite end the "south pole." By stroking a piece of steel with a "natural magnet," as these iron ore fragments were called, it was found that the magnetic properties of the natural magnet could be transferred or "induced" into the steel.

Steel which will retain magnetic properties for an extended period of time after being subjected to a strong magnetic field are called "permanent magnets;" iron or steel that loses such magnetic properties soon after being subjected to a magnetic field are called "temporary magnets." Soft iron will lose magnetic properties almost immediately after being removed from a magnetic field, and so is used where this property is desirable.

The area affected by a magnet is called a "field of force." The extent of this field of force is related to the strength of the magnet and can be determined by use of a compass. In practice, it is common to illustrate the field of force surrounding a magnet by lines as shown in Fig. 3-1 and the field of force is usually called "lines of force" or "flux." Actually, there are no "lines;" however, this is a convenient method of illustrating the presence of the invisible magnetic forces and if a certain magnetic force is defined as a "line of force," then all magnetic forces may be measured by comparison. The number of "lines of force" making up a strong magnetic field is enormous.

Most materials when placed in a magnetic field are not attracted by the magnet, do not change the magnitude or direction of the magnetic field, and so are called "non-magnetic materials." Materials such as iron, cobalt, nickel or their alloys, when placed in a magnetic field will concentrate the field of force and hence are magnetic conductors or "magnetic materials." There are no materials known in which magnetic fields will not penetrate and magnetic lines of force can be deflected only by magnetic materials or by another magnetic field.

Alnico, an alloy containing aluminum, nickel and cobalt, retains magnetic properties for a very long period of time after being subjected to a strong magnetic field and is extensively used as a permanent magnet. Soft iron, which loses magnetic properties quickly, is used to concentrate magnetic fields as in Fig. 3-1.

ELECTRICITY. Electricity, like magnetism, is an invisible physical force whose effects may be more readily explained than the theory of what electricity consists of. All of us are familiar with the property of electricity to produce light, heat and mechanical power. What must be explained for the purpose of understanding ignition system operation is the inter-relationship of magnetism and electricity and how the ignition spark is produced.

Electrical current may be defined as a flow of energy in a conductor which, in some ways, may be compared to flow of water in a pipe. For electricity to flow, there must be a pressure (voltage) and a complete circuit (closed path) through which the electrical energy may return, a comparison being a water pump and a pipe that receives water from the outlet (pressure) side of the pump and returns the water to the inlet side of the pump. An electrical circuit may be completed by electricity flowing through the earth (ground), or through the metal framework of an engine or other equipment ("grounded" or "ground" connections). Usually, air is an insulator through which electrical energy will not flow. However, if the force (voltage) becomes great, the resistance of air to the flow of electricity is broken down and a current will flow, releasing energy in the form of a spark. By high voltage electricity breaking down the resistance of the air gap between the spark plug electrodes, the ignition spark is formed.

ELECTRO-MAGNETIC INDUCTION. The principle of electro-magnetic induction is as follows: When a wire (conductor) is moved through a field of magnetic force so as to cut across the lines of force (flux), a potential voltage or electromotive force (emf) is induced in the wire. If the wire is a part of a completed electrical circuit, current will flow through the circuit as illustrated in Fig. 3-2. It should be noted that the movement of the wire through the lines of magnetic force is a relative motion; that is, if the lines of force of a moving magnetic field cut across a wire, this will also induce an emf to the wire.

The direction of an induced current is related to the direction of magnetic force and also to the direction of movement of the wire through the lines of force, or flux. The voltage of an induced

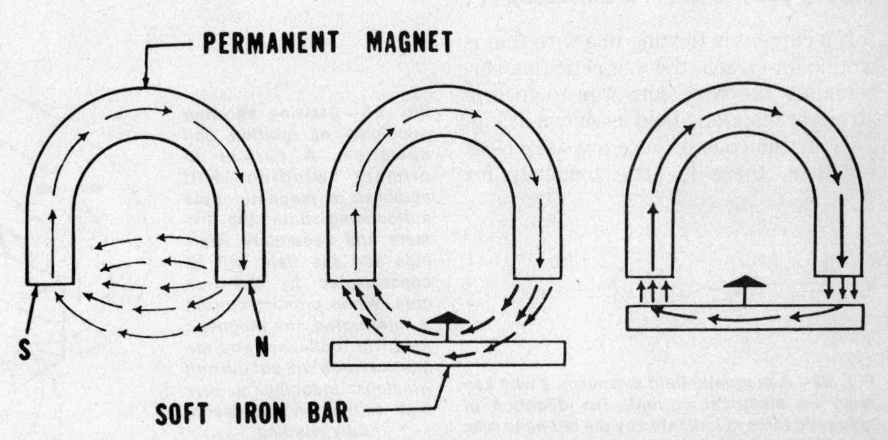

Fig. 3-1 — In left view, field of force of permanent magnet is illustrated by arrows showing direction of magnetic force from north pole (N) to south pole (S). In center view, lines of magnetic force are being attracted by soft iron bar that is being moved into the magnetic field. In right view, the soft iron bar has been moved to the magnet and the field of magnetic force is concentrated within the bar.

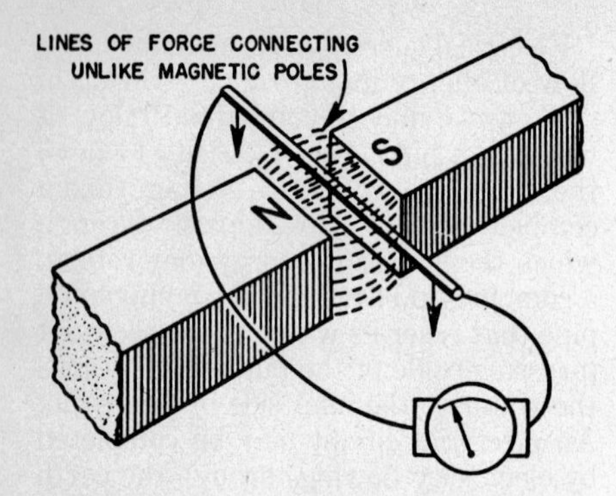

Fig. 3-2 — When a conductor is moved through a magnetic field so as to cut across lines of force, a potential voltage will be induced in the conductor. If the conductor is a part of a completed electrical circuit, current will flow through the circuit as indicated by the gage.

current is related to the strength, or concentration of lines of force, of the magnetic field and to the rate of speed at which the wire is moved through the flux. If a length of wire is wound into a coil and a section of the coil is moved through magnetic lines of force, the voltage induced will be proportional to the number of turns of wire in the coil.

ELECTRICAL MAGNETIC FIELDS. When a current is flowing in a wire, a magnetic field is present around the wire as illustrated in Fig. 3–3. The direction of lines of force of this magnetic field is related to the direction of current in the wire. This is known as the left hand rule and is stated as follows: If a wire carrying a current is grasped in the left hand with thumb pointing in direction electrons are moving, the curved fingers will point the direction of lines of magnetic force (flux) encircling the wire.

NOTE: The currently used electron theory explains the movement of electrons from negative to positive. Be sure to use the LEFT HAND RULE with the thumb pointing the direction electrons are moving (the positive end of a conductor).

If a current is flowing in a wire that is wound into a coil, the magnetic flux surrounding the wire converge to form a stronger magnetic field as shown in Fig. 3–4. If the coils of wire are very close together, there is little tendency for

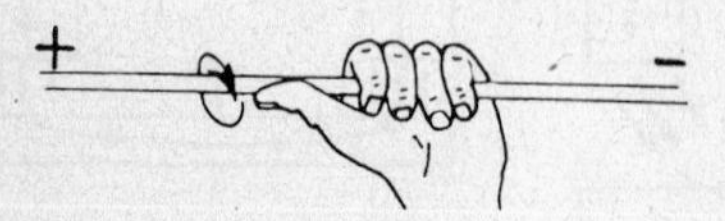

Fig. 3-3 — A magnetic field surrounds a wire carrying an electrical current. The direction of magnetic force is indicated by the left hand rule; that is, if thumb of left hand points in direction that electrical current is flowing in conductor, fingers of the left hand will indicate direction of magnetic force.

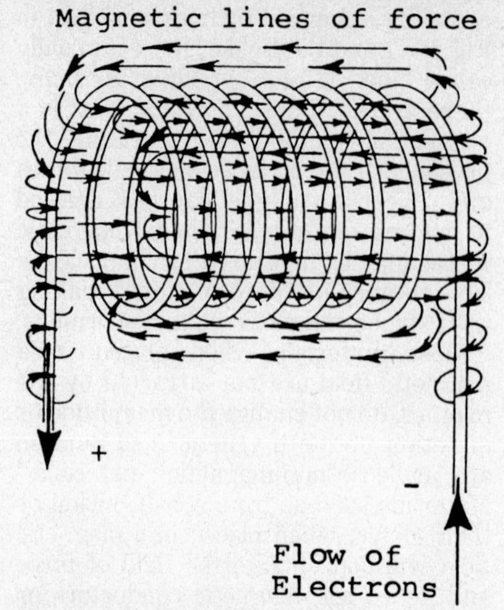

Fig. 3-4 — When a wire is wound in a coil, the magnetic force created by a current in the wire will tend to converge in a single strong magnetic field as illustrated. If the loops of the coil are wound closely together, there is little tendency for lines of force to surround individual loops of the coil.

magnetic flux to surround individual loops of the coil and a strong magnetic field will surround the entire coil. The strength of this field will vary with the current flowing through the coil.

STEP-UP TRANSFORMERS (IGNITION COILS). In both battery and magneto ignition systems, it is necessary to step-up, or transform, a relatively low primary voltage to the 15,000 to 20,000 volts required for the ignition spark. This is done by means of an ignition coil which utilizes the interrelationship of magnetism and electricity as explained in preceding paragraphs.

Basic ignition coil design is shown in Fig. 3–5. The coil consists of two separate coils of wire which are called the primary coil winding and the secondary coil winding, or simply the primary winding and the secondary winding. The primary winding as indicated by the heavy, black line is of larger diameter wire and has a smaller number of turns when compared to the secondary winding indicated by the light line.

A current passing through the primary winding creates a magnetic field (as indicated by the "lines of force") and this field, concentrated by the soft iron core, surrounds both the primary and secondary windings. If the primary winding current is suddenly interrupted, the magnetic field will collapse and the lines of force will cut through the coil windings. The resulting induced voltage in the secondary winding is greater than the voltage of the current that was flowing in the primary winding and is related to the number of turns of wire in each winding. Thus:

$$\text{Induced secondary voltage} = \text{primary voltage} \times \frac{\text{No. of turns in secondary winding}}{\text{No. of turns in primary winding}}$$

For example, if the primary winding of an ignition coil contained 100 turns of wire and the secondary winding contained 10,000 turns of wire, a current having an emf of 200 volts flowing in the primary winding, when suddenly interrupted, would result in and emf of:

$$200 \text{ Volts} \times \frac{10{,}000 \text{ turns of wire}}{100 \text{ turns of wire}} = 20{,}000 \text{ volts}$$

SELF-INDUCTANCE. It should be noted the collapsing magnetic field resulting from interrupted current in the primary winding will also induce a current in the primary winding. This effect is termed "self-inductance." This self-induced current is such as to oppose any interruption of current in the primary winding, slowing the collapse of

Fig. 3-5 — Drawing showing principles of ignition coil operation. A current in primary winding will establish a magnetic field surrounding both the primary and secondary windings and the field will be concentrated by the iron core. When primary current is interrupted, the magnetic field will "collapse" and the lines of force will cut the coil windings inducing a very high voltage in the secondary winding.

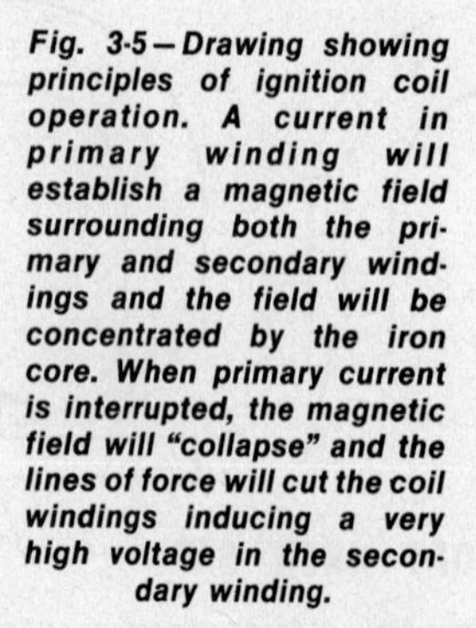

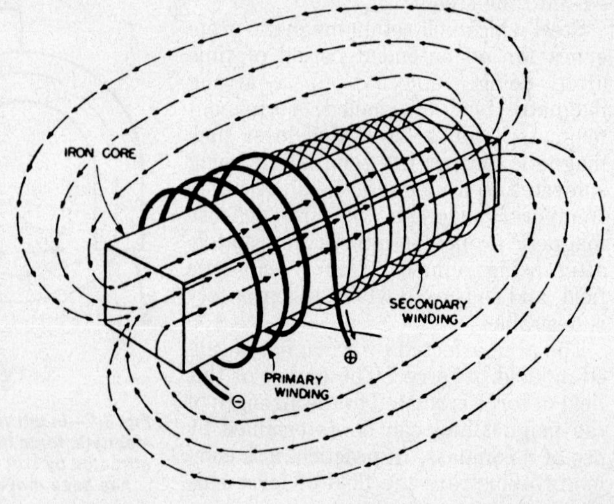

the magnetic field and reducing the efficiency of the coil. The self-induced primary current flowing across the slightly open breaker switch, or contact points, will damage the contact surfaces due to the resulting spark.

To momentarily absorb, then stop the flow of current across the contact points, a capacitor or, as commonly called, a condenser is connected in parallel with the contact points. A simple condenser is shown in Fig. 3–6; however, the capacity of such a condenser to absorb current (capacitance) is limited by the small surface area of the plates. To increase capacity to absorb current, the condenser used in ignition systems is constructed as shown in Fig. 3–9.

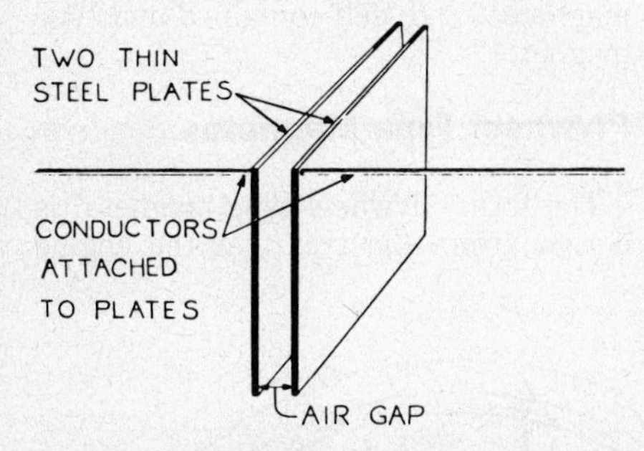

Fig. 3-6—Drawing showing construction of a simple condenser. Capacity of such a condenser to absorb current is limited due to the relatively small surface area. Also, there is a tendency for current to arc across the air gap. Refer to Fig. 3-9 for construction of typical ignition system condenser.

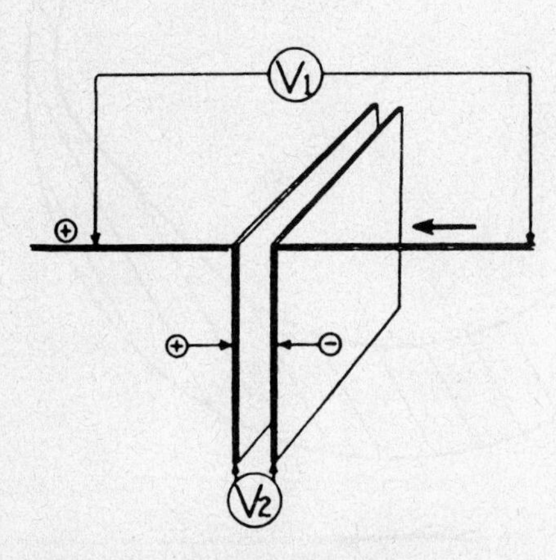

Fig. 3-7—A condenser in an electrical circuit will absorb electrons until an opposing voltage (V2) is built up across condenser plates which is equal to the voltage (V1) of the electrical current.

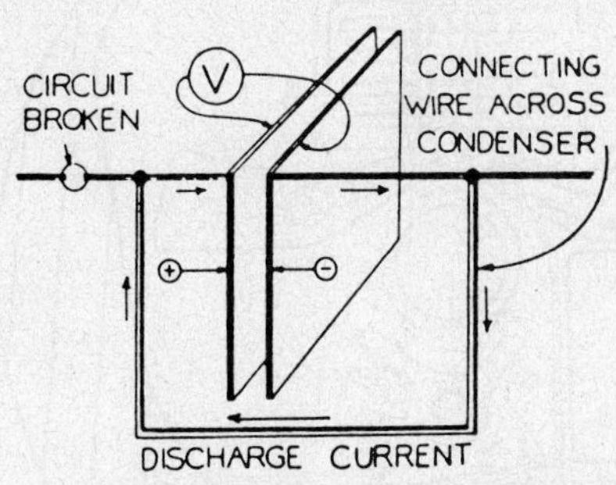

Fig. 3-8—When a circuit containing a condenser is interrupted (circuit broken), the condenser will retain a potential voltage (V). If a wire is connected across the condenser, a current will flow in reverse direction of charging current until condenser is discharged (voltage across condenser plates is zero).

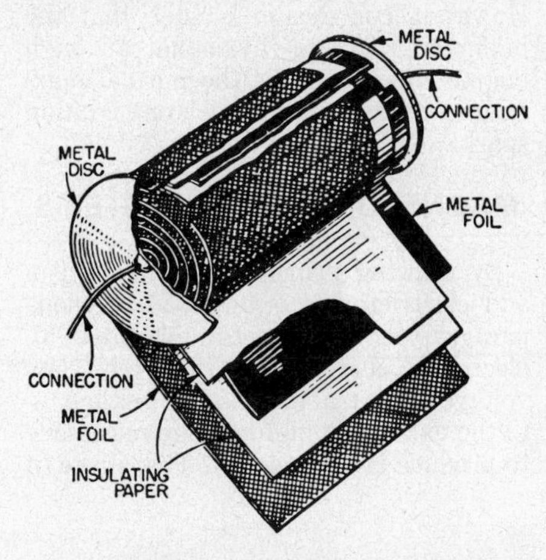

Fig. 3-9—Drawing showing construction of typical ignition system condenser. Two layers of metal foil, insulated from each other with paper, are rolled tightly together and a metal disc contacts each layer, or strip of foil. Usually, one disc is grounded through the condenser shell.

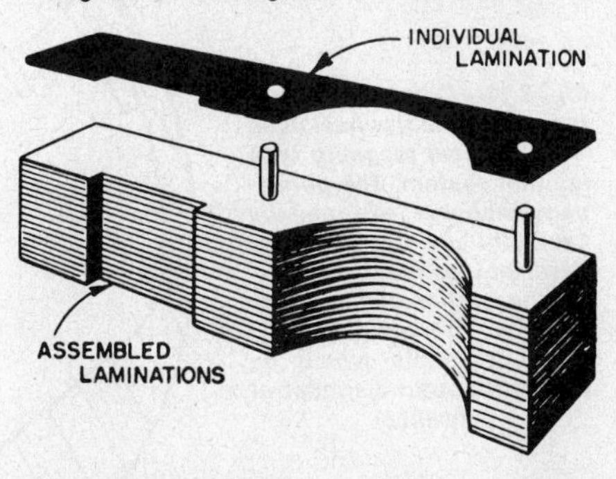

Fig. 3-10—To prevent formation of "eddy currents" within soft iron cores used to concentrate magnetic fields, core is assembled of plates or "laminations" that are insulated from each other. In a solid iron core, there is a tendency for counteracting magnetic forces to build up from stray currents induced in the core.

EDDY CURRENTS. It has been found that when a solid soft iron bar is used as a core for an ignition coil, stray electrical currents are formed in the core. These stray, or "eddy currents" create opposing magnetic forces causing the core to become hot and also decrease efficiency of the coil. As a means of preventing excessive formation of eddy currents within the core, or other magnetic field carrying parts of a magneto, a laminated plate construction as shown in Fig. 3–10 is used instead of solid material. The plates, or laminations, are insulated from each other by a natural oxide coating formed on the plate surfaces or by coating the plates with varnish. The cores of some ignition coils are constructed of soft iron wire instead of plates and each wire is insulated by a varnish coating. This type of construction serves the same purpose as laminated plates.

BATTERY IGNITION SYSTEMS

Some engines are equipped with a battery ignition system. A schematic diagram of a typical battery ignition system for a single cylinder engine is shown in Fig. 3–11. Designs of battery to location of breaker points and method for actuating the points; however, all operate on the same basic principles.

BATTERY IGNITION SYSTEM PRINCIPLES. Refer to the schematic diagram in Fig. 3–11. When the timer cam is turned so the contact points are

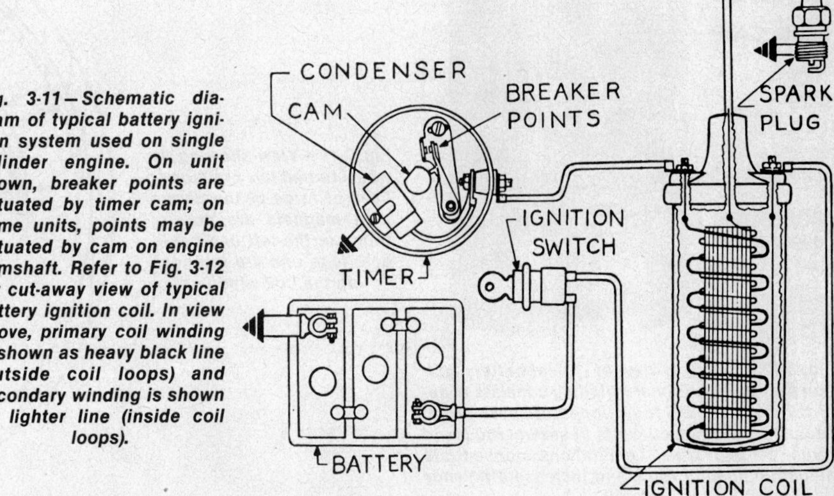

Fig. 3-11—Schematic diagram of typical battery ignition system used on single cylinder engine. On unit shown, breaker points are actuated by timer cam; on some units, points may be actuated by cam on engine camshaft. Refer to Fig. 3-12 for cut-away view of typical battery ignition coil. In view above, primary coil winding is shown as heavy black line (outside coil loops) and secondary winding is shown by lighter line (inside coil loops).

closed, a current is established on the primary circuit by the emf of the battery. This current flowing through the primary winding of the ignition coil establishes a magnetic field concentrated in the core laminations and surrounding the windings. A cut-away view of a typical ignition coil is shown in Fig. 3–12. At the proper time for the ignition spark, contact points are opened by the timer cam and primary ignition circuit is interrupted. The condenser, wired in parallel with breaker contact points between timer terminal and ground, absorbs self-induced current in the primary circuit for an instant and brings the flow of current to a quick, controlled stop. The magnetic field surrounding the coil rapidly cuts the primary and secondary windings creating an emf as high as 250 volts in the primary winding and up to 25,000 volts in the secondary winding. Current absorbed by the condenser is discharged as breaker points close, grounding the condenser lead wire.

Due to resistance of the primary winding, a certain period of time is required for maximum primary current flow after breaker contact points are closed. At high engine speeds, points remain closed for a smaller interval of time, hence primary current does not build up to maximum and secondary voltage is somewhat less than at low engine speed. However, coil design is such that the minimum voltage available at high engine speed exceeds the normal maximum voltage required for ignition spark.

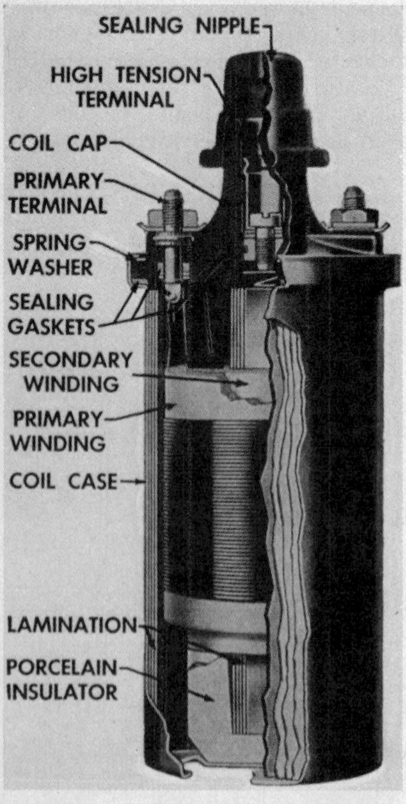

Fig. 3-12 — Cut-away view of typical battery ignition system coil. Primary winding consists of approximately 200-250 turns (loops) of heavier wire; secondary winding consists of several thousand turns of fine wire. Laminations concentrate magnetic lines of force and increase efficiency of coil.

MAGNETO IGNITION SYSTEMS

By utilizing principles of magnetism and electricity as outlined in previous paragraphs, a magneto generates an electrical current of relatively low voltage, then transforms this voltage into the extremely high voltage necessary to produce ignition spark. This surge of high voltage is timed to create the ignition spark and ignite the compressed fuel-air mixture in the engine cylinder at the proper time in the Otto cycle as described in the paragraphs on fundamentals of engine operation principles.

Two different types of magnetos are used on air-cooled engines and, for discussion in this section of the manual, will be classified as "flywheel type magnetos" and "self-contained unit type magnetos."

Flywheel Type Magnetos

The term "flywheel type magneto" is derived from the fact that the engine

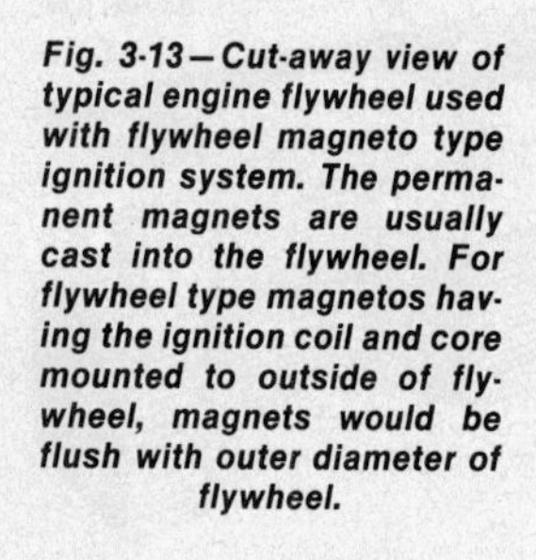

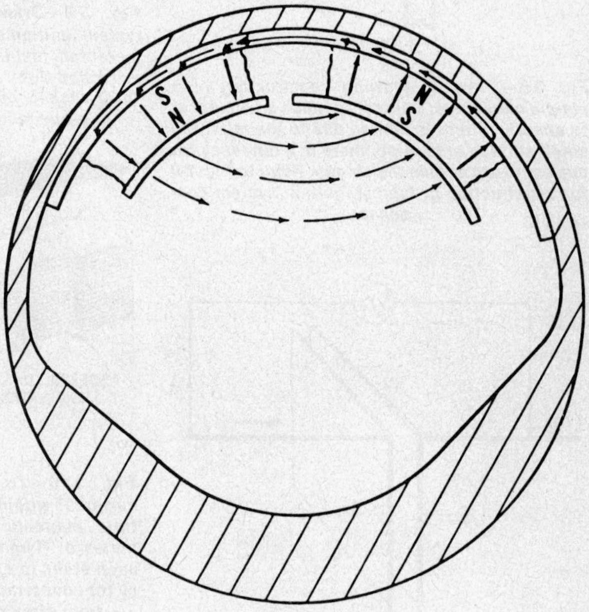

Fig. 3-13 — Cut-away view of typical engine flywheel used with flywheel magneto type ignition system. The permanent magnets are usually cast into the flywheel. For flywheel type magnetos having the ignition coil and core mounted to outside of flywheel, magnets would be flush with outer diameter of flywheel.

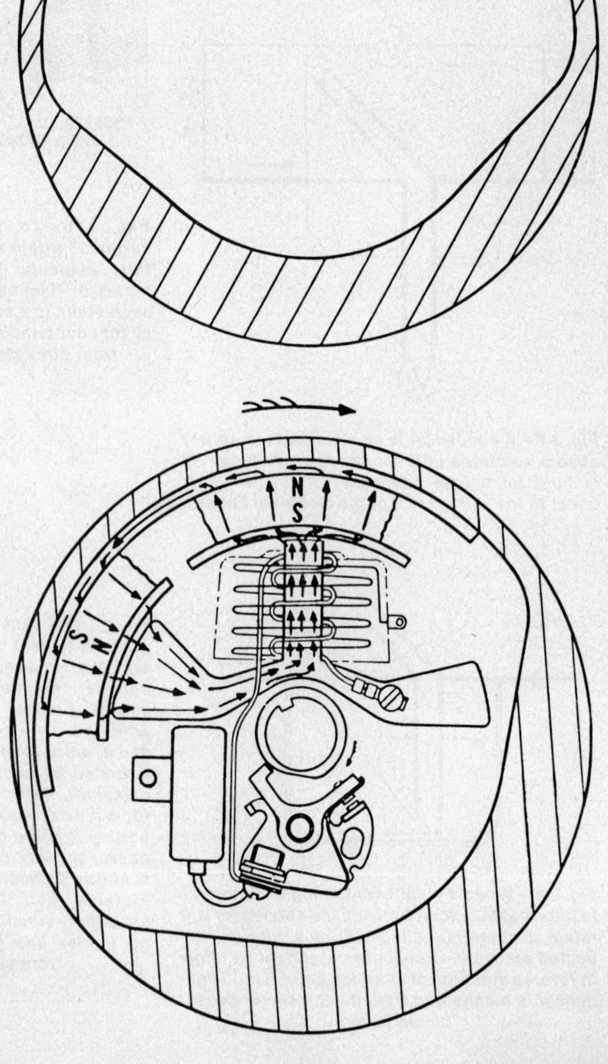

Fig. 3-14 — View showing flywheel turned to a position so lines of force of the permanent magnets are concentrated in the left and center core legs and are interlocking the coil windings.

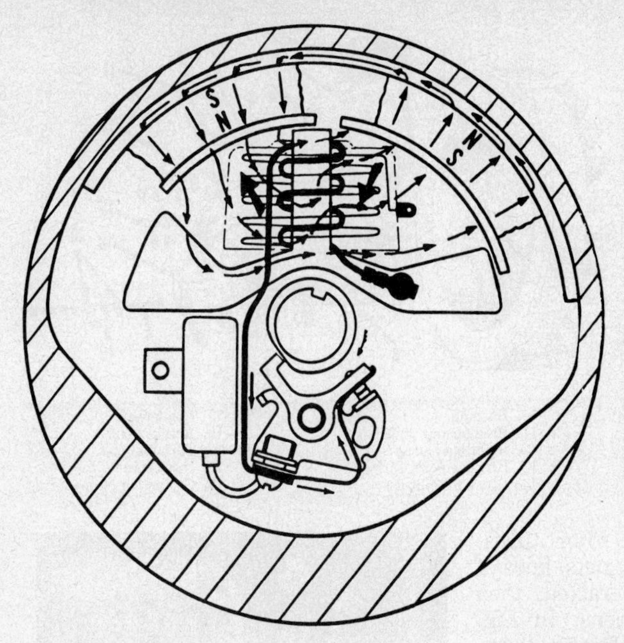

Fig. 3-15 — View showing flywheel turned to a position so lines of force of the permanent magnets are being withdrawn from the left and center core legs and are being attracted by the center and right core legs. While this event is happening, the lines of force are cutting up through the coil windings section between the left and center legs and are cutting down through the section between the right and center legs as indicated by the heavy black arrows. As the breaker points are now closed by the cam, a current is induced in the primary ignition circuit as the lines of force cut through the coil windings.

flywheel carries the permanent magnets and is the magneto rotor. In some similar systems, magneto rotor is mounted on engine crankshaft as is the flywheel, but is a part separate from flywheel.

FLYWHEEL MAGNETO OPERATING PRINCIPLES. In Fig. 3 – 13, a cross-sectional view of a typical engine flywheel (magneto rotor) is shown. The arrows indicate lines of force (flux) of the permanent magnets carried by the flywheel. As indicated by arrows, direction of force of magnetic field is from north pole (N) of left magnet to south pole (S) of right magnet.

Figs. 3 – 14, 3 – 15, 3 – 16 and 3 – 17 illustrate operational cycle of flywheel type magneto. In Fig. 3 – 14, flywheel magnets have moved to a position over left and center legs of armature (ignition coil) core. As magnets moved into this position, their magnetic field was attracted by armature core as illustrated in Fig. 3 – 1 and a potential voltage (emf) was induced in coil windings. However, this emf was not sufficient to cause current to flow across spark plug electrode gap in high tension circuit and points were open in primary circuit.

In Fig. 3 – 15, flywheel magnets have moved to a new position to where their magnetic field is being attracted by center and right legs of armature core, and is being withdrawn from left and center legs. As indicated by heavy black arrows, lines of force are cutting up through the section of coil windings between left and center legs of armature and are cutting down through coil windings section between center and right legs. If the left hand rule, as explained in a previous paragraph, is applied to the lines of force cutting through coil sections, it is seen resulting emf induced in primary circuit will cause a current to flow through primary coil windings and breaker points which have now been closed by action of the cam.

At instant movement of lines of force cutting through coil winding sections is at maximum rate, maximum flow of current is obtained in primary circuit. At this time, cam opens breaker points interrupting primary circuit and, for an instant, flow of current is absorbed by condenser as illustrated in Fig. 3 – 16. An emf is also induced in secondary coil windings, but voltage is not sufficient to cause current to flow across spark plug gap.

Flow of current in primary windings created a strong electromagnetic field surrounding coil windings and up through center leg of armature core as shown in Fig. 3 – 17. As breaker points were opened by cam, interrupting primary circuit, magnetic field starts to collapse cutting coil windings as indicated by heavy black arrows. The emf induced in primary circuit would be sufficient to cause a flow of current across opening breaker points were it not for the condenser absorbing flow of current and bringing it to a controlled stop. This allows electromagnetic field to collapse at such a rapid rate to induce a very high voltage in coil high tension or secondary

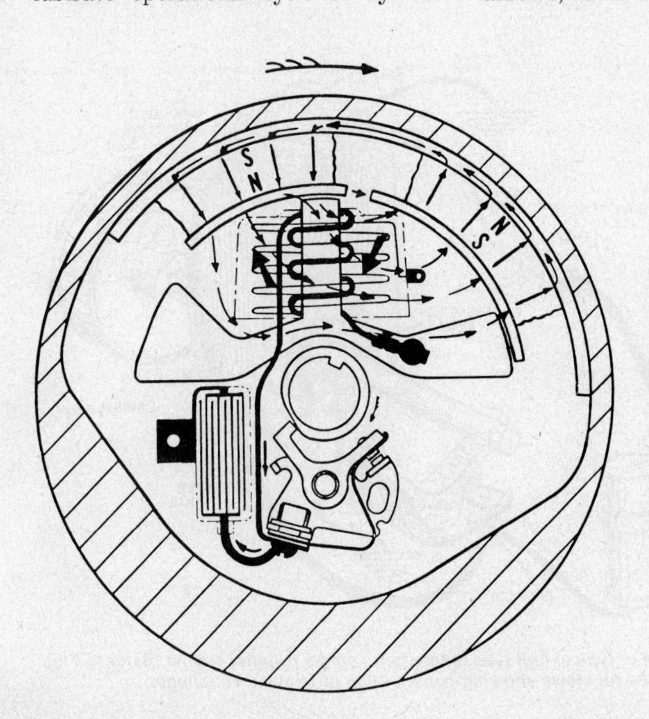

Fig. 3-16 — The flywheel magnets have now turned slightly past the position shown in Fig. 3-15 and the rate of movement of lines of magnetic force cutting through the coil windings is at the maximum. At this instant, the breaker points are opened by the cam and flow of current in the primary circuit is being absorbed by the condenser, bringing the flow of current to a quick, controlled stop. Refer now to Fig. 3-17.

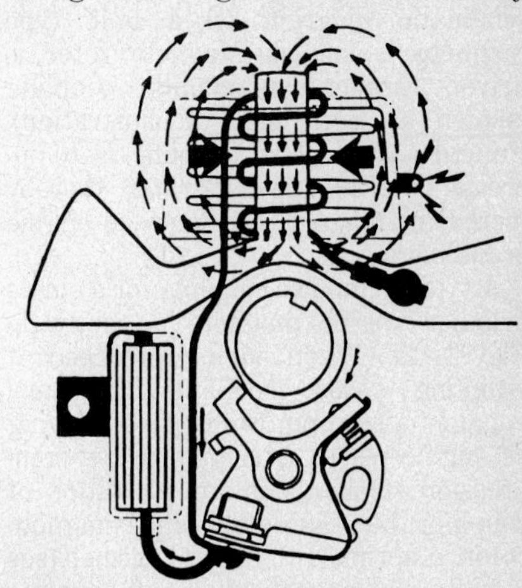

Fig. 3-17 — View showing magneto ignition coil, condenser and breaker points at same instant as illustrated in Fig. 3-16; however, arrows shown above illustrate lines of force of the electromagnetic field established by current in primary coil windings rather than the lines of force of permanent magnets. As the current in the primary circuit ceases to flow, the electromagnetic field collapses rapidly, cutting the coil windings as indicated by heavy arrows and inducing a very high voltage in the secondary coil winding resulting in the ignition spark.

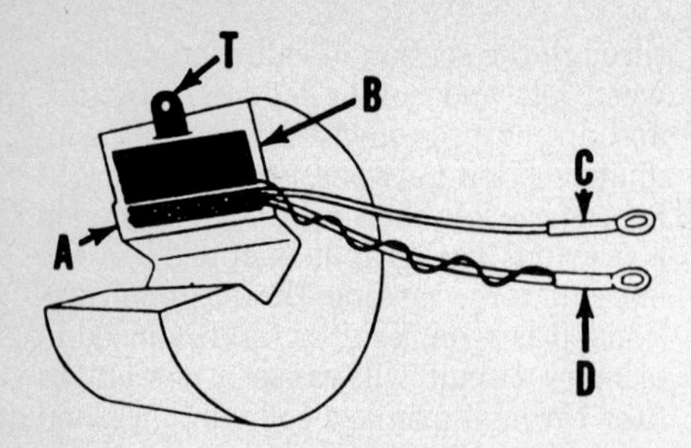

Fig. 3-18—Drawing showing construction of a typical flywheel magneto ignition coil. Primary windings (A) consists of about 200 turns of wire. Secondary winding (B) consists of several thousand turns of fine wire. Coil primary and secondary ground connection is (D); primary connection to breaker point and condenser terminal is (C); and coil secondary (high tension) terminal is (T).

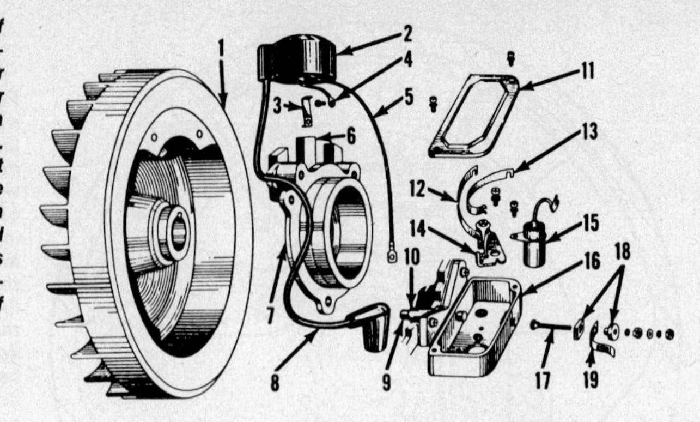

Fig. 3-19—Exploded view of a typical flywheel type magneto used on single cylinder engines in which the breaker points (14) are actuated by a cam on engine camshaft. Push rod (9) rides against cam to open and close points. In this type unit, an ignition spark is produced only on alternate revolutions of the flywheel as the camshaft turns at one-half speed.

1. Flywheel
2. Ignition coil
3. Coil clamps
4. Coil ground lead
5. Breaker point lead
6. Armature core (laminations)
7. Crankshaft bearing retainer
8. High tension lead
9. Push rod
10. Bushing
11. Breaker box cover
12. Point lead strap
13. Breaker point spring
14. Breaker point assy.
15. Condenser
16. Breaker box
17. Terminal bolt
18. Insulators
19. Grounding (stop) spring

windings. This voltage, in the order of 15,000 to 25,000 volts, is sufficient to break down the resistance air gap between spark plug electrodes and a current will flow across gap. This creates ignition spark which ignites compressed fuel-air mixture in engine cylinder.

Self-Contained Unit Type Magnetos

Some four-stroke cycle engines are equipped with a magneto which is a self-contained unit as shown in Fig. 3–20. This type magneto is driven from engine timing gears via a gear or coupling. All components of the magneto are enclosed in one housing and magneto can be removed from engine as a unit.

UNIT TYPE MAGNETO OPERATING PRINCIPLES. In Fig. 3–21, a schematic diagram of a unit type magneto is shown. Magneto rotor is driven through an impulse coupling (shown at right side of illustration). Function of impluse coupling is to increase rotating speed of rotor, thereby increasing magneto efficiency, at engine cranking speeds.

A typical impulse coupling for a single cylinder engine magneto is shown in Fig. 3–22. When engine is turned at cranking speed, coupling hub pawl engages a stop pin in magneto housing as engine piston is coming up on compression stroke. This stops rotation of coupling hub assembly and magneto rotor. A spring within coupling shell (see Fig. 3–23) connects shell and coupling hub; as engine continues to turn, spring winds up until pawl kickoff contacts pawl and disengages it from stop pin. This occurs at the time an ignition spark is required to ignite compressed fuel-air mixture in engine cylinder. As pawl is released, spring connecting coupling shell and hub unwinds and rapidly spins magneto rotor.

Magneto rotor (see Fig. 3–21) carries permanent magnets. As rotor turns, alternating position of magnets, lines of force of magnets are attracted, then withdrawn from laminations. In Fig. 3–21, arrows show magnetic field concentrated within laminations, or armature core. Slightly further rotation of magnetic rotor will place magnets to where laminations will have greater attraction for opposite poles of magnets. At this instant, lines of force as indicated by arrows will suddenly be withdrawn and an opposing field of force will be established in laminations. Due to this rapid movement of lines of force, a current will be induced in primary magneto circuit as coil windings are cut by lines of force. At instant maximum current is induced in primary windings, breaker points are opened by a cam on magnetic rotor shaft interrupting primary circuit. The lines of magnetic force established by primary current (refer to Fig. 3–5) will cut through secondary windings at such a rapid rate to induce a very high voltage in secondary (or high tension) circuit. This voltage will break down

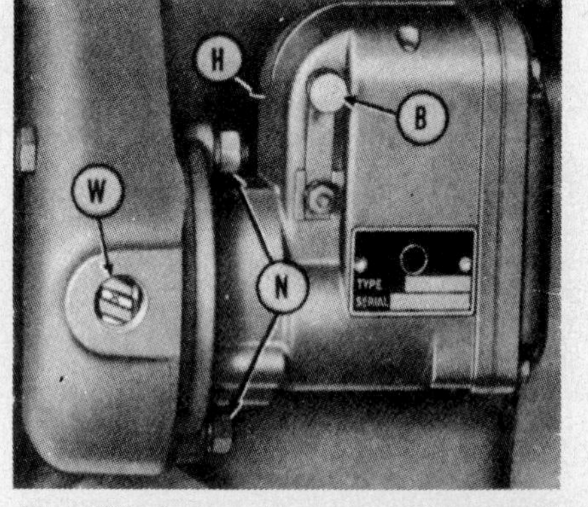

Fig. 3-20—Some engines are equipped with a unit type magneto having all components enclosed in a single housing (H). Magneto is removable as a unit after removing retaining nuts (N). Stop button (B) grounds out primary magneto circuit to stop engine. Timing window is (W).

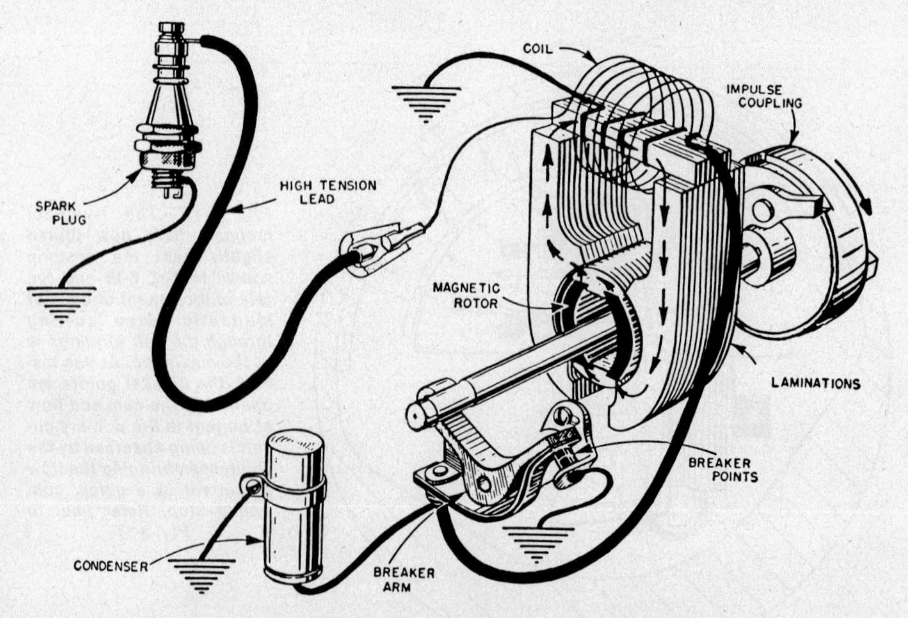

Fig. 3-21—Schematic diagram of typical unit type magneto for single cylinder engine. Refer to Figs. 3-22, 3-23, and 3-24 for views showing construction of impulse couplings.

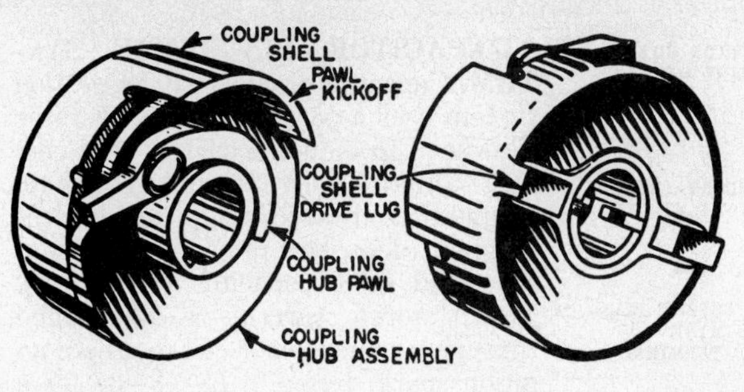

Fig. 3-22—Views of typical impulse coupling for magneto driven by engine shaft with slotted drive connection. Coupling drive spring is shown in Fig. 3-23. Refer to Fig. 3-24 for view of combination magneto drive gear and impulse coupling used on some magnetos.

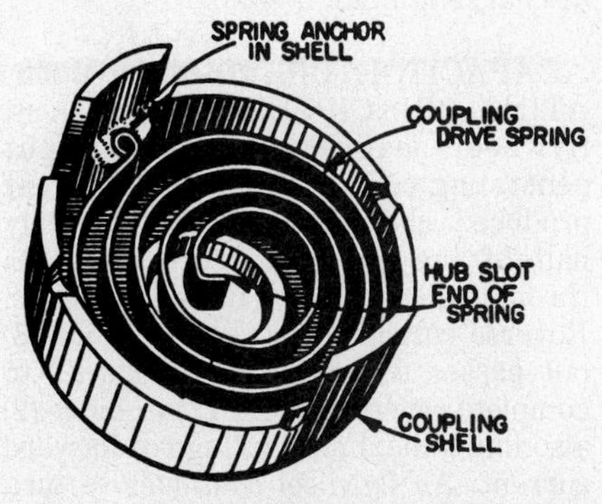

Fig. 3-23—View showing impulse coupling shell and drive spring removed from coupling hub assembly. Refer to Fig. 3-22 for views of assembled unit.

resistance of spark plug electrode gap and a spark across electrodes will result.

At engine operating speeds, centrifugal force will hold impulse coupling hub pawl (See Fig. 3–22) in a position so it cannot engage stop pin in magneto housing and magnetic rotor will be driven through spring (Fig. 3–23) connecting coupling shell to coupling hub. The impulse coupling retards ignition spark, at cranking speeds, as engine piston travels closer to top dead center while magnetic rotor is held stationary by pawl and stop pin. The difference in degrees of impluse coupling shell rotation between position of retarded spark and normal running spark is known as impulse coupling lag angle.

SOLID STATE IGNITION SYSTEM

BREAKERLESS MAGNETO SYSTEM. Solid state (breakerless) magneto ignition system operates somewhat on the same basic principles as conventional type flywheel magneto previously described. The main difference is breaker contact points are replaced by a solid state electronic Gate Controlled Switch (GCS) which has no moving parts. Since, in a conventional system breaker points are closed over a longer period of crankshaft rotation than is the "GCS", a diode has been added to the circuit to provide the same characteristics as closed breaker points.

BREAKERLESS MAGNETO OPERATING PRINCIPLES. The same basic principles for electro-magnetic induction of electricity and formation of magnetic fields by electrical current as outlined for conventional flywheel type magneto also apply to solid state magneto. Therefore principles of different components (diode and GCS) will complete operating principles of solid state magneto.

The diode is represented in wiring diagrams by the symbol shown in Fig. 3–25. The diode is an electronic device that will permit passage of electrical current in one direction only. In electrical schematic diagrams, current flow is opposite direction the arrow part of symbol is pointing.

The symbol shown in Fig. 3–26 is used to represent gate controlled switch (GCS) in wiring diagrams. The GCS acts as a switch to permit passage of current from cathode (C) terminal to anode (A) terminal when in "ON" state and will not permit electric current to flow when in "OFF" state. The GCS can be turned "ON" by a positive surge of electricity at gate (G) terminal and will remain "ON" as long as current remains positive at gate terminal or as long as current is flowing through GCS from cathode (C) terminal to anode (A) terminal.

The basic components and wiring diagram for solid state breakerless magneto are shown schematically in Fig. 3–27. In Fig. 3–28, magneto rotor (flywheel) is turning and ignition coil magnets have just moved into position so their lines of force are cutting ignition coil windings and producing a negative surge of current in primary windings. The diode allows current to flow opposite to the direction of diode symbol arrow and action is same as conventional magneto with breaker points closed. As rotor (flywheel) continues to turn as shown in Fig. 3–29, direction of magnetic flux lines will reverse in armature center leg. Direction of current will change in primary coil circuit and previously conducting diode will be shut off. At this point, neither diode is conducting. As voltage begins to build up as rotor continues to turn, condenser acts

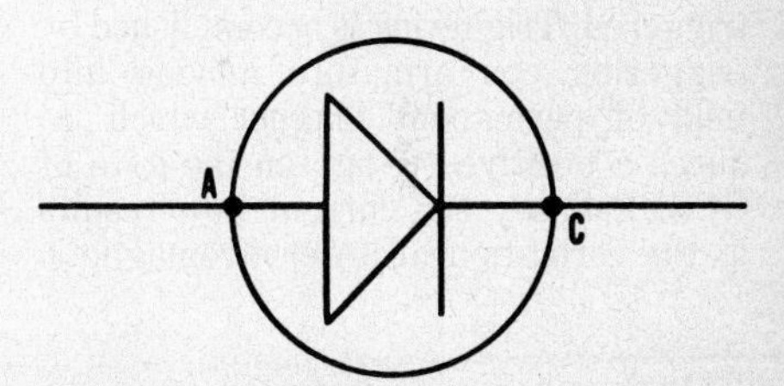

Fig. 3-25—In a diagram of an electrical circuit, the diode is represented by the symbol shown above. The diode will allow current to flow in one direction only, from cathode (C) to anode (A).

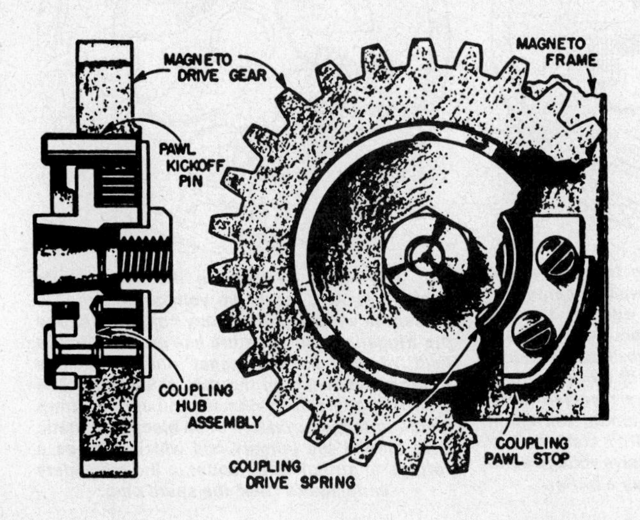

Fig. 3-24—Views of combination magneto drive gear and impulse coupling used on some magnetos.

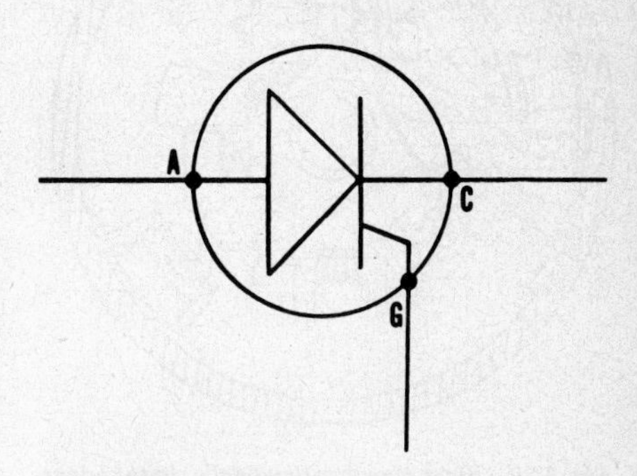

Fig. 3-26—The symbol used for a Gate controlled Switch (GCS) in an electrical diagram is shown above. The GCS will permit current to flow from cathode (C) to anode (A) when "turned on" by a positive electrical charge at gate (G) terminal.

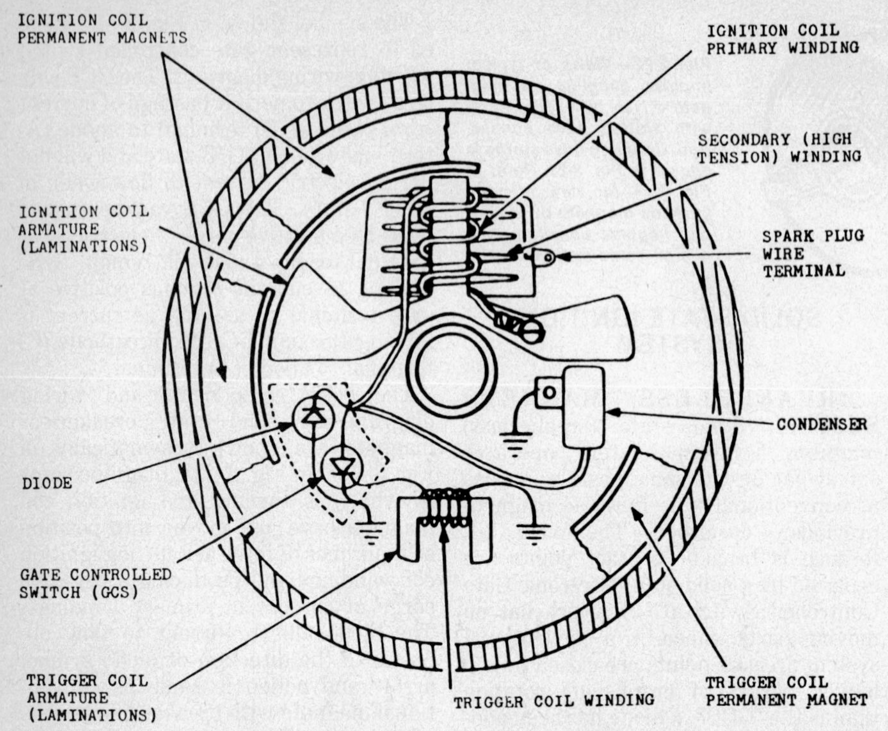

Fig. 3-27 – Schematic diagram of typical breakerless magneto ignition system. Refer to Figs. 3-28, 3-29 and 3-30 for schematic views of operating cycle.

as a buffer to prevent excessive voltage build up at GCS before it is triggered.

When rotor reaches approximate position shown in Fig. 3–30, maximum flux density has been achieved in center leg of armature. At this time the GCS is triggered. Triggering is accomplished by triggering coil armature moving into field of permanent magnet which induces a positive voltage on the gate of GCS. Primary coil current flow results in the formation of an electromagnetic field around primary coil which induces a voltage of sufficient potential in secondary coil windings to "fire" spark plug.

When rotor (flywheel) has moved magnets past armature, GCS will cease to conduct and revert to "OFF" state until it is triggered. The condenser will discharge during time GCS was conducting.

CAPACITOR DISCHARGE SYSTEM. Capacitor discharge (CD) ignition system uses a permanent magnet rotor (flywheel) to induce a current in a coil, but unlike conventional flywheel magneto and solid state breakerless magneto described previously, current is stored in a capacitor (condenser). Then, stored current is discharged through a transformer coil to create ignition spark. Refer to Fig. 3–31 for a schematic of a typical capacitor discharge ignition system.

CAPACITOR DISCHARGE OPERATING PRINCIPLES. As permanent flywheel magnets pass by input generating coil (1–Fig. 3–31), current produced charges capacitor (6). Only half of the generated current passes through diode (3) to charge capacitor. Reverse current is blocked by diode (3) but passes through Zener diode (2) to complete reverse circuit. Zener diode (2) also limits maximum voltage of forward current. As flywheel continues to turn and magnets pass trigger coil (4), a small amount of electrical current is generated. This current opens gate controlled switch (5) allowing capacitor to discharge through pulse transformer (7). The rapid voltage rise in transformer primary coil induces a high voltage secondary current which forms ignition spark when it jumps spark plug gap.

THE SPARK PLUG

In any spark ignition engine, the spark plug (See Fig. 3–32) provides means for

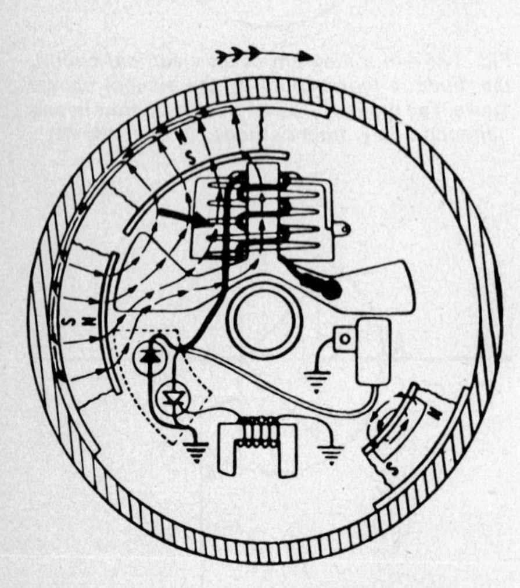

Fig. 3-28 – View showing flywheel of breakerless magneto system at instant of rotation where lines of force of ignition coil magnets are being drawn into left and center legs of magneto armature. The diode (see Fig. 3-25) acts as a closed set of breaker points in completing the primary ignition circuit at this time.

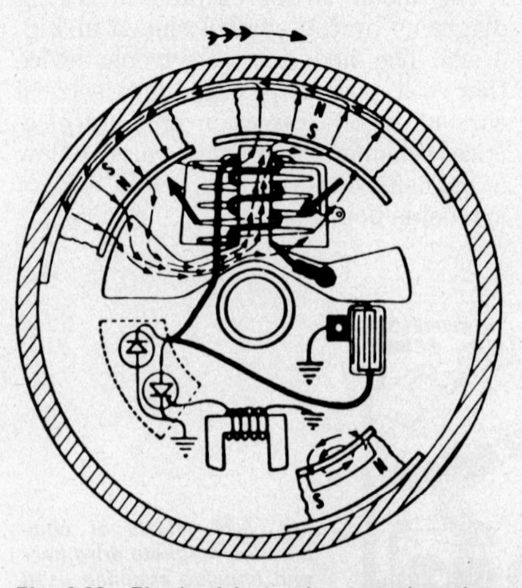

Fig. 3-29 – Flywheel is turning to point where magnetic flux lines through armature center leg will reverse direction and current through primary coil circuit will reverse. As current reverses, diode which was previously conducting will shut off and there will be no current. When magnetic flux lines have reversed in armature center leg, voltage potential will again build up, but since GCS is in "OFF" state, no current will flow. To prevent excessive voltage build up, the condenser acts as a buffer.

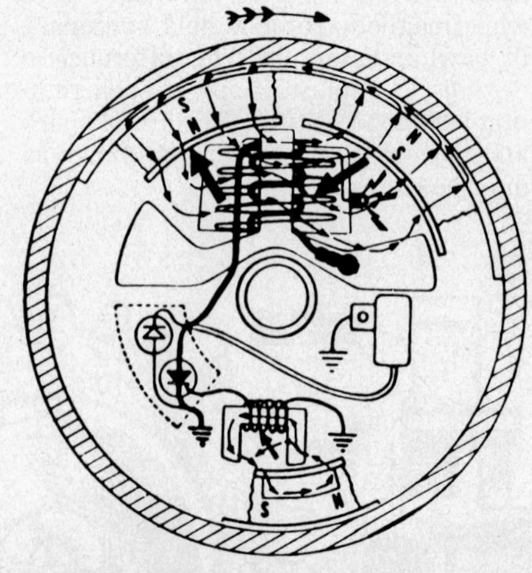

Fig. 3-30 – With flywheel in the approximate position shown, maximum voltage potential is present in windings of primary coil. at this time the triggering coil armature has moved into the field of permanent magnet and a positive voltage is induced on the gate of the GCS. The GCS is triggered and primary coil current flows resulting in the formation of an electromagnetic field around the primary coil which induces a voltage of sufficient potential in the secondary windings to "fire" the spark plug.

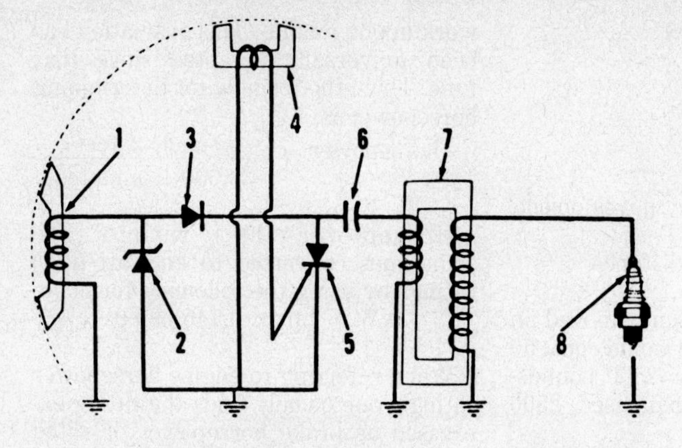

Fig. 3-31—Schematic diagram of typical capacitor discharge ignition system.

1. Generating coil
2. Zener diode
3. Diode
4. Trigger coil
5. Gate controlled switch
6. Capacitor
7. Pulse transformer (coil)
8. Spark plug

igniting compressed fuel-air mixture in cylinder. Before an electric charge can move across an air gap, intervening air must be charged with electricity, or ionized. If spark plug is properly gapped and system is not shorted, not more than 7,000 volts may be required to initiate a spark. Higher voltage is required as spark plug warms up, or if compression pressures or distance of air gap is increased. Compression pressures are highest at full throttle and relatively slow engine speeds, therefore, high voltage requirements or a lack of available secondary voltage most often shows up as a miss during maximum acceleration from a slow engine speed.

There are many different types and sizes of spark plugs which are designed for a number of specific requirements.

THREAD SIZE. The threaded, shell portion of the spark plug and the attaching hole in cylinder are manufactured to meet certain industry established standards. The diameter is referred to as "Thread Size." Those commonly used are: 10mm, 14mm, 18mm, ⅞-inch and ½-inch pipe.

REACH. The length of thread, and thread depth in cylinder head or wall are also standardized throughout the industry. This dimension is measured from gasket seat of plug to cylinder end of thread. See Fig. 3-33. Four different reach plugs commonly used are: ⅜-inch, 7/16-inch, ½-inch and ¾-inch.

HEAT RANGE. During engine operation, part of the heat generated during combustion is transferred to the spark plug, and from plug to cylinder through shell threads and gasket. The operating temperature of spark plug plays an important part in engine operation. If too much heat is retained by plug, fuel-air mixture may be ignited by contact with heated surface before ignition spark occurs. If not enough heat is retained, partially burned combustion products (soot, carbon and oil) may build up on plug tip resulting in "fouling" or shorting out of plug. If this happens, secondary current is dissipated uselessly as it is generated instead of bridging plug gap as a useful spark, and engine will misfire.

Operating temperature of plug tip can be controlled, within limits, by altering length of the path heat must follow to reach threads and gasket of plug. Thus, a plug with a short, stubby insulator around center electrode will run cooler than one with a long, slim insulator. Refer to Fig. 3-34. Most plugs in more popular sizes are available in a number of heat ranges which are interchangeable within the group. The proper heat range is determined by engine design and type of service. Refer to SPARK PLUG SERVICING FUNDAMENTALS for additional information on spark plug selection.

SPECIAL TYPES. Sometimes, engine design features or operating conditions call for special plug types designed for a particular purpose.

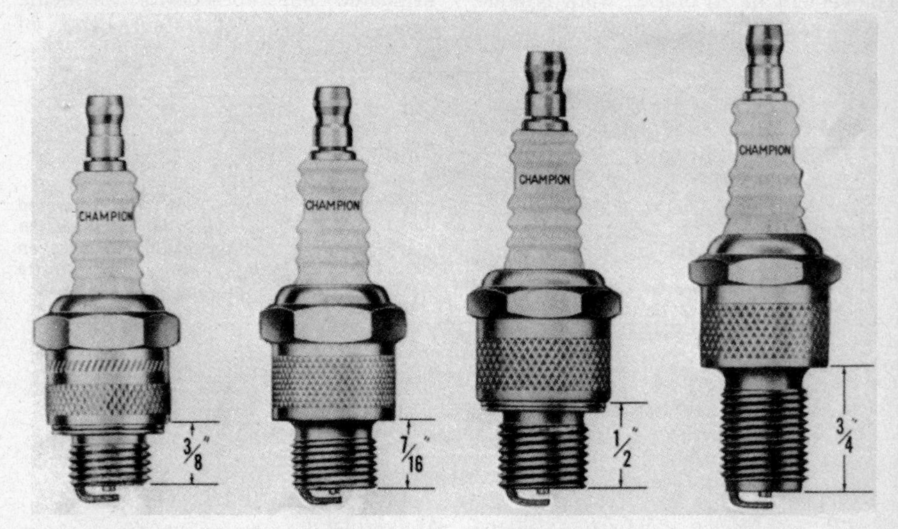

Fig. 3-33—Views showing spark plugs with various "reaches" available. A ⅜-inch reach spark plug measures ⅜-inch from firing end of shell to gasket surface of shell.

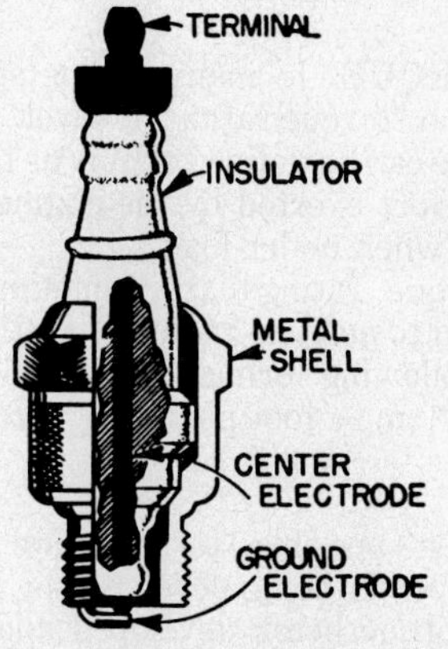

Fig. 3-32—Cross-sectional drawing of spark plug showing construction and nomenclature.

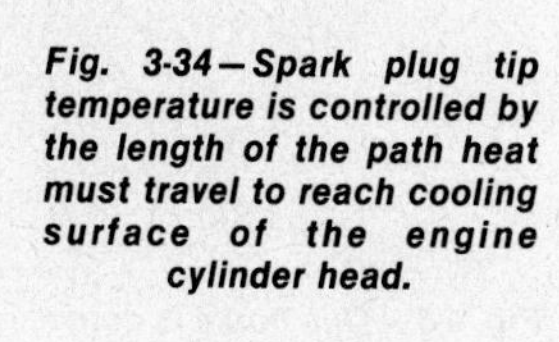

Fig. 3-34—Spark plug tip temperature is controlled by the length of the path heat must travel to reach cooling surface of the engine cylinder head.

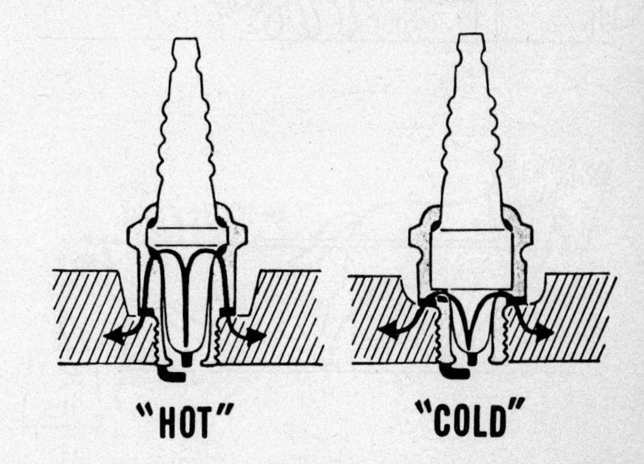

ENGINE POWER AND TORQUE RATNGS

The following paragraphs discuss terms used in expressing engine horsepower and torque ratings and explains methods for determining different ratings. Some engine repair shops are now equipped with a dynamometer for measuring engine torque and/or horsepower and the mechanic should be familiar with terms, methods of measurement and how actual power developed by an engine can vary under different conditions.

GLOSSARY OF TERMS

FORCE. Force is an action against an object that tends to move the object from a state of rest, or to accelerate movement of an object. For use in calculating torque or horsepower, force is measured in pounds.

WORK. When a force moves an object from a state of rest, or accelerates movement of an object, work is done. Work is measured by multiplying force applied by distance force moves object, or:

$$\text{work} = \text{force} \times \text{distance}$$

Thus, if a force of 50 pounds moved an object 50 feet, work done would equal 50 pounds times 50 feet, or 2500 pounds-feet (or as it is usually expressed, 2500 foot-pounds).

POWER. Power is the rate at which work is done; thus, if:

$$\text{work} = \text{force} \times \text{distance},$$

then:

$$\text{power} = \frac{\text{force} \times \text{distance.}}{\text{time}}$$

From the above formula, it is seen that power must increase if time in which work is done decreases.

HORSEPOWER. Horsepower is a unit of measurement of power. Many years ago, James Watt, a Scotsman noted as inventor of the steam engine, evaluated one horsepower as being equal to doing 33,000 foot-pounds of work in one minute. This evaluation has been universally accepted since that time. Thus, the formula for determining horsepower is:

$$\text{horsepower} = \frac{\text{pounds} \times \text{feet}}{33{,}000 \times \text{minutes}}$$

Horsepower (hp.) ratings are sometimes converted to kilowatt (kW) ratings by using the following formula:

$$\text{kW} = \text{hp.} \times 0.745\ 699\ 9$$

When referring to engine horsepower ratings, one usually finds the rating expressed as brake horsepower or rated horsepower, or sometimes as both.

BRAKE HORSEPOWER. Brake horsepower is the maximum horsepower available from an engine as determined by use of a dynamometer, and is usually stated as maximum observed brake horsepower or as corrected brake horsepower. As will be noted in a later paragraph, observed brake horsepower of a specific engine will vary under different conditions of temperature and atmospheric pressure. Corrected brake horsepower is a rating calculated from observed brake horsepower and is a means of comparing engines tested at varying conditions. The method for calculating corrected brake horsepower will be explained in a later paragraph.

RATED HORSEPOWER. An engine being operated under a load equal to the maximum horsepower available (brake horsepower) will not have reserve power for overloads and is subject to damage from overheating and rapid wear. Therefore, when an engine is being selected for a particular load, the engine's brake horsepower rating should be in excess of the expected normal operating load. Usually, it is recommended that the engine not be operated in excess of 80% of the engine maximum brake horsepower rating; thus, the "rated horsepower" of an engine is usually equal to 80% of maximum horsepower the engine will develop.

TORQUE. In many engine specifications, a "torque rating" is given. Engine torque can be defined simply as the turning effort exerted by the engine output shaft when under load.

Torque ratings are sometimes converted to newton-meters (N•m) by using the following formula:

$$\text{N}\cdot\text{m} = \text{foot pounds of torque} \times 1.355\ 818$$

It is possible to calculate engine horsepower being developed by measuring torque being developed and engine output speed. Refer to the following paragraphs.

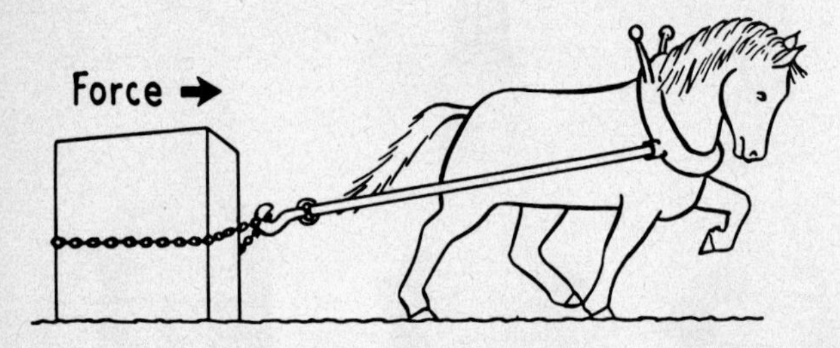

Fig. 4-1 — A force, measured in pounds, is defined as an action tending to move an object or to accelerate movement of an object.

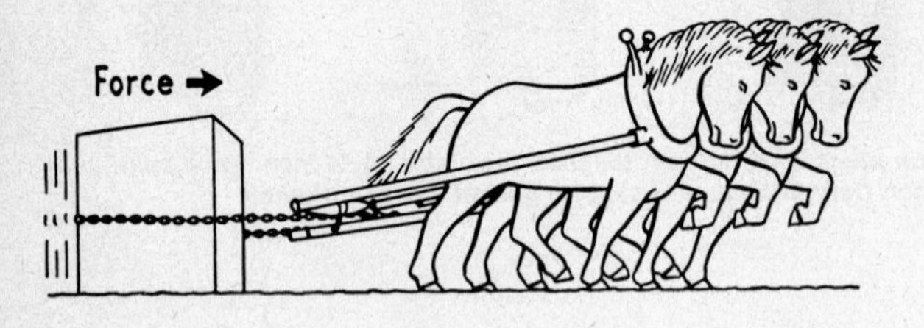

Fig. 4-2 — If a force moves an object from a state of rest or accelerates movement of an object, then work is done.

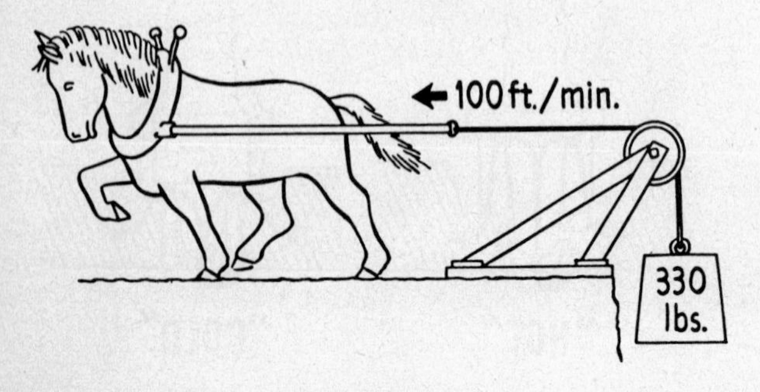

Fig. 4-3 — This horse is doing 33,000 foot-pounds of work in one minute, or one horsepower.

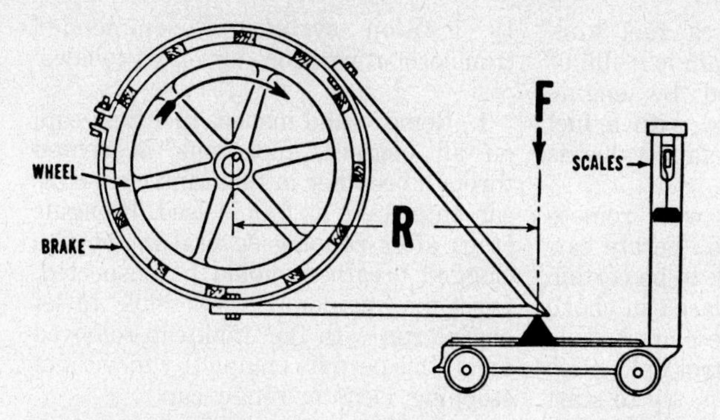

Fig. 4-4—Diagram showing a prony brake on which the torque being developed by an engine can be measured. By also knowing the rpm of the engine output shaft, engine horsepower can be calculated.

MEASURING ENGINE TORQUE AND HORSEPOWER

PRONY BRAKE. The prony brake is the most simple means of testing engine performance. Refer to diagram in Fig. 4–4. A torque arm is attached to a brake on wheel mounted on engine output shaft. The torque arm, as the brake is applied, exerts a force (F) on scales. Engine torque is computed by multiplying force (F) times length of torque arm radius (R), or:

$$\text{engine torque} = F \times R.$$

If, for example, torque arm radius (R) is 2 feet and force (F) being exerted by torque arm on scales is 6 pounds, engine torque would be 2 feet × 6 pounds, or 12 foot-pounds.

To calculate engine horsepower being developed by use of the prony brake, we must also count revolutions of engine output shaft for a specific length of time. In formula for calculating horsepower:

$$\text{horsepower} = \frac{\text{feet} \times \text{pounds}}{33{,}000 \times \text{minutes}};$$

Feet in formula will equal circumference transcribed by torque arm radius multiplied by number of engine output shaft revolutions. Thus:

$$\text{feet} = 2 \times 3.14 \times \text{radius} \times \text{revolutions}.$$

Pounds in formula will equal force (F) of torque arm. If, for example, force (F) is 6 pounds, torque arm radius is 2 feet and engine output shaft speed is 3300 revolutions per minute, then:

$$\text{horsepower} = \frac{2 \times 3.14 \times 2 \times 3300 \times 6}{33{,}000 \times 1}$$

or,

horsepower × 7.54

DYNAMOMETERS. Some commercial dynamometers for testing small engines are now available, although the cost may be prohibitive for all but larger repair shops. Usually, these dynamometers have a hydraulic loading device and scales indicating engine speed and load; horsepower is then calculated by use of a slide rule type instrument. For further information on commercial dynamometers, refer to manufacturers listed in special service tool section of this manual.

HOW ENGINE HORSEPOWER OUTPUT VARIES

Engine efficiency will vary with the amount of air taken into the cylinder on each intake stroke. Thus, air density has a considerable effect on horsepower output of a specific engine. As air density varies with both temperature and atmospheric pressure, any change in air temperature, barometric pressure, or elevation will cause a variance in observed engine horsepower. As a general rule, engine horsepower will:

A. Decrease approximately 3% for each 1000 foot increase above 1000 ft. elevation;

B. Decrease approximately 3% for each 1 inch drop in barometric pressure; or,

C. Decrease approximately 1% for each 10° rise in temperature (Farenheit).

Thus, to fairly compare observed horsepower readings, observed readings should be corrected to standard temperature and atmospheric pressure conditions of 60°F., and 29.92 inches of mercury. The correction formula specified by the Society of Automotive Engineers is somewhat involved; however for practical purposes, the general rules stated above can be used to approximate corrected brake horsepower of an engine when observed maximum brake horsepower is known.

For example, suppose the engine horsepower of 7.54 as found by use of the prony brake was observed at an altitude of 3000 feet and at a temperature of 100°F. At standard atmospheric pressure and temperature conditions, we could expect an increase of 4% due to temperature (100° − 60° × 1% per 10°) and an increase of 6% due to altitude (3000 ft. − 1000 ft. × 3% per 1000 ft.) or a total increase of 10%. Thus, corrected maximum horsepower from this engine would be approximately 7.54 + .75, or approximately 8.25 horsepower.

TROUBLESHOOTING

When servicing an engine to correct a specific complaint, such as engine will not start, is hard to start, etc., a logical step-by-step procedure should be followed to determine cause of trouble before performing any service work. This procedure is "TROUBLESHOOTING."

Of course, if an engine is received in your shop for a normal tune up or specific repair work is requested, troubleshooting procedure is not required and work should be performed as requested. It is wise, however, to fully check the engine before repairs are made and recommend any additional repairs or adjustments necessary to ensure proper engine performance.

The following procedures, as related to a specific complaint or trouble, have proven to be a satisfactory method for quickly determining cause of trouble in a number of engine repair shops.

NOTE: It is not suggested the troubleshooting procedure as outlined in following paragraphs be strictly adhered to at all times. In many instances, customer's comments on when trouble was encountered will indicate cause of trouble. Also, the mechanic will soon develop a diagnostic technique that can only come with experience. In addition to the general troubleshooting procedure, reader should also refer to special notes following this section and to the information included in engine, carburetor and magneto servicing fundamentals sections.

If Engine Will Not Start—Or Is Hard To Start

1. If engine is equipped with a rope or crank starter, turn engine slowly. As engine piston is coming up on compression stroke, a definite resistance to turning should be felt on rope or crank. This resistance should be noted every other crankshaft revolution on a single cylinder engine, on every revolution of a two cylinder engine and on every ½-revolution of a four cylinder engine. If correct cranking resistance is noted, engine compression can be considered as not the cause of trouble at this time.

NOTE: Compression gages for gasoline engines are available and are of value in troubleshooting engine service problems.

Where available from engine manufacturer, specifications will be given for engine compression pressure in engine service sections of this manual. On engines having electric starters, remove spark plug and check engine compression with gage; if gage is not available, hold thumb so spark plug hole is partly covered. An alternating blowing and suction action should be noted as engine is cranked.

If very little or no compression is noted, refer to appropriate engine repair section for repair of engine. If check indicates engine is developing compression, proceed to step 2.

2. Remove spark plug wire and hold wire terminal about 1/8-inch (3.18 mm) away from cylinder (on wires having rubber spark plug boot, insert a small screw or bolt in terminal).

NOTE: A test plug with 1/8-inch (3.18 mm) gap is available or a test plug can be made by adjusting the electrode gap of a new spark plug to 0.125 inch (3.18 mm).

While cranking engine, a bright blue spark should snap across the 1/8-inch (3.18 mm) gap. If spark is weak or yellow, or if no spark occurs while cranking engine, refer to **IGNITION SYSTEM SERVICE FUNDAMENTALS** for information on appropriate type system.

If spark is satisfactory, remove and inspect spark plug. Refer to **SPARK PLUG SERVICE FUNDAMENTALS.** If in doubt about spark plug condition, install a new plug.

NOTE: Before installing plug, make certain electrode gap is set to proper dimension shown in engine repair section of this manual. Refer also to Fig. 5-1.

If ignition spark is satisfactory and engine will not start with new plug, proceed with step 3.

3. If engine compression and ignition spark are adequate, trouble within the fuel system should be suspected. Remove and clean or renew air cleaner or cleaner element. Check fuel tank (Fig. 5-2) and make certain it is full of fresh fuel as prescribed by engine manufacturer. If equipped with a fuel shut-off valve, make certain valve is open.

If engine is equipped with remote throttle controls that also operate carburetor choke plate, check to be certain that when controls are placed in choke position, carburetor choke plate is fully closed. If not, adjust control linkage so choke will fully close; then, try to start engine. If engine does not start after several turns, remove air cleaner assembly; carburetor throat should be wet with gasoline. If not, check for reason fuel is not getting to carburetor. On models with gravity feed from fuel tank to carburetor (fuel tank above carburetor), disconnect fuel line at carburetor to see that fuel is flowing through the line. If no fuel is flowing, remove and clean fuel tank, fuel line and any fuel filters or shut-off valve.

On models having a fuel pump separate from carburetor, remove fuel line at carburetor and crank engine through several turns; fuel should spurt from open line. If not, disconnect fuel line from tank to fuel pump at pump connection. If fuel will not run from open line, remove and clean fuel tank, line and if so equipped, fuel filter and/or shut-off valve. If fuel runs from open line, remove and overhaul or renew the fuel pump.

After making sure clean, fresh fuel is available at carburetor, again try to start engine. If engine will not start, refer to recommended initial adjustments for carburetor in appropriate engine repair section of this manual and adjust carburetor idle and/or main fuel needles.

If engine will not start when compression and ignition test within specifications and clean, fresh fuel is available to carburetor, remove and clean or overhaul carburetor as outlined in CARBURETOR SERVICING FUNDAMENTALS section of this manual.

4. The preceding troubleshooting techniques are based on the fact that to run, an engine must develop compression, have an ignition spark and receive proper fuel-air mixture. In some instances, there are other factors involved. Refer to special notes following this section for service hints on finding common causes of engine trouble that may not be discovered in normal troubleshooting procedure.

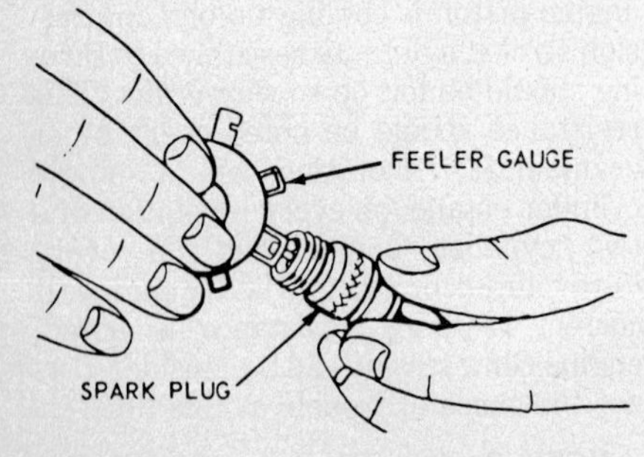

Fig. 5-1 — Be sure to check spark plug electrode gap with proper size feeler gage and adjust gap to specification recommended by manufacturer.

If Engine Starts, Then Stops

This complaint is usually due to fuel starvation, but may be caused by a faulty ignition system. Recommended troubleshooting procedure is as follows:

1. Remove and inspect fuel tank cap; on all engines, fuel tank is vented through breather in fuel tank cap so air can enter tank as fuel is used. If engine stops after running several minutes, a clogged breather should be suspected. On some engines, it is possible to let engine run with fuel tank cap removed and if this permits engine to run without stopping, clean or renew cap.

CAUTION: Be sure to observe safety precautions before attempting to run engine without fuel tank cap in place. If there is any danger of fuel being spilled on engine or spark entering open tank, *do not* attempt to run engine without fuel tank cap in place. If in doubt, try a new cap.

2. If clogged breather in fuel tank cap is eliminated as cause of trouble, a partially clogged fuel filter or fuel line should be suspected. Remove and clean fuel tank and line and if so equipped, clean fuel shut-off valve and/or fuel tank filter. On some engines, a screen or felt type fuel filter is located in carburetor fuel inlet; refer to engine repair section for appropriate engine make and model for carburetor construction.

3. After cleaning fuel tank, line, filter, etc., if trouble is still encountered, a sticking or faulty carburetor inlet needle valve or float may be cause of trouble. Remove, disassemble and clean carburetor using data in engine repair section and in CARBURETOR SERVICE FUNDAMENTALS as a guide.

4. If fuel system is eliminated as cause of trouble by performing procedure outlined in steps 1, 2 and 3, check magneto or battery ignition coil on

Fig. 5-2 — Condensation can cause water and rust to form in fuel tank even though only clean fuel has been poured into tank.

tester if such equipment is available. If not, check for ignition spark immediately after engine stops. Renew coil, condenser and breaker points if no spark is noted. Also, check for engine compression immediately after engine stops; trouble may be caused by sticking intake or exhaust valve or cam followers (tappets). If no or little compression is noted immediately after engine stops, refer to ENGINE SERVICE FUNDAMENTALS section and to engine repair data in appropriate engine repair section of this manual.

Engine Overheats

When air cooled engines overheat, check for:

1. Air inlet screen in blower housing plugged with grass, leaves, dirt or other debris.

2. Remove blower housing and shields and check for dirt or debris accumulated on or between cooling fins on cylinder.

3. Missing or bent shields or blower housing. (Never attempt to operate an air cooled engine without all shields and blower housing in place.)

4. A too lean main fuel-air adjustment of carburetor.

5. Improper ignition spark timing. Check breaker point gap, and on engine with unit type magneto, check magneto to engine timing. On battery ignition units with timer or distributor, check for breaker points opening at proper time.

6. Engines being operated under loads in excess of rated engine horsepower or at extremely high ambient (surrounding) air temperatures may overheat.

Engine Surges When Running

Trouble with an engine surging is usually caused by improper carburetor adjustment or improper governor adjustment.

1. Refer to CARBURETOR paragraphs in appropriate engine repair section and adjust carburetor as outlined.

2. If adjusting carburetor did not correct surging condition, refer to GOVERNOR paragraph and adjust governor linkage.

3. If any wear is noted in governor linkage and adjusting linkage did not correct problem, renew worn linkage parts.

4. If trouble is still not corrected, remove and clean or overhaul carburetor as necessary. Also check for any possible air leaks between the carburetor to engine gaskets or air inlet elbow gaskets.

Special Notes on Engine Troubleshooting

ENGINES WITH COMPRESSION RELEASE. Several different makes of four-stroke cycle engines now have a compression release that reduces compression pressure at cranking speeds, thus making it easier to crank the engine. Most models having this feature will develop full compression when turned in a reverse direction with throttle in wide open position. Refer to the appropriate engine repair section in this manual for detailed information concerning compression release used on different makes and models.

IGNITION SYSTEM SERVICE

The fundamentals of servicing ignition systems are outlined in the following paragraphs. Refer to appropriate heading for type ignition system being inspected or overhauled.

BATTERY IGNITION SERVICE FUNDAMENTALS

Usually all components are readily accessible and while use of test instruments is sometimes desirable, condition of the system can be determined by simple checks. Refer to following paragraphs.

GENERAL CONDITION CHECK. Remove spark plug wire and if terminal is rubber covered, insert small screw or bolt in terminal. Hold uncovered end of terminal or bolt inserted in terminal about 1/8-inch (3.18 mm) away from engine or connect spark plug wire to test plug. Crank engine while observing gap between spark plug wire terminal and engine; if a bright blue spark snaps across gap, condition of system can be considered satisfactory. However, ignition timing may have to be adjusted. Refer to timing procedure in appropriate engine repair section.

VOLTAGE, WIRING AND SWITCH CHECK. If no spark, or a weak yellow-orange spark occurred when checking system as outlined in preceding paragraph, proceed with following check:

Test battery condition with hydrometer or voltmeter. If check indicates a dead cell, renew battery; recharge battery if a discharged condition is indicated.

NOTE: On models with electric starter or starter-generator unit, battery can be assumed in satisfactory condition if the starter cranks the engine freely.

If battery checks within specifications, but starter unit will not turn engine, a faulty starter unit is indicated and ignition trouble may be caused by excessive current draw of such a unit. If battery and starting unit, if so equipped, are in satisfactory condition, proceed as follows:

Remove battery lead wire from ignition coil and connect a test light of same voltage as battery between disconnected lead wire and engine ground. Light should go on when ignition switch is in "on" position and go off when switch is in "off" position. If not, renew switch and/or wiring and recheck for satisfactory spark. If switch and wiring are funtioning properly, but no spark is obtained, proceed as follows:

BREAKER POINTS AND CONDENSER. Remove breaker box cover and, using small screwdriver, separate and inspect breaker points. If burned or deeply pitted, renew breaker points and condenser. If point contacts are clean to grayish in color and are only slightly pitted, proceed as follows: Disconnect condenser and ignition coil lead wires from breaker point terminal and connect a test light and battery between terminal and engine ground. Light should go on when points are closed and should go out when points are open. If light fails to go out when points are open, breaker arm insulation is defective and breaker points must be renewed. If light does not go on when points are in closed position, clean or renew breaker points. In some instances, new breaker point contact surfaces may have an oily or wax coating or have foreign material between the surfaces so proper contact is prevented. Check ignition timing and breaker point gap as outlined in appropriate engine repair section of this manual.

Connect test light and battery between condenser lead and engine ground; if light goes on, condenser is shorted out and should be renewed. Capacity of condenser can be checked if test instrument is available. It is usually good practice to renew condenser whenever new breaker points are being installed if tester is not available.

IGNITION COIL. If a coil tester is available, condition of coil can be checked. However, if tester is not available, a reasonably satisfactory performance test can be made as follows:

Disconnect high tension wire from spark plug. Turn engine so cam has allowed breaker points to close. With ignition switch on, open and close points with small screwdriver while holding high tension lead about 1/8 to 1/4-inch (3.18 to 6.35 mm) away from engine ground. A bright blue spark should snap across gap between spark plug wire and ground each time points are opened. If no spark occurs, or spark is weak and yellow-orange, renewal of ignition coil is indicated.

Sometimes, an ignition coil may perform satisfactorily when cold, but fail after engine has run for some time and coil is hot. Check coil when hot if this condition is indicated.

FLYWHEEL MAGNETO SERVICE FUNDAMENTALS

In servicing a flywheel magneto ignition system, the mechanic is concerned with troubleshooting, service adjustments and testing magneto components. The following paragraphs outline basic steps in servicing a flywheel type magneto. Refer to appropriate engine section for adjustment and test specifications for a particular engine.

Troubleshooting

If engine will not start and malfunction of ignition system is suspected, make the following checks to find cause of trouble.

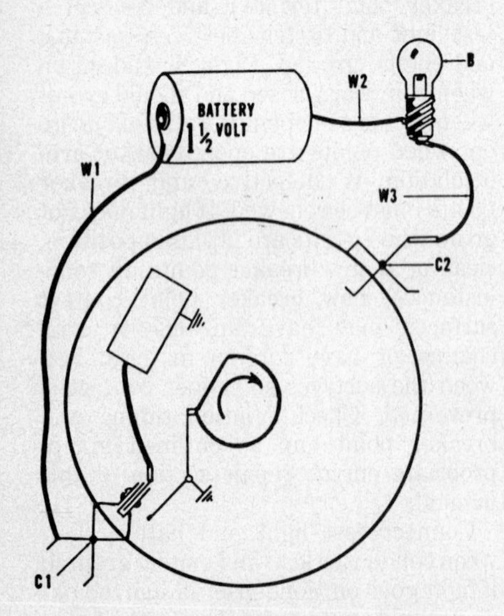

Fig. 5-3 – Drawing showing a simple test lamp for checking ignition timing and/or breaker point opening.

B. 1½ volt bulb
C1. Spring clamp
C2. Spring clamp
W1. Wire
W2. Wire
W3. Wire

Check to be sure ignition switch (if so equipped) is in "On" or "Run" position and the insulation on wire leading to ignition switch is in good condition. Switch can be checked with timing and test light as shown in Fig. 5–3. Disconnect lead from switch and attach one clip of test light to switch terminal and remaining clip to engine. Light should go on when switch is in "Off" or "Stop" position, and should go off when switch is in "On" or "Run" position.

Inspect high tension (spark plug) wire for worn spots in insulation or breaks in wire. Frayed or worn insulation can be repaired temporarily with plastic electrician's tape.

If no defects are noted in ignition switch or ignition wires, remove and inspect spark plug as outlined in SPARK PLUG SERVICING section. If spark plug is fouled or is in questionable condition, connect a spark plug of known quality to high tension wire, ground base of spark plug to engine and turn engine rapidly with starter. If spark across electrode gap of spark plug is a bright blue, magneto can be considered in satisfactory condition.

NOTE: Some engine manufacturers specify a certain type spark plug and a specific test gap. Refer to appropriate engine service section; if no specific spark plug type or electrode gap is recommended for test purposes, use spark plug type and electrode gap recommended for engine make and model.

If spark across gap of test plug is weak or orange colored, or no spark occurs as engine is cranked, magneto should be serviced as outlined in the following paragraphs.

Magneto Adjustments

BREAKER CONTACT POINTS. Adjustment of breaker contact points affects both ignition timing and magneto edge gap. Therefore, breaker contact point gap should be carefully adjusted according to engine manufacturer's specifications. Before adjusting breaker contact gap, inspect contact points and renew if condition of contact surfaces is questionable. It is sometimes desirable to check condition of points as follows: Disconnect condenser and primary coil leads from breaker point terminal. Attach one clip of a test light (See Fig. 5–3) to breaker point terminal and remaining clip of test light to magneto ground. Light should be out when contact points are open and should go on when engine is turned to close breaker contact points. If light stays on when points are open, insulation of breaker contact arm is defective. If light does not go on when points are closed, contact surfaces are dirty, oily or are burned.

Adjust breaker point gap as follows unless manufacturer specifies adjusting breaker gap to obtain correct ignition timing. First, turn engine so points are closed to be sure contact surfaces are in alignment and seat squarely. Then, turn engine so breaker point opening is maximum and adjust breaker gap to manufacturer's specification. A wire type feeler gage is recommended for checking and adjusting the breaker contact gap. Be sure to recheck gap after tightening breaker point base retaining screws.

IGNITION TIMING. On some engines, ignition timing is non-adjustable and a certain breaker point gap is specified. On other engines, timing is adjustable by changing position of magneto stator plate with a specified breaker point gap or by simply varying breaker point gap to obtain correct timing. Ignition timing is usually specified either in degrees of engine (crankshaft) rotation or in piston travel before piston reaches top dead center position. In some instances, a specification is given for ignition timing even though timing may be non-adjustable; if a check reveals timing is incorrect on these engines, it is an indication of incorrect breaker point adjustment or excessive wear of breaker cam. Also, on some engines, it may indicate a wrong breaker can has been installed or cam has been installed in a reversed position on engine crankshaft.

Some engines may have a timing mark or flywheel locating pin to locate flywheel at proper position for ignition spark to occur (breaker points begin to open). If not, it will be necessary to measure piston travel as illustrated in Fig. 5–4 or install a degree indicating device on engine crankshaft.

A timing light as shown in Fig. 5–3 is a valuable aid in checking or adjusting

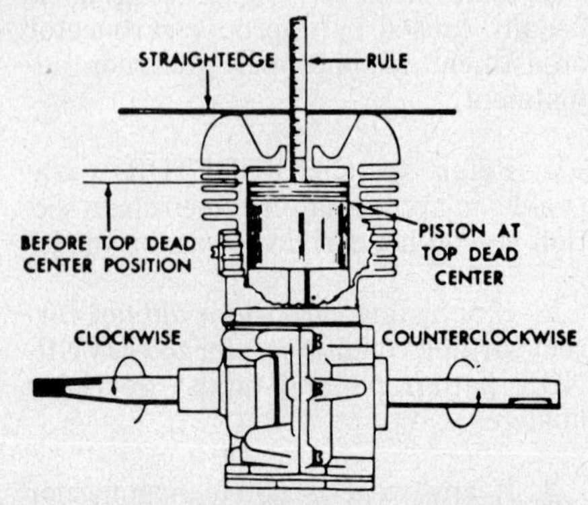

Fig. 5-4 – On some engines, it will be necessary to measure piston travel with rule, dial indicator or special timing gage when adjusting or checking ignition timing.

engine timing. After disconnecting ignition coil lead from breaker point terminal, connect leads of timing light as shown. If timing is adjustable by moving magneto stator plate, be sure breaker point gap is adjusted as specified. Then, to check timing, slowly turn engine in normal direction of rotation past point at which ignition spark should occur. Timing light should be on, then go out (breaker points open) just as correct timing location is passed. If not, turn engine to proper timing location and adjust timing by relocating magneto stator plate or varying breaker contact gap as specified by engine manufacturer. Loosen screws retaining stator plate or breaker points and adjust position of stator plate or points so points are closed (timing light is on). Then, slowly move adjustment until timing light goes out (points open) and tighten retaining screws. Recheck timing to be sure adjustment is correct.

ARMATURE AIR GAP. To fully concentrate magnetic field of flywheel magnets within armature core, it is necessary that flywheel magnets pass as closely to armature core as possible without danger of metal to metal contact. Clearance between flywheel magnets and legs of armature core is called armature air gap.

On magnetos where armature and high tension coil are located outside of the flywheel rim, adjustment of armature air gap is made as follows: Turn engine so flywheel magnets are located directly under legs of armature core and check clearance between armature core and flywheel magnets. If measured clearance is not within manufacturers specifications, loosen armature core mounting screws and place shims of thickness equal to minimum air gap specification between magnets and armature core (Fig. 5–5). The magnets will pull armature core against shim stocks. Tighten armature core mounting screws, remove shim stock and turn engine through several revolutions to be sure flywheel does not contact armature core.

Where armature core is located under or behind flywheel, the following methods may be used to check and adjust armature air gap: On some engines, slots or openings are provided in flywheel through which armature air gap can be checked. Some engine manufacturers provide a cut-away flywheel that can be installed temporarily for checking armature air gap. A test flywheel can be made out of a discarded flywheel (See Fig. 5–6), or out of a new flywheel if service volume on a particular engine warrants such expenditure. Another method of checking armature air gap is to remove flywheel and place a layer of plastic tape equal to minimum specified air gap over legs of armature core. Reinstall flywheel and turn engine through several revolutions and remove flywheel; no evidence of contact between flywheel magnets and plastic tape should be noticed. Then cover legs of armature core with a layer of tape of thickness equal to maximum specified air gap; then, reinstall flywheel and turn engine through several revolutions. Indication of the flywheel magnets contacting plastic tape should be noticed after flywheel is again removed. If magnets contact first thin layer of tape applied to armature core legs, or if they do not contact second thicker layer of tape, armature air gap is not within specifications and should be adjusted.

NOTE: Before loosening armature core mounting screws, scribe a mark on mounting plate against edge of armature core so that adjustment of air gap can be gaged.

In some instances, it may be necessary to slightly enlarge armature core mounting holes before proper air gap adjustment can be made.

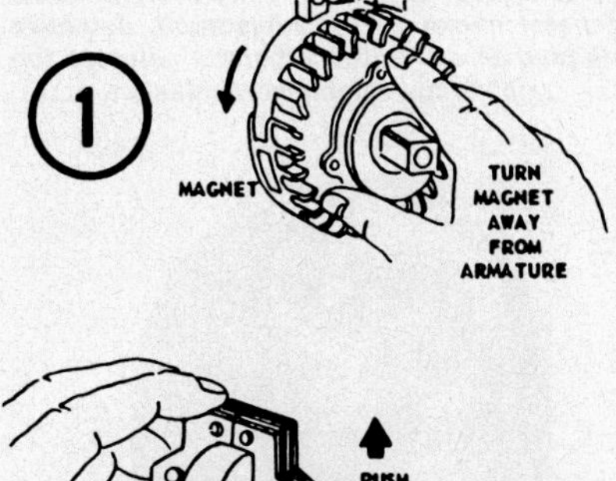

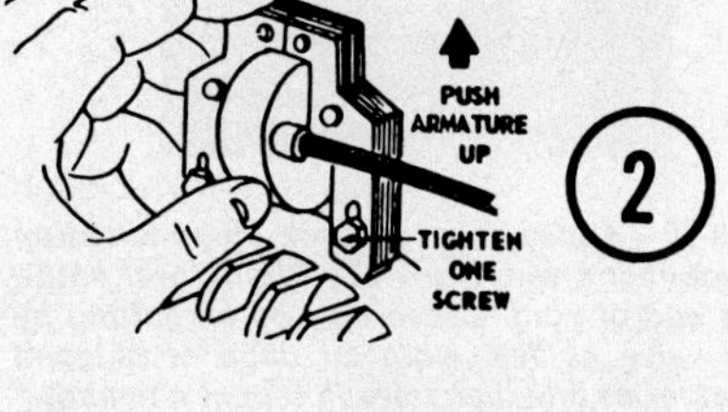

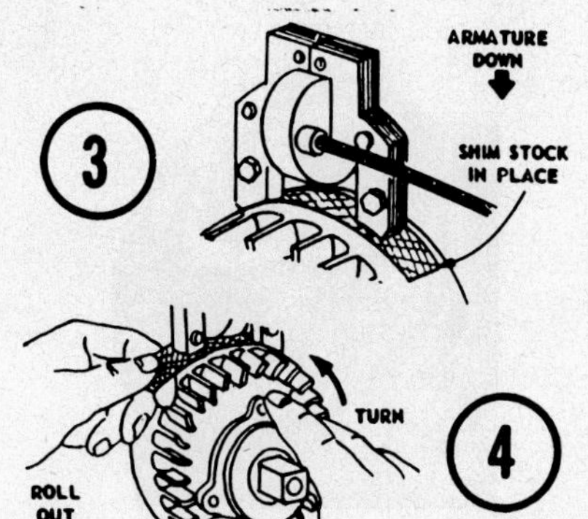

Fig. 5-5—Views showing adjustment of armature air gap when armature is located outside flywheel. Refer to Fig. 5-6 for engines having armature located inside flywheel.

MAGNETO EDGE GAP. The point of maximum acceleration of movement of flywheel magnetic field through high tension coil (and therefore, the point of maximum current induced in primary coil windings) occurs when trailing edge of flywheel magnet is slightly past left hand leg of armature core. The exact point of maximum primary current is determined by using electrical measuring devices, distance between trailing edge of flywheel magnet and leg of armature core at this point is measured and becomes a service specification. This distance, which is stated either in thousandths of an inch or in degrees of flywheel rotation, is called the Edge Gap or "E" Gap.

For maximum strength of ignition spark, breaker points should just start to open when flywheel magnets are at specified edge gap position. Usually, edge gap is non-adjustable and will be maintained at proper dimension if contact breaker points are adjusted to recommended gap and correct breaker cam is installed. However magneto edge gap can change (and spark intensity thereby reduced) due to the following:

a. Flywheel drive key sheared
b. Flywheel drive key worn (loose)
c. Keyway in flywheel or crankshaft worn (oversized)
d. Loose flywheel retaining nut which can also cause any above listed difficulty
e. Excessive wear on breaker cam
f. Breaker cam loose on crankshaft
g. Excessive wear on breaker point rubbing block or push rod so points cannot be properly adjusted.

Unit Type Magneto Service Fundamentals

Improper functioning of carburetor, spark plug or other components often

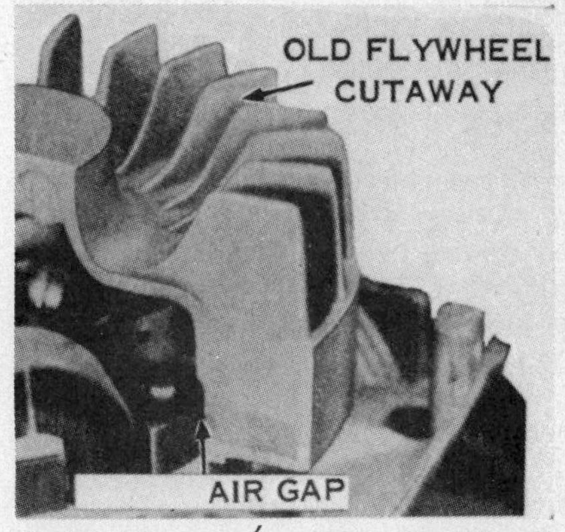

Fig. 5-6—Where armature core is located inside flywheel, check armature gap by using a cut-away flywheel unless other method is provided by manufacturer; refer to appropriate engine repair section. Where possible, an old discarded flywheel should be used to cut-away section for checking armature gap.

causes difficulties that are thought to be an improperly functioning magneto. Since a brief inspection will often locate other causes for engine malfunction, it is recommended one be certain magneto is at fault before opening magneto housing. Magneto malfunction can easily be determined by simple tests as outlined in following paragraph.

Troubleshooting

With a properly adjusted spark plug in good condition, ignition spark should be strong enough to bridge a short gap in addition to actual spark plug gap. With engine running, hold end of spark plug wire not more than 1/16-inch (1.59 mm) away from spark plug terminal. Engine should not misfire.

To test magneto spark if engine will not start, remove ignition wire from magneto end cap socket. Bend a short piece of wire so when it is inserted in end cap socket, other end is about ⅛-inch (3.18 mm) from engine casting. Crank engine slowly and observe gap between wire and engine; a strong blue spark should jump gap the instant impulse coupling trips. If a strong spark is observed, it is recommended magneto be eliminated as source of engine difficulty and spark plug, ignition wire and terminals be thoroughly inspected.

If, when cranking engine, impulse coupling does not trip, magneto must be removed from engine and coupling overhauled or renewed. It should be noted that if impulse coupling will not trip, a weak spark will occur.

Magneto Adjustments and Service

BREAKER POINTS. Breaker points are accessible for service after removing magneto housing end cap. Examine point contact surfaces for pitting or pyramiding (transfer of metal from one surface to the other); a small tungsten file or fine stone may be used to resurface points. Badly worn or badly pitted points should be renewed. After points are resurfaced or renewed, check breaker point gap with rotor turned so points are opened maximum distance. Refer to MAGNETO paragraph in appropriate engine repair section for point gap specifications.

When installing magneto end cap, both end cap and housing mating surfaces should be throughly cleaned and a new gasket be installed.

CONDENSER. Condenser used in unit type magneto is similar to that used in other ignition systems. Refer to MAGNETO paragraph in appropriate engine repair section for condenser test specifications. Usually, a new condenser should be installed whenever breaker points are being renewed.

COIL. The ignition coil can be tested without removing the coil from housing. Instructions provided with coil tester should have coil test specifications listed.

ROTOR. Usually, service on magneto rotor is limited to renewal of bushings or bearings, if damaged. Check to be sure rotor turns freely and does not drag or have excessive end play.

MAGNETO INSTALLATION. When installing a unit type magneto on an engine, refer to MAGNETO paragraph in appropriate engine repair section for magneto to engine timing information.

SOLID STATE IGNITION SERVICE FUNDAMENTALS

Because of differences in solid state ignition construction, it is impractical to outline a general procedure for solid state ignition service. Refer to specific engine section for testing, overhaul notes and timing of solid state ignition systems.

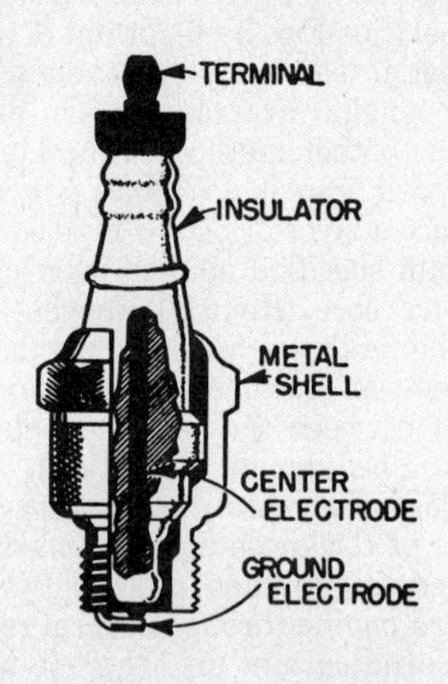

Fig. 5-7—Cross-sectional drawing of spark plug showing construction and nomenclature.

Fig. 5-8—Normal plug appearance in four-stroke cycle engine. Insulator is light tan to gray in color and electrodes are not burned. Renew plug at regular intervals as recommended by engine manufacturer.

SPARK PLUG SERVICING

ELECTRODE GAP. Spark plug electrode gap should be adjusted by bending the ground electrode. Refer to Fig. 5-7. Recommended gap is listed in SPARK PLUG paragraph in appropriate engine repair section of this manual.

CLEANING AND ELECTRODE CONDITIONING. Spark plugs are most usually cleaned by abrasive action commonly referred to as "sand blasting." Actually, ordinary sand is not used, but a special abrasive which is nonconductive to electricity even when melted, thus the abrasive cannot short

Fig. 5-9—Appearance of spark plug indicating cold fouling. Cause of cold fouling may be use of a too-cold plug, excessive idling or light loads, carburetor choke out of adjustment, defective spark plug wire or boot, carburetor adjusted too "rich" or low engine compression.

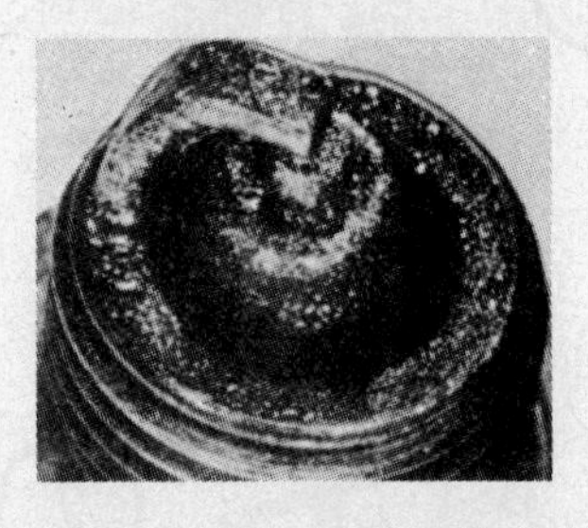

Fig. 5-10—Appearance of spark plug indicating wet fouling; a wet, black oily film is over entire firing end of plug. Cause may be oil getting by worn valve guides, worn oil rings or plugged breather or breather valve in tappet chamber.

Fig. 5-11—Appearance of spark plug indicating overheating. Check for plugged cooling fins, bent or damaged blower housing, engine being operated without all shields in place or other causes of engine overheating. Also can be caused by too lean a fuel-air mixture or spark plug not tightened properly.

out plug current. Extreme care should be used in cleaning plugs after sand blasting, however, as any particles of abrasive left on plug may cause damage to piston rings, piston or cylinder walls. Some engine manufacturers recommend spark plug be renewed rather than cleaned because of possible engine damage from cleaning abrasives.

After plug is cleaned by abrasive, and before gap is set, electrode surfaces between grounded and insulated electrodes should be cleaned and returned as nearly as possible to original shape by filing with a point file. Failure to properly dress electrodes can result in high secondary voltage requirements, and misfire of the plug.

PLUG APPEARANCE DIAGNOSIS. Appearance of a spark plug will be altered by use, and an examination of plug tip can contribute useful information which may assist in obtaining better spark plug life. Figs. 5-8 and 5-11 are provided by Champion Spark Plug Company to illustrate typical observed conditions. Listed in captions are probable causes and suggested corrective measures.

CARBURETOR SERVICING FUNDAMENTALS

The bulk of carburetor service consists of cleaning, inspection and adjustment. After considerable service it may become necessary to overhaul the carburetor and renew worn parts to restore original operating efficiency. Although carburetor condition affects engine operating economy and power, ignition and engine compression must also be considered to determine and correct causes of poor performance.

Before dismantling carburetor for cleaning or overhaul, clean all external surfaces and remove accumulated dirt and grease. Refer to appropriate engine repair section for carburetor exploded or cross-sectional views. Dismantle carburetor and note any discrepancies to assure correction during overhaul. Thoroughly clean all parts and inspect for damage or wear. Wash jets and passages and blow clear with clean, dry compressed air. Do not use a drill or wire to clean jets as possible enlargement of calibrated holes will disturb operating balance. Measurement of jets to determine extent of wear is difficult and new parts are usually installed to assure satisfactory results.

Carburetor manufacturers provide for many of their models an assortment of gaskets and other parts usually needed to do a correct job of cleaning and overhaul. These assortments are usually catalogued as Gasket Kits and Overhaul Kits respectively.

On float type carburetors, inspect float pin and needle valve for wear and renew if necessary. Check metal floats for leaks and where a dual type float is installed, check alignment of float sections. Check cork floats for loss of protective coating and absorption of fuel.

NOTE: Do not attempt to recoat cork floats with shellac or varnish or to resolder leaky metal floats. Renew part if defective.

Check fit of throttle and choke valve shafts. Excessive clearance will cause improper valve plate seating and will permit dust or grit to be drawn into engine. Air leaks at throttle shaft bores due to wear will upset carburetor calibration and contribute to uneven engine operation. Rebush valve shaft holes where necessary and renew dust seals. If rebushing is not possible, renew body part supporting shaft. Inspect throttle and choke valve plates for proper installation and condition.

Power or idle adjustment needles must not be worn or grooved. Check condition of needle seal packing or "O" ring and renew packing or "O" ring if necessary.

Reinstall or renew jets, using correct size listed for specific model. Adjust power and idle settings as described for specific carburetors in engine service section of manual.

It is important that carburetor bore at idle discharge ports and in vicinity of throttle valve be free of deposits. A partially restricted idle port will produce a "flat spot" between idle and mid-range rpm. This is because the restriction makes it necessary to open throttle wider than the designed opening to obtain proper idle speed. Opening throttle wider than the design specified amount will uncover more of the port than was intended in calibration of carburetor. As a result an insufficient amount of the port will be available as a reserve to cover transition period (idle to mid-range rpm) when the high speed system begins to function.

When reassembling float type carburetors, be sure float position is properly adjusted. Refer to CARBURETOR paragraph in appropriate engine repair section for float level adjustment specifications.

ENGINE SERVICE

DISASSEMBLY AND ASSEMBLY

Special techniques must be developed in repair of engines of aluminum alloy or magnesium alloy construction. Soft threads in aluminum or magnesium castings are often damaged by carelessness in overtightening fasteners or in attempting to loosen or remove seized fasteners. Manufacturer's recommended torque values for tightening screw fasteners should be followed closely.

NOTE: If damaged threads are encountered, refer to following paragraph, "REPAIRING DAMAGED THREADS."

A given amount of heat applied to aluminum or magnesium will cause it to expand a greater amount than will steel under similar conditions. Because of different expansion characteristics, heat is usually recommended for easy installation of bearings, pins, etc., in aluminum or magnesium castings. Sometimes, heat can be used to free parts that are seized or where an interference fit is used. Heat, therefore, becomes a service tool and the application of heat one of the required service techniques. An open flame is not usually advised because it destroys paint and other protective coatings and because a uniform and controlled temperature with open flame is difficult to obtain. Methods commonly used are heating in oil or water, with a heat lamp, electric hot plate, electric hot air gun, or in an oven or kiln. The use of water or oil gives a fairly accurate temperature control but is somewhat limited as to the size and type of part that can be handled. Thermal crayons are available which can be used to determine temperature of a heated part. These crayons melt when the part reaches a specified temperature, and a number of crayons for different temperatures are available. Temperature indicating crayons are usually available at welding equipment supply houses.

Use only specified gaskets when reassembling, and use an approved gasket cement or sealing compound unless otherwise stated. Seal all exposed threads and repaint or retouch with an approved paint.

REPAIRING DAMAGED THREADS

Damaged threads in castings can be renewed by use of thread repair kits which are recommended by a number of equipment and engine manufacturers. Use of thread repair kits is not difficult, but instructions must be carefully

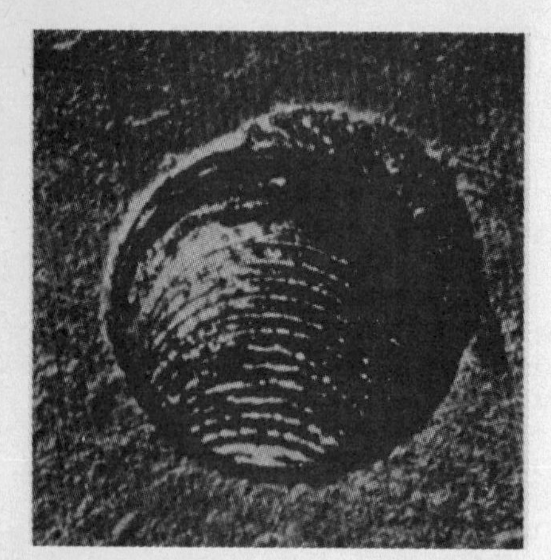

Fig. 5-12—Damaged threads in casting before repair. Refer to Figs. 5-13, 5-14 and 5-15 for steps in installing thread insert. (Series of photos provided by Heli-Coil Corp., Danbury, Conn.)

Fig. 5-13—First step in repairing damaged threads is to drill out old threads using exact size drill recommended in instructions provided with thread repair kit. Drill all the way through an open hole or all the way to bottom of blind hole, making sure hole is straight and that centerline of hole is not moved in drilling process.

Fig. 5-14—Special drill taps are provided in thread repair kit for threading drilled hole to correct size for outside of thread insert. A standard tap cannot be used.

followed. Refer to Figs. 5–12 through 5–15 which illustrate the use of Heli-Coil thread repair kits that are manufactured by the Heli-Coil Corporation, Danbury, Connecticut.

Heli-Coil thread repair kits are available through parts departments of most engine and equipment manufacturers; thread inserts are available in all National Coarse (USS) sizes from #4 to 1½ inch and National Fine (SAE) sizes from #6 to 1½ inch. Also, sizes for repairing 14mm and 18mm spark plug ports are available.

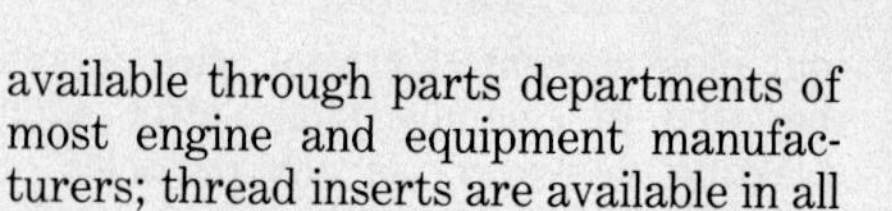

Fig. 5-15—A thread insert and a completed repair are shown above. Special tools are provided in thread repair kit for installation of thread insert.

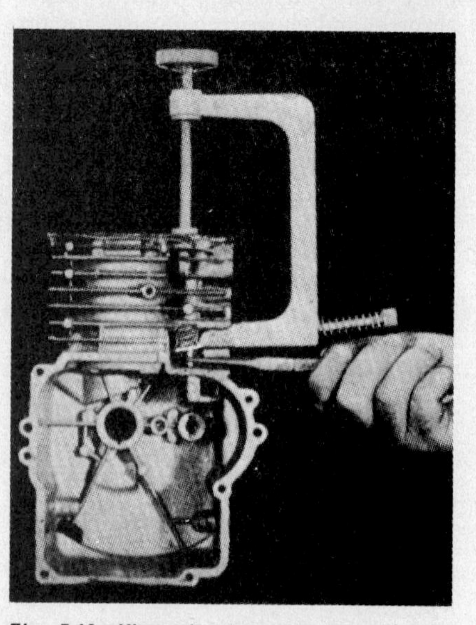

Fig. 5-16—View showing one type of valve spring compressor being used to remove keeper. (Block is cut-away to show valve spring.)

VALVE SERVICE FUNDAMENTALS

When overhauling engines, obtaining proper valve sealing is of primary importance. The following paragraphs cover fundamentals of servicing intake and exhaust valves, valve seats and valve guides.

REMOVING AND INSTALLING VALVES. A valve spring compressor, one type of which is shown in Fig. 5–16, is a valuable aid in removing and installing intake and exhaust valves. This tool is used to hold spring compressed while removing or installing pin, collars or retainer from valve stem. Refer to Fig. 5–17 for views showing some of the different methods of retaining valve spring to valve stem.

VALVE REFACING. If valve face (See Fig. 5–18) is slightly worn, burned or pitted, valve can usually be refaced providing proper equipment is available. Many shops will usually renew valves, however, rather than invest in somewhat costly valve refacing tools.

Before attempting to reface a valve, refer to specifications in appropriate engine repair section for valve face angle. On some engines, manufacturer recommends grinding the valve face to an angle of ½° to 1° less than that of the valve seat. Refer to Fig. 5–19. Also, nominal valve face angle may be either 30° or 45°.

After valve is refaced, check thickness of valve "margin" (See Fig. 5–18). If margin is less than manufacturer's minimum specification (refer to specifications in appropriate engine repair section), or is less than one-half the margin of a new valve, renew valve. Valves having excessive material removed in refacing operation will not give satisfactory service.

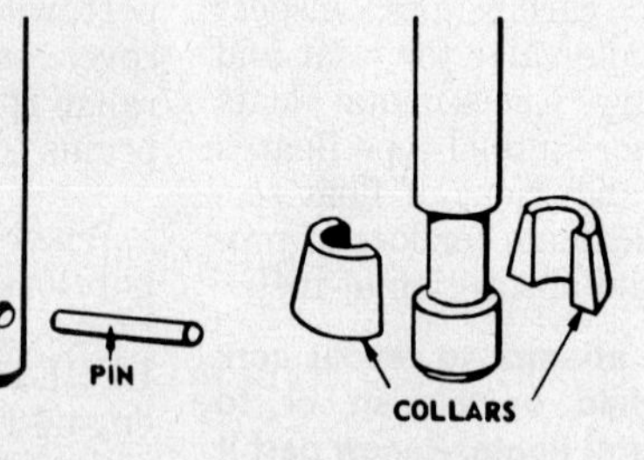

Fig. 5-17—Drawing showing three types of valve spring keepers used.

When refacing or renewing a valve, the seat should also be reconditioned, or in engines where valve seat is renewable, a new seat should be installed. Refer to following paragraph "RESEATING OR RENEWING VALVE SEATS." Then, the seating surfaces should be lapped in using a fine valve grinding compound.

RESEATING OR RENEWING VALVE SEATS. On engines having valve seat machined in cylinder block casting, seat can be reconditioned by using a correct angle seat grinding stone or valve seat cutter. When reconditioning valve seat, care should be taken that only enough material is removed to provide a good seating on valve contact surface. The width of seat should then be measured (See Fig. 5–20) and if width exceeds manufacturer's maximum specifications, seat should be narrowed by using one stone or cutter with an angle 15° greater than seat angle and a second stone or cutter with an angle 15° less than seat angle. When narrowing seat, coat seat lightly with Prussian blue and check where seat contacts valve face by inserting valve in guide and rotating valve lightly against seat. Seat should contact approximately center of valve face. By using only narrow angle seat narrowing stone or cutter, seat contact will be moved towards outer edge of valve face.

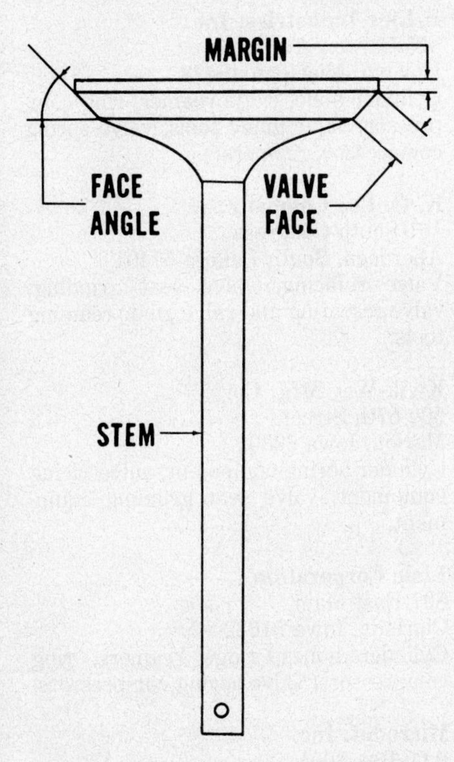

Fig. 5-18 – Drawing showing typical four-stroke cycle engine valve. Face angle is usually 30° or 45°. On some engines, valve face is ground to an angle of ½ or 1 degree less than seat angle.

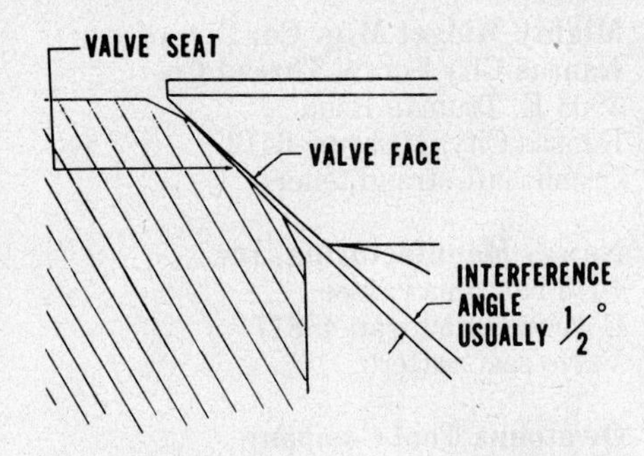

Fig. 5-19 – Drawing showing line contact of valve face with valve seat when valve face is ground at smaller angle than valve seat; this is specified on some engines.

On engines having renewable valve seats, refer to appropriate engine repair section in this manual for recommended method of removing old seat and installing new seat. Refer to Fig. 5–21 for one method of installing new valve seats. Seats are retained in cylinder block bore by an interference fit; that is, seat is slightly larger than bore in block. It sometimes occurs that valve seat will become loose in bore, especially on engines with aluminum crankcase. Some manufacturers provide oversize valve seat inserts (insert O.D. larger than standard part) so that if standard size insert fits loosely, bore can be cut oversize and a new insert be tightly installed. After installing valve seat insert in engines of aluminum construction, metal around seat should be peened as shown in Fig. 5–22. Where a loose insert is encountered and an oversize insert is not available, loose insert can usually be tightened by center-punching cylinder block material at three equally spaced points around insert, then peening completely around insert as shown in Fig. 5–22.

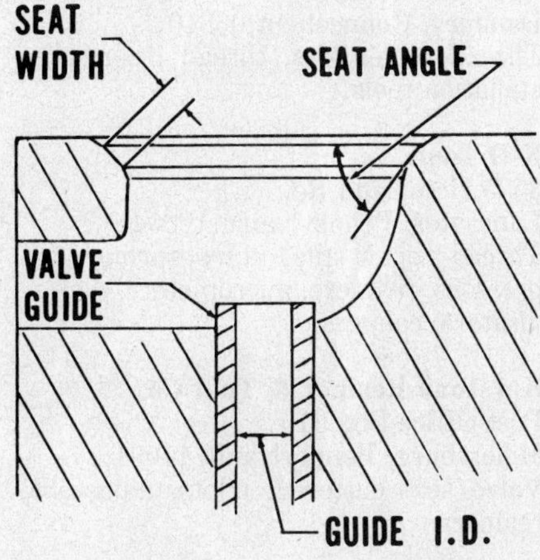

Fig. 5-20 – Cross-sectional drawing of typical valve seat and valve guide as used on some engines. Valve guide may be integral part of cylinder block; on some models so constructed, valve guide I.D. may be reamed out and an oversize valve stem installed. On other models, a service guide may be installed after counterboring cylinder block.

For some engines with cast iron cylinder blocks, a service valve seat insert is available for reconditioning valve seat, and is installed by counterboring cylinder block to specified dimensions, then driving insert into place. Refer to appropriate engine repair section in this manual for information on availability and installation of service valve seat inserts for cast iron engines.

INSTALLING OVERSIZE PISTON AND RINGS

Some engine manufacturers have over-size piston and ring sets available for use in repairing engines in which cylinder bore is excessively worn and standard size piston and rings cannot be

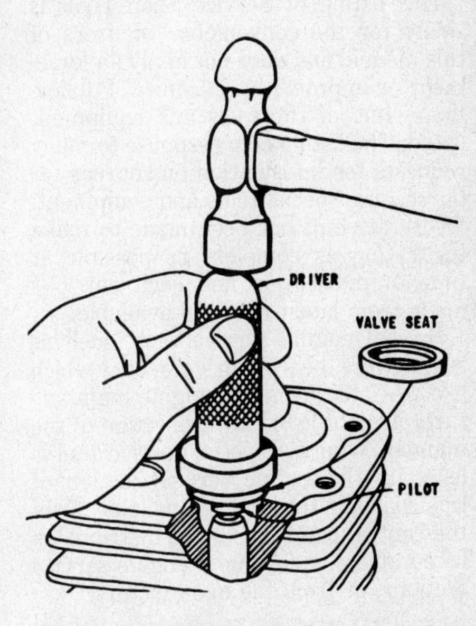

Fig. 5-21 – View showing one method used to install valve seat insert. Refer to appropriate engine repair section for manufacturer's recommended method.

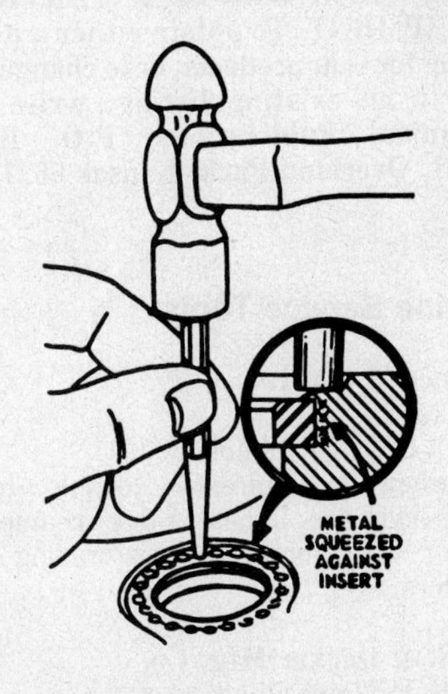

Fig. 5-22 – It is usually recommended that on aluminum block engines, metal be peened around valve seat insert after insert is installed.

used. If care and approved procedure are used in oversizing cylinder bore, installation of an oversize piston and ring set should result in a highly satisfactory overhaul.

Cylinder bore may be oversized by using either a boring bar or a hone; however, if a boring bar is used, it is usually recommended cylinder bore be finished with a hone. Refer to Fig. 5-23.

Where oversize piston and rings are available, it will be noted in appropriate engine repair section of this manual. Also, the standard bore diameter will also be given. Before attempting to rebore or hone the cylinder to oversize, carefully measure the cylinder bore to be sure standard size piston and rings will not fit within tolerance. Also, it may be possible cylinder is excessively worn or damaged and reboring or honing to largest oversize will not clean up worn or scored surface.

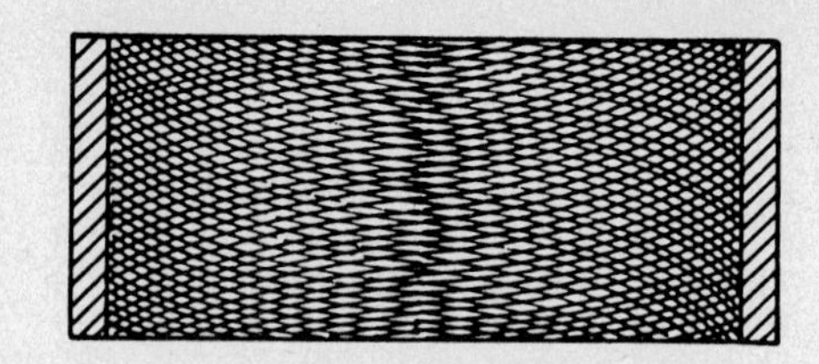

Fig. 5-23—A cross-hatch pattern as shown should be obtained when honing cylinder. Pattern is obtained by moving hone up and down cylinder bore as it is being turned by slow speed electric drill.

SERVICE SHOP TOOL BUYER'S GUIDE

This listing of Service Shop Tools is solely for the convenience of users of this manual and does not imply endorsement or approval by Technical Publications, Inc. of the tools and equipment listed. The listing is in response to many requests for information on sources for purchasing special tools and equipment. Every attempt has been made to make the listing as complete as possible at time of publication and each entry is made from latest material available.

Special engine service tools such as seal drivers, bearing drivers, etc., which are available from the engine manufacturer are not listed in this section of the manual. Where a special service tool is listed in the engine service section of this manual, the tool is available from the central parts or service distributors listed at the end of most engine service sections, or from the manufacturer.

NOTE TO MANUFACTURERS AND NATIONAL SALES DISTRIBUTORS OF ENGINE SERVICE TOOLS AND RELATED SERVICE EQUIPMENT. To obtain either a new listing for your products, or to change or add to an existing listing, write to Technical Publications, P.O. Box 12901, Overland Park, Kansas 66212.

Engine Service Tools

Ammco Tools, Inc.
Wacker Park
North Chicago, Illinois 60064
Valve spring compressor, torque wrenches, cylinder hones, ridge reamers, piston ring compressors, piston ring expanders.

Black & Decker Mfg. Co.
701 East Joppa Road
Towson, Maryland 21204
Valve grinding equipment.

Bloom, Inc.
Route Four, Hiway 20 West
Independence, Iowa 50644
Engine repair stand with crankshaft straightening attachment.

Brush Research Mfg. Inc.
4642 East Floral Drive
Los Angeles, California 90022
Cylinder hones.

E-Z Lok
P.O. Box 2069
Gardena, California 90247
Thread repair insert kits for metal, wood and plastic.

Frederick Mfg. Co., Inc.
1400 C., Agnes Avenue
Kansas City, Missouri 64127
Crankshaft straightener.

Heli-Coil Products Division
Heli-Coil Corporation
Shelter Rock Lane
Danbury, Connecticut 06810
Thread repair kits, thread inserts, installation tools.

K-D Tools
3575 Hempland Rd.
Lancaster, Pennsylvania 17604
Thread repair kits, valve spring compressors, reamers, micrometers, dial indicators, calipers.

Keystone Reamer & Tool Co.
Post Office Box 310
Millersburg, Pennsylvania 17061
Valve seat cutter & pilots, adjustable reamers.

Ki-Sol Corporation
100 Larkin Williams Ind. Court
Fenton, Missouri 63026
Cylinder hone, ridge reamer, ring compressor, ring expander, ring groove cleaner, torque wrenches, valve spring compressor, valve refacing equipment.

K-Line Industries, Inc.
315 Garden Avenue
Holland, Michigan 49423
Cylinder hone, ridge reamer, ring compressor, valve guide tools, valve spring compressor, reamers.

K. O. Lee Company
101 South Congress
Aberdeen, South Dakota 57401
Valve refacing, valve seat grinding, valve reseating and valve guide reaming tools.

Kwik-Way Mfg. Co.
500 57th Street
Marion, Iowa 52302
Cylinder boring equipment, valve facing equipment, valve seat grinding equipment.

Lisle Corporation
807 East Main
Clarinda, Iowa 51632
Cylinder hones, ridge reamers, ring compressors, valve spring compressors.

Microdot, Inc.
P.O. Box 3001
800 So. State College Blvd.
Fullerton, California 92631
Thread repair insert kits.

Mighty Midget Mfg. Co. Div. of Kansas City Screw Thread Co.
2908 E. Truman Road
Kansas City, Missouri 64127
Crankshaft straightener.

Neway Manufacturing, Inc.
1013 No. Shiawassee
Corunna, Michigan 48817
Valve seat cutters.

Owatonna Tool Company
436 Eisenhower Drive
Owatonna, Minnesota 55060
Valve tools, spark plug tools, piston ring tools, cylinder hones.

Precision Manufacturing & Sales Co., Inc.
2140 Range Rd., P.O. Box 149
Clearwater, Florida 33517
Cylinder boring equipment, measuring instruments, valve equipment, hones, porting, hand tools, test equipment, threading, presses, parts washers, milling machines, lathes, drill presses, glass beading machines, dynos, safety equipment.

Sunnen Product Company
7910 Manchester Avenue
Saint Louis, Missouri 63143
Cylinder hones, rod reconditioning, valve guide reconditioning.

Rexnord
Specialty Fastener Div.
3000 W. Lomita Blvd.
Torrance, California 90505
Thread repair insert kits (Keenserts) and installation tools.

Vulcan Tools Div. of TRW, Inc.
2300 Kenmore Avenue
Buffalo, New York 14207
Cylinder hones, reamers, ridge removers, valve spring compressors, valve spring testers, ring compressor, ring groove cleaner.

Waters-Dove Manufacturing
Post Office Box 40
Skiatook, Oklahoma 74070
Crankshaft straightener, oil seal remover.

Test Equipment and Gages

Allen Test Products
2101 North Pitcher Street
Kalamazoo, Michigan 49007
Coil and condenser testers, compression gages.

Applied Power, Inc., Auto Div.
P.O. Box 27207
Milwaukee, Wisconsin 53227
Compression gage, condenser tester, tachometer, timing light, ignition analyzer.

AW Dynamometer, Inc.
131½ East Main Street
Colfax, Illinois 61728
Engine test dynamometer.

B. C. Ames Company
131 Lexington
Waltham, Massachusetts 02254
Micrometer dial gages and indicators.

The Bendix Corporation
Engine Products Div.
Delaware Avenue
Sidney, New York 13838
Condenser tester, magneto test equipment, timing light.

Burco
208 Delaware Ave.
Delmar, New York 12054
Coil and condenser tester, compression gage, carburetor tester, grinders, Pow-R-Arms, chain breakers, rivet spinners.

Dixson, Inc.
Post Office Box 1449
Grand Junction, Colorado 81501
Tachometer, compression gage, timing light.

Fox Valley Instrument Co.
Route 5, Box 390
Cheboygan, Michigan 49721
Coil and condenser tester, ignition actuated tachometer.

Graham-Lee Electronics, Inc.
4220 Central Avenue NE
Minneapolis, Minnesota 55421
Coil and condenser tester.

K-D Tools
3575 Hempland Rd.
Lancaster, Pennsylvania 17604
Diode tester & installation tools, compression gage, timing light, timing gages.

Ki-Sol Corporation
100 Larkin Williams Ind. Court
Fenton, Missouri 63026
Micrometers, telescoping gages, compression gages, cylinder gages.

K-Line Industries, Inc.
315 Garden Avenue
Holland, Michigan 49423
Compression gage, tachometers.

Merc-O-Tronic Instruments Corporation
215 Branch Street
Almont, Michigan 48003
Ignition analyzers for conventional, solid state and magneto systems, electric tachometers, electronic tachometer and dwell meter, power timing lights, ohmmeters, compression gages, mechanical timing devices.

Nichoff Automotive Div. TRW Inc.
4925 Lawrence Avenue
Chicago, Illinois 60630
Timing light, compression gage, vacuum gage, tachometer, ignition analyzer.

Owatonna Tool Company
436 Eisenhower Drive
Owatonna, Minnesota 55060
Feeler gages, hydraulic test gages.

Prestolite Electronics Div.
An Allied Company
Post Office Box 931
Toledo, Ohio 43694
Magneto Test Plug

Simpson Electric Company
853 Dundee Avenue
Elgin, Illinois 60120
Electrical and electronic test equipment.

L.S. Starrett Company
1-165 Crescent Street
Athol, Massachusetts 01331
Micrometers, dial gages, bore gages, feeler gages.

Stevens Instrument Company, Inc.
Post Office Box 193
Waukegan, Illinois 60085
Ignition analyzers, timing lights, tachometers, volt-ohmmeters, spark checkers, CD ignition testers.

Stewart-Warner Corporation
1826 Diversey Parkway
Chicago, Illinois 60614
Compression gage, ignition tachometer, timing light ignition analyzer.

P. A. Sturtevant Co.,
Div. Dresser Ind.
3201 North Wolf Road
Franklin Park, Illinois 60131
Torque wrenches, torque multipliers, torque analyzers.

Sun Electric Corporation
Instrument Products Div.
1560 Trimble Road
San Jose, California 95131
Compression gage, hydraulic test gages, coil and condenser tester, ignition tachometer, ignition analyzer.

Westberg Mfg. Inc.
3400 Westach Way
Sonoma, California 95476
Ignition tachometer, magneto test equipment, ignition system analyzer.

Shop Tools and Equipment

A-C Delco Div., General Motors Corp.
400 Renaissance Center
Detroit, Michigan 48243
Spark plug cleaning and test equipment

Applied Power, Inc., Auto. Div.
Box 27207
Milwaukee, Wisconsin 53227
Arc and gas welding equipment, welding rods, accessories, battery service equipment.

Black & Decker Mfg. Co.
701 East Joppa Road
Towson, Maryland 21204
Air and electric powered tools.

Bloom, Inc.
Route 4, Hiway 20 West
Independence, Iowa 50644
Lawn mower repair bench with built-in engine cranking mechanism.

Campbell Chain Division
McGraw-Edison Co.
Post Office Box 3056
York, Pennsylvania 17402
Chain type engine and utility slings.

Champion Pneumatic Machinery Co.
1301 No. Euclid Ave.
Princeton, Illinois 61356
Air compressors.

Champion Spark Plug Co.
Post Office Box 910
Toledo, Ohio 43661
Spark plug cleaning and testing equipment, gap tools and wrenches.

Chicago Pneumatic Tool Co.
2200 Bleecker St.
Utica, New York 13503
Air impact wrenches, air hammers, air drills and grinders, nutrunners, speed ratchets.

Clayton Manufacturing Company
415 North Temple City Boulevard
El Monte, California 91731
Steam cleaning equipment.

E-Z Lok
P.O. Box 2069
240 E. Rosecrans Ave.
Gardena, California 90247
Thread repair insert kits for metal, wood and plastic.

G & H Products, Inc.
Post Office Box 770
St. Paris, Ohio 43027
Motorized lawnmower stand.

General Scientific Equipment Company
Limekiln Pike & Williams Ave.
Box 27309
Philadelphia, Pennsylvania 19118-0255
Safety equipment.

Graymills Corporation
3705 North Lincoln Avenue
Chicago, Illinois 60613
Parts washing stand.

Heli-Coil Products Div.,
Heli-Coil Corp.
Shelter Rock Lane
Danbury, Connecticut 06810
Thread repair kits, thread inserts and installation tools.

Ingersoll-Rand
253 E. Washington Ave.
Washington, New Jersey 07882
Air and electric impact wrenches, electric drills and screwdrivers.

Jaw Manufacturing Co.
39 Mulberry Street, P.O. Box 213
Reading, Pennsylvania 19603
Files for renewal of damaged threads, hex nut size rethreader dies, flexible shaft drivers and extensions, screw extractors, impact drivers.

Jenny Div. of Homestead Ind., Inc.
Box 348
Coraopolis, Pennsylvania 15108-0348
Steam cleaning equipment, pressure washing equipment.

Keystone Reamer & Tool Co.
Post Office Box 310
Millersburg, Pennsylvania 17061
Adjustable reamers, twist drills, tape, dies, etc.

K-Line Industries, Inc.
315 Garden Avenue
Holland, Michigan 49423
Air and electric impact wrenches.

Microdot, Inc.
P.O. Box 3001
800 So. State College Blvd.
Fullerton, California 92631
Thread repair insert kits.

Ingersoll-Rand Co.
Deerfield Industrial Park
South Deerfield, Massachusetts 01373
Impact wrenches, portable electric tools, battery powered ratchet wrench.

Owatonna Tool Company
436 Eisenhower Drive
Owatonna, Minnesota 55060
Bearing and gear pullers, hydraulic shop presses.

Proto Tool Div., Ingersoll-Rand
2600 East Nutwood Avenue
Fullerton, California 92631
Torque wrenches, gear & bearing pullers.

Shure Manufacturing Corp.
1601 South Hanley Road
Saint Louis, Missouri 63144
Steel shop benches, desks, engine overhaul stand.

Sioux Tools, Inc.
2801-2999 Floyd Blvd.
Sioux City, Iowa 51102
Portable air and electric tools.

Rexnord Specialty Fastener Div.
3000 W. Lomita Blvd.
Torrance, California 90505
Thread repair insert kits (Keenserts) and installation tools.

Vulcan Tools
2300 Kenmore Avenue
Buffalo, New York 14207
Air and electric impact wrenches.

Sharpening and Maintenance Equipment for Small Engine Powered Implements

Bell Industries, Saw & Machine Div.
Post Office Box 2510
Eugene, Oregon 97402
Saw chain grinder.

Desa Industries, Inc.
25000 South Western Ave.
Park Forest, Illinois 60466
Saw chain breakers and rivet spinners, chain files, file holder and filing guide.

Foley-Belsaw Company
6301 Equitable Rd.
P.O. Box 593
Kansas City, Missouri 64141
Circular, band and hand saw filers, heavy duty grinders, saw setters, retoothers, circular saw vises, lawn mower sharpener, saw chain sharpening and repair equipment.

Granberg Industries
200 S. Garrard Blvd.
Richmond, California 94804
Saw chain grinder, file guides, chain breakers and rivet spinners, chain saw lumber cutting attachments.

Ki-Sol Corporation
100 Larkin Williams Ind. Court
Fenton, Missouri 63026
Mower blade balancer.

Magna-Matic Div.,
A. J. Karrels Co.
Box 348
Port Washington, Wisconsin 53074
Rotary mower blade balancer, "track" checking tool.

Omark Industries, Inc.
4909 International Way
Portland, Oregon 97222
Saw chain and saw bars, maintenance equipment, to include file holders, filing vises, rivet spinners, chain breakers, filing depth gages, bar groove gages, electric chain saw sharpeners.

S.I.P. Grinding Machines
American Marsh Pumps
P.O. Box 23038
722 Porter St.
Lansing, Michigan 48909
Lawn mower sharpeners, saw chain grinders, lapping machine.

Specialty Motors Mfg.
641 California Way, P.O. Box 157
Longview, Washington 98632
Chain saw bar rebuilding equipment.

Mechanic's Hand Tools

Channellock, Inc.
1306 South Main Street
Meadville, Pennsylvania 16335

John H. Graham & Company, Inc.
617 Oradell Avenue
Oradell, New Jersey 07649

Jaw Manufacturing Company
39 Mulberry Street
Reading, Pennsylvania 19603

K-D Tools
3575 Hempland Rd.
Lancaster, Pennsylvania 17604

K-Line Industries, Inc.
315 Garden Avenue
Holland, Michigan 49423

Millers Falls Div., Ingersoll-Rand
Deerfield Industrial Park
South Deerfield, Massachusetts 01373

New Britain Tool Company
Division of Litton Industrial Products
P.O. Box 12198
Research Triangle Park, N.C. 27709

Owatonna Tool Company
436 Eisenhower Drive
Owatonna, Minnesota 55060

Proto Tool Div., Ingersoll-Rand
2600 East Nutwood Avenue
Fullerton, California 92631

Snap-On Tools
2801 80th Street
Kenosha, Wisconsin 53140

Triangle Corporation—Tool Division
Cameron Road
Orangeburg, South Carolina 29115

Vulcan Tools Div. of TRW, Inc.
2300 Kenmore Avenue
Buffalo, New York 14207

J. H. Williams Div. of TRW, Inc.
400 Vulcan Street
Buffalo, New York 14207

Shop Supplies (Chemicals, Metallurgy Products, Seals, Sealers, Common Parts Items, etc.)

ABEX Corp.
Amsco Welding Products
Fulton Industrial Park
3-9610-14
P.O. Box 258
Wauseon, OH 43567
Hardfacing alloys.

Atlas Tool & Manufacturing Co.
7100 S. Grand Ave.
Saint Louis, Missouri 63111
Rotary mower blades.

Bendix Automotive Aftermarket
1094 Bendix Drive, Box 1632
Jackson, Tennessee 38301
Cleaning chemicals.

CR Industries
900 North State Street
Elgin, Illinois 60120
Shaft & bearing seals.

Clayton Manufacturing Company
415 North Temple City Blvd.
El Monte, California 91731
Steam cleaning compounds and solvents.

E-Z Lok
P.O. Box 2069, 240 E. Rosecrans Ave.
Gardena, California 90247
Thread repair insert kits for metal, wood and plastic.

Eutectic Welding Alloys Corp.
40-40 172nd Street
Flushing, New York 11358
Specialized repair and maintenance welding alloys.

Frederick Manufacturing Co., Inc.
1400 C Agnes Street
Kansas City, Missouri 64127
Throttle controls and parts.

Heli-Coil Products Division
Heli-Coil Corporation
Shelter Rock Lane
Danbury, Connecticut 06810
Thread repair kits, thread inserts and installation tools.

King Cotton Cordage
617 Oradell Ave.
Oradell, New Jersey 07649
Starter rope, nylon cord.

Loctite Corporation
705 North Moutain Road
Newington, Connecticut 06111
Thread locking compounds, bearing mounting compounds, retaining compounds and sealants, instant and structural adhesives.

McCord Gasket Division
Ex-Cell-O Corporation
2850 West Grand Boulevard
Detroit, Michigan 48202
Gaskets, seals.

Microdot, Inc.
P.O. Box 3001
800 So. State College Blvd.
Fullerton, California 92631
Thread repair insert kits.

Niehoff Automotive Div. TRW Inc.
4925 Lawrence Ave.
Chicago, Illinois 60630
Ignition parts and ignition test equipment.

Permatex Industrial
705 N. Mountain Rd.
Newington, Connecticut 06111
Cleaning chemicals, gasket sealers, pipe sealants, adhesives, lubricants.

Radiator Specialty Co.
Box 34689
Charlotte, North Carolina 28234
Cleaning chemicals (Gunk), and solder seal.

Rexnord Specialty Fastener Div.
3000 Lomita Blvd.
Torrance, California 90506
Thread repair insert kits (Keenserts) and installation tools.

Union Carbide Coporation
Home & Auto Products Division
Old Ridgebury Rd.
Danbury, CT 06817
Cleaning chemicals, engine starting fluid.

BRIGGS & STRATTON

BRIGGS & STRATTON CORPORATION
Milwaukee, Wisconsin 53201

BRIGGS & STRATTON ENGINE IDENTIFICATION INFORMATION

In order to obtain correct service parts for Briggs & Stratton engines it is necessary to correctly identify engine model or series and provide engine serial number.

Briggs & Stratton model or series number also provides information concerning important mechanical features or optional equipment.

Refer to Fig. B1 for an explanation of each digit in relation to engine identification or description of mechanical features and options.

As an example, a 401417 series model number is broken down in the following manner.

40 – Designates 40 cubic inch displacement.
1 – Designates design series 1.
4 – Designates horizontal shaft, Flo-Jet carburetor and mechanical governor.
1 – Designates flange mounting with plain bearings.
7 – Designates electric starter, 12 volt gear drive with alternator.

	FIRST DIGIT AFTER DISPLACEMENT	SECOND DIGIT AFTER DISPLACEMENT	THIRD DIGIT AFTER DISPLACEMENT	FOURTH DIGIT AFTER DISPLACEMENT
CUBIC INCH DISPLACEMENT	BASIC DESIGN SERIES	CRANKSHAFT, CARBURETOR GOVERNOR	BEARINGS, REDUCTION GEARS & AUXILIARY DRIVES	TYPE OF STARTER
6	0	0 -	0 - Plain Bearing	0 - Without Starter
8	1	1 - Horizontal Vacu-Jet	1 - Flange Mounting Plain Bearing	1 - Rope Starter
9	2	2 - Horizontal Pulsa-Jet	2 - Ball Bearing	2 - Rewind Starter
10	3	3 - Horizontal Flo-Jet (Pneumatic Governor)	3 - Flange Mounting Ball Bearing	3 - Electric - 110 Volt, Gear Drive
11	4	4 - Horizontal Flo-Jet (Mechanical Governor)	4 -	4 - Elec. Starter-Generator - 12 Volt, Belt Drive
13	5	5 - Vertical Vacu-Jet	5 - Gear Reduction (6 to 1)	5 - Electric Starter Only - 12 Volt, Gear Drive
14	6	6 -	6 - Gear Reduction (6 to 1) Reverse Rotation	6 - Alternator Only *
17	7	7 - Vertical Flo-Jet	7 -	7 - Electric Starter, 12 Volt Gear Drive, with Alternator
19	8	8 -	8 - Auxiliary Drive Perpendicular to Crankshaft	8 - Vertical-pull Starter
20	9	9 - Vertical Pulsa-Jet	9 - Auxiliary Drive Parallel to Crankshaft	
23				
24				
25				
30				
32				

* Digit 6 formerly used for "Wind-Up" Starter on 60000, 80000 and 92000 Series

Fig. B1 – Explanation of model or series number of Briggs & Straton engines.

A. First one or two digits indicate CUBIC INCH DISPLACEMENT.
B. First digit after displacement indicates BASIC DESIGN SERIES, relating to cylinder construction, ignition, general configuration, etc.
C. Second digit after displacement indicates POSITION OF CRANKSHAFT AND TYPE OF CARBURETOR.
D. Third digit after displacement indicates TYPE OF BEARINGS and whether or not engine is equipped with REDUCTION GEAR or AUXILIARY DRIVE.
E. Last digit indicates TYPE OF STARTER.

BRIGGS & STRATTON

BRIGGS & STRATTON CORPORATION
Milwaukee, Wisconsin 53201

Model	No. Cyls.	Bore	Stroke	Displacement	Power Rating
170000, 171000	1	3 in. (76.2 mm)	2.375 in. (60.3 mm)	16.8 cu. in. 275.1 cc)	7 hp. (5.2 kW)
190000, 191000	1	3 in. (76.2 mm)	2.75 in. (69.85 mm)	19.44 cu. in. (318.5 cc)	8 hp. (6 kW)
220000	1	3.438 in. (87.3 mm)	2.375 in. (60.3 mm)	22.04 cu. in. (361.1 cc)	10 hp. (7.5 kW)
250000, 251000, 252000, 253000	1	3.438 in. (87.3 mm)	2.625 in. (66.68 mm)	24.36 cu. in. (399.1 cc)	11 hp. (8.2 kW)

Engines in this section are four-cycle, one cylinder engines and both vertical and horizontal crankshaft models are covered. Crankshaft is supported in main bearings which are an integral part of crankcase and cover or ball bearings which are a press fit on crankshaft. Cylinder block and bore are a single aluminum casting.

Connecting rod in all models rides directly on crankpin journal. Horizontal crankshaft models are splash lubricated by an oil dipper attached to connecting rod cap and vertical crankshaft models are lubricated by an oil slinger wheel located on governor gear.

Early models use a magneto type ignition system with points and condenser located underneath flywheel. Late models use a breakerless (Magnetron) ignition system.

A float type carburetor is used on all models and a fuel pump is available as optional equipment for some models.

Refer to **BRIGGS & STRATTON ENGINE IDENTIFICATION INFORMATION** section for engine identification and always give engine model and serial number when ordering parts or service material.

MAINTENANCE

SPARK PLUG. Recommended spark plug is Champion J8 or equivalent. To decrease radio interference use Champion XJ8 or equivalent spark plug. Electrode gap for all models is 0.030 inch (0.762 mm).

CARBURETOR. One of two different Flo-Jet carburetors will be used. Carburetor may be identified as Flo-Jet I (Fig. B2) or Flo-Jet II (Fig. B3). Refer to appropriate figure for model being serviced.

FLO-JET I CARBURETOR. Initial adjustment procedure varies according to horsepower rating and crankshaft type.

For initial carburetor adjustment for 7, 8 and 10 horsepower horizontal crankshaft models, open idle mixture screw 1½ turns and main fuel mixture screw 2½ turns.

For initial carburetor adjustment for 11 horsepower horizontal crankshaft models, open idle mixture screw 1 turn and main fuel mixture screw 1½ turns.

For initial carburetor adjustment for all vertical crankshaft models, open idle mixture screw 1¼ turns and main fuel mixture screw 1½ turns.

Make final adjustment with engine at normal operating temperature and running. Place engine under load and adjust main fuel mixture screw for leanest mixture that will allow satisfactory acceleration and steady governor operation. Set engine at idle speed, no load and adjust idle mixture screw to obtain smoothest idle operation.

As each adjustment affects the other, adjustment procedure may have to be repeated.

FLO-JET II CARBURETOR. For initial adjustment open idle mixture screw 1 turn and main fuel mixture screw 1½ turns. Make final adjustment with engine at normal operating temperature and running. Place engine under load and adjust main fuel mixture screw for leanest mixture that will allow satisfactory acceleration and steady governor operation. Set engine at idle speed, no load and adjust idle mixture screw to obtain smoothest idle operation.

As each adjustment affects the other, adjustment procedure may have to be repeated.

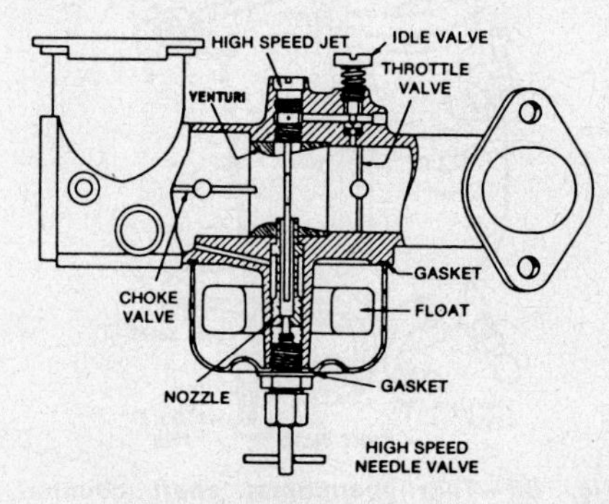

Fig. B2—Cross-sectional view of Flo-Jet I carburetor.

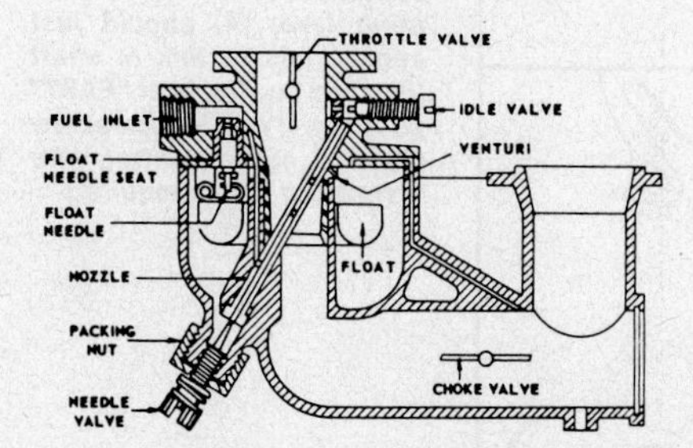

Fig. B3—Cross-sectional view of Flo-Jet II carburetor. Before separating upper and lower body sections, loosen packing nut and power needle valve as a unit and use special screwdriver (tool number 19062) to remove nozzle.

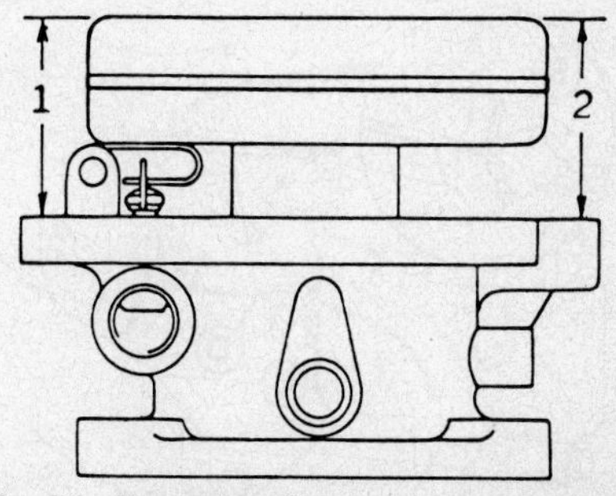

Fig. B4—Dimension (2) must be the same as dimension (1) plus or minus 1/32-inch (0.79 mm).

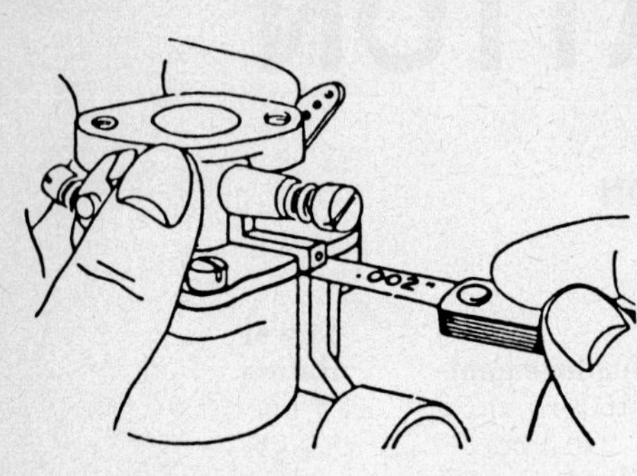

Fig. B5—Check upper body of Flo-Jet II carburetor for warpage as outlined in text.

To check float level for all models, invert carburetor body and float assembly. Refer to Fig. B4 for proper float level dimensions. Adjust by bending float lever tang that contacts inlet valve.

Check Flo-Jet II carburetor for upper body warpage using an 0.002 inch (0.0508 mm) feeler gage as shown in Fig. B5. If upper body is warped more than 0.002 inch (0.0508 mm) it must be renewed.

CHOKE-A-MATIC CARBURETOR CONTROLS. Engines may be equipped with a control unit with which carburetor choke, throttle and magneto grounding swtich are operated from a single lever (Choke-A-Matic carburetors).

To check operation of Choke-A-Matic controls, move control lever to "CHOKE" position; carburetor choke slide or plate must be completely closed. Move control lever to "STOP" position; magneto grounding switch should be making contact. With control lever in "RUN", "FAST" or "SLOW" position, carburetor choke should be completely open. On units with remote controls, synchronize movement of remote lever to carburetor control lever by loosening screw (C–Fig. B6) and moving control wire housing (D) as required. Tighten screw to clamp housing securely. refer to Fig. B7 to check remote control wire movement.

AUTOMATIC CHOKE (THERMOSTAT TYPE). A thermostat operated choke is used on some models equipped with Flo-Jet II carburetor. To adjust choke linkage, hold choke shaft so thermostat lever is free. At room temperature, stop screw in thermostat collar should be located midway between thermostat stops. If not, loosen stop screw, adjust collar and tighten stop screw. Loosen set screw (S–Fig. B8) on thermostat lever. Slide lever on shaft to insure free movement of choke unit. Turn thermostat shaft clockwise until stop screw contacts thermostat stop. While holding shaft in this position, move shaft lever until choke is open exactly 1/8-inch (3 mm) and tighten lever set screw. Turn thermostat shaft counterclockwise until stop screw contacts thermostat stop as shown in Fig. B9. Manually open choke valve until it stops against top of choke link opening. At this time choke should be open at least 3/32-inch (2.38 mm), but not more than 5/32-inch (4 mm). Hold choke valve in wide open position and check position of counterweight lever. Lever should be in a horizontal position with free end towards right.

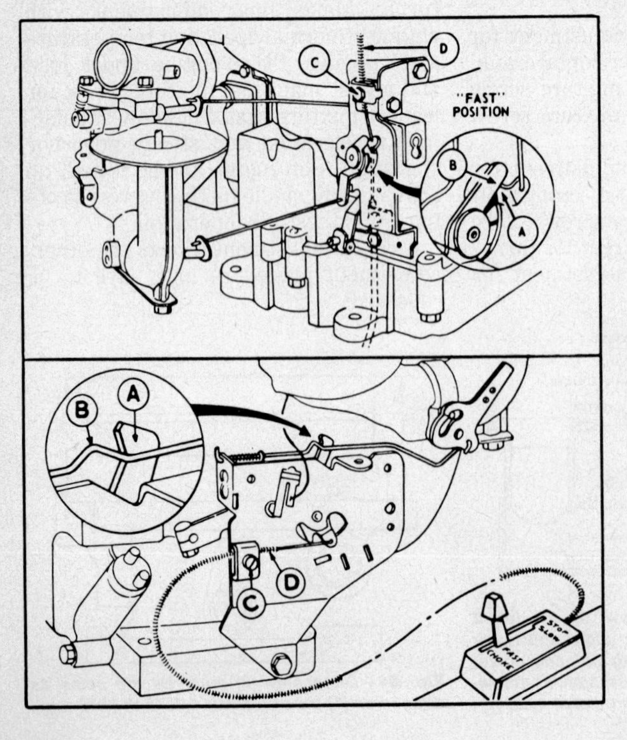

Fig. B6—On Choke-A-Matic controls shown, choke actuating lever (A) should just contact choke link or shaft (B) when control is at "FAST" position. If not, loosen screw (C) and move control wire housing (D) as required.

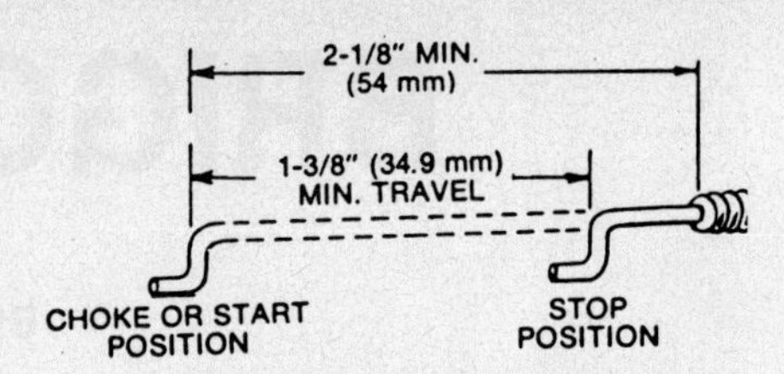

Fig. B7—For proper operation of Choke-A-Matic controls, remote control wire must extend to dimension shown and have a minimum travel of 1 3/8 inches (34.9 mm).

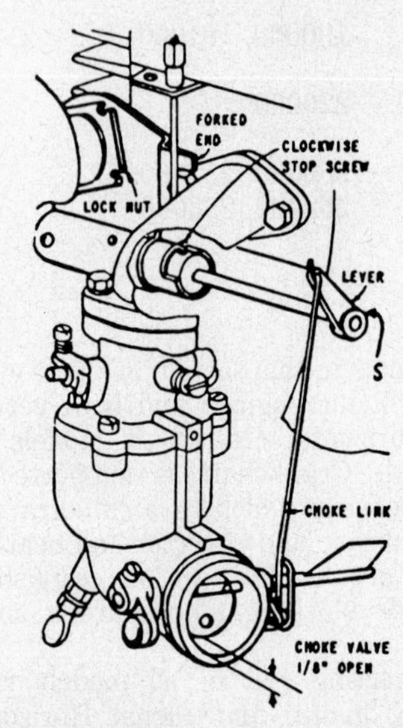

Fig. B8—Automatic choke used on some models equipped with Flo-Jet II carburetor showing unit in "HOT" position.

FUEL TANK OUTLET. Some models are equipped with a fuel tank outlet as shown in Fig. B10. On other engines, a fuel sediment bowl is incorporated with fuel tank outlet as shown in Fig. B11.

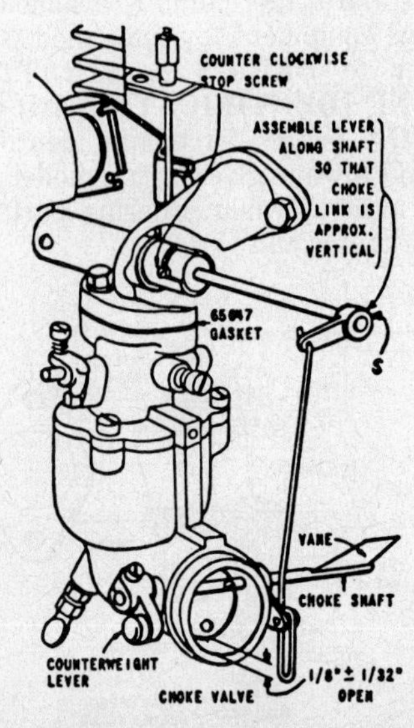

Fig. B9—Turn thermostat shaft counterclockwise until stop screw contacts thermostat stop as shown.

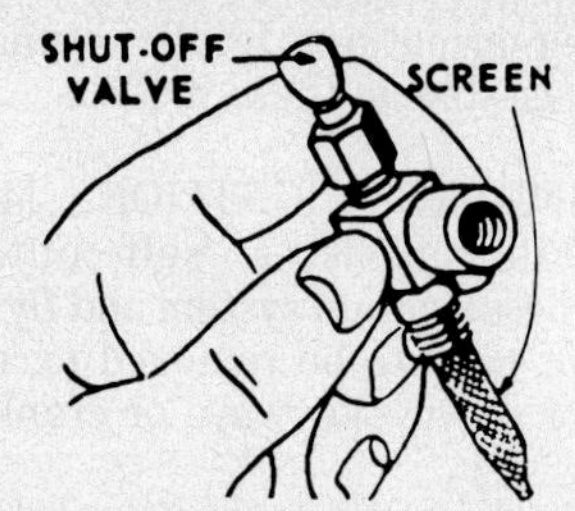

Fig. B10—Fuel tank outlet used on some B&S models.

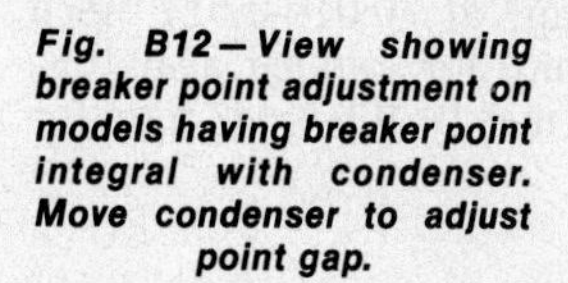

Fig. B12—View showing breaker point adjustment on models having breaker point integral with condenser. Move condenser to adjust point gap.

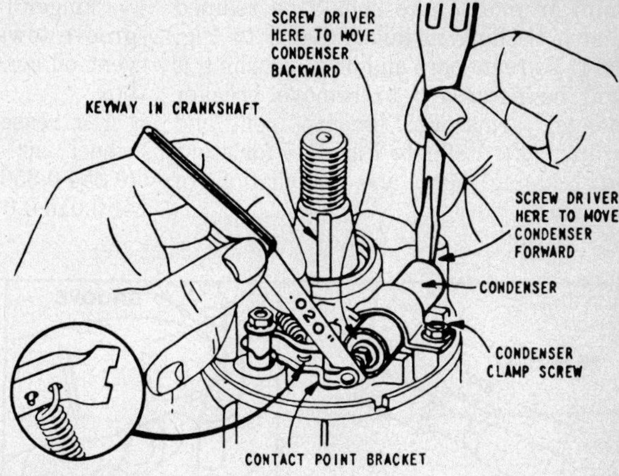

Clean any lint and dirt from tank outlet screens with a brush. Varnish or other gasoline deposits may be removed by use of a suitable solvent. Tighten packing nut or remove nut and shut-off valve, then renew packing if leakage occurs around shut-off valve stem.

FUEL PUMP. A fuel pump is available as optional equipment on some models. Refer to **SERVICING BRIGGS & STRATTON ACCESSORIES** section for service information.

GOVERNOR. Engines are equipped with a gear driven mechanical governor and governor unit is enclosed within engine crankcase and is driven from camshaft gear. Lubrication oil slinger is an integral part of governor unit on all vertical crankshaft engines. All binding or slack due to wear must be removed from governor linkage to prevent "hunting" or unsteady operation. To adjust carburetor to governor linkage, loosen clamp bolt on governor lever. Move link end of governor lever until carburetor throttle shaft is in wide open position. Using a screwdriver, rotate governor lever shaft clockwise as far as possible and tighten clamp bolt.

Governor gear and weight unit can be removed when engine is disassembled. Refer to exploded views of engines in Fig. B21, B22, B23 and B28. Remove governor lever, cotter pin and washer from outer end of governor lever shaft. Slide governor lever out of bushing towards inside of engine. Governor gear and weight unit can now be removed. Renew governor lever shaft bushing in crankcase, if necessary and ream new bushing after installation to 0.2385-0.239 inch (6.05-6.07 mm).

IGNITION SYSTEM. Early models use a magneto system which incorporates a breaker point set and condenser. Late models use Magnetron breakerless ignition system. Refer to appropriate paragraph for model being serviced.

MAGNETO IGNITION. All models use breaker points and condenser located under flywheel.

One of two different types of ignition points as shown in Fig. B12 and B13 are used. Breaker point gap is 0.020 inch (0.508 mm) for all models with magneto ignition.

On each type, breaker contact arm is actuated by a plunger in a bore in engine crankcase which rides against a cam on engine crankshaft. Plunger can be removed after removing breaker points. Renew plunger if worn to a length of 0.870 inch (22.098 mm) or less. If breaker point plunger bore in crankcase is worn, oil will leak past plunger. Check bore with B&S gage, tool number 19055. If plug gage will enter bore ¼-inch (6.35

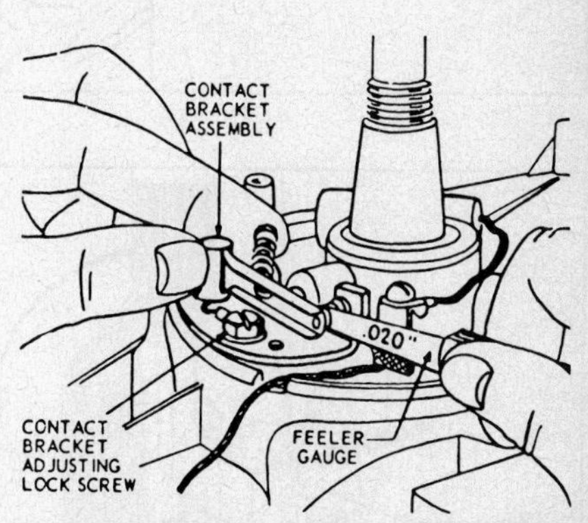

Fig. B13—Adjusting breaker point gap on models having breaker points separate from condenser.

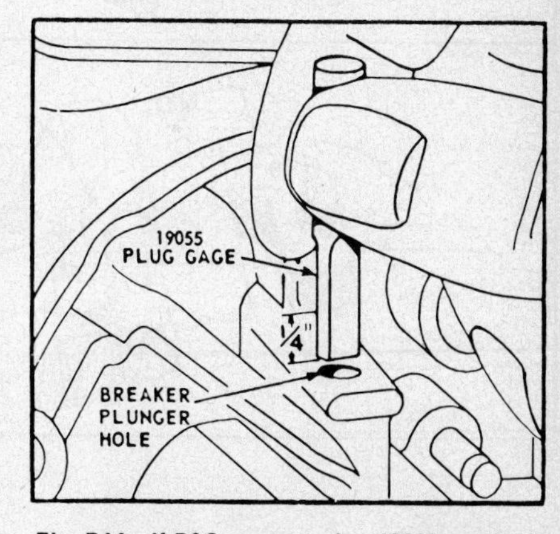

Fig. B14—If B&S gage number 19055 can be inserted in plunger bore ¼-inch (6.35 mm) or more, bore is worn and must be rebushed.

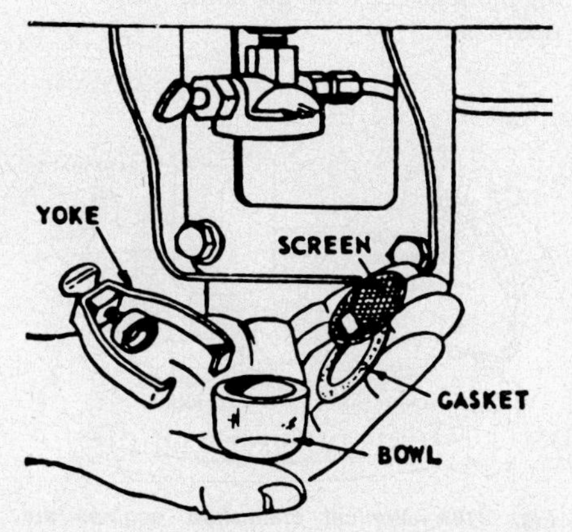

Fig. B11—Fuel sediment bowl and tank outlet used on some models.

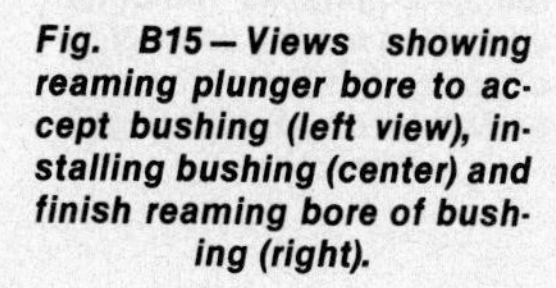

Fig. B15—Views showing reaming plunger bore to accept bushing (left view), installing bushing (center) and finish reaming bore of bushing (right).

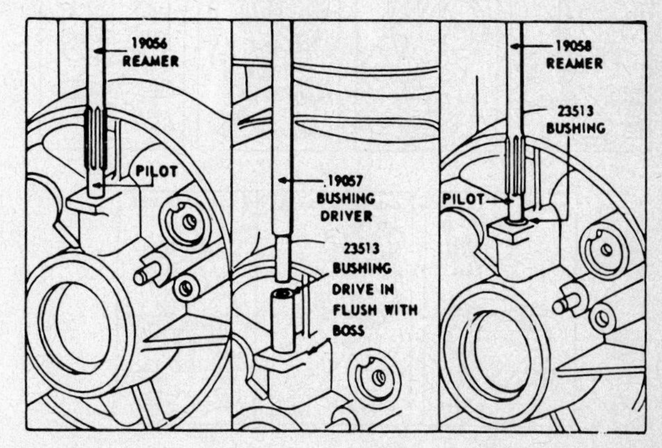

mm) or more, bore should be reamed and a bushing installed. Refer to Fig. B14. To ream bore and install bushing it will be necessary to remove breaker points, armature, ignition coil and crankshaft. Refer to Fig. B15 for steps in reaming bore and installation of bushing.

Plunger must be reinstalled with groove toward top (Fig. B15A) to prevent oil contamination in breaker point box.

For reassembly set armature to flywheel air gap at 0.010-0.014 inch (0.254-0.356 mm) for two-leg armature or 0.016-0.019 inch (0.406-0.483 mm) for three-leg armature. Ignition timing is non-adjustable on these models.

MAGNETRON IGNITION. Magnetron ignition is a self-contained breakerless ignition system and flywheel does not need to be removed except to check or service keyways or crankshaft key.

To check spark, remove spark plug and connect spark plug cable to B&S tester, part number 19051 and ground remaining tester lead to engine cylinder head. Spin engine at 350 rpm or more. If spark jumps the 0.166 inch (4.2 mm) tester gap, system is functioning properly.

To remove armature and Magnetron module, remove flywheel shroud and armature retaining screws. Use a 3/16-inch (4.76 mm) diameter pin punch to release stop switch wire from module. To remove module, unsolder wires, push module retainer away from laminations and push module off. See Fig. B15B.

Resolder wires for reinstallation and use Permatex or equivalent to hold ground wires in position.

Adjust armature air gap to 0.010-0.014 inch (0.25-0.36 mm).

Fig. B15A — Insert plunger into bore with groove toward top.

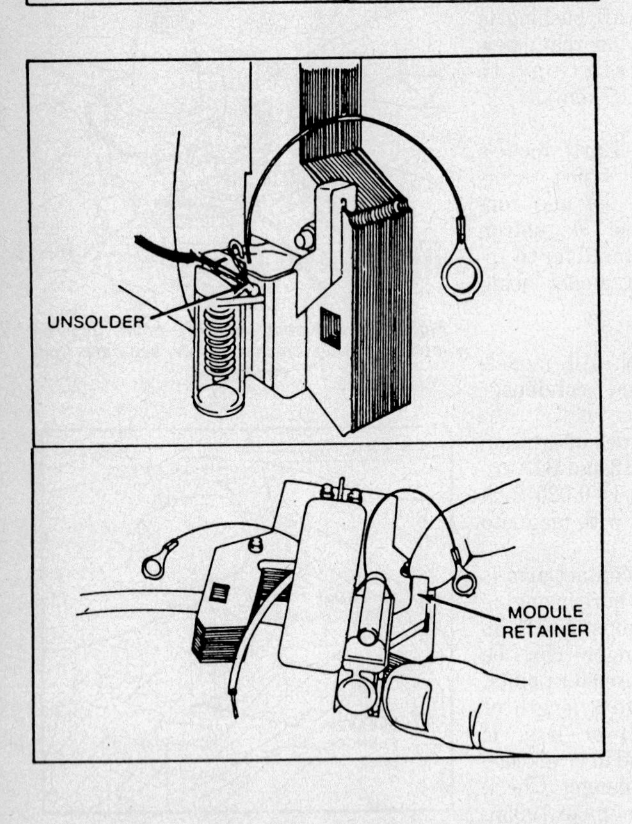

Fig. B15B — Wires must be unsoldered to remove Magnetron module.

LUBRICATION. Horizontal crankshaft engines have a splash lubrication system provided by an oil dipper attached to connecting rod. Refer to Fig. B16 for view of various types of dippers used.

Vertical crankshaft engines are lubricated by an oil slinger wheel on governor gear which is driven by camshaft gear. See Fig. B16A.

Oils approved by manufacturer must meet requirements of API service classification SC, SD, SE or SF.

Use SAE 10W-40 oil for temperatures above 20°F (-7°C) and SAE 5W-30 oil for temperatures below 20°F (-7°C).

Check oil level at five hour intervals and maintain at bottom edge of filler plug. **DO NOT** overfill.

Recommended oil change interval for all models is every 25 hours of normal operation.

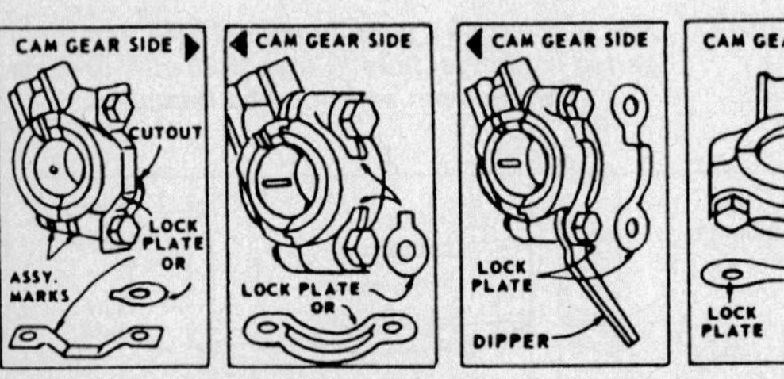

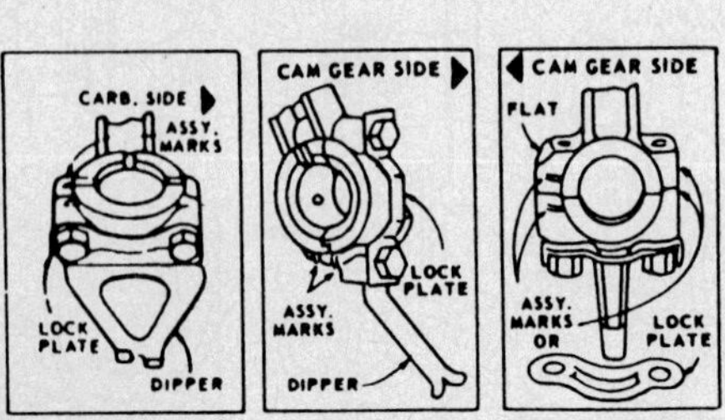

Fig. B16 — Install connecting rod in engine as indicated according to type used. Note dipper installation on horizontal crankshaft engine connecting rod.

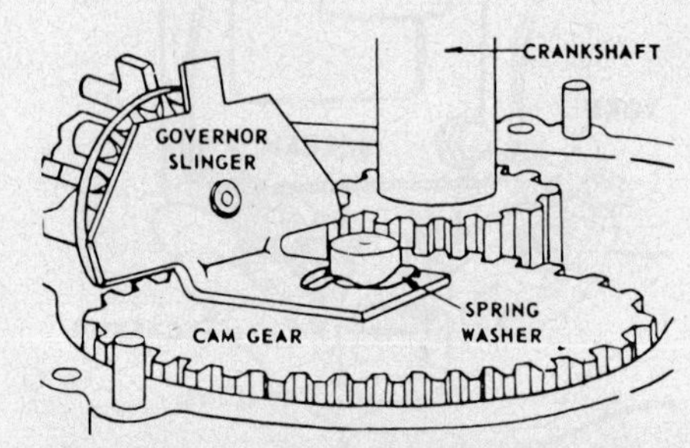

Fig. B16A — Vertical crankshaft engines are lubricated by an oil slinger mounted on governor gear assembly.

Crankcase oil capacity for 16.8 and 19.44 cubic inch (275.1 and 318.5 cc) engines is 2¼ pints (1.1 L) for vertical crankshaft models and 2¾ pints (1.3 L) for horizontal crankshaft models.

Crankcase oil capacity for 22.04 and 24.36 cubic inch (361.1 and 399.1 cc) engines is 3 pints (1.4 L) for both vertical and horizontal crankshaft models.

CRANKCASE BREATHER. A crankcase breather is built into engine valve cover. A partial vacuum must be maintained in crankcase to prevent oil from being forced out past oil seals and gaskets or past breaker point plunger or piston rings. Air can flow out of crankcase through breather, but a one-way valve blocks return flow, maintaining necessary vacuum. Breather mounting holes are offset one way. A vent tube connects breather to carburetor air horn for extra protection against dusty conditions.

REPAIRS

CYLINDER HEAD. When removing cylinder head note location from which different length cap screws are removed as they must be reinstalled in their original positions.

Always use a new gasket when reinstalling cylinder head. Do not use sealer on gasket. Lubricate cylinder head bolt threads with graphite grease, install in correct locations and tighten in several even steps in sequence shown in Fig. B17 until 165 in.-lbs. (19 N•m) torque is obtained. Start and run engine until normal operating temperature is reached, stop engine and retighten head bolts as outlined.

It is recommended carbon and lead deposits be removed at 100 to 300 hour intervals, or whenever cylinder head is removed.

CONNECTING ROD. Connecting rod and piston are removed from cylinder head end of block as an assembly. Aluminum alloy connecting rod rides directly on induction hardened crankshaft crankpin journal. Rod should be rejected if crankpin bore is scored or out-of-round more than 0.0007 inch (0.0178 mm) or if piston pin bore is scored or out-of-round more than 0.0005 inch (0.0127 mm). Renew connecting rod if either crankpin journal or piston pin bore is worn to, or larger than, sizes given in chart.

REJECT SIZES FOR CONNECTING ROD

Model	Crankpin bore	Pin bore*
170000, 171000	1.0949 in. (27.81 mm)	0.674 in. (17.12 mm)
190000, 191000	1.1265 in. (28.61 mm)	0.674 in. (17.12 mm)
All other models	1.252 in. (31.8 mm)	0.802 in. (20.37 mm)

*Piston pins of 0.005 inch (0.127 mm) oversize are available for service. Piston pin bore in rod can be reamed to this size of crankpin bore is within specifications.

Refer to Fig. B16, locate style rod used for model being serviced and install in engine as indicated.

Tighten connecting rod cap screws to 165 in.-lbs. (19 N•m) torque on 170000, 171000, 190000 and 191000 models and to 190 in.-lbs. (22 N•m) torque on all other models.

PISTON, PIN AND RINGS. Chrome plated aluminum piston used in aluminum bore (Kool-Bore) engines should be renewed if top ring side clearance in groove exceeds 0.007 inch (0.18 mm) or if piston is visibly scored or damaged. Piston should also be renewed, or pin bore reamed for 0.005 inch (0.127 mm) oversize pin, if pin bore is 0.673 inch (17.09 mm) or larger for 170000, 171000, 190000 and 191000 models or 0.801 inch (20.32 mm) or larger for all other models.

Renew piston pin if pin is 0.0005 inch (0.0127 mm) or more out-of-round or if pin is worn to a diameter of 0.671 inch (17.04 mm) or smaller for 170000, 171000, 190000 or 191000 models or 0.799 inch (20.29 mm) or smaller for all other models.

Ring end gap should be 0.010-0.025 inch (0.254-0.635 mm) and compression ring should be renewed if end gap is greater than 0.035 inch (0.89 mm) and oil control ring should be renewed if end gap is greater than 0.045 inch (1.14 mm).

Pistons used in 220000 and 250000 engines have a notch and a letter "F" stamped in piston (Fig. B17A) which must face flywheel side of engine after installation.

Pistons and rings are available in a variety of oversizes as well as standard.

A chrome ring set is available for slightly worn standard bore cylinders. Refer to note in **CYLINDER** section.

CYLINDER. Standard cylinder bore diameter is 2.999-3.000 inches (76.175-76.230 mm) for 170000, 171000, 190000 and 191000 models and 3.4365-3.4375 inches (87.287-87.313 mm) for all other models.

If cylinder is worn 0.003 inch (0.076 mm) or more, or out-of-round 0.0025 inch (0.0635 mm) or more, it should be

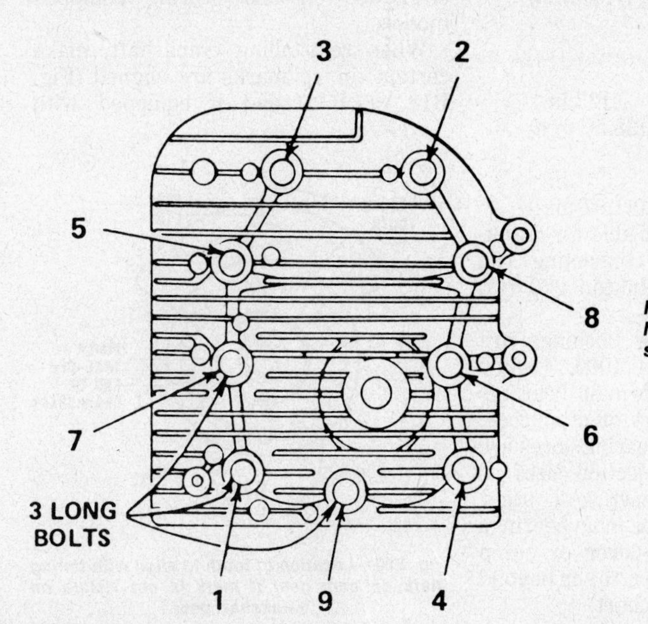

Fig. B17—Tighten cylinder head bolts in sequence shown. Note location of three long bolts.

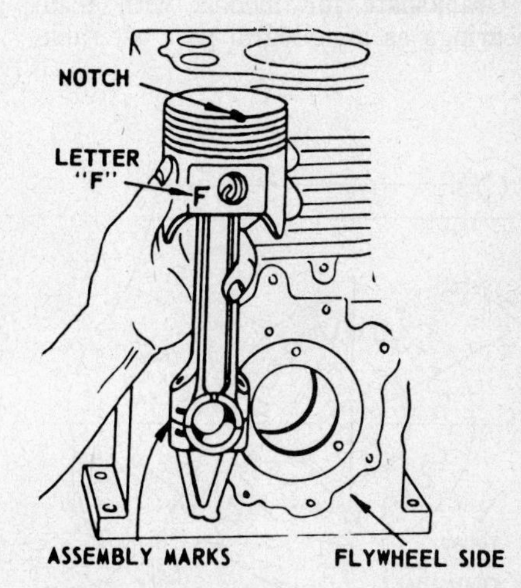

Fig. B17A—Notch and letter "F" stamped in piston of 220000 and 250000 engines must face flywheel side of engine after installation.

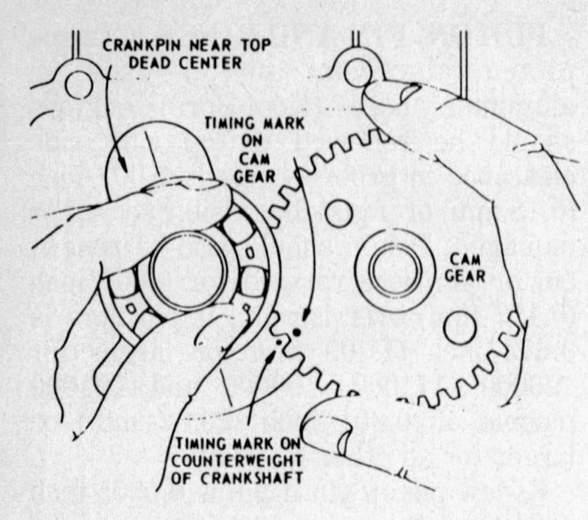

Fig. B18—Align timing mark on cam gear with mark on crankshaft counterweight on ball bearing equipped models.

resized to nearest oversize for which piston and rings are available.

NOTE: A chrome piston ring set is available for slightly worn standard bore cylinders. No honing or cylinder deglazing is required for these rings. Cylinder bore can be a maximum of 0.005 inch (0.127 mm) oversize when using chrome rings.

It is recommended a hone be used for resizing cylinders. Operate hone at 300-700 rpm with an up and down movement which will produce a 45° angle cross-hatch pattern. Clean cylinder after honing with oil or soap and water.

Approved hone for aluminum bore is Ammco No. 3956 for rough and finishing or Sunnen AN200 for rough and Sunnen AN500 for finishing.

CRANKSHAFT AND MAIN BEARINGS. Crankshaft may be supported at each end in main bearings which are an integral part of crankcase, cover or sump or ball bearing mains which are a press fit on crankshaft and fit into machined bores in crankcase, cover or sump.

Crankshaft for models with main bearings as an integral part of crankcase, cover or sump, should be renewed or reground if main bearing journals are worn to, or beyond crankshaft rejection specifications as listed in following chart.

CRANKSHAFT REJECTION SIZES

Model	Magneto end journal	Drive end journal
170000, 190000	0.9975 in. (25.34 mm)	1.1790 in. (29.95 mm)
171000 Synchro-balanced 191000 Synchro-balanced	1.1790 in. (29.95 mm)	1.1790 in. (29.95 mm)
All other models	1.376 in. (34.95 mm)	1.376 in. (34.95 mm)

Crankshaft for models with ball bearing main bearings should be renewed if new bearings are loose on journals. Bearings should be a press fit.

Crankshaft for all models should be renewed or reground if connecting rod crankpin journal diameter is worn to, or beyond crankshaft rejection specifications as listed in following chart.

CRANKSHAFT REJECTION SIZES

Model	Crankpin journal
170000, 171000 Synchro-balanced	1.090 in. (27.69 mm)
190000, 191000 Synchro-balanced	1.122 in. (28.50 mm)
All other models	1.247 in. (31.67 mm)

Connecting rod is available for crankshaft which has had connecting rod crankpin journal reground to 0.020 inch (0.508 mm) undersize.

Service main bearing bushings are available for 170000, 171000, 190000 and 191000 models with main bearings cast as an integral part of crankcase, cover or sump if main bearing bores are worn to, or beyond rejection sizes as listed in following chart. All other models with integral type main bearings must have crankcase, cover or sump renewed if they are worn to, or beyond rejection sizes listed in chart.

MAIN BEARING REJECT SIZES

Model	Magneto end bearing	Drive end bearing
170000, 171000, 190000, 191000	1.0036 in. (25.491 mm)	1.185 in. (30.099 mm)
Synchro-balanced models	1.185 in. (30.099 mm)	1.185 in. (30.099 mm)
All other models	1.383 in. (35.128 mm)	1.383 in. (35.128 mm)

To install service main bushings it is necessary to ream main bearing bores to correct size. Main bearing tool kit, part number 19184, is available and contains all necessary reamers, guides, drivers and supports. Make certain all metal cuttings are removed before pressing bushings into place and align oil notches or holes as bushings are installed.

Ball bearing mains are a press fit on crankshaft and must be removed by pressing crankshaft out of bearing. Renew ball bearing if worn or rough. Expand new bearing by heating in oil and install on crankshaft with shield side towards crankpin journal.

Crankshaft end play should be 0.002-0.008 inch (0.0508-0.2032 mm). At least one 0.015 inch thick cover or sump gasket must be used. Additional cover gaskets in a variety of thicknesses are available if end play is greater than 0.008 inch (0.2032 mm) metal shims are available for use on crankshaft.

Place metal shims between crankshaft gear and cover or sump on models with integral type main bearings or between magneto end of crankshaft and crankcase on ball bearing equipped models.

When reinstalling crankshaft, make certain timing marks are aligned (Fig. B18 or B19) and if equipped with

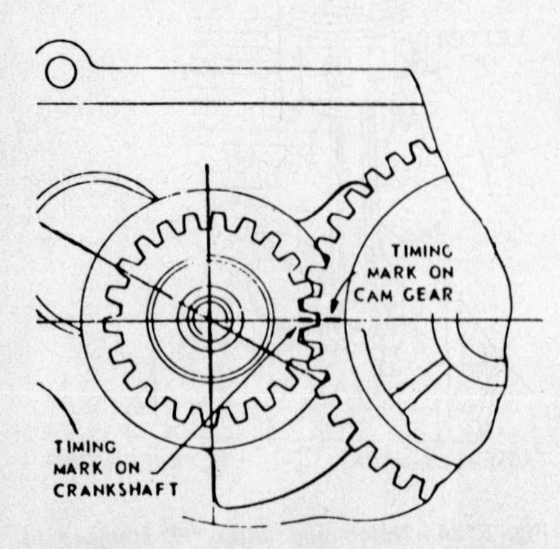

Fig. B19—Align timing marks on cam gear and crankshaft gear on plain bearing models.

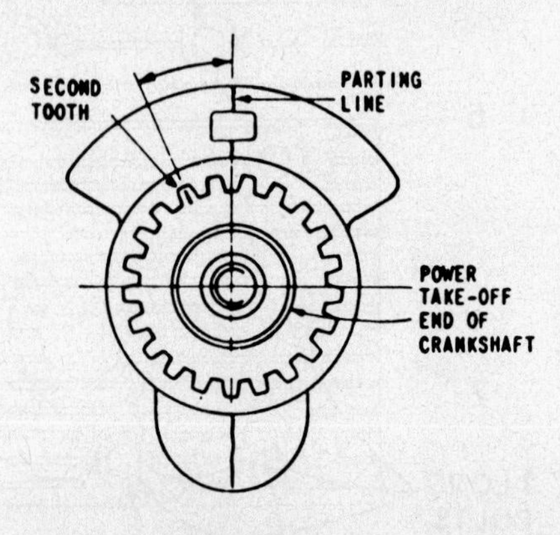

Fig. B20—Location of tooth to align with timing mark on cam gear if mark is not visible on crankshaft gear.

1. Cylinder block
2. Head gasket
3. Cylinder head
4. Connecting rod
5. Rod bolt lock
6. Rings
7. Piston
8. Rotocoil (exhaust valve)
9. Retainer clips
10. Piston pin
11. Intake valve
12. Exhaust valve
13. Retainers
14. Crankcase cover
15. Oil seal
16. Crankcase gasket
17. Main bearings
18. Key
19. Crankshaft
20. Camshaft
21. Tappet
22. Governor gear
23. Governor crank
24. Governor lever
25. Ground wire
26. Governor control plate
27. Spring
28. Governor rod
29. Spring
30. Nut

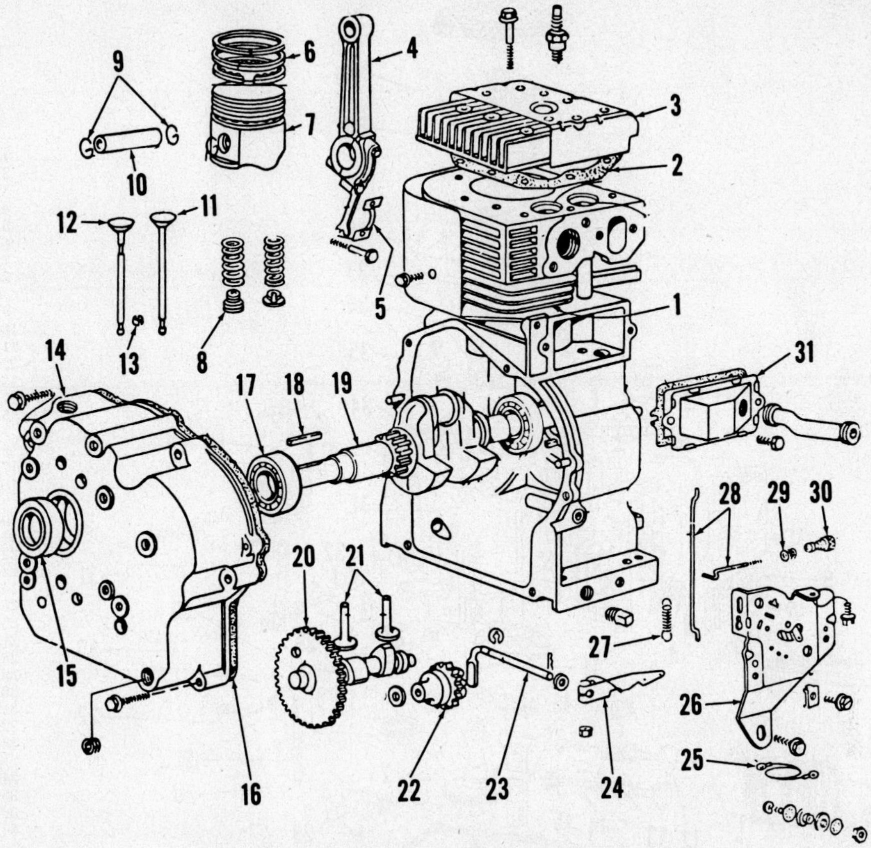

Fig. B21—Exploded view of 220000 engine assembly.

1. Cylinder head
2. Head gasket
3. Cylinder block
4. Rod bolt lock
5. Connecting rod
6. Rings
7. Piston
8. Piston pin
9. Retainer clip
10. Dipstick assy.
11. Crankcase cover
12. Crankcase gasket
13. Oil seal
14. Counterweight & bearing assy.
15. Retainer
16. Key
17. Crankshaft
18. Camshaft assy.
19. Tappet
20. Governor gear
21. Governor crank
22. Governor lever
23. Governor nut & spring
24. Governor control rod
25. Ground wire
26. Governor control plate
27. Drain plug
28. Spring
29. Governor link
30. Choke link
31. Breather assy.
32. Rotocoil (exhaust valve)
33. Valve springs
34. Retainer
35. Exhaust valve
36. Intake valve

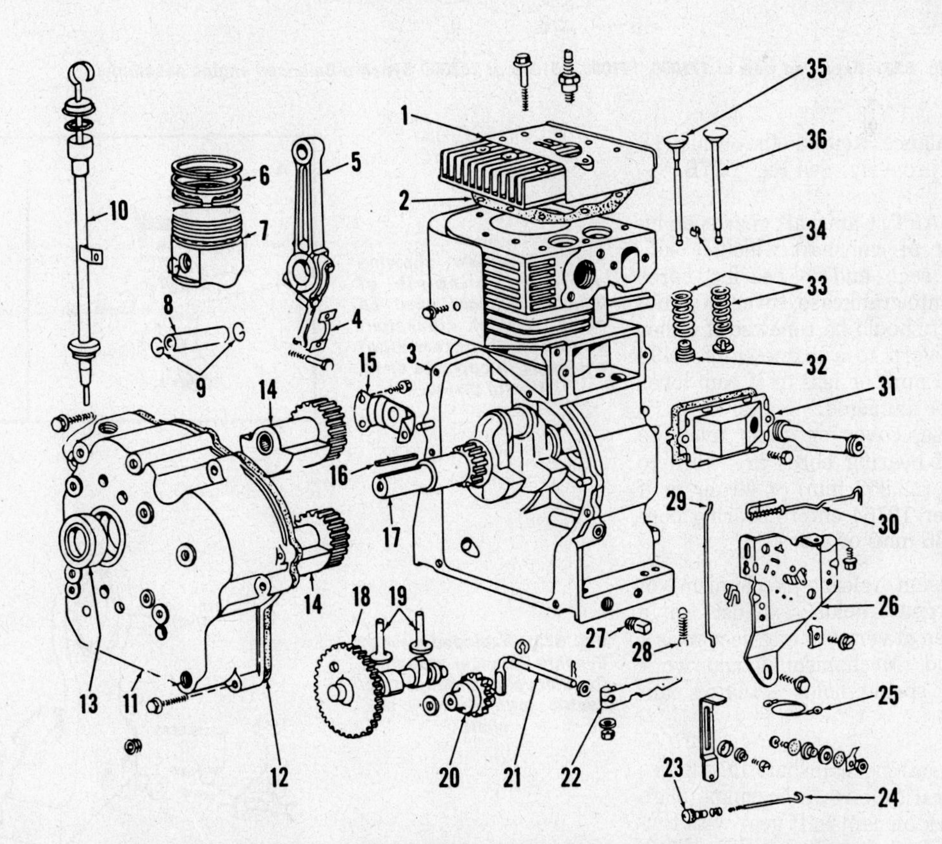

Fig. B22—Exploded view of 170000, 190000, 251000, 252000 or 253000 engine assembly.

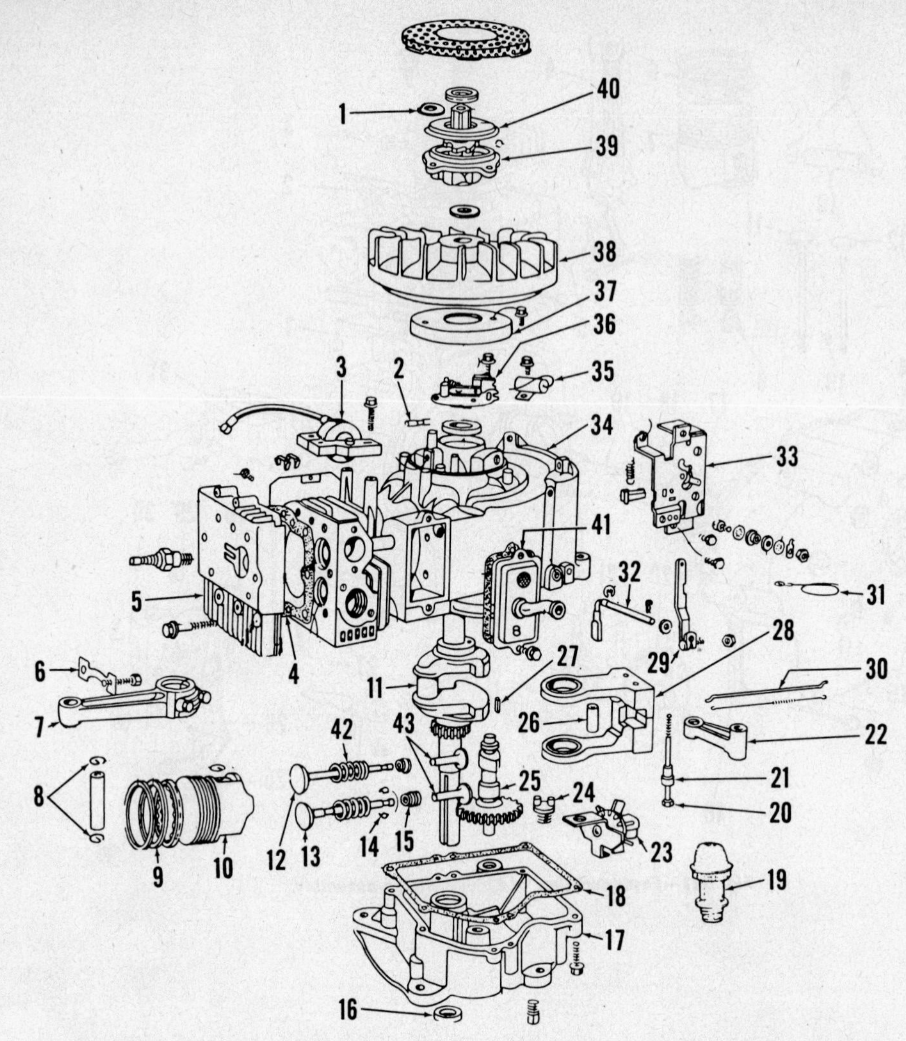

1. Thrust washer
2. Breaker point plunger
3. Armature assy.
4. Head gasket
5. Cylinder head
6. Rod screw lock
7. Connecting rod
8. Piston pin & retaining clips
9. Rings
10. Piston
11. Crankshaft
12. Intake valve
13. Exhaust valve
14. Retainer
15. Rotocoil (exhaust valve)
16. Oil seal
17. Oil sump (engine base)
18. Crankcase gasket
19. Oil minder
20. Cap screw (2 used)
21. Spacer (2 used)
22. Link
23. Governor & oil slinger
24. Plug
25. Cam gear assy.
26. Dowel pin (2 used)
27. Key
28. Counterweight assy.
29. Governor lever
30. Governor link
31. Ground wire
32. Governor crank
33. Choke-A-Matic control
34. Cylinder assy.
35. Condenser
36. Breaker points
37. Cover
38. Flywheel assy.
39. Clutch housing
40. Rewind starter clutch
41. Breather assy.
42. Valve spring
43. Tappets

Fig. B23—Exploded view of 171000, 191000, 251000 or 252000 Synchro-Balanced engine assembly.

counter balance weights they must be positioned properly. See Fig. B27B.

CAMSHAFT. Camshaft gear is an integral part of camshaft which is supported at each end in bearing bores machined into crankcase, cover or sump.

Camshaft should be renewed if either journal is worn to a diameter of 0.498 inch (12.66 mm) or less or if cam lobes are worn or damaged.

Crankcase, cover or sump must be renewed if bearing bores are worn to 0.506 inch (12.852 mm) or larger or if tool number 19164 enters bearing bore ¼-inch (6.35 mm) or more.

Compression release mechanism on camshaft gear holds exhaust valve slightly open at very low engine rpm as a starting aid. Mechanism should work freely and spring holds actuator cam against pin.

When installing camshaft in engines with integral type main bearings, align timing mark on camshaft gear with timing mark on crankshaft gear (Fig. B19).

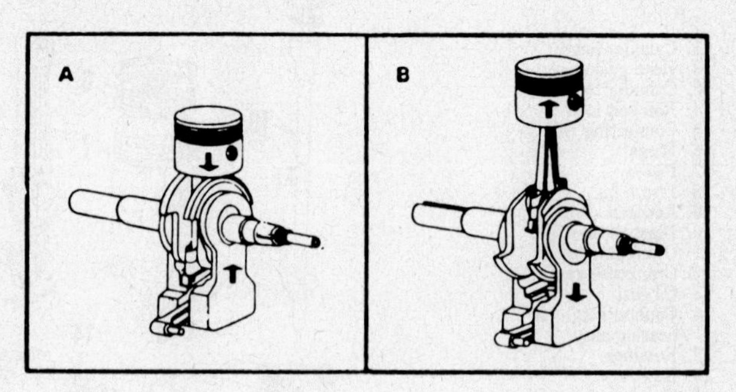

Fig. B24—View showing operating principle of Synchro-Balancer used on some vertical crankshaft engines. Counterweight oscillates in opposite direction of piston.

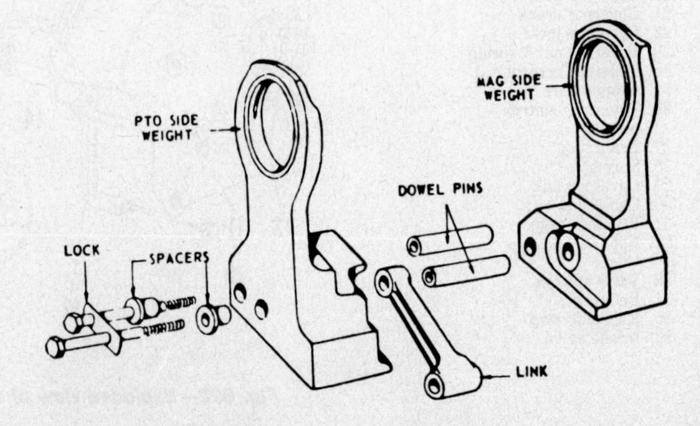

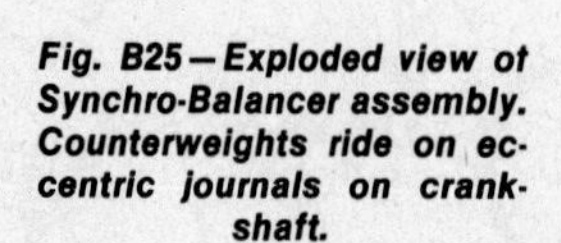

Fig. B25—Exploded view of Synchro-Balancer assembly. Counterweights ride on eccentric journals on crankshaft.

When installing camshaft in engines with ball bearing mains, align timing mark on camshaft gear with timing mark on crankshaft counterweight (Fig. B18).

If timing mark is not visible on crankshaft gear, align camshaft gear timing mark with second tooth to the left of crankshaft counterweight parting line (Fig. B20).

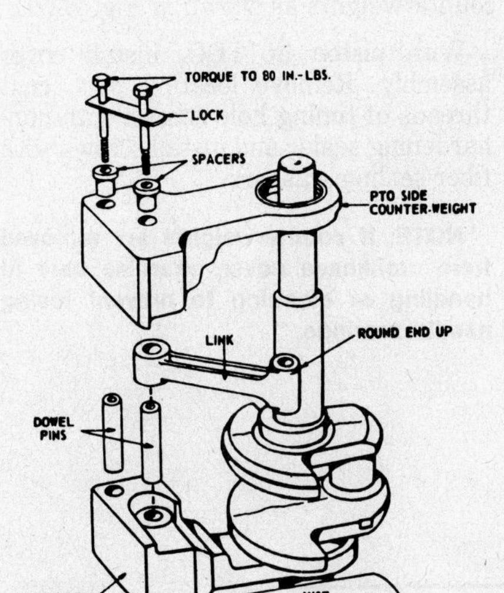

Fig. B26—Assemble balance units on crankshaft as shown. Install link with rounded edge on free end toward pto end of crankshaft.

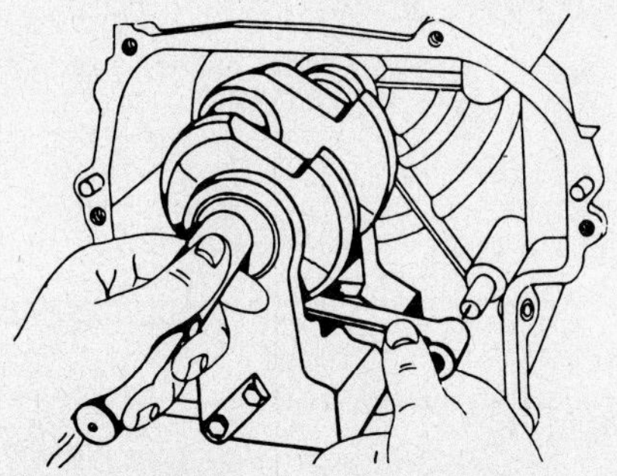

Fig. B27—When installing crankshaft and balancer assembly, place free end of link on anchor pin in crankcase.

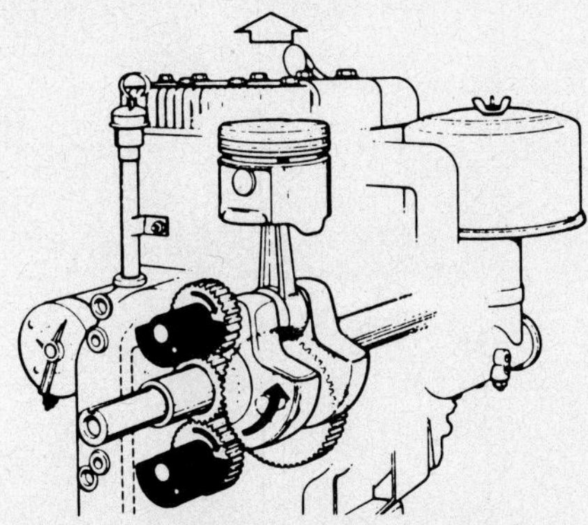

Fig. B27A—View of rotating counterbalance system used on some models. Counterweight gears are driven by crankshaft.

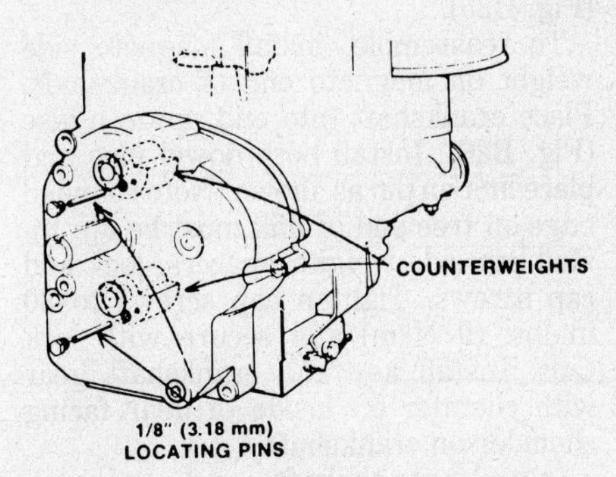

Fig. B27B—To properly align counterweights, remove two small screws from crankcase cover and insert ⅛-inch (3.18 mm) diameter locating pins.

VALVE SYSTEM. Valve seats are machined directly into cylinder block and are ground at a 45° angle. Seat width should be 3/64 to 1/16-inch (1.19 to 1.58 mm).

Valve face is ground at 45° angle and valve should be renewed if margin is 1/64-inch (0.4 mm) or less after refacing.

Valve guides should be checked for wear using valve guide gage number 19151. If gage enters valve guide 5/16-inch (7.9 mm) or more, guides should be reamed using reamer number 19183 and bushing number 19192 should be installed.

Briggs & Stratton also has a tool kit, part number 19232, available for removing factory or field installed guide bushings so new bushing, part number 231218 may be installed.

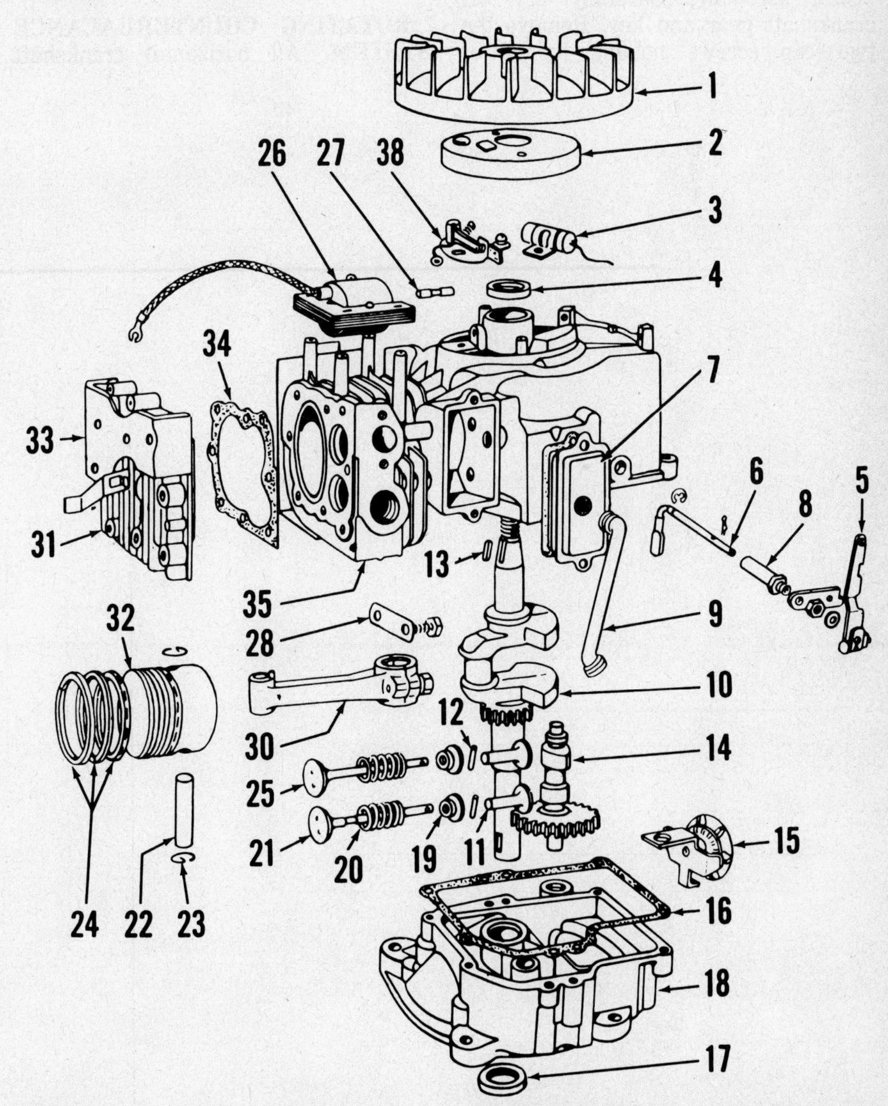

Fig. B28—Exploded view of 170000, 190000 and 220000 series vertical crankshaft engine assembly.

1. Flywheel
2. Breaker point cover
3. Condenser
4. Oil seal
5. Governor lever
6. Governor crank
7. Breather & valve
8. Bushing
9. Breather vent tube
10. Crankshaft
11. Tappets
12. Valve retaining pins
13. Flywheel key
14. Camshaft and gear
15. Governor & oil slinger assy.
16. Gasket
17. Oil seal
18. Oil sump (engine base)
19. Valve spring retainer
20. Valve spring
21. Exhaust valve
22. Piston pin
23. Retaining rings
24. Piston rings
25. Intake valve
26. Armature & coil assy.
27. Breaker plunger
28. Rod bolt lock
30. Connecting rod
31. Cylinder head
32. Piston
33. Air baffle
34. Head gasket
35. Cylinder block
38. Breaker points

Valve tappet gap (cold) for all models is 0.005-0.007 inch (0.13-0.18 mm) for intake valve and 0.009-0.011 inch (0.23-0.28 mm) for exhaust valve.

To adjust valves, piston must be at "top dead center" on compression stroke. Grind end of valve stem off squarely to obtain clearance.

SYNCHRO–BALANCER. All vertical crankshaft engines, except 220000 models, may be equipped with an oscillating Synchro-Balancer. Balance weight assembly rides on eccentric journals on crankshaft and moves in opposite direction of piston (Fig. B24).

To disassemble balancer unit, first remove flywheel, engine base, cam gear, cylinder head and connecting rod and piston assembly. Carefully pry off crankshaft gear and key. Remove the two cap screws holding halves of counterweight together. Separate weights and remove link, dowel pins and spacers. Slide weights from crankshaft (Fig. B25).

To reassemble, install magneto side weight on magneto end of crankshaft. Place crankshaft (pto end up) in a vise (Fig. B26). Install both dowel pins and place link on pin as shown. Note rounded edge on free end of link must be up. Install pto side weight, spacers, lock and cap screws. Tighten cap screws to 80 in.-lbs. (9 N•m) and secure with lock tabs. Install key and crankshaft gear with chamfer on inside of gear facing shoulder on crankshaft.

Install crankshaft and balancer assembly in crankcase, sliding free end of link on anchor pin as shown in Fig. B27. Reassemble engine.

ROTATING COUNTERBALANCE SYSTEM. All horizontal crankshaft engines, except 220000 models, may be equipped with two gear driven counterweights in constant mesh with crankshaft gear. Gears, mounted in crankcase cover, rotate in opposite direction of crankshaft (Fig. B27A).

To properly align counterweights when installing cover, remove two small screws from cover and insert 1/8-inch (3.18 mm) diameter locating pins through holes and into holes in counterweights as shown in Fig. B27B.

With piston at TDC, install cover assembly. Remove locating pins, coat threads of timing hole screws with non-hardening sealer and install screws with fiber sealing washers.

NOTE: If counterweights are removed from crankcase cover, exercise care in handling or cleaning to prevent losing needle bearings.

BRIGGS & STRATTON

BRIGGS & STRATTON CORPORATION
Milwaukee, Wisconsin 53201

Model	No. Cyls.	Bore	Stroke	Displacement	Power Rating
19, 19D, 191000, 193000	1	3 in. (76.2 mm)	2.625 in. (66.675 mm)	18.56 cu. in. (304.1 cc)	7.25 hp. (5.5 kW)
200000	1	3 in. (76.2 mm)	2.875 in. (73.025 mm)	20.32 cu. in. (333 cc)	8 hp. (6 kW)
23,23A, 23C, 23D, 23100, 233000	1	3 in. (76.2 mm)	3.25 in. (82.55 mm)	22.97 cu. in (376.5 cc)	9 hp. 6.7 kW)
243000	1	3.0625 in. (77.79 mm)	3.25 in. (82.55 mm)	23.94 cu. in. (392.3 cc)	10 hp. (7.5 kW)
300000, 301000	1	3.4375 in. (87.31 mm)	3.25 in. (82.55 mm)	30.16 cu. in. (494.2 cc)	12 hp. (9 kW)
302000	1	3.4375 in. (87.31 mm)	3.25 in. (82.55 mm)	30.16 cu. in. (494.2 cc)	13 hp. (9.7 kW)
320000	1	3.5625 in. (90.5 mm)	3.25 in. (82.55 mm)	32.4 cu. in. (531 cc)	14 hp. (10.4 kW)
325000	1	3.5625 in. (90.5 mm)	3.25 in. (82.55 mm)	32.4 cu. in. (531 cc)	15 hp. (11.2 kW)
326000	1	3.5625 in. (90.5 mm)	3.25 in. (82.55 mm)	32.4 cu. in. (531 cc)	16 hp. (11.9 kW)

Engines in this section are four-cycle, one cylinder horizontal crankshaft models. Crankshaft is supported at each end in bushing type main bearings which are an integral part of main bearing support plate or by ball bearings which are pressed onto crankshaft and fit into machined bores of main bearing support plates. Cylinder block and bore are a single cast iron casting.

Connecting rod in all models rides directly on crankpin journal and are splash lubricated by an oil dipper attached to connecting rod cap.

Early models use a variety of magneto ignition systems with points and condenser either mounted externally or underneath flywheel. Late production models use Briggs & Stratton Magnetron breakerless ignition system.

A float type carburetor is used on all models and a fuel pump is available as optional equipment for some models.

Refer to **BRIGGS & STRATTON ENGINE IDENTIFICATION INFORMATION** section for engine identification and always give engine model and serial number when ordering parts or service material.

MAINTENANCE

SPARK PLUG. Recommended spark plug is Champion J8 or equivalent. To decrease radio interference use Champion XJ8 or equivalent spark plug. Electrode gap for all models is 0.030 inch (0.762 mm).

CARBURETOR. All models are equipped with a two-piece float type carburetor. Refer to Fig. B30 for a cross-sectional view of carburetor and location of mixture adjustment screws.

For initial carburetor adjustment, open idle mixture screw 1 turn and main fuel mixture screw 1½ turns.

Make final adjustment with engine at normal operating temperature and running. Place engine under load and adjust main fuel mixture screw for leanest mixture that will allow satisfactory acceleration and steady governor operation. Set engine idle mixture screw to obtain smoothest idle operation.

As each adjustment affects the other, adjustment procedure may have to be repeated.

To check float level for all models, invert carburetor body and float assembly. Refer to Fig. B34 for proper float level dimensions. Adjust by bending float lever tang that contacts inlet valve.

Check carburetor for upper body warpage using a 0.002 inch (0.0508 mm)

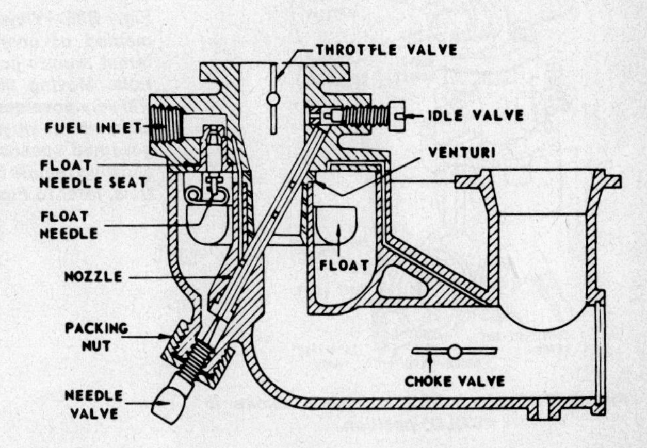

Fig. B30—Cross-sectional view of typical two-piece float type carburetor. Before separating upper and lower carburetor bodies, loosen packing nut and unscrew needle valve and packing nut. Use special screwdriver, part number 19062 to remove nozzle.

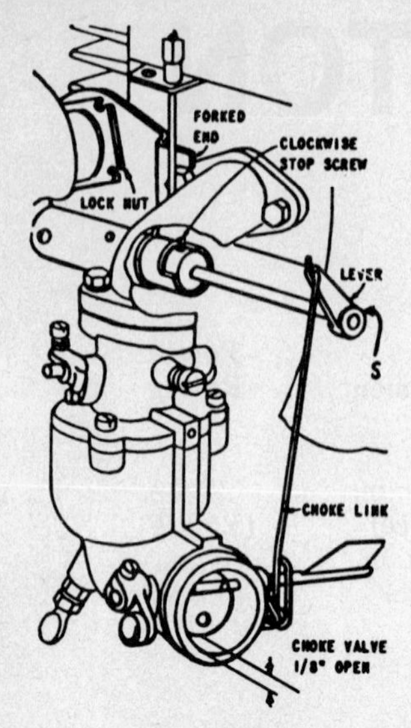

Fig. B31—Typical B&S automatic choke unit in "HOT" position.

feeler gage as shown in Fig. B33. If upper body is warped more than 0.002 inch (0.0508 mm) it should be renewed.

AUTOMATIC CHOKE. A thermostat operated choke is used on some models. To adjust choke linkage, hold choke shaft so thermostat lever is free. At room temperature, stop screw in thermostat collar should be located midway between thermostat stops (Fig. B31 & B32). If not, loosen stop screw, adjust collar and tighten stop screw. Loosen set screw (S–Fig. B31) on thermostat lever. Slide lever on shaft to insure free movement of choke unit. Turn thermostat shaft clockwise until stop screw contacts thermostat stop. While holding shaft in this position, move shaft lever

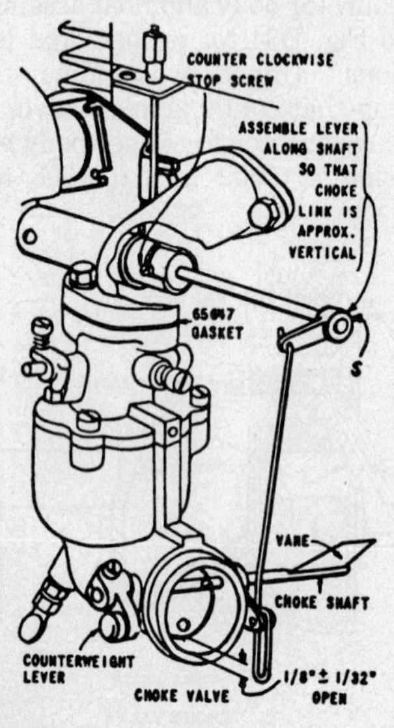

Fig. B32—Typical B&S automatic choke in "COLD" position.

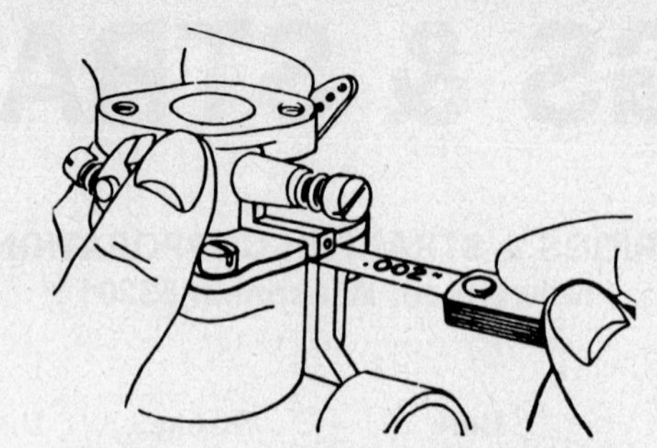

Fig. B33—Check carburetor upper body for warpage. If a 0.002 inch (0.0508 mm) feeler gage can be inserted between upper and lower bodies as shown, renew upper body.

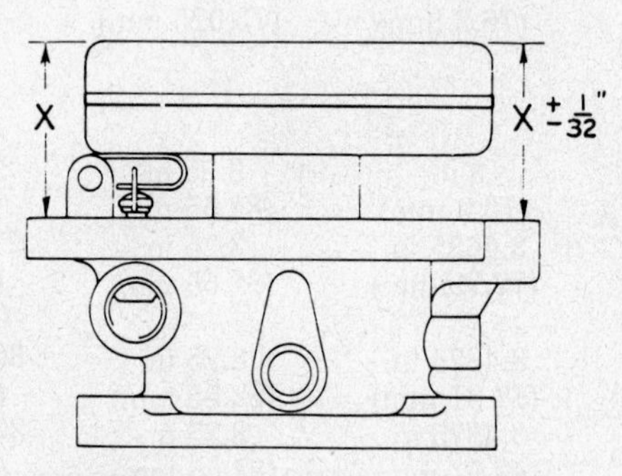

Fig. B34—Check carburetor float setting as shown. Bend tang if necessary to adjust float level.

until choke is open exactly 1/8-inch (3.18 mm) and tighten lever set screw. Turn thermostat shaft counter-clockwise until stop screw contacts thermostat stop as shown in Fig. B32. Manually open choke valve until it stops against top of choke link opening. At this time choke should be open at least 3/32-inch (2.25 mm), but not more than 5/32-inch (4 mm). Hold choke valve in wide open position and check position of counterweight lever. Lever should be in a horizontal position with free end towards right.

FUEL PUMP. A diaphragm type fuel pump is available on some models as optional equipment. Refer to **SERVICING BRIGGS & STRATTON ACCESSORIES** section for service information.

GOVERNOR. All models are equipped with a gear driven mechanical governor. Governor weight unit is enclosed within engine and is driven by camshaft gear. Refer to Fig. B62, B63 or B64 for appropriate exploded view of model being serviced.

To adjust governor, loosen screw holding governor lever to governor crank (46–Fig. B62, B63 or B64). Push governor lever as far clockwise as it will go. Tighten screw holding governor lever to crank until it is snug, but not tight. Push governor lever as far counter-clockwise as it will go and tighten screw securely.

Governor gear and weight unit can be removed when engine is disassembled. Remove governor lever, cotter pin and washer from outer end of governor crank. Slide crank out of bushing towards inside of engine. Governor gear and weight unit will then slide off of shaft. Shaft can be removed from crankcase if necessary. Renew bushing in crankcase for governor crank if bushing is worn and ream new bushing, after installation, to 0.2385-0.239 inch (6.06-6.07 mm).

Refer to Figs. B35 through B38 for governor control and carburetor linkage hook-up and adjustments.

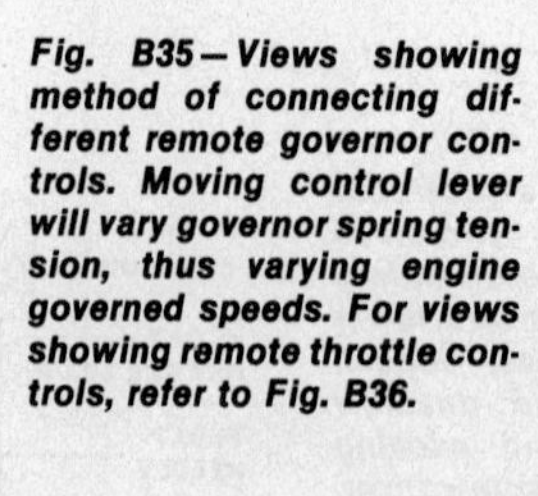

Fig. B35—Views showing method of connecting different remote governor controls. Moving control lever will vary governor spring tension, thus varying engine governed speeds. For views showing remote throttle controls, refer to Fig. B36.

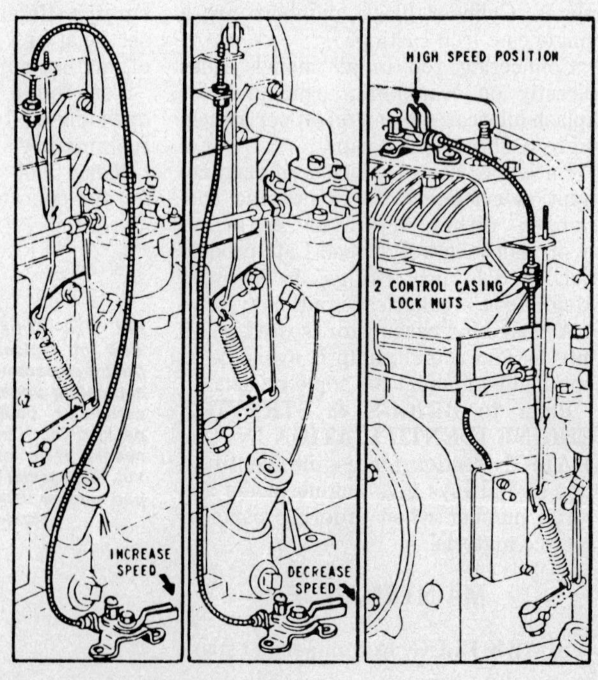

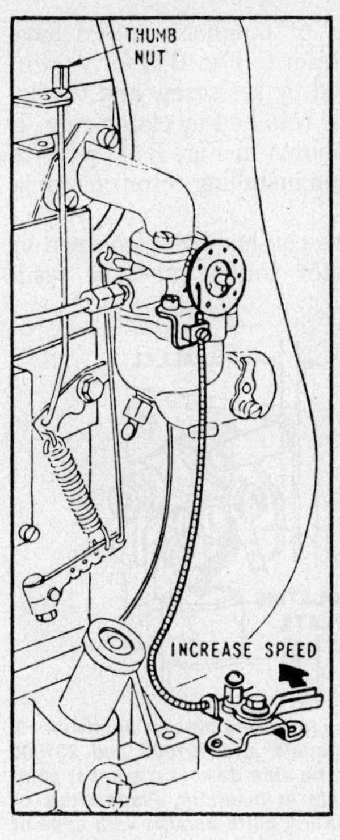

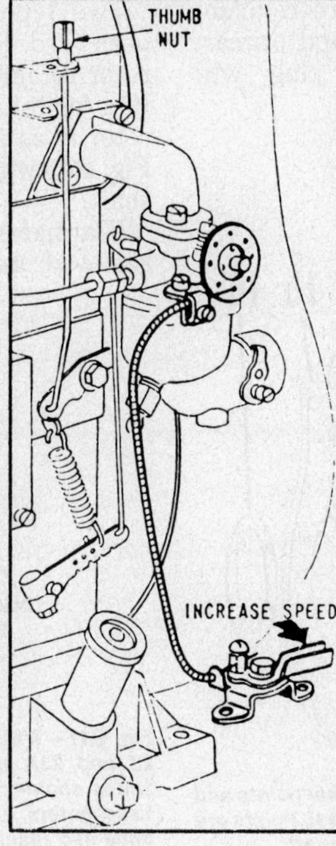

Fig. B36—Views showing methods of connecting remote throttle controls. When control lever is in high speed position, governor controls speed of engine and governed speed is adjusted by turning thumb nut. Moving control lever to slow speed position moves carburetor throttle shaft stop to decrease engine speed.

IGNITION SYSTEM. Early models use a variety of magneto type ignition systems and late production models use Briggs & Stratton Magnetron ignition system. Refer to appropriate paragraph for model being serviced.

MODEL 23C MAGNETO. Refer to Fig. B39 for exploded view of magneto used on 23C model engine. Condenser and breaker points are mounted on crankshaft bearing support plate (17) and are accessible after removing flywheel and magneto cover (7).

Breaker point gap is 0.020 inch (0.508 mm) and condenser capacity is 0.18-0.24 mfd.

Breaker point plunger can be removed from bore in crankshaft bearing support when breaker points are removed. Plunger should be renewed if worn to a length of 0.870 inch (22.098 mm) or less. Plunger bore diameter may be checked using B&S plunger bore gage, part number 19055. If gage enters plunger bore ¼-inch (6.35 mm) or more, install service bushing, part number 23513. To install bushing, remove bearing support from engine, then ream bore using B&S reamer, part number 19056, drive bushing in flush with outer end of plunger bore with B&S driver, part number 19057 and use B&S reamer, part number 19058, to finish ream inside of bushing.

When reinstalling flywheel, inspect the soft metal key and renew if damaged in any way.

NOTE: Renew key with correct B&S part. DO NOT substitute a steel key.

After flywheel is installed and retaining nut tightened, check armature air gap and adjust as necessary to 0.022-0.026 inch (0.5588-0.6604 mm).

MODELS 19, 23 & 23A AND SERIES 191000 & 231000 MAGNETO. Refer to Fig. B40 for exploded view of magneto used on these models.

Condenser and breaker points are mounted externally in a breaker box located on carburetor side of engine and are accessible after removing breaker box cover (18–Fig. B40).

Breaker point gap is 0.020 inch (0.508 mm) and condenser capacity is 0.18-0.24 mfd.

When renewing breaker points, or if oil leak is noted, breaker shaft oil seal (16) should be renewed. To renew points and/or seal, turn engine so breaker point gap is at maximum. Remove terminal and breaker spring screws and loosen breaker arm retaining nut (19) so it is flush with end of shaft (29). Tap loosened nut lightly to free breaker arm (21) from taper on shaft, then remove breaker arm, breaker plate (22), pivot (23), insulating plate (24) and eccentric (17). Pry oil seal out and press new oil seal in with metal side out. Place breaker plate on insulating plate with dowel on breaker plate entering hole in insulator. Then, install unit with edges of plates parallel with breaker box as shown in Fig. B41. Turn breaker shaft clockwise as far as possible and install breaker arm while holding shaft in this position.

Breaker box can be removed without removing points or condenser. Refer to Fig. B42. Disassembly of unit is evident after removal from engine.

To renew ignition coil, engine flywheel must be removed. Disconnect coil ground and primary wires and pull spark plug wire from hole in back plate. Disengage clips (8–Fig. B40) retaining coil core (9) to armature (14) and push

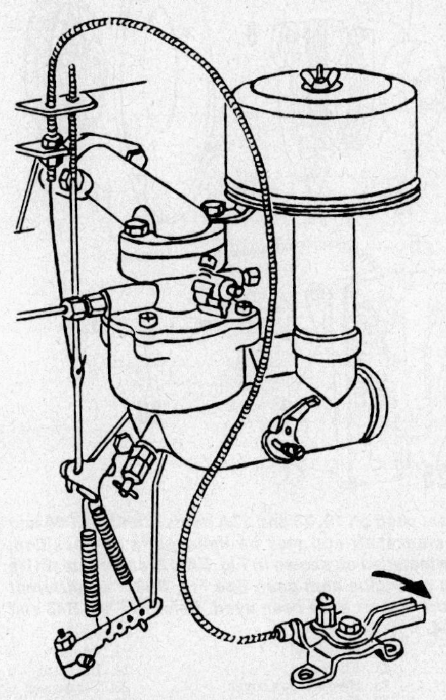

Fig. B37—View showing governor linkage, springs and levers properly installed on 243000, 300000, 301000, 302000, 320000, 325000 and 326000 series engines. Refer to Fig. B38 for remote control hook-up.

Fig. B38—Remote governor control installation on 243000, 300000, 301000, 302000, 320000, 325000 and 326000 series engines.

core from coil. Insert core in new coil with rounded side of core towards spark plug wire. Place coil and core on armature with retainer (7) between coil and armature. Reinstall core retaining clips, connect coil ground and primary wires and insert spark plug wire through hole in back plate.

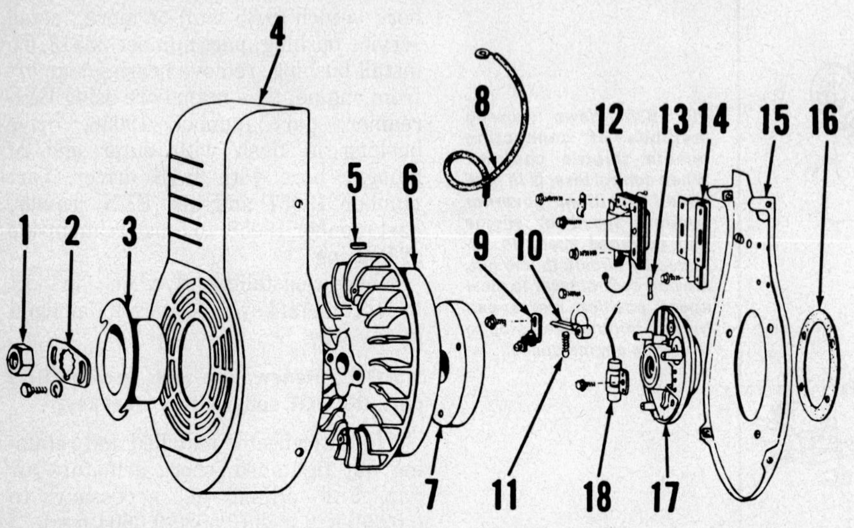

Fig. B39 – Exploded view of magneto ignition system used on 23C model engines. Breaker points and condenser are mounted on bearing support (17) and are enclosed by cover (7) and flywheel. Points are actuated by plunger (13) which rides against breaker cam machined on engine crankshaft.

1. Flywheel nut
2. Nut retainer
3. Starter pulley
4. Blower housing
5. Flywheel key
6. Flywheel
7. Breaker cover
8. Spark plug wire
9. Breaker point base
10. Breaker point arm
11. Breaker spring
12. Coil & armature
13. Plunger
14. Armature support
15. Back plate
16. Shim gasket
17. Bearing support
18. Condenser

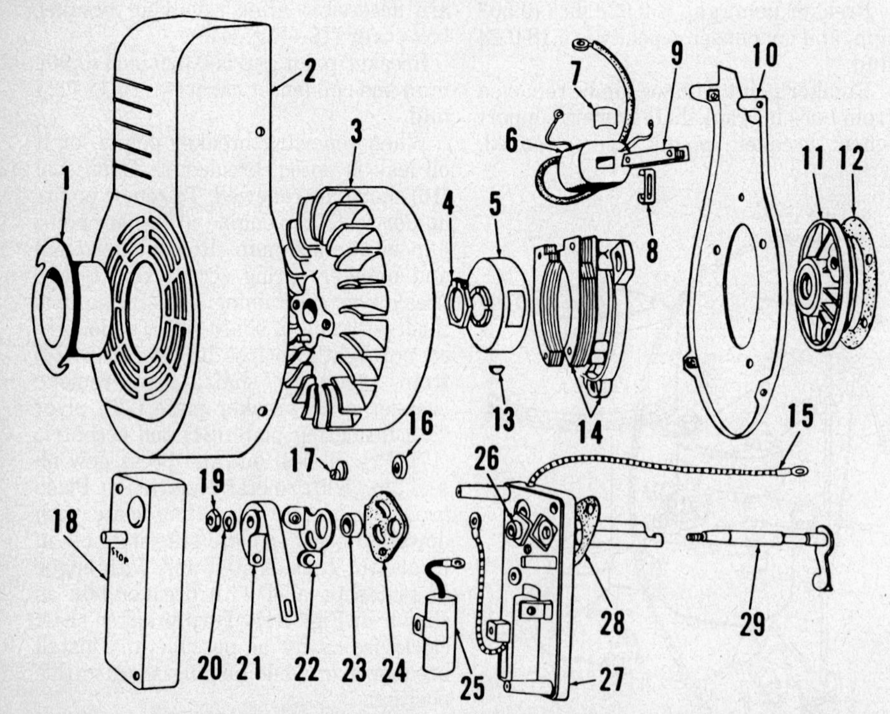

Fig. B40 – Exploded view of magneto ignition system used on 19, 23 and 23A models and 191000 and 231000 series engines. Flywheel is not keyed to crankshaft and may be installed in any position; however, on crank start models, flywheel should be installed as shown in Fig. B47. Breaker arm (21) is mounted on shaft (29) which is actuated by a cam on engine cam gear. See Fig. B48. Two different methods of attaching magneto rotor (5) to engine crankshaft have been used. Refer to Figs. B43 and B44.

1. Starter pulley
2. Blower housing
3. Flywheel
4. Rotor clamp
5. Magneto rotor
6. Ignition coil
7. Coil clip
9. Coil core
10. Back plate
11. Bearing support
12. Shim gasket
13. Rotor key
14. Armature
15. Primary coil lead
16. Shaft seal
17. Eccentric
18. Breaker box cover
19. Nut
20. Washer
21. Breaker point arm
22. Breaker point base
23. Pivot
24. Insulator
25. Condenser
26. Seal retainer
27. Breaker box
28. Gasket
29. Shaft

Two types of magnetic rotors have been used. Refer to Fig. B43 for view of rotor retained by set screw and to Fig. B44 for rotor retained by clamp ring. If rotor is as shown in Fig. B44, refer to Fig. B45 when installing rotor on crankshaft.

If armature coil has been loosened or removed, rotor timing must be read-

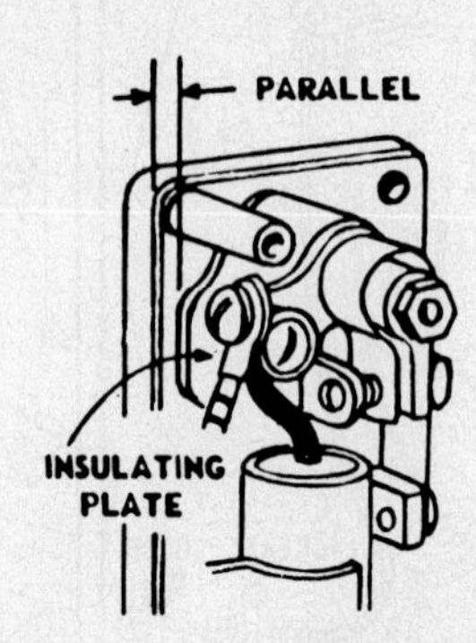

Fig. B41 – When installing breaker points on 19, 23 and 23A models and 191000 and 231000 series engines, be sure dowel on breaker point base enters hole in insulator. Place sides of base and insulating plate parallel with edge of breaker box.

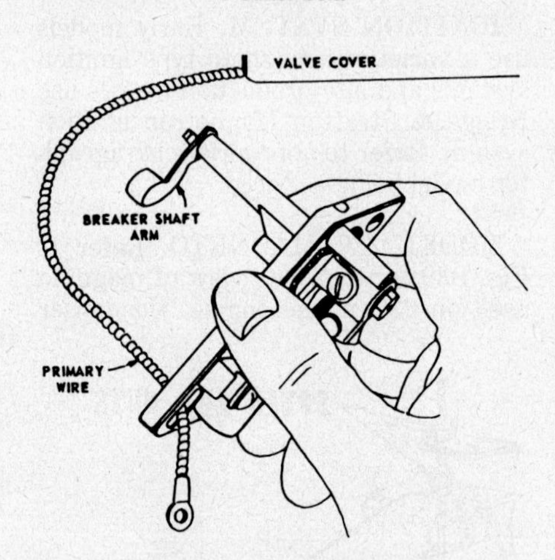

Fig. B42 – To remove breaker box on 19, 23 and 23A models and 191000 and 231000 series engines it is not necessary to remove points and condenser as unit may be removed as an assembly.

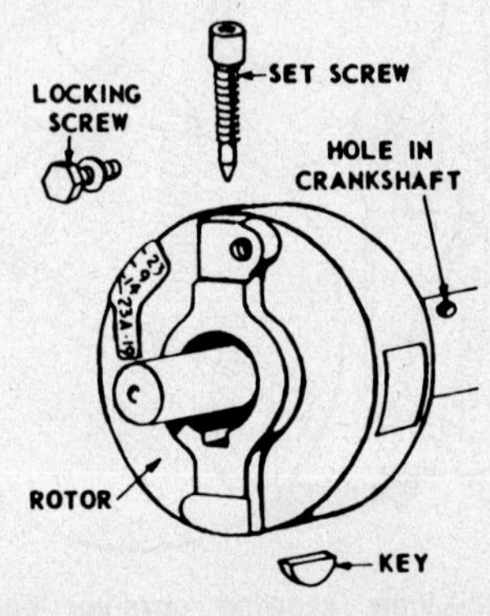

Fig. B43 – On some models, magneto rotor is fastened to crankshaft by a set screw which enters hole in shaft. Set screw is locked inplace by a second screw. Refer to Fig. B44 also.

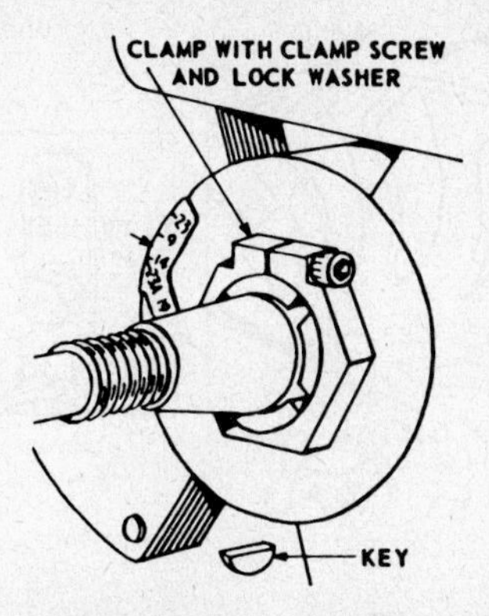

Fig. B44—View showing magneto rotor fastened to crankshaft with clamp. Make certain split in clamp is between two slots in rotor as shown and check clearance between rotor and shoulder on crankshaft as shown in Fig. B45.

justed. With point gap adjusted to 0.020 inch (0.508 mm), connect a static timing light from breaker point terminal to ground (coil primary wire must be disconnected) and turn engine in normal direction of rotation until light goes on. Then, turn engine very slowly in same direction until light just goes out (breaker points start to open). At this time, engine model number on magneto rotor should be aligned with arrow on armature as shown in Fig. B46. If not, loosen armature core retaining cap screws and turn armature in slotted mounting holes so arrow is aligned with appropriate engine model number on rotor. Tighten armature mounting screws.

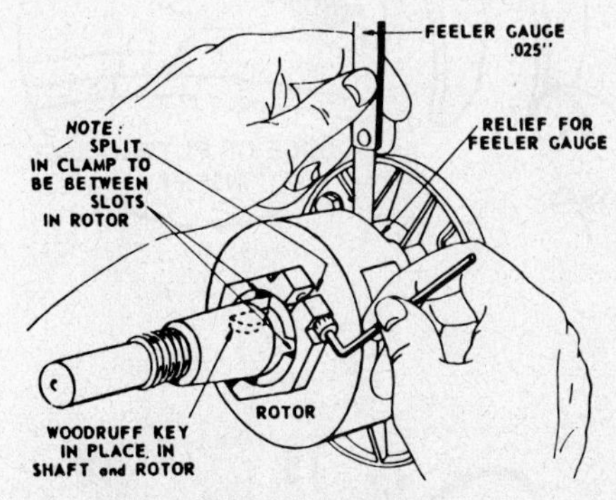

Fig. B45—Models having magneto rotor clamped to crankshaft, position rotor 0.025 inch (0.635 mm) from shoulder on shaft, then tighten clamping screw.

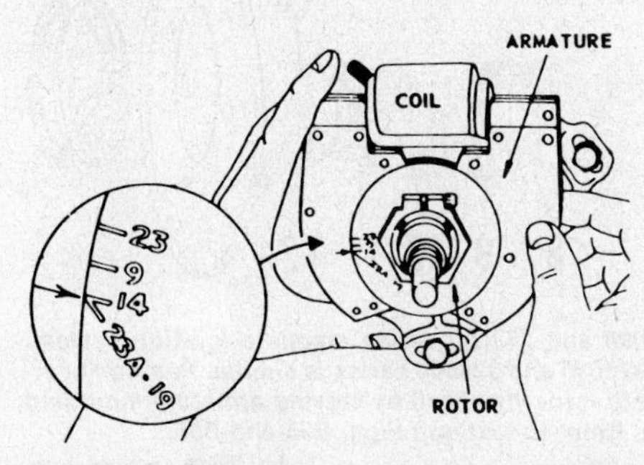

Fig. B46—With engine turned so breaker points are just starting to open, align model number line on rotor with arrow on armature by rotating armature in slotted mounting holes.

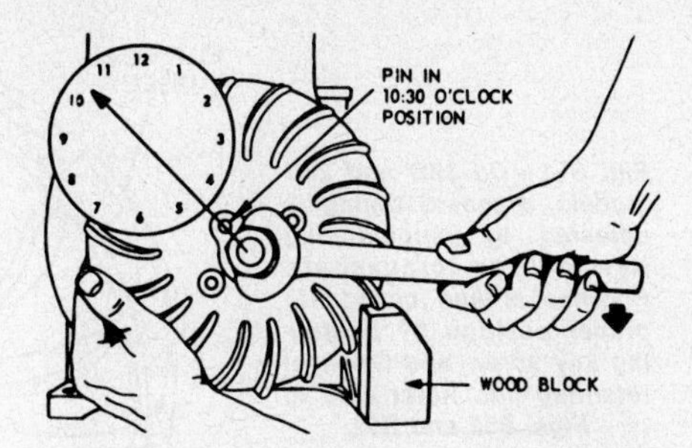

Fig. B47—On 19, 23 and 23A models and 191000 and 231000 series engines equipped with crank starter, mount flywheel with pin in position shown with armature timing marks aligned.

To install flywheel on models with crank starter, place flywheel on crankshaft as shown in Fig. B47 with magneto timing marks aligned as in previous paragraph. On models not having a crank starter, flywheel may be installed in any position.

Magneto breaker points are actuated by a cam on a centrifugal weight mounted on engine camshaft (Fig. B48). When engine is being overhauled, or cam gear is removed, check action of advance spring and centrifugal weight unit by holding cam gear in position shown and pressing weight down. When weight is released, spring should return weight to its original position. If weight does not return to its original position, check weight for binding and renew spring.

MODELS 19D & 23D MAGNETO. Refer to Fig. B49 for exploded view of magneto system.

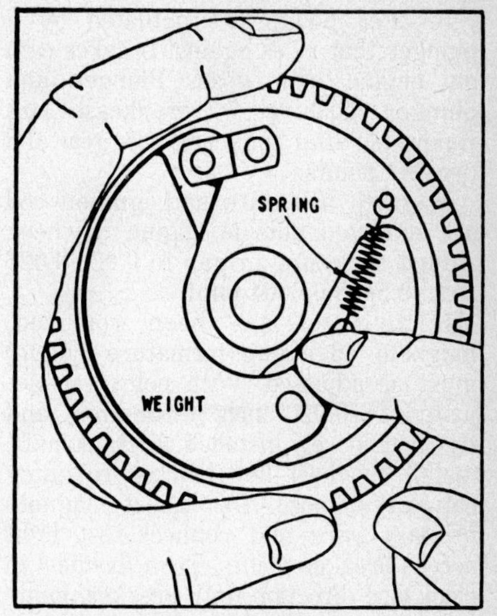

Fig. B48—Check timing advance weight and spring on 19, 23 and 23A models and 191000 and 231000 series engines. Refer to text.

Condenser and breaker points are mounted externally in a breaker box located on carburetor side of engine and are accessible after removing cover (8–Fig. B49).

Breaker point gap is 0.020 inch (0.508 mm) and condenser capacity is 0.18-0.24 mfd.

Installation of new breaker points is made easier by turning engine so points are open to their widest gap before removing old points. For method of adjusting breaker point gap, refer to Fig. B50.

NOTE: When installing points, apply Permatex or equivalent sealer to retaining screw threads to prevent engine oil from leaking into breaker box.

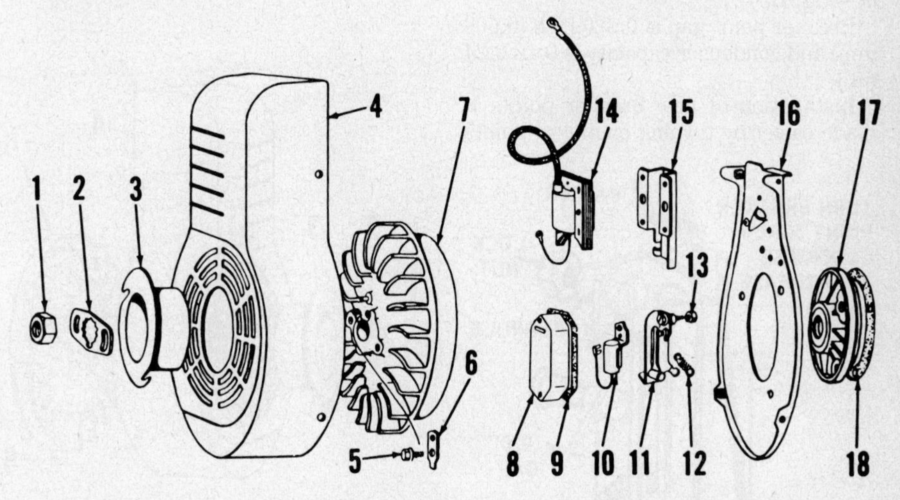

Fig. B49—Exploded view of magneto ignition system used on 19D and 23D models. Magneto rotor (flywheel) to armature timing is adjustable as shown in Figs. B51 and B52. Adjust breaker points as in Fig. B50.

1. Flywheel nut
2. Nut retainer
3. Starter pulley
4. Blower housing
5. Key cap screw
6. Flywheel key
7. Flywheel
8. Breaker box cover
9. Gasket
10. Condenser
11. Breaker points
12. Breaker spring
13. Lock nut
14. Coil & armature assy.
15. Armature mounting bracket
16. Back plate
17. Bearing support
18. Shim gasket

Breaker points are actuated by a plunger that rides against breaker cam on engine cam gear. Plunger and plunger bushing in crankcase are renewable after removing cam gear and breaker points.

Magneto armature and ignition coil are mounted outside engine flywheel. Adjust armature air gap to 0.022-0.026 inch (0.5588-0.6604 mm).

If flywheel has been removed, magneto edge gap (armature timing) must be adjusted. With point gap adjusted to 0.020 inch (0.508 mm) and flywheel loosely installed on crankshaft, install flywheel key leaving retaining cap screw loose. Disconnect magneto primary wire and connect test light across breaker points. Turn flywheel in clockwise direction until breaker points just start to open (timing light goes out). Now, while making sure engine crankshaft does not turn, turn flywheel back slightly in counter-clockwise direction so edge of flywheel insert lines up with edge of armature as shown in Fig. B53. Tighten flywheel key screw, then tighten flywheel retaining nut to 110-118 ft.-lbs. (149-160 N·m) torque for 19D model and to 138-150 ft.-lbs. (187-203 N·m) torque on 23D model. Readjust armature air gap as necessary.

SERIES 193000, 200000, 233000, 243000, 300000, 301000, 302000, 320000, 325000 & 326000 MAGNETO. Refer to Fig. 54 for exploded view of magneto system.

Condenser and breaker points are mounted externally in a breaker box located on carburetor side of engine and are accessible after removing cover (8–Fig. B54).

Breaker point gap is 0.020 inch (0.508 mm) and condenser capacity is 0.18-0.24 mfd.

Installation of new breaker points is made easier by turning engine so points

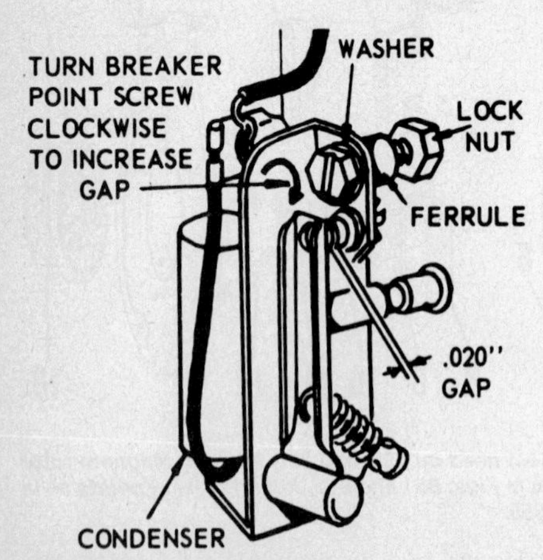

Fig. B50—On all models equipped with this type of breaker points, loosen lock nut and turn adjusting screw to obtain 0.020 inch (0.508 mm) gap.

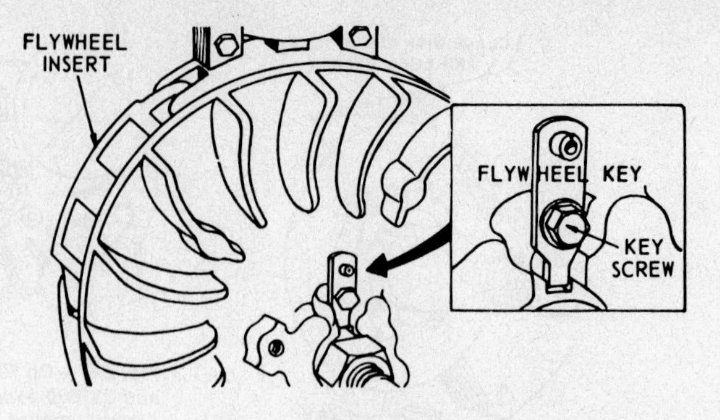

Fig. B51—On 19D and 23D models, magneto timing is adjusted by repositioning flywheel on crankshaft. Flywheel is then locked into proper position by tightening key screw and flywheel retaining nut. Refer also to Figs. B52 and B53.

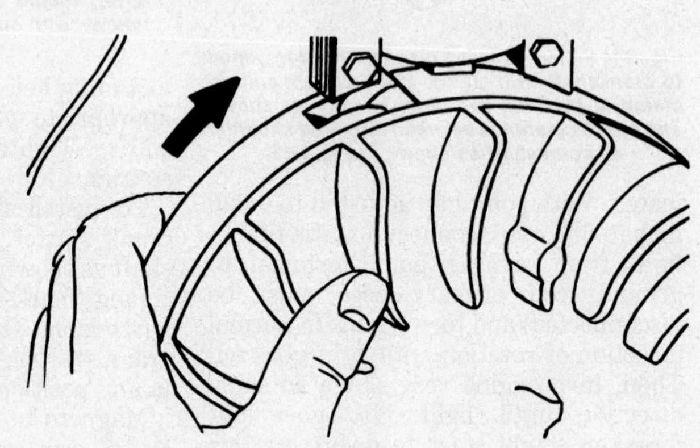

Fig. B52—Turning engine in normal direction of rotation with flywheel and flywheel key loose. Turn engine slowly until breaker points are just starting to open. Refer to text and Figs. B51 and B53.

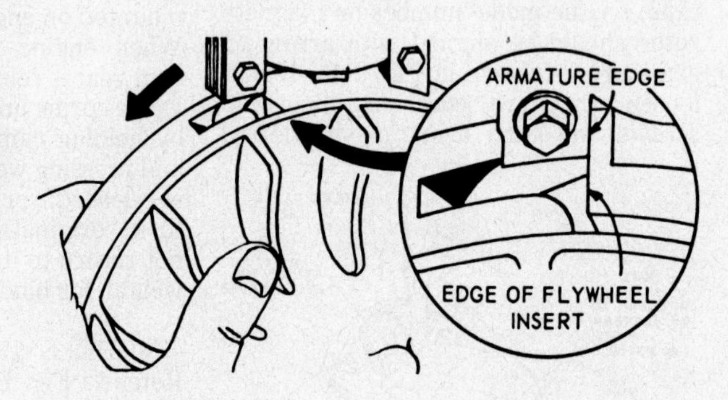

Fig. B53—Turn flywheel counter-clockwise on crankshaft until edge of flywheel insert is aligned with edge of armature as shown in insert. Then, tighten flywheel key screw and retaining nut.

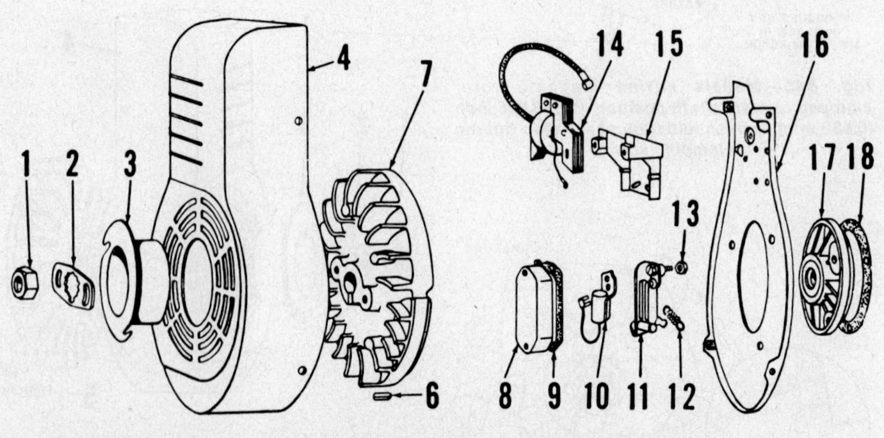

Fig. B54—Exploded view of 193000, 200000, 233000 and 243000 series magneto ignition system. Magneto used on 300000, 301000, 302000, 320000, 325000 and 326000 series is similar. Position of armature is adjustable to time armature with magneto rotor (flywheel) by moving armature mounting bracket (15) on slotted mounting holes. Refer to text and Figs. B55 and B56.

1. Flywheel nut
2. Nut retainer
3. Starter pulley
4. Blower housing
6. Flywheel key
7. Flywheel
8. Breaker box cover
9. Gasket
10. Condenser
11. Breaker points
12. Breaker spring
13. Lock nut
14. Coil & armature assy.
15. Armature mounting bracket
16. Back plate
17. Bearing support
18. Shim gasket

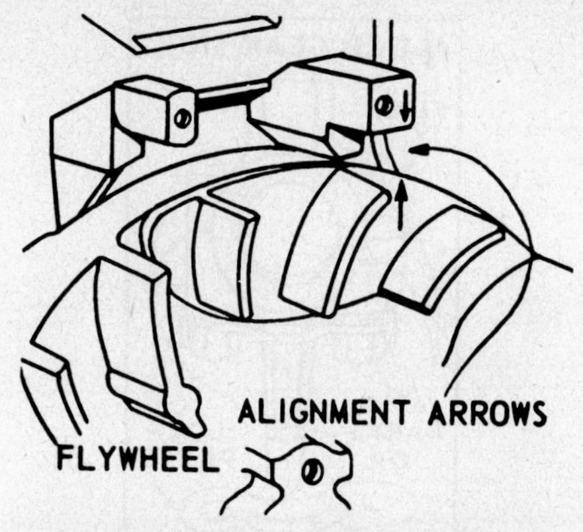

Fig. B55—On 193000, 200000, 233000, 243000, 300000, 301000, 302000, 320000, 325000 and 326000 series, time magneto by aligning armature core support so arrow on support is aligned with arrow on flywheel when breaker points are just starting to open. Refer to text for procedure.

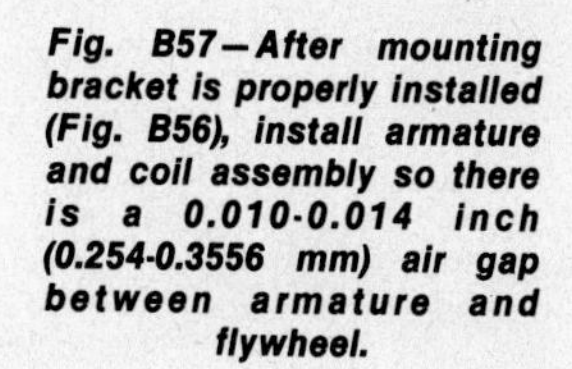

Fig. B57—After mounting bracket is properly installed (Fig. B56), install armature and coil assembly so there is a 0.010-0.014 inch (0.254-0.3556 mm) air gap between armature and flywheel.

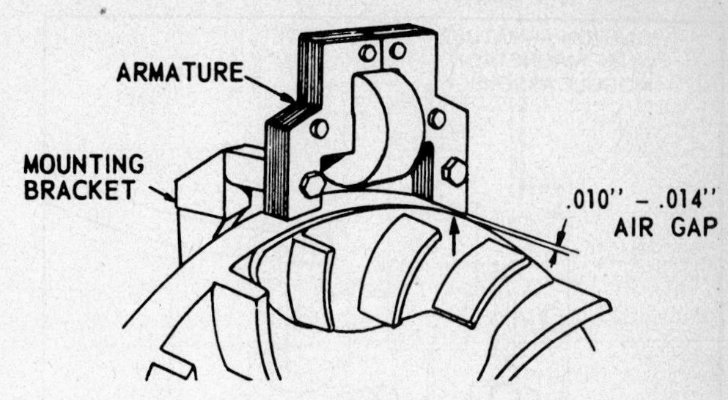

are open to their widest gap before removing old points. For method of adjusting breaker point gap, refer to Fig. B50.

NOTE: When installing points, apply Permatex or equivalent sealer to retaining screw threads to prevent engine oil from leaking into breaker box.

Breaker points are actuated by a plunger that rides against breaker cam on engine cam gear. Plunger (43–Fig. B62) and plunger bushing (PB) are renewable after removing engine cam gear and breaker points. On 300000, 301000, 302000, 320000, 325000 and 326000 series engines breaker plunger and plunger bushing are similar to those shown in Fig. B62.

Magneto armature and ignition coil are mounted outside engine flywheel. Adjust armature air gap to 0.010-0.014 inch (0.254-0.3556 mm).

If flywheel has been removed, magneto edge gap (armature timing) must be adjusted. With point gap adjusted to 0.020 inch (0.508 mm) and armature ignition coil assembly removed from bracket connect a test light across breaker points. Disconnect coil primary wire and slowly turn flywheel in a clockwise direction until light just goes out (breaker points start to open). At this time, arrow on flywheel should be exactly aligned with arrow on armature mounting bracket (Fig. B55). If not, mark position of bracket, remove flywheel and shift bracket on slotted mounting holes (Fig. B56) to bring arrows into alignment.

Reinstall flywheel, make certain arrows are aligned and tighten flywheel nut to 110-118 ft.-lbs. (149-160 N•m) torque on 193000 and 200000 series engines and 138-150 ft.-lbs. (187-203 N•m) torque on all remaining models. Readjust armature air gap as necessary.

MAGNETRON IGNITION. Magnetron ignition is a self-contained breakerless ignition system. Flywheel does not need to be removed except to change timing by moving armature bracket or to service crankshaft key or keyway.

To check spark, remove spark plug and connect spark plug cable to B&S tester, part number 19051 and ground remaining tester lead to engine cylinder head. Spin engine at 350 rpm or more. If spark jumps the 0.166 inch (4.2 mm) tester gap, system is functioning properly.

To remove armature and Magnetron module, remove flywheel shroud and armature retaining screws. Use a 3/16-inch diameter pin punch to release stop switch wire from module. To remove module, unsolder wires, push module retainer away from laminations and push module off. See Fig. B57A.

Resolder wires for reinstallation and use Permatex or equivalent gasket sealer to hold ground wires in position.

To set timing for gasoline fuel operation, install armature bracket so mounting screws are centered in slots (Fig. B57B).

To set timing for kerosene fuel operation, install armature bracket as far to left as possible (Fig. B57B).

Adjust armature air gap to 0.010-0.014 inch (0.25-0.36 mm).

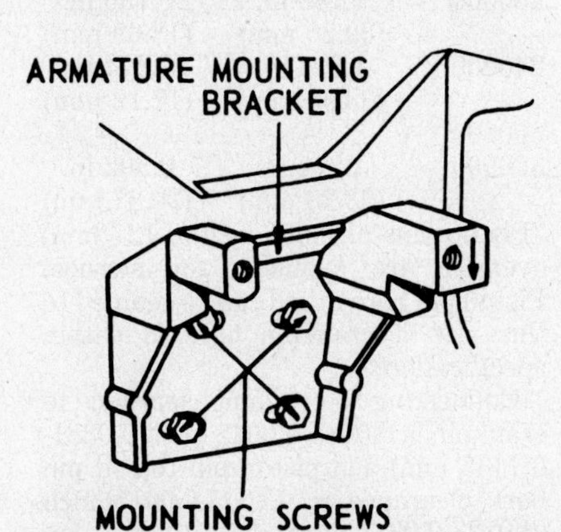

Fig. B56—View with flywheel removed on 193000, 200000, 233000, 243000, 300000, 301000, 302000, 320000, 325000 and 326000 series engines. Magneto is timed by shifting armature mounting bracket on slotted mounting holes. Refer to Fig. B55 also.

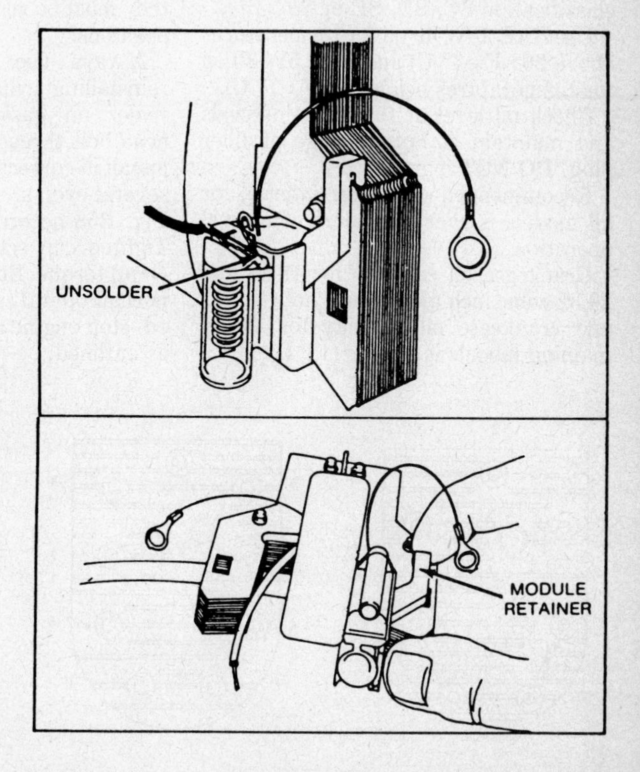

Fig. B57A—Wires must be unsoldered to remove Magnetron module.

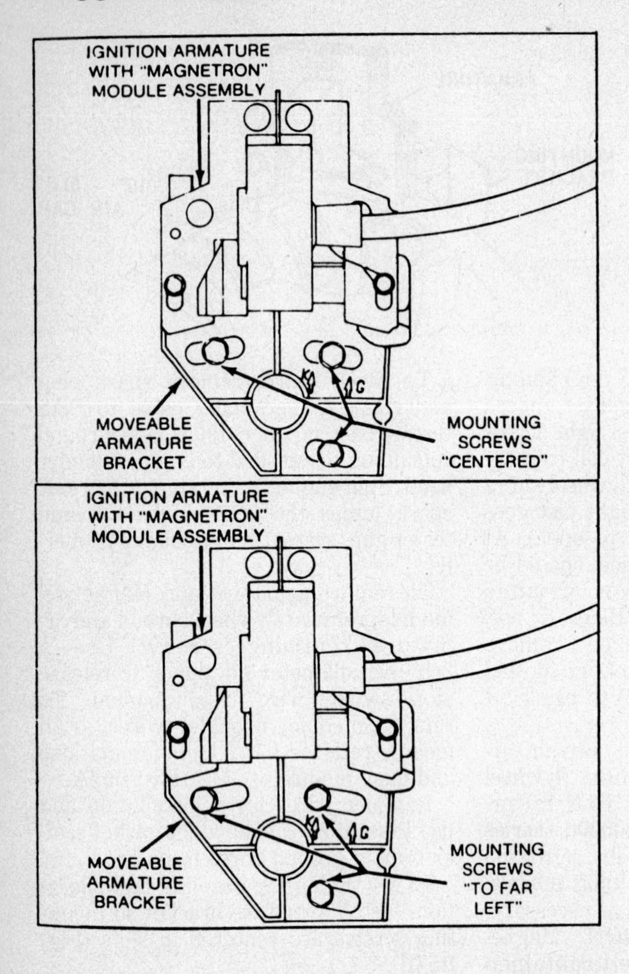

Fig. B57B — Upper view shows position of armature bracket for gasoline fuels and lower view shows position of armature bracket for kerosene fuels for correct ignition timing.

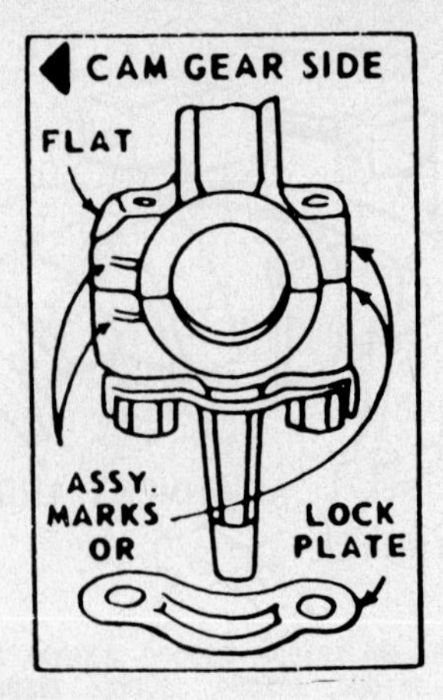

Fig. B59 — Drawing of lower end of connecting rod showing clearance flat and assembly marks.

LUBRICATION. All models are splash lubricated by an oil dipper attached to connecting rod.

Oils approved by manufacturer must meet requirements of API service classification SC, SD, SE or SF.

Use SAE 10W-40 oil for temperatures above 20° F (-7° C) and SAE 5W-30 oil for temperatures below 20° F (-7° C).

Check oil level at five hour intervals and maintain at bottom edge of filler plug. **DO NOT** overfill.

Recommended oil change interval for all models is every 25 hours of normal operation.

Crankcase oil capacity for 18.56 and 20.32 cubic inch models is 3 pints (1.4 L) and crankcase oil capacity for all remaining models is 4 pints (1.9 L).

REPAIRS

CYLINDER HEAD. When removing cylinder head note location from which different cap screws are removed as they must be reinstalled in their original positions.

Always use a new gasket when reinstalling cylinder head. Do not use sealer on gasket. Lubricate cylinder head bolt threads with graphite grease, install in correct locations and tighten in several even steps in sequence shown in Fig. B58 according to type head used. Tighten cap screws to 190 in.-lbs. (22 N·m) torque. Start and run engine until normal operating temperature is reached, stop engine and retighten head bolts as outlined.

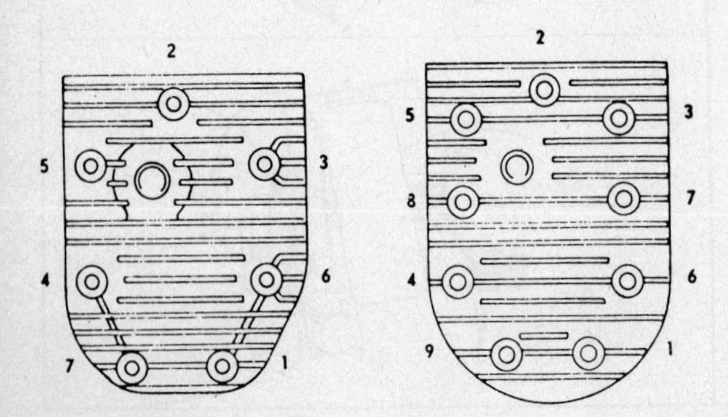

Fig. B58 — View of two different cylinder heads used. Tighten cylinder head cap screws in sequence shown to 190 in.-lbs. (22 N·m) torque. Refer to text for tightening procedure.

It is recommended carbon and lead deposits be removed at 100 to 300 hour intervals, or whenever cylinder head is removed.

CONNECTING ROD. Connecting rod and piston are removed from cylinder head end of block as an assembly. Aluminum alloy connecting rod rides directly on induction hardened crankshaft crankpin journal. Rod should be renewed if crankpin bore is scored or out-of-round more than 0.0007 inch (0.0178 mm) or if piston pin bore is scored or out-of-round more than 0.0005 inch (0.0127 mm).

REJECT SIZES FOR CONNECTING ROD

Model	Crankpin bore	Pin bore*
19, 19D, 190000	1.001 in. (25.43 mm)	0.674 in. (17.12 mm)
200000	1.127 in. (28.63 mm)	0.674 in. (17.12 mm)
23, 23A, 23C, 23D, 230000	1.189 in. (30.20 mm)	0.736 in. (18.69 mm)
240000	1.314 in. (33.38 mm)	0.674 in. (17.12 mm)
300000, 320000	1.314 in. (33.38 mm)	0.802 in. (20.37 mm)

*Piston pins of 0.005 inch (0.127 mm) oversize are available for service. Piston pin bore in rod can be reamed to this size if crankpin bore is within specification.

Connecting rod running clearance to crankpin is 0.0015-0.0045 inch (0.0381-0.1143 mm) and piston pin to rod pin bore clearance is 0.0005-0.0015 inch (0.0127-0.0381 mm).

On 300000, 301000, 302000, 320000, 325000 or 326000 series engines, notch on top of piston and letter "F" on side of piston must be on the same side as assembly marks on connecting rod

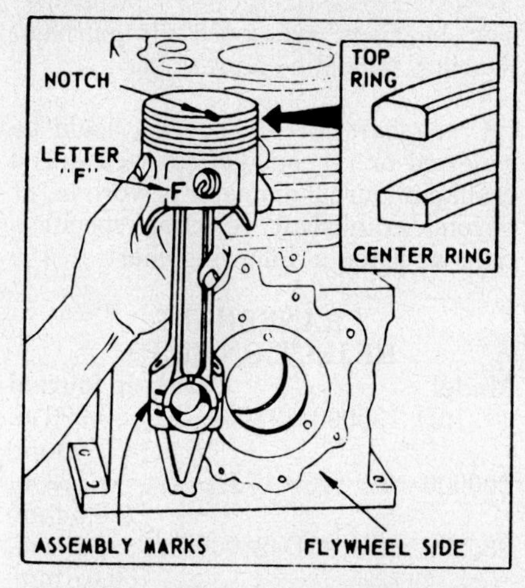

Fig. B60—On 300000, 301000, 302000, 320000, 325000 and 326000 series, assemble piston to connecting rod with notch and stamped letter "F" on piston to same side as assembly marks on rod. Install assembly in cylinder with assembly marks to flywheel side of crankcase.

(Fig. B60). Install assembly in cylinder with assembly marks to flywheel side of crankcase.

On all remaining models connecting rod position in crankcase is determined by clearance flat on rod as shown in Fig. B59. Assemble cap to rod by matching assembly marks.

Tighten connecting rod cap screws on all models to 190 in.-lbs. (22 N·m) torque.

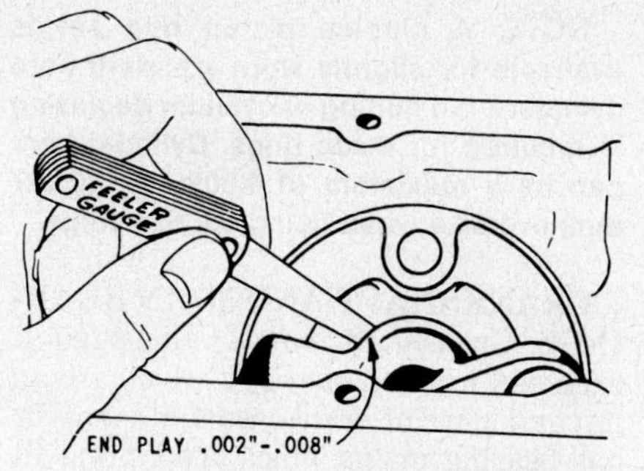

Fig. B61—Crankshaft end play can be checked as shown on models with main bearings which are an integral part of bearing support plates.

PISTON, PIN AND RINGS. Aluminum alloy piston has two compression and one oil control ring and piston should be renewed if top ring side clearance in groove exceeds 0.007 inch (0.178 mm) or if piston is visibly scored or damaged. Piston should also be renewed or pin bore reamed for 0.005 inch (0.127 mm) oversize pin if pin bore is worn to, or greater than rejection diameter shown in chart.

PISTON PIN BORE REJECTION SIZES

Model	Pin bore
19, 19D, 191000, 193000, 200000, 243000	0.673 in. (17.09 mm)
23, 23A, 23C, 23D, 231000, 233000	0.736 in. (18.69 mm)
300000, 301000, 302000, 320000, 325000, 326000	0.801 in. (20.35 mm)

Renew piston pin if pin is 0.0005 inch (0.0127 mm) out-of-round or if pin is worn to, or below, rejection diameter shown in chart.

PISTON PIN REJECTION SIZES

Model	Pin diameter
19, 19D, 191000, 193000, 200000, 243000	0.671 in. (17.04 mm)
23, 23A, 23C, 23D, 231000, 233000	0.734 in. (18.64 mm)
300000, 301,000, 302,000, 320000, 325000, 326000	0.799 in. (20.30 mm)

Ring end gap should be 0.010-0.025 inch (0.245-0.635 mm) and compression ring should be renewed if end gap is greater than 0.030 inch (0.75 mm). Oil control ring should be renewed if end gap is greater than 0.035 inch (0.90 mm).

Pistons used in 300000, 301000, 302000, 320000, 325000 and 326000 series engines have a notch and a letter "F" stamped in piston (Fig. B60) which

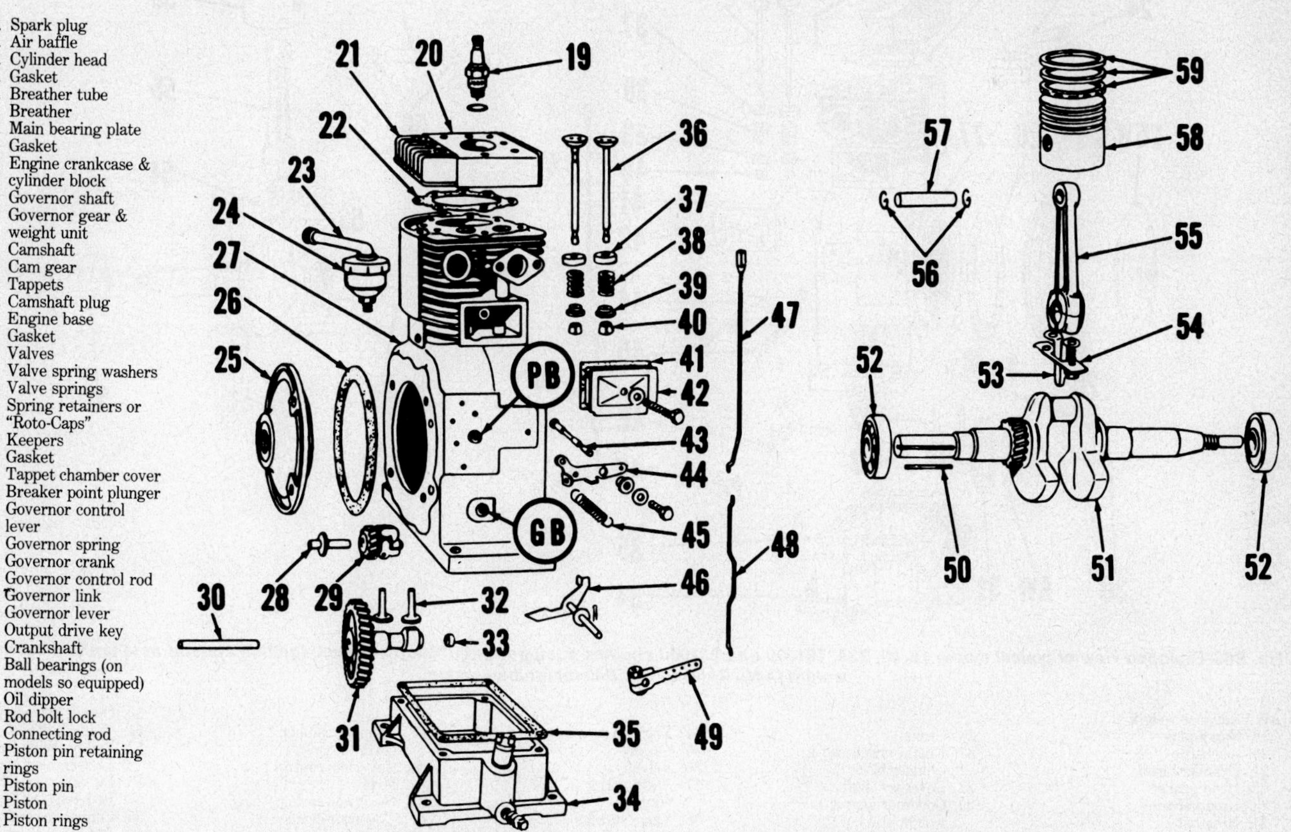

Fig. B62—Exploded view of typical 19D, 23D, 193000, 200000, 233000 and 243000 engines. Breaker plunger bushing (PB) and governor crank bushing (GB) in engine crankcase (27) are renewable. Breaker plunger (43) rides against a lobe on cam gear (31).

must face flywheel side of engine after installation.

Pistons and rings are available in a variety of oversizes as well as standard.

CYLINDER. Standard cylinder bore diameters are listed in following chart.

STANDARD CYLINDER BORE SIZES

Model	Standard bore
19, 19D, 191000, 193000, 23, 23A, 23C, 23D, 200000 231000, 233000	2.999-3.000 in. (76.18-76.23 mm)
243000	3.0615-3.0625 in. (76.79-76.82 mm)
300000, 301000, 302000	3.4365-3.4375 in. (87.287-87.313 mm)
320000, 325000, 326000	3.5615-3.5625 in. (90.462-90.488 mm)

If cylinder is worn 0.003 inch (0.076 mm) or more, or out-of-round 0.0025 inch (0.0635 mm) or more, it should be resized to nearest oversize for which piston and rings are available.

NOTE: A chrome piston ring set is available for slightly worn standard bore cylinders. No honing or cylinder deglazing is required for these rings. Cylinder bore can be a maximum of 0.005 inch (0.127 mm) oversize when using chrome rings.

CRANKSHAFT AND MAIN BEARINGS. Crankshaft may be supported at each end in main bearings which are an integral part of crankcase and sump or ball bearing mains which are a press fit on crankshaft and fit into machined bores in main bearing support plates.

Crankshaft for models with main bearings as an integral part of main bearing support plates should be renewed or reground if journals are out-of-round 0.0007 inch (0.0178 mm) or more or if main bearing journals are worn to a diameter of 1.179 inch (29.95 mm) or less for 19, 19D models and 191000, 193000 and 200000 series engines or 1.3759 inch (34.95 mm) or less for 23, 23A, 23C, 23D models and 231000 and 233000 series engines.

Crankshaft for models with ball bearing main bearings should be renewed if new bearings are loose on journals. Bearings should be a press fit.

Crankshaft for all models should be renewed or reground if connecting rod crankpin journal diameter is worn to, or beyond crankshaft rejection specifications as listed in following chart.

CRANKSHAFT REJECTION SIZES

Model	Crankpin journal
19, 19D, 190000, 193000	0.996 in. (25.30 mm)
200000	1.122 in. (28.5 mm)
23, 23A, 23C, 23D, 230000	1.184 in. (30.07 mm)
243000, 300000, 301000, 302000, 320000, 325000, 326000	1.309 in. (33.25 mm)

Connecting rod is available for crankshaft which has had connecting rod crankpin journal reground to 0.020 inch (0.508 mm) undersize.

Crankshaft main bearing support plates should be renewed if integral type

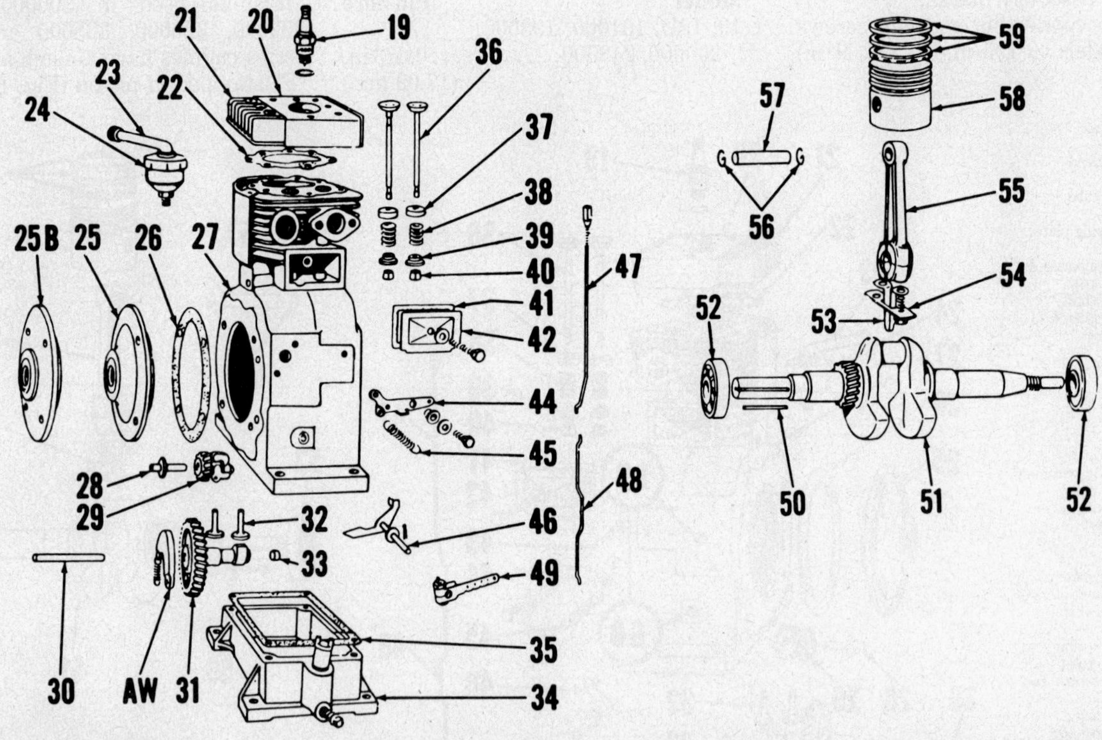

Fig. B63–Exploded view of typical model 19, 23, 23A, 191000 and 231000 engines equipped with "Magna-Matic" ignition system; note ignition advance weight (AW). Refer to Fig. B40 for ignition system.

AW. Advance weight
19. Spark plug
20. Air baffle
21. Cylinder head
22. Head gasket
23. Breather tube
24. Breather
25. Bearing plate (plain bushing)
25B. Bearing plate (ball bearing)
26. Gasket
27. Engine crankcase & cylinder block
28. Governor shaft
29. Governor gear & weight unit
30. Camshaft
31. Cam gear
32. Tappets
33. Camshaft plug
34. Engine base
35. Gasket
36. Valves
37. Valve spring washers
38. Valve springs
39. Spring retainers or "Roto-Caps"
40. Keepers
41. Gasket
42. Tappet chamber cover
44. Governor control lever
45. Governor spring
46. Governor crank
47. Governor control rod
48. Governor link
49. Governor lever
50. Output drive key
51. Crankshaft
52. Ball bearings (on models so equipped)
53. Oil dipper
54. Rod bolt lock
55. Connecting rod
56. Retaining rings
57. Piston pin
58. Piston
59. Piston rings

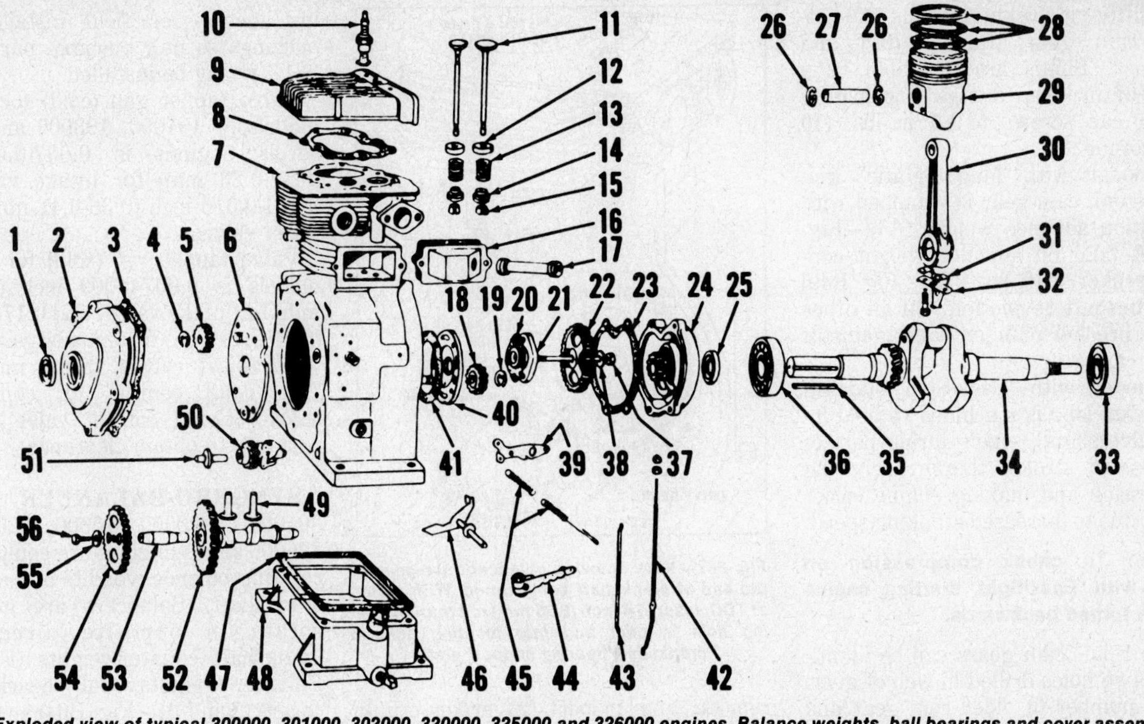

Fig. B64—Exploded view of typical 300000, 301000, 302000, 320000, 325000 and 326000 engines. Balance weights, ball bearings and cover assemblies (2 and 24) are serviced only as assemblies.

1. Oil seal
2. Cover & balance assy. (pto end)
3. Gasket
4. "E" ring
5. Idler gear
6. Bearing support plate
7. Cylinder block
8. Head gasket
9. Cylinder head
10. Spark plug
11. Valves
12. Spring caps
13. Valve springs
14. Spring retainer or rotator
15. Keepers
16. Breather assy.
17. Breather tube
18. Idler gear
19. "E" ring
20. Shim
21. Cam bearing
22. Balancer drive gear
23. Gasket
24. Cover & balance assy. (magneto end)
25. Oil seal
26. Retaining rings
27. Piston pin
28. Piston rings
29. Piston
30. Connecting rod
31. Oil dipper
32. Rod bolt lock
33. Ball bearing
34. Crankshaft
35. Key
36. Ball bearing
37. Drive gear bolt
38. Belleville washer
39. Governor control lever
40. Bearing support plate
41. Shim
42. Control rod
43. Link
44. Governor lever
45. Governor springs
46. Governor crank assy.
47. Gasket
48. Engine base
49. Valve lifters
50. Governor gear & weight unit
51. Governor shaft
52. Cam gear
53. Camshaft
54. Balancer drive gear
55. Belleville washer
56. Drive gear bolt

main bearing bores are 0.0007 inch (0.0178 mm) or more out-of-round or if bearing bore diameter is 1.185 inches (30.1 mm) or more for 19, 19D models or 191000, 193000 and 200000 series engines, 1.382 inches (34.1 mm) or more for 23, 23A, 23C, 23D models or 243000, 300000, 301000, 302000, 320000, 325000 and 326000 series engines. No service bushings for main bearings are available.

Ball bearing mains are a press fit on crankshaft and must be removed by pressing crankshaft out of bearing. Renew ball bearing if worn or rough. Expand new bearing by heating in oil and install on crankshaft with shield side towards crankpin journal.

Crankshaft end play should be 0.002-0.008 inch (0.0508-0.2032 mm) and is adjusted by varying thickness of gaskets between flywheel main bearing support plate and crankcase. Gaskets are available in a variety of thicknesses.

When reinstalling crankshaft, make certain timing marks are aligned (Fig. B66 or B67).

CAM GEAR AND CAMSHAFT. On 19, 19D, 23, 23A, 23C, 23D models or 191000, 193000, 200000, 233000 and 243000 series engines timing gear and camshaft lobes are cast as an integral part and are referred to as a "cam gear". Cam gear turns on a stationary shaft referred to as a "camshaft". Reject camshaft if it is worn to, or less than, a diameter of 0.4968 inch (12.62 mm).

On all remaining models cam gear and lobes are an integral part referred to as a "cam gear". Camshaft (53–Fig. B64) rotates with cam gear (52). Camshaft rides in a bearing in cylinder block on pto end of engine and journal on camgear rides in renewable cam bearing (21) on magneto end of engine.

Renew camshaft if worn to 0.6145 inch (15.6 mm) or less and renew cam gear (52) if journal diameter is 0.8105 inch (20.6 mm) or less or if lobes are worn to a diameter of 1.184 inches (30.1 mm) or less for 300000, 301000 and 302000 series engines or 1.215 inches (30.9 mm) or less for 320000, 325000 and 326000 series engines.

When installing cam gear in 300000, 301000, 302000, 320000, 325000 and 326000 series engines, cam gear end play should be 0.002-0.008 inch (0.0508-0.2032 mm) and is controlled by

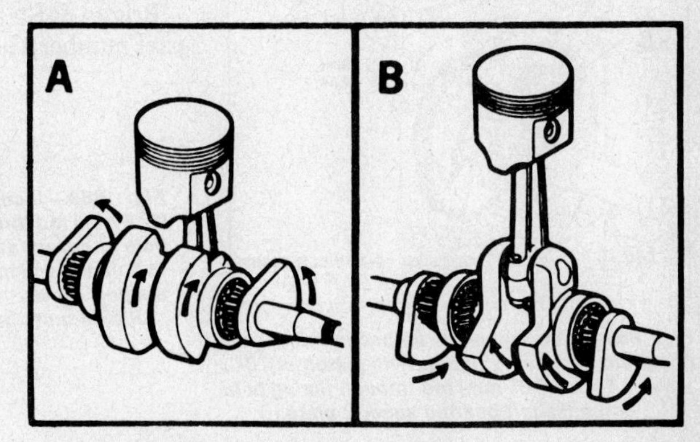

Fig. B65—Synchro-Balance weights rotate in opposite direction of crankshaft counterweights on 300000, 301000, 302000, 320000, 325000 and 326000 series engines.

using different thickness shims (20) between cam gear bearing (21) and crankcase. Shims are available in a variety of thicknesses. Tighten cam gear bearing cap screws to 85 in.-lbs. (10 N•m) torque.

On models with "Magna-Matic" ignition system, cam gear is equipped with an ignition advance weight (AW–Fig. B63). A tang on advance weight contacts breaker shaft lever (29–Fig. B40) each camshaft revolution. On all other models, breaker plunger rides against a lobe on cam gear.

On models with "Easy-Spin" starting intake cam lobe is machined to hold intake valve slightly open during part of compression stroke, thereby relieving compression and making engine easier to start due to increased cranking speed.

NOTE: To check compression on models with "Easy-Spin" starting, engine must be turned backwards.

"Easy-Spin" cam gears can be identified by two holes drilled in web of gear. If part number of older cam gear and "Easy-Spin" cam gear are the same except for an "E" following new part number, gears are interchangeable.

On all models, align timing marks on crankshaft gear and cam gear when reassembling engine.

GOVERNOR WEIGHT UNIT. Tangs on governor weights should be square and smooth and weights should operate freely. If not, renew gear and weight assembly (29–Figs. B62 and B63) or (50–Fig. B64). Renew governor shaft if worn or scored.

GOVERNOR CRANK. With governor weight and gear unit removed,

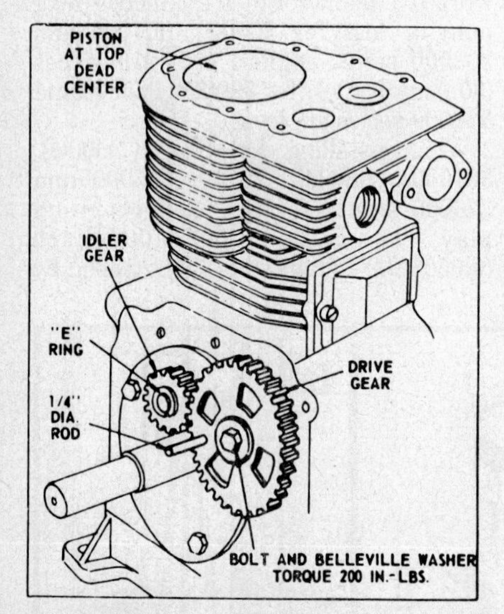

Fig. B66 – View showing balancer drive gear (magneto end) being timed. With piston at TDC, insert 1/4-inch (6.35 mm) rod through timing hole in crankshaft bearing support plate.

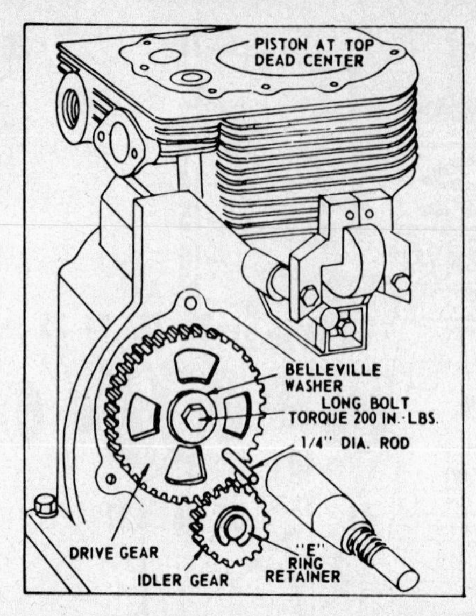

Fig. B67 – View showing balancer drive gear on pto end of crankshaft being timed. With piston at TDC, insert 1/4-inch (6.35 mm) rod through timing hole in gear and into locating hole in crankshaft bearing support plate.

remove and inspect governor crank; renew if worn. Governor crank should be free fit in bushing in engine crankcase with minimum bushing to crank clearance. Renew bushing if new crank fits loosely. Bushing should be finish reamed to 0.2385-0.239 inch (6.0579-6.0706 mm) after installation.

VALVE SYSTEM. All models are equipped with renewable exhaust valve seat inserts and intake valve seats are machined directly into cylinder block. Seats are ground at a 45° angle and seat width should be 3/64 to 1/16-inch (1.2 to 1.6 mm). Service exhaust and intake valve seat inserts are available and are installed using special tools available from Briggs & Stratton.

Valve face is ground at a 45° angle and valve should be renewed if margin is 1/64-inch (0.4 mm) or less after refacing.

Valve guides should be checked for wear using valve guide gage number 19151. If gage enters valve guide 5/16-inch (7.9 mm) or more, guides should be reamed using reamer number 19183 and bushing number 19192 should be installed.

Briggs & Stratton also had a tool kit, part number 19232, available for removing factory or field installed guide bushings so new bushing, part number 231218 may be installed.

Valve tappet gap (cold) for 19, 19D models or 191000, 193000 and 200000 series engines is 0.007-0.009 inch (0.18-0.23 mm) for intake valves and 0.014-0.016 inch (0.36-0.41 mm) for exhaust valves.

Valve tappet gap (cold) for all other models is 0.007-0.009 inch (0.18-0.23 mm) for intake valves and 0.17-0.19 inch (0.43-0.48 mm) for exhaust valves.

To adjust valves, piston must be at "top dead center" on compression stroke. Grind end of valve stem off squarely to obtain clearance.

SYNCHRO-BALANCER. 300000, 301000, 302000, 320000, 325000 and 326000 series engines are equipped with rotating balance weights at each end of crankshaft. Balancers are geared to rotate in opposite direction of crankshaft counterweights (Fig. B65). Balance weights, ball bearings and covers (2 and 24–Fig. B64) are serviced as assemblies only.

Balancers are driven from idler gears (5 and 18) which are driven by gears (22 and 54). Drive gears (22 and 54) are bolted to camshaft. To time balancers, first remove cover and balancer assemblies (2 and 24). Position piston at TDC. Loosen bolts (37 and 56) until drive gears will rotate on cam gear and camshaft. Insert a ¼-inch (6.35 mm) rod through timing hole in each drive gear and into locating holes in main bearing support plates as shown in Figs. B66 and B67. With piston at TDC and ¼-inch (6.35 mm) rods in place, tighten drive gear bolts to a torque of 200 in.-lbs. (23 N•m). Remove ¼-inch (6.35 mm) rods. Remove timing hole screws (Fig. B68) and insert ⅛-inch (3.18 mm) rods through timing holes and into hole in balance weights. Then, with piston at TDC, carefully slide cover assemblies into position. Tighten cap screws in pto end cover to 200 in.-lbs. (23 N•m) torque and tighten magneto end cover cap screws to 120 in.-lbs. (14 N•m) torque. Remove ⅛-inch (3.18 mm) rods. Coat threads of timing hole screws with Permatex or equivalent and install screws with fiber sealing washers.

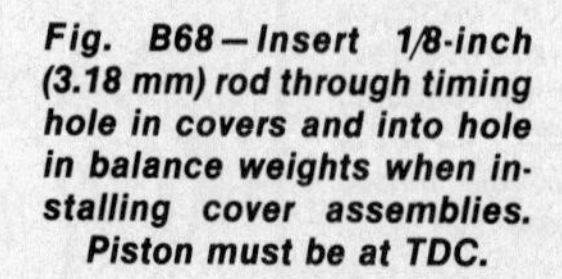

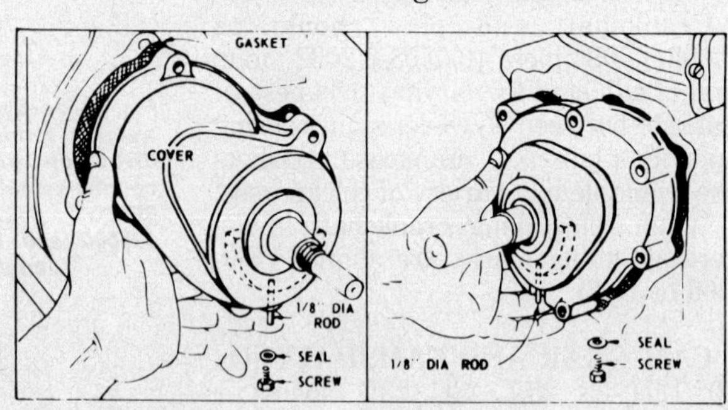

Fig. B68 – Insert 1/8-inch (3.18 mm) rod through timing hole in covers and into hole in balance weights when installing cover assemblies. Piston must be at TDC.

BRIGGS & STRATTON

BRIGGS & STRATTON CORPORATION
Milwaukee, Wisconsin 53201

Model	No. Cyls.	Bore	Stroke	Displacement	Power Rating
400400	2	3.44 in. (87.3 mm)	2.16 in. (54.8 mm)	40 cu. in. (656 cc)	14 hp. (10.4 kW)
400700	2	3.44 in. (87.3 mm)	2.16 in. (54.8 mm)	40 cu. in. (656 cc)	14 hp. (10.4 kW)
401400	2	3.44 in. (87.3 mm)	2.16 in. (54.8 mm)	40 cu. in. (656 cc)	16 hp. (11.9 kW)
401700	2	3.44 in. (87.3 mm)	2.16 in. (54.8 mm)	40 cu. in. (656 cc)	16 hp. (11.9 kW)
402400	2	3.44 in. (87.3 mm)	2.16 in. (54.8 mm)	40 cu. in. (656 cc)	16 hp. (11.9 kW)
402700	2	3.44 in. (87.3 mm)	2.16 in. (54.8 mm)	40 cu. in. (656 cc)	16 hp. (11.9 kW)
421400	2	3.44 in. (87.3 mm)	2.28 in. (57.9 mm)	42.33 cu. in. (694 cc)	18 hp. (13.4 kW)
421700	2	3.44 in. (87.3 mm)	2.28 in. (57.9 mm)	42.33 cu. in. (694 cc)	18 hp. (13.4 kW)
422400	2	3.44 in. (87.3 mm)	2.28 in. (57.9 mm)	42.33 cu. in. (694 cc)	18 hp. (13.4 kW)
422700	2	3.44 in. (87.3 mm)	2.28 in. (57.9 mm)	42.33 cu. in. (694 cc)	18 hp. (13.4 kW)

Engines in this section are four cycle, two cylinder opposed, horizontal or vertical crankshaft engines. Crankshaft is supported at each end in bearings which are an integral part of crankcase and sump or cover, DU type bearings or ball bearings which fit into machined bores of crankcase and sump or cover. Cylinder block and crankcase are a single aluminum casting, however, some models are equipped with cast iron cylinder liners which are an integral part of aluminum casting.

Connecting rods for all models ride directly on crankpin journals. Vertical crankshaft models are lubricated by a gear driven oil slinger and horizontal crankshaft models are splash lubricated by an oil dipper attached to number one cylinder connecting rod cap.

Early models are equipped with a flywheel magneto ignition with points, condenser and coil mounted externally on engine. Late models are equipped with "Magnetron" breakerless ignition.

All models use a float type carburetor with an integral fuel pump.

Always give engine model and serial number when ordering parts or service material.

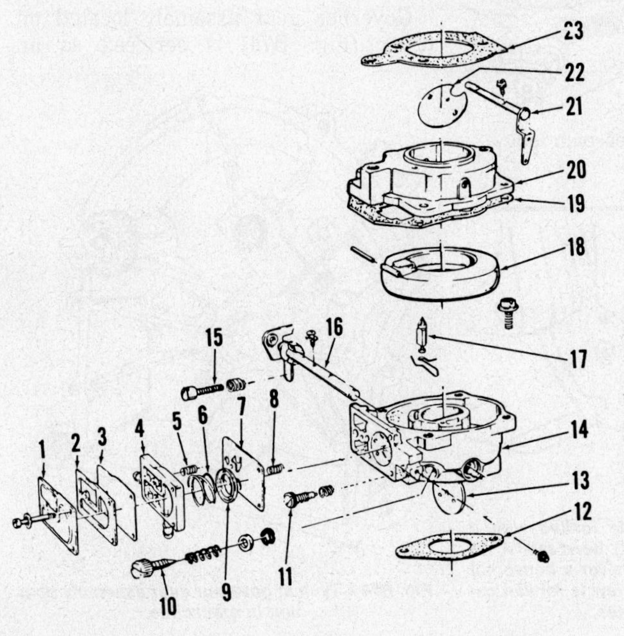

Fig. B70—Exploded view of downdraft Flo-Jet with integral fuel pump.

1. Diaphragm cover
2. Gasket
3. Damping diaphragm
4. Pump body
5. Spring
6. Pump spring
7. Diaphragm
8. Spring
9. Spring cap
10. Needle valve assy.
11. Idle valve assy.
12. Mounting gasket
13. Throttle valve
14. Lower carburetor body
15. Throttle adjustment screw
16. Throttle shaft
17. Fuel inlet valve
18. Float
19. Carburetor body gasket
20. Upper carburetor body
21. Choke shaft
22. Choke valve
23. Air cleaner gasket

MAINTENANCE

SPARK PLUG. Recommended spark plug for all models is Champion RJ12. Electrode gap is 0.030 inch (0.762 mm).

CARBURETOR. A downdraft float type carburetor with integral fuel pump is used. Refer to Fig. B70.

For initial carburetor adjustment, open idle mixture screw (11–Fig. B70) and main fuel mixture screw (10) 1½-turns each.

Make final adjustments with engine at operating temperature and running. Place governor speed control lever in idle position and hold throttle lever against idle stop. Turn idle speed adjusting screw (15) to obtain 1400 rpm. Adjust idle mixture screw to obtain smoothest idle operation. Hold throttle shaft in closed position and readjust idle speed adjusting screw (15) to obtain 900 rpm. Release throttle. Move remote control to a position where a ⅛-inch (3.18

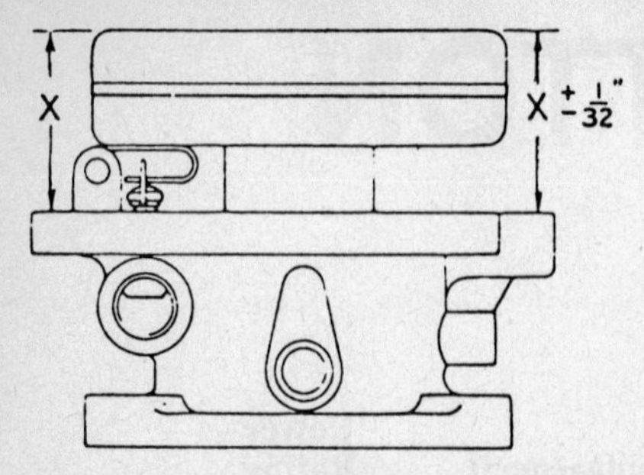

Fig. B70A—Check carburetor float setting as shown. Bend tang if necessary to adjust float level.

mm) diameter pin can be inserted through two holes in governor control plate (Fig. B71). With remote control in governed idle position, bend tab "A", Fig. B72, to obtain 1400 rpm. Remove pin. Place governor speed control lever in fast position and adjust main fuel mixture screw for leanest mixture that will allow satisfactory acceleration and steady governor operation.

To disassemble carburetor, remove idle and main fuel mixture screws. Remove fuel pump body and upper carburetor body. Remove float assembly and fuel inlet valve. Inlet valve seat is a press fit in upper carburetor body. Use a self threading screw to remove seat. New seat should be pressed into upper carburetor body until flush with body. Remainder of carburetor disassembly is evident after inspection and reference to Fig. B70.

If necessary to renew throttle shaft bushings, use a 1/4 x 20 tap to remove old bushings. Press new bushings in using a vise and ream with a 7/32-inch drill if throttle shaft binds.

To check float level, invert carburetor body and float assembly. Refer to Fig. B70A for proper float level dimensions. Adjust by bending float lever tang that contacts inlet valve. Reassemble carburetor by reversing disassembly procedure.

Correct choke and speed control operation is dependent upon proper adjustment of remote controls. To adjust choke, place control lever in "CHOKE" position. Loosen casing clamp screw. Move casing and wire until choke is completely closed and tighten screw.

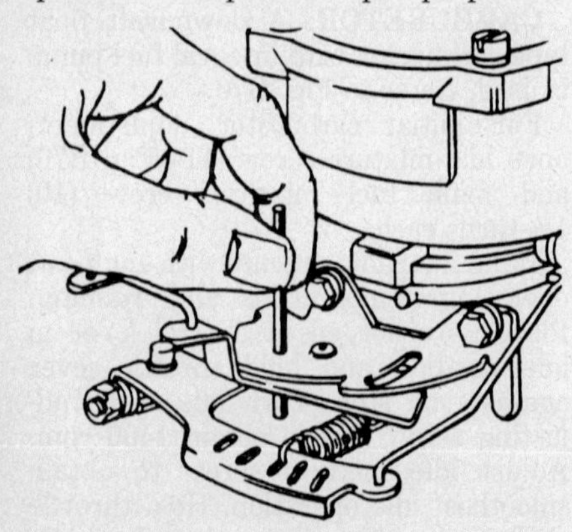

Fig. B71—Insert a 1/8-inch (3.18 mm) diameter pin through the two holes in governor control plate to correctly set governed idle position.

FUEL PUMP. All parts of the vacuum-diagram type pump are serviced separately. When disassembling pump, care must be taken to prevent damage to pump body (plastic housing) and diaphragm. Inspect diaphragm for punctures, wrinkles or wear. All mounting surfaces must be free of nicks, burrs and debris.

To assemble pump, position diaphragm on carburetor. Place spring and cup on top of diaphragm. Install flapper valve springs. Carefully place pump body, remaining diaphragm, gasket and cover plate over carburetor casting and install mounting screws. Tighten screws in a staggered sequence to avoid distortion.

GOVERNOR. All models are equipped with a gear driven mechanical governor. Governor gear and weight assembly is enclosed within the engine and is driven by the camshaft gear. Refer to Fig. B74.

To adjust governor, loosen nut holding governor lever to governor shaft. Push governor lever counter-clockwise until throttle is wide open. Hold lever in this position while rotating governor shaft counter-clockwise as far as it will go. Tighten governor lever nut to a torque of 100 in.-lbs. (11 N•m).

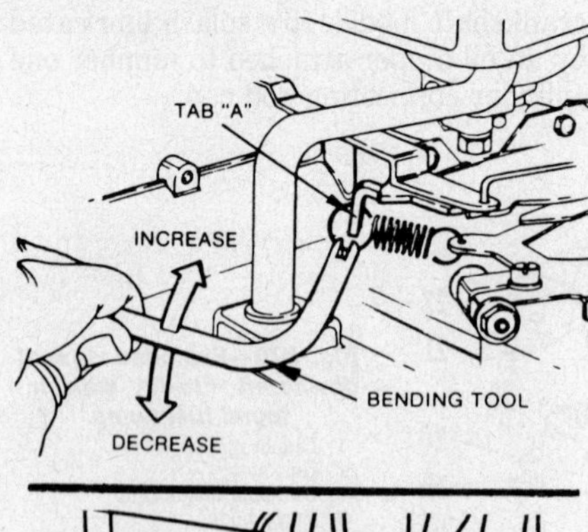

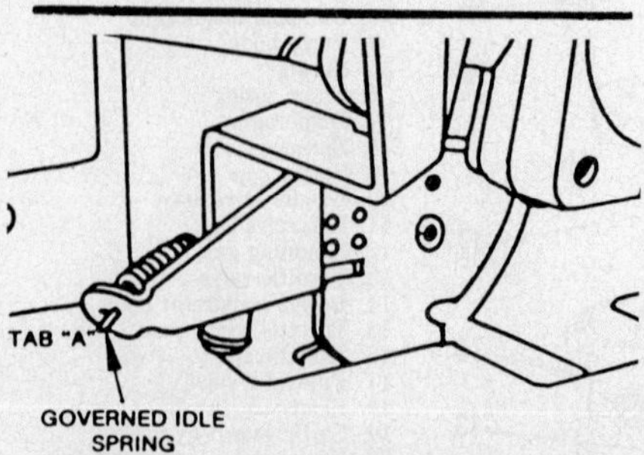

Fig. B72—With governor plate locked with a 1/8-inch (3.18 mm) pin (Fig. B71), bend tab "A" to obtain 1400 rpm. Upper view is for a horizontal crankshaft engine and lower view is for vertical crankshaft engines.

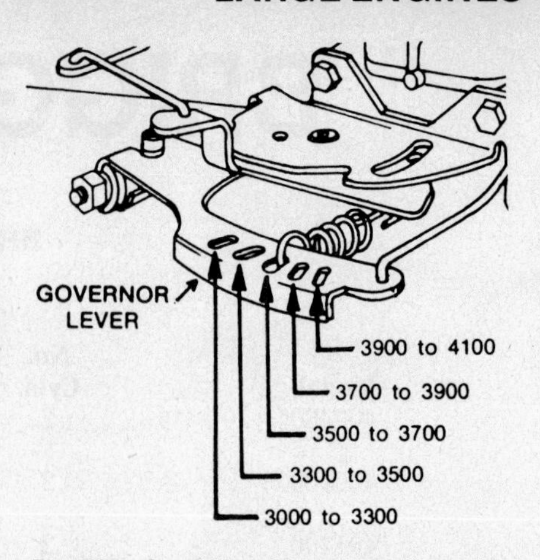

Fig. B73—Governor spring should be installed with end loops as shown. Install loop in appropriate governor lever hole for engine speed (rpm) desired.

To adjust top governed speed, first adjust carburetor and governed idle as previously outlined. Install governor spring end loop in appropriate hole in governor lever for desired engine rpm as shown in Fig. B73. Check engine top governed speed using an accurate tachometer.

Governor gear and weight assembly can be removed when engine is disassembled. Loosen nut and remove governor lever. To obtain maximum clearance, rotate crankshaft until timing mark on crankshaft gear is at approximately 10 o'clock position. Remove "E" ring and thick washer on outer end of governor shaft. Carefully slide shaft down past crankshaft.

CAUTION: Be careful not to bind shaft against crankshaft as governor lower bearing could be damaged.

To reassemble governor shaft, refer to Fig. B75 and reverse disassembly procedure.

Governor gear assembly located on cover (Fig. B74) is serviced as an

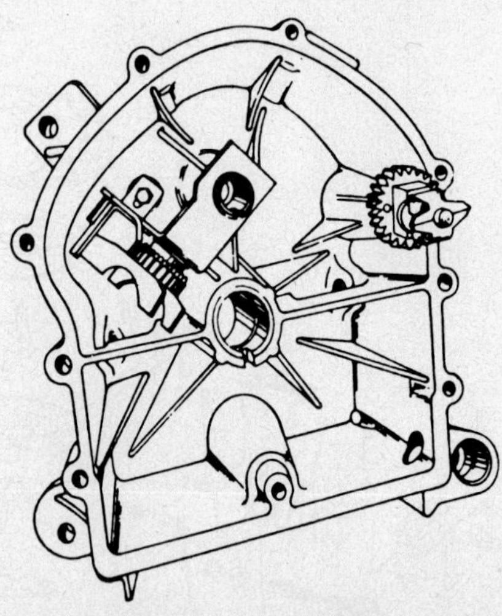

Fig. B74—Typical governor gear assembly position in crankcase.

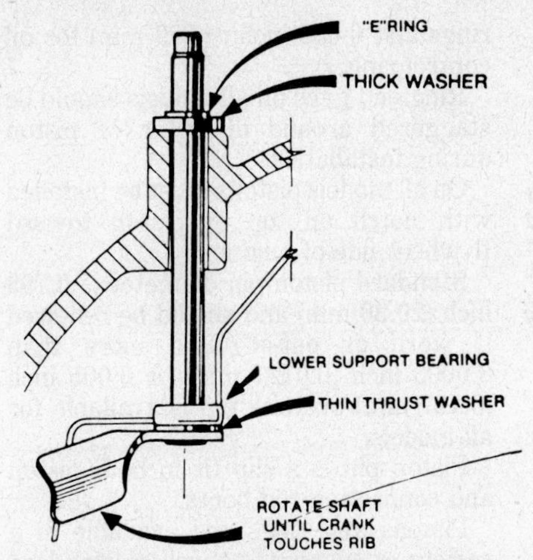

Fig. B75—Cross-sectional view of governor shaft assembly. To disassemble, remove "E" ring and washer. Carefully guide shaft down past crankshaft.

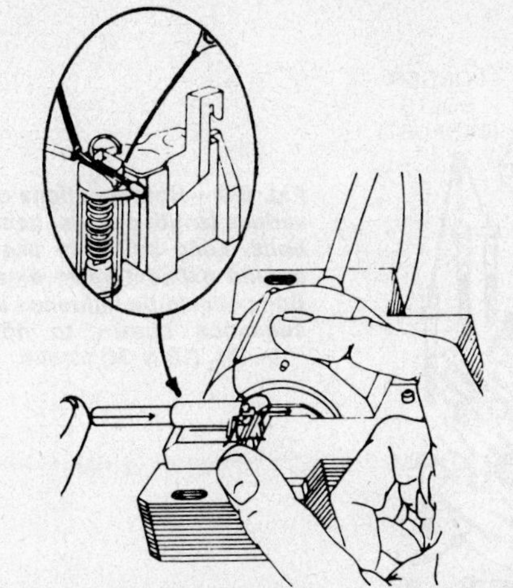

Fig. B77—Use a 5/32-inch (3.97 mm) diameter rod to release wires from "Magnetron" module. Refer to text.

assembly only. Note that there is a thrust washer installed between cover and governor gear assembly.

IGNITION SYSTEM. Early production engines were equipped with a flywheel type magneto ignition with points, condenser and coil located externally on engine and late production engines are equipped with "Magnetron" breakerless ignition system.

Refer to appropriate paragraph for model being serviced.

FLYWHEEL MAGNETO IGNITION. Flywheel magneto system consists of a permanent magnet cast into flywheel, armature and coil assembly, breaker points and condenser.

Breaker points and condenser are located under or behind intake manifold and are protected by a metal cover which must be sealed around edges and at wire entry location to prevent entry of dirt or moisture.

Breaker point gap should be 0.020 inch (0.508 mm) for all models.

Breaker points are actuated by a plunger (Fig. B76) which is installed with the smaller diameter end toward breaker points. Renew plunger if length

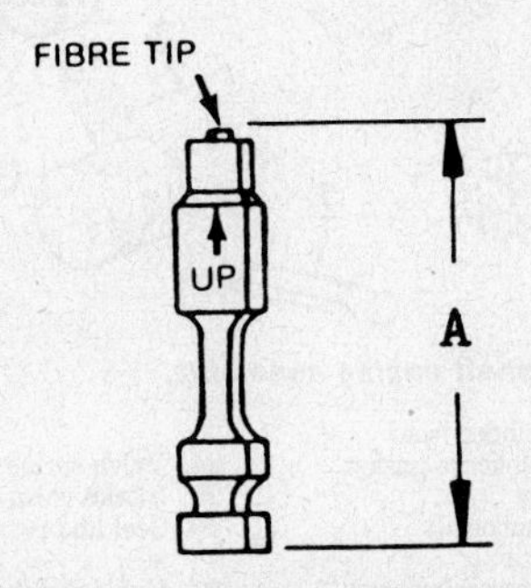

Fig. B76—Plunger must be renewed if plunger length (A) is worn to 1.115 inch (28.32 mm) or less.

is 1.115 inch (28.32 mm) or less. Renew plunger seal by installing seal on plunger (make certain it is securely attached) and installing seal and plunger assembly into plunger bore. Slide seal over plunger boss until seated against casting at base of boss.

Armature air gap should be 0.010-0.014 inch (0.25-0.36 mm) and is adjusted by loosening armature retaining bolts and moving armature as necessary on slotted holes. Tighten armature retaining bolts.

MAGNETRON IGNITION. "Magnetron" ignition consists of permanent magnets cast in flywheel and a self-contained transistor module mounted on ignition armature.

To check ignition, attach B&S tester number 19051 to each spark plug lead and ground tester to engine. Spin flywheel rapidly. If spark jumps the 0.166 inch (4.2 mm) tester gap, ignition system is operating satisfactorily.

Armature air gap should be 0.008-0.012 inch (0.20-0.30 mm) and is adjusted by loosening armature retaining bolts and moving armature as necessary on slotted holes. Tighten armature bolts.

Flywheel does not need to be removed to service "Magnetron" ignition except to check condition of flywheel key or keyway.

To remove Magnetron module from armature, remove stop switch wire, module primary wire and armature primary wire from module by using a 5/32 inch (3.97 mm) rod to release spring and retainer (Fig. B77). Remove spring and retainer clip. Unsolder wires. Remove module by pulling out on module retainer while pushing down on module until free of armature laminations.

During reinstallation, use 60/40 rosin core solder and make certain all wires are held firmly against coil body with tape or "Permatex" number 2 or equivalent gasket sealer.

LUBRICATION. Vertical crankshaft models are splash lubricated by a gear driven oil slinger (5–Fig. B79) and horizontal crankshaft models are splash lubricated by an oil dipper attached to number one connecting rod.

Oils approved by manufacturer must meet requirements of API service classification SC, SD, SE or SF.

Use SAE 10W-40 oil for temperatures above 20°F (–7°C) and SAE 5W-20 oil for temperatures below 20° F (-7° C).

Check oil at regular intervals and maintain at "FULL" mark on dipstick. Dipstick should be pushed or screwed in completely for accurate measurement. **DO NOT** overfill.

Recommended oil change interval for all models is every 25 hours of normal operation.

Crankcase oil capacity for early production engines is 3.5 pints (1.65 L) and for late production engines oil capacity is 3 pints (1.42 L). Check oil level with dipstick. **DO NOT** overfill.

CRANKCASE BREATHER. Crankcase breathers are built into engine valve covers. Horizontal crankshaft models have a breather valve in each cover assembly and vertical crankshaft models have only one breather in cover of number one cylinder.

Breathers maintain a partial vacuum in crankcase to prevent oil from being forced out past oil seals and gaskets or past breaker point plunger or piston rings.

Fiber disc of breather assembly must not be stuck or binding. A 0.045 inch (1.14 mm) wire gage **SHOULD NOT** enter space between fiber disc valve and body. Check with gage at 90° intervals around fiber disc.

When installing breathers make certain side of gasket with notches is toward crankshaft.

REPAIRS

CYLINDER HEADS. When removing cylinder heads, note locations from which different length bolts are removed as they must be reinstalled in their original positions.

Always use a new gasket when reinstalling cylinder head. Do not use sealer on gasket. Lubricate cylinder head bolt threads with graphite grease, install in correct locations and tighten in several even steps in sequence shown in Fig. B78 to 160 in.-lbs. (18 N·m) torque.

It is recommended carbon and lead deposits be removed at 100 to 300 hour

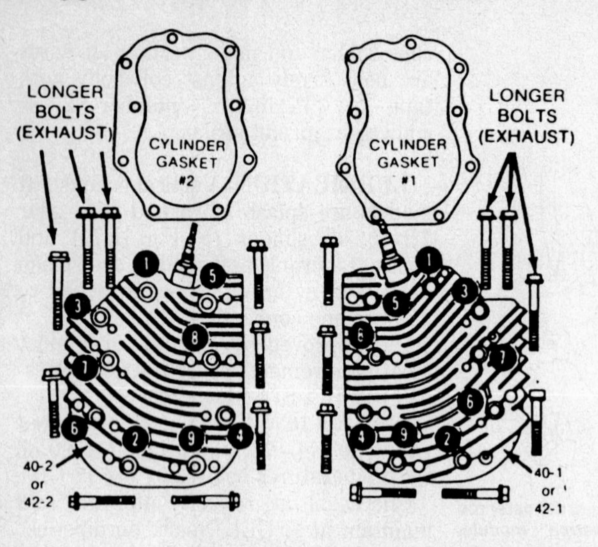

Fig. B78—Note locations of various length cylinder head bolts. Long bolts are used around exhaust valve area. Bolts should be tightened in sequence shown to 160 in.-lbs. (18 N·M) torque.

intervals, or whenever cylinder head is removed.

CONNECTING RODS. Connecting rods and pistons are removed from cylinder head end of block as an assembly. Aluminum alloy connecting rods ride directly on crankpin journals.

Connecting rod should be renewed if crankpin bearing bore measures 1.627 inches (41.33 mm) or more, if bearing surfaces are scored or damaged or if pin bore measures 0.802 inch (20.37 mm) or more.

NOTE: A 0.005 inch (0.127 mm) oversize piston pin is available.

Connecting rod should be installed on piston so oil hole in connecting rod is toward cam gear side of engine when notch on top of piston is toward flywheel with piston and rod assembly installed. Make certain match marks on connecting rod and cap are aligned and if equipped with an oil dipper (horizontal crankshaft models), it should be installed on number one rod. Install special washers and nuts and tighten to 190 in.-lbs. (22 N·m) torque for all models.

PISTONS, PINS AND RINGS. Pistons used in engines with cast iron cylinder liners (Series 400400, 400700, 402400, 402700, 422400 and 422700) have an "L" on top of piston. A chrome plated aluminum piston is used in aluminum bore (Kool-Bore) cylinders. Due to the different cylinder bore materials, pistons **WILL NOT** interchange.

Pistons for all models should be renewed if they are scored or damaged, or if a 0.007 inch (0.178 mm) feeler gage can be inserted between a new top ring and ring groove. If piston pin bore measures 0.801 inch (20.35 mm) or more, piston should be renewed or pin bore reamed for 0.005 inch (0.127 mm) oversize pin.

Piston ring end gaps for 401400, 401700, 421400 and 421700 models should be 0.035 inch (0.889 mm) for compression rings and 0.045 inch (1.14 mm) for oil control ring. Piston ring end gaps for 400400, 400700, 402400, 402700, 422400 and 422700 models should be 0.030 inch (0.762 mm) for compression rings and 0.035 inch (0.889 mm) for oil control ring.

Ring end gaps on all models should be staggered around diameter of piston during installation.

On all models pistons must be installed with notch on top of piston toward flywheel side of engine.

Standard piston pin diameter is 0.799 inch (20.30 mm) and should be renewed if worn or out-of-round more than 0.0005 inch (0.0127 mm). A 0.005 inch (0.127 mm) oversize pin is available for all models.

Piston pin is a slip fit in both piston and connecting rod bores.

Pistons and rings are available in a variety of oversizes as well as standard and as with pistons, ring sets for aluminum bore (Kool-Bore) engines and ring sets for engines with cast iron cylinder liners should not be interchanged.

CYLINDERS. Cylinder bores may be either aluminum, or a cast iron liner which is an integral part of the cylinder block casting. Pistons and rings for aluminum cylinders and cast iron

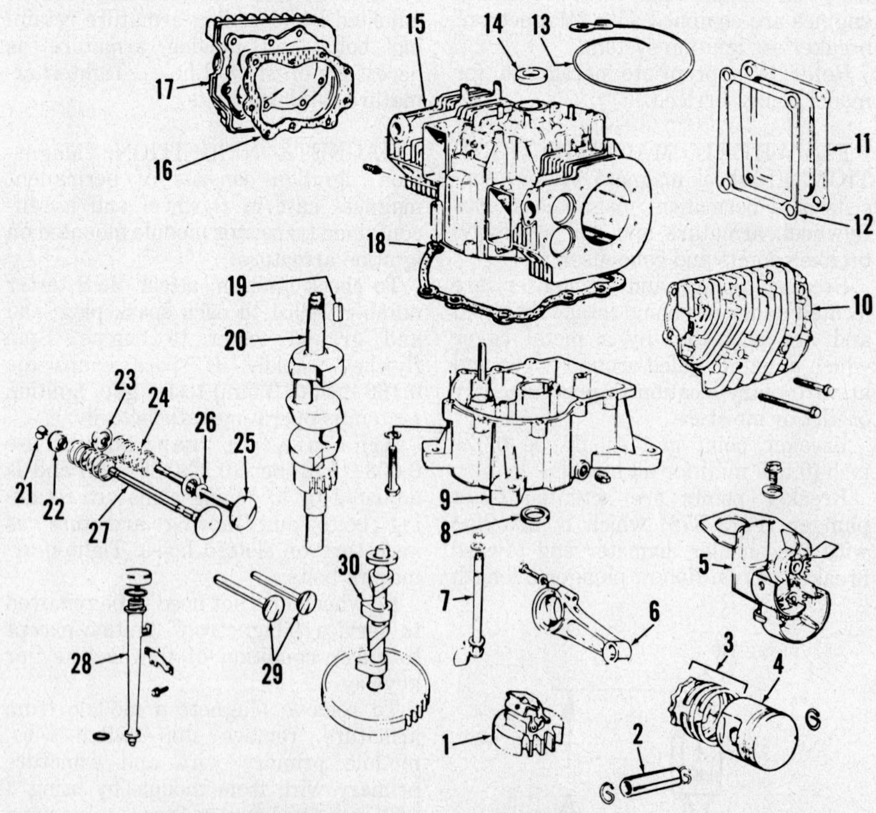

Fig. B79—Exploded view of vertical crankshaft engine assembly.

1. Governor gear
2. Piston pin & clips
3. Piston rings
4. Piston
5. Oil slinger assy.
6. Connecting rod
7. Governor shaft assy.
8. Oil seal
9. Crankcase assy.
10. Cylinder head
11. Crankcase cover plate
12. Crankcase gasket
13. Ground wire
14. Oil seal
15. Cylinder assy.
16. Head gasket
17. Cylinder head
18. Crankcase gasket
19. Key
20. Crankshaft
21. Retainer
22. Rotocoil (exhaust valve)
23. Retainer (intake valve)
24. Valve spring (2)
25. Intake valve
26. Seal and retainer assy.
27. Exhaust valve
28. Oil dipstick assy.
29. Valve tappet (2)
30. Camshaft assy.

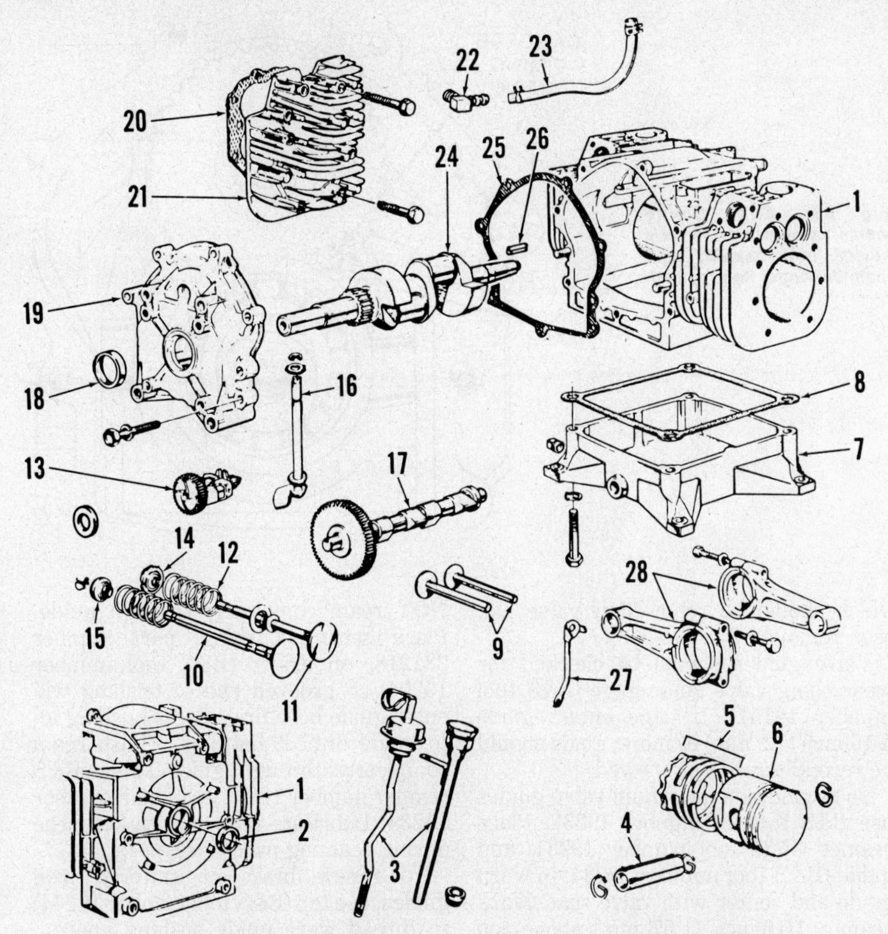

Fig. B80—Exploded view of horizontal crankshaft engine assembly.

1. Cylinder assy.
2. Oil seal
3. Dipstick assy.
4. Piston pin & retainer clips
5. Piston rings
6. Piston
7. Crankcase
8. Crankcase gasket
9. Valve tappets
10. Exhaust valve
11. Intake valve
12. Valve spring
13. Governor gear assy.
14. Intake valve retainer
15. Rotocoil (exhaust valve)
16. Governor shaft
17. Camshaft
18. Oil seal
19. Crankcase cover
20. Head gasket
21. Cylinder head
22. Elbow connector
23. Fuel line
24. Crankshaft
25. Crankcase cover gasket
26. Key
27. Oil dipper
28. Connecting rods

cylinders **SHOULD NOT** be interchanged. Series 400400, 400700, 402400, 402700, 422400 and 422700 have cast iron cylinder liners as an integral part of cylinder block casting and Series 401400, 401700, 421400 and 421700 have aluminum cylinder bores (Kool-Bore). Standard cylinder bore for all models is 3.4365-3.4375 inches (87.29-87.31 mm) and should be resized using a suitable hone (B&S part number 19205 for aluminum bore or 19211 for cast iron bore) if more than 0.003 inch (0.0762 mm) oversize or 0.0015 inch (0.0381 mm) out-of-round for cast iron liner engines or 0.003 inch (0.0762 mm) oversize or 0.0025 inch (0.0635 mm) out-of-round for aluminum bore (Kool-Bore) engines. Resize to nearest oversize for which piston and rings are available.

CRANKSHAFT AND MAIN BEARINGS. Crankshaft may be supported at each end in main bearings which are an integral part of crankcase, cover or sump, DU type bearings or ball bearing mains which are a press fit on crankshaft and fit into machined bores in crankcase, cover or sump.

To remove crankshaft from engines with integral type or DU type main bearings, remove necessary air shrouds. Remove flywheel and front gear cover or sump. Remove cam gear making certain valve tappets clear camshaft lobes. Remove crankshaft.

To reinstall crankshaft, reverse removal procedure making certain timing marks are aligned as shown in Fig. B81.

To remove crankshaft from engines with ball bearing main bearing, remove any necessary air shrouds and remove flywheel. Remove front gear cover or sump. Compress exhaust and intake valve springs on number two cylinder to provide clearance for camshaft lobes. Remove crankshaft and camshaft together.

To reinstall crankshaft, reverse removal procedure making certain timing marks are aligned as shown in Fig. B82.

Crankshaft for models with integral type or DU type main bearings should be renewed if main bearing journals measure 1.376 inches (34.95 mm) or less. Crankshaft should be renewed or reground if crankpin journal measures 1.622 inches (41.15 mm) or less. Connecting rods for 0.020 inch (0.508 mm) undersize journal is available.

Ball bearing main bearing is a press fit on crankshaft and must be removed by pressing crankshaft out of bearing. Renew ball bearing if worn or rough. Expand new bearing by heating in oil and install with shield side towards crankpin.

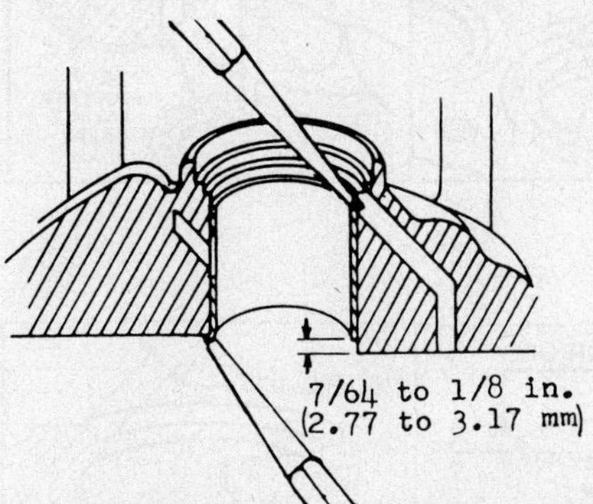

Fig. B80A—Press DU type bearings in until 7/64 to 1/8-inch (2.77 to 3.17 mm) from thrust surface. Make certain oil holes are aligned and stake bearing as shown. Refer to text.

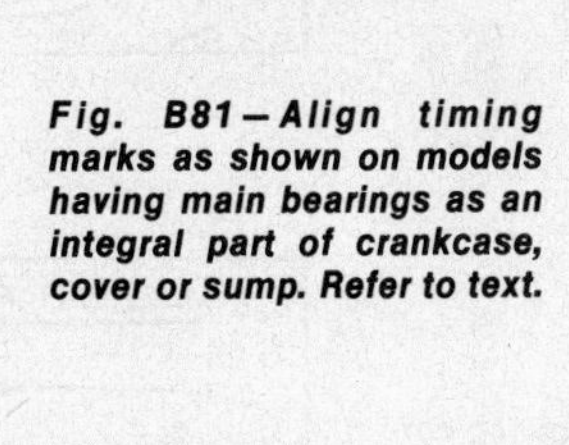

Fig. B81—Align timing marks as shown on models having main bearings as an integral part of crankcase, cover or sump. Refer to text.

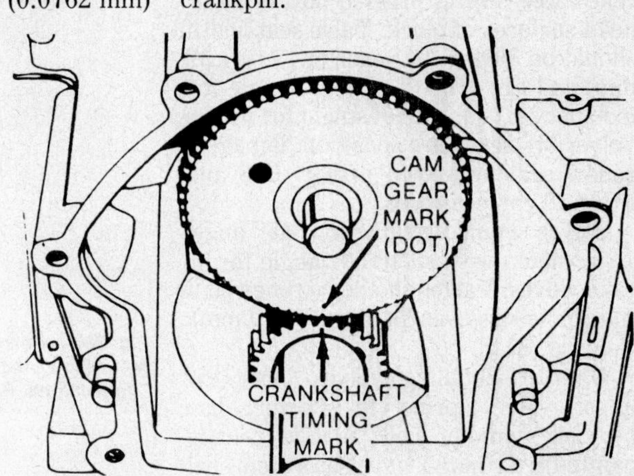

Integral type main bearings should be reamed out and service bushings installed if 0.0007 inch (0.0178 mm) or more out-of-round or if they measure 1.383 inches (35.13 mm) or more in diameter. Special reamers are available from Briggs & Stratton.

DU type main bearings should be renewed if 0.0007 inch (0.0178 mm) or more out-of-round or if they measure 1.383 inches (35.13 mm) or more in diameter.

Worn DU type bearings are pressed out of bores using B&S cylinder support number 19227 and driver number 19226. Make certain oil holes in bearings will align with holes in block and cover or sump and press bearings in until they are 7/64 to ⅛-inch (2.77 to 3.17 mm) below thrust face. Stake bearings into place. See Fig. B80A.

If ball bearing is loose in crankcase, cover or sump bores, crankcase, cover or sump must be renewed.

Crankshaft end play for all models should be 0.002-0.008 inch (0.050-0.200 mm). At least one 0.015 inch cover or sump gasket must be used. Additional cover gaskets of 0.005 and 0.009 inch thickness are available if end play is less than 0.002 inch (0.050 mm). If end play is over 0.008 inch (0.200 mm), metal shims are available for use on crankshaft between crankshaft gear and cover or sump.

NOTE: Thrust washer can not be used on double ball bearing engines.

CAMSHAFT. Camshaft and camshaft gear are an integral part which ride in journals at each end of camshaft.

To remove camshaft, refer to appropriate **CRANKSHAFT AND MAIN BEARING** section for model being serviced.

Camshaft should be renewed if gear teeth or lobes are worn or damaged or if bearing journals measure 0.623 inches (15.82 mm) or less.

VALVE SYSTEM. Valves seat in renewable inserts pressed into cylinder head surfaces of block. Valve seat width should be 3/64 to 1/16-inch (1.17 to 1.57 mm) and are ground at a 45° angle for exhaust seats and a 30° angle for intake valves. If seats are loose or damaged, renew seat as shown in Fig. B83 and grind to correct angle.

Valves should be refaced at 45° angle for exhaust valves and 30° angle for intake valves. Valves should be renewed if margin is less than 1/64-inch (0.10 mm). See Fig. 84.

When reinstalling valves, note exhaust valve spring is shorter, has heavier diameter coils and is usually painted red. Intake valve has a stem seal which should be renewed if valve has been removed.

Valve guides should be checked for wear using valve guide gage (B&S tool number 19151). If gage enters guide 5/16-inch (7.9 mm) or more, guide should be reconditioned or renewed.

To recondition aluminum valve guides use B&S tool kit number 19232. Place reamer (B&S tool number 19231) and guide (B&S tool number 19234) in worn guide and center with valve seat. Mark reamer 1/16-inch (1.57 mm) above top edge of service bushing (Fig. B85). Ream worn guide until mark on reamer is flush with top of guide bushing. **DO NOT** ream completely through guide. Place service bushing, part number 231218, on driver (B&S tool number 19204) so grooved end of bushing will enter guide bore first. Press bushing into guide until it bottoms. Finish ream completely through guide using B&S reamer number 19233 and guide number 19234. Lubricate reamer with kerosene during reaming procedure.

To renew brass or sintered iron guides, use tap (B&S tool number 19264) to thread worn guide bushing approximately ½-inch (12.7 mm) deep. **DO NOT** thread more than 1 inch (25.4 mm) deep. Install puller washer (B&S tool

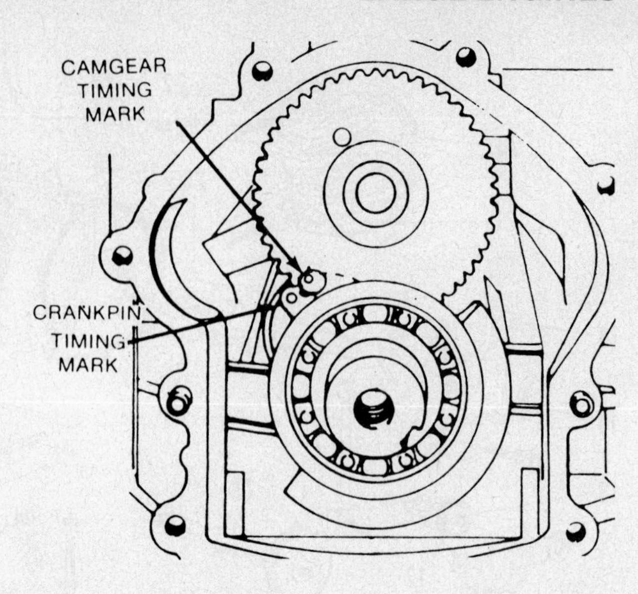

Fig. B82—Align timing marks as shown on models having ball bearing type main bearings. Refer to text.

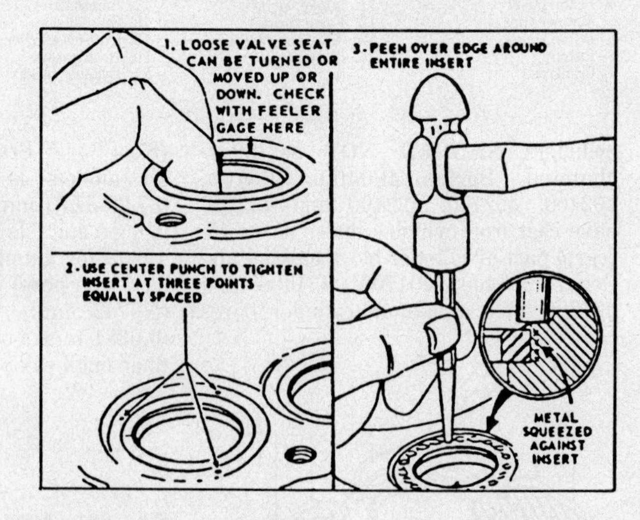

Fig. B83—Loose inserts can be tightened or renewed as shown. If a 0.005 inch (0.127 mm) feeler gage can be inserted between seat and seat bore, renew cylinder.

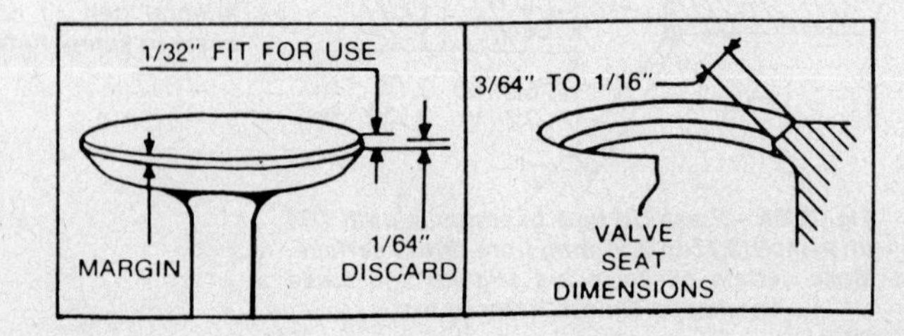

Fig. B84—View showing correct valve face and seat dimensions. Refer to text.

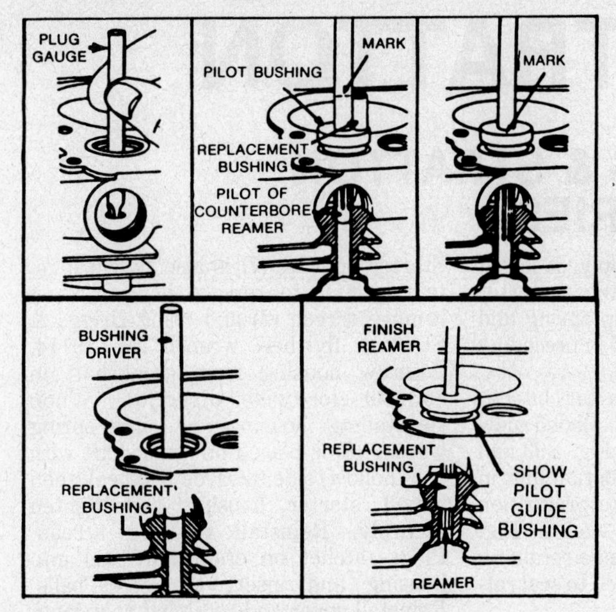

Fig. B85—View showing correct procedure for reconditioning valve guides. Refer to text.

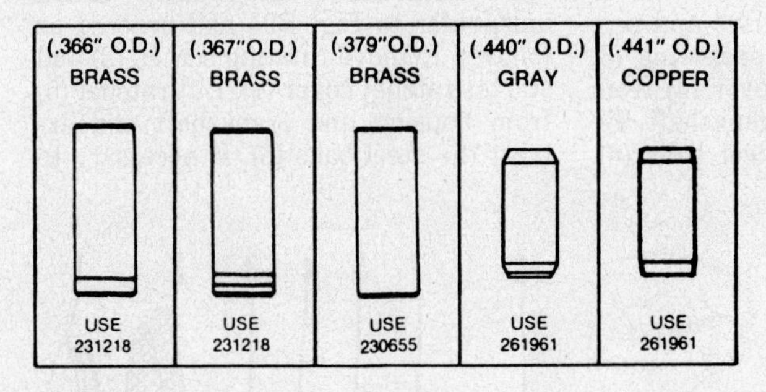

Fig. B86—Use correct bushing as indicated when renewing valve guide bushings.

number 19240) on puller screw (B&S tool number 19238) and thread screw and washer assembly into worn guide. Center washer on valve seat and tighten puller nut (B&S tool number 19239) against washer, continue to tighten while keeping threaded screw from turning (Fig. B85) until guide has been removed. Identify guide using Fig. B86 and find appropriate replacement. Place correct service bushing on driver (B&S tool number 19024) so the two grooves on service bushings number 231218 are down. Remaining bushing types can be installed either way. Press bushing in until it bottoms. Finish ream with B&S reamer number 19233 and reamer guide number 19234. Lubricate reamer with kerosene and ream completely through new service bushing.

To adjust valves for all models, cylinder to be adjusted must be at "top dead center" on compression stroke. If springs are installed, stem end clearance should be 0.007-0.009 inch (0.18-0.23 mm) for exhaust valves and 0.004-0.006 inch (0.10-0.15 mm) for intake valves. Without springs, stem end clearance should be 0.009-0.0011 inch (0.23-0.28 mm) for exhaust valves and 0.006-0.008 inch (0.15-0.23 mm) for intake valves. If clearance is less than specified, grind end of stem as necessary. If clearance is excessive, grind valve seat deeper as necessary.

BRIGGS & STRATTON

SERVICING BRIGGS & STRATTON ACCESSORIES

REWIND STARTER

OVERHAUL. To renew broken rewind spring, proceed as follows: Grasp free outer end of spring (S–Fig. B90) and pull broken end from starter housing. With blower housing removed, remove tangs (T) and remove starter pulley from housing. Untie knot in rope (R) and remove rope and inner end of broken spring from pulley. Apply a small amount of grease on inner face of pulley, thread inner end of spring through notch in starter housing, engage inner end of spring in pulley hub and place bar in pulley hub and turn pulley approximately 13½ turns in a counter-clockwise direction as shown in Fig. B91. Tie wrench to blower housing with wire to hold pulley so hole (H) in pulley is aligned with rope guide (G) in housing as shown in Fig. B92. Hook a wire in inner end of rope and thread rope through guide and hole in pulley; then, tie a knot in rope and release the pulley allowing spring to wind rope into pulley groove.

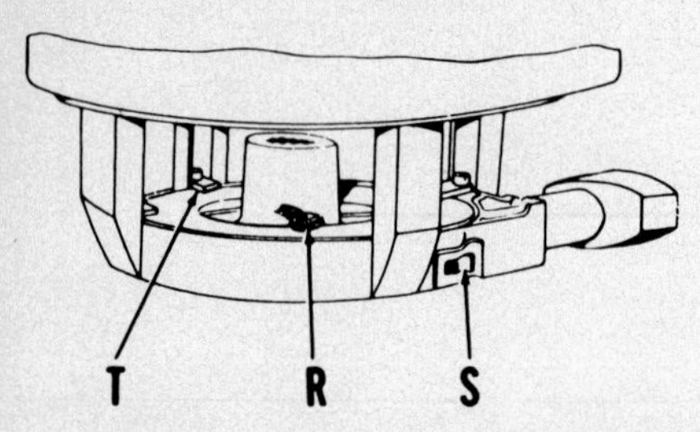

Fig. B90 – View of rewind starter showing rope (R), spring end (S) and retaining tangs (T).

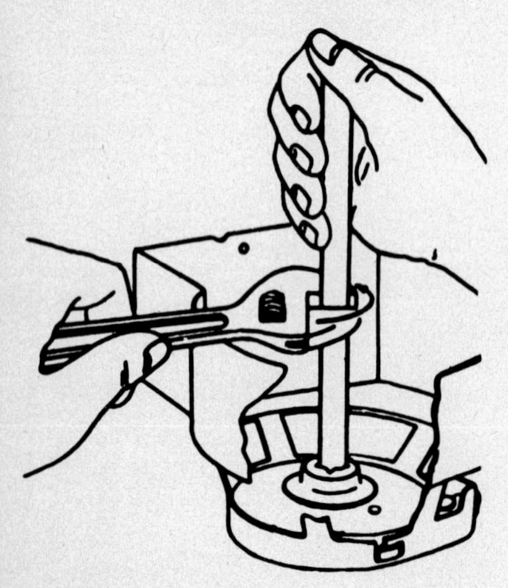

Fig. B91 – Using square shaft and wrench to wind up the rewind starter spring. Refer to text.

To renew starter rope only, it is not generally necessary to remove starter pulley and spring. Wind up spring and install new rope as outlined in preceding paragraph.

Two different types of starter clutches have been used; refer to exploded view of early production unit in Fig. B93 and exploded view of late production unit in Fig. B94. Outer end of late production ratchet (refer to cutaway view in Fig. B95) is sealed with a felt and a retaining plug and a rubber ring is used to seal ratchet to ratchet cover.

To disassemble early type starter clutch unit, refer to Fig. B93 and proceed as follows: Remove snap ring (3) and lift ratchet (5) and cover (4) from starter housing (7) and crankshaft. Be careful not to lose the steel balls (6).

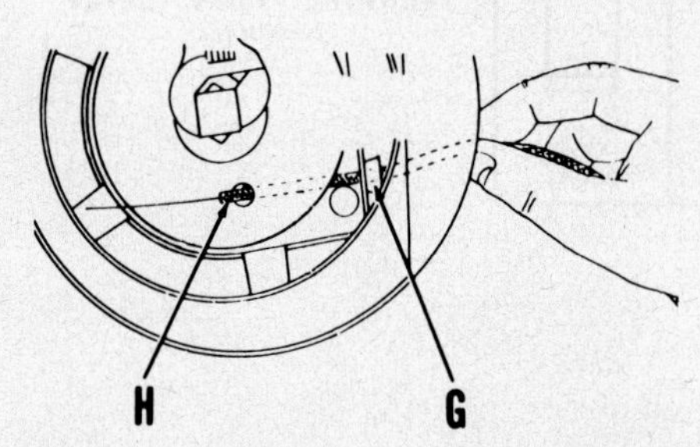

Fig. B92 – Threading starter rope through guide (G) in blower housing and hole (H) in starter pulley with wire hooked in end of rope.

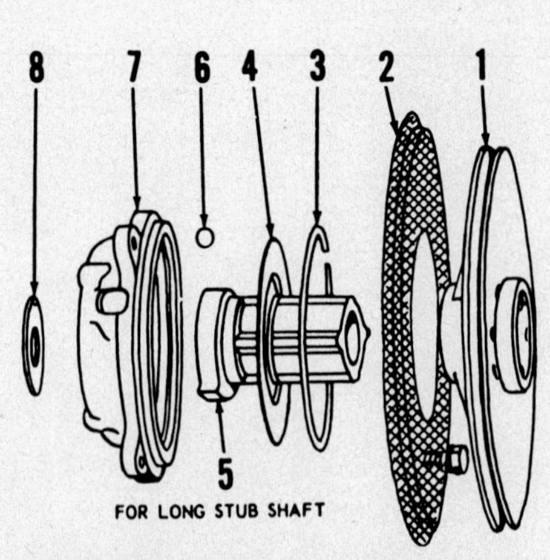

Fig. B93 – Exploded view of early production starter clutch unit; refer to Fig. B96 for "long stub shaft". A late type unit (Fig. B94) should be installed when renewing "long" crankshaft with "short" (late production) shaft.

1. Starter rope pulley
2. Rotating screen
3. Snap ring
4. Ratchet cover
5. Starter ratchet
6. Steel balls
7. Clutch housing (flywheel nut)
8. Spring washer

Starter housing (7) is also flywheel retaining nut; to remove housing, first remove screen (2) and using Briggs & Stratton flywheel wrench No. 19114, unscrew housing from crankshaft in counter-clockwise direction. When reinstalling housing, be sure spring washer (8) is placed on crankshaft with cup (hollow) side towards flywheel, then install starter housing and tighten securely. Reinstall rotating screen. Place ratchet on crankshaft and into housing and insert the steel balls. Reinstall cover and retaining snap ring.

To disassemble late starter clutch unit, refer to Fig. B94 and proceed as follows: Remove rotating screen (2) and starter ratchet cover (4). Lift ratchet (5) from housing and crankshaft and extract the steel balls (6). If necessary to

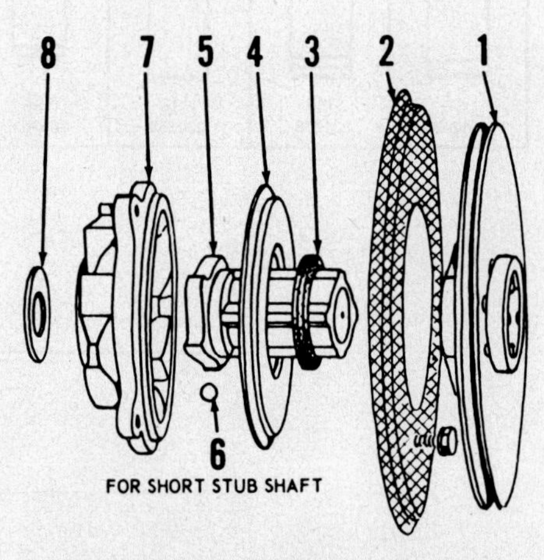

Fig. B94 – Exploded view of late production sealed starter clutch unit. Late unit can be used with "short stub shaft" only; refer to Fig. B96.

1. Starter rope pulley
2. Rotating screen
3. Snap ring
4. Ratchet cover
5. Starter ratchet
6. Steel balls
7. Clutch housing (flywheel nut)
8. Spring washer

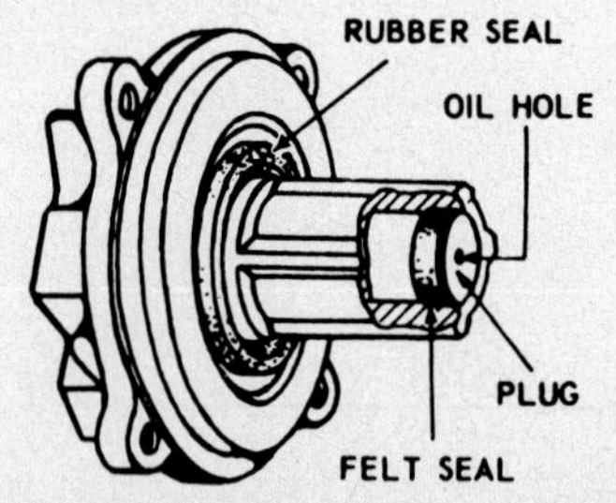

Fig. B95 – Cut-away showing felt seal and plug in end of late production starter ratchet (5 – Fig. B94).

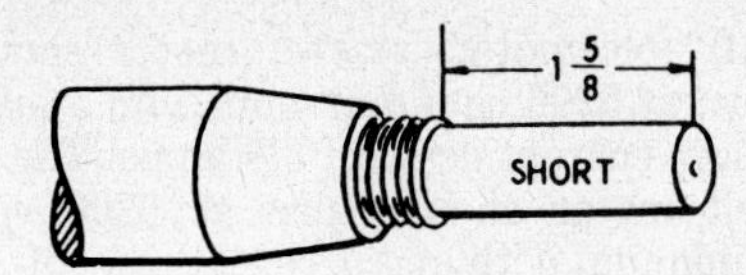

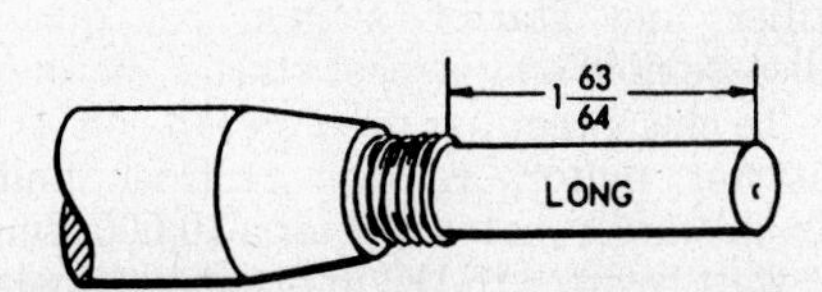

Fig. B96 — Crankshaft with short, 1⅝-inch (31.275 mm) stub (top view) must be used with late production starter clutch assembly. Early crankshaft (bottom view) can be modified by cutting off the 1-63/64 inch (50.4 mm) stub end to the dimension shown in top view and beveling end of shaft to allow installation of late type clutch unit.

remove housing (7), hold flywheel and unscrew housing in counter-clockwise direction using Briggs & Stratton flywheel wrench No. 19114. When installing housing, be sure spring washer (8) is in place on crankshaft with cup (hollow) side towards flywheel; then, tighten housing securely. Inspect felt seal and plug in outer end of ratchet; renew ratchet if seal or plug are damaged as these parts are not serviced separately. Lubricate the felt with oil and place ratchet on crankshaft. Insert the steel balls and install ratchet cover, rubber seal and rotating screen.

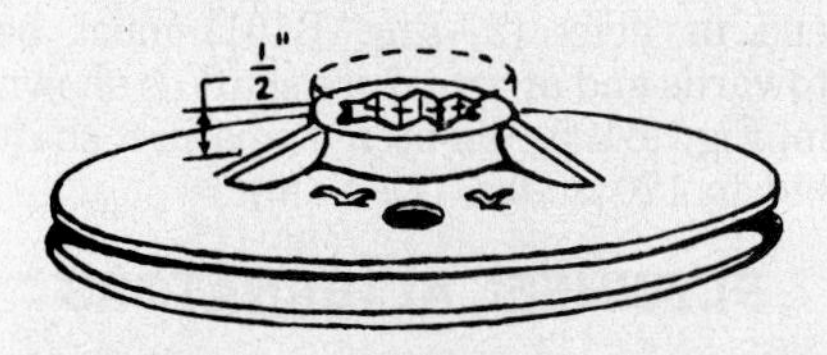

Fig. B97 — When installing late type starter clutch unit as a replacement for early type, either install new starter rope pulley or cut hub of old pulley to ½-inch (12.7 mm) as shown.

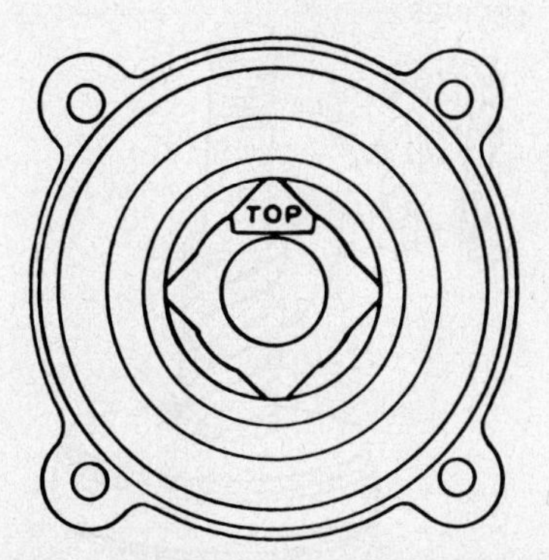

Fig. B98 — When installing blower housing and starter assembly, turn starter ratchet so word "TOP" stamped on outer end of ratchet is towards engine cylinder head.

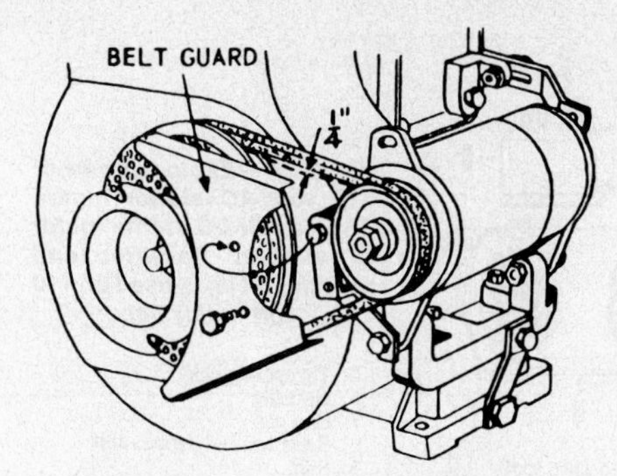

Fig. B99 — View showing starter generator belt adjustment on models so equipped. Refer to text.

NOTE: Crankshafts used with early and late starter clutches differ; refer to Fig. B96.

If renewing early (long) crankshaft with late (short) shaft, also install late type starter clutch unit. If renewing early starter clutch with late type unit, crankshaft must be shortened to dimension shown for short shaft in Fig. B96; also, hub or starter rope pulley must be shortened to ½-inch (12.7 mm) dimension shown in fig. B97. Bevel end of crank-shaft after removing the approximate ⅜-inch (9.52 mm) from shaft.

When installing blower housing and starter assembly, turn starter ratchet so word "TOP" on ratchet is toward engine cylinder head.

12-VOLT STARTER-GENERATOR UNITS

Combination starter-generator functions as a cranking motor when starting switch is closed. When engine is operating and with starting switch open, unit operates as a generator. Generator output and circuit voltage for battery and various operating requirements are controlled by a current-voltage regulator. On units where voltage regulator is mounted separately from generator unit, do not mount regulator with cover down as regulator will not function in this position. To adjust belt tension, apply approximately 30 pounds (13.6 kg) pull on generator adjusting flange and tighten mounting bolts. Belt tension is correct when a pressure of 10 pounds (44.5 N) applied midway between pulleys will deflect belt ¼-inch (6.35 mm). See Fig. B99. On units equipped with two drive belts, always renew belts in pairs. A 50-ampere capacity battery is recommended. Starter-generator units are intended for use in temperatures above 0° F (-18° C). Refer to Fig. B100 for exploded view of starter-generator. Parts and service on starter-generator are available at authorized Delco-Remy service stations.

GEAR-DRIVE STARTERS

Two types of gear drive starters may be used, a 110 volt AC starter or a 12 volt DC starter. Refer to Figs. B101 and B102 for exploded views of starter motors. A properly grounded receptacle should be used with power cord connected to 110 volt AC starter motor. A 32 ampere hour capacity battery is recommended for use with 12 volt DC starter motor.

To renew a worn or damaged flywheel ring gear, drill out retaining rivets using a 3/16-inch drill. Attach new ring gear using screws provided with new ring gear.

To check for correct operation of starter motor, remove starter motor from engine and place motor in a vise or other holding fixture. Install a 0-5 amp ammeter in power cord to 110 volt AC starter motor. On 12 volt DC motor, connect a 12 volt battery to motor with a 0-50 amp ammeter in series with positive line from battery to starter motor. Connect a tachometer to drive end of starter. With starter activated on

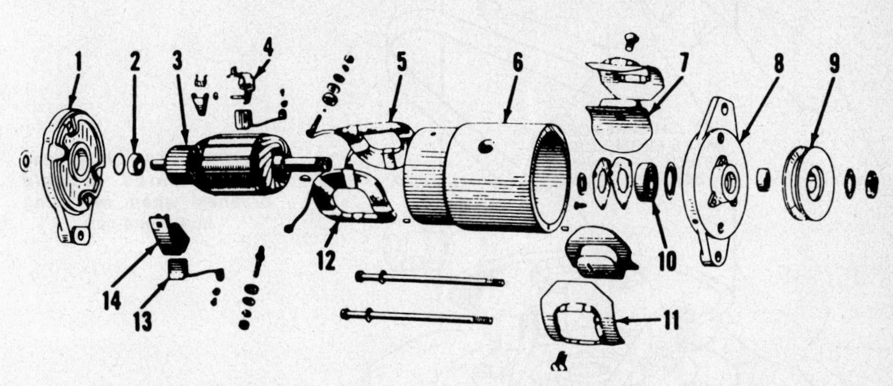

Fig. B100 — Exploded view of Delco-Remy starter-generator unit used on some B&S engines.

1. Commutator end frame
2. Bearing
3. Armature
4. Ground brush holder
5. Field coil L.H.
6. Frame
7. Pole shoe
8. Drive end frame
9. Pulley
10. Bearing
11. Field coil insulator
12. Field coil R.H.
13. Brush
14. Insulated brush holder

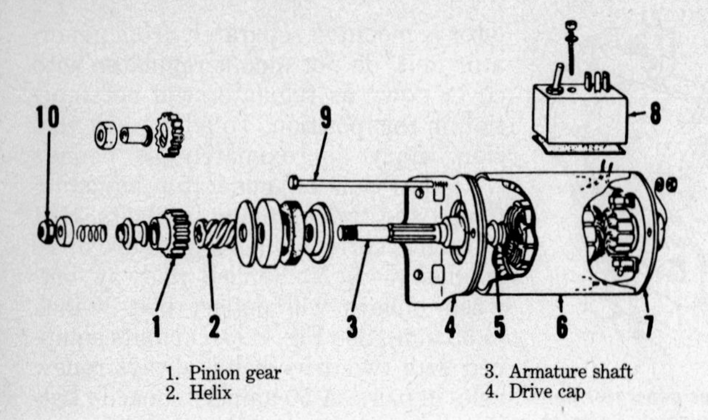

Fig. B101 — Exploded view of 110 volt AC starter motor. Twelve volt DC starter motor is similar. Rectifier and switch unit (8) is used on 110 volt motor only.

1. Pinion gear
2. Helix
3. Armature shaft
4. Drive cap
5. Thrust washer
6. Housing
7. End cap
8. Rectifier and switch unit
9. Bolt
10. Nut

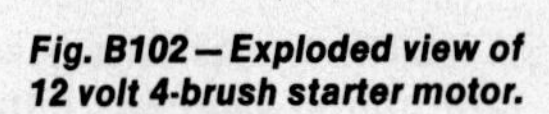

Fig. B102 — Exploded view of 12 volt 4-brush starter motor.

1. Cap
2. Roll pin
3. Retainer
4. Pinion spring
5. Spring cup
6. Starter gear
7. Clutch assy.
8. Drive end cap assy.
9. Armature
10. Housing
11. Spring
12. Brush assy.
13. Battery wire terminal
14. Commutator end cap assy.

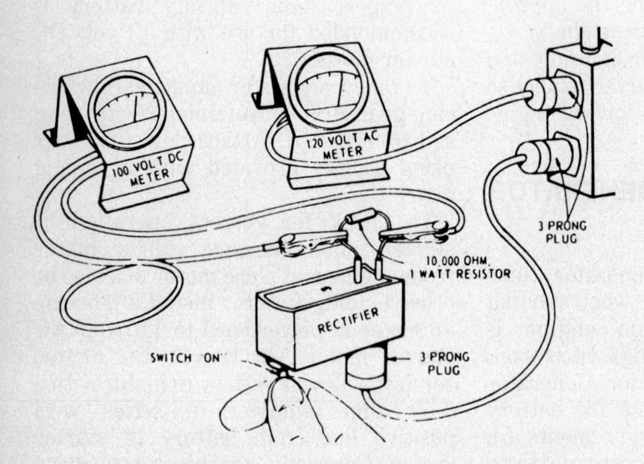

Fig. B103 — View of test connections for 110 volt rectifier. Refer to text for procedure.

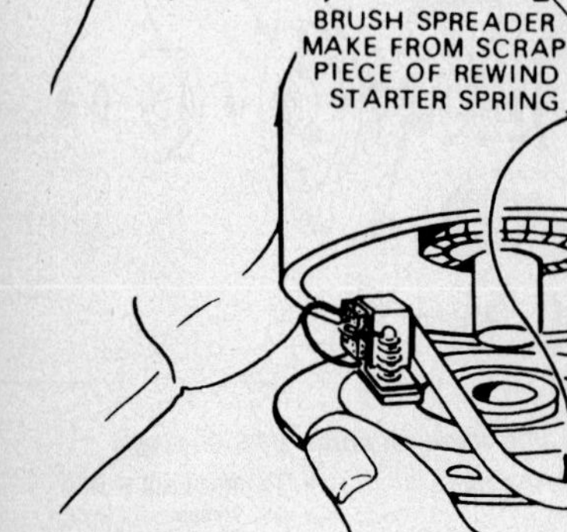

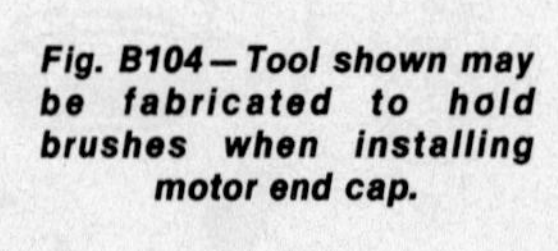

Fig. B104 — Tool shown may be fabricated to hold brushes when installing motor end cap.

110 volt motor, starter motor should turn at 5200 rpm minimum with a maximum current draw of 3½ amps. The 12 volt motor should turn at 6200 rpm minimum with a current draw of 16 amps maximum. If starter motor does not operate satisfactorily, check operation of rectifier or starter switch. If rectifier and starter switch are good, disassemble and inspect starter motor.

To check rectifier used on 110 volt AC starter motor, remove rectifier unit from starter motor. Solder a 10,000 ohm 1 watt resistor to DC internal terminals of rectifier as shown in Fig. B103. Connect a 0-100 range DC voltmeter to resistor leads. Measure voltage of AC outlet to be used. With starter swtich in "OFF" position, a zero reading should be shown on DC voltmeter. With starter switch in "ON" position, DC voltmeter should show a reading that is 0-14 volts lower than AC line voltage measured previously. If voltage drop exceeds 14 volts, renew rectifier unit.

Disassembly of starter motor is self-evident after inspection of unit and referral to Fig. B101 and B102. Note position of bolts during disassembly so they can be installed in their original positions during reassembly. When reassembling motor lubricate end cap bearings with SAE 20 oil. Be sure to match drive cap keyway to stamped key in housing when sliding armature into motor housing. Brushes may be held in their holders during installation by making a brush spreader tool from a piece of metal as shown in Fig. B104. Splined end of helix (2 – Fig. B101) must be towards end of armature shaft as shown in Fig. B105. Tighten armature shaft nut to 170 in.-lbs. (19 N·m).

FLYWHEEL ALTERNATORS

4 Amp. Non-Regulated Alternator

Some engines are equipped with the 4 ampere non-regulated flywheel alter-

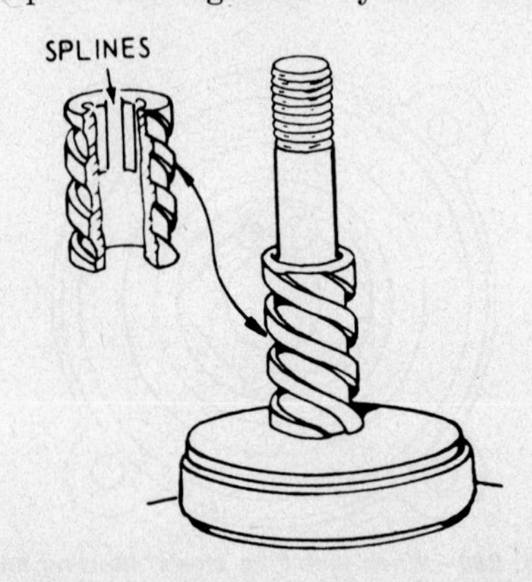

Fig. 105 — Install helix on armature so splines of helix are toward outer end of shaft as shown.

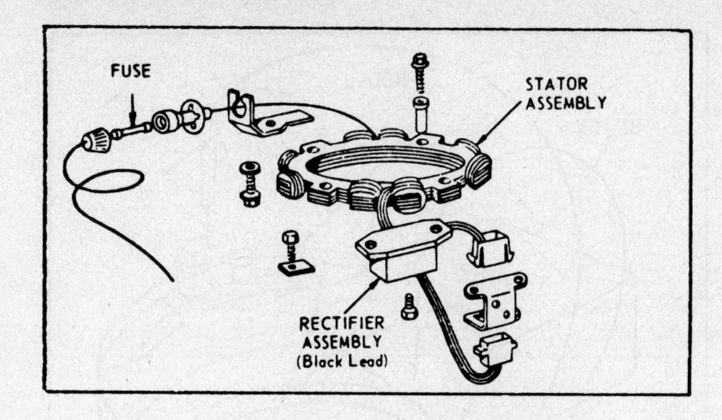

Fig. B106—Stator and rectifier assemblies used on the 4 ampere non-regulated flywheel alternator. Fuse is 7½ amp AGC or 3AG.

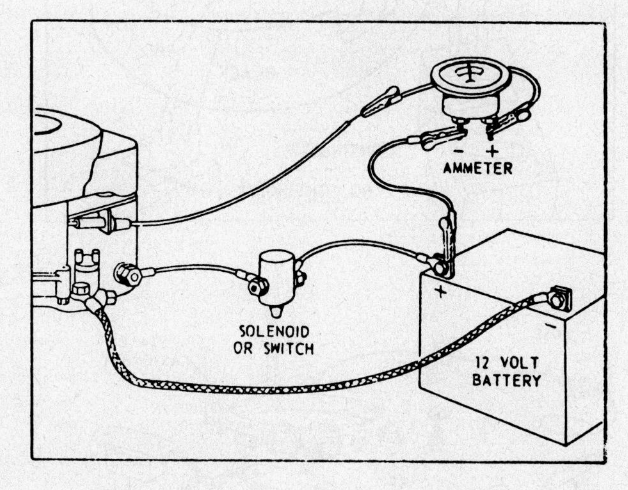

Fig. B107—Install ammeter as shown for output test.

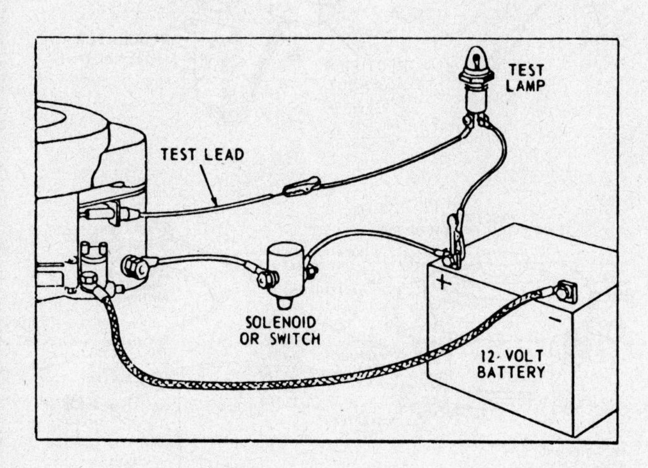

Fig. B108—Connect a test lamp as shown to test for shorted stator or defective rectifier. Refer to text.

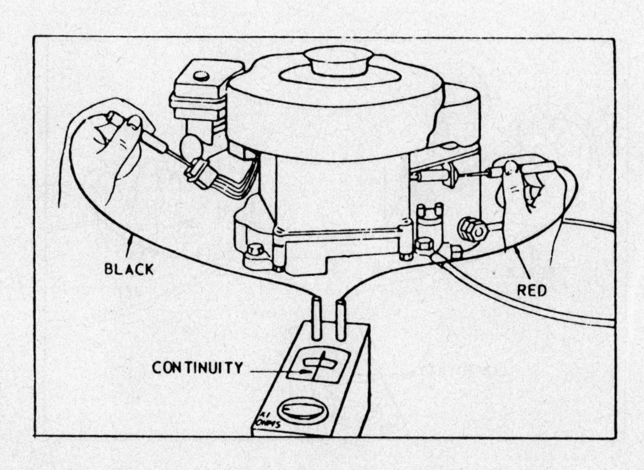

Fig. B109—Use an ohmmeter to check condition of stator. Refer to text.

nator shown in Fig. B106. A solid state rectifier and 7½ amp fuse is used with this alternator.

If battery is run down and no output from alternator is suspected, first check the 7½ amp fuse. If fuse is good, clean and tighten all connections. Disconnect charging lead and connect an ammeter as shown in Fig. B107. Start engine and check for alternator output. If ammeter shows no charge, stop engine, remove ammeter and install a test lamp as shown in Fig. B108. Test lamp should not light. If it does light, stator or rectifier is defective. Unplug rectifier plug under blower housing. If test lamp does not go out, stator is shorted.

If shorted stator is indicated, use an ohmmeter and check continuity as follows: Touch one test lead to lead inside of fuse holder as shown in Fig. B109. Touch remaining test lead to each of the four pins in rectifier connector. Unless ohmmeter shows continuity at each of the four pins, stator winding is open and stator must be renewed.

If defective rectifier is indicated, unbolt and remove flywheel blower housing with rectifier. Connect one ohmmeter test lead to blower housing and remaining test lead to the single pin connector in rectifier connector. See Fig. B110. Check for continuity, then reverse leads and again test for continuity. If tests show no continuity in either direction or continuity in both directins, rectifier is faulty and must be renewed.

7 Amp. Regulated Alternator

A 7 amp regulated flywheel alternator is used witn 12 volt gear drive starter motor on some models. Alternator is equipped with a solid state rectifier and regulator. An isolation diode is also used on most models.

If engine will not start, using electric start system and trouble is not in starting motor, install an ammeter in circuit as shown in Fig. B112. Start engine manually. Ammeter should indicate charge. If ammeter does not show battery charging taking place, check for defective wiring and if necessary proceed with troubleshooting.

If battery charging occurs with engine running, but battery does not retain charge, then isolation diode may be defective. The isolation diode is used to prevent battery drain if alternator circuit malfunctions. After troubleshooting diode, remainder of circuit should be inspected to find reason for excessive battery drain. To check operation of diode, disconnect white lead of diode from fuse holder and connect a test lamp from the diode white lead to negative terminal of battery. Test lamp should not light. If test lamp lights, diode is defective. Disconnect test lamp and disconnect red lead of diode. Test continuity of diode with ohmmeter by connecting leads of ohmmeter to leads of diode then reverse lead connections. Ohmmeter should show continuity in one direction and an open circuit in the other direction. If readings are incorrect, then diode is defective and must be renewed.

To troubleshoot alternator assembly, proceed as follows: Disconnect white lead of isolation diode from fuse holder and connect a test lamp between positive terminal of battery and fuse holder on engine. Engine must not be started. With connections made, test lamp should not light. If test lamp does light, stator, regulator or rectifier is defective. Unplug rectifier-regulator plug under blower housing. If lamp remains lighted, stator is grounded. If lamp goes out, regulator or rectifier is shorted.

If previous test indicated stator is grounded, check stator leads for defects and repair if necessary. If shorted leads are not found, renew stator. Check stator for an open circuit as follows: Using a ohmmeter, connect positive lead to fuse holder as shown in Fig. B113 and negative lead to one of the pins in rectifier and regulator connector. Check each of the four pins in connector. Ohmmeter should show continuity at each pin, if not, then there is an open in stator and stator must be renewed.

To test rectifier, unplug rectifier and regulator connector plug and remove blower housing from engine. Using an ohmmeter check for continuity between connector pins connected to black wires and blower housing as shown in Fig. B114. Be sure good contact is made with metal of blower housing. Reverse ohmmeter leads and check continuity again. Ohmmeter should show a continuity reading for one direction only on each plug. If either pin shows a continuity reading for both directions, or if either pin shows no continuity for either direction, then rectifier must be renewed.

To test regulator unit, repeat procedure used to test rectifier unit except connect ohmmeter lead to pins connected to red wire and white wire. If ohmmeter shows continuity in either direction for red lead pin, regulator is defective an must be renewed. White lead pin should read as an open on ohmmeter in one direction and a weak reading in the other direction. Otherwise, regulator is defective and must be renewed.

10 Amp Regulated Alternator

Engines may be equipped with a 10 ampere flywheel alternator and a solid state rectifier-regulator. To check charging system, disconnect charging lead from battery. Connect a DC

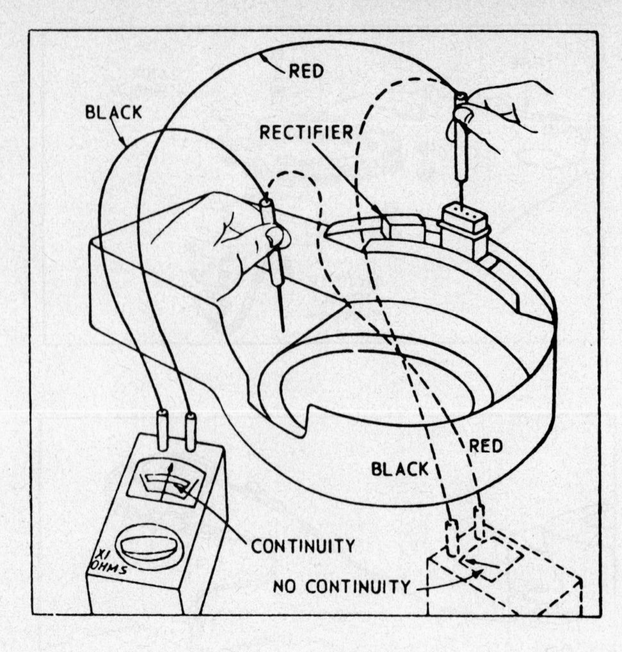

Fig. B110—If ohmmeter shows continuity in both directions or in neither direction, rectifier is defective.

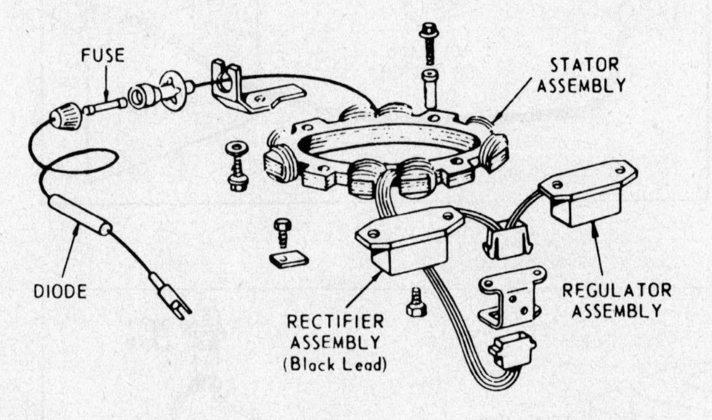

Fig. B111—Stator, rectifier and regulator assemblies used on the 7 ampere regulated flywheel alternator.

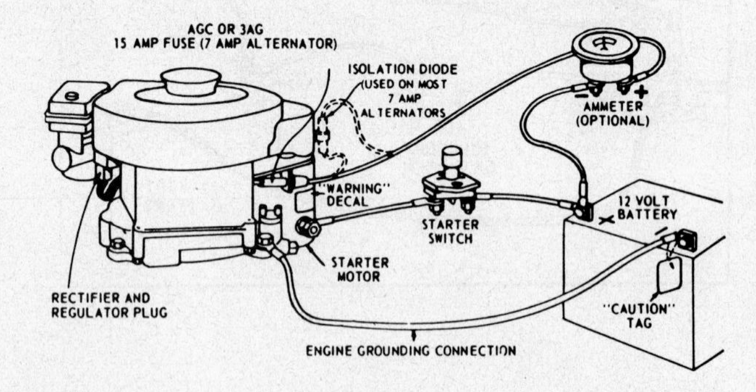

Fig. B112—Typical wiring used on engines equipped with 7 ampere flywheel alternator.

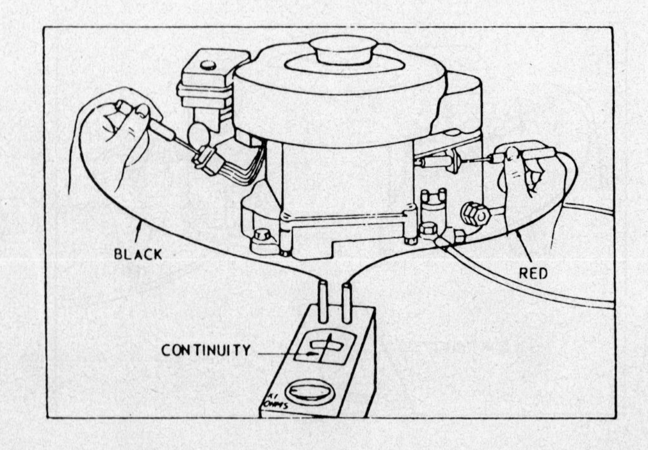

Fig. B113—Use an ohmmeter to check condition of stator. Refer to text.

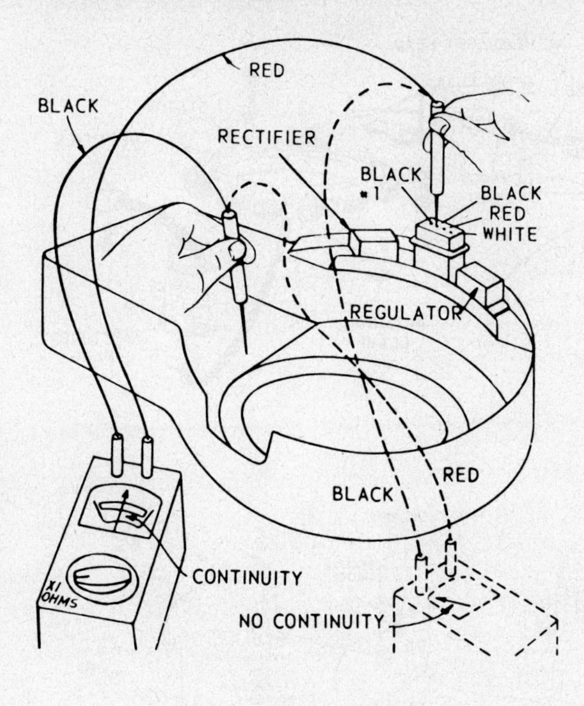

Fig. B114—Be sure good contact is made between ohmmeter test lead and metal cover when checking rectifier and regulator.

voltmeter between charging lead and ground as shown in Fig. B116. Start engine and operate at 3600 rpm. A voltmeter reading of 14 volts or above indicates alternator is functioning. If reading is less than 14 volts, stator or rectifier-regulator is defective.

To test stator, disconnect stator plug from rectifier-regulator. Operate engine at 3600 rpm and connect AC voltmeter leads to AC terminals in stator plug as shown in Fig. B117. Voltmeter reading above 20 volts indicates stator is good. A reading less than 20 volts indicates stator is defective.

To test rectifier-regulator, make certain charging lead is connected to battery and stator plug is connected to rectifier-regulator. Check voltage across battery terminals with DC voltmeter (Fig. B118). If voltmeter reading is 13.8 volts or higher, reduce battery voltage by connecting a 12 volt load lamp across battery terminals. When battery voltage is below 13.5 volts, start engine and operate at 3600 rpm. Voltmeter reading should rise. If battery is fully charged,

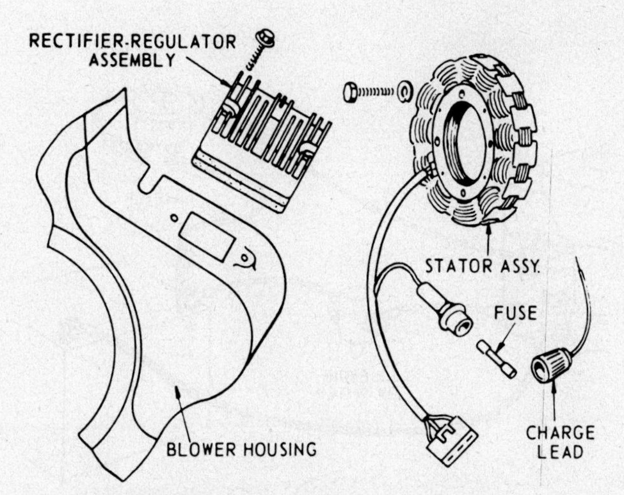

Fig. B115—View of 10 ampere regulated flywheel alternator stator and rectifier-regulator used on some engines.

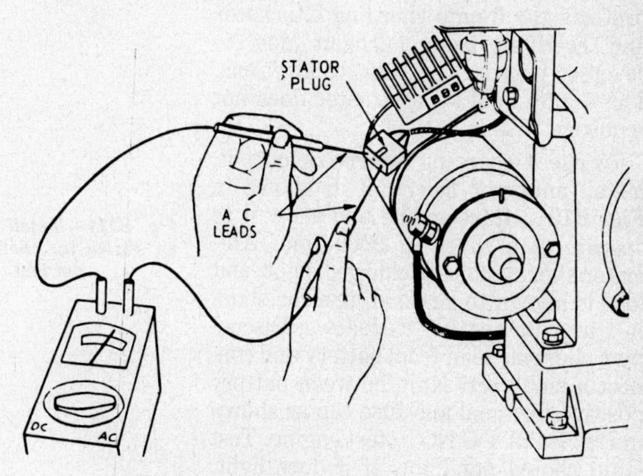

Fig. B117—AC voltmeter is used to test stator.

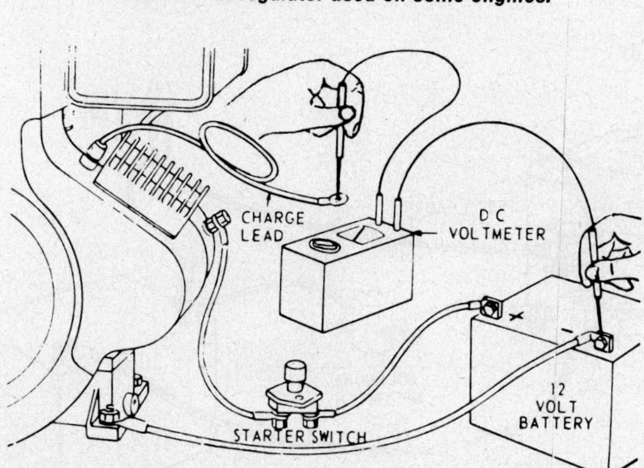

Fig. B116—DC voltmeter is used to determine if alternator is functioning. Refer to text.

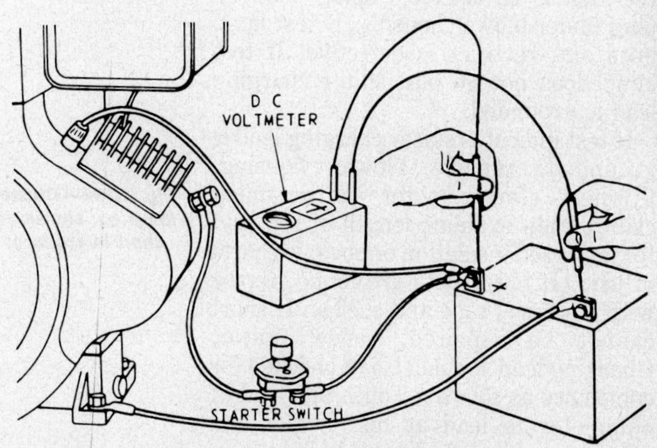

Fig. B118—Check battery voltage with DC voltmeter. Refer to text for rectifier-regulator test.

reading should rise above 13.8 volts. If voltage does not increase or if voltage reading rises above 14.7 volts, rectifier-regulator is defective and must be renewed.

Dual Circuit Alternator

A dual circuit alternator is used on some models. This system operates as two separate alternators. A single ring of magnets inside the flywheel supplies magnetic field for both sets of windings on stator. One alternator uses a solid state rectifier and provides 2 amperes at 2400 rpm or 3 amperes at 3600 rpm for battery charging current. The other alternator feeds alternating current directly to lights. Since the two are electrically independent, use of lights does not reduce charge going into battery.

Current for lights is available only when engine is operating. Twelve volt lights with a total rating of 60 to 100 watts may be used. With a rating of 70 watts, voltage rises from 8 volts at 2400 rpm to 12 volts at 3600 rpm. Since output depends on engine speed, brightness of lights changes with engine speed.

Battery charging current connection is made through a 7½ amp fuse mounted in fuse holder. See Fig. B120. Current for lights is available at plastic connector below fuse holder. The 7½ amp fuse protects the 3 amp charging alternator and rectifier from burn-out due to reverse polarity battery connections. The 5 amp lighting alternator does not require a fuse.

To check charging alternator output, install ammeter in circuit as shown in Fig. B121. Start engine and allow it to operate at a speed of 3000 rpm. Ammeter should indicate charge. If not, and fuse is known to be good, test for short in stator or rectifier as follows: Disconnect charging lead from battery and connect a small test lamp between battery positive terminal and fuse cap as shown in Fig. B122. DO NOT start engine. Test lamp should not light. If it does light, stator's charging lead is grounded or rectifier is defective. Unplug rectifier plug under blower housing. If test lamp goes out, rectifier is defective. If test lamp does not go out, stator charging lead is grounded.

If test indicates stator charging lead is grounded, remove blower housing, flywheel, starter motor and retaining clamp, then examine length of red lead for damaged insulation or obvious shorts in lead. If bare spots are found, repair with electrical tape and shellac. If short cannot be repaired, renew stator. Charging lead should also be checked for continuity as follows: Touch one lead of ohmmeter to lead at fuse holder and other ohmmeter lead to red lead pin in connector as shown in Fig. B123. If

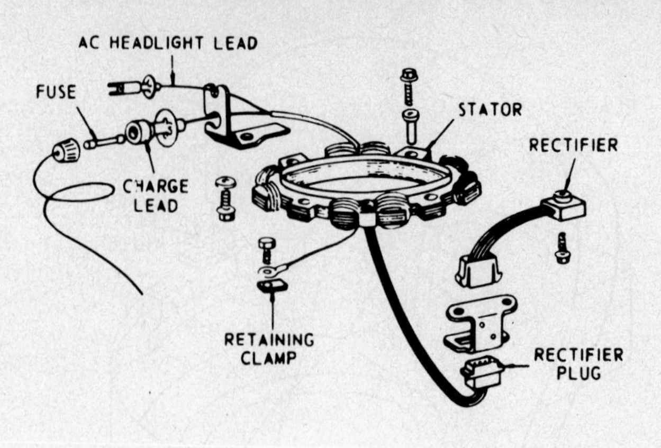

Fig. B119—Stator and rectifier assemblies used on the dual circuit alternator. Fuse is 7½ amp AGC or 3AG.

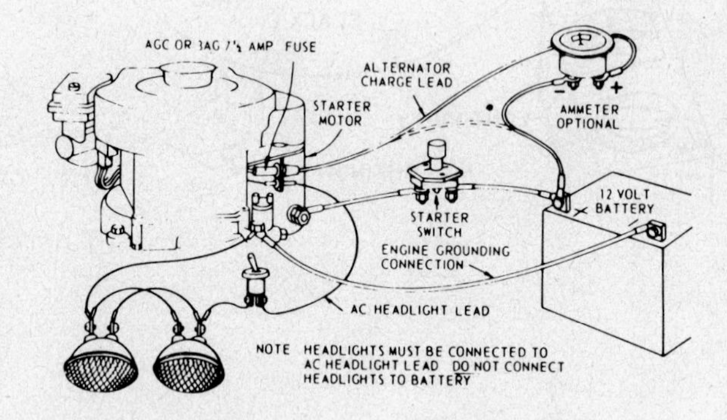

Fig. B120—Typical wiring used on engines equipped with the dual circuit flywheel alternator.

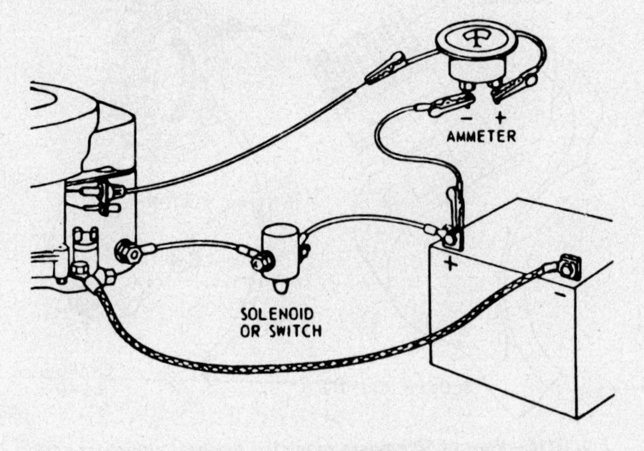

Fig. B121—Install ammeter as shown for charging output test.

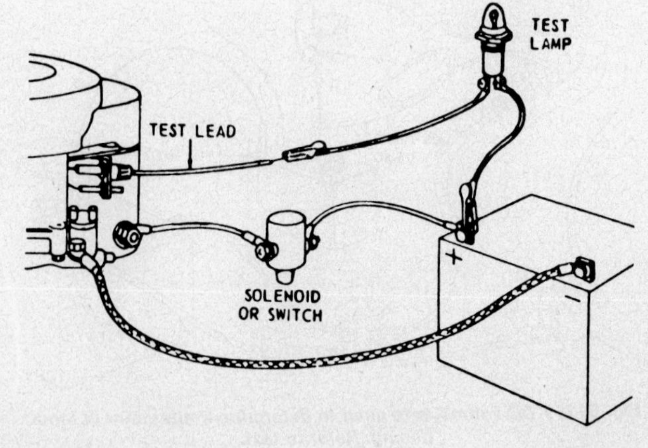

Fig. B122—Connect a test lamp as shown to test for short in stator or rectifier.

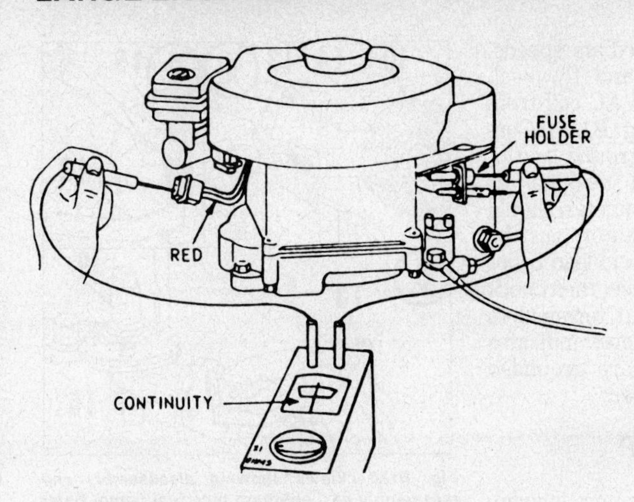

Fig. B123—Use an ohmmeter to check charging lead for continuity. Refer to text.

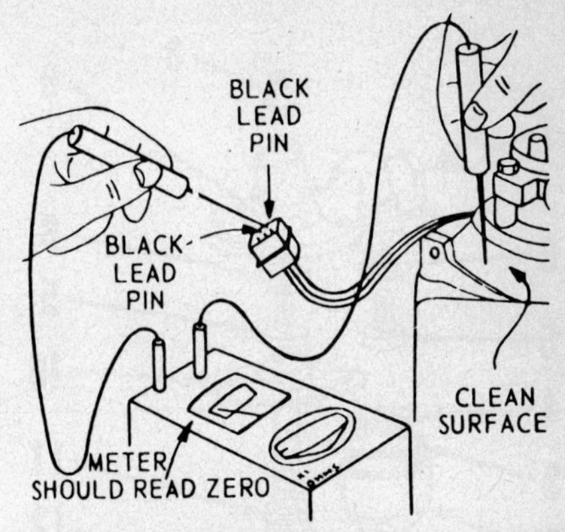

Fig. B125—Checking for grounded charging coils. Refer to text.

ohmmeter does not show continuity, charging lead is open and stator must be renewed. Charging coils should be checked for continuity as follows: Touch ohmmeter test leads to the two black lead pins as shown in Fig. B124. If ohmmeter does not show continuity, charging coils are defective and stator must be renewed. Test for grounded charging coils by touching one test lead of ohmmeter to a clean ground surface on the engine and the other test lead to each of the black lead pins as shown in Fig. B125. If ohmmeter shows continuity, charging coils are grounded and stator must be renewed.

To test rectifier, use an ohmmeter and check for continuity between each of the three lead pin sockets and blower housing. See Fig. B126. Reverse ohmmeter leads and check continuity again. Ohmmeter should show a continuity reading for one direction only on each lead socket. If any pin socket shows continuity reading in both directions or neither direction, rectifier is defective and must be renewed.

To test AC lighting alternator circuit, connect a load lamp to AC output plug

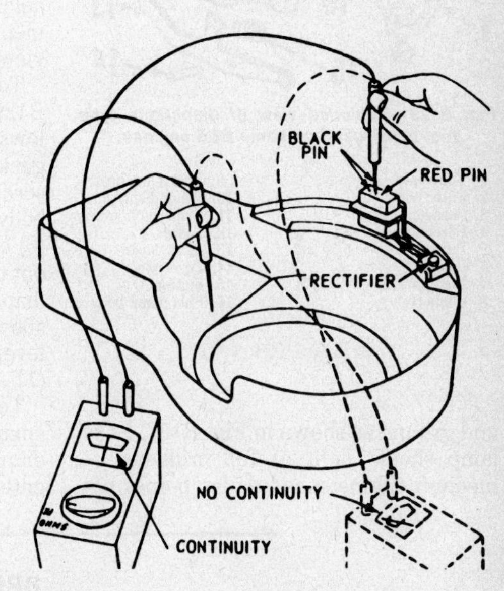

Fig. B126—If ohmmeter shows continuity in both directions or neither direction, rectifier is defective.

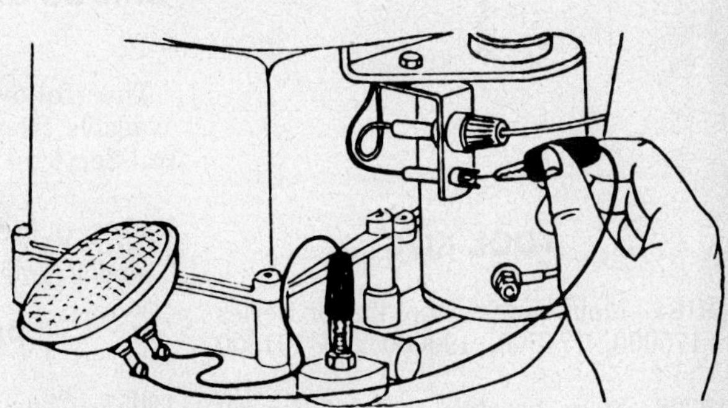

Fig. B127—Load lamp (GE 4001 or equivalent) is used to test AC lighting circuit output.

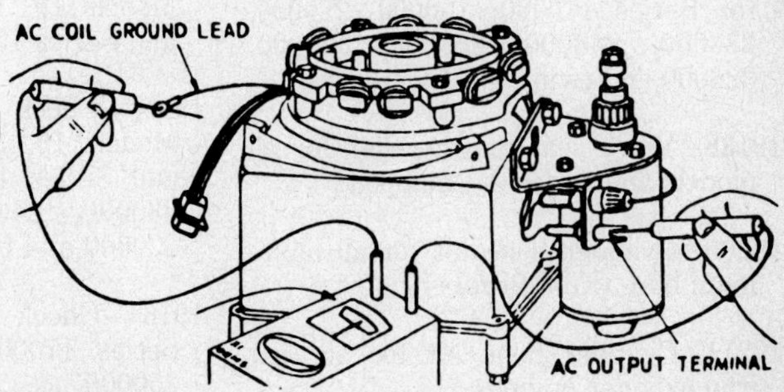

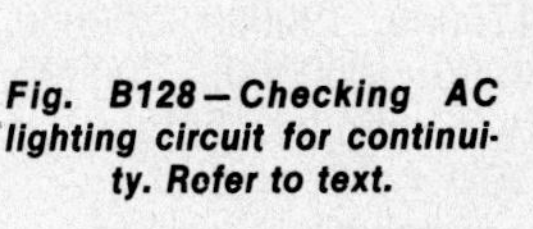

Fig. B128—Checking AC lighting circuit for continuity. Refer to text.

Fig. B124—Checking charging coils for an "open". Meter should show continuity.

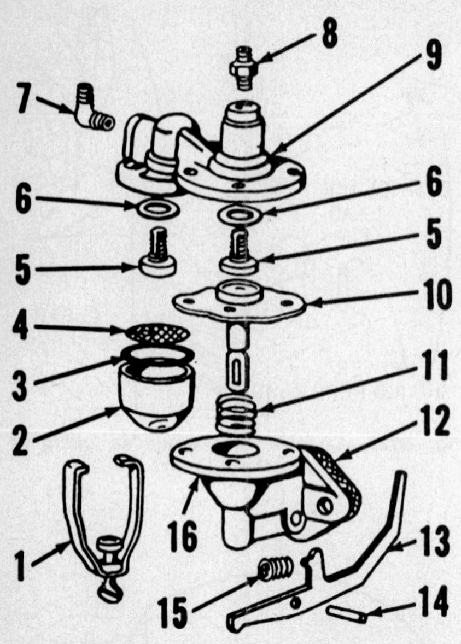

Fig. B129—Exploded view of diaphragm type fuel pump used on some B&S engines.

1. Yoke assy.
2. Filter bowl
3. Gasket
4. Filter screen
5. Pump valves
6. Gaskets
7. Elbow fitting
8. Connector
9. Fuel pump head
10. Pump diaphragm
11. Diaphragm spring
12. Gasket
13. Pump lever
14. Lever pin
15. Spring
16. Fuel pump body

and ground as shown in Fig. B127. Load lamp should light at full brilliance at medium engine speed. If lamp does not light or is very dim at medium speeds, remove blower housing and flywheel. Disconnect ground end of AC coil from retaining clamp screw (Fig. B119). Connect ohmmeter between ground lead of AC coil and AC output terminal as shown in Fig. B128. Ohmmeter should show continuity. If not, stator must be renewed. Be sure AC ground lead is not touching a grounded surface, then check continuity from AC output terminal to engine ground. If ohmmeter indicates continuity, lighting coils are grounded and stator must be renewed.

FUEL PUMP

A diaphragm type fuel pump is available on many models as optional equipment. Refer to Fig. B129 for exploded view of pump.

To disassemble pump, refer to Figs. B129 and B130; then, proceed as follows: Remove clamp (1), fuel bowl (2), gasket (3) and screen (4). Remove screws retaining upper body (9) to lower body (16). Pump valves (5) and gaskets (6) can now be removed. Drive pin (14) out to either side of body (16), then press diaphragm (10) against spring (11) as shown in view A, Fig. B130, and remove lever (13). Diaphragm and spring (11–Fig. B129) can now be removed.

To reassemble, place diaphragm spring in lower body and place diaphragm on spring, being sure spring enters cup on bottom side of diaphragm and slot in shaft is at right angle to pump lever. Then, compressing diaphragm against spring as in view A, Fig. B130, insert hooked end of lever with hole in lower body and drive pin into place. Then, insert lever spring (15) into body and push outer end of spring into place over hook on arm of lever as shown in view B. Hold lever downward as shown in view C while tightening screws holding upper body to lower body. When installing pump on engine, apply a liberal amount of grease on lever (13) at point where it contacts groove in crankshaft.

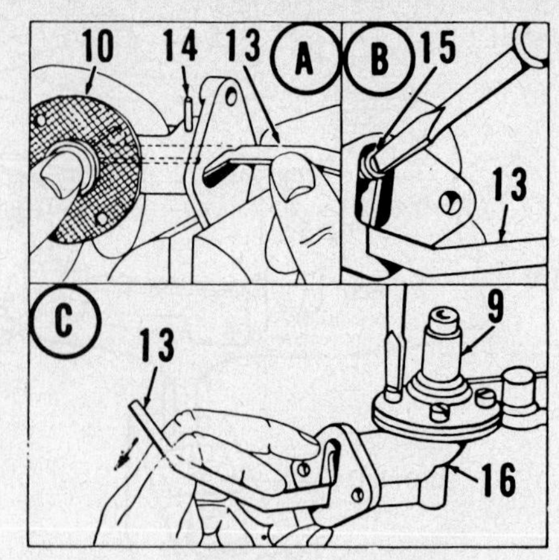

Fig. B130—Views showing disassembly and reassembly of diaphragm type fuel pump. Refer to text for procedure and to Fig. B129 for exploded view of pump and for legend.

BRIGGS & STRATTON SPECIAL TOOLS

The following special tools are available from Briggs & Stratton Central Service Distributors.

TOOL KITS

19184–Main bearing tool kit for Series 170000, 171000, 190000 and 191000

19228–Main bearing tool kit for all twin-cylinder engines with integral and DU type main bearings

19232–Valve guide puller/reamer kit for Series 170000, 190000, 220000, 233000, 243000, 250000, 300000, 326000 and twin-cylinder engines

19138–Valve seat insert puller for all models and series so equipped

19205–Cylinder hone kit for all aluminum bore (Kool-Bore) engines

19211–Cylinder hone kit for all cast iron cylinder engines

19237–Valve seat cutter kit for all models and series

PLUG GAGES

19055–Check breaker plunger bore on Model 23C and Series 170000, 190000, 220000 and 251000.

19117–Check main bearing bore on Models 19, 19D, 23, 23A, 23C, 23D and Series 190000,200000 and 230000

19151–Check valve guide bores on Models 19, 19D, 23, 23A, 23C, 23D and Series 170000, 190000, 200000, 230000, 240000, 250000, 300000, 320000 and twin-cylinder engines

19164–Check camshaft bearings on Series 170000, 190000, 220000 and 250000

19178–Check main bearing bore on Series 170000, 171000, 190000 and 191000

19219–Check integral or DU type main bearing bore on twin-cylinder engines

REAMERS

19056–Breaker plunger bushing reamer for 170000, 190000, 220000 and 251000

19058–Finish reamer for breaker plunger bushing for Series 170000, 190000, 220000 and 251000

19172–Counter bore reamer for main bearings for Series 170000 and 190000

19173–Finish reamer for main bearings for Series 170000 and 190000

19174 – Counter bore reamer for main bearings for Series 170000, 171000, 190000 and 191000

19175 – Finish reamer for main bearings for Series 170000, 171000, 190000 and 191000

19183 – Valve guide reamer for bushing installation for Models 19, 19D, 23, 23A, 23C, 23D and Series 170000, 190000, 200000, 220000, 230000, 240000, 251000, 300000 and 320000

19231 – Valve guide bushing reamer for Models 19 and 23 and Series 170000, 190000, 200000, 220000, 230000, 240000, 250000, 300000, 320000 and twin-cylinder engines

19233 – Valve guide bushing finish reamer for Models 19 and 23 and Series 170000, 190000, 200000, 220000, 230000, 240000, 250000, 300000, 320000 and twin-cylinder engines

PILOTS

19096 – Main bearing reamer pilot for Series 170000, 171000, 190000 and 191000

19127 – Expansion pilot for valve seat counterbore cutter for Models 19 and 23 and Series 170000, 190000, 200000, 220000, 230000, 240000, 300000, 320000 and twin-cylinder engines

19130 – "T" handle for 19127 pilot

REAMER GUIDE BUSHING

19168 – Guide bushing for main bearing reaming for Series 170000 and 190000

19169 – Guide bushing for main bearing reaming for Series 170000 and 190000

19170 – Guide bushing for main bearing reaming for Series 170000 and 190000

19171 – Guide bushing for main bearing reaming for Series 170000 and 190000

19192 – Guide bushing for valve guide reaming for Models 19 and 23 and Series 170000, 190000, 200000, 220000, 230000, 240000, 251000, 300000 and 320000

19201 – Guide bushing for main bearing reaming for Series 171000 and 191000

19222 – Guide bushing for tool kit 19228 for twin-cylinder engines

19234 – Guide bushing for tool kit 19232 for twin-cylinder engines

PILOT GUIDE BUSHINGS

19168 – Pilot guide bushing for main bearing reaming for Series 170000 and 190000

19169 – Pilot guide bushing for main bearing reaming for Series 170000 and 190000

19220 – Pilot guide bushing for main bearing reaming for tool kit 19228 for twin-cylinder engines

COUNTERBORE CUTTERS

19131 – Counterbore valve seat cutter for Models 19, 19D, 23, 23A, 23D and Series 190000, 200000, 230000, 240000 and 243000

CRANKCASE SUPPORT

19123 – To support crankcase when removing or installing main bearings on Series 170000 and 190000

19227 – To support crankcase when removing or installing DU type main bearings on twin-cylinder engines

DRIVERS

19057 – To install breaker plunger bushing on Series 170000, 190000, 220000 and 251000

19136 – To install valve seat inserts on all models and series

19179 – To install main bearings on Series 170000 and 190000

19226 – To install main bearing on twin-cylinder engines

FLYWHEEL PULLERS

19068 – Flywheel removal on Models 19, 19D, 23, 23A, 23C, and 23D and Series 190000, 200000, 230000, 240000, 300000 and 320000

19165 – Flywheel removal on Series 170000, 190000 and 250000

19203 – Flywheel removal on Models 19 and 23 and Series 190000, 200000, 220000, 230000, 240000, 250000, 251000, 300000, 320000 and twin-cylinder engines

VALVE SPRING COMPRESSOR

19063 – Valve spirng compressor for all models and series

PISTON RING COMPRESSOR

19230 – Ring compressor for all models and series

STARTER WRENCH

19114 – All models and series with rewind starter

19161 – All models and series with rewind starter (to be used with ½-inch drive torque wrench)

CLUTCH WRENCH

19244 – To remove, install and torque rewind starter clutches

BENDING TOOL

19229 – Governor tang bending tool

SPARK TESTER

19051 – Test ignition spark on all models and series

BRIGGS & STRATTON CENTRAL SERVICE DISTRIBUTORS

(Arranged Alphabetically by States)

These franchised firms carry extensive stocks of repair parts. Contact them for name of the nearest service distributor who may have the parts you need.

BEBCO, Incorporated
2221 Second Avenue, South
Birmingham, Alabama 35233

Power Equipment Company
3141 North 35th Avenue, Unit 101
Phoenix, Arizona 85107

Pacific Power Equipment Company
1565 Adrain Road
Burlingame, California 94010

Power Equipment Company
1045 Cindy Lane
Carpinteria, California 93013

Pacific Power Equipment Company
5000 Oakland Street
Denver, Colorado 80239

Spencer Engine Incorporated
1114 West Cass Street
Tampa, Florida 33606

Sedco Incorporated
1414 Red Plum Road
Norcross, Georgia 30093

Small Engine Clinic, Incorporated
98019 Kam Highway
Aiea, Hawaii 96701

Midwest Engine Warehouse
515 Romans Road
Elmhurst, Illinois 60126

Medart Engines & Parts of Kansas
15500 West 109th Street
Lenexa, Kansas 66215

Commonwealth Engine, Incorporated
1361 South 15th Street
Louisville, Kentucky 40210

Suhren Engine Company, Inc.
8330 Earhart Boulevard
New Orleans, Louisiana 70118

Grayson Company of Louisiana, Inc.
100 Fannin Street
Shreveport, Louisiana 71101

W. J. Connell Company
65 Green Street
Foxboro, Massachusetts 02035

Carl A. Anderson Inc. of Minnesota
2737 West Service Road
Eagan, Minnesota 55121

Medart Engines & Parts
100 Larkin Williams Ind. Ct.
Fenton, Missouri 63026

Original Equipment, Incorporated
905 Second Avenue, North
Billings, Montana 59101

Carl A. Anderson Inc. of Nebraska
7410 "L" Street
Omaha, Nebraska 68127

W. J. Connel Co., Bound Brook Div.
Chimney Rock Road, Route 22
Bound Brook, New Jersey 08805

Power Equipment Company
7209 Washington Street, North East
Alburquerque, New Mexico 87109

AEA, Incorporated
700 West 28th Street
Charlotte, North Carolina 28206

Gardner, Incorporated
1150 Chesapeake Avenue
Columbus, Ohio 43212

American Electric Ignition Company
124 North West Eighth
Oklahoma City, Oklahoma 73101

Brown & Wiser, Incorporated
9991 South West Avery Street
Tualatin, Oregon 97062

Pitt Auto Electric Company
2900 Stayton Street
Pittsburgh, Pennsylvania 15212

Automotive Electric Corporation
3250 Millbranch Road
Memphis, Tennessee 38116

American Electric Ignition Co., Inc.
618 Jackson Street
Amarillo, Texas 79101

Grayson Company, Incorporated
1234 Motor Street
Dallas, Texas 75207

Engine Warehouse, Incorporated
7415 Empire Central Drive
Houston, Texas 77040

Frank Edwards Company
110 South 300 West
Salt Lake City, Utah 84101

RBI Corporation
101 Cedar Ridge Drive
Ashland, Virginia 23005

Bitco Western, Incorporated
4030 1st Avenue, South
Seattle, Washington 98134

Air Cooled Engine Supply
South 127 Walnut Street
Spokane, Washington 99204

Wisconsin Magnetic Incorporated
4727 North Teutonia Avenue
Milwaukee, Wisconsin 53209

CANADIAN DISTRIBUTORS

Canadian Curtiss-Wright, Ltd.
2-3571 Viking Way
Richmond, British Columbia V6V 1W1

Canadian Curtiss-Wright, Ltd.
89 Paramount Road
Winnipeg, Manitoba R2X 2W6

Canadian Curtiss-Wright, Ltd.
1815 Sismet Road
Mississauga, Ontario L4W 1P9

Canadian Curtiss-Wright, Ltd.
8100-N Trans Canada Hwy.
St. Laurent, Quebec H4S 1M5

CLINTON

CLINTON ENGINES CORPORATION
Maquoketa, Iowa

CLINTON ENGINE IDENTIFICATION INFORMATION

In order to obtain correct service replacement parts when overhauling Clinton engines, it is important engine be properly identified as to:
1. Model number
2. Variation number
3. Type letter

A typical nameplate from model number series engines prior to 1961 is shown in Fig. CL1. In this example, the following information is noted from nameplate:
1. Model number—**B-760**
2. Variation number—**AOB**
3. Type letter—**B**

In some cases, model number may be shown as in following example:

D-790-2124

Thus, "D-790" would be model number of a D-700-2000 series engine in which digits "2124" would be the variation number.

In late 1961, identification system for Clinton engines was changed to be acceptable for use with IBM inventory record systems. A typical nameplate from one of these models is shown in Fig. CL2.

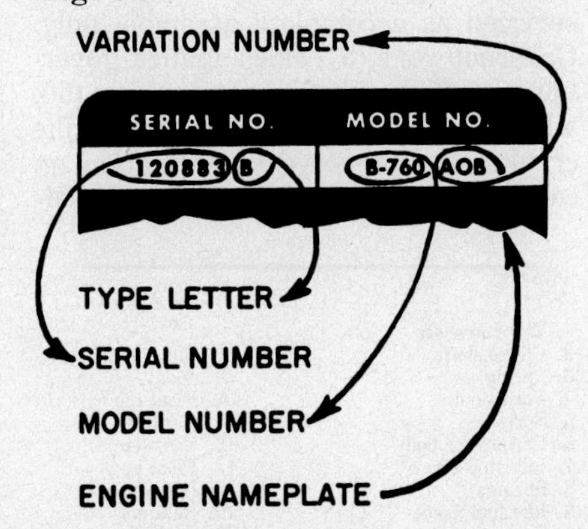

Fig. CL1—Typical nameplate from Clinton engine manufactured prior to late 1961.

In this example, the following information is noted from nameplate:
1. Model number—405-0000-000
2. Variation number—070
3. Type letter—D

In addition, the following information may be obtained from model number on nameplate:

First digit—identifies type of engine, i.e., 4 means four-cycle engine and 5 means two-cycle engine.

Second and third digits—completes basic identification of engine. Odd numbers will be used for vertical shaft engines and even numbers for horizontal shaft engines, i.e., 405 would indicate a four-cycle vertical shaft engine and 500 would indicate a two-cycle horizontal shaft engine.

Fourth digit—indicates type of starter as follows:
- 0—Recoil starter
- 1—Rope starter
- 2—Impulse starter
- 3—Crank starter
- 4—12-Volt electric starter
- 5—12-Volt starter-generator
- 6—110-Volt starter
- 7—12-Volt generator
- 8—Unassigned to date
- 9—Short block assembly

Fifth digit—indicates bearing type as follows:
- 0—Standard bearing
- 1—Aluminum or bronze sleeve bearing with flange mounting surface and pilot diameter on engine mounting face for mounting equipment concentric to crankshaft center line.
- 2—Ball or roller bearing
- 3—Ball or roller bearing with flange mounting surface and pilot diameter on engine mounting face for mounting equipment concentric to crankshaft center line.
- 4—Numbers 4 through 9 are unassigned to date.

Sixth digit—indicates auxiliary power take-off and speed reducers as follows:
- 0—Without auxiliary pto or speed reducer
- 1—Auxiliary power take-off
- 2—2:1 speed reducer
- 3—Not assigned to date
- 4—4:1 speed reducer
- 5—Not assigned to date
- 6—6:1 speed reducer
- 7—Numbers 7 through 9 are unassigned to date

Seventh digit—if other than "0" will indicate a major design change.

Eighth, ninth and tenth digits—identifies model variations.

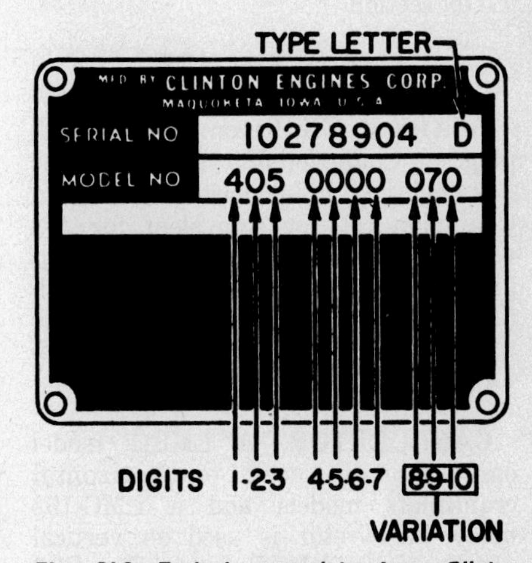

Fig. CL2—Typical nameplate from Clinton engine after model identification system was changed in late 1961. First seven digits indicate basic features of engine. Engines with "Mylar" (plastic) nameplate have engine model and serial numbers stamped on cylinder air deflector next to nameplate.

CLINTON

CLINTON ENGINES CORPORATION
Maquoketa, Iowa

Horizontal Crankshaft Engines

Model	No. Cyls.	Bore	Stroke	Displacement	Power Rating
412	1	2.813 in. (71.5 mm)	2.5 in. (63.5 mm)	15.5 cu. in. (254 cc)	7 hp. (5.22 kW)

Vertical Crankshaft Engines

Model	No. Cyls.	Bore	Stroke	Displacement	Power Rating
413	1	2.813 in. (71.5 mm)	2.5 in. (63.5 mm)	15.5 cu. in. (254 cc)	7 hp. (5.22 kW)

Engines in this section are one cylinder, four-cycle engines with either a vertical or a horizontal crankshaft. All horizontal crankshaft models have bushing type main bearing at flywheel end of crankshaft and either a bushing type or tapered roller bearing main at output end. All vertical crankshaft models use bushing type main bearings at each end of crankshaft.

Connecting rod on all models rides directly on crankpin journal. Horizontal crankshaft models are splash lubricated while vertical crankshaft models are lubricated by a gear type pump driven from lower end of camshaft.

All models incorporate a magneto type ignition system with breaker points and condenser located underneath flywheel.

All models are equipped with a side draft carburetor.

Engine identification is explained in **CLINTON ENGINE IDENTIFICATION** section.

MAINTENANCE

SPARK PLUG. Recommended spark plug is a Champion J8 or equivalent for 412 horizontal crankshaft models and a Champion H11 or equivalent for 413 vertical crankshaft models. Spark plug gap for all models is 0.025-0.028 inch (0.635-0.711 mm) and plug should be tightened to 23-25 ft.-lbs. (31-34 N·m) torque.

CARBURETOR. A LME-1 model carburetor is used on horizontal crankshaft models and a LMG-162 model carburetor is used on vertical crankshaft models. Refer to Fig. CL3 for exploded view of typical "LM" series carburetor.

For initial adjustment, open both the main fuel mixture and the idle fuel mixture screws 1½ turns. Make final adjustment with engine at operating temperature and running. Place engine under load and adjust main fuel needle to leanest mixture that will allow satisfactory acceleration and steady governor operation.

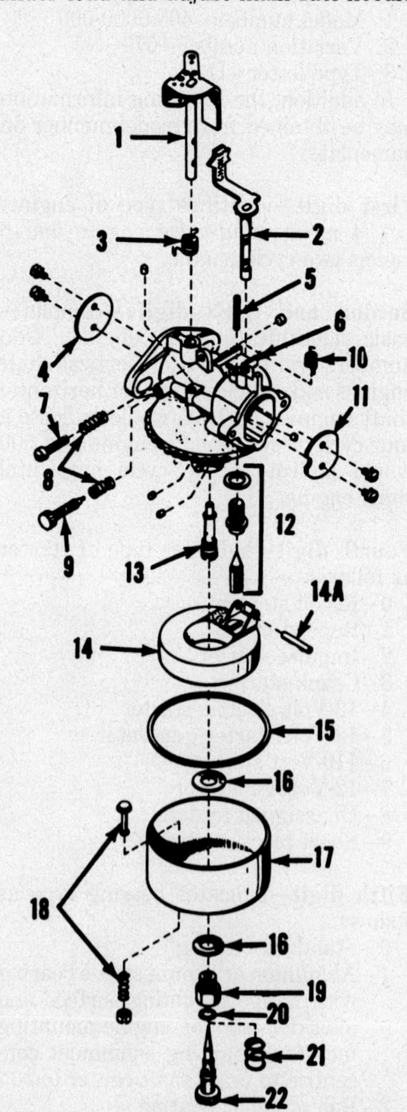

Fig. CL3—Exploded view of typical "LM" series float type carburetor as used on both horizontal and vertical shaft models.

1. Throttle shaft
2. Choke shaft
3. Spring
4. Throttle disc
5. Spring
6. Carburetor body
7. Idle stop screw
8. Springs
9. Idle fuel needle
10. Spring
11. Choke disc
12. Inlet needle
13. Main nozzle
14. Float
14A. Float pin
15. Gasket
16. Gaskets
17. Float bowl
18. Drain valve
19. Retainer
20. Seal
21. Spring
22. Main fuel needle

Run engine at idle speed, no load and adjust idle mixture screw for smoothest idle operation. As each adjustment affects the other, adjustment procedure may have to be repeated.

To check float level, invert carburetor throttle body and float assembly. There should be 5/32 to 11/64-inch (4-4.4 mm) clearance between machined surface of body casting and free end of float. Adjust float by bending float lever tang that contacts inlet valve.

Float drop should not exceed 3/16-inch (4.5 mm). Bend vertical tang at rear of hinge on float to adjust.

GOVERNOR. Horizontal crankshaft 412 models and vertical crankshaft 413 models are equipped with a mechanical (flyweight) governor. Refer to Fig. CL5 for exploded view of horizontal crankshaft model and to Fig. CL8 for vertical crankshaft model. Both governors are similar.

Governor ring (58 – Fig. CL5), weights (57), weight pins (54) and retainer (55) are attached to engine camshaft (53) by rivets (56). Governor and camshaft are serviced as a complete assembly only. Governor yoke (51) rides against governor ring (58) and pivots in bushing (50) which is pressed into bore in engine crankcase. For service information on camshaft-governor unit, refer to CAM-

SHAFT paragraph. Governor yoke may be removed and bushing in crankcase renewed after engine camshaft and governor assembly is removed.

To adjust governor, loosen screw (35) retaining governor arm (23) to governor adjusting bracket (24). Move speed control to hold carburetor throttle in wide open position. Turn adjusting bracket and yoke so as to close governor weights, then while holding bracket in this position tighten screw (35).

MAGNETO. Magneto assembly used on both horizontal crankshaft models and vertical crankshaft models is shown in exploded view of horizontal crankshaft model in Fig. CL5. Refer to inset (21). Magneto armature back plate is piloted to engine crankcase (73) by adapter ring (20) and armature mounting holes are not slotted. Therefore, magneto air gap and ignition timing are fixed and non-adjustable. For check purposes, air gap must be from 0.007-0.017 inch (0.1778-0.4318 mm) and edge gap from 0.1094-0.2500 inch (2.78-6.35 mm). A layer of plastic tape 0.008-0.009 inch (0.2032-0.2286 mm) laid over coil laminations will serve to show clearance of flywheel magnets. It may be necessary to renew armature (13) or flywheel (19) if gap is excessive.

Magneto is accessible after removing engine flywheel and breaker box cover. Adjust breaker point gap to 0.018-0.021 inch (0.4572-0.5334 mm). Condenser capacity is 0.15-0.19 mfd.

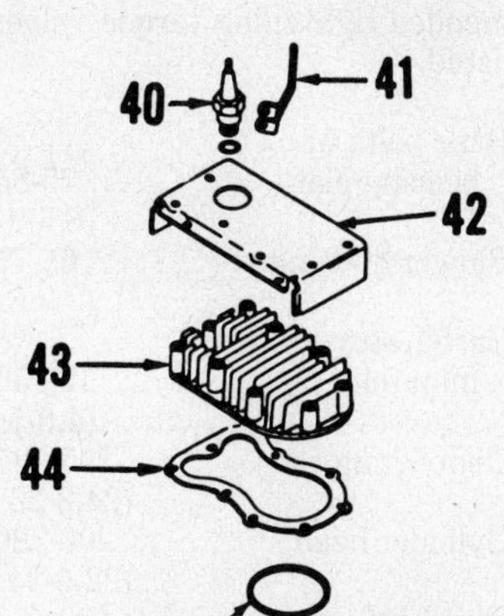

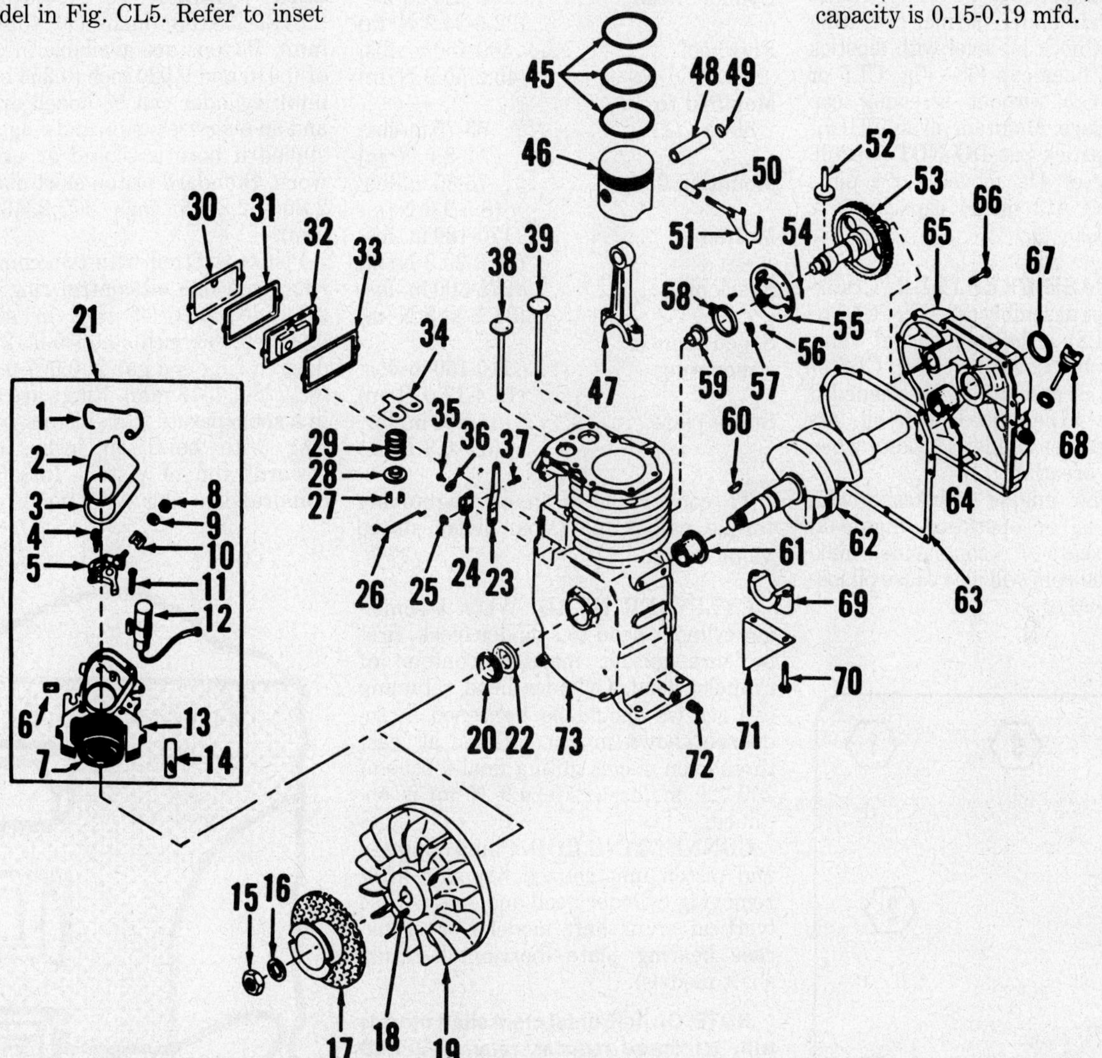

Fig. CL5—Exploded view of horizontal crankshaft model with sleeve type bearings (bushings). On some models, a roller type bearing is used at output end of crankshaft instead of sleeve bearing (64). Also, speed reducer units with ratios of 2:1, 4:1 and 6:1 are available. Refer to Fig. CL37 in SERVICING CLINTON ACCESSORIES section. Lock for connecting rod cap screws (70) is not shown. Refer to item (28) in Fig. CL8.

1. Spring
2. Breaker box cover
3. Gasket
4. Screw
5. Breaker points
6. Lubrication felt
7. Ignition coil
8. Nut
9. Washer
10. "Jamtite" terminal
11. Screw
12. Condenser
13. Armature & stator plate assy.
14. Coil clip
15. Flywheel nut
16. Washer
17. Starter pulley & screen
18. Pin
19. Flywheel
20. Magneto adapter ring
21. Magneto assy.
22. Oil seal
23. Governor arm
24. Adjusting bracket
25. Nut
26. Governor link
27. Valve keepers
28. Spring retainers
29. Valve springs
30. Cover
31. Gasket
32. Breather assy.
33. Gasket
34. Valve spring retainer plate
35. Screw
36. Washer
37. Bolt
38. Intake valve
39. Exhaust valve
40. Spark plug
41. Grounding switch
42. Air baffle
43. Cylinder head
44. Gasket
45. Piston rings
46. Piston
47. Connecting rod (includes 69)
48. Piston pin
49. Pin retainers (2)
50. Bushing for 51
51. Governor yoke assy.
52. Tappets
53. Camshaft & governor assy.
54. Pins
55. Retainer
56. Rivets
57. Governor weights
58. Governor ring
59. Camshaft bushing
60. Flywheel key
61. Flange bushing (main bearing)
62. Crankshaft
63. Gasket
64. Sleeve bushing (main bearing)
65. Bearing plate
66. Bolts
67. Oil seal
68. Oil dipstick
69. Connecting rod cap
70. Connecting rod cap screw
71. Oil dipper
72. Crankcase drain plug
73. Crankcase & cylinder casting

LUBRICATION. Horizontal crankshaft engines are splash lubricated by an oil dipper (71–Fig. CL5) attached to connecting rod cap. Vertical crankshaft models are lubricated by a gear type oil pump (36–Fig. CL8) which is driven by a pin in lower end of engine camshaft.

Oils approved by manufacturer must meet requirements of API service classification SB, SC, SD or SE.

For operation in temperatures above 32° F (0° C), use SAE 30 oil. For temperatures between 32° F (0° C) and -10° F (-23° C), use SAE 10 oil and for temperatures below -10° F (-23° C), SAE 5 oil is recommended.

It is recommended oil be changed at 25 hour intervals under normal operating conditions. Check oil level with dipstick attached to filler cap (68–Fig. CL5 or 46–Fig. CL8) without screwing cap down into place. Maintain oil at "FULL" mark on dipstick but **DO NOT** overfill. Oil capacity of 412 model is 2½ pints (1.18 L) and 413 model capacity is 2 pints (0.95 L).

CRANKCASE BREATHER. Crankcase breather assembly (32–Fig. CL5 or 58–Fig. CL8) is located behind valve tappet chamber cover (30–Fig. CL5 or 56–Fig. CL8) and should be cleaned if difficulty is experienced with oil loss through breather. Renew gaskets when reinstalling breather.

Over filling engine crankcase with lubricating oil or operating engine at speeds in excess of recommended maximum of 3600 rpm will also cause oil loss through breather.

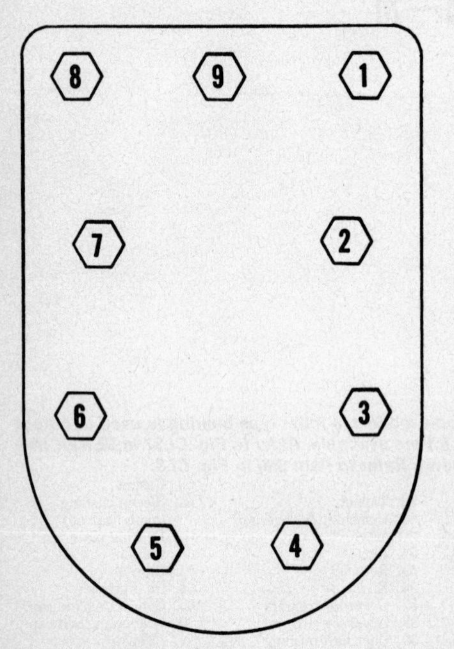

Fig. CL6—When installing cylinder head, tighten retaining cap screws in order shown. Refer to tightening torque table.

REPAIRS

TIGHTENING TORQUES. Recommended tightening torque values are as listed.

Base bolts or bearing plate	75-85 in.-lbs. (8.5-9.6 N·m)
Blower housing	65-75 in.-lbs. (7.3-8.5 N·m)
Carburetor to manifold	35-50 in.-lbs. (4.0-5.7 N·m)
Connecting rod	215-235 in.-lbs. (24.3-26.6 N·m)
Cylinder head	200-220 in.-lbs. (22.6-24.9 N·m)
Flywheel	400-450 in.-lbs. (45.2-50.8 N·m)
Manifold to block	65-75 in.-lbs. (7.3-8.5 N·m)
Mounting flange	75-85 in.-lbs. (8.5-9.6 N·m)
Muffler	170-180 in.-lbs. (19.2-20.3 N·m)
Spark plug	275-300 in.-lbs. (31.1-33.9 N·m)
Speed reducer mounting	110-150 in.-lbs. (12.4-17.0 N·m)
Stator plate	50-60 in.-lbs. (5.7-6.8 N·m)

To convert listed in.-lbs. tightening torque values to ft.-lbs. divide stated value by 12.

CYLINDER HEAD. When assembling cylinder head to cylinder block, first be sure gasket matches contour of cylinder head. Cylinder head retaining cap screws should be tightened in sequence shown in Fig. CL6 in at least three even stages until a final torque of 200-220 in.-lbs. (22.6-24.9 N–m) is obtained.

CONNECTING ROD. Connecting rod and piston unit can be removed after removing cylinder head and engine base (vertical crankshaft models) or crankcase bearing plate (horizontal crankshaft models).

NOTE: On horizontal crankshaft models with 2:1 speed reducer, refer to SPEED REDUCERS paragraph in Servicing Clinton Accessories section of this manual for information on removing bearing plate.

The aluminum alloy connecting rod rides directly on crankshaft crankpin journal. Recommended rod to crankpin clearance is 0.001-0.002 inch (0.0254-0.0508 mm). Renew rod if clearance is excessive or if rod is scored or otherwise shows signs of damage.

Connecting rod to piston pin clearance is 0.0002-0.0011 inch (0.0051-0.0279 mm). Renew piston pin and/or connecting rod if clearance is 0.002 inch (0.0508 mm) or more.

When installing rod and piston assembly, install with oil hole or chamfer of connecting rod towards flywheel and with embossments on rod and cap aligned. Tighten cap screws to a torque of 215-235 in.-lbs. (24.3-26.6 N·m) and bend tabs of lock against screw heads.

PISTON, PIN AND RINGS. Recommended piston skirt to cylinder bore clearance is 0.007-0.009 inch (0.1778-0.2286 mm). Renew piston if scored or if clearance is excessive. Standard cylinder bore diameter is 2.8125-2.8135 inch (71.438-71.4629 mm). Pistons are available in oversizes of 0.010 and 0.020 inch (0.254 and 0.508 mm); cylinder can be honed or rebored and an oversize piston and ring set be installed if bore is scored or excessively worn. Standard piston skirt diameter is 2.8045-2.8055 inch (71.2343-71.2597 mm).

Piston is fitted with two compression rings and one oil control ring. Desired side clearance of ring in groove is 0.0025-0.005 inch (0.0635-0.1270 mm); desired ring end gap is 0.007-0.017 inch (0.1778-0.4318 mm). Rings are available in a set separate from piston. Install top ring with bevel on inside diameter towards top of piston. Install second compression ring with notch on outer

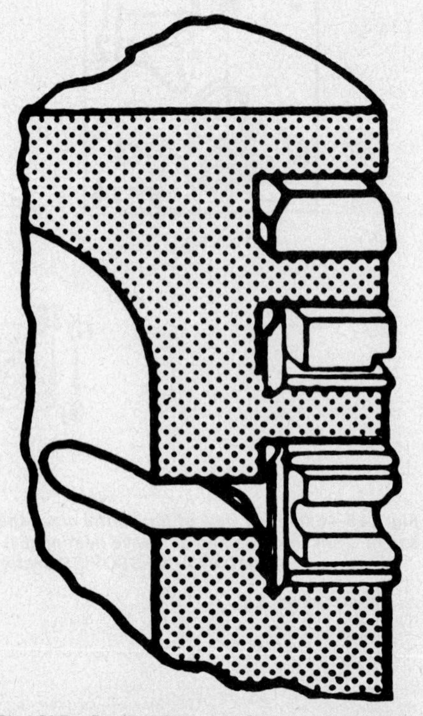

Fig. CL7—Profile view showing arrangement of piston rings. Note inside bevel of compression ring is at top and notch of scraper ring is down. Install expanders and rails as shown only with chromium rings.

face downward. Oil control ring can be installed with either side up. See Fig. CL7.

Piston pin diameter is 0.6726-0.6728 inch (17.08-17.09 mm), and pin is available in standard size only. Pin should be "light press fit" which is 0.0001 inch (0.0025 mm) interference fit to 0.0004 inch (0.0102 mm) loose in piston.

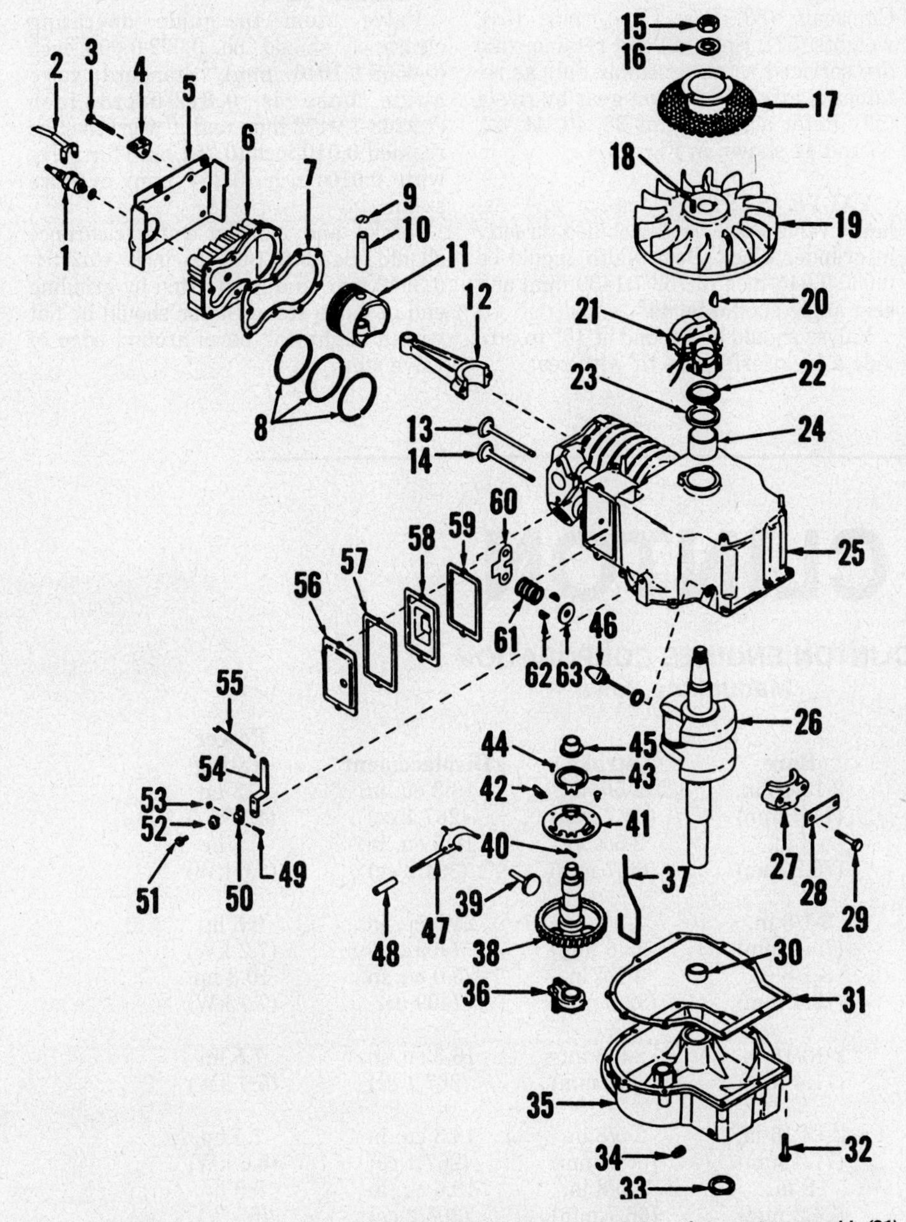

Fig. CL8—Exploded view of vertical crankshaft model. For exploded view of magneto assembly (21), refer to Fig. CL5. Gear type oil pump (36) is driven by pin in lower end of camshaft (38).

1. Spark plug
2. Grounding switch
3. Cylinder head bolts
4. Lifting bracket
5. Air baffle
6. Cylinder head
7. Gasket
8. Piston rings
9. Pin retainers
10. Piston pin
11. Piston
12. Connecting rod (includes 27)
13. Intake valve
14. Exhaust valve
15. Flywheel nut
16. Washer
17. Starter pulley & screen assy.
18. Pin
19. Flywheel
20. Flywheel key
21. Magneto assy.
22. Magneto adapter ring
23. Oil seal
24. Flange bushing (main bearing)
25. Crankcase & cylinder casting
26. Crankshaft
27. Connecting rod cap
28. Rod cap screw lock
29. Connecting rod cap screw
30. Sleeve bushing (main bearing)
31. Gasket
32. Cap screws
33. Oil seal
34. Oil drain plug
35. Engine base
36. Oil pump
37. Oil tube
38. Camshaft
39. Tappets
40. Pins
41. Retainer
42. Weights
43. Governor ring
44. Rivets
45. Camshaft bushing
46. Oil dipstick
47. Governor yoke assy.
48. Bushing
49. Bolt
50. Adjusting bracket
51. Screw
52. Washer
53. Nut
54. Governor arm
55. Governor link
56. Cover
57. Gasket
58. Breather assy.
59. Gasket
60. Valve spring retainer
61. Valve springs
62. Valve keepers
63. Spring retainers

NOTE: Special chrome ring sets are available in both standard and 0.020 inch (0.508 mm) oversize for use in cylinder bores having up to 0.010 inch (0.254 mm) wear, taper and/or out-of-round. Installation instructions are packaged with each ring set.

Some models are factory equipped with chrome ring set, stellite exhaust valve and rotocap and a standard exhaust valve in place of intake valve.

CYLINDER AND CRANKCASE. Cylinder and crankcase are an integral unit of cast iron construction.

Standard cylinder bore diameter is 2.8125-2.8135 inch (71.438-71.4629 mm). Cylinder can be bored to 0.010 or 0.020 inch (0.254 or 0.508 mm) oversize for installation of oversize piston and ring set.

CRANKSHAFT, MAIN BEARINGS AND SEALS. All vertical crankshaft models are equipped with bushing type main bearings. All horizontal crankshaft models have a bushing type main bearing at flywheel end of crankshaft and either a bushing or tapered roller bearing at output end. Bushing used at flywheel end of crankshaft on all models is a flanged bronze bushing. A sleeve bushing is used at output end on models so equipped.

Crankshaft main bearing journal diameter is 1.2510-1.2515 inches (31.7754-31.7881 mm) at each end. Bushing inside diameter is 1.2525-1.2535 inches (31.81-31.84 mm) resulting in a journal to bushing clearance of 0.001-0.0025 inch (0.0254-0.0635 mm). Renew bushings and/or crankshaft if clearance is excessive or if bushings or journals are scored. Renew crankshaft if crankpin is 0.001 inch (0.0254 mm) or more out-of-round or excessively worn or scored. Crankpin diameter is 1.2493-1.250 inches (31.73-31.75 mm). Connecting rod to crankpin clearance should be 0.001-0.002 inch (0.0254-0.0508 mm).

Crankshaft end play for models with bushing type main bearings at each end of crankshaft should be 0.002-0.006 inch (0.0508-0.1524 mm). For models with tapered roller bearing at output end of crankshaft, end play should be 0.001-0.006 inch (0.0254-0.1524 mm). End play is adjusted by varying thickness of gasket between crankcase and engine base of bearing plate.

Horizontal crankshaft models equipped with speed reducer should have 0.001-0.006 inch (0.0254-0.1524 mm) crankshaft end play, which is controlled by varying thickness of gasket between speed reducer gear housing and engine bearing plate. No oil seal is used at output end of crankshaft as oil level in gear box is maintained from engine lubricating oil supply.

CAMSHAFT. Camshaft is supported in a bushing (59–Fig. CL5 or 45–Fig. CL8) in cylinder block and rides directly in unbushed bore of engine bearing plate (65–Fig. CL5) or engine base (35–Fig. CL8). Desired clearance of camshaft journal to bore in bearing plate of horizontal crankshaft models is 0.001-0.004 inch (0.0254-0.1016 mm). For vertical crankshaft models, clearance of journal to bore in engine base should be 0.0005-0.002 inch (0.0127-0.0508 mm). On all models, clearance of camshaft journal to bushing in engine cylinder block should be 0.001-0.003 inch (0.0254-0.0762 mm). Camshaft lobe maximum diameter should be 1.2085 inches (30.6959 mm).

Inspect the three governor weights in camshaft gear for wear and be sure they pivot freely on their pins. Check governor ring for wear and free movement. Camshaft (53–Fig. CL5), pins (54), weights (57), ring (58) and retainer (55) are serviced as an assembly only as retainer is attached to cam gear by rivets (56). Refer also to items 38, 40, 41, 42, 43 and 44 shown in Fig. CL8.

VALVE SYSTEM. Intake and exhaust valve seats are machined directly in cylinder block. Seat width should be 0.030-0.045 inch (0.762-1.1430 mm) and seat angle should be 44°.

Valves should be ground at 45° to provide a 1° interference fit with seat.

Stellite exhaust valve and rotocap are used on some models and can be installed in place of standard exhaust valve. If stellite exhaust valve is used, install a standard exhaust valve in place of standard intake valve.

Valve stem to guide operating clearance should be 0.002-0.004 inch (0.0508-0.1016 mm). Standard valve guide bore is 0.312-0.313 inch (7.9248-7.9472 mm) and if worn may be reamed 0.010 inch (0.254 mm) for valve with 0.010 inch (0.254 mm) oversize stem.

Intake and exhaust valve clearance should be 0.011-0.012 inch (0.2794-0.3048 mm) and is adjusted by grinding end of valve stem. Stems should be flat with a slight 45° bevel around edge of valve stem.

CLINTON

CLINTON ENGINES CORPORATION
Maquoketa, Iowa

Model	No. Cyls.	Bore	Stroke	Displacement	Power Rating
1600, A-1600	1	2-13/16 in. (71.4 mm)	2-5/8 in. (66.7 mm)	16.3 cu. in. (267.1 cc)	6.3 hp (4.7 kW)
18	1	3 in. (76.2 mm)	2-5/8 in. (66.7 mm)	18.6 cu. in. (304.2 cc)	7.5 hp (5.6 kW)
2500, A-2500, B-2500	1	3-1/8 in. (79.4 mm)	3-1/4 in. (82.6 mm)	25.0 cu. in. (409 cc)	9.6 hp (7.2 kW)
2790	1	3-1/8 in. (79.4 mm)	3-1/4 in. (82.6 mm)	25.0 cu. in. (409 cc)	10.3 hp (7.7 kW)
414-1300-000, 414-1301-000	1	2-13/16 in. (71.4 mm)	2-5/8 in. (66.7 mm)	16.3 cu. in. (267.1 cc)	7.5 hp (5.6 kW)
416-1300-000, 418-1300-000	1	2-13/16 in. (71.4 mm)	2-5/8 in. (66.7 mm)	16.3 cu. in. (267.1 cc)	7.5 hp (5.6 kW)
418-1301-000	1	3 in. (76.2 mm)	2-5/8 in. (66.7 mm)	18.6 cu. in. (304.2 cc)	8.0 hp (6 kW)
420-1300-000, 420-1301-000	1	3-1/8 in. (79.4 mm)	3-1/4 in. (82.6 mm)	25.0 cu. in. (409 cc)	9.6 hp (7.2 kW)
422-1300-000, 422-1301-000	1	3-1/8 in. (79.4 mm)	3-1/4 in. (82.6 mm)	25.0 cu. in. (409 cc)	10.3 hp (7.7 kW)

Engines covered in this section are one cylinder, four-cycle, horizontal crankshaft engines in which crankshaft is supported at each end by either bushing type main bearings or tapered roller bearings.

Connecting rod on all models rides directly on crankpin journal. All models are splash lubricated by an oil distributor (dipper) attached to connecting rod cap.

A magneto type ignition system with externally located breaker points and condenser is used on all models.

Float type carburetors of various manufacture are used according to model and application.

Engine identification is explained in **CLINTON ENGINE IDENTIFICATION** section.

MAINTENANCE

SPARK PLUG. Recommended spark plug is a Champion H10 or equivalent. Spark plug gap for all models should be 0.025-0.028 inch (0.635-0.711 mm) and plug should be tightened to 23-25 ft.-lbs. (31-34 N·m) torque.

CARBURETOR. Carter "UT" series or Clinton "HEW" series float type carburetors are used according to model and application. Refer to appropriate paragraph for adjustment and assembly information.

HEW FLOAT TYPE CARBURETOR. Refer to Fig. CL9 for exploded view of typical HEW carburetor.

For initial adjustment, open idle mixture and main fuel mixture screws 1½ turns each. Make final adjustment with engine running and at operating temperature. Place engine under load and adjust main fuel mixture screw for leanest mixture that will allow satisfactory acceleration and steady governor operation. Set engine at idle speed, no load and adjust idle mixture screw to obtain smoothest idle operation.

As each adjustment affects the other, adjustment procedure may have to be repeated.

To check float level, invert carburetor throttle body and float assembly. There should be 3/16-inch (4.8 mm) clearance between free side of float and machined surface of body casting. To adjust, carefully bend float lever tang that contacts inlet valve to provide correct float level measurement.

Float drop should not exceed 3/16-inch (4.5 mm). Bend vertical tang at rear of hinge on float to adjust.

Fig. CL9—Exploded view of "HEW" series float type carburetor.

1. Float bowl
2. Throttle body
3. Oil cup
5. Air cleaner
7. Float
9. Gasket
10. Nozzle tip gasket
11. Gasket
13. Filter housing
14. Idle fuel needle
15. Main fuel needle
16. Fuel inlet needle & seat
17. Nozzle
19. "O" ring
20. Adapter plate
21. Drain plug
22. Plugs
25. Idle speed screw
28. Float pin
29. Choke shaft
30. Throttle shaft
31. Choke stop spring
32. Lower spring
33. Upper spring
35. Throttle disc
36. Choke disc
37. Venturi
38. Washer

"UT" CARTER FLOAT TYPE CARBURETOR. Refer to Fig. CL10 for exploded view of typical "UT" series float type carburetor. Identification number for carburetor is stamped on metal tag and attached to carburetor body by one of the bowl retaining screws. Make certain tag is reinstalled when reassembling carburetor.

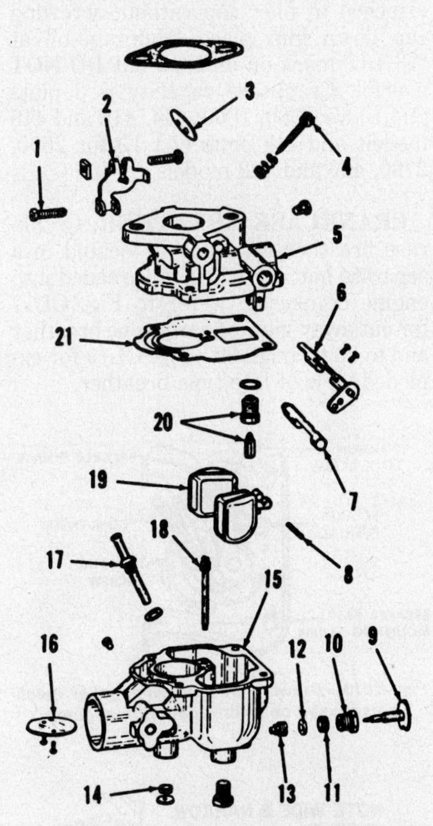

Fig. CL10—Exploded view of typical Carter "UT" float type carburetor. Venturi is not shown.

1. Idle speed stop screw
2. Throttle lever
3. Throttle disc
4. Idle fuel needle
5. Throttle body
6. Choke shaft
7. Throttle shaft
8. Float pin
9. Main fuel needle
10. Nut
11. Packing
12. Washer
13. Main jet
14. Dust seal
15. Lower body
16. Choke disc
17. Fuel nozzle
18. Idle jet
19. Float
20. Inlet needle & seat
21. Gasket

For initial adjustment, open main fuel mixture screw 1¼ turns and idle mixture screw 1½ turns. Make final adjustment with engine running and at operating temperature. Place engine under load and adjust main fuel needle for leanest setting that will allow satisfactory acceleration and steady governor operation. Set engine at idle speed, no load and adjust idle mixture screw to obtain smoothest idle operation.

As each adjustment affects the other, adjustment procedure may have to be repeated.

To check float level, invert carburetor throttle body and float assembly. Match number stamped on tag attached to carburetor with carburetor identification number listed to determine correct float setting which is measured from float seam to carburetor body. Carefully bend float lever tang that contacts inlet valve to provide correct float level measurement.

Carburetor Identification Number	Carburetor Float Level
2217-S	11/64-in. (4.36 mm)
2230-S	17/64-in. (6.75 mm)
2336-S	1/4-in. (6.35 mm)
2337-S	1/4-in. (6.35 mm)
2398-S	*1/4-in. *(6.35 mm)
2712-S	19/64-in. (7.54 mm)
2713-S	19/64-in. (7.54 mm)
2714-S	1/4-in. (6.35 mm)

*9/32-inch (7.14 mm) when non-metallic inlet needle valve seat is used.

GOVERNOR. All models are equipped with a mechanical governor which consists of a governor gear and weight unit (Fig. CL11) which rotates on a shaft pressed into engine crankcase. Governor is driven by cam gear.

Governor weight unit can be serviced after removing engine base. See Fig. CL18.

To adjust governor travel, loosen governor lever on shaft (66–Fig. CL18 or 76–Fig. CL19), turn shaft as far counter-clockwise as possible and hold in this position. Turn carburetor throttle shaft to wide open position and tighten governor lever on shaft.

MAGNETO AND TIMING. Magneto consists of an armature and coil assembly located under engine flywheel and breaker points and condenser which are located in a breaker box attached to outside of engine crankcase and are accessible after removing box cover. Refer to Fig. CL12 and CL13 for details of different breaker assemblies used.

Breaker cam is located on camshaft and ignition timing is automatically advanced by flyweight and spring on cam assembly as engine speed increases. Ignition timing is adjustable to a slight degree by varying breaker point gap from 0.028-0.030 inch (0.7112-0.7620 mm). Ignition should occur at 19° to 21° BTDC when timing is fully advanced. Retarded timing is 1° to 3° BTDC on early engines and 6° to 8° ATDC on later engines or early engines fitted with late style cam assemblies.

Breaker points are actuated by a push rod located between cam and external breaker box. Refer to Figs. CL14 and CL15 for details of push rod assembly. Make certain small coil of spring is seated in narrow groove of push rod as shown in Fig. CL14. Push rod shield (S–Fig. CL15) was installed about 1966 and is available for service on earlier models and Clinton recommends it be installed whenever possible.

Condenser capacity is 0.15-0.19 mfd. and armature air gap should be 0.012-0.020 inch (0.3048-0.508 mm).

Refer to Fig. CL16 for method of securing high tension lead to backplate if flywheel rubbing against lead is encountered.

LUBRICATION. All models are splash lubricated by an oil distributor (dipper) (30–Fig. CL18 or 43–Fig. CL19) attached to connecting rod cap.

Oils approved by manufacturer must meet requirements of API service classification SB, SC, SD or SE.

For operation in temperatures above 32° F (0° C), use SAE 30 oil. For temperatures between 32° F (0° C) and -10° F (-23° C use SAE 10 oil and for temperatures below -10° F (-23° C) SAE 5 oil is recommended.

It is recommended oil be changed at 25 hour intervals under normal operating conditions. Check oil level with dipstick attached to filler cap without screwing cap down into place. Maintain oil at "FULL" mark on dipstick but **DO NOT** overfill. Crankcase capacity is 3 pints (1.4 L) for 1600, 1800, 414, 416 and 418 models and 4½ pints (2.1 L) for 2500, 2790, 420 and 422 models.

CRANKCASE BREATHER. Crankcase breather assembly is located in a separate housing which is threaded into engine crankcase. Refer to Fig. CL17 for cut-away view of early type breather and to 26 through 32–Fig. CL19 for exploded view of late type breather.

Breather should be removed and cleaned after each 100 hours of normal engine operation. Refer to Figs. CL17 and CL19 for proper reassembly of breather.

Fig. CL11—Mechanical governor can be serviced after removing base from engine crankcase.

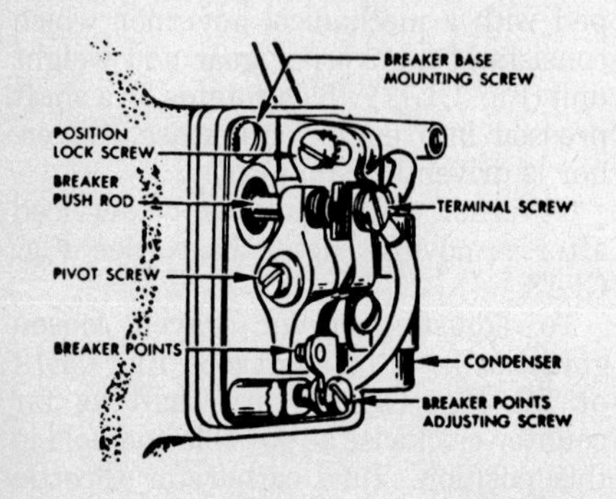

Fig. CL12—Drawing of ignition breaker mechanism used on early production engines.

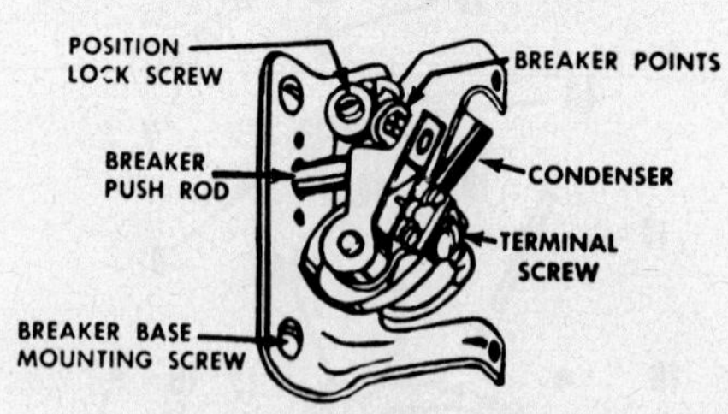

Fig. CL13—Drawing of ignition breaker mechanism used on later production engines.

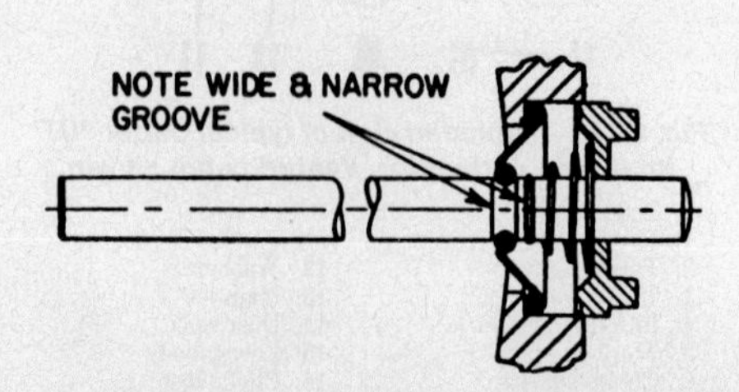

Fig. CL14—Be sure seal is installed in wide groove and return spring is seated in narrow groove of breaker point push rod. Pins formed on spring retainer fit in holes in breaker box.

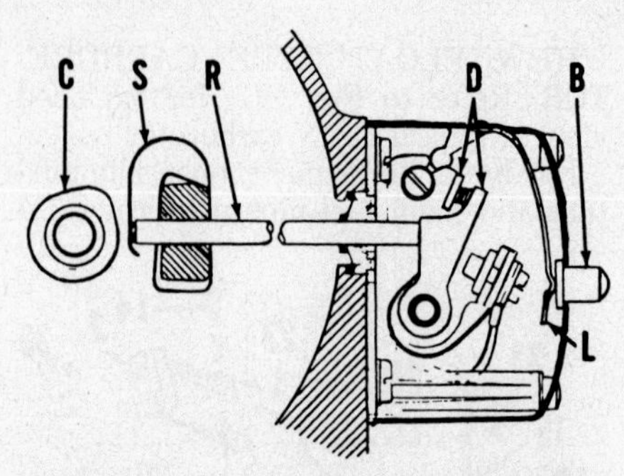

Fig. CL15—Cross-sectional view of camshaft and ignition points assembly. Spring-type push rod shield (S) was added about 1966 and can be installed on earlier units.

B. Kill button
C. Camshaft
D. Breaker points
L. Spring leaf
R. Push rod
S. Shield

Fig. CL16—Anchor high tension lead as shown if difficulty is encountered with flywheel rubbing insulation from wire.

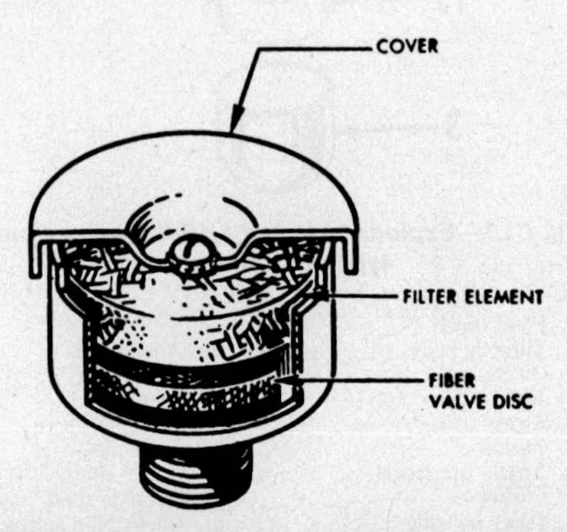

Fig. CL17—Cut-away view of breather assembly used on early models. Refer to (26-Fig. CL19) for exploded view of breather used on later engines. Breather should be serviced after each 100 hours of engine operation.

REPAIRS

TIGHTENING TORQUES. Recommended torque values are as follows:

Base bolts . . . . 12-13 ft.-lbs. (16-18 N•m)
Bearing plate . . . . 13-15 ft.-lbs. (18-20 N•m)
Blower housing . . . . 5-6 ft.-lbs. (7-8 N•m)
Carburetor or manifold . . . . 5-5.4 ft.-lbs. (7-7.3 N•m)
Connecting rod . . . . 18-20 ft.-lbs. (24-27 N•m)
Cylinder head . . . . 17-18 ft.-lbs. (23-25 N•m)
Flywheel . . . . 8-10 ft.-lbs. (11-14 N•m)
Spark plug . . . . 23-25 ft.-lbs. (31-34 N•m)
Stator plate . . . . 7-8 ft.-lbs. (9-11 N•m)

CONNECTING ROD. Connecting rod and piston assembly can be removed from engine after removing cylinder head and engine base.

Recommended connecting rod to crankshaft clearance for 1600, 1800, 414, 416 and 418 models is 0.0015-0.0025 inch (0.0381-0.0635 mm) and should be 0.0010-0.0018 inch (0.0254-0.0457 mm) for 2500, 2790, 420 and 422 models.

Recommended connecting rod to piston pin clearance for all models should be 0.0002-0.0011 inch (0.0051-0.0279 mm).

Connecting rod is available in standard size only. When reassembling, make certain embossments on connecting rod and cap are aligned as shown in Fig. CL20. The "clearance side" of connecting rod must be installed toward camshaft side of engine. Refer to Fig. CL21. Make certain connecting rod, rod

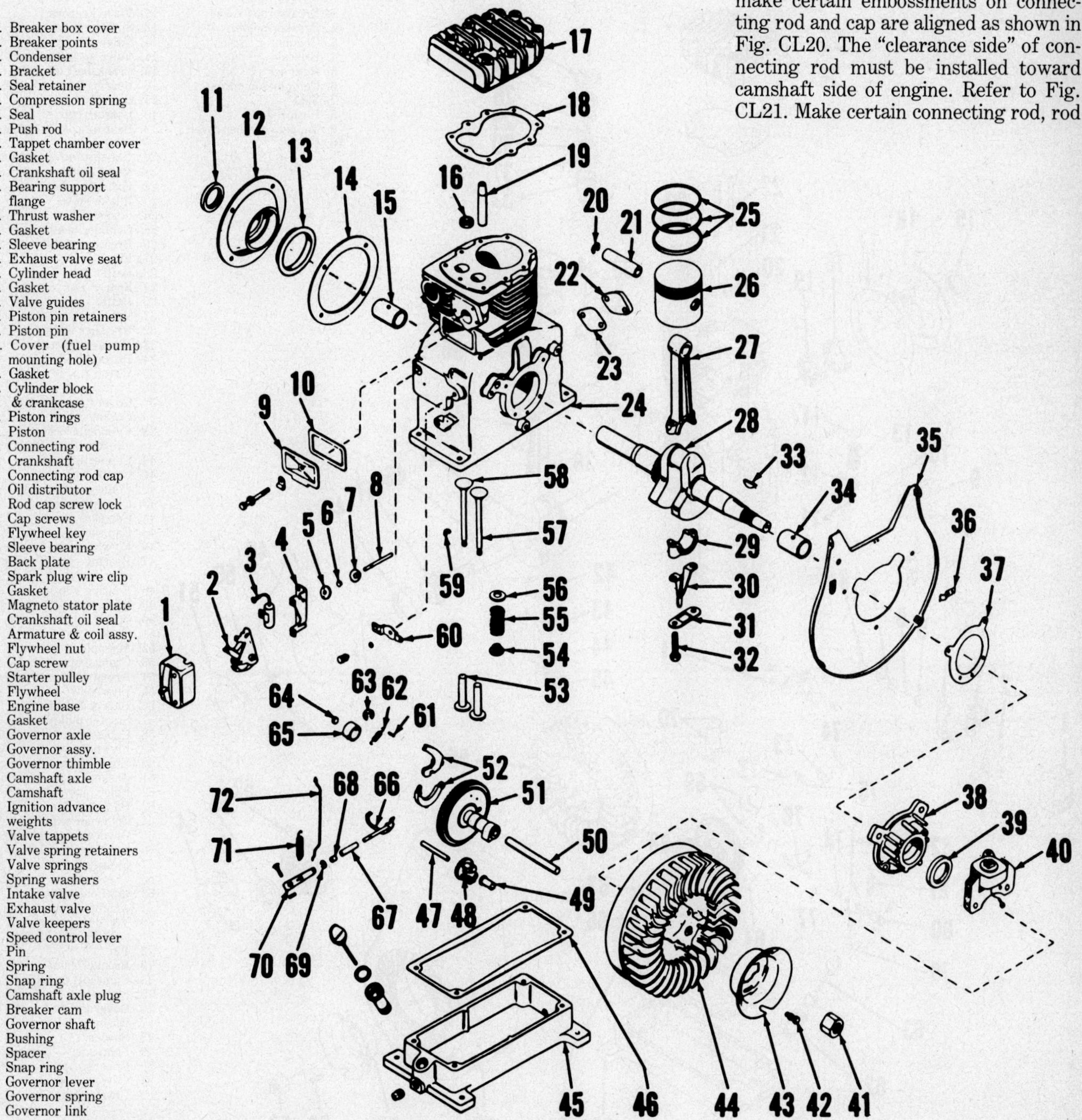

Fig. CL18 – Exploded view of model with bushing type main bearings (15 and 34). Note flange thrust washer (13) which controls crankshaft end play; end play may be adjusted by thickness of gasket (14) used between bearing support (12) and cylinder block. Gasket (14) is available in a variety of thicknesses; gasket (37) is available in one thickness only.

locks and oil distributor clear camshaft after assembly.

PISTON, PIN AND RINGS. Recommended piston skirt to cylinder bore clearance is as follows:

Model	Bore	Clearance
1600, 414,416, 418	2.8125 in. (71.4 mm)	0.007-0.009 in. (0.1778-0.2286 mm)
18,418-1301	3.0 in. (76.2 mm)	0.007-0.0085 in. (0.1778-0.2159 mm)
2500, B-2500, 420	3.125 in. (79.4 mm)	0.0065-0.0085 in. (0.1651-0.2159 mm)
A-2500, 2790, 422	3.125 in. (79.4 mm)	0.005-0.007 in. (0.1270-0.1778 mm)

Piston is fitted with two compression rings and one oil control ring. Recommended ring side clearance in groove is 0.0025-0.005 inch (0.0635-0.1270 mm). Piston is available in oversizes of 0.010 and 0.020 inch (0.254 and 0.508 mm) as well as standard.

Piston pin should be a "light press fit", 0.0002 inch (0.0051 mm) tight to 0.0003 inch (0.0076 mm) loose in piston. Pin is available in standard size only.

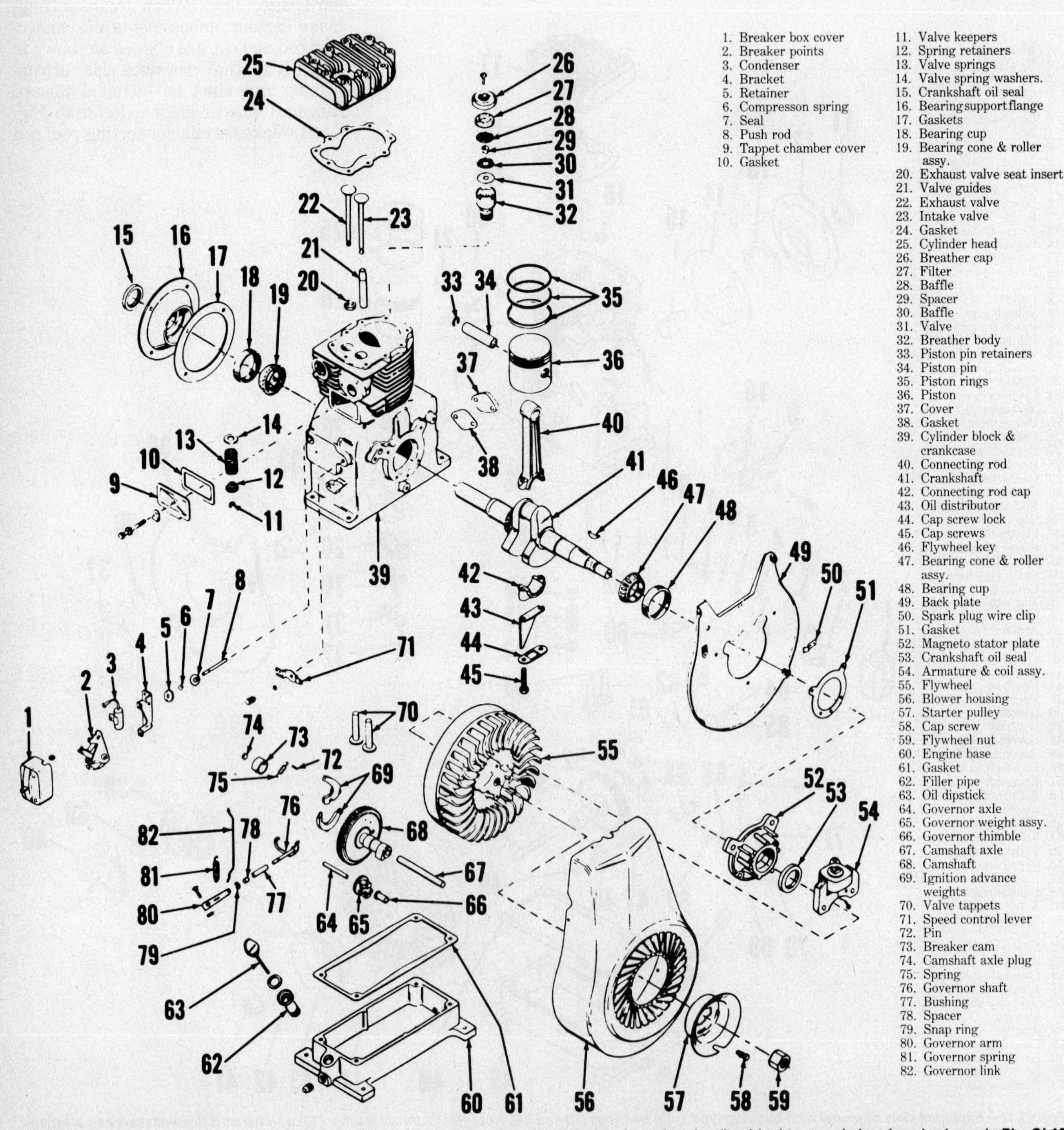

Fig. CL19—Exploded view of model having tapered rollered main bearings. Model with sleeve bearing (bushing) type main bearings is shown in Fig. CL18. Refer to Fig. CL23 for installation of bearing plate (16) on models equipped with roller type main bearings.

Recommended piston ring end gap is as follows:

Cylinder Bore	Clearance
2.8125 in. (71.4 mm)	0.007-0.017 in. (0.1778-0.4318 mm)
3.0 in. (76.2 mm)	0.007-0.017 in. (0.1778-0.4318 mm)
3.125 in. (79.4 mm)	0.010-0.020 in. (0.254-0.508 mm)

Piston rings are available in oversizes of 0.010 and 0.020 inch (0.254 and 0.508 mm) as well as in standard size. Special oversize ring sets of 0.005, 0.030 and 0.040 inch (0.1270, 0.762 and 1.016 mm) have been available for some models. Chrome ring sets come as standard and 0.020 inch (0.508 mm) oversize.

NOTE: Before altering diameter of cylinder bore by boring or honing, check local supplier for available oversizes of replacement parts.

Refer to Fig. CL22 for correct place ment of piston rings and make certain ring end gaps are staggered.

CYLINDER AND CRANKCASE. Cylinder and crankcase are an integral casting of cast iron. Standard cylinder bore diameter for A-1600, 1600, 414, 416 and 418 models is 2.8125-2.8135 inches (71.4-71.7 mm), for 18 and 418-1301 models bore is 2.9995-3.0005 inches (76.187-76.213 mm) and for 420, 422, 2790, A-2500, B-2500 and 2500 models bore is 3.1245-3.1255 inches (79.362-79.388 mm).

If engine bore is within recommended tolerances it can be returned to service after removing any "glaze" on cylinder walls using appropriate hone or glaze-breaker. Leave a cross-hatch pattern for proper seating of new rings.

CRANKSHAFT AND MAIN BEARINGS. All current production models are equipped with ball, needle or tapered roller main bearings. Earlier production, A-1600, 1800 and B-2500 engines were equipped with bushing (sleeve) type main bearings. See Fig. CL18 or CL19 for comparison.

Sleeve type main bearing inside diameter should be measured by use of a micrometer bore gage and readings compared to measured outer diameter of crankshaft main journals. Models A-1600, 1800 and B-2500 engines equipped with sleeve bearings call for a minimum clearance of 0.0018 inch (0.0457 mm) and a maximum of 0.0035 inch (0.0889 mm). Bearings must be renewed if clearance exceeds 0.005 inch (0.1270 mm). New sleeve bearings must be pressed approximately 1/32-inch (0.794 mm) below flush from crankshaft thrust face of bearing plate or crankcase. Align bearing oil holes with oil passages. Allowable crankshaft end play for sleeve bearing equipped engines is 0.006-0.020 inch (0.1524-0.508 mm).

Engines equipped with tapered roller, ball or needle bearings call for crankshaft end play of 0.001-0.006 inch (0.0254-0.1524 mm). Adjustment is made by varying gasket thickness between pto end bearing plate and engine crankcase. Standard gasket thickness is 0.015 inch with a variety of thicknesses available.

When installing new needle bearings, pressure should be exerted against lettered side of bearing only.

When installing new ball bearings, shielded side faces inward, toward crankcase. Oil drain hole of pto end bearing plate (Fig. CL23) should be installed downward, toward oil sump.

Crankpin diameter for 1600, A-1600, 1800, 414, 416 and 418 models should be 1.1243-1.1250 inches (28.5572-28.5850 mm) if standard and connecting rod clearance should be 0.0015-0.0025 inch (0.0381-0.0635 mm).

Crankpin diameter for 2500, A-2500, B-2500, 2790, 420 and 422 models should be 1.2495-1.2500 inches (31.7373-31.7500 mm) if standard. Connecting rod clearance should be 0.001-0.0018 inch (0.0254-0.0457 mm).

On all models if clearances are exceeded or if crankpin is over 0.001 inch (0.0254 mm) out-of-round, renewal of crankshaft and/or connecting rod is required. Manufacturer recommends great care in assembly of cap to connecting rod and strict adherence to specified bolt torque values to prevent distortion. Refer to appropriate preceding section.

CAMSHAFT. A hollow camshaft and gear assembly rotates on a stationary shaft on all models. Recommended shaft to camshaft bore clearance for all models is 0.0015-0.0035 inch (0.0381-0.0889 mm). Renew shaft and/or camshaft if clearance is 0.0055 inch (0.1397 mm) or more.

Ignition breaker cam and advance mechanism are mounted on cam gear and camshaft. Ignition cam, advance

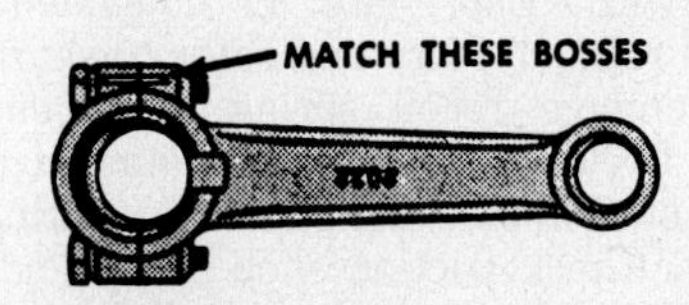

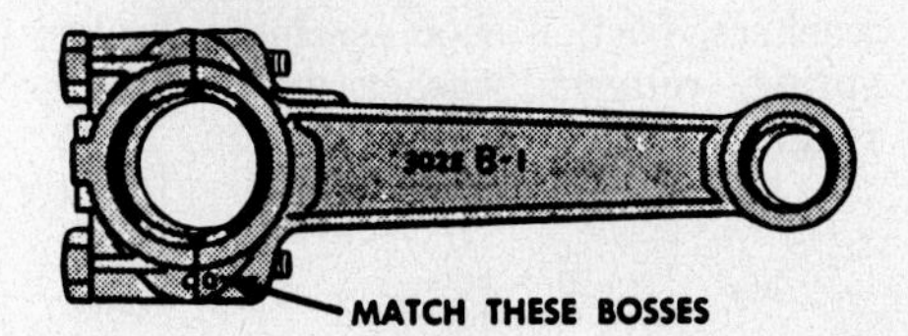

Fig. CL20—When assembling cap to connecting rod, be sure embossments on rod and cap are aligned as shown.

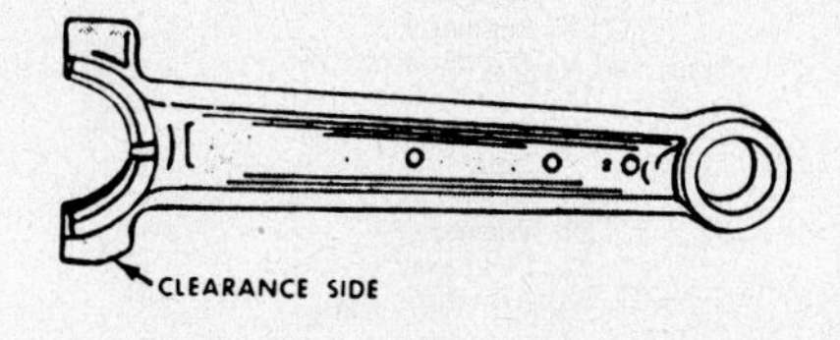

Fig. CL21—Connecting rods of some models have a clearance side which must be installed toward camshaft.

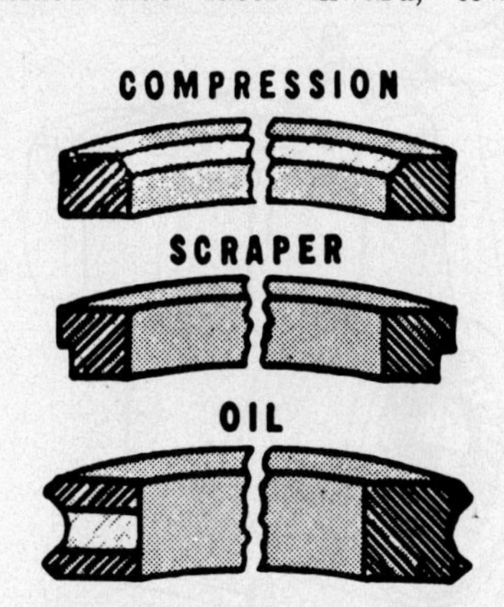

Fig. CL22—Drawing showing proper placement of piston rings. Note bevel at top of ring on inside diameter on top compression ring and notch in lower side of outside diameter of second compression (scraper) ring. Oil ring may be installed with either side up.

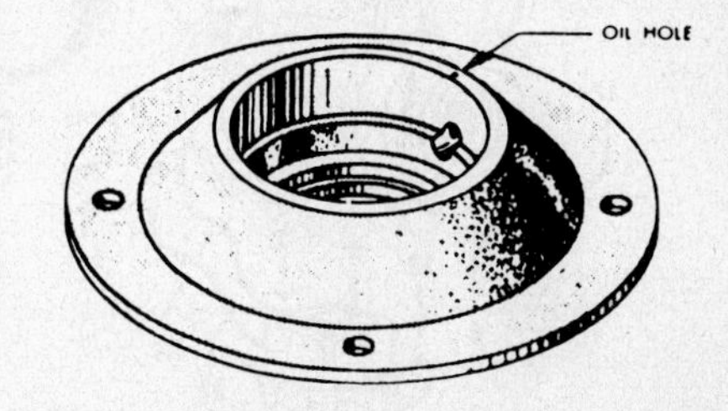

Fig. CL23—Main bearing (pto end) plate on ball bearing equipped models should be installed with oil drain hole towards bottom of engine.

Fig. CL24—Valves are properly timed when single marked tooth of cam gear is meshed between the two marked teeth on crankshaft gear. Crankshaft gear may have diagonal marks instead of dots as shown.

weights, pivot pins, rivets and springs are renewable separately from camshaft.

VALVE SYSTEM. Valve seats may be machined directly into cylinder casting or renewable inserts may be installed according to model and application. Inserts may be renewed or installed on models not so equipped. Seat width for both valves should be 0.030-0.045 inch (0.762-1.1430 mm) and seat angle should be 44°.

Valves should be ground to 45° to provide a 1° interference fit with seat.

Stellite exhaust valve and rotocap are used on some models and can be installed in place of standard exhaust valve. If stellite exhaust valve is used, install a standard exhaust valve in place of standard intake valve.

Valve stem to guide operating clearance should be 0.002-0.004 inch (0.0508-0.1016 mm). Renew valves and/or guides if clearance is 0.006 inch (0.1524 mm) or more. New guides should be installed so upper ends are 1-1/4 inches (31.75 mm) below cylinder head surface of engine block.

Valves are operated by mushroom type tappets which ride in unbushed bores in engine crankcase. Recommended tappet to bore clearance is 0.002-0.004 inch (0.0508-0.1016 mm). Refer to Fig. CL24 for location of valve timing marks.

CLINTON

SERVICING CLINTON ACCESSORIES

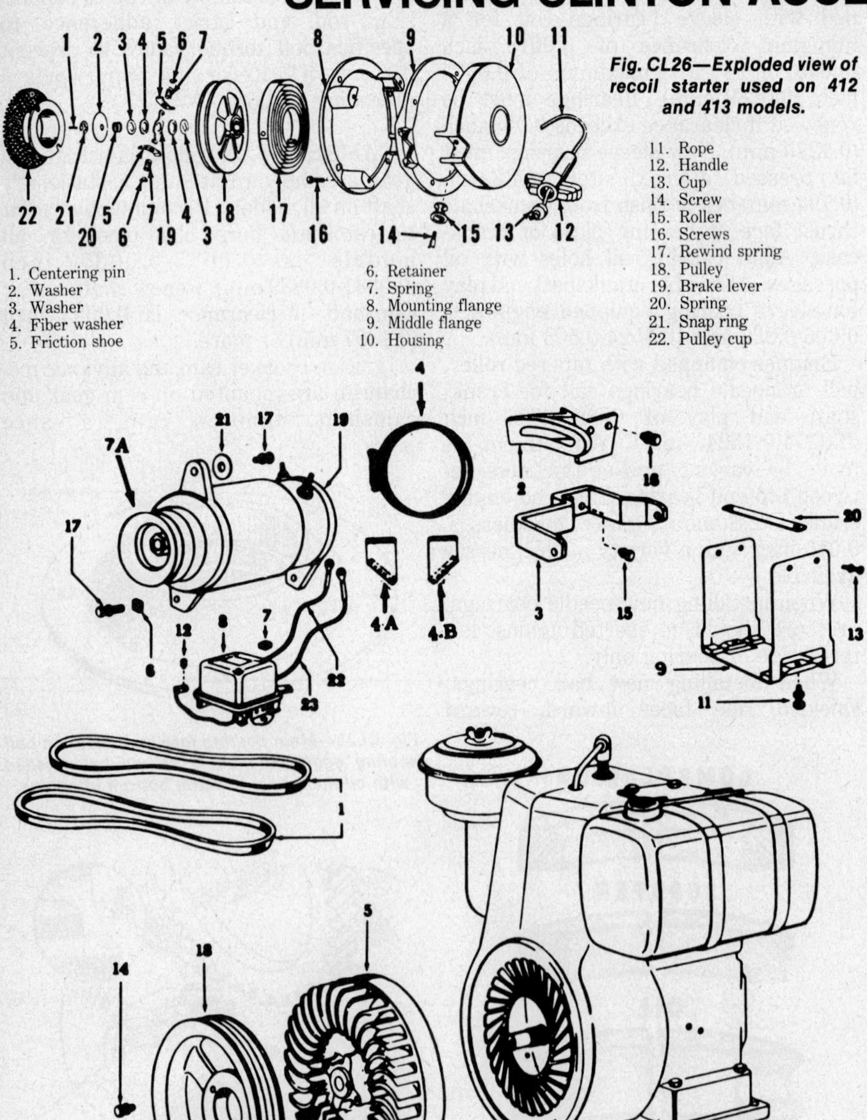

Fig. CL26—Exploded view of recoil starter used on 412 and 413 models.

1. Centering pin
2. Washer
3. Washer
4. Fiber washer
5. Friction shoe
6. Retainer
7. Spring
8. Mounting flange
9. Middle flange
10. Housing
11. Rope
12. Handle
13. Cup
14. Screw
15. Roller
16. Screws
17. Rewind spring
18. Pulley
19. Brake lever
20. Spring
21. Snap ring
22. Pulley cup

Fig. CL27—Delco-Remy starter-generator used on larger displacement horizontal crankshaft engine. See Fig. CL28 for wiring arrangement. Refer to text.

1. Drive belts
2. Adjusting bracket
3. Mounting bracket
4. Clamp
4A. Regulator clip
4B. Regulator clip
5. Flywheel
6. Nut
7. Nut
7A. Pulley
8. Regulator
9. Regulator bracket
18. Flywheel sheave
19. Starter-generator unit.
20. Shock mount strap.
21. Washer
22. Field assy.
23. Armature assy.

RETRACTABLE STARTERS

Fairbanks-Morse retractable starters are used on some models. To disassemble starter, remove retainer ring, retainer washer, brake spring, friction washer, friction shoe assembly and second friction washer as shown in Fig. CL26. Hold rope handle in one hand and cover in the other and allow rotor to rotate to unwind recoil spring preload. Lift rotor from cover, shaft and recoil spring. Note winding direction of recoil spring and rope for aid in reassembly. Remove recoil spring from cover and unwind rope from rotor.

When reassembling unit, lubricate recoil spring, cover shaft and its bore in rotor with "Lubriplate" or equivalent. Install rope on rotor and rotor to shaft, then engage recoil spring inner end hook. Preload recoil spring four turns and install middle flange and mounting flange. Check friction shoe sharp ends and renew if necessary. Install friction washers, friction shoe assembly, brake spring, retainer washer and retainer ring. Make certain friction shoe assembly is installed properly for correct starter rotation. If properly installed, sharp ends of friction shoe plates will extend when rope is pulled.

Starter operation can be reversed by winding rope and recoil spring in opposite direction and turning friction shoe assembly upside down.

12-VOLT STARTER-GENERATOR

A combination 12-volt starter-generator manufactured by Delco-Remy is used on some models. Starter-generator functions as a cranking motor when starting switch is closed. When engine is operating and with starting switch open, unit operates as a generator. Generator output and circuit voltage for battery and various operating requirements are controlled by a current-voltage regulator.

Refer to Fig. CL28 for typical wiring arrangement and obtain service parts from authorized Delco-Remy service center.

STARTER-ALTERNATOR-MAGNETO

Models 412 and 413 may be equipped with an American Bosch starter installed as shown in Fig. CL29 or Fig. CL30. Exploded view of starter drive assembly is shown in Fig. CL30A.

Because of size limitations, this electric starter motor will not stand up to extended operation under full load. It should not be operated for more than ten seconds at a time and at least one minute should be allowed between attempts to start engine. If starter turns engine over rapidly but engine fails to start, check for fuel, ignition or mechanical problem such as a stuck valve.

Fig. CL28—Wiring arrangement for Delco-Remy starter-generator. Starter and battery cables, A, B and C must be at least No. 4 AWG, up to No. 1 AWG. Other wiring should be at least No. 14 AWG.

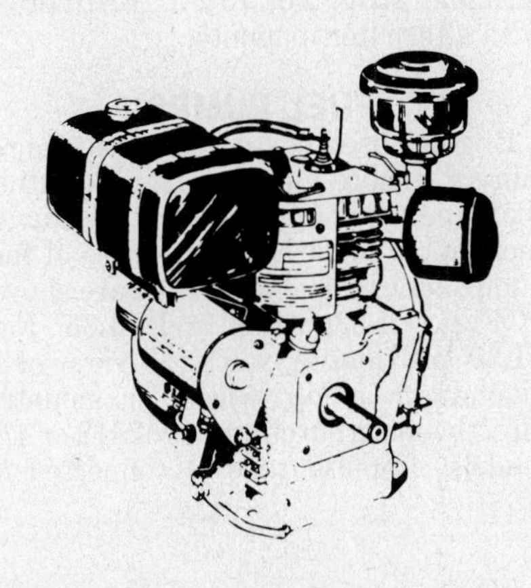

Fig. CL29—American-Bosch starter mounted on 412 model engine. AC rectifier and terminal strip are visible below starter body.

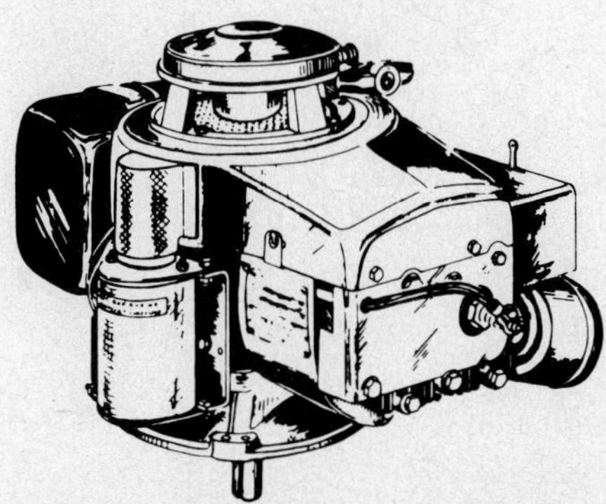

Fig. CL30—View of American-Bosch starter as mounted on 413 model vertical crankshaft engine. See Fig. CL30A for view of starter drive.

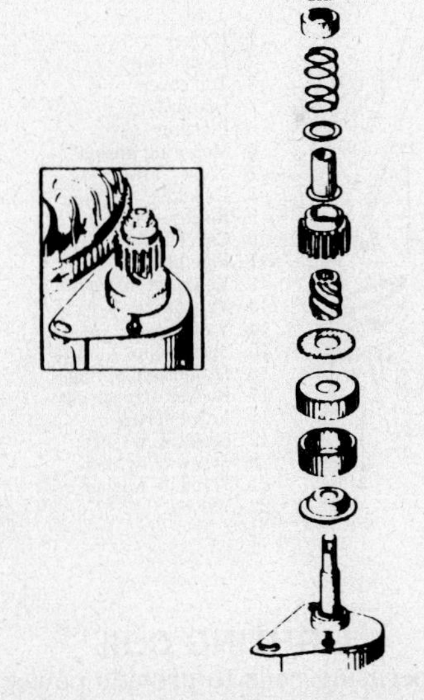

Fig. CL30A—Exploded view of starter drive components of 12 V American-Bosch starter. Be sure to allow a slight amount of backlash between starter pinion and flywheel ring gear.

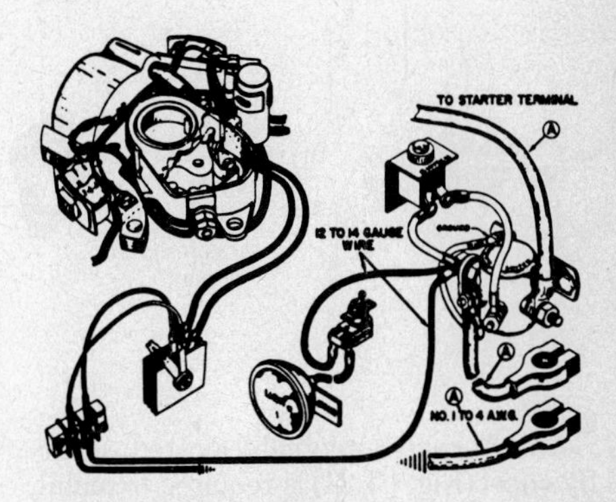

Fig. CL31—Simplified view of alternator magneto wiring layout used with American-Bosch starter. Note rectifier and lighting connections.

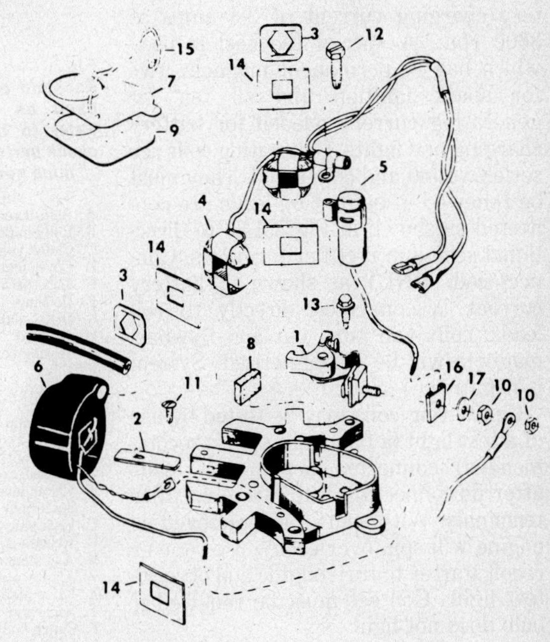

CL32—Exploded view of magneto stator plate to show arrangement of ignition and charging (generating) components.

1. Breaker points
2. Coil retainer clip
3. Retainer
4. Generating coils
5. Condenser
6. Ignition coil
7. Dust cover
8. Felt strip
9. Dust cover gasket
10. Nut (shorting wire)
11. Screw
12. Condenser screw
13. Points screw
14. Shield (4)
15. Dust cover spring
16. Condenser terminal
17. Condenser washer

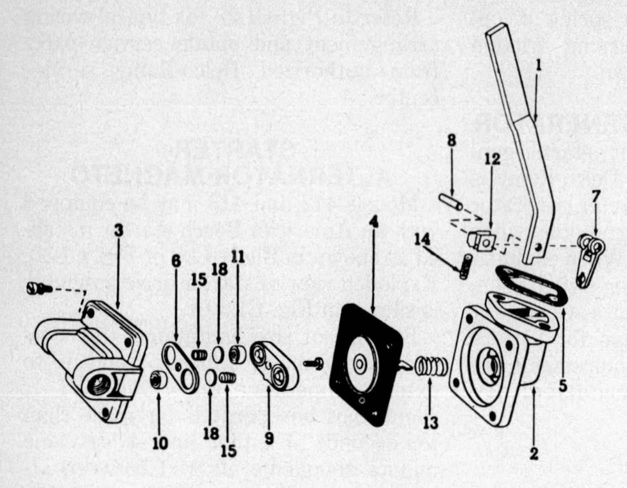

Fig. CL33—Exploded view of mechanical fuel pump used on some engines.

1. Rocker arm
2. Body
3. Cover
4. Diaphragm
5. Gasket
6. Gasket
7. Linkage
8. Pin
9. Valve retainer
10. Valve seat
11. Valve seat
12. Clip
13. Spring
14. Spring
15. Springs
18. Valves

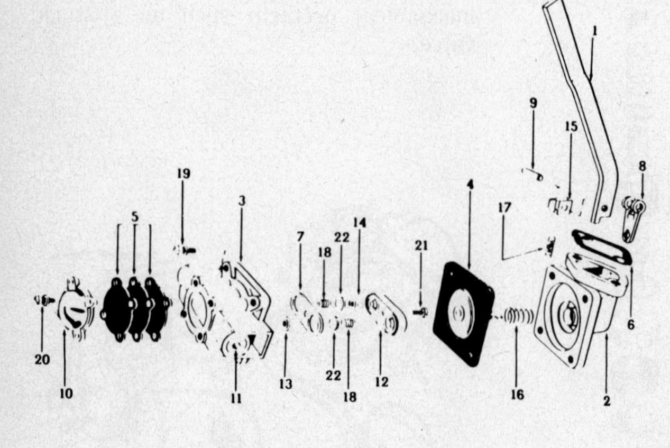

Fig. CL34—Exploded view of mechanical fuel pump used on some engines of larger displacement. Note pulsator installed in top cover.

1. Rocker arm
2. Lower body
3. Top cover
4. Diaphragm
5. Pulsator
6. Mounting gasket
7. Valve gasket
8. Linkage
9. Rocker pin
10. Cover plate
11. Pipe plug
12. Valve retainer
13. Valve seat
14. Valve seat
15. Spring clip
16. Diaphragm spring
17. Rocker arm spring
18. Valve spring
19. Screw & washer
20. Screw & washer
21. Screw & washer
22. Valves

An alternator-magneto located under flywheel (Fig. CL31) is required to maintain 12-volt battery charge in systems with Bosch starter. System component parts are coil, condenser, points and a pair of generating coils to provide battery charging current of 2-4 amps at 3600 rpm. A special flywheel is used which has 10 permanent magnets, two for spark ignition and all ten for generating current needed for battery charging and lights. Generating coils are series-wound and connected. They must be renewed in pairs. Coil leads are connected as shown in Fig. CL31 to directional selenium rectifier terminals. Connect coils **ONLY** as shown. If battery current is connected directly to coil leads, coils will burn out and flywheel magnets will be demagnetized. System is negative (-) ground.

Generating coils may be tested by use of a test light (a flashlight bulb is recommended) connected across coil leads after disconnecting leads from rectifier terminals. With spark plug removed so engine will spin over easily, use rope or recoil starter to turn engine and observe test light. Coil set must be renewed if bulb does not light.

LIGHTING COIL

Generating coils to provide power for lights only may be mounted on some engines equipped for manual (recoil and rope) starting. Lighting coils are mounted beneath flywheel on ignition stator plate. See Fig. CL32. Flywheel on these models is also furnished with additional magnets and testing of coils is performed exactly as in preceding section. Lighting coil leads are connected to terminal strip, not to AC rectifier as with alternator-magneto.

IMPULSE LINE CONNECTOR

FUEL OUTLET

FUEL INLET

Fig. CL35—Diaphragm type fuel pump used on 412 and 413 models. This style pump attaches directly to carburetor inlet with impulse tube connected to vacuum source at intake manifold. In case of failure, entire pump must be renewed.

FUEL PUMPS

Engines used in some applications require a remote fuel tank be used rather than engine-mounted tank with gravity fuel supply to carburetor. Types of fuel pumps which have been used are shown in Figs. CL33, CL34 and CL35. Fig. CL35 provides a sectional view of a diaphragm pump which is mounted directly on carburetor inlet of 412 or 413 models. Impulse tube is connected to

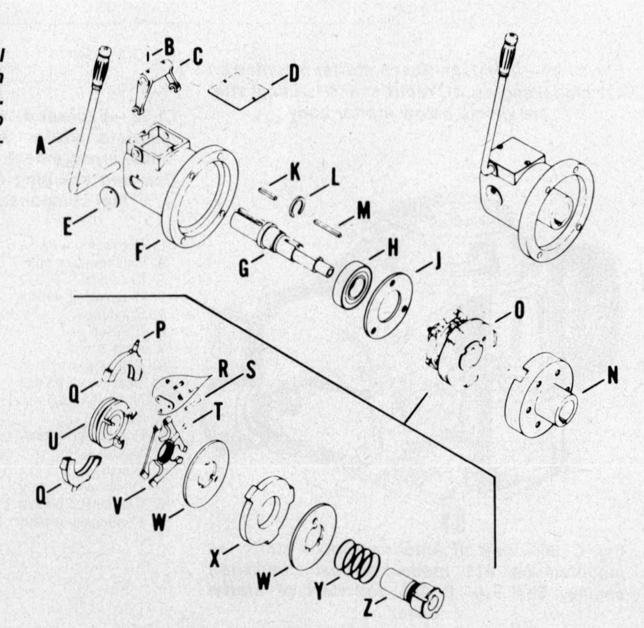

Fig. CL36—Typical Rockford over-center clutch used on these engines. Refer to text. Be sure to check parts source to determine need for adapter.

A. Shift lever
B. Roll pin (2)
C. Clutch yoke
D. Cover plate
E. Lube fitting cover
F. Housing
G. Drive shaft
H. Bearing
J. Bearing retainer
K. Key
L. Snap ring
M. Key
N. Drive cup
O. Clutch assy.
P. Grease fitting
Q. Release bearing
R. Clutch lever (6)
S. Spider pin (3)
T. Lock plug
U. Release sleeve
V. Adjusting spider
W. Clutch plate (2)
X. Drive plate
Y. Separator spring
Z. Clutch body

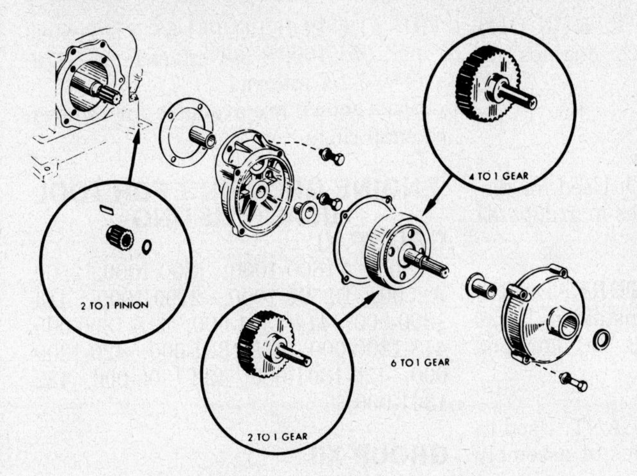

Fig. CL37 – Reduction gear unit available for engines which exceed five horsepower. Optional gear sets are shown in insets. If sleeve type bearing as shown supports engine crankshaft, finish reaming is required. In later production, tapered roller bearings are used.

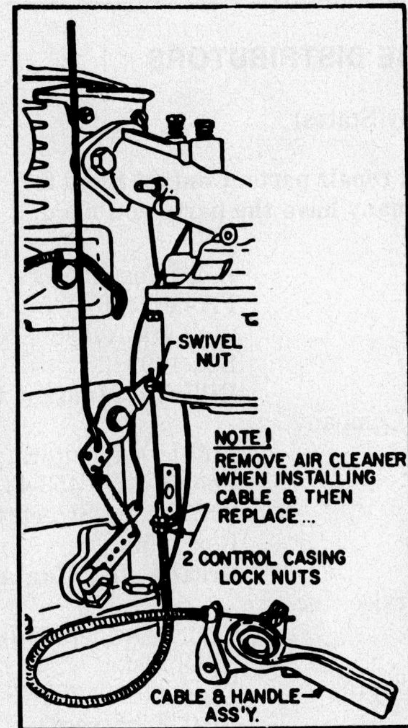

Fig. CL38 – Installed view of remote control cables. Refer to text. Take care maximum engine speed is not exceeded.

engine intake manifold. Mechanical, cam-operated pumps shown in other views are used on larger engines which may require greater lift.

CLUTCH

Some models may be equipped with a dry disc type over-center clutch manufactured by Rockford Clutch Division of Borg-Warner Corporation.

For adjustment remove cover plate (D – Fig. CL36) and loosen set screw on spider (V). Clutch is tightened by rotating spider in a clockwise direction. When properly adjusted it will require a 20 pound (9 kg) pulling force at end of shift lever (A) to engage clutch.

Occasional lubrication of bronze throw-out (clutch release) bearing (Q) is required and should be done working through opening in housing under cover (E) at grease fitting (P). Take care to not over lubricate bearing.

REDUCTION GEAR

Gear reduction sets which are mounted over pto shaft end and bolted to engine block at bearing plate position are available for some models. Reduction gears are available in 2:1 or 4:1 ratio with clockwise rotation or 6:1 ratio with counter-clockwise rotation.

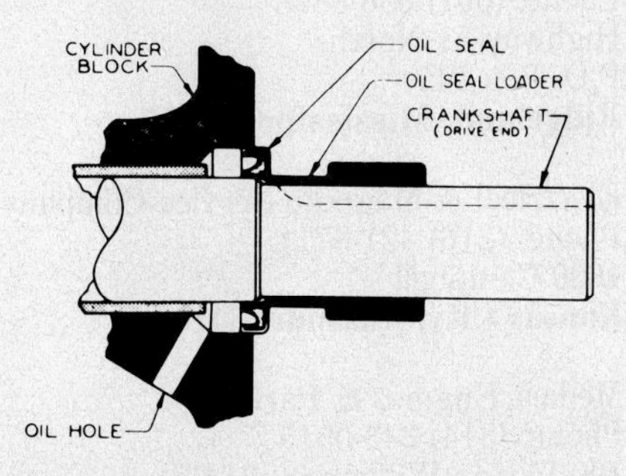

Fig. CL39 – Technique for use of oil seal loader when fitting new seals over shafts.

Gearcase should be filled to check plug level with oil of same weight and rating as oil used in engine crankcase. Oil transfers from crankcase to reduction case via a drilled passage through inner half of gearcase. Oil in reduction case should be changed at least every 100 hours of normal operation.

Bronze sleeve bearing shown in Fig. CL37 which supports end of crankshaft within gear housing has been supplanted in later models by a tapered roller bearing. Be sure to check unit before ordering renewal parts.

REMOTE CONTROLS

Refer to Fig. CL38 when servicing Bowden cable used for remote control of engine speed.

Adjust remote control with engine running by the following procedure:

1. Loosen screw in swivel nut on control lever.
2. Move control lever to high speed position.
3. Move control wire through control casing and swivel nut until maximum desired speed is obtained. (Do not exceed maximum of 3600 rpm). Lever should remain in high speed position.
4. Tighten screw in swivel nut to hold control wire.
5. Move control lever to slow speed position to be sure engine can slow down to idle speed.

CLINTON SERVICE TOOLS

The following special tool list indicates the CLINTON part number, the tool name and usage by engine group. Following the tool list is the engine group listing.

Refer to list of Clinton Central Warehouse Distributors as a source of these special tools.

Oil Seal Loaders

For loading (installing) oil seals over shafts. See Fig. CL39.

951-40 ....for 1 3/8 inch diameter shaft
951-49 ...for 1 3/16 inch diameter shaft
951-47 ......for 1/2 inch diameter shaft
951-118 ...for 1 1/4 inch diameter shaft

Valve Tools

951-32 VALVE SPRING COMPRESSOR. For engines in group XII.

951-67 VALVE SPRING COMPRESSOR. For engines in groups XI, XII.

951-38 VALVE SEATING PILOT. For use with Valve Seat Cutter 951-37 on groups XI & XII.

951-37 VALVE SEAT CUTTER. For cutting valve seats, groups XI & XII. Replacement cutters: 951-131.

951-53 ROLLING TOOL. Used to peen metal around margin of valve seat insert after installation, groups XI & XII.

Piston Ring Tools

951-34 PISTON RING COMPRESSOR Used for compressing piston rings up to 3 1/8 inch diameter on engines in groups XI & XII.

Flywheel & Crankshaft Tools

951-42 FLYWHEEL HOLDER. Used to hold flywheel during removal of flywheel nut.

951-64 CRANKSHAFT RUN-OUT GAGE. Used to check engines for bent crankshafts.

Miscellaneous Tools

951-65 ENGINE STAND. Used for display or repair of engines in groups XI & XII.

951-236 HELI-COIL THREAD REPAIR SET. Use for installing 1/4-20 and 5/16-18 Heli-Coils in damaged threads of castings.

951-234 LOCTITE SEALANT. Used to assist holding properties of assembly screws.

TRU-ARC PLIERS. 951-68, 5 5/8 external, 951-100, 8 3/8 internal, 951-132-500, 6 1/2 internal.

Sizes shown are available for removal of snap rings.

ENGINE GROUPING FOR TOOL USAGE LISTING

GROUP XI

1600, A1600-1000, 1800-1000, 2500, A2500, B2500-1000, 2790-1000, 414-1300-000, 414-1301-000, 416-1300-000, 418-1300-000, 418-1301-000, 420-1300-000, 420-1301-000, 422-1300-000, 422-1301-000.

GROUP XII

412-0000-000, 413-0000-000

CLINTON CENTRAL WAREHOUSE DISTRIBUTORS

(Arranged Alphabetically by States)

These franchised firms carry extensive stocks of repair parts. Contact them for name and address of nearest distributor who many have the parts you need.

Charlie C. Jones Battery & Electric Company
Phone: (602) 272-5621
2440 West McDowell Road
P.O. Box 6654
Phoenix, Arizona 85005

Garden Equipment Company
Phone: (213) 633-8105
6600 Cherry Avenue
Long Beach, California 90805

Air Cooled Engines
Phone: (408) 295-4790
1076 North 10th
San Jose, California 95112

Radco Distributors Company
Phone: (904) 733-7956
4909 Victor Street
P.O. Box 5459
Jacksonville, Florida 32207

Power Products Distributors, Incorporated
Phone: (813) 748-0734
1523 8th Avenue
P.O. Box 7
Palmetto, Florida 33561

Rogers Engines
Phone: (305) 689-8300
3330 West 45th Street
West Palm Beach, Florida 33407

Lanco Engines Services Incorporated
Phone: (808) 845-2244
720 Moowoa Street
Honolulu, Hawaii 96817

Midwest Engine Warehouse
Phone: (312) 883-1200
515 Romans Road
Elmhurst, Illinois 60126

Mid-East Power Equipment Company
Phone: (606) 253-0688
185 Lisle Road
P.O. Box 40505
Lexington, Kentucky 40505

Springfield Auto Electric Service, Inc.
Phone: (413) 736-3684
117 West Gardner Street
Springfield, Massachusetts 01105

Automotive Products Company
Phone: (906) 863-2651
520 First Street
Menominee, Michigan 49858

Engine Parts Supply Company
Phone: (612) 338-7086
1220-24 Harmon Place
Minneapolis, Minnesota 55403

Johnson Big Wheel Mowers, Inc.
Phone: (601) 856-8848
Highway 51 North
P.O. Box 717
Ridgeland, Mississippi 39157

Electrical & Magneto Service Company
Phone: (816) 421-3711
1600 Campbell
Kansas City, Missouri 64108

Medart Engines & Parts
Phone: (314) 343-0515
100 Larkin Williams Ind. Court
Fenton, Missouri 63026

A & I Distributors
Phone: (406) 245-6443
2112-4th Avenue, North
Box 1999
Billings, Montana 59103

A & I Distributors
Phone: (406) 453-4314
317 Second Street, South
Box 1159
Great Falls, Montana 59403

Carl A. Anderson Incorporated of Nebraska
Phone: (402) 339-4944
7510 "L" Street
Box 27139
Omaha, Nebraska 68127

Central Motive Power Incorporated
Phone: (505) 345-8335
3740 Princeton Drive, North East
P.O. Box 1924
Albuquerque, New Mexico 87103

Wayne Auto Electric Incorporated
Phone: (315) 255-1111
26 East Genesee Street
Auburn, New York 13021

H. G. S. Power House
Phone: (516) 423-1348
68-70 West Jericho Turnpike
(West of Route 116)
Huntington Station, New York 11746

E. J. Smith & Sons Company
Phone: (704) 394-3361
4250 Golf Acres Drive
P.O. Box 668887
Charlotte, North Carolina 28266

United Power Equipment
Phone: (701) 255-2450
2031 East Lee Avenue
P.O. Box 1077
Bismark, North Dakota 58502

V. E. Petersen Company
Phone: (419) 838-5911
28101 East Broadway
Walbridge (Toledo), Ohio 43465

Victory Motors Incorporated
Phone: (918) 863-6675
605 Cherokee
Muskogee, Oklahoma 74001

Brown & Wisner Incorporated
Phone: (503) 692-0330
9991 South West Avery Street
Box 1109
Tualatin, Oregon 97062

Fry's Power Equipment
Phone: (814) 944-6259
4th Street & Bell Avenue
Altoona, Pennsylvania 16602

McCullough Distributing Company
Phone: (215) 288-9700
5613-19 Tulip Street
Philadelphia, Pennsylvania 19124

Locke Auto Electric Service Inc.
Phone: (605) 336-2780
231 North Dakota Avenue
Box 1165
Sioux Falls, South Dakota 57101

RCH Distributors Incorporated
Phone: (901) 345-2200
3140 Carrier Street
Box 173
Memphis, Tennessee 38101

Automobile Electric Service
Phone: (615) 255-7515
1008 Charlotte Avenue
Nashville, Tennessee 37203

Wes Kern's Repair Service
Phone: (915) 532-2455
2423 East Missouri
El Paso, Texas 79901

McCoy Sales & Service
Phone: (817) 838-6338
4045 East Belknap Street
Fort Worth, Texas 76111

Yazoo of Texas Incorporated
Phone: (713) 923-5979
1409 Telephone Road
Houston, Texas 77023

A-1 Engine & Mower Company
Phone: (801) 364-3653
437 East 9th, South
Salt Lake City, Utah 84111

Hunt Auto Supply
Phone: (804) 583-4553
1300 Monticello Avenue
Norfolk, Virginia 23510

R. B. I. Corporation
Phone: (804) 798-1541
101 Cedar Run Drive
P.O. Box 9318
Richmond, Virginia 23227

Fremont Electric Company
Phone: (206) 633-2323
744 North 34th Street
P.O. Box 31640
Seattle, Washington 98103

CANADIAN DISTRIBUTORS

Loveseth, Ltd.
Phone: (403) 436-8250
9570-58th Avenue
Edmonton, Alberta T6E 0B6

Western Air Cooled Engines, Ltd.
Phone: (604) 254-8121
1205 East Hastings
Vancouver, British Columbia V6A 1S5

Yetman's Ltd.
Phone: (204) 586-8046
949 Jarvis Avenue
Winnipeg, Manitoba R2X 0A1

Longwood Equipment Company
Phone: (416) 438-3710
1940 Ellesmere Road
Unit 8
Scarborough, Ontario M1H 2V7

CRAFTSMAN

SEARS ROEBUCK & CO.
Chicago, Ill. 60607

Craftsman engines are manufactured by the Tecumseh Products Company. Except for various parts, Craftsman engines are comparable to Tecumseh models of the same size with the same equipment. Refer to the Tecumseh engine section in this manual for service information. The following cross reference listing may be helpful, but is intended for service information only. The complete Craftsman model and serial number is neccessary when ordering parts.

Craftsman Model Number	Basic Tecumseh Model Number
143.186022	V70
143.186032	V70
143.186042	V70
143.186072	V70
143.196042	V70
143.196052	V70
143.196062	V70
143.196072	V70
143.206022	V70
143.216012	V70
143.216022	V70
143.216032	V70
143.216072	V70
143.216082	V70
143.216092	V70
143.216132	V80
143.216152	V80
143.216162	V80
143.216172	V80
143.226072	V70
143.226082	V70
143.226092	V80
143.226102	V80
143.226112	V80
143.226122	V80
143.226192	V80
143.226202	V70
143.226212	V70
143.226272	V70
143.226282	V70
143.226292	V80
143.226302	V70
143.226312	V80
143.226342	V70
143.226352	V80
143.236022	VM80
143.236032	VM80
143.236062	VM80
143.236072	VM80
143.236092	V70
143.236122	V70
143.236142	VM80
143.236162	V70
143.236172	V70
143.246022	VM80
143.246032	VM80
143.246052	V70
143.246062	V70
143.246072	V70
143.246082	VM80
143.246092	VM80
143.246102	V70
143.246112	V70
143.246122	VM80
143.246132	V70
143.246142	V70
143.246152	VM80
143.246162	VM80
143.246172	VM80
143.246182	VM80
143.246192	VM80
143.246202	VM80
143.246212	VM80
143.246222	V70
143.246232	VM80
143.246242	V70
143.246252	VM80
143.246262	V70
143.246272	VM80
143.246282	VM80
143.246292	VM80
143.246302	V70
143.246312	V70
143.246322	VM80
143.246332	VM80
143.246342	V70
143.246362	VM100
143.246372	VM80
143.246382	VM100
143.249012	VH100
143.249022	VH100
143.256012	VM80
143.256032	V70
143.256042	VM80
143.256062	VM80
143.256072	VM80
143.256102	V70
143.256112	VM80
143.256132	V70
143.259012	VH100
143.259022	VH100
143.266012	VM80
143.266022	VM80
143.266042	V70
143.266052	V70
143.266092	V70
143.266102	V70
143.266112	V70
143.266122	V70
143.266132	V70
143.266142	VM80
143.266162	VM80
143.266172	VM80
143.266182	VM80
143.266202	VM80
143.266212	VM80
143.266222	VM80
143.266232	VM80
143.266242	VM80
143.266302	V70
143.266312	VM80
143.266322	VM80
143.266332	V70
143.266342	VM80
143.266352	V70
143.266362	VM80
143.266422	VM80
143.266462	VM100
143.266472	VM100
143.266482	VM80
143.269012	VH100
143.269022	VH100
143.276022	VM100
143.276032	VM80
143.276042	VM80
143.276052	VM100
143.276062	VM80
143.276072	VM80
143.276092	V70
143.276102	V70
143.276112	V70
143.276122	VM80
143.276132	VM100
143.276142	VM80
143.276152	VM80
143.276162	VM80
143.276172	V70
143.276192	V70
143.276212	V70
143.276222	V70
143.276232	VM80
143.276242	VM80
143.276262	VM80
143.276272	VM80
143.276282	V70
143.276292	VM80
143.276302	VM80
143.276322	V70
143.276332	V70
143.276342	V70
143.276352	VM80
143.276362	VM100

Craftsman Model Number	Basic Tecumseh Model Number
143.276372	V70
143.276382	V70
143.276392	V70
143.276402	VM100
143.276422	V70
143.276432	VM80
143.276442	VM80
143.276452	VM80
143.276462	VM80
143.276472	VM80
143.276482	VM100
143.279012	VH100
143.286032	V70
143.286042	VM100
143.286052	VM100
143.286062	V70
143.286072	VM80
143.286082	VM80
143.286092	VM80
143.286102	VM100
143.286112	VM80
143.286122	VM80
143.286132	V70
143.286142	VM80
143.286152	VM80
143.286162	VM80
143.286172	VM80
143.286182	VM80
143.286192	VM100
143.286202	VM80
143.286212	VM100
143.286222	V70
143.286232	VM100
143.286242	VM100
143.286252	V70
143.286262	VM100
143.286272	VM100
143.286312	VM80
143.286322	V70
143.286332	V70
143.286362	VM100
143.296012	VM80
143.296022	VM100
143.296032	VM80
143.296042	VM100
143.296052	VM80
143.296062	VM80
143.296072	VM100
143.296082	VM80
143.296092	VM80
143.296102	VM100
143.296112	VM100
143.296122	VM100
143.296132	VM100
143.296142	VM80
143.296152	VM80
143.296162	VM80
143.296172	VM80
143.296182	VM80
143.296192	VM80
143.296202	VM100
143.296212	VM100
143.296222	VM80
143.296232	VM100
143.296242	VM80
143.296252	VM80
143.296262	VM80
143.306012	VM80
143.306022	VM100
143.306032	VM100
143.306042	VM100
143.316012	V70
143.316022	VM80
143.316042	VM70
143.316052	VM80
143.316062	VM80
143.316082	VM80
143.316092	VM100
143.316102	VM80
143.316112	VM80
143.316122	VM100
143.316132	VM100
143.316142	VM100
143.316152	VM80
143.316162	VM100
143.316172	VM70
143.316182	VM80
143.316192	VM100
143.316202	VM100
143.316222	VM80
143.316232	VM80
143.316242	VM80
143.316252	VM80
143.316262	VM100
143.316272	VM100
143.316282	VM100
143.316292	VM80
143.316302	VM100
143.316312	VM80
143.326012	TVM195
143.326022	TVM195
143.326032	TVM195
143.326042	TVM195
143.326052	TVM195
143.326062	TVM195
143.326072	TVM195
143.326082	TVM195
143.326092	TVM195
143.326102	TVM195
143.326112	TVM195
143.326122	TVM220
143.326132	TVM220
143.326142	TVM220
143.326152	TVM220
143.326162	TVM220
143.326172	TVM220
143.326182	TVM220
143.326192	V70
143.326202	V70
143.326212	V70
143.326222	V70
143.326232	V70
143.326242	V70
143.326252	V70
143.326262	V70
143.326272	V70
143.326312	V70
143.326322	TVM170
143.326332	TVM195
143.326342	TVM195
143.326372	TVM170
143.336022	TVM220
143.336032	TVM220
143.336042	TVM220
143.558012	HH80
143.558022	HH80
143.558032	HH80
143.558052	HH80
143.559012	HH100
143.559022	HH100
143.559032	HH100
143.559042	HH100
143.562012	HH120
143.562022	HH120
143.562032	HH120
143.568032	HH80
143.569022	HH100
143.569032	HH100
143.569042	HH100
143.569052	HH100
143.569082	HH100
143.572012	HH120
143.572022	HH120
143.572032	HH120
143.572042	HH120
143.572052	HH120
143.572062	HH120
143.572092	HH120
143.572102	HH120
143.578012	HH80
143.578022	HH80
143.578052	HH80
143.578062	HH80
143.578072	HH80
143.579012	HH100
143.579022	HH100
143.579032	HH100
143.579042	HH100
143.579052	HH100
143.579062	HH100
143.579072	HH100
143.579082	HH100
143.579092	HH100
143.579102	HH100
143.579112	HH100
143.579132	HH100
143.582012	HH120
143.582022	HH120
143.582032	HH120
143.582042	HH120
143.582052	HH120
143.582062	HH120
143.582072	HH120
143.582082	HH120
143.582092	HH120
143.582102	HH120
143.582112	HH120
143.582122	HH120
143.582132	HH120
143.582142	HH120
143.582172	HH120
143.586052	H70
143.586062	H70
143.586112	H70
143.586122	H70

Craftsman Model Number	Basic Tecumseh Model Number
143.586132	H70
143.586142	H70
143.586162	H70
143.586252	H70
143.588012	HH80
143.588032	HH80
143.589012	HH100
143.589022	HH100
143.589032	HH100
143.589042	HH100
143.589052	HH100
143.589072	HH100
143.592012	HH120
143.592022	HH120
143.592032	HH120
143.592052	HH120
143.592062	HH120
143.592072	HH120
143.592082	HH120
143.596012	H70
143.596022	H70
143.596052	H70
143.598012	HH80
143.599012	HH100
143.599022	HH100
143.599042	HH100
143.599052	HH100
143.599062	HH100
143.602012	HH120
143.602022	HH120
143.602032	HH120
143.602052	HH120
143.602062	HH120
143.602072	HH120
143.602082	HH120
143.602092	HH120
143.602102	HH120
143.602112	HH120
143.602122	HH120
143.606012	H70
143.606022	H70
143.606032	H70
143.606042	H70
143.606052	H70
143.606102	H70
143.608012	HH80
143.608022	HH80
143.608032	HH80
143.609022	HH100
143.609032	HH100
143.609042	HH100
143.609052	HH100
143.609072	HH100
143.612012	HH120
143.612022	HH120
143.616012	H70
143.616122	H70
143.619012	HH100
143.622012	HH120
143.622022	HH120
143.622032	HH120
143.622042	HH120
143.622052	HH120
143.622062	HH120
143.622072	HH120
143.622082	HH120
143.622092	HH120
143.622102	HH120
143.626012	H70
143.626032	H70
143.626052	H70
143.626062	H70
143.626082	H70
143.626092	H70
143.626102	H70
143.626122	H70
143.626142	H70
143.626152	H70
143.626172	H70
143.626192	H70
143.626212	H70
143.626282	HM80
143.626292	H70
143.626312	H70
143.626322	H70
143.628012	HH80
143.628022	HH80
143.629012	HH100
143.629022	HH100
143.629032	HH100
143.629042	HH100
143.629052	HH100
143.629062	HH100
143.629072	HH100
143.630012	HH160
143.630022	HH160
143.632012	HH120
143.632022	HH120
143.632032	HH120
143.636012	H80
143.636022	H80
143.636032	H70
143.636042	HM80
143.636062	H70
143.636072	HM80
143.639012	HH100
143.640012	HH160
143.640022	HH160
143.640032	HH160
143.640052	HH160
143.642012	HH120
143.642022	HH120
143.642032	HH120
143.642042	HH120
143.646012	H70
143.646022	H70
143.646032	H70
143.646042	HM80
143.646052	HM80
143.646072	H70
143.646082	H70
143.646092	H70
143.646102	H70
143.646122	H70
143.646132	H70
143.646142	HM80
143.646152	H70
143.646162	HM80
143.646172	H70
143.646182	H70
143.646202	H70
143.646212	HM80
143.646222	HM80
143.646232	HM80
143.649012	HH100
143.649022	HH100
143.650012	HH160
143.650022	HH140
143.650032	HH160
143.652012	HH120
143.652022	HH120
143.652032	HH120
143.652042	HH120
143.652052	HH120
143.652062	HH120
143.652072	HH120
143.656062	H70
143.656072	HM80
143.656082	HM80
143.656102	H70
143.656122	H70
143.656132	H70
143.656152	H70
143.656192	H70
143.656212	HM80
143.656222	HM80
143.656232	H70
143.656262	H70
143.656272	H70
143.656282	HM80
143.659012	HH100
143.659022	HH100
143.659032	HH100
143.660012	HH160
143.660022	HH160
143.662012	HH120
143.666012	H70
143.666022	H70
143.666032	HM80
143.666042	H70
143.666052	H70
143.666062	H70
143.666072	H70
143.666082	HM80
143.666092	HM80
143.666152	HM80
143.666162	HM80
143.666222	H70
143.666232	HM80
143.666252	H70
143.666282	H70
143.666302	H70
143.666312	H70
143.666322	HM80
143.666332	HM100
143.666342	H70
143.666352	HM80
143.666362	HM100
143.666382	H70
143.669012	HH100
143.670022	OH160
143.670032	OH160
143.670042	OH140
143.670052	OH160
143.670062	OH160

Craftsman Model Number	Basic Tecumseh Model Number
143.670072	OH160
143.670082	OH140
143.670092	OH140
143.672012	HH120
143.672022	HH120
143.672032	HH120
143.672042	HH120
143.672052	HH120
143.672062	HH120
143.672072	HH120
143.676012	HM80
143.676022	HM80
143.676032	H70
143.676042	HM80
143.676052	HM80
143.676062	HM100
143.626072	HM100
143.676082	HM100
143.676092	HM100
143.676102	H70
143.676122	H70
143.676142	HM80
143.676152	HM100
143.676162	HM100
143.676172	H70
143.676182	HM80
143.676192	H70
143.676202	HM80
143.676212	HM100
143.676222	HM80
143.676252	HM80
143.676262	HM100
143.679012	HH100
143.679022	HH100
143.679032	HH100
143.680012	OH140
143.680022	OH140
143.686012	HM70
143.686022	HM70
143.686032	HM80
143.686042	HM80
143.686052	HM80

Craftsman Model Number	Basic Tecumseh Model Number
143.686062	H70
143.686082	HM80
143.686092	HM80
143.686102	HM100
143.686132	H70
143.686142	HM80
143.686152	HM100
143.686172	HM100
143.696032	HM80
143.696052	H70
143.696062	HM80
143.696072	HM100
143.696092	HM80
143.696102	H70
143.696112	HM100
143.696132	HM80
143.700012	OH180
143.706032	HM80
143.706042	HM80
143.706052	HM80
143.706072	HM80
143.706082	HM80
143.706092	H70
143.706112	H70
143.706122	HM80
143.706132	HM100
143.706152	HM100
143.706162	HM80
143.706172	H70
143.706222	HM80
143.706232	HM100
143.710012	OH160
143.710022	OH140
143.712012	HH120
143.712032	HH120
143.712042	HH120
143.716012	HM70
143.716022	HM80
143.716052	HM70
143.716062	HM70
143.716072	HM80
143.716082	HM80

Craftsman Model Number	Basic Tecumseh Model Number
143.716092	HM100
143.716102	HM100
143.716112	H70
143.716122	H70
143.716132	H70
143.716142	H70
143.716152	HM80
143.716172	HM80
143.716182	HM80
143.716192	HM80
143.716202	H70
143.716212	HM100
143.716222	HM80
143.716252	H70
143.716262	HM100
143.716312	H70
143.716322	H70
143.716332	HM100
143.716342	HM70
143.716362	H70
143.716372	HM100
143.716382	HM100
143.716392	HM80
143.716432	HM100
143.726032	HM80
143.726042	H70
143.726052	H70
143.726102	HM100
143.726152	HM80
143.726182	H70
143.726192	H70
143.726202	H70
143.726212	H70
143.726232	HM100
143.726252	H70
143.726282	H70
143.726292	HM100
143.726312	HM100
143.726322	HM100
143.730012	OH160
143.736042	HM80
143.736082	HM80

KOHLER

KOHLER COMPANY
Kohler, Wisconsin 53044

Model	No. Cyls.	Bore	Stroke	Displacement	Power Rating
K-141*	1	2.875 in. (73.025 mm)	2.5 in. (63.5 mm)	16.22 cu. in. (266 cc)	6.25 hp. (4.7 kW)
K-141**	1	2.9375 in. (74.613 mm)	2.5 in. (63.5 mm)	16.9 cu. in. (277.7 cc)	6.25 hp. (4.7 kW)
K-160	1	2.875 in. (73.025 mm)	2.5 in. (63.5 mm)	16.22 cu. in. (266 cc)	7 hp. (5.2 kW)
K-161*	1	2.875 in. (73.025 mm)	2.5 in. (63.5 mm)	16.22 cu. in. (266 cc)	7 hp. (5.2 kW)
K-161**	1	2.9375 in. (74.613 mm)	2.5 in. (63.5 mm)	16.9 cu. in. (277.7 cc)	7 hp. (5.2 kW)
K-181	1	2.9375 in. (74.613 mm)	2.75 in. (69.85 mm)	18.6 cu. in. (305.4 cc)	8 hp. (6 kW)
KV-181	1	2.9375 in. (74.613 mm)	2.75 in. (69.85 mm)	18.6 cu. in. (305.4 cc)	8 hp. (6 kW)
K-241	1	3.25 in. (82.55 mm)	2.875 in. (73.025 mm)	23.9 cu. in. (390.8 cc)	10 hp. (7.5 kW)
K-301	1	3.375 in. (85.725 mm)	3.25 in. (82.55 mm)	29.07 cu. in. (476.5 cc)	12 hp. (8.9 kW)
K-321	1	3.5 in. (88.9 mm)	3.25 in. (82.55 mm)	31.27 cu. in. (512.4 cc)	14 hp. (10.4 kW)
K-341	1	3.75 in. (95.25 mm)	3.25 in. (82.55 mm)	35.89 cu. in. (588.2 mm)	16 hp. (11.9 kW)

* Designates early production (before 1970) engines.

** Designates late production (after 1970) engines.

All engines in this section are one cylinder, four-cycle engines. All models except KV-181 have horizontal crankshafts. Model KV-181 has a vertical crankshaft.

Model K-141 has a ball bearing main at pto end of crankshaft and a bushing type main bearing at flywheel end. Model KV-181 has a ball bearing main at pto end of crankshaft and a needle roller bearing main at flywheel end. All remaining models have ball bearing mains at each end of crankshaft.

Connecting rod on all models rides directly on crankpin journal. All models except KV-181 are splash lubricated. Model KV-181 has an oil circulating system which lubricates upper main bearing and crankpin.

Various models may be equipped with either a battery type ignition system, a magneto type system, each with externally mounted breaker points or a solid state breakerless ignition system.

Either a side draft or an updraft carburetor is used depending on model and application.

Special engine accessories or equipment is noted by a suffix letter on the engine model number. Key to suffic letters is as follows:

C – Over-center lever operated clutch.
P – Crankcase and crankshaft machined for direct drive applications.
R – Reduction drive unit.
S – Starter-generator unit or gear drive.
T – Rewind starter.

MAINTENANCE

SPARK PLUG. Recommended spark plug for K-241, K-301, K-321 and K-341 engines is Champion H10 or equivalent. All other models use Champion J8 or equivalent. Spark plug gap should be 0.025 inch (0.635 mm) for all models.

If radio noise reduction is needed use Champion EH10 or equivalent for Model K-241, K-301, K-321 and K-341 engines and Champion EJ8 or equivalent in remaining models. Spark plug gap should be 0.020 inch (0.508 mm) for all models.

CARBURETOR. Model K-141 engines are equipped with a Tillotson "E" series updraft carburetor. Early production K-161, K-181, K-241 and K-301 engines are equipped with Carter "N" model side draft carburetors. Late production K-161, K-181, K-241 and K-301 and all KV-181, K-321 and K-341 engines are equipped with Kohler side draft carburetors. Refer to appropriate paragraph for carburetor service information.

CARTER CARBURETOR. Refer to Fig. KO1 for exploded view of typical "N" model Carter carburetor.

For initial adjustment, open idle fuel needle 1½ turns and open main fuel needle 2 turns. Make final adjustment with engine at normal operating temperature and running. Place engine under load and adjust main fuel needle for leanest setting that will allow satisfactory acceleration and steady governor operation. Set engine at idle speed, no load and adjust idle mixture screw to obtain smoothest idle operation.

As each adjustment affects the other, adjustment procedure may have to be repeated.

To check float level, invert carburetor throttle body and float assembly. There should be 13/64-inch (5.159 mm) clearance between free side of float and machined surface of body casting. Carefully bend float lever tang that contacts inlet valve as necessary to provide correct measurement.

KOHLER CARBURETOR. Refer to Fig. KO2 for exploded view of Kohler carburetor. For initial adjustment, open main fuel needle 2 turns and open idle fuel needle 1¼ turns. Make final adjustment with engine at normal operating temperature and running. Place engine under load and adjust main fuel needle to leanest mixture that will allow satisfactory acceleration and steady governor operation.

Adjust idle speed stop screw to maintain an idle speed of 1000 rpm, then adjust idle fuel needle for smoothest idle operation. As each adjustment affects the other, adjustment procedure may have to be repeated.

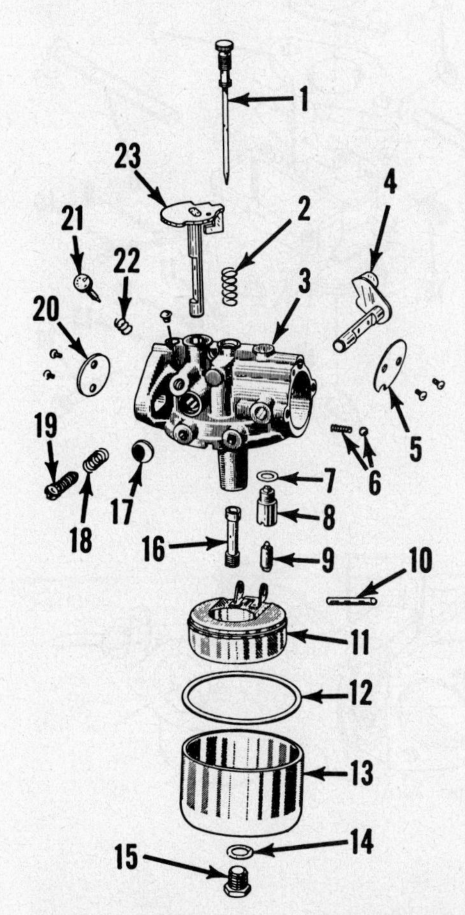

Fig. KO1 – Exploded view of typical Carter "N" model carburetor used on early production K-160, K-161, K-181, K-241 and K-301 engines.

1. Main fuel needle
2. Spring
3. Carburetor body
4. Choke shaft
5. Choke disc
6. Choke detent
7. Sealing washer
8. Inlet valve seat
9. Inlet valve
10. Float pin
11. Float
12. Gasket
13. Float bowl
14. Sealing washer
15. Retainer
16. Main jet
17. Plug
18. Spring
19. Idle stop screw
20. Throttle disc
21. Idle fuel needle
22. Spring
23. Throttle shaft

To check float level, invert carburetor body and float assembly. There should be 11/64-inch (4.366 mm) clearance between machined surface of body casting and free end of float. Carefully bend float lever tang that contacts inlet valve as necessary to provide correct measurement.

TILLOTSON CARBURETOR. Refer to Fig. KO3 for exploded view of Tillotson "E" series carburetor similar to that used on Model K-141 engine. Design of choke shaft lever will differ from that shown in exploded view.

For initial adjustment, open idle fuel mixture needle ¾-turn and open main fuel mixture needle 1 turn. Make final adjustment with engine at normal operating temperature and running. Place engine under load and adjust main fuel needle to leanest mixture that will allow satisfactory acceleration and steady governor operation. Slow engine to idle speed and adjust idle mixture screw to obtain smoothest idle operation. As each adjustment affects the other, adjustment procedure may have to be repeated.

To check float level, invert carburetor cover and float assembly. There should be 1-5/64 inches (27.384 mm) clearance between free side of float and machined surface of cover. See Fig. KO4. Carefully bend float lever tang that contacts inlet valve as necessary to provide correct measurement.

AUTOMATIC CHOKE. Some models equipped with Kohler or Carter carburetors are also equipped with an automatic choke. "Thermostatic" type shown in Fig. KO5 may be used on either electric start or manual start models. "Electric-Thermostatic" type shown in Fig. KO6 can be used only on electric start models. Automatic choke adjustment should be made with engine cold.

THERMOSTATIC TYPE. To adjust "Thermostatic" type, loosen lock screw (Fig. KO5) and rotate adjustment bracket as necessary to obtain correct amount of choking. Tighten lock screw. In cold temperature, choke should be closed. At 70° F. (21° C) choke should be partially open and choke lever should be in vertical position. Start engine and allow it to run until normal operating temperature is reached. Choke should be fully open when engine is at normal operating temperature.

ELECTRIC-THERMOSTATIC TYPE. To adjust "Electric-Thermostatic" type, refer to Fig. KO6 and move choke lever arm until hole in brass cross shaft is aligned with slot in bearings. Insert a No. 43 (0.089 inch) drill through shaft until drill is engaged in notch in base of choke. Loosen clamp bolt on lever arm and move link end of lever arm upward until choke disc is closed to desired position. Tighten clamp bolt while holding lever arm in this position. Remove drill. Choke should be fully closed when starting a cold engine and should be fully open when engine is running at normal operating temperature.

GOVERNOR. All models are equipped with a gear driven flyweight governor that is located inside engine crankcase. Maximum recommended governed engine speed is 3800 rpm on KV-181 engines and 3600 rpm for all other engines. Recommended idle speed of 1000 rpm is controlled by adjustment of throttle stop screw on carburetor.

Before attempting to adjust engine governed speed, synchronize governor linkage as follows: On Models K-141,

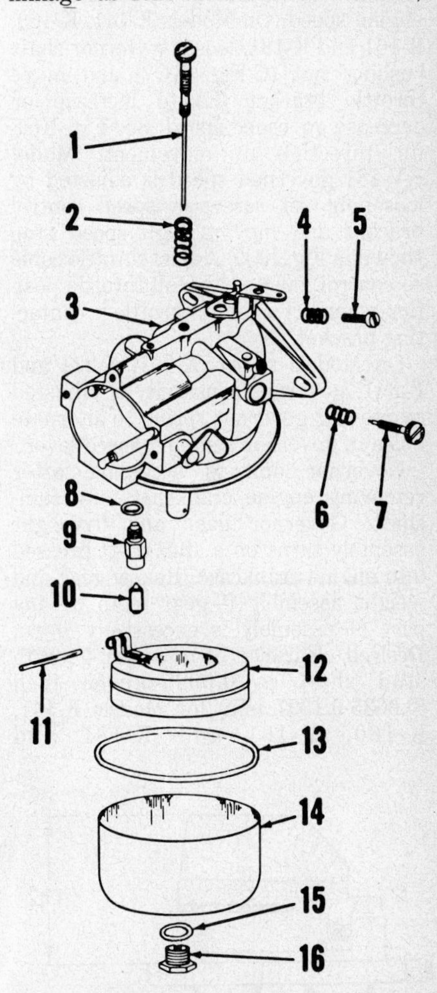

Fig. KO2 – Exploded view of Kohler carburetor used on late production K-161, K-181, K-241 and K-301 and all KV-181, K-321 and K-341 engines.

1. Main fuel needle
2. Spring
3. Carburetor body assy.
4. Spring
5. Idle speed stop screw
6. Spring
7. Idle fuel needle
8. Sealing washer
9. Inlet valve seat
10. Inlet valve
11. Float pin
12. Float
13. Gasket
14. Float bowl
15. Gasket
16. Bowl retainer

K-160, K-161 and K-181, loosen bolt clamping governor arm (G–Fig. KO7) to governor cross shaft (F) and turn governor cross shaft counter-clockwise as far as possible. While holding cross shaft in this position, move governor arm away from carburetor to limit of linkage travel and tighten clamping bolt.

To synchronize governor linkage on KV-181, follow procedure for K-141, K-160, K-161 and K-181 and refer to Fig. KO9.

On Models K-241, K-301, K-321 and K-341, refer to Fig. KO10 and loosen governor arm hex nut. Rotate governor cross shaft counter-clockwise as far as possible, move governor arm away from carburetor to limit of linkage travel, then tighten hex nut on arm.

To adjust maximum governed speed on Models K-241, K-301, K-321 and K-341, adjust position of stop so speed control lever contacts stop at desired engine speed. On Models K-141, K-160, K-161 and K-181, loosen governor shaft bushing nut (C-Fig. KO7) and move throttle bracket (E) to increase or decrease governed speed. See Fig. KO8 for direction of movement. Model KV-181 governed speed is adjusted by loosening set screw in speed control bracket and moving high speed stop shown in Fig. KO9. Adjust throttle cable so control handle is in full throttle position when drive pin in throttle is contacting bracket.

On Models K-241, K-301, K-321 and K-341, governor sensitivity is adjusted by moving governor spring to alternate holes in governor arm and speed lever.

Governor unit is accessible after removing engine crankshaft and camshaft. Governor gear and flyweight assembly turns on a stub shaft pressed into engine crankcase. Renew gear and weight assembly if gear teeth or any part of assembly is excessively worn. Desired clearance of governor gear to stud shaft is 0.0025-0.0055 inch (0.0635-0.1397 mm) for Models K-141, K-160, K-161 and K-181 and

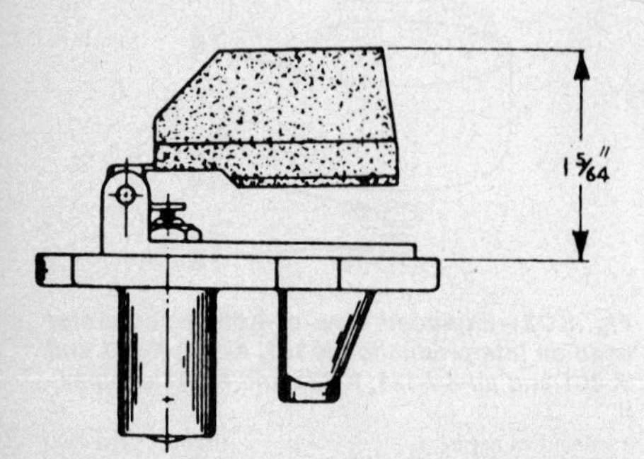

Fig. KO4—Check float setting on Tillotson carburetor by inverting throttle body and float assembly and measuring distance from free end of float to edge of cover as shown. Float setting is correct if distance is 1-5/64 inch (27.384 mm).

Fig. KO3—Exploded view of Tillotson "E" series carburetor used on K-141 engine: Refer to Fig. KO4 for method of checking float level.

1. Float bowl cover
2. Gasket
3. Gasket
4. Inlet valve assy.
5. Float
6. Gasket
7. Carburetor body
8. Choke shaft
9. Spring
10. Idle speed stop screw
11. Packing
12. Washer
13. Spring
14. Main fuel needle
15. Screw plug
16. Choke disc
17. Idle fuel needle
18. Spring
19. Throttle shaft
20. Spring
21. Choke shaft detent
22. Plug
23. Throttle disc
24. Float pin

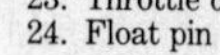

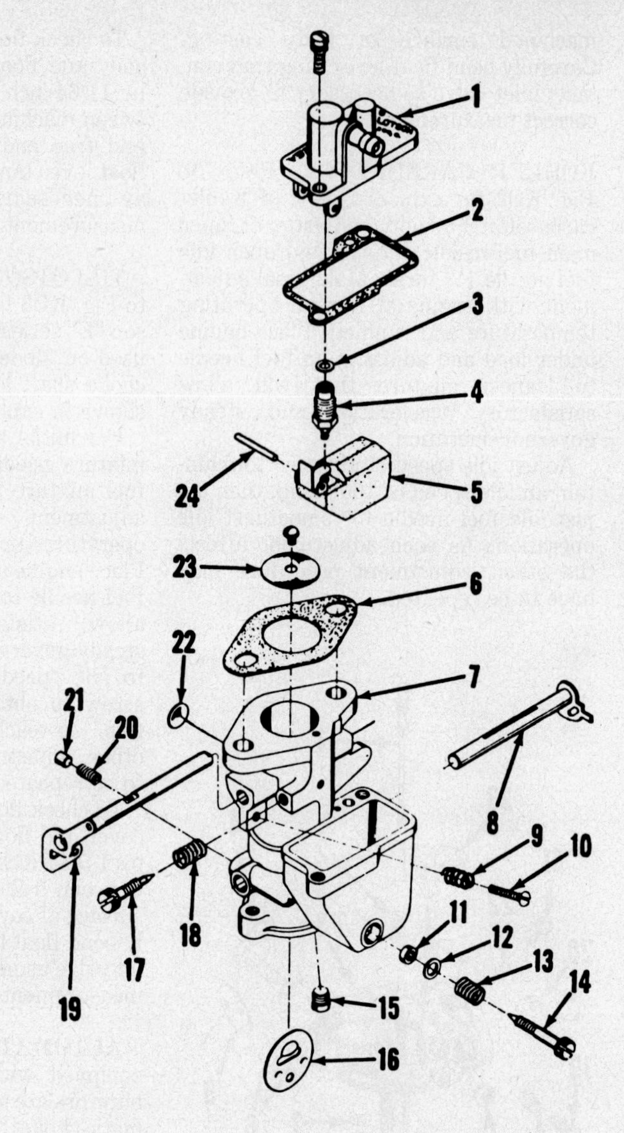

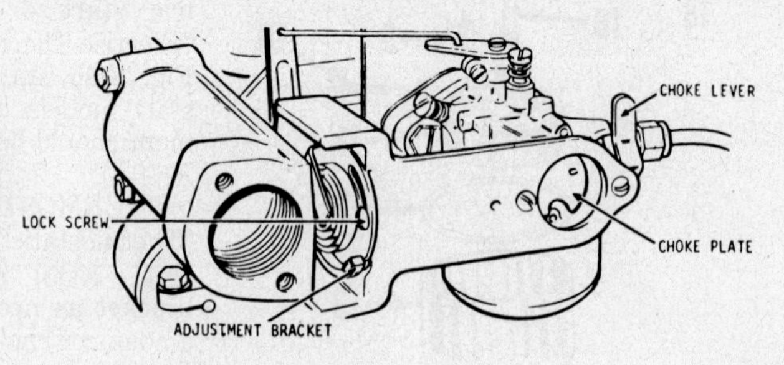

Fig. KO5—View showing "Thermostatic" automatic choke used on some engines equipped with Kohler or Carter carburetors.

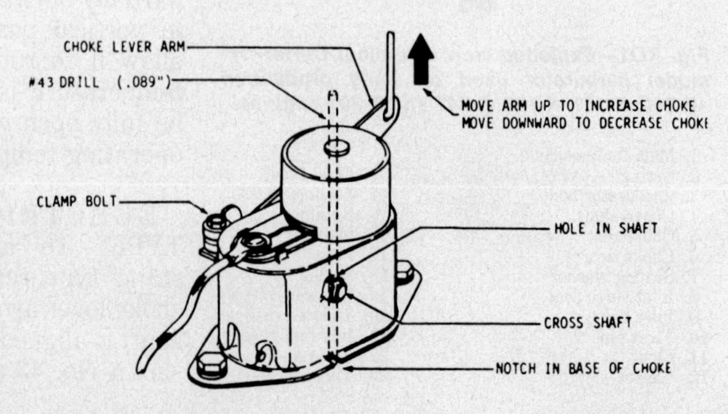

Fig. KO6—View of "Electric-Thermostatic" automatic choke used on some engines equipped with Kohler or Carter carburetor.

0.0005-0.002 inch (0.0127-0.0508 mm) for Models KV-181, K-241, K-301, K-321 and K-341.

IGNITION AND TIMING. Three types of ignition systems used are as follows: Battery ignition, flywheel magneto ignition and solid state breakerless ignition. Refer to appropriate paragraphs for timing and service procedures.

BATTERY AND MAGNETO IGNITION. On engines equipped with either battery or magneto ignition systems, breaker points are located externally on engine crankcase as shown in Fig. KO11. Breaker points are actuated by a cam through a push rod.

On early models, breaker cam is driven by camshaft gear through an automatic advance mechanism as shown in Fig. KO12. At cranking speed, ignition occurs at 3° BTDC on Models K-141, K-160, K-161 and K-181 or 3° ATDC on Models K-241 and K-301. At operating speeds, centrifugal force of the advance weights overcomes spring

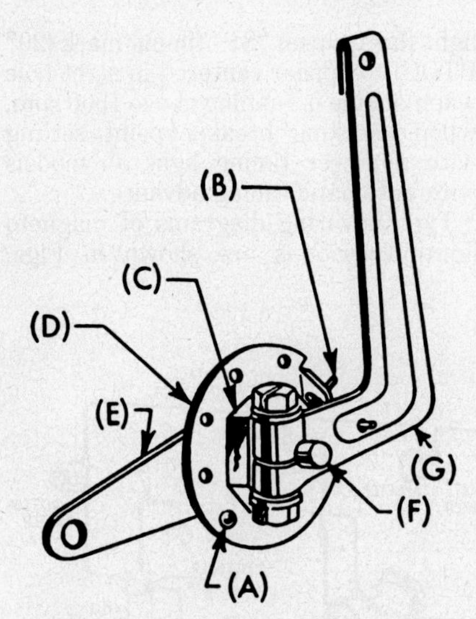

Fig. KO7—Drawing showing governor lever and related parts on K-141, K-160, K-161 and K-181 models.

A. Drive pin
B. Governor spring
C. Bushing nut
D. Speed control disc
E. Throttle bracket
F. Governor Shaft
G. Governor lever

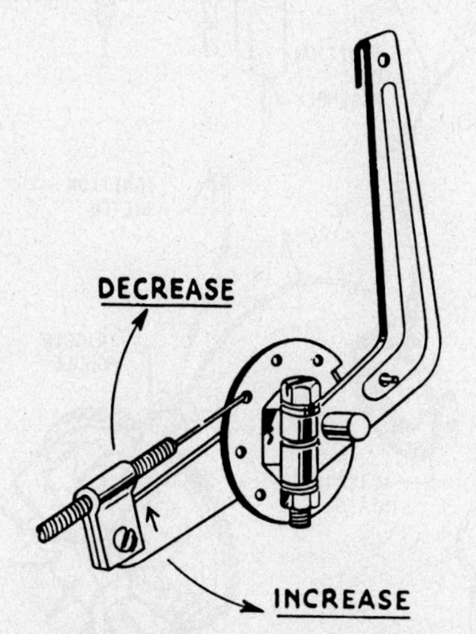

Fig. KO8—On K-141, K-160, K-161 and K-181, loosen bushing nut (C—Fig. KO7) and move throttle bracket as shown to change governed speed of engine.

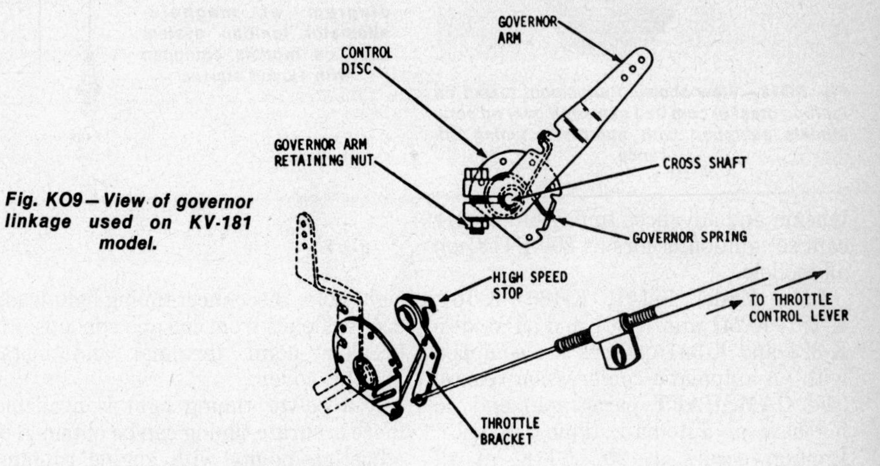

Fig. KO9—View of governor linkage used on KV-181 model.

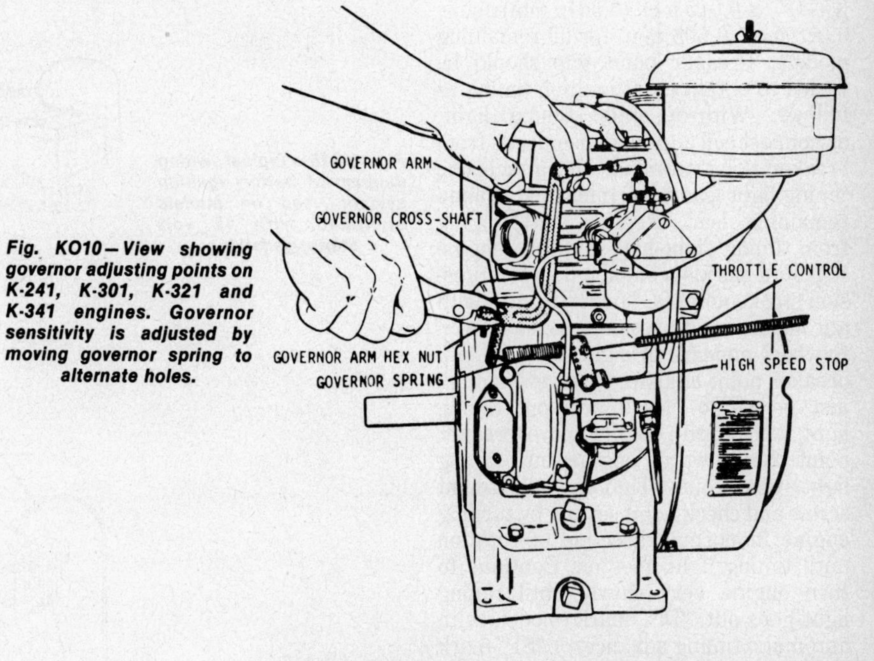

Fig. KO10—View showing governor adjusting points on K-241, K-301, K-321 and K-341 engines. Governor sensitivity is adjusted by moving governor spring to alternate holes.

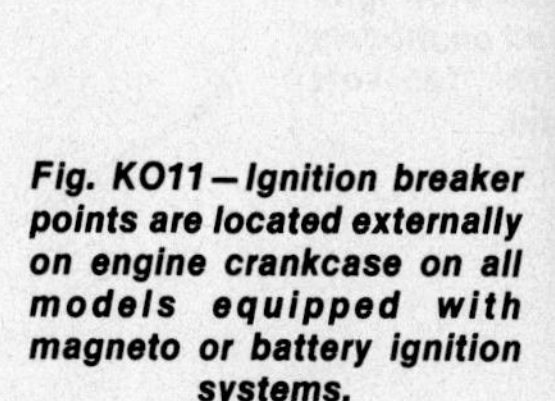

Fig. KO11—Ignition breaker points are located externally on engine crankcase on all models equipped with magneto or battery ignition systems.

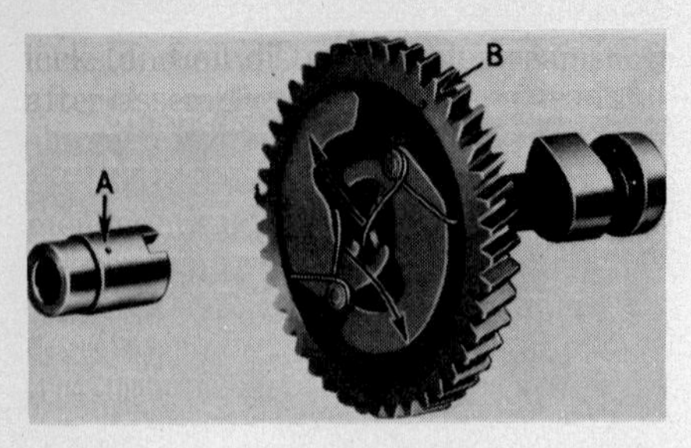

Fig. KO12—View showing alignment marks on ignition breaker cam and camshaft gear on early models equipped with automatic timing advance.

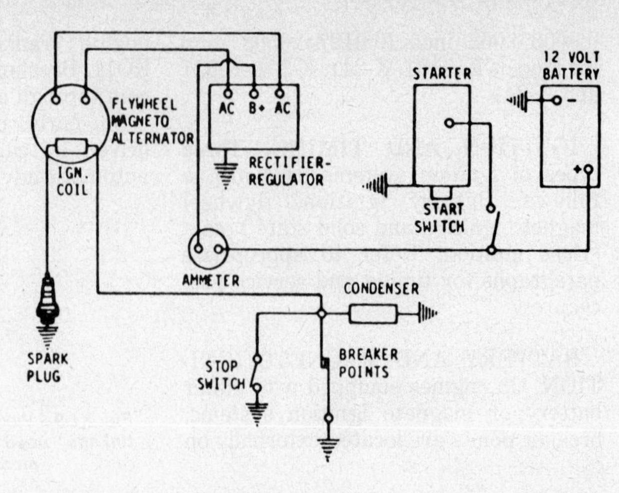

Fig. KO14—Typical wiring diagram of magneto-alternator ignition system used on models equipped with 12 volt starter.

tension and advances timing of breaker cam so ignition occurs at 20° BTDC on all models.

Late Model K-141, K-160, K-161, K-181, K-241 and K-301 and all Models K-321 and K-341 engines are equipped with an automatic compression release (see **CAMSHAFT** paragraph) and do not have an automatic timing advance. Ignition occurs at 20° BTDC at all engine speeds.

Initial breaker point gap on Model KV-181 is 0.015 inch (0.3810 mm) and is 0.020 inch (0.508 mm) for all remaining models. Breaker point gap should be varied to obtain exact ignition timing as follows: With a static timing light, disconnect coil and condenser leads from breaker point terminal and attach one timing light lead to terminal and ground remaining lead. Remove button plug from timing sight hole and turn engine so piston has just completed its compression stroke and "DC" mark (models with automatic compression release) on flywheel appears in sight hole. Loosen breaker point adjustment screw and adjust points to closed position (timing light will be on); slowly move breaker point base towards engine until timing light goes out. Tighten adjustment screw and check point setting by turning engine in normal direction of rotation until timing light goes on. Continue to turn engine very slowly until timing light goes out. "DC" mark (models with automatic timing advance) or "SP" mark (models with automatic compression release) should now be in register with sight hole. Disconnect timing light leads, connect leads from coil and condenser to breaker point terminal and install breaker cover.

If a power timing light is available, more accurate timing can be obtained by adjusting points with engine running. Breaker point gap should be adjusted so light flash causes "SP" timing mark (20° BTDC) to appear centered in sight hole when engine is running above 1500 rpm, when checking breaker point setting with a power timing light on models with automatic timing advance.

Typical wiring diagrams of magneto ignition systems are shown in Figs.

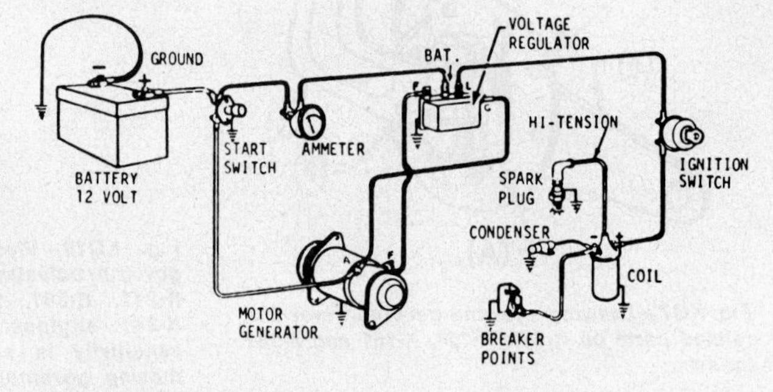

Fig. KO15—Typical wiring diagram of battery ignition system used on models equipped with 12 volt starter-generator.

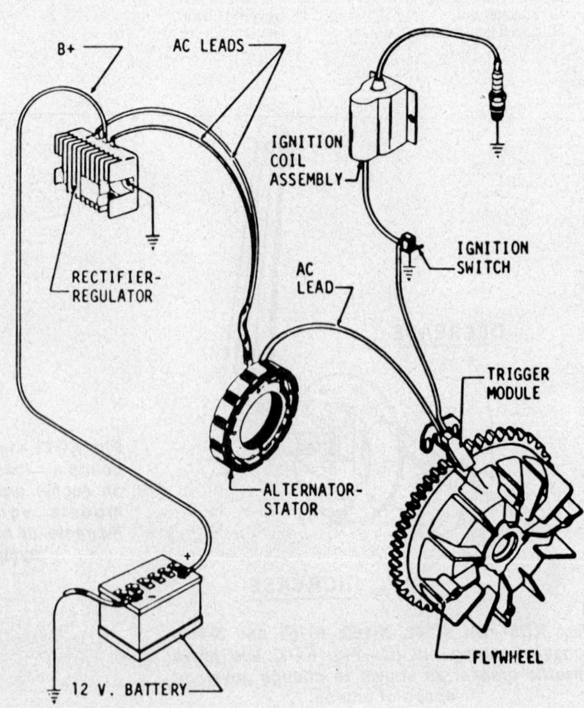

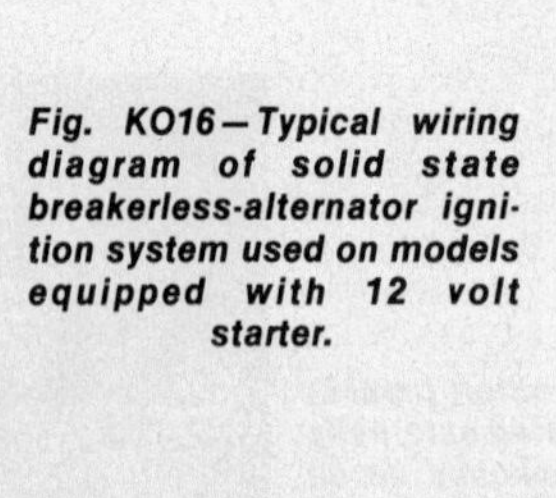

Fig. KO16—Typical wiring diagram of solid state breakerless-alternator ignition system used on models equipped with 12 volt starter.

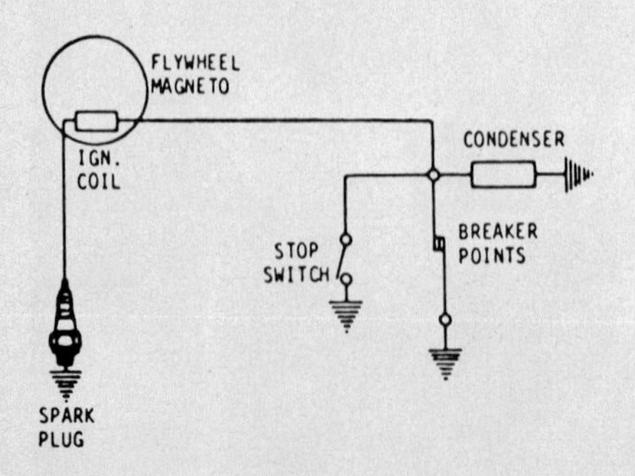

Fig. KO13—Typical wiring diagram of magneto ignition system used on manual start engines.

KO13 and KO14. Fig. KO15 is typical wiring diagram of battery ignition system.

SOLID STATE BREAKERLESS IGNITION. Breakerless-alternator ignition system uses solid state devices which eliminate need for mechanically operated breaker points. Ignition timing is non-adjustable. The only adjustment is trigger module to flywheel trigger projection air gap. To adjust air gap, rotate flywheel until projection is adjacent to trigger module. Loosen trigger retaining screws and move trigger until an air gap of 0.005-0.010 inch (0.127-0.254 mm) is obtained. Make certain flat surfaces on trigger and projection are parallel, then tighten retaining screws. Refer to Fig. KO16 for typical wiring diagram of breakerless ignition system.

The four main components of a breakerless ignition system are ignition winding (on alternator stator), trigger module, ignition coil assembly and special flywheel with trigger projection. System also includes a conventional spark plug, high tension lead and ignition switch. Ignition winding is separate from battery charging AC windings on alternator stator.

Trigger module includes three diodes, a resistor, a sensing coil and magnet plus the SCR (electronic switch). Trigger module has two clip-on type terminals. See Fig. KO17. Terminal marked "A" must be connected to ignition winding on alternator stator. "I" terminal must be connected to ignition switch. Improper connection will cause damage to electronic devices.

Ignition coil assembly includes capacitor and a pulse transformer arrangement similar to a conventional high tension coil. Flywheel has a special projection for triggering ignition.

If ignition trouble exists and spark plug is known to be good, the following tests should be made to determine which component is at fault. A flashlight type continuity tester can be used to test ignition coil assembly and trigger module.

To test coil assembly, remove high tension lead (spark plug wire) from coil. Insert one tester lead in coil terminal and remaining tester lead to coil mounting bracket. Tester light should be on. Connect one tester lead to coil mounting bracket and remaining tester lead to ignition switch wire of coil. Tester light should be out. Renew ignition coil if either test shows wrong results.

To test trigger module, connect one tester lead to AC inlet (A – Fig. KO17) and remaining lead to ignition lead terminal (I). Check for continuity. Reverse tester leads and again check for continuity. Test light must be on in one test only. The second test is made by connecting one tester lead to trigger module mounting bracket and other lead to AC inlet (A). Check for continuity. Reverse tester leads and again check for continuity. Again, test light must be on in one test only. The third test is made by connecting POSITIVE lead of tester to (I) terminal of trigger and connect remaining tester lead to trigger module mounting bracket. Test light should be off. Rotate flywheel in normal direction of rotation. When trigger projection on flywheel passes trigger module, test light should turn on. Renew trigger module if any of the three tests show wrong results.

If ignition trouble still exists after ignition coil assembly and trigger module tests show they are good, ignition winding on alternator stator is faulty. In this event, renew stator.

LUBRICATION. All models are splash lubricated except Model KV-181 which has an oil circulating system. See Fig. KO18. This vertical crankshaft engine has two oil sumps. Outer sump (7) is oil reservoir and inner sump (13) is a low level oil drain back sump. Oil is drawn from outer sump (7) through main gallery (6), oil passage (5) in crankcase and passage (4) in bearing plate. After lubricating upper main bearing (3), oil flows through thrust bearing (2) which acts as a centrifugal oil pump and is forced through tapered hole in crankpin (9) to lubricate connecting rod. Throw-off oil lubricates piston, camshaft, gears and lower main bearing (14). This drain back oil is then taken from inner sump (13) through oil pick-up assembly (10) and reed valve (11) and returned via gallery (12) to outer sump (7).

Maintain crankcase oil level at full mark on dipstick, but do not overfill. High quality detergent oil having API classification "SC" is recommended.

Use SAE 30 oil in temperatures above 32° F. (0° C), SAE 10W-30 oil in temperatures between 32° F (0° C) and 0° F (-18° C) and SAE 5W-20 oil in temperatures below 0° F (-18° C).

CRANKCASE BREATHER. Refer to Figs. KO19 and KO20. A reed valve assembly is located in valve spring compartment to maintain a partial vacuum in crankcase and thus reduce leakage of oil at bearing seals. If a slight amount of crankcase vacuum (5-10 inches water or ½-1 inch mercury) is not present, reed valve is faulty or engine has excessive blow-by past rings and/or valves or worn oil seals.

AIR CLEANER. Engine may be

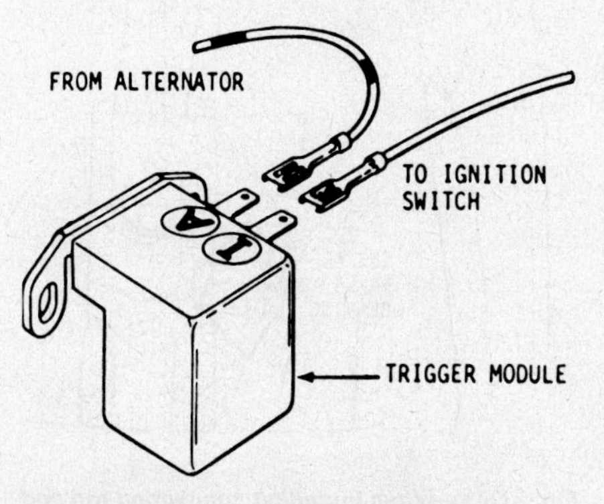

Fig. KO17 – Wires must be connected to trigger module as shown. Reversing connections will damage electronic devices.

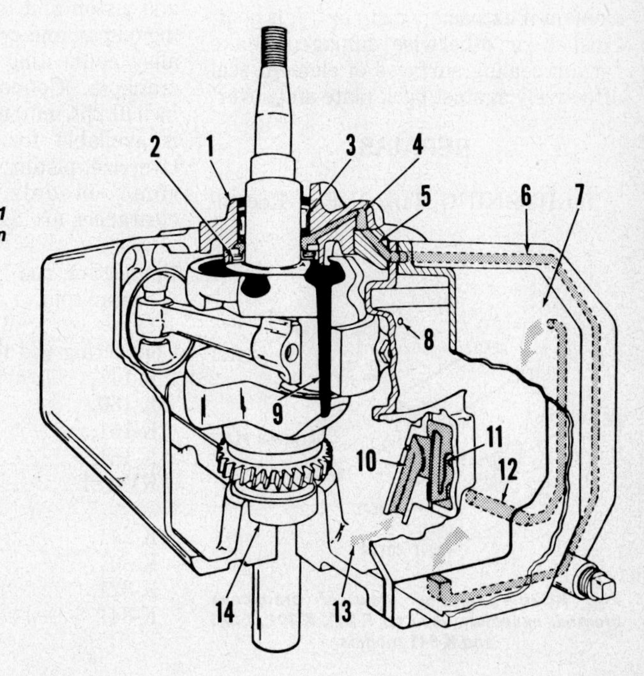

Fig. KO18 – View of KV-181 engine showing lubrication system.

1. Bearing plate
2. Thrust bearing
3. Upper main bearing (needle)
4. Oil passage in bearing plate
5. Oil passage in crankcase
6. Main oil gallery
7. Outer sump
8. Vent hole (0.030 inch)
9. Tapered hole in crankpin
10. Oil pick-up assy.
11. Reed valve
12. Oil return gallery
13. Inner sump
14. Lower main bearing (ball)

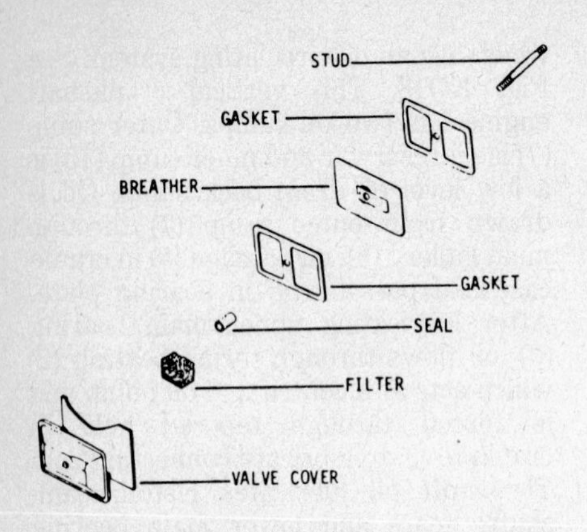

Fig. KO19—Exploded view of crankcase breather assembly used on K-141, K-160, K-161, K-181 and KV-181 models.

equipped with either an oil bath type air cleaner shown in Fig. KO21 or a dry element type shown in Fig. KO22.

Oil bath air cleaner should be serviced every 25 hours of normal operation or more often if operating in extremely dusty conditions. To service oil bath type, remove complete cleaner assembly from engine. Remove cover and element from bowl. Empty used oil from bowl, then clean cover and bowl in solvent. Clean element in solvent and allow to drip dry. Lightly re-oil element. Fill bowl to oil level marked on bowl using same grade and weight oil as used in engine crankcase. Renew gaskets as necessary when reassembling and reinstalling unit.

Dry element type air cleaner should be cleaned every 100 hours of normal operation or more frequently if operating in dusty conditions. Remove dry element and tap element lightly on a flat surface to remove surface dirt. Do not wash element or attempt to clean element with compressed air. Renew element if extremely dirty or if it is bent, crushed or otherwise damaged. Make certain sealing surfaces of element seal effectively against back plate and cover.

REPAIRS

TIGHTENING TORQUES. Recommended tightening torques are as follows:

Fig. KO21-View of oil bath type air cleaner.

Spark plug ....22 ft.-lbs. (30 N·m)

Connecting rod
cap screws* .....See CONNECTING ROD section

Cylinder head
cap screws*
K-141,
K-160,
K-161,
K-181,
KV-181 ....20 ft.-lbs. (27 N·m)
K-241,
K-301,
K-321,
K-341 ....30 ft.-lbs. (41 N·m)

Flywheel retaining nut
Models with 5/8-inch nut ....60 ft.-lbs. (81 N·m)
Models with 3/4-inch nut ...100 ft.-lbs. (136 N·m)

*With threads lubricated.

CONNECTING ROD. Connecting rod and piston unit is removed after removing oil pan and cylinder head. Aluminum alloy connecting rod rides directly on crankpin. Connecting rod with 0.010 inch (0.254 mm) undersize crankpin bore is available for reground crankshaft. Oversize piston pins are available on some models. Desired running clearances are as follows:

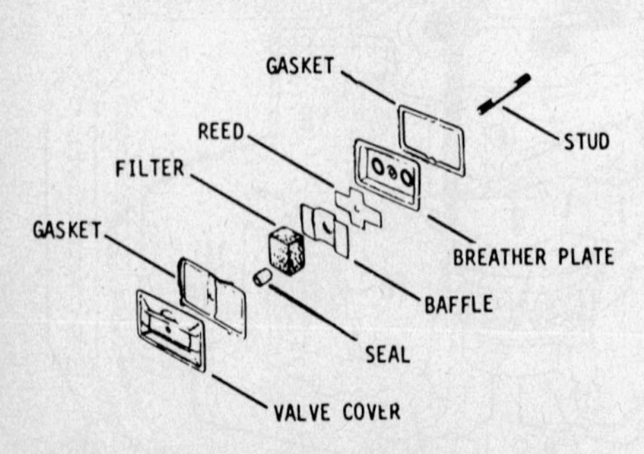

Fig. KO20—Exploded view of crankcase breather assembly used on K-241, K-301, K-321 and K-341 models.

Connecting rod
to crankpin ....0.001-0.002 in. (0.0254-0.0508 mm)

Connecting rod to piston pin,
K-141,
K-160,
K-161,
K-181,
KV-181 ....0.0006-0.0011 in. (0.0152-0.0279 mm)
K-241,
K-301,
K-321,
K-341 ....0.0003-0.0008 in. (0.0076-0.0203 mm)

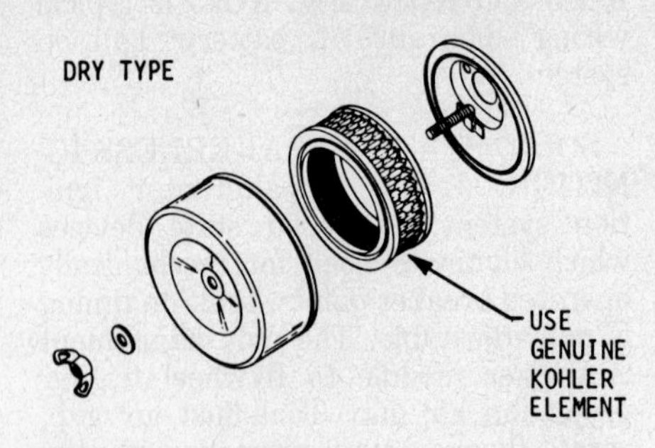

Fig. KO22—Exploded view of dry element type air cleaner.

Rod side play
on crankpin
K-141,
K-160,
K-161,
K-181,
KV-181 ....0.005-0.016 in. (0.1270-0.4064 mm)
K-241,
K-301,
K-321,
K-341 ....0.007-0.016 in. (0.1778-0.4064 mm)

Standard crankpin diameter is 1.1855-1.1860 inches (30.1117-30.1244 mm) on Models K-141, K-160, K-161, K-181 and KV-181 and 1.4995-1.5000 inches (38.0873-38.1000 mm) on Models K-241, K-301, K-321 and K-341.

Assemble piston to rod so arrow on piston (when so marked) faces away from valves and match marks (Fig. KO23) on rod and cap are aligned and towards camshaft side of engine. Always use new retainer rings to secure piston pin in piston.

Kohler recommends connecting rod cap screws be overtorqued to one specification, loosened, then tightened

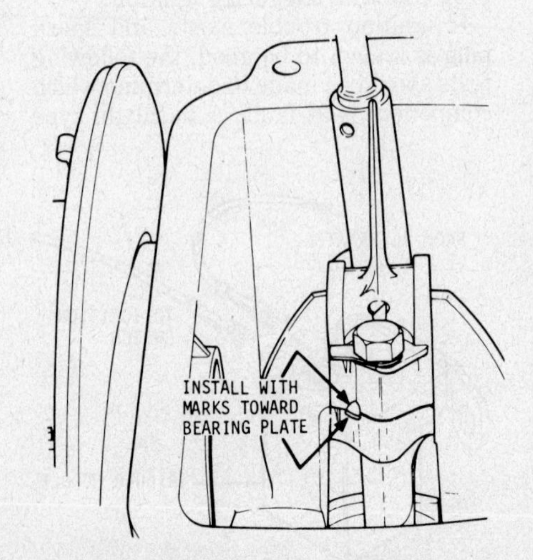

Fig. KO23—When installing connecting rod and piston unit, be sure marks on rod and cap are aligned and are facing toward flywheel side of engine.

to correct torque specification. Tightening specifications are as follows:

K-141,
K-160,
K-161,
K-181,
KV-181 (Overtorque) . . . . 20 ft.-lbs. (27 N·m)
(Final torque) . . . . 17 ft.-lbs. (23 N·m)

K-241,
K-301,
K-321,
K-341 (Overtorque) . . . . 28 ft.-lbs. (38 N·m)
(Final torque) . . . . 24 ft.-lbs. (32 N·m)

NOTE: Torque specifications for connecting rod cap screws are with lubricated threads.

PISTON, PIN AND RINGS. Aluminum alloy piston is fitted with two compression rings and one oil control ring. Renew piston if scored or if side clearance of new ring in piston top groove exceeds 0.005 inch (0.1270 mm). Pistons and rings are available in oversizes of 0.010, 0.020 and 0.030 inch (0.254, 0.508 and 0.762 mm) as well as standard sizes.

Piston pin fit in piston bore should be from 0.0001 inch (0.0025 mm) interference to 0.0003 inch (0.0076 mm) loose on Models K-141, K-160, K-161, K-181 and KV-181 and 0.0002-0.0003 inch (0.0051-0.0076 mm) loose on Models K-241, K-301, K-321 and K-341. Standard piston pin diameter is 0.6248 inch (15.8699 mm) on Models K-141, K-160, K-161, K-181 and KV-181, 0.8592 inch (21.8237 mm) on Model K-241 and 0.8753 inch (22.2326 mm) on Models K-301, K-321 and K-341. Piston pins are available in oversize of 0.005 inch (0.127 mm) on some models. Always renew piston pin retaining rings.

Recommended piston to cylinder bore clearance measured at thrust side at bottom of skirt is as follows:

K-141, K-160,
K-161, K-181,
KV-181 . . . . 0.0045-0.0065 in. (0.1143-0.1651 mm)
K-241, K-301 . . . . 0.003-0.004 in. (0.0762-0.1016 mm)
K-321, K-341 . . . . 0.0035-0.0045 in. (0.0889-0.1143 mm)

Recommended piston to cylinder bore clearance measured at thrust side just below oil ring is as follows:

K-141, K-160,
K-161, K-181,
KV-181 . . . . 0.006-0.0075 in. (0.1524-0.1905 mm)
K-241 . . . . 0.0075-0.0085 in. (0.1905-0.2159 mm)
K-301 . . . . 0.0065-0.0095 in. (0.1651-0.2413 mm)
K-321,
K-341 . . . . 0.007-0.010 in. (0.1778-0.254 mm)

Minimum piston diameters measured just below oil ring 90° from piston pin are as follows:

K-141*
K-160
K-161* . . . . 2.866 in. (72.7964 mm)
K-181
K-141**
K-161**
KV-181 . . . . 2.9275 in. (74.3585 mm)
K-241 . . . . 3.2400 in. (82.296 mm)
K-301 . . . . 3.3640 in. (85.4456 mm)
K-321 . . . . 3.4885 in. (88.6079 mm)
K-341 . . . . 3.7385 in. (94.9579 mm)

*Designates early production (before 1970) engines.

**Designates late production (after 1970) engines.

Renew any piston which does not meet specifications.

Kohler recommends piston rings always be renewed whenever they are removed. Piston ring specifications are as follows:

Ring end gap,
K-141, K-160,
K-161, K-181,
KV-181 . . . . 0.007-0.017 in. (0.1778-0.4318 mm)
K-241, K-301,
K-321, K-341 . . . . 0.010-0.020 in. (0.254-0.508 mm)

Ring side clearance (compression rings),
K-141, K-160,
K-161, K-181,
KV-181 . . . . 0.0025-0.004 in. (0.0635-0.1016 mm)

Ring side clearance (oil control ring),
K-141, K-160,
K-161, K-181,
KV-181 . . . . 0.001-0.0025 in. (0.0254-0.0635 mm)
K-241, K-301,
K-321, K-341 . . . . 0.001-0.003 in. (0.0254-0.0762 mm)

If compression ring has a groove or bevel on outside surface, install ring with groove or bevel down. If groove or bevel is on inside surface of compression ring, install ring with groove or bevel up. Oil control ring can be installed either side up.

CYLINDER BLOCK. If cylinder wall is scored or bore is tapered more than 0.0025 inch (0.0635 mm) on Models K-141, K-160, K-161, K-181 and KV-181 or 0.0015 inch (0.0381 mm) on Models K-241, K-301, K-321 and K-341, or out-of-round more than 0.005 inch (0.1270 mm), cylinder should be honed to nearest suitable oversize of 0.010, 0.020 or 0.030 inch (0.254, 0.508 or 0.762 mm) for which piston and rings are available.

Standard cylinder bore diameters are as follows:

K-141*
K-160
K-161* . . . . 2.875 in. (73.025 mm)
K-141**
K-161**
K-181
KV-181 . . . . 2.9375 in. (74.613 mm)
K-241 . . . . 3.251 in. (82.58 mm)
K-301 . . . . 3.375 in. (85.725 mm)
K-321 . . . . 3.500 in. (88.90 mm)
K-341 . . . . 3.750 in. (95.725 mm)

*Designates early production (before 1970) engines.
**Designates late production (after 1970) engines.

CYLINDER HEAD. Always use a new gasket when installing cylinder head. Cylinder head should be checked for warpage by placing cleaned head on flat surface. If clearance between head and plate exceeds 0.003 inch (0.0762 mm) when checked with feeler gage, renew head. Tighten cylinder head cap screws evenly and in steps using correct sequence shown in Fig. KO24, KO25 or KO25A.

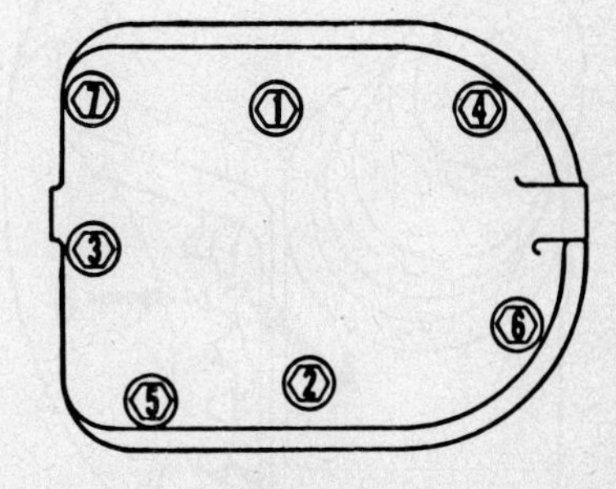

Fig. KO24—On K-141, K-160, K-161, K-181 and KV-181 models, tighten cylinder head cap screws evenly, in sequence shown, to 15-20 ft.-lbs. (20-27 N·m) torque.

CRANKSHAFT. Crankshaft on standard K-141 engine is supported by ball bearing in crankcase at pto end of shaft and a bushing type bearing in bearing plate at flywheel end. Model KV-181 has a roller bearing at flywheel end of crankshaft and a ball bearing at opposite end. Crankshaft on all other models is supported in two ball bearings.

On Model K-141 engine with bushing type main bearing, renew crankshaft and/or bearing plate if crankshaft journal and bearing are excessively worn or scored. Recommended crankshaft journal to bushing running clearance is 0.001-0.0025 inch (0.0254-0.0635 mm).

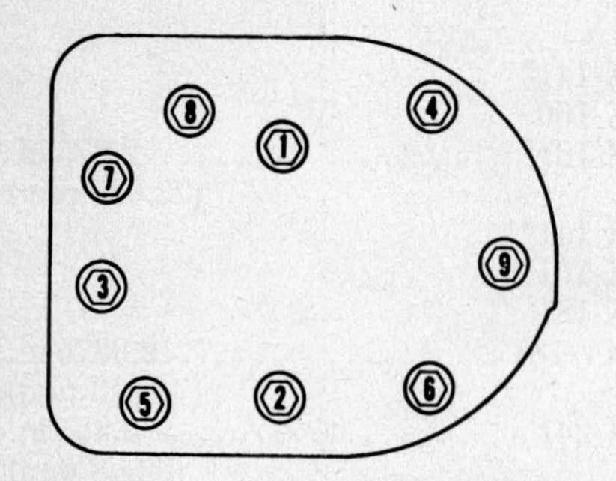

Fig. KO25—On K-241, K-301 and K-321 models, tighten cylinder head cap screws evenly, in sequence shown, to 25-30 ft.-lbs. (34-41 N·m) torque.

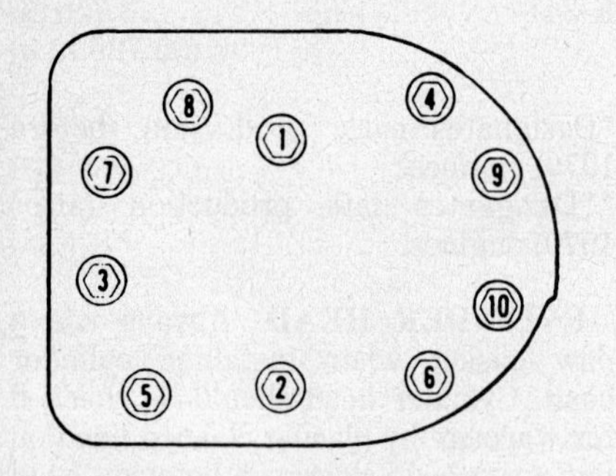

Fig. KO25A—On K-341 models, tighten cylinder head cap screws evenly, in sequence shown, to 25-30 ft.-lbs. (34-41 N·m) torque.

On all models, renew ball bearing type mains if excessively loose or rough. Crankshaft end play should be 0.003-0.028 inch (0.0762-0.7112 mm) on KV-181 models, 0.003-0.020 inch (0.0762-0.508 mm) on K-241, K-301, K-321 and K-341 models, and 0.002-0.023 inch (0.0508-0.5842 mm) on all other models. End play is controlled by thickness of bearing plate gaskets. Gaskets are available in thicknesses of 0.010 and 0.020 inch. Install ball bearing mains with sealed side towards crankpin.

Crankpin journal may be reground to 0.010 inch (0.254 mm) undersize for use of undersize connecting rod if journal is scored or out-of-round. Standard crankpin diameter is 1.1855-1.1860 inch (30.1117-30.1244 mm) on Models K-141, K-160, K-161, K-181 and KV-181 and 1.4995-1.500 inch (38.0873-38.10 mm) on Models K-241, K-301, K-321 and K-341.

When installing crankshaft, align timing marks on crankshaft and camshaft gear as shown in Fig. KO26.

NOTE: On Models K-241, K-301, K-321 and K-341 equipped with dynamic balancer, refer to DYNAMIC BALANCER paragraph for installation and timing of balancer gears.

On all models, Kohler recommends crankshaft seals be installed in crankcase and bearing plate after crankshaft and bearing plate are installed. Carefully work oil seals over crankshaft and drive seals into place with hollow driver that contacts outer edge of seals.

CAMSHAFT. The hollow camshaft and integral camshaft gear turn on a pin that is a slip fit in flywheel side of crankcase and a drive fit in closed side of crankcase. Remove and install pin from open side (bearing plate side) of crankcase. Desired camshaft to pin running clearance is 0.0005-0.003 inch (0.0127-0.0762 mm) on Models K-141, K-160, K-161, K-181 and KV-181 and 0.001-0.0035 inch (0.0254-0.0889 mm) on all other models. Desired camshaft end play of 0.005-0.010 inch (0.1270-0.254 mm) is controlled by use of 0.005 and 0.010 inch thick spacer washers between camshaft and cylinder block at bearing plate side of crankcase.

On early models, ignition spark advance weights, springs and weight pivot pins are renewable separately from camshaft gear. When reinstalling camshaft in engine, be sure breaker cam is cor-

Fig. KO27—Views showing operation of camshaft with automatic compression release. In view 1, spring (C) has moved control lever (D) which moves cam lever (B) upward so tang (T) is above exhaust cam lobe. This tang holds exhaust valve open slightly on a portion of compression stroke to relieve compression while cranking engine. At engine speeds of 650 rpm or more, centrifugal force moves control lever (D) outward allowing tang (T) to move below lobe surface as shown in view 2.

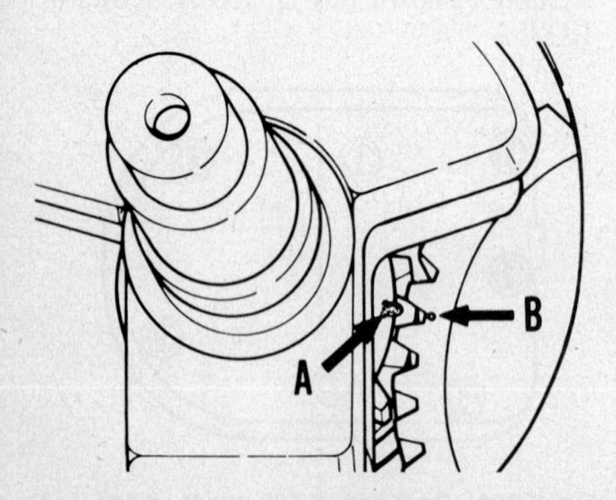

Fig. KO26—When installing crankshaft, make certain timing mark (A) on crankshaft is aligned with timing mark (B) on camshaft gear.

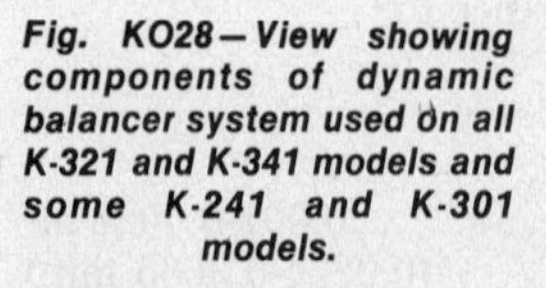

Fig. KO28—View showing components of dynamic balancer system used on all K-321 and K-341 models and some K-241 and K-301 models.

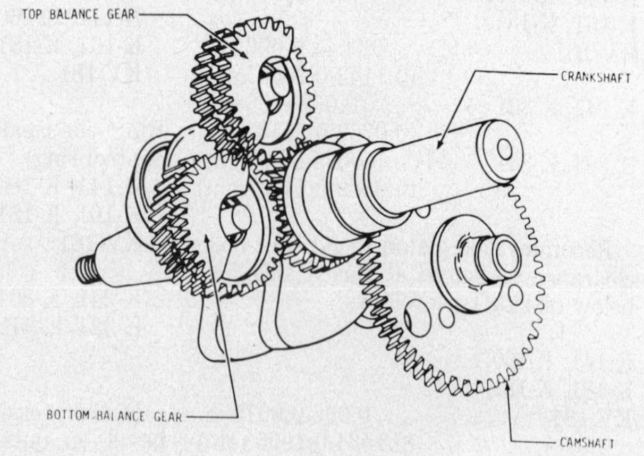

rectly installed. Spread springs as shown in Fig. KO12 and install breaker cam on tangs of flyweights with timing mark on cam and camshaft gear aligned.

Late production camshafts do not have automatic timing advance mechanism shown in Fig. KO12, but are equipped with automatic compression release mechanism shown in Fig. KO27. The automatic compression release mechanism holds exhaust valve slightly open during first part of compression stroke, reducing compression pressure and allowing easier cranking of engine. Refer to Fig. KO27 for operational details. At speeds above 650 engine rpm, compression release mechanism is inactive. Service procedures remain the same as for early production camshaft units except for the difference in timing advance and compression release mechanisms. Service kits are available for adding automatic compression release to early production engines.

To check compression on engine equipped with automatic compression release, engine must be cranked at 650 rpm or higher to overcome compression release action. A reading can also be obtained by rotating flywheel in reverse direction with throttle wide open. Compression reading should be 110-120 psi (758-827kPa) on an engine in top mechanical condition. When compression reading falls below 100 psi (689 kPa), it indicates leaking rings or valves.

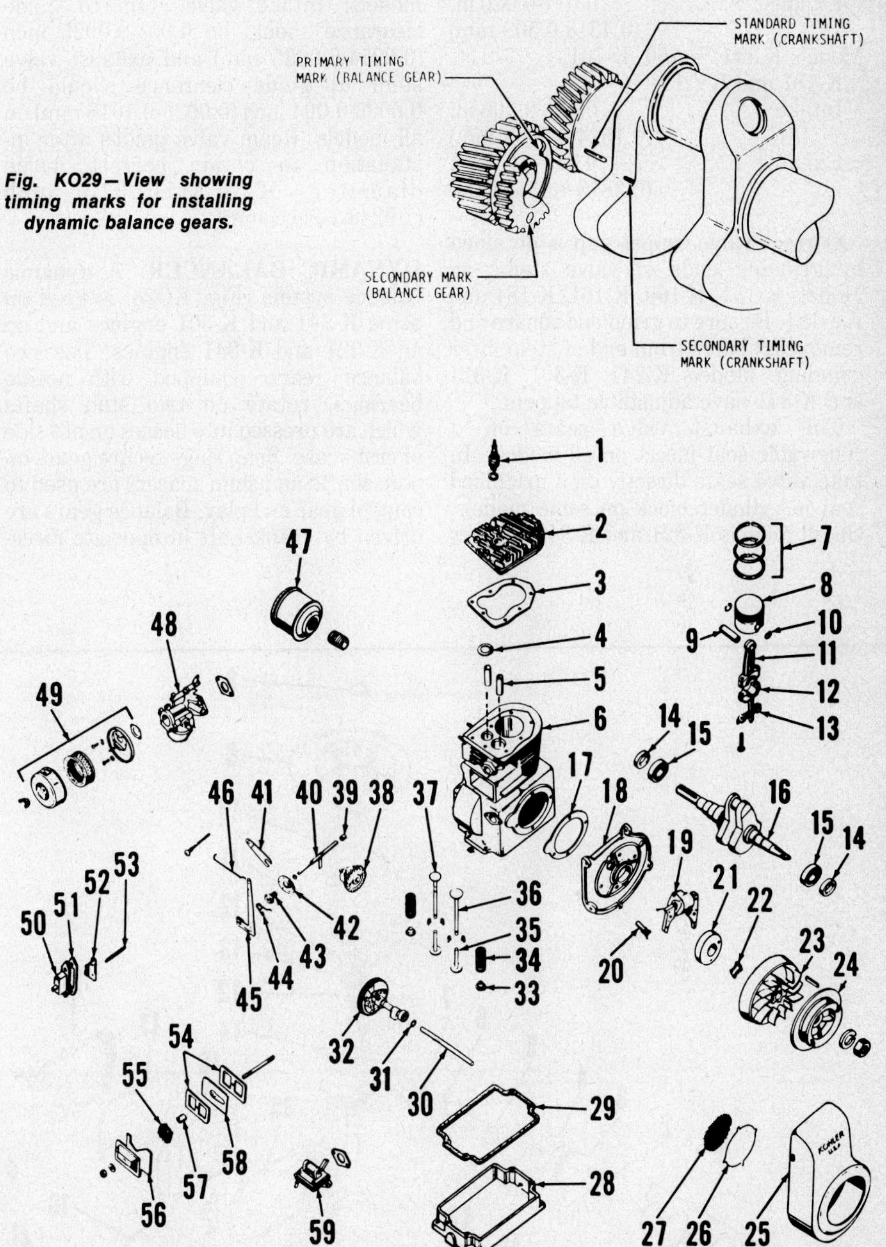

Fig. KO29 — View showing timing marks for installing dynamic balance gears.

Fig. KO30 — Exploded view of K-181 model basic engine assembly. Models K-141, K-160 and K-161 are similar. Standard K-141 engine is equipped with a bushing type main bearing in bearing plate (18).

1. Spark plug
2. Cylinder head
3. Head gasket
4. Exhaust valve seat
5. Valve guide
6. Cylinder block
7. Piston rings
8. Piston
9. Piston pin
10. Retaining rings
11. Connecting rod
12. Rod cap
13. Rod bolt lock
14. Oil seal
15. Ball bearing
16. Crankshaft
17. Gasket
18. Bearing plate
19. Magneto
20. Condenser
21. Magneto rotor
22. Wave washer
23. Flywheel
24. Pulley
25. Shroud
26. Screen retainer
27. Screen
28. Oil pan
29. Gasket
30. Camshaft pin
31. Shim washer
32. Camshaft
33. Spring retainer
34. Valve spring
35. Valve tappet
36. Intake valve
37. Exhaust valve
38. Governor gear & weight assy.
39. Needle bearing
40. Governor shaft
41. Bracket
42. Speed disc
43. Bushing
44. Governor spring
45. Governor lever
46. Link
47. Muffler
48. Carburetor
49. Air cleaner assy.
50. Breaker cover
51. Gasket
52. Breaker points
53. Push rod
54. Gaskets
55. Filter
56. Valve cover
57. Breather seal
58. Reed plate
59. Fuel pump

11/16" NOMINAL - PRESS DEPTH .735"
STUB SHAFT
ABOVE MAIN BEARING BOSS
STUB SHAFT BOSS
7/16"
IF SHAFT PROTRUDES ABOUT 11/16" ABOVE STUB SHAFT BOSS, DO NOT USE 3/8" SPACER

Fig. KO28A — If stub shaft boss is approximately 7/16-inch (10.113 mm) above main bearing boss, press new shaft in until it is 0.735 inch (18.669 mm) above stub shaft boss.

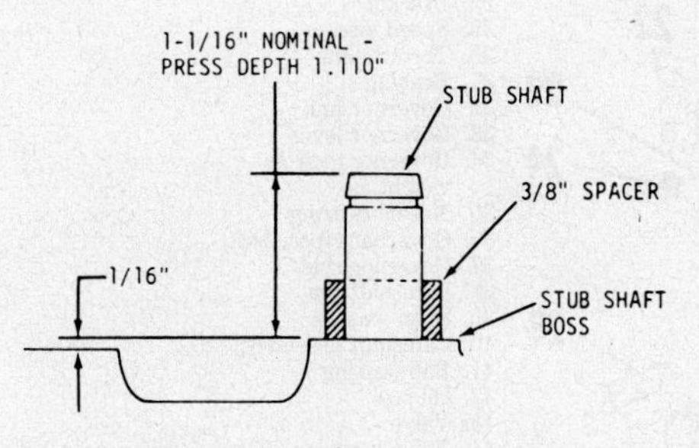

Fig. KO28B — If stub shaft boss is approximately 1/16-inch (1.588 mm) above main bearing boss, press new shaft in until it is 1.110 inch (28.194 mm) above stub shaft boss and use a 3/8-inch spacer between block and gear.

VALVE SYSTEM. Valve tappet gap (cold) is as follows:

Models K-241, K-301, K-321 and K-341
Intake ....0.008-0.010 in. (0.2032-0.254 mm)
Exhaust ....0.017-0.020 in. (0.4318-0.508 mm)

Models K-141, K-160, K-161, K-181 and KV-181
Intake ....0.006-0.008 in. (0.1524-0.2032 mm)
Exhaust ....0.015-0.017 in. (0.3810-0.4318 mm)

Correct valve tappet gap is obtained by grinding ends of valve stems on Models K-141, K-160, K-161, K-181 and KV-181. Be sure to grind end square and remove all burrs from end of stem after grinding. Models K-241, K-301, K-321 and K-341 have adjustable tappets.

The exhaust valve seats on a renewable seat insert on all models. In take valve seats directly on a machined seat in cylinder block on some models. On all Models K-321 and K-341 engines and some other models, a renewable intake valve seat insert is used. Valve face and seat angle is 45°. Desired seat width is 1/32 to 1/16-inch (0.794 to 1.588 mm).

Renewable valve guides are used on all models. Intake valve stem to guide clearance should be 0.001-0.0025 inch (0.0254-0.0635 mm) and exhaust valve stem to guide clearance should be 0.0025-0.004 inch (0.0635-0.1016 mm) on all models. Ream valve guides after installation to obtain correct inside diameter of 0.312-0.313 inch (7.9248-7.9502 mm).

DYNAMIC BALANCER. A dynamic balance system (Fig. KO28) is used on some K-241 and K-301 engines and on all K-321 and K-341 engines. The two balance gears, equipped with needle bearings, rotate on two stub shafts which are pressed into bosses on pto side of crankcase. Snap rings secure gears on stub shafts and shim spacers are used to control gear end play. Balance gears are driven by crankshaft in opposite direction of crankshaft rotation. Use following procedure to install and time dynamic balancer components.

To renew stub shafts, press old shafts out and discard. If stub shaft boss is approximately 7/16-inch (10.113 mm) above main bearing boss (see Fig. KO28A), press new shaft in until it is 0.735 inch (18.669 mm) above stub shaft boss. On blocks where stub shaft boss is approximately 1/16-inch (1.588 mm) above main bearing boss (Fig. KO28B), press new shaft in until it is 1.110 inch (27.19 mm) above stub shaft boss and use a 3/8-inch (9.525 mm) spacer between block and gear.

To install top balance gear-bearing assembly, first place one 0.010 inch shim on stub shaft, install top gear assembly on shaft making certain timing marks are facing flywheel side of crankcase. Install one 0.005 inch, one 0.010 inch and one 0.020 inch shim spacers in this order and install snap ring. Using a feeler gage, check gear end play. Proper end play of balance gear is 0.005-0.010 inch.

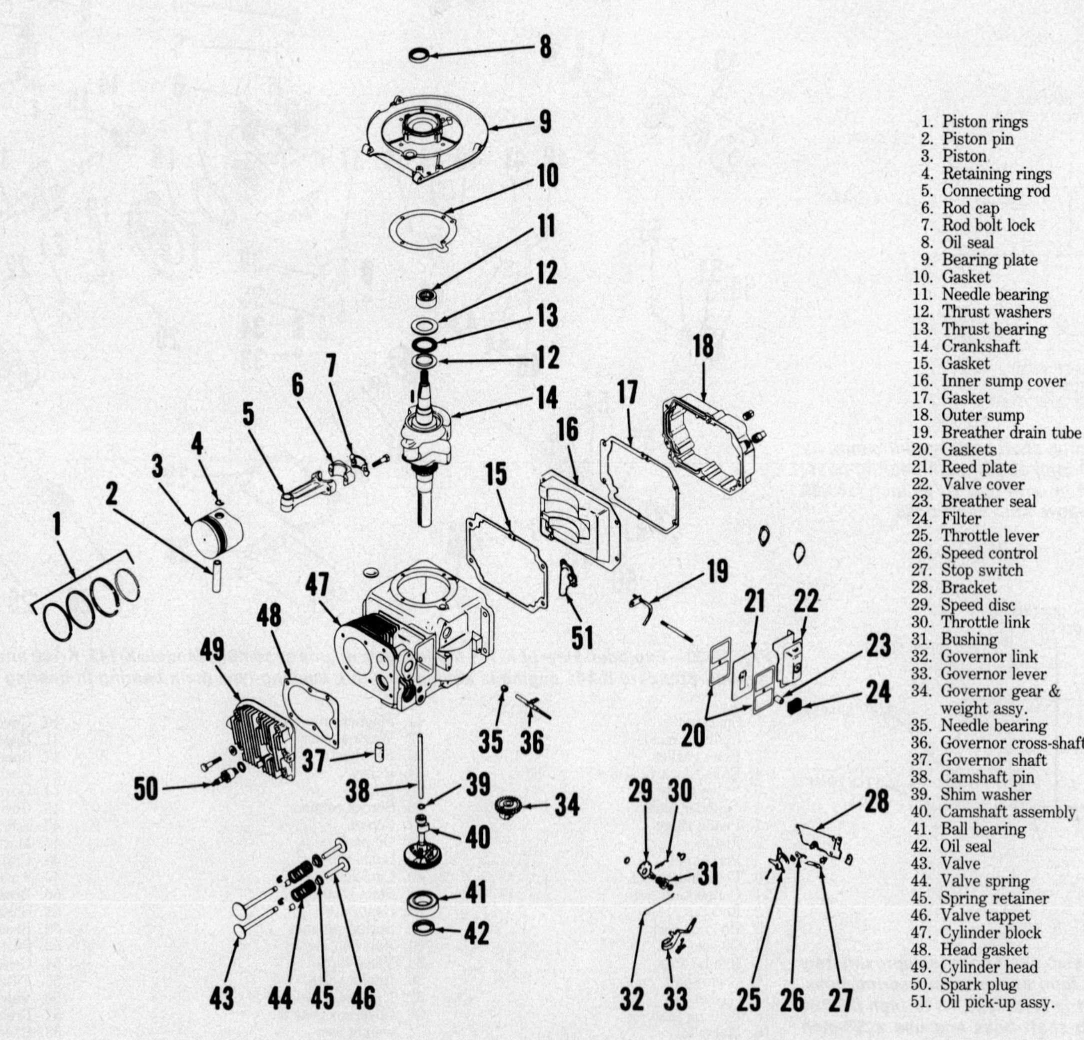

Fig. KO31 – Exploded view of KV-181 model basic engine assembly.

Add or remove 0.005 inch thick spacers as necessary to obtain correct end play.

NOTE: Always install the 0.020 inch thick spacer next to snap ring.

Install crankshaft in crankcase and align primary timing mark on top balance gear with standard timing mark on crankshaft. See Fig. KO29. With primary timing marks aligned, engage crankshaft gear 1/16-inch into narrow section of top balance gear and rotate crankshaft to align timing marks on camshaft gear and crankshaft as shown in Fig. KO26. Press crankshaft into crankcase until it is seated firmly into ball bearing main.

Rotate crankshaft until crankpin is approximately 15° past bottom dead center. Install one 0.010 inch shim on stub shaft, align secondary timing mark on bottom balance gear with secondary timing mark on crankshaft counterweight. See Fig. KO29. Install gear assembly onto stub shaft. If properly timed, secondary timing mark on bottom balance gear will be aligned with standard timing mark on crankshaft after gear is fully on stub shaft. Install one 0.005 inch and one 0.020 inch shim, then install snap ring. Check bottom balance gear end play and add or remove 0.005 inch thick spacers as required to obtain proper end play of 0.005-0.010 inch. Make certain the 0.020 inch shim is used against the snap ring.

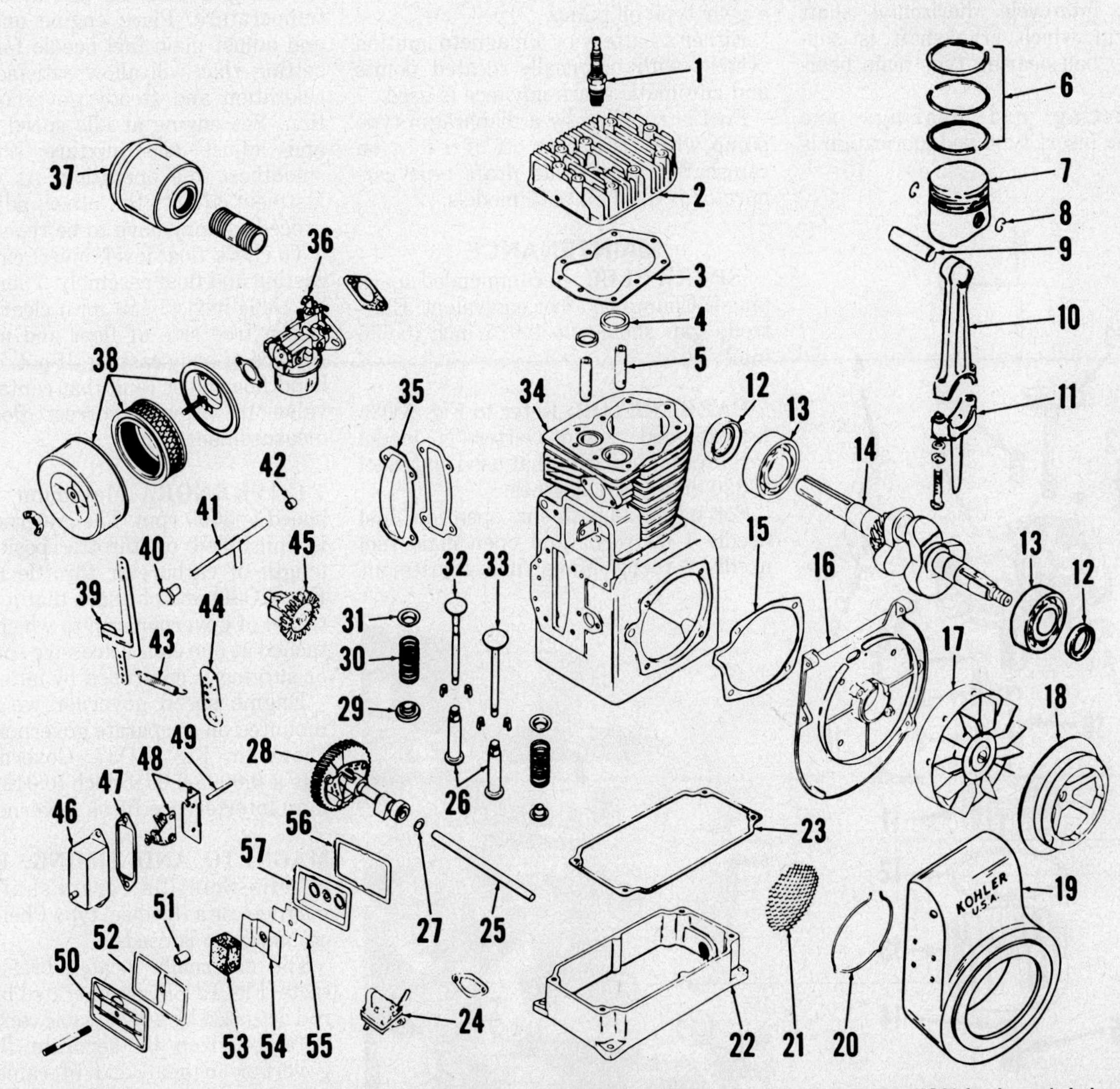

Fig. KO32 — Exploded view of K-241 or K-301 basic engine assembly. Models K-321 and K-341 are similar. Refer to Fig. KO28 for dynamic balancer used on all K-321 and K-301 models.

1. Spark plug
2. Cylinder head
3. Head gasket
4. Valve seat insert
5. Valve guide
6. Piston rings
7. Piston
8. Retaining rings
9. Piston pin
10. Connecting rod
11. Rod cap
12. Oil seal
13. Ball bearing
14. Crankshaft
15. Gasket
16. Bearing plate
17. Flywheel
18. Pulley
19. Shroud
20. Screen retainer
21. Screen
22. Oil pan
23. Gasket
24. Fuel pump
25. Camshaft pin
26. Valve tappets
27. Shim washer
28. Camshaft
29. Valve rotator
30. Valve spring
31. Spring retainer
32. Exhaust valve
33. Intake valve
34. Cylinder block
35. Camshaft cover
36. Carburetor
37. Muffler
38. Air cleaner assy.
39. Governor lever
40. Bushing
41. Governor shaft
42. Needle bearing
43. Governor spring
44. Speed lever
45. Governor gear & weight unit
46. Breaker cover
47. Gasket
48. Breaker point assy.
49. Push rod
50. Valve cover
51. Breather seal
52. Gasket
53. Filter
54. Baffle
55. Reed
56. Gasket
57. Breather plate

KOHLER

KOHLER COMPANY
Kohler, Wisconsin 53044

Model	No. Cyls	Bore	Stroke	Displacement	Power Rating
K-330, K-331	1	3.625 in. (92.08 mm)	3.25 in. (82,55 mm)	33.6 cu. in. (549.7 cc)	12.5 hp (9.32 kW)

Model K-330 and K-331 are one cylinder, four-cycle, horizontal shaft engines in which crankshaft is supported by ball bearing type main bearings.

Connecting rod bearings are renewable insert type and lubrication is provided by a full pressure system with a gear type oil pump.

Either a battery or a magneto ignition system with externally located points and automatic spark advance is used.

Fuel is supplied by a diaphragm type pump which is driven off of a lobe on camshaft. Carter side draft type carburetor is used on both models.

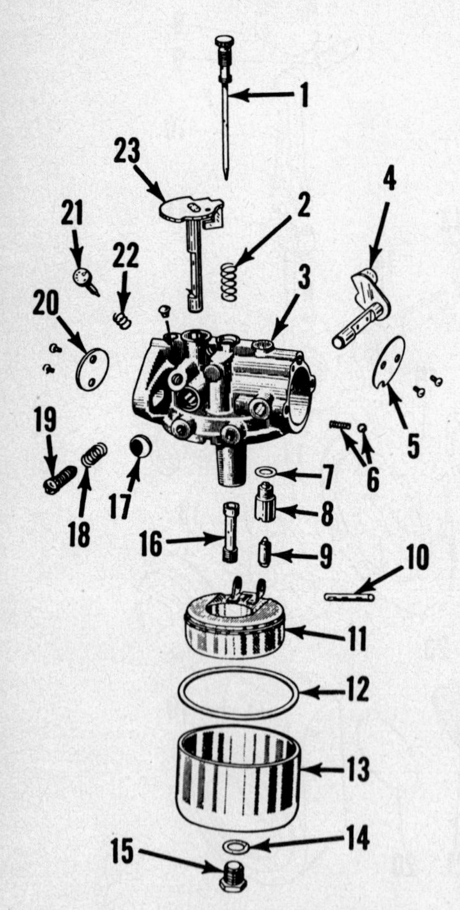

Fig. KO35—Exploded view of typical Carter "N" model carburetor. Refer to Fig. KO36 for location of idle fuel adjustment needle on K-330 and K-331 models.

1. Main fuel needle
2. Spring
3. Carburetor body
4. Choke shaft
5. Choke disc
6. Choke detent
7. Sealing washer
8. Inlet valve seat
9. Inlet valve
10. Float pin
11. Float
12. Gasket
13. Float bowl
14. Sealing washer
15. Bowl retainer
16. Main jet
17. Plug
18. Spring
19. Idle stop screw
20. Throttle disc
21. Idle fuel needle
22. Spring
23. Throttle shaft

MAINTENANCE

SPARK PLUG. Recommended spark plug is Champion J8 or equivalent. Electrode gap should be 0.025 inch (0.635 mm).

CARBURETOR. Refer to Fig. KO35 for exploded view of Carter "N" model carburetor similar to that used on Model K-330 and K-331 engines.

For initial adjustment, open idle fuel needle 1-1/2 turns and open main fuel needle 2 turns. Make final adjustment with engine running and at operating temperature. Place engine under load and adjust main fuel needle for leanest setting that will allow satisfactory acceleration and steady governor operation. Set engine at idle speed, no load and adjust idle mixture screw for smoothest idle operation. As each adjustment affects the other, adjustment procedure may have to be repeated.

Fig. KO36—View of carburetor adjustment points and variable speed governor linkage.

A. Sensitivity adjustment
B. Main fuel needle
C. Idle stop screw
D. Idle fuel needle
E. Choke lever
TR. Throttle rod

To check float level, invert carburetor casting and float assembly. There should be 13/64-inch (5.159 mm) clearance between free side of float and machined surface of body casting. If not, carefully bend float lever tang that contacts inlet valve to provide correct float level measurement.

GOVERNOR. Maximum no-load speed is 3600 rpm. The governed speed is controlled by throttle position, but length of carburetor throttle rod (TR-Fig. KO36) must be such that it matches travel of governor arm to which it is attached at one end. Excessive speed drop or surging is controlled by nuts (A).

Engine speed governor weights are mounted on a separate governor gear as shown in Fig. KO37. Governor gear has a 0.0005-0.0015 inch (0.0127-0.0381 mm) interference fit on governor shaft.

MAGNETO AND TIMING. Either a Bendix-Scintilla crankshaft type magneto or a flywheel type Phelon (Repco) magneto is used.

The externally located breaker contacts (Fig. KO38) are operated by a push rod actuated by a removable cam located in and driven by separate flyweight governor in gear end of camshaft as

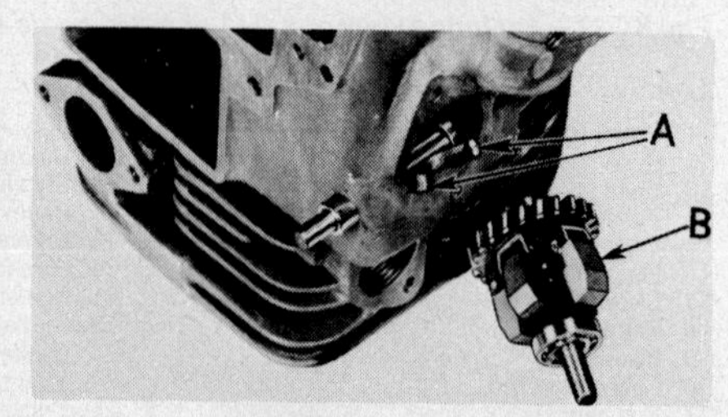

Fig. KO37—View showing governor gear and flyweight assembly (B) removed from engine. Governor yoke is shown at (A).

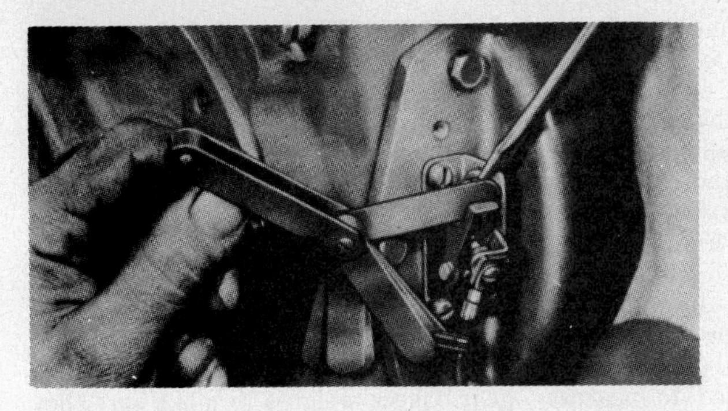

Fig. KO38 – Ignition breaker points are located externally on engine crankcase for either magneto or battery ignition.

shown in Fig. KO39. This governor varies angular position of breaker cam in accordance with engine speed to provide automatic advance of ignition timing. Note assembly marks "A" and "B" must be in register when assembling breaker cam to governor.

A timing inspection port is provided in bearing plate and there are two timing marks on flywheel. Satisfactory timing is obtained by adjusting breaker contact gap to 0.018-0.020 inch (0.457-0.508 mm). For precision timing, use a timing light and adjust breaker gap until the first or "SP" timing mark is centered in inspection port with engine running at high idle speed.

BATTERY IGNITION. Battery ignition is generally used on electric starting engines. An automotive type Delco-Remy coil and condenser are used with this system. However, since breaker points and adjustments are the same as magneto ignition, refer to previous paragraph for adjustment procedure.

Fig. KO39 – View showing alignment marks on ignition breaker cam and camshaft gear. Spread springs as indicated by arrows and insert cam on drive tangs of weights.

LUBRICATION. Engine is pressure lubricated by a gear type oil pump. High quality detergent type oil having API classification "MS" or "SC" should be used. Use SAE 30 oil in temperatures above 32° F (0° C), SAE 10W oil in temperatures between 32° F (0° C) and 0°F (-18°C) and SAE 5W oil in below 0°F (-18°C) temperatures. Maintain oil level at full mark on dipstick, but do not overfill.

CRANKCASE BREATHER. A reed valve assembly is located in valve spring compartment to maintain a partial vacuum in crankcase and thus reduce leakage of oil at bearing seals. If a slight amount of crankcase vacuum is not present, reed valve is faulty or engine has excessive blow-by past rings and/or valves.

AIR CLEANER. Engines may be equipped with either an oil bath type air cleaner shown in Fig. KO40 or a dry element type shown in Fig. KO41.

Oil bath air cleaner should be serviced every 25 hours of normal operation or more often if operating in extremely dusty conditions. To service oil bath type, remove complete cleaner assembly from engine. Remove cover and element from bowl. Empty used oil from bowl, then clean cover and bowl in solvent. Clean element in solvent and allow to drip dry. Lightly re-oil element. Fill bowl using same grade and weight oil as used in engine crankcase. Renew gaskets as necessary when reassembling and reinstalling unit.

Dry element type air cleaner should be cleaned every 100 hours of normal operation or more frequently if operating in dusty conditions. Remove dry element and tap element lightly on a flat surface to remove surface dirt. Do not wash element or attempt to clean element with compressed air. Renew element if extremely dirty or if it is bent, crushed or otherwise damaged. Make certain sealing surfaces of element seal effectively against back plate and cover.

REPAIRS

TIGHTENING TORQUES. Recommended tightening torque values with threads lightly lubricated are as follows:

Spark plug . . . . 22 ft.-lbs. (30 N·m)
Connecting rod . . . . See CONNECTING ROD paragraph
Cylinder head . . . . 40 ft.-lbs. (54 N·m)
Flywheel nut . . . . 100 ft.-lbs. (136 N·m)

CONNECTING ROD. Connecting rod and piston unit is removed from above after removing cylinder head and oil pan (engine base). Aluminum alloy connecting rod is fitted with renewable crankpin bearing inserts, but rides directly on piston pin. Refer to the following specifications:

Desired clearances –
Rod to crankpin . . . . 0.0003-0.0023 in. (0.0076-0.0584 mm)
Rod to piston pin . . . 0.0003-0.0008 in. (0.0076-0.0203 mm)
Rod side play on crankpin . . . . 0.007-0.011 in. (0.178-0.279 mm)

Crankpin diameter is 1.873 inches (47.68 mm). Crankpin bearing inserts

OIL BATH TYPE

OIL LEVEL

FILL TO LEVEL MARK WITH SAME OIL AS ENGINE

Fig. KO40 – View of oil bath type air cleaner.

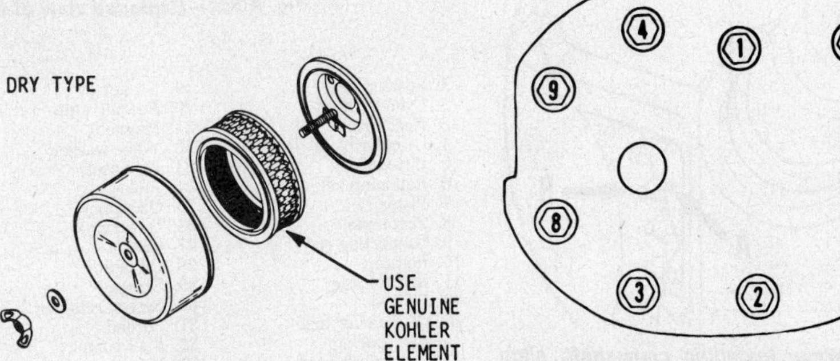

Fig. KO41 – Exploded view of dry element type air cleaner.

Fig. KO42 – Tighten cylinder head cap screws evenly to 40 ft.-lbs. (54 N·m) torque using sequence shown.

for connecting rod are available for 0.002, 0.010 and 0.020 inch (0.0508, 0.0254 and 0.508 mm) undersize crankshaft as well as standard size. If crankpin is only moderately worn, 0.002 inch (0.0508 mm) oversize bearing inserts can be installed. If crankpin is excessively worn, out-of-round or scored, crankpin can be reground for use with 0.010 or 0.020 inch (0.254 or 0.508 mm) oversize bearing inserts. Piston pin is available in standard size and 0.005 inch (0.127 mm) oversize.

When reinstalling piston and connecting rod assembly, piston may be installed either way on connecting rod; but match marks on connecting rod and cap must be aligned and be installed towards flywheel side of engine. Tighten connecting rod cap screws to a torque of 40 ft.-lbs. (54 N·m), then loosen cap screws and re-tighten to a torque of 35 ft.-lbs. (48 N·m).

PISTON, PIN AND RINGS. Two 3/32-inch (2.381 mm) compression rings and one 3/16-inch (4.763 mm) oil control ring are fitted with end gap limits of 0.007-0.017 inch (0.178-0.432 mm). Side clearance of compression rings in grooves should be 0.0025-0.0045 inch (0.0635-0.1143 mm); side clearance of oil ring 0.002-0.0035 inch (0.0508-0.0889 mm). Rings of 0.010, 0.020 and 0.030 inch (0.254, 0.508 and 0.762 mm) oversize are available.

The floating piston pin is retained in piston by a snap ring at each end. Recommended clearance of pin in unbushed aluminum rod is 0.0003-0.0008 inch (0.0076-0.0203 mm); in piston bosses, 0.0001-0.0003 inch (0.0025-0.0076 mm). Oversize pin is available.

Aluminum alloy piston should have a clearance of 0.0005-0.0015 inch (0.0127-0.0381 mm) in cylinder bore measured at top of skirt thrust face below oil ring. Reject piston if a new piston ring has 0.005 inch (0.127 mm) or more side clearance in top ring groove. Pistons of 0.010, 0.020 and 0.030 inch (0.254, 0.508 and 0.762 mm) oversize as well as standard size are available.

CYLINDER HEAD. Always use new head gasket when installing cylinder head. Tighten cylinder head cap screws evenly to a torque of 40 ft.-lbs. (54 N·m) using sequence shown in Fig. KO42.

CYLINDER BLOCK. If walls are scored or bore is tapered or out-of-round more than 0.005 inch (0.127 mm), cylinder should be resized to nearest suitable oversize of 0.010, 0.020 or 0.030 inch (0.254, 0.508 or 0.762 mm). Standard cylinder bore diameter is 3.625 inches (92.075 mm).

CAMSHAFT ASSEMBLY. Camshaft and gear assembly should have 0.001-0.0025 inch (0.0254-0.0635 mm) running clearance on camshaft support pin. Similar clearance should exist between breaker cam and camshaft pin. Desired camshaft end play of 0.005-0.010 inch (0.127-0.254 mm) is controlled by spacer washers (44-Fig. KO44) which are available in a variety of thicknesses. Ignition advance cam weights are mounted on face of cam gear. Check weights to make sure they work freely through their full travel. Ignition breaker cam should be installed with mark (A-Fig. KO39) in register with mark (B) on cam gear.

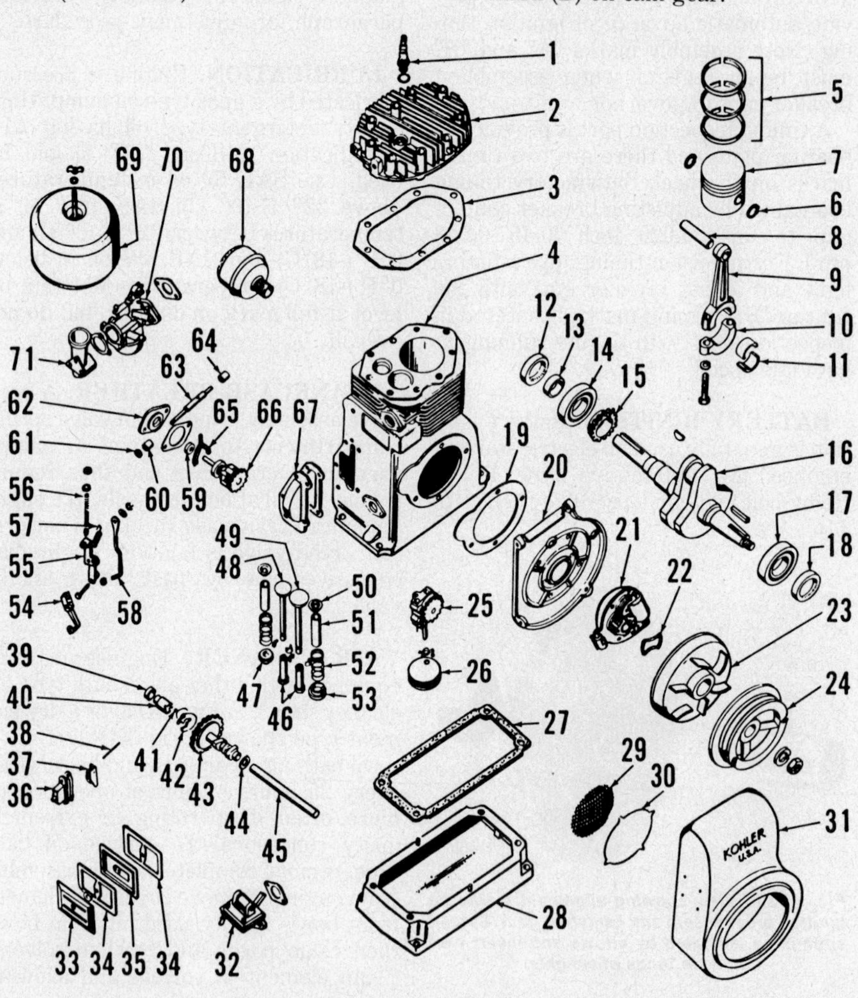

Fig. KO44—Exploded view of K-330 or K-331 basic engine assembly.

1. Spark plug
2. Cylinder head
3. Head gasket
4. Cylinder block
5. Piston rings
6. Retaining rings
7. Piston
8. Piston pin
9. Connecting rod
10. Rod cap
11. Rod bearing
12. Oil seal
13. Oil transfer tube
14. Ball bearing
15. Crankshaft gear
16. Crankshaft
17. Ball bearing
18. Oil seal
19. Gasket
20. Bearing plate
21. Magneto
22. Wave washer
23. Flywheel
24. Pulley
25. Oil pump
26. Oil strainer
27. Gasket
28. Oil pan
29. Screen
30. Screen retainer
31. Shroud
32. Fuel pump
33. Valve cover
34. Gaskets
35. Breather plate
36. Breaker cover
37. Breaker points
38. Push rod
39. Spacer
40. Breaker cam
41. Advance spring
42. Advance weight
43. Camshaft
44. Shim spacer
45. Camshaft pin
46. Valve tappets
47. Valve rotator
48. Exhaust valve
49. Intake valve
50. Spring seat
51. Valve guide
52. Valve spring
53. Spring retainer
54. Throttle arm
55. Control arm
56. Link
57. Governor spring
58. Throttle extension
59. Bushing
60. Needle bearing
61. Oil seal
62. Governor cover
63. Governor shaft
64. Needle bearing
65. Yoke
66. Governor gear & weight unit
67. Camshaft cover
68. Muffler
69. Air cleaner
70. Carburetor
71. Air intake elbow

Fig. KO43—When installing crankshaft, align timing mark (A) on crankshaft with timing mark (B) on camshaft gear.

Camshaft pin is a slip fit in open (flywheel) side of crankcase and a press fit in closed (pto) side of crankcase. Remove and install pin from open side of crankcase.

CRANKSHAFT. Crankshaft is supported in two ball bearing mains. Crankshaft end play should be 0.003-0.008 inch (0.076-0.203 mm) on pump models and 0.005-0.011 inch (0.127-0.279 mm) on other engines. End play is controlled by thickness of gasket used between crankcase and bearing plate (closure plate). Gaskets are available in a variety of thicknesses.

When installing crankshaft, align timing mark (A-Fig. KO43) on crankshaft with mark (B) on camshaft gear.

Crankshaft gear (15-Fig. KO44) should have an interference fit of 0.001-0.0015 inch (0.0254-0.0381 mm) on crankshaft. Install crankshaft main bearings in crankcase and bearing plate with shielded side of ball bearings toward inside of crankcase. Install oil seals with lips to inside of crankcase. Use a tapered protective sleeve or tape over crankshaft keyways to prevent damage to seals.

Oil transfer sleeve (13-Fig. KO44) is a press fit in crankcase between oil seal (12) and main bearing (14) at pto end of crankshaft. Oil clearance between transfer sleeve and crankshaft should be 0.001-0.0035 inch (0.0254-0.0889 mm). Oil holes in sleeve must be aligned with oil passage in crankcase. Purpose of transfer sleeve is to conduct oil to drilled passage in crankshaft for lubrication of crankpin bearing.

VALVE SYSTEM. Valve tappet gap (cold) is 0.008 inch (0.2032 mm) for intake valve and 0.020 inch (0.508 mm) for exhaust valve. Adjustable tappets are used.

The exhaust valve on all engines is equipped with a valve rotator and seats in a renewable hardened alloy steel insert. On some engines, a renewable intake valve seat insert is used although usually the intake valve seats directly on machined surface in crankcase and cylinder casting. Desired valve seat width is 1/32-inch (0.794 mm). If valve seat width exceeds 1/16-inch (1.588 mm), narrow the seat using 30° and 60° cutters or stones; then renew seating surface with 45° cutter or stone. Valve face and seat angle is 45°.

Desired clearance for intake valve stem in guide is 0.0005-0.002 inch (0.0127-0.0508 mm) and for exhaust valve stem in guide is 0.002-0.0035 inch (0.0508-0.0889 mm). Renew valve guides if clearance of new valve stem in guide is excessive. Clearance of tappets in bores of crankcase should be 0.0008-0.0023 inch (0.0203-0.0584 mm).

Free length of intake valve spring should be 2-1/4 inches (57.15 mm) and free length of exhaust valve spring should be 1-13/16 inches (46.038 mm). Renew spring if free length is not approximately equal to specified free length.

OIL PUMP. A gear type oil pump is used to supply lubricating oil under pressure to connecting rod. A sudden drop in oil pressure may be caused by dirt or foreign particles in the oil pump. Sometimes it is possible to disconnect oil line to oil pressure gage and dislodge foreign material by forcing compressed air into pump. Pump pressure is adjusted at factory. Should it be necessary to readjust oil pressure, turn screw in pump body to left to decrease pressure or to right to increase pressure. Seal screw with permatex or equivalent sealer when adjustment is completed.

When installing pump assembly, check pump drive gear for backlash and alignment with camshaft gear.

KOHLER

KOHLER COMPANY
Kohler, Wisconsin 503044

Model	No. Cyls.	Bore	Stroke	Displacement	Power Rating
K-361	1	3.75 in. (95.25 mm)	3.25 in. (82.55 mm)	35.89 cu. in. (588.2 cc)	18 hp (13.4 kW)

Model K-361 is a one cylinder, four-cycle, horizontal crankshaft engine in which crankshaft is supported by ball bearing type main bearings.

Connecting rod rides directly on crankpin journal and is lubricated by oil splash system.

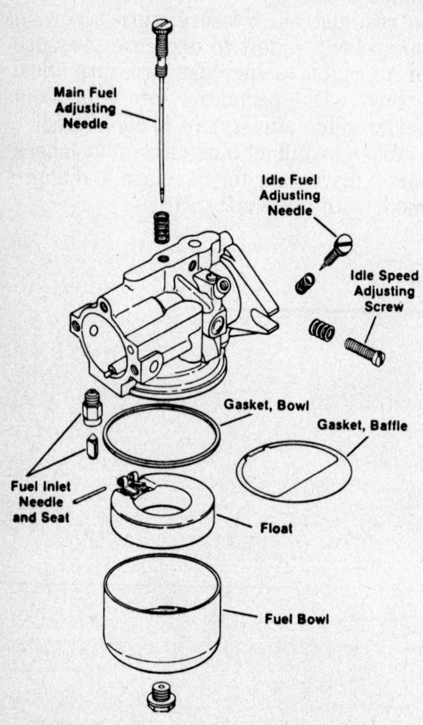

Fig. KO45 – Exploded view of Kohler side draft carburetor used on K-361 engines.

A battery type ignition system with externally mounted breaker points is used.

Engines are equipped with Kohler side draft carburetors.

Special engine accessories or equipment is noted by a suffix letter on engine model number. Key to suffix letters is as follows:

A – Oil pan type
C – Clutch model
P – Pump model
Q – Quiet model
S – Electric start
T – Retractable start

MAINTENANCE

SPARK PLUG. Recommended spark plug is Champion J8 or equivalent and for radio noise reduction, Champion EH10 or equivalent. Electrode gap is 0.035 inch (0.889 mm).

CARBURETOR. Refer to Fig. KO45 for exploded view of Kohler side draft carburetor. For initial adjustment, open main fuel needle 2-1/2 turns and open idle fuel needle 3/4-1 turn. Make final adjustment with engine at operating temperature and running. Place engine under load and adjust main fuel needle to leanest mixture that will allow satisfactory acceleration and steady governor operation.

Adjust idle speed stop screw so engine idles at 1725-1875 rpm, then adjust idle fuel needle for smoothest idle operation.

As each adjustment affects the other, adjustment procedure may have to be repeated.

To check float level, invert carburetor body and float assembly. There should be 11/64-inch. (4.366 mm) clearance plus or minus 1/32-inch (0.794 mm) between machined surface of body casting and free end of float. Adjust float by bending float lever tang that contacts inlet valve. See Fig. KO46.

AUTOMATIC CHOKE. Some models are equipped with a Thermo-Electric automatic choke and fuel shutdown solenoid (Fig. KO47).

When installing or adjusting choke unit, position unit leaving mounting screw slightly loose. Hold choke plate wide open and rotate choke unit clockwise with slight pressure until it can no longer be rotated. Hold choke unit and

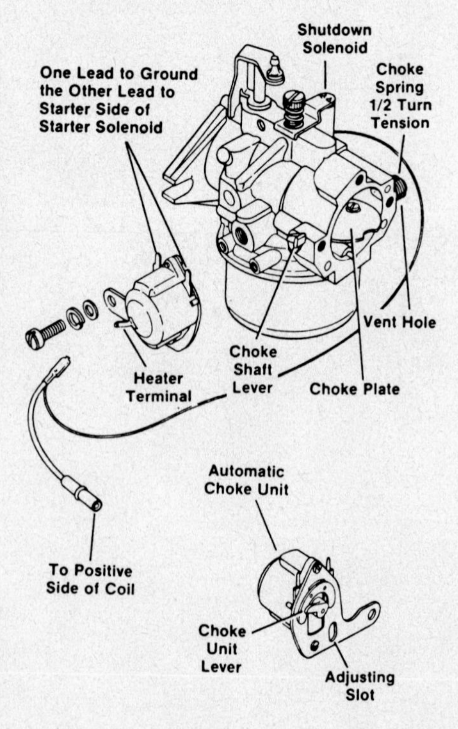

Fig. KO47 – Typical carburetor with Thermo-Electric automatic choke and shutdown control. Refer to text for details.

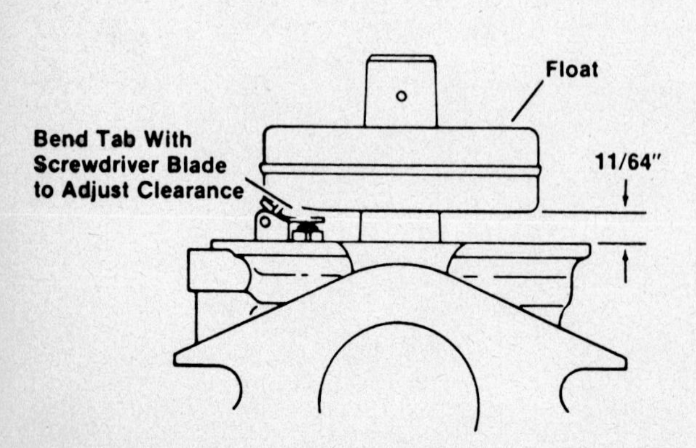

Fig. KO46 – Check float setting by inverting carburetor throttle body and float assembly and measuring distance from free end of float to machined surface of casting. Correct clearance is 11/64-inch (4.366 mm).

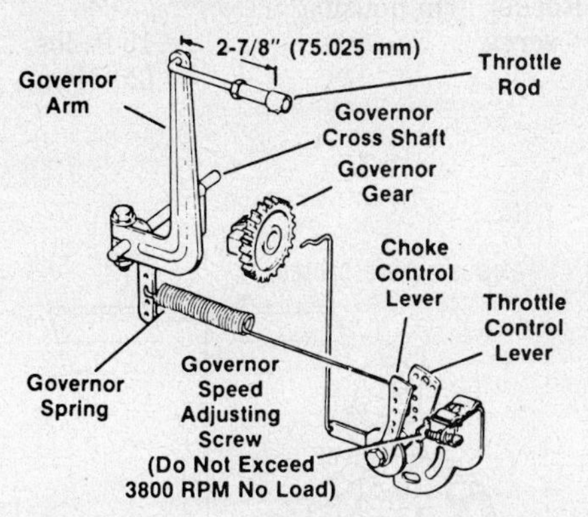

Fig. KO48 – View of typical governor linkage used on K-361 model.

tighten mounting screws. Choke valve should close 5°-10° at 75°F (23.8°C).

With ignition on, shutdown solenoid plunger raises and opens float bowl vent to air horn. With ignition off, solenoid plunger closes vent and stops fuel flow. To check solenoid, remove solenoid and plunger from carburetor body. With lead wire connected, pull plunger approximately 1/4-inch from solenoid and ground casing to engine surface. Turn ignition switch on. Renew solenoid unit if solenoid does not draw plunger in. After renewing solenoid, reset main fuel adjusting screw.

GOVERNOR. Model K-361 is equipped with a centrifugal flyweight mechanical governor. Governor flyweight mechanism is mounted within crankcase and driven by camshaft. Maximum no-load speed is 3800 rpm.

Governor sensitivity can be adjusted by repositioning governor spring in governor arm and throttle control lever. If too sensitive, surging will occur with change of load. If a big drop in speed occurs when normal load is applied, sensitivity should be increased. Normal spring position is in third hole from bottom on governor arm and second hole from top on throttle control lever. To increase sensitivity, move spring end upward on governor arm. Refer to Fig. KO48.

Governor unit is accessible after removing crankshaft and camshaft. Gear and flyweight assembly turn on a stub shaft pressed into crankcase. To remove gear assembly, unscrew governor stop pin and slide gear off stub shaft. Remove governor cross-shaft by unscrewing governor bushing nut and removing shaft from inside. Renew gear assembly, cross-shaft or stub shaft if excessively worn or broken. If stub shaft must be renewed, press new shaft into block until it protrudes 3/8-inch (9.525 mm) above boss. Governor gear to stub shaft clearance is 0.0005-0.002 inch (0.0127-0.0508 mm).

IGNITION TIMING. A battery ignition system with externally mounted breaker points (Fig. KO49) is actuated by camshaft through a push rod. System is equipped with an automatic compression release located on camshaft and does not have an automatic timing advance. Ignition occurs at 20° BTDC at all engine speeds.

Initial breaker point gap is 0.020 inch (0.508 mm) but should be varied to obtain exact ignition timing as follows:

For static timing, disconnect spark plug lead from coil to prevent engine starting. Remove breaker point cover and rotate engine, by hand, in normal direction of rotation (clockwise, from flywheel side). Points should just begin to open when "S" mark appears in center of timing sight hole. Continue rotating engine until points are fully open. Check gap with a feeler gage. Gap setting can vary from 0.017-0.022 inch (0.4318-0.5588 mm), set to achieve smoothest running.

To set timing, using a power timing light, connect light to spark plug and with engine running at 1200-1800 rpm adjust points so "S" mark on flywheel is centered in timing sight hole.

LUBRICATION. A splash lubrication system is used. Oils approved by manufacturer must meet requirements of API service classification SE, SC, CC or SD.

For operation in temperatures above 32°F (0°C), use SAE 30 or SAE 10W-30 oil. For temperatures below 32°F (0°C) use SAE 5W-20 or 5W-30 oil.

It is recommended oil be changed at 25 hour intervals under normal operating conditions. Maintain oil level at "FULL" mark on dipstick but **DO NOT** overfill. Oil capacity of K-361 model is 2 quarts

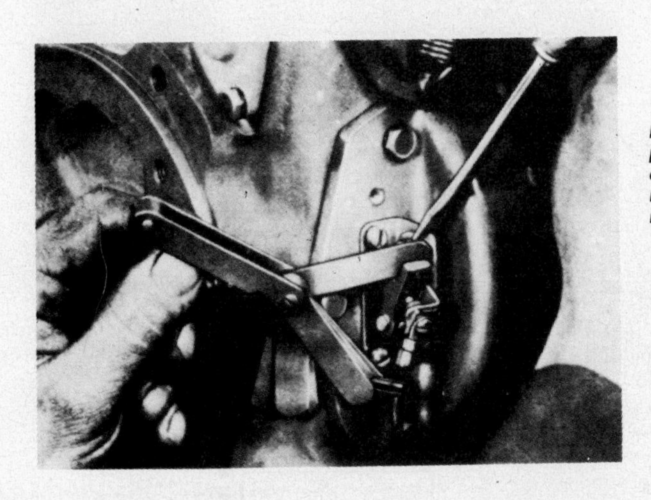

Fig. KO49 – Ignition breaker points are located externally on engine crankcase on all models equipped with magneto or battery ignition systems.

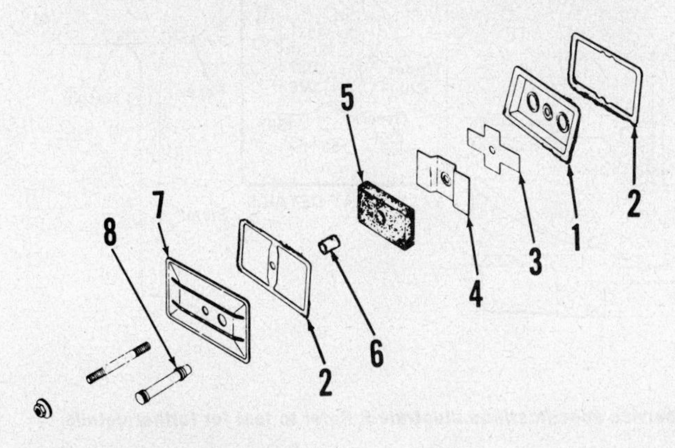

Fig. KO50 – Exploded view of crankcase breather assembly.

1. Breather plate
2. Gasket
3. Reed
4. Baffle
5. Filter
6. Seal
7. Valve cover
8. Breather hose

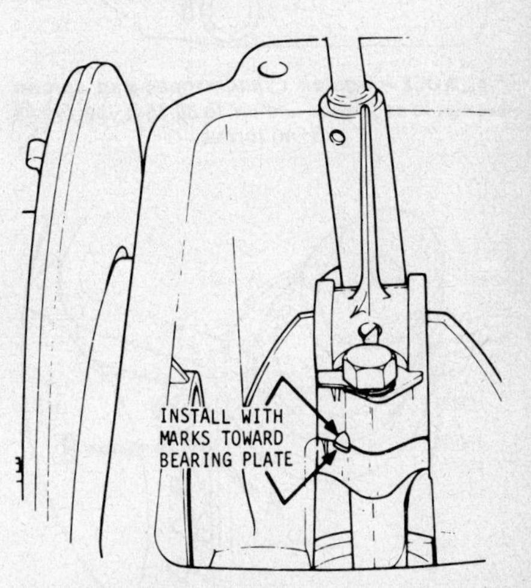

Fig. KO51 – When installing connecting rod and piston unit, be sure marks on rod and cap are aligned and are facing toward flywheel side of engine.

(1.9 L), K-361A model capacity is 1.5 quarts (1.4 L).

AIR CLEANER. Dry element type air cleaner should be cleaned every 50 hours of normal operation or more frequently if operating in dusty conditions. Remove dry element and tap element lightly on a flat surface to remove surface dirt. Do not wash element or attempt to clean element with compressed air. Renew element if extremely dirty or if it is bent, crushed or otherwise damaged. Make certain sealing surfaces of element seal effectively against back plate and cover.

REPAIRS

TIGHTENING TORQUES. Recommended tightening torques, with lightly lubricated threads, are as follows:

Spark plug .... 18-22 ft.-lbs. (24-30 N·m)
Connecting rod cap screws .... 25 ft.-lbs. (34 N·m)
Cylinder head cap screws .... 30-35 ft.-lbs. (40-48 N·m)
Flywheel retainer nut .... 60-70 ft.-lbs. (81-95 N·m)
Fuel pump screws (plastic pump) .... 5.8 ft.-lbs. (7.9 N·m)
Rocker arm housing screw .... 15 ft.-lbs. (20 N·m)

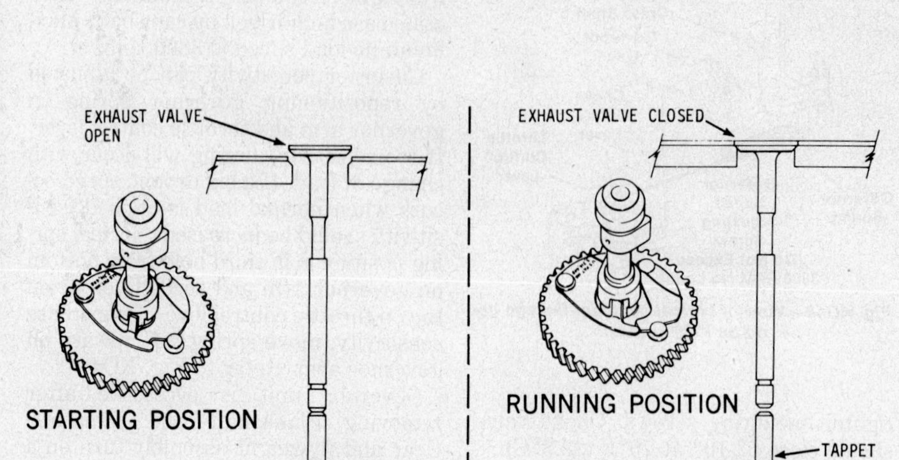

Fig. KO54—Views showing operation of camshaft with automatic compression release. In starting position, spring has moved control lever which moves cam lever upward so tang is above exhaust cam lobe. This tang holds exhaust valve open slightly on a portion of the compression stroke to relieve compression while cranking engine. At engine speeds of 650 rpm or more, centrifugal force moves control lever outward allowing tang to move below lobe surface.

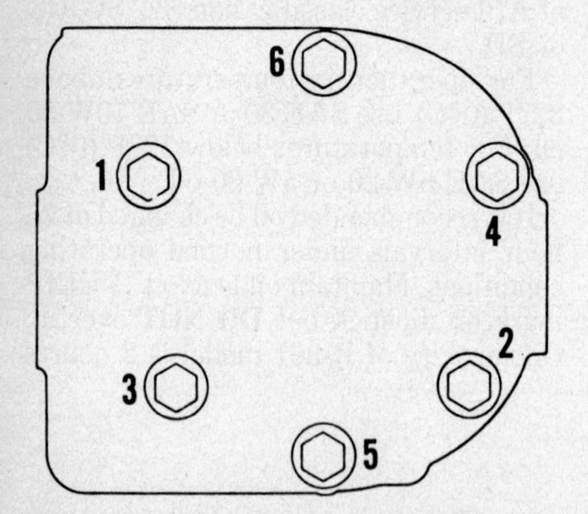

Fig. KO52—Tighten cylinder head cap screws evenly, in sequence shown, to 30-35 ft.-lbs. (40-48 N·m) torque.

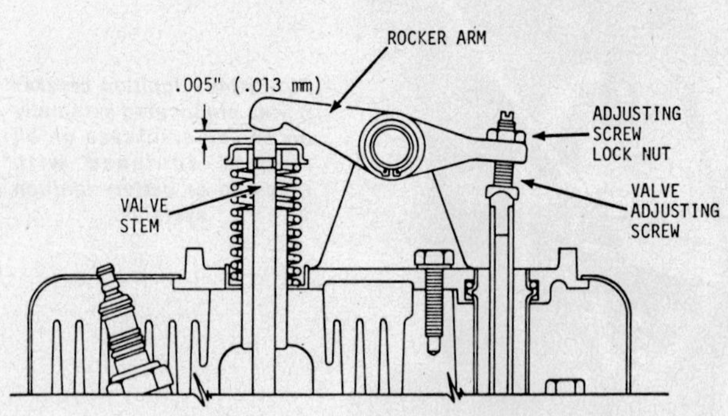

Fig. KO55—View of adjusting points on K-361 valve systems. Both intake and exhaust valve tappet clearance should be adjusted cold, to 0.005 inch (0.127 mm).

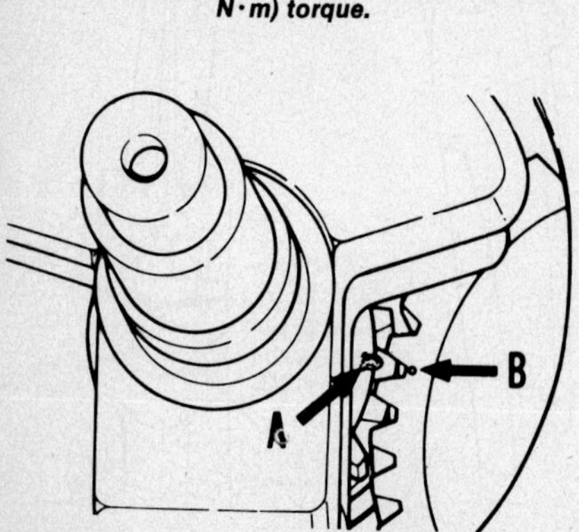

Fig. KO53—When installing crankshaft, make certain timing mark (A) on crankshaft is aligned with timing mark (B) on camshaft gear.

Fig. KO56—Principal valve service specifications illustrated. Refer to text for further details.

CONNECTING ROD. Connecting rod and piston assembly is removed after removing oil pan and cylinder head. Aluminum alloy connecting rod rides directly on crankpin. Connecting rods are available in standard size as well as one for 0.010 inch (0.254 mm) undersize crankshafts. Desired running clearances are as follows:

Connecting rod to
crankpin . . . . 0.001-0.002 in.
(0.0254-0.0508 mm)
Connecting rod to
piston pin . . . . 0.0003-0.0008 in.
(0.0076-0.0203 mm)
Rod side play
on crankpin . . . . 0.007-0.0175 in.
(0.1778-0.4445 mm)

Standard crankpin diameter is 1.4995-1.500 inch (38.09-38.10 mm).

When reinstalling connecting rod and piston assembly, piston can be installed either way on rod, but make certain match marks (Fig. KO51) on rod and cap are aligned and are towards flywheel side of engine. Kohler recommends connecting rod cap screws be torqued to 30 ft.-lbs. (40 N•m), loosened and re-torqued to 25 ft.-lbs (34 N•m). Cap screw threads should be lightly lubricated before installation.

PISTON, PIN AND RINGS. Aluminum alloy piston is fitted with two 0.093 inch (2.3622 mm) wide compression rings and one 0.187 inch (4.7498 mm) wide oil control ring. Renew piston if scored or if side clearance of new ring in piston top groove exceeds 0.005 inch (0.127 mm). Pistons and rings are available in oversizes of 0.010, 0.020 and 0.030 inch (0.254, 0.508 and 0.762 mm) as well as standard size. Piston pin fit in piston bore should be from 0.0001 inch (0.0025 mm) interference to 0.0003 inch (0.0076 mm) loose. Standard piston pin diameter is 0.8753 inch (22.23 mm).

Recommended piston to cylinder bore clearance measured at thrust side at bottom of skirt is 0.003-0.0045 inch (0.0762-0.1143 mm). Recommended clearance when measured at thrust side just below oil ring is 0.007-0.0095 inch (0.1778-0.2413 mm).

Kohler recommends piston rings always be renewed if they are removed.

Piston ring specifications are as follows:

Ring end gap . . . . 0.010-0.020 in.
(0.254-0.508 mm)
Ring side clearance–
Compression
ring . . . . 0.002-0.004 in.
(0.0508-0.1016 mm)
Oil control
ring . . . . 0.001-0.003 in.
(0.0254-0.0762 mm)

If compression ring has a groove or bevel on outside surface, install ring with groove or bevel down. If groove or bevel is on inside surface of compression ring, install ring with groove or bevel up. Oil control ring can be installed either side up.

CYLINDER BLOCK. If cylinder wall is scored or bore is tapered more than 0.0015 inch (0.0381 mm) or out-of-round more than 0.005 inch (0.127 mm), cylinder should be bored to nearest suitable oversize of 0.010, 0.020 or 0.030 inch (0.254, 0.508 or 0.762 mm). Standard cylinder bore is 3.750 inches (95.25 mm).

CYLINDER HEAD. Always use a new head gasket when installing cylinder head. Lightly lubricate cylinder head cap screws and tighten them evenly and in equal steps using sequence shown in Fig. KO52.

CRANKSHAFT. Crankshaft is supported by two ball bearings. Bearings have an interference fit with cylinder block of 0.0006-0.0022 inch (0.0152-0.0559 mm) and with bearing plate of 0.0012-0.0028 inch (0.0305-0.0711 mm). Ball bearing to crankshaft clearance is 0.0004 inch (0.0102 mm) interference to 0.0005 inch (0.0127 mm) loose.

Renew ball bearings if excessively loose or rough. Crankshaft end play should be 0.003-0.020 inch (0.0762-0.508 mm) and is controlled by varying thickness of bearing plate gaskets. Bearing plate gaskets are available in a variety of thicknesses.

Standard crankpin journal diameter is 1.4995-1.500 inches (38.09-38.10 mm) and may be reground to 0.010 inch (0.254 mm) undersize.

When installing crankshaft, align timing marks on crankshaft and camshaft gears as shown in Fig. KO53. Refer to **DYNAMIC BALANCER** paragraph for installation and timing of balancer gears.

Kohler recommends crankshaft seals be installed in crankcase and bearing plate after crankshaft and bearing plate are installed. Carefully work oil seals over crankshaft and drive seals into place with hollow driver that contacts outer edge of seals.

CAMSHAFT. The hollow camshaft and integral camshaft gear turn on a pin that is a slip fit in flywheel side of crankcase and a drive fit in closed side of crankcase. Remove and install pin from open side (bearing plate side) of crankcase. Desired camshaft to pin running clearance is 0.001-0.0035 inch (0.0254-0.0889 mm). Desired camshaft

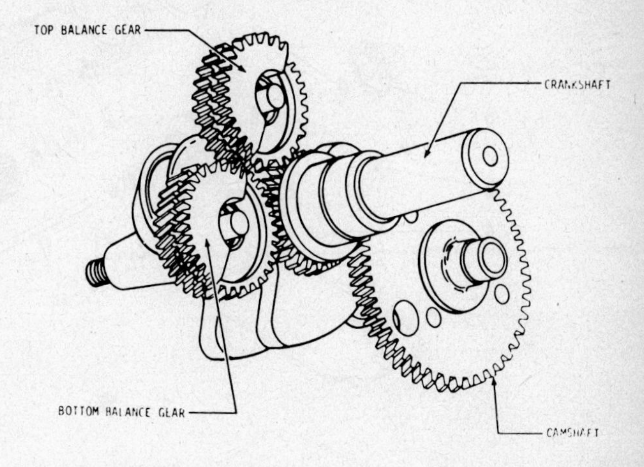

Fig. KO57—View showing components of dynamic balancer system.

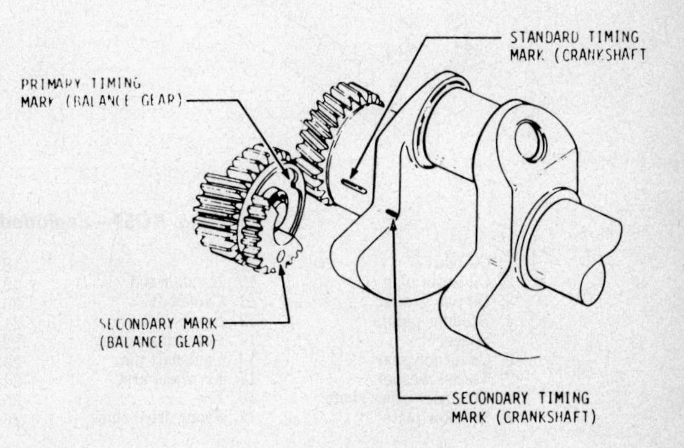

Fig. KO58—View showing timing marks for installing dynamic balance gears.

end play of 0.005-0.010 inch (0.127-0.254 mm) is controlled by use of 0.005 and 0.010 inch (0.127 and 0.254 mm) thick spacer washers between camshaft and cylinder block at bearing plate side of crankcase.

Camshaft is equipped with automatic compression release shown in Fig. KO54. Automatic compression release mechanism holds exhaust valve open during first part of compression stroke, reducing compression pressure and allowing easier cranking. Refer to Fig. KO54 for operational details. At speeds above 650 rpm, release mechanism is inactive. Release mechanism weights and weight pivot pins are not renewable separately from camshaft gear, but weight spring is available for service.

To check compression on engine equipped with automatic compression release, engine must be cranked at 650 rpm or higher to overcome release mechanism. A reading can also be obtained by rotating flywheel in reverse directon with throttle wide open. Compression reading for engine in good condition should be 110-120 psi (758-827 kPa). When reading falls below 100 psi (690 kPa), it indicates leaking rings or valves.

VALVE SYSTEM. Valve seats are hardened steel inserts cast into head. If seats become worn or damaged entire cylinder head must be renewed. Refer to Fig. KO56 for valve seat and guide details.

Valve stem to guide operating clearance should be 0.0029-0.0056 inch (0.0762-0.1422 mm) for exhaust valve and 0.001-0.0027 inch (0.0254-0.0686 mm) for intake valve.

Intake and exhaust valve clearance should be adjusted to 0.005 inch (0.127

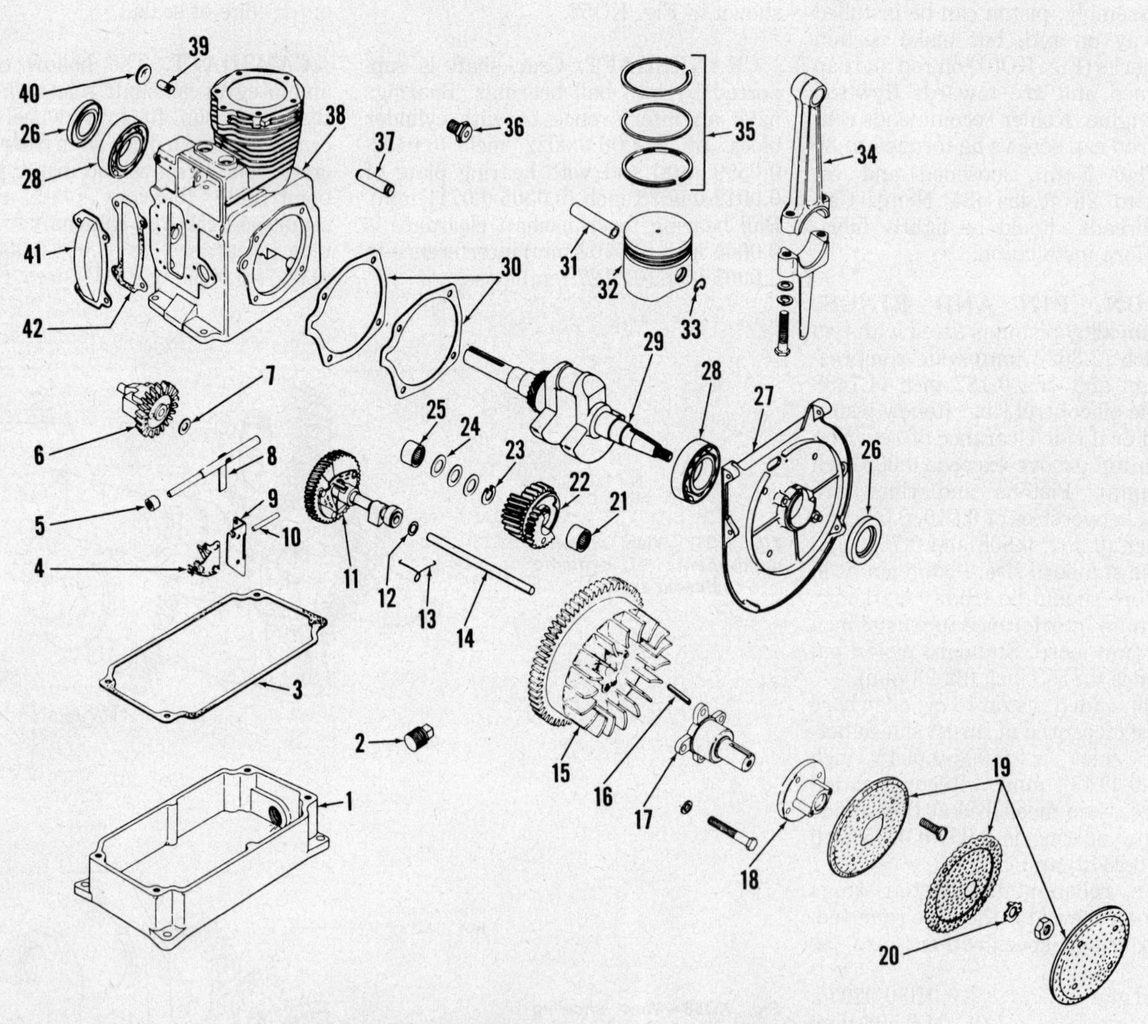

Fig. KO59—Exploded view of K-361 basic engine assembly.

1. Oil pan
2. Oil drain plug
3. Oil pan gasket
4. Breaker points
5. Spacer
6. Governor gear
7. Thrust washer
8. Governor cross-shaft
9. Breaker plate
10. Breaker rod
11. Camshaft
12. Spacer
13. Actuating spring
14. Camshaft pin
15. Flywheel assy.
16. Key
17. Front drive shaft
18. Front drive adapter
19. Grass screen
20. Flywheel washer
21. Needle bearing
22. Balancer gear (2 used)
23. Retainer
24. Spacer
25. Needle bearing
26. Oil seal
27. Bearing plate
28. Main ball bearing
29. Crankshaft
30. Gasket
31. Piston pin
32. Piston
33. Retainer
34. Connecting rod assy.
35. Piston rings
36. Bushing
37. Shaft
38. Cylinder block assy.
39. Governor shaft
40. Expansion plug
41. Camshaft cover
42. Camshaft cover gasket

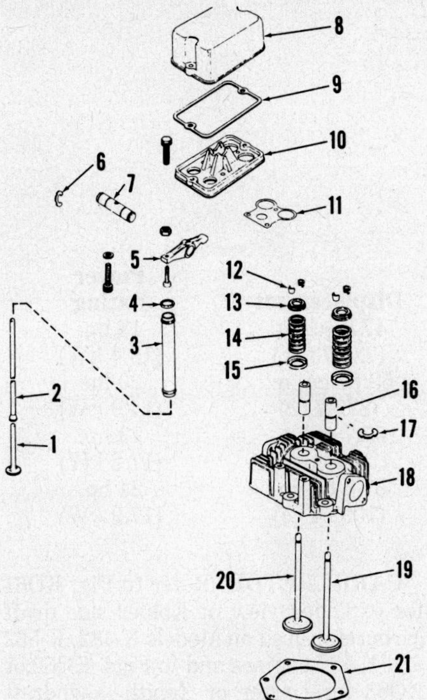

Fig. KO60—Exploded view of overhead valve system used on K-361 engines.

1. Tappet
2. Push rod
3. Push rod tube
4. "O" rings
5. Rocker arm
6. Retainer ring
7. Rocker arm shaft
8. Valve cover
9. Gasket
10. Rocker arm housing
11. Housing gasket
12. Valve keepers
13. Valve rotators
14. Valve springs
15. Spring retainers
16. Valve guides
17. Retainer ring
18. Cylinder head
19. Intake valve
20. Exhaust valve
21. Head gasket

mm) while cold. To adjust, disconnect spark plug wire and rotate engine by hand until both valves are seated and piston is at TDC. Loosen adjusting screw locknut (Fig. KO55) and turn adjusting screw in or out to obtain correct clearance. Tighten locknuts and recheck clearance.

DYNAMIC BALANCER. A dynamic balance system (Fig. KO57) is used on all K-361 engines. The two balance gears, equipped with needle bearings, rotate on two stub shafts which are pressed into bosses on pto side of crankcase. Snap rings secure gears on stub shafts and shim spacers are used to control gear end play. Balancer gears are driven by crankshaft in opposite direction of crankshaft rotation. Use following procedure to install and time dynamic balancer components.

To install new stub shafts, press shafts into special bosses in crankcase until they protrude 1.110 inches (28.194 mm) above thrust surface of bosses.

To install top balance gear-bearing assembly first place one 3/8-inch spacer and one 0.010 inch shim spacer on stub shaft, then slide top gear assembly on shaft. Timing marks must face flywheel side of crankcase. Install one 0.005, one 0.010 and one 0.020 inch shim spacers in this order, then install snap ring. Using a feeler gage, check gear end play. Correct end play of balance gear is 0.002-0.010 inch (0.0508-0.254 mm). Add or remove 0.005 inch thick spacers as necessary to obtain correct end play. Always make certain a 0.020 inch thick spacer is next to snap ring.

Install crankshaft in crankcase and align primary timing mark on top balance gear with standard timing mark on crankshaft. See Fig. KO58. With primary timing marks aligned, engage crankshaft gear 1/16-inch (1.588 mm) into narrow section of top balancer gear. Rotate crankshaft to align timing marks on camshaft gear and crankshaft as shown in Fig. KO53. Press crankshaft into crankcase until it is seated firmly into ball bearing main.

Rotate crankshaft until crankpin is approximately 15° past bottom dead center. Install one 3/8-inch spacer and one 0.010 inch shim spacer on stub shaft. Align secondary timing mark on bottom balance gear with secondary timing mark on crankshaft. See Fig. KO58. Slide gear assembly into position on stub shaft. If properly timed, secondary timing mark on bottom balance gear will be aligned with standard timing mark on crankshaft after gear is fully on stub shaft. Install one 0.005 inch and one 0.020 inch shim spacer, then install snap ring. Check bottom balance gear end play and add or remove 0.005 inch thick spacers as required to obtain proper end play of 0.002-0.010 inch (0.0508-0.254 mm).

Always make certain a 0.020 inch thick spacer is next to snap ring.

KOHLER

KOHLER COMPANY
Kohler, Wisconsin 53044

Model	No. Cyls.	Bore	Stroke	Displacement	Power Rating
K-482	2	3.250 in.	2.875 in.	47.8 cu. in.	18 hp.
		(82.55 mm)	(73.03 mm)	(781.73 cc)	(13.4 kW)
K-532	2	3.375 in.	3.0 in.	53.67 cu. in.	20 hp.
		(85.73 mm)	(76.2 mm)	(879.7 cc)	(14.9 kW)
K-582	2	3.5 in.	3.0 in.	57.7 cu. in.	23 hp.
		(88.9 mm)	(76.2 mm)	(946 cc)	(17.2 kW)
K-662	2	3.625 in.	3.250 in.	67.2 cu. in.	24 hp.
		(92.08 mm)	(82.55 mm)	(1099.4 cc)	(17.9 kW)

All engines in this section are four cycle, two cylinder opposed, horizontal crankshaft engines. Crankshaft for K-482, K-532 and K-582 models is supported by a ball bearing at output end and a sleeve type bearing at flywheel end of crankshaft. Crankshaft for K-662 model is supported at each end by tapered roller bearings.

Connecting rod for K-482, K-532 and K-582 models rides directly on crankpin journal and K-662 model connecting rod is equipped with a renewable type precision bearing insert. Pressure lubrication is provided by a gear type oil pump driven by crankshaft gear.

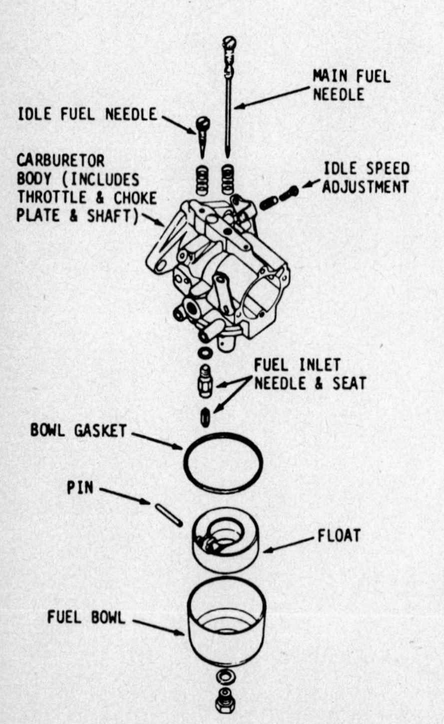

Fig. KO61—Exploded view of Kohler carburetor used on K-482, K-532 and K-582 models.

Either a battery type or magneto type ignition system is used according to model and application.

A side draft, float type carburetor is used for K-482, K-532 and K-582 models and a down draft, float type carburetor is used on K-662 model.

Engine model, serial and specification numbers must be given when ordering parts or service material.

MAINTENANCE

SPARK PLUG. Recommended spark plug is Champion H10 or equivalent for K-482, K-532 and K-582 models or Champion J8 or equivalent for K-662 model. Electrode gap is 0.025 inch (0.635 mm) for all models except K-582. Electrode gap for K-582 is 0.035 inch (0.889 mm).

CARBURETOR. Refer to Fig. KO61 for exploded view of Kohler side draft carburetor used on Models K-482, K-532 and K-582 engines and to Figs. KO62 or KO63 for Carter or Zenith downdraft carburetors used on Model K-662 engines. For initial adjustment on all models, open idle fuel needle 1-1/4 turns and open main fuel needle 2 turns. Make final adjustment with engine at operating temperature and running. Operate engine under load and adjust main fuel needle for leanest mixture that will allow satisfactory acceleration and steady governor operation.

Adjust idle speed stop screw to maintain a low idle speed of 1000 rpm. Then, adjust idle mixture needle for smoothest idle operation.

Since main fuel and idle fuel adjustments affect each other, recheck

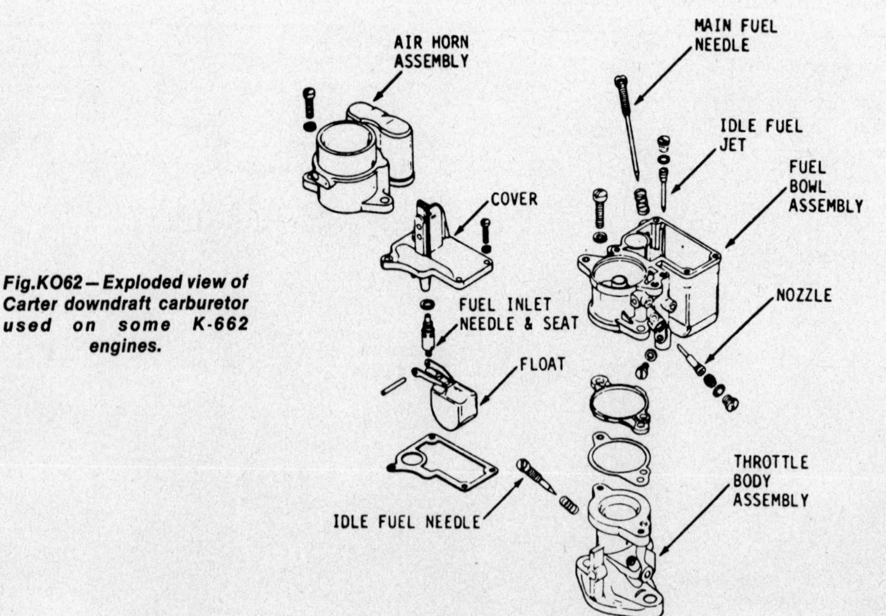

Fig.KO62—Exploded view of Carter downdraft carburetor used on some K-662 engines.

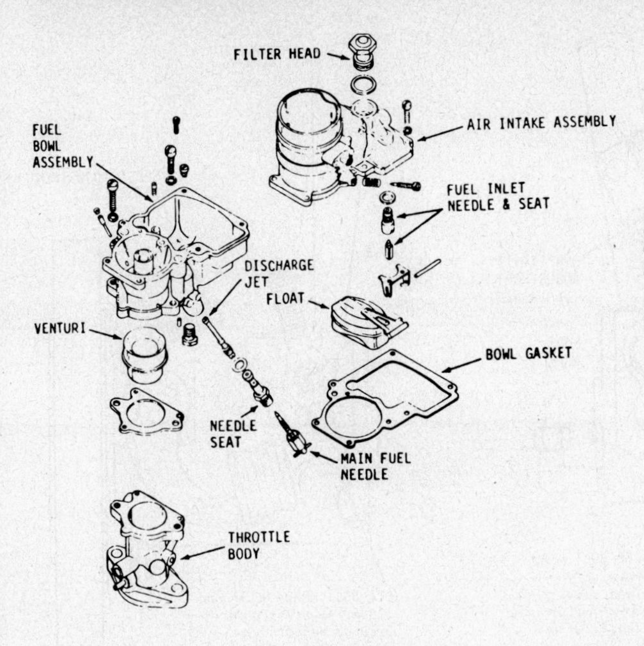

Fig. KO63—Exploded view of Zenith downdraft carburetor used on some K-662 engines.

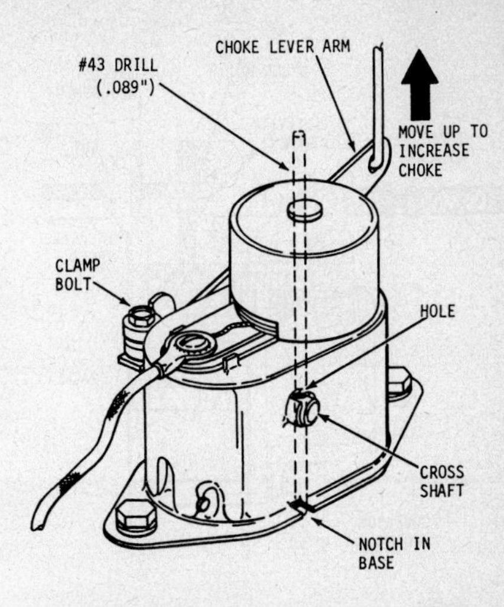

Fig. KO63B—To adjust choke, remove air cleaner and insert round rod (number 43 drill) through cross-shaft until it engages notch in base. Loosen clamp bolt on choke lever and adjust choke plate to desired setting. Tighten clamp bolt.

engine operation and readjust fuel needles as necessary.

To check float level, invert carburetor body and float assembly. There should be 11/64-inch (4.366 mm) clearance between machined surface of body casting and free end of float on Kohler carburetor used on Models K-482, K-532 and K-582. On Carter or Zenith carburetors used on Model K-662, there should be a distance of 1 1/2 inches (38.1 mm) between machined surface of casting and top of float. Adjust as necessary by bending float lever tang that contacts inlet valve.

AUTOMATIC CHOKE. One of two different types automatic choke may be used according to model and application. One is an integral part of the downdraft carburetor while the other is mounted on exhaust manifold and choke plate is controlled through external linkage.

To adjust type which is integral with carburetor, loosen the three screws (Fig. KO63A) and rotate housing in clockwise direction to close, or counterclockwise to open.

To adjust exhaust manifold mounted type, remove air cleaner and move choke arm (Fig. KO63B) until hole in brass shaft aligns with slot in bearing and insert a small round rod as shown. Loosen clamp bolt on choke lever and adjust choke plate to desired position. Tighten clamp bolt and remove rod.

The electrical lead on this choke is connected so current flows to thermostatic element when ignition is switched on. Tension of thermostatic spring should be set to allow full choke at starting. Current through heating element controls tension of thermostatic spring, which gradually opens choke as engine warms up.

Choke on all models should be fully open when engine has reached operating temperature.

GOVERNOR. An externally mounted centrifugal flyweight type governor is used on all engines. Governor is driven by the camshaft gear and is lubricated by an external oil line connected to engine lubrication system.

GOVERNOR INSTALLATION AND TIMING (K-482, K-532 and K-582). Ignition breaker points are mounted on governor housing on Models K-482,

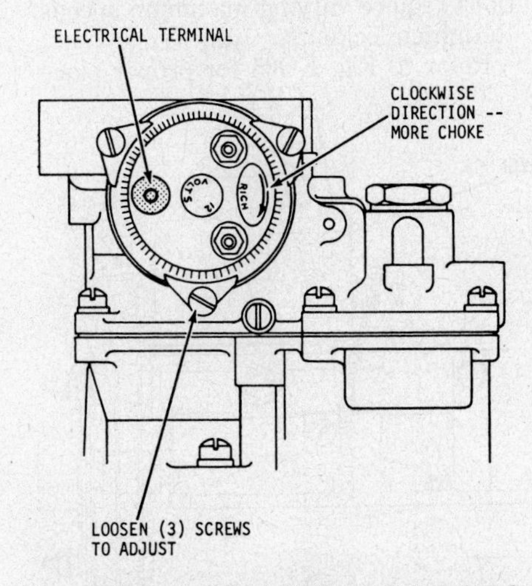

Fig. KO63A—To adjust choke which is integral part of carburetor, loosen the three screws and rotate housing clockwise to increase or counterclockwise to decrease choke action.

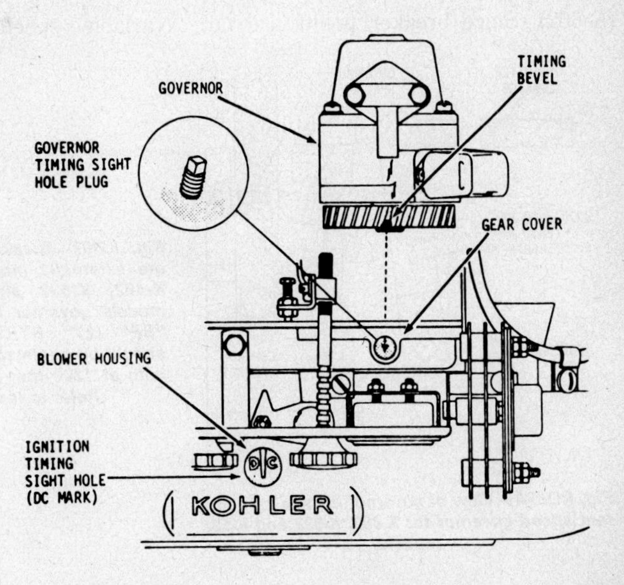

Fig. KO64—Top view showing location of sight holes and timing marks used when installing governor assembly on K-482, K-532 or K-582 models.

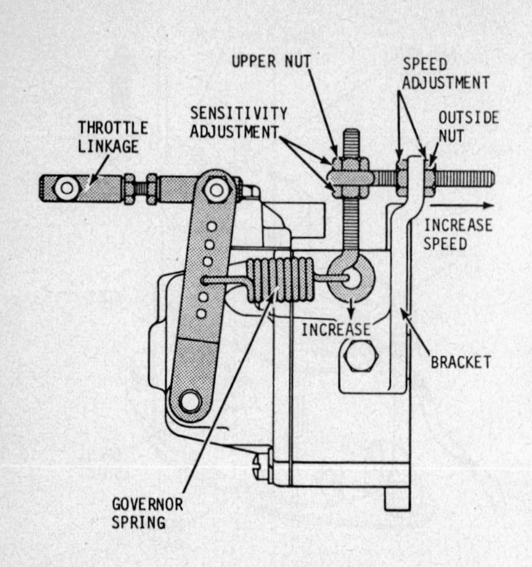

Fig. KO65 — View of governor linkage, maximum speed stop set screw and governor spring used on K-482, K-532 and K-582 models. Governor sensitivity is adjusted by moving governor spring to alternate holes.

K-532 and K-582 and are activated by a push rod which rides on a cam on governor drive shaft. Therefore, governor must be timed to engine.

To install governor assembly, first uncover ignition timing sight hole in top center of blower housing. See Fig. KO64. Rotate engine crankshaft until "DC" mark on flywheel is centered in sight hole. Remove governor timing sight hole plug from engine gear cover. Install governor assembly so tooth with special timing bevel on governor gear is centered in governor timing sight hole. Install governor retaining cap screws. Recheck to make certain both the "DC" mark on flywheel and beveled tooth on governor gear are centered in sight holes, then install sight hole plug and cover.

After governor is installed, readjust ignition timing as outlined in **IGNITION TIMING** paragraph.

GOVERNOR INSTALLATON (K-662). Since breaker points are not

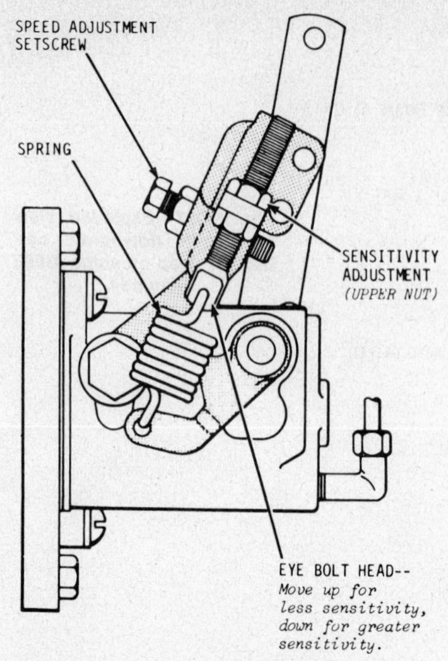

Fig. KO66 — View of governor spring, maximum speed adjustment set screw and sensitivity adjustment on K-662 model governor.

mounted on or driven off governor on Model K-662, governor does not have to be timed to the engine. Installation of governor assembly is obvious.

GOVERNOR ADJUSTMENT. Different governor adjustment procedures must be followed for models running at constant speed than for models running at variable speed. Model K-662 has a slightly different adjustment procedure due to the use of a different governor assembly. Make certain correct procedure is followed for model being serviced and set idle and high speed adjustments to manufacturers recommendation for application for which engine is used.

To adjust governor on all models with variable speed application use the

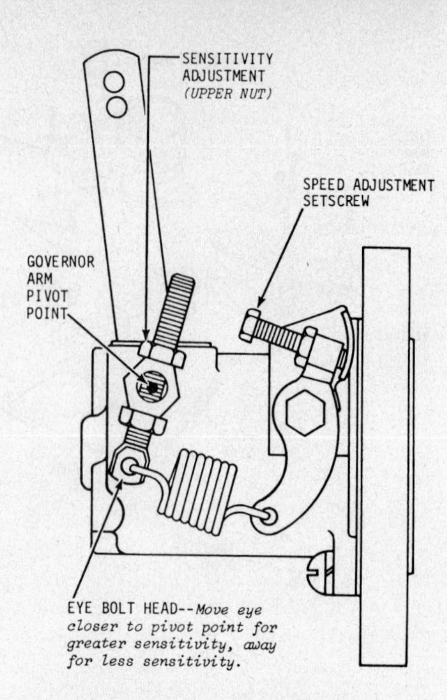

Fig. KO66A — View of K-662 governor linkage showing various adjustment points.

following procedure: First move governor arm forward to high idle position and check to see that throttle linkage moves carburetor throttle shaft to wide open position. If not, adjust length of throttle linkage as necessary. Make certain throttle linkage moves freely.

Start engine and move speed control lever to high idle position. Using a tachometer, check engine speed. Loosen locknuts and adjust high speed stop set screw on speed control bracket (Fig. KO65 or KO66) to obtain maximum no-load speed shown in operators manual and make certain allowable maximum speed for each different engine application is not exceeded, varying applications require varying maximum speeds. Retighten locknuts.

Refer to Fig. KO65 for proper place-

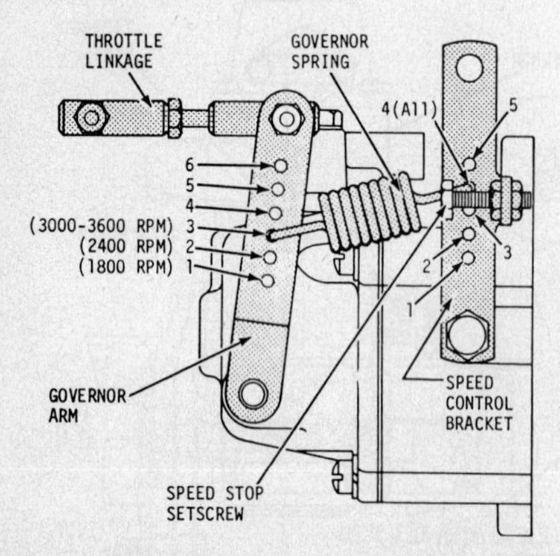

Fig. KO65A — View of governor linkage on constant speed governor for K-482, K-532 and K-582 models.

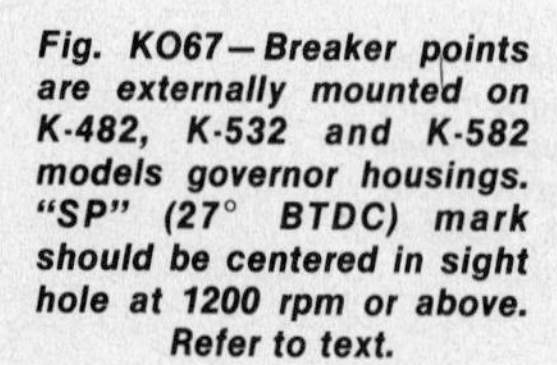

Fig. KO67 — Breaker points are externally mounted on K-482, K-532 and K-582 models governor housings. "SP" (27° BTDC) mark should be centered in sight hole at 1200 rpm or above. Refer to text.

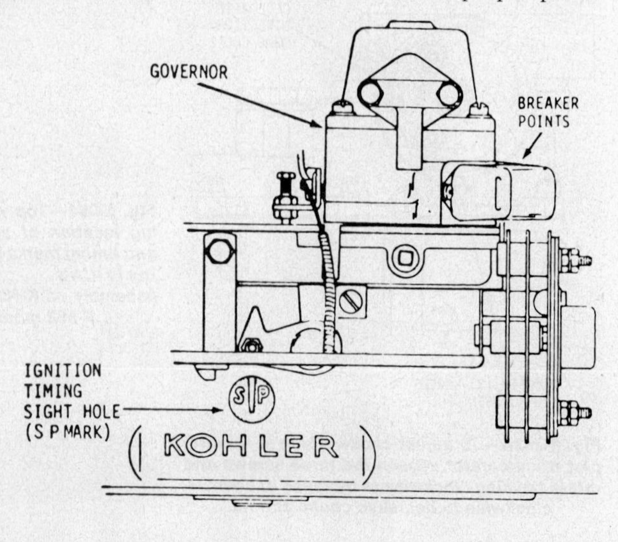

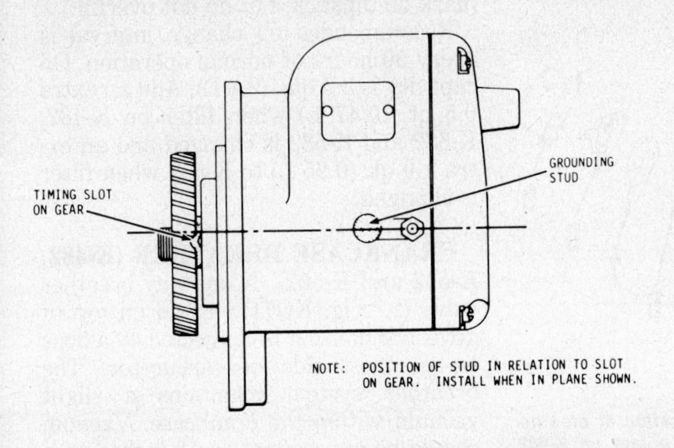

Fig. KO68—View of standard (simultaneous firing) magneto used on some K-662 engines.

ment of governor spring in hole of governor arm and control bracket. If governor surging or hunting occurs, governor is too sensitive and governor spring should be moved to a set of holes closer together. If a big drop in speed occurs when normal load is applied, governor should be set for greater sensitivity. Increase spring tension by placing spring in holes spaced further apart. Move spring one hole at a time and recheck engine operation and high idle rpm after each adjustment.

On Model K-662, governor sensitivity is adjusted by loosening one nut and tightening opposite nut on spring eye bolt. See Fig. KO66. If governor hunting or surging occurs, governor is too sensitive. To correct, loosen bottom nut and tighten top nut. To increase sensitivity, loosen top nut and tighten bottom nut. After adjustment, recheck engine high idle rpm as changing sensitivity will affect high idle speed.

To adjust governor on all models except K-662, constant speed application, use the following procedure: Remove air cleaner and note throttle shaft must be a certain length to maintain a fixed operating speed such as 1800 rpm for 60 cycle electric generators. Any changes in length will result in changes in generator output, therefore only slight readjustment of speed is recommended and always check generator for correct output after adjustments. To increase speed, loosen speed adjusting nuts (Fig. KO65A) and draw eye of bolt closer to the bracket. Reverse procedure to decrease speed. Make certain all adjusting nuts are tight after final adjustments.

Sensitivity is varied by changing length of sensitivity eye bolt. Move bolt downward to increase sensitivity and upward to decrease sensitivity.

On K-662 models, refer to Fig. KO66A. To increase sensitivity draw eye of bolt closer to pivot point and to decrease sensitivity move eye of bolt away from pivot point.

On all models recheck speeds and check output of generators if applicable.

IGNITION TIMING (K-482, K-532 and K-582). Ignition points are externally mounted on governor housing and are activiated by a push rod which rides on a cam on governor drive shaft. An automatic spark advance-retard mechanism is incorporated in the governor which allows engine to start at 8° BTDC and run (1200 rpm or above) at 27° BTDC.

To adjust ignition timing, first remove breaker point cover and adjust breaker contact gap to 0.020 inch (0.508 mm). Install breaker point cover, then remove cover from ignition timing sight hole in top center of blower housing. Attach timing light to number "1" spark plug (cylinder nearest to flywheel). Start engine and operate at 1200 rpm or above. Aim timing light at ignition timing sight hole. The "SP" mark on flywheel should be centered in sight hole when light flashes. See Fig. KO67. If "SP" mark is not centered in sight hole, loosen governor mounting cap screws and rotate governor assembly until the "SP" (27° BTDC) mark is centered as light flashes. Retighten governor mounting cap screws.

If slotted holes in governor flange will not allow enough rotation to center the timing mark, check governor timing as outlined in **GOVERNOR INSTALLATION** and **TIMING** paragraph.

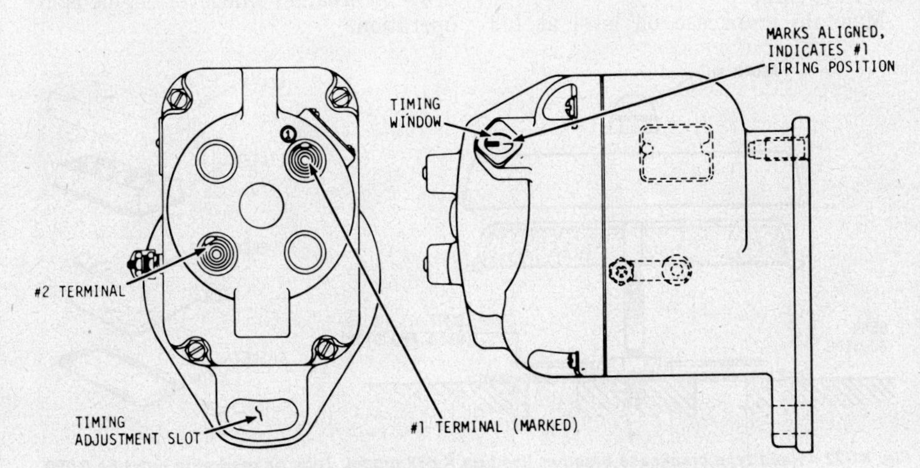

Fig. KO69—View of distributor type magneto used on some K-662 engines.

IGNITION TIMING (K-662). Two types of external, self-contained magnetos have been used on Model K-662 engines. Standard magneto (Fig. KO68) is referred to as a simultaneous firing magneto. This type fires both spark plugs at the same time. Ignition occurs only in cylinder in which piston is on compression stroke. At this time, piston in opposite cylinder is on exhaust stroke and that spark is ineffective.

The other magneto (Fig. KO69) is classified as a distributor type magneto. This magneto is equipped with a distributor rotor which directs ignition voltage to the proper spark plug at appropriate time.

Breaker point gap on either type magneto is 0.015 inch (0.381 mm).

To install and time standard type magneto, first crank engine over slowly by hand until "DC" mark on flywheel is centered in timing sight hole (A–Fig. KO70). Rotate magneto drive gear until impulse coupling trips. Then turn gear backwards until timing slot in gear is aligned with grounding stud on case as shown in Fig. KO68. Using a new flange gasket, carefully install magneto assembly. Tighten retaining cap screws securely. Slowly crank engine one revolution until impulse coupling trips. At this time, "DC" mark on flywheel should be just past center of timing sight hole. To

Fig. KO70—Location of ignition timing sight hole (A) on K-662 model engine.

check and adjust running timing, attach a timing light, start engine and operate at 1200 rpm or above. Aim timing light at timing sight hole (A – Fig. KO70). The "SP" mark on flywheel should be centered in sight hole when timing light flashes. If not, loosen magneto mounting cap screws and rotate magneto assembly until the "SP" (22° BTDC) mark is centered in sight hole. Tighten mounting cap screws.

To install and time distributor type magneto (Fig. KO69), first remove spark plug from No. 1 cylinder (cylinder closest to flywheel). Slowly crank engine over until No. 1 piston is on compression stroke and "M" mark on engine flywheel is centered in ignition sight hole (A – Fig. KO70). Turn magneto drive gear counter-clockwise (facing gear) until white mark is centered in timing window on magneto. See Fig. KO69. Using a new flange gasket, carefully install magneto assembly. Tighten retaining cap screws securely. Slowly crank engine in normal direction of rotation until impulse coupling trips. At this time, "DC" mark on flywheel should be just in center of timing sight hole. To check and adjust running timing, attach a timing light, start engine and operate at 1200 rpm or above. Aim timing light at timing sight hole (A-Fig. KO70). The "SP" (22° BTDC) mark should be centered in sight hole. Tighten mounting cap screws.

COMPRESSION TEST. Results of a compression test can be used to partially determine the condition of the engine. To check compression, remove spark plugs, make certain air cleaner is clean and set throttle and choke in wide open position. Crank engine with starting motor to a speed of about 1000 rpm. Insert gage in spark plug hole and take several readings on both cylinders. Consistent readings in the 110-120 psi (758-827 kPa) range indicate good compression. When compression reading falls below 100 psi (690 kPa) on either cylinder, valve leakage or excessive wear in cylinder, or ring wear are indicated.

If compression reading is higher than 120 psi (827 kPa), it indicates excessive carbon deposits have built up in combustion chamber. Remove cylinder heads and clean carbon from heads.

LUBRICATION. A gear type oil pump supplies oil through internal galleries to front main bearing, camshaft bearings, connecting rods and other wear areas. A full flow, spin-on type oil filter, located on crankcase in front of number "2" cylinder on Models K-482, K-532 and K-582, or just to the rear of number "1" cylinder on Model K-662, or a remote mounted cannister type oil filter connected to engine via two pressure hoses located on left side of engine block, are used in the oil pressure system. Normal oil pressure for all models when operating at normal temperature should be 25 psi (172 kPa) at 1200 rpm, 30-50 psi (207-345 kPa) at 1800 rpm, 35-55 psi (241-379 kPa) at 2200 rpm and 45-65 psi (310-448 kPa) at 3200-3600 rpm. On Models K-482, K-532 and K-582, an oil pressure relief valve is located on crankcase just forward of number "1" cylinder (cylinder nearest to flywheel). To adjust oil pressure, loosen locknut and turn adjusting screw clockwise to increase pressure or counter-clockwise to decrease pressure. Retighten locknut.

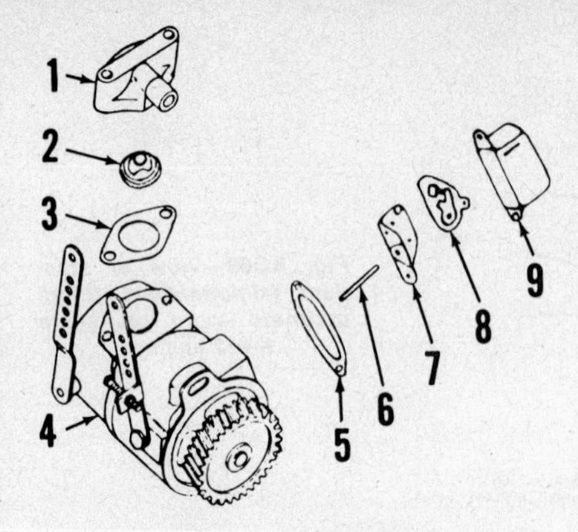

Fig. KO71 – View showing location of breather valve assembly and breaker points on K-482, K-532 or K-582 models governor assembly. Breather valve (2) and valve housing (1) are available as an assembly.

1. Valve housing
2. Breather valve
3. Gasket
4. Governor assy.
5. Gasket
6. Breaker push rod
7. Bracket
8. Breaker points
9. Cover

High quality detergent type oil having API classification "SC", "SD", "SE" or "SF" are recommended for use in Kohler engines. Use SAE 30 oil if temperature is above 32° F. (0° C) and use SAE 5W-20 or 5W-30 if temperature is below 32° F. (0° C).

Maintain crankcase oil level at full mark on dipstick, but do not overfill.

Recommended oil change interval is every 50 hours of normal operation. Oil capacity is 3.0 qts. (2.8 L). Add an extra 0.5 qt. (0.47 L) when filter on K-482, K-532 and K-582 is changed and an extra 1.0 qt. (0.95 L) to K-662 when filter is changed.

CRANKCASE BREATHER (K-482, K-532 and K-582). A one-way breather valve (2 – Fig. KO71) located on top of governor housing is connected by a hose to air inlet side of carburetor. The breather system maintains a slight vacuum within the crankcase. Vacuum should be approximately 12 inches on a water manometer with engine operating at 3600 rpm. If manometer shows crankcase pressure, breather valve is faulty, oil seals or gaskets are worn or damaged or engine has excessive blow-by due to worn rings.

Breather valve is pressed into valve housing (1) and is available only as an assembly with the housing.

CRANKCASE BREATHER (K-662). A reed type breather valve (Fig. KO72) located on top of gear cover maintains a partial vacuum within engine crankcase. Vacuum should be approximately 16 inches on a water manometer with engine operating at 3200 rpm. If manometer shows crankcase pressure, breather valve is faulty, oil seals or gaskets are worn or damaged, or engine has excessive blow-by due to worn rings.

Breather components must be correctly installed to function properly. See Fig. KO72. Reed must be free of rust, dents or cracks and must lay flat on breather ports in gear cover. Tabs on reed stop must be 0.020 inch (0.508 mm) above reed.

A closed system was also used on some K-662 models and all parts are the same except for a special housing which is connected to air inlet side of carburetor by a tube and provides a positive draw on breather whenever engine is in operation.

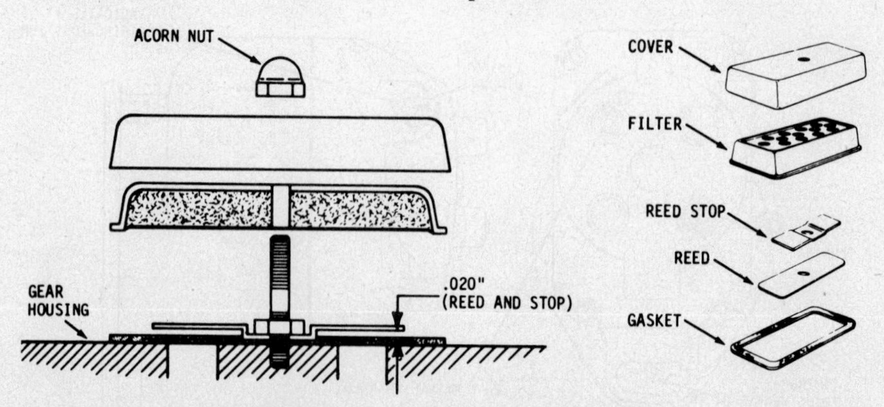

Fig. KO72 – Reed type crankcase breather used on K-662 model. Tabs on reed stop must be 0.020 inch (0.508 mm) above reed as shown.

REPAIRS

TIGHTENING TORQUES. Recommended tightening torques with lightly lubricated threads are as follows:

Spark plug .... 22 ft.-lbs. (30 N·m)
Camshaft nut–
K-482, K-532 & K-582...... 40 ft.-lbs. (54 N·m)
K-662 .... 25 ft.-lbs. (34 N·m)
Connecting rod cap screws–
K-482, K-532 & K-582...... 25 ft.-lbs. (34 N·m)
K-662 .... 35 ft.-lbs. (48 N·m)
Cylinder head cap screws–
K-482, K-532 & K-582...... 35 ft.-lbs. (48 N·m)
K-662 .... 40 ft.-lbs. (54 N·m)
Closure plate cap screws–
K-482, K-532 & K-582...... 30 ft.-lbs. (40 N·m)
K-662 .... 50 ft.-lbs. (68 N·m)
Flywheel nut–
K-482, K-532 & K-582..... 115 ft.-lbs. (156 N·m)
K-662 .... 130 ft.-lbs. (176 N·m)
Oil pan cap screws–
K-482, K-532 & K-582...... 30 ft.-lbs. (40 N·m)
K-662 .... 45 ft.-lbs. (61 N·m)

CONNECTING RODS. Connecting rod and piston assemblies can be removed after removing oil pan (engine base) and cylinder heads. Identify each rod and piston assembly so they can be reinstalled in their original cylinders and do not intermix connecting rod caps.

On Models K-482, K-532 and K-582, connecting rods ride directly on the crankpins. Connecting rods 0.010 inch (0.254 mm) oversize are available for undersize reground crankshafts. Standard crankpin diameter is 1.6245-1.6250 inches (41.262-41.275 mm). Desired running clearances are as follows:

Connecting rods to
crankpins .... 0.001-0.0035 in. (0.0254-0.0889 mm)
Connecting rods to
piston pins .... 0.0003-0.0008 in. (0.0076-0.0203 mm)
Rod side play on
crankpins .... 0.005-0.014 in. (0.1270-0.3556 mm)

On K-662 model connecting rods are equipped with renewable insert type bearings and a renewable piston pin bushing. Bearings are available for 0.002, 0.010 and 0.020 inch (0.0508, 0.254 and 0.508 mm) undersize crankshafts. Standard crankpin diameter is 1.8745-1.8750 inches (47.612-47.625 mm). Desired running clearances are as follows:

Connecting rods to
crankpins .... 0.0003-0.0035 in. (0.0076-0.0889 mm)
Connecting rods to
piston pins .... 0.0001-0.0006 in. (0.0025-0.0152 mm)
Rod side play on
crankpins .... 0.007-0.011 in. (0.1778-0.2794 mm)

On all models, when installing connecting rod and piston units, make certain raised match marks on rods and rod caps are aligned and are towards flywheel side of engine. Kohler recommends connecting rod cap screws of K-482, K-532 and K-582 be tightened evenly to a torque of 30 ft.-lbs. (40 N·m), then loosened and retorqued to 25 ft.-lbs. (34 N·m). Kohler recommends connecting rod cap screws on K-662 engines be tightened to 40 ft.-lbs. (54 N·m), then loosened and retorqued to 35 ft.-lbs. (48 N·m).

On all models connecting rod cap screw threads should be lightly lubricated before installation.

PISTONS, PINS AND RINGS. Aluminum alloy pistons are fitted with two compression rings and one oil control ring. Renew pistons if scored or if side clearance of new ring in piston top ring groove exceeds 0.006 inch (0.1524 mm). Pistons are available in oversizes of 0.010, 0.020 and 0.030 inch (0.254, 0.508 and 0.762 mm) as well as standard size. Recommended piston to cylinder bore clearance when measured just below oil ring at right angle to pin is as follows:

K-482 .... 0.007-0.009 in. (0.1778-0.2286 mm)
K-532 .... 0.0065-0.0095 in. (0.1651-0.2413 mm)
K-582 .... 0.007-0.010 in. (0.178-0.254 mm)
K-662 .... 0.001-0.003 in. (0.0254-0.0762 mm)

Piston pin fit in piston boss on K-482, K-532 and K-582 should be from 0.000 (light interference) to 0.0005 inch (0.0127 mm) loose and for K-662 should be 0.0001-0.0003 inch (0.0025-0.0076 mm) loose.

Piston pin fit in rod on K-482, K-532 and K-582 should be 0.0003-0.0008 inch (0.0076-0.0203 mm) loose. Model K-662 connecting rod has a renewable bushing for piston pin and piston fit should be 0.0001-0.0006 inch (0.0025-0.0152 mm) loose. Always renew piston pin retaining rings.

Kohler recommends piston rings should always be renewed when they are removed.

Piston ring end gap for all models should be 0.010-0.020 inch (0.254-0.508 mm) for new bores and a maximum of 0.030 inch (0.762 mm) for used bores.

Piston ring specifications are as follows:

Ring width
Compression rings .... 0.093 in. (2.36 mm)
Oil ring .... 0.187 in. (4.750 mm)
Ring side clearance (K-482, K-532 and K-582)
Top ring .... 0.002-0.004 in. (0.0508-0.1016 mm)
2nd ring.......... 0.0015-0.0035 in. (0.0381-0.0889 mm)
Oil ring .... 0.001-0.003 in. (0.0254-0.0762 mm)
Ring side clearance (K-662)
Top ring .... 0.0025-0.0045 in. (0.0635-0.1143 mm)
2nd ring .... 0.0025-0.0045 in. (0.0635-0.1143 mm)
Oil ring .... 0.002-0.0035 in. (0.0508-0.0889 mm)

Piston rings are available in standard size and oversizes of 0.010, 0.020 and 0.030 inch (0.254, 0.508 and 0.762 mm).

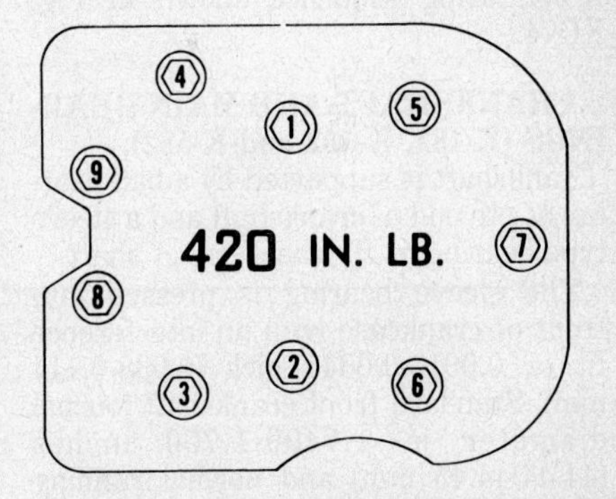

Fig. KO73–Tighten cylinder head cap screws on K-482, K-532 and K-582 models to 35 ft.-lbs. (48 N·m) torque using sequence shown.

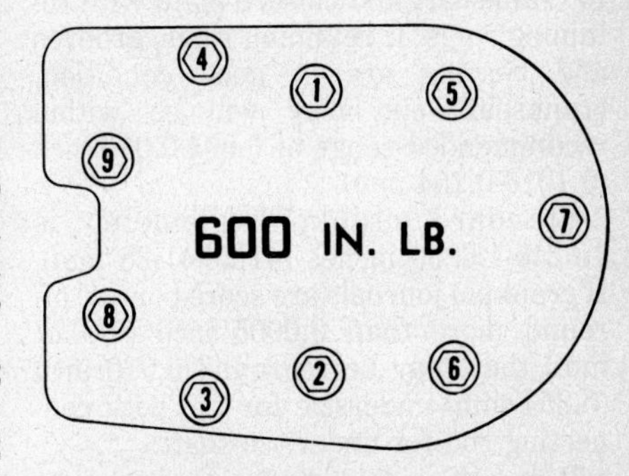

Fig. KO74–Tighten cylinder head cap screws on K-662 model to 50 ft.-lbs. (68 N·m) torque using sequence shown.

Fig. KO75—View showing timing marks on camshaft gear and crankshaft gear on K-482, K-532 and K-582 models. Model K-662 is similar.

Fig. KO76—Front crankshaft seal on K-662 model rotates with crankshaft and seals against gear cover.

CYLINDERS. If cylinder walls are scored or bores are tapered or out-of-round more than 0.005 inch (0.127 mm) cylinders should be bored to nearest suitable oversize of 0.010, 0.020 or 0.030 inch (0.254, 0.508 or 0.762 mm). Standard cylinder bore for K-482 is 3.250 inches (82.55 mm), for K-532 is 3.375 inches (85.73 mm), for K-582 is 3.5 inches (88.9 mm) and for K-662 is 3.625 inches (92.08 mm).

CYLINDER HEADS. Always use new head gaskets when installing cylinder heads. Tighten cylinder head cap screws evenly to a torque of 35 ft.-lbs. (48 N·m) on Models K-482, K-532 and K-582 using sequence shown in Fig. KO73 or 50 ft.-lbs. (68 N·m) on Model K-662 using sequence shown in Fig. KO74.

CRANKSHAFT AND MAIN BEARINGS (K-482, K-532 and K-582). Crankshaft is supported by a ball bearing at pto end of crankshaft and a sleeve type bearing at flywheel end of shaft.

The sleeve bearing is pressed into front of crankcase with an interference fit of 0.0015-0.0045 inch (0.038-0.114 mm). Standard front crankshaft journal diameter is 1.7490-1.750 inches (44.44-44.45 mm) and normal running clearance is 0.002-0.0035 inch (0.0508-0.0889 mm) in sleeve bearing.

The rear main (ball bearing) is secured to crankshaft and closure plate with retaining rings. If retaining rings, grooves and bearing are in good condition, crankshaft end play will be within recommended range of 0.004-0.010 inch (0.1016-0.254 mm).

Standard crankpin diameter is 1.6245-1.6250 inches (41.26-41.28 mm). If crankpin journals are scored or out-of-round more than 0.0005 inch (0.0127 mm) they may be reground 0.010 inch (0.254 mm) undersize for use with connecting rod for undersize shaft.

Renewable crankshaft gear is a light press fit on crankshaft. When installing crankshaft gear, align single punch mark on crankshaft gear with two punch marks on camshaft gear as shown in Fig. KO75.

Crankshaft oil seals can be installed after gear cover and closure plate are installed. Carefully work oil seals over crankshaft and drive seals into place

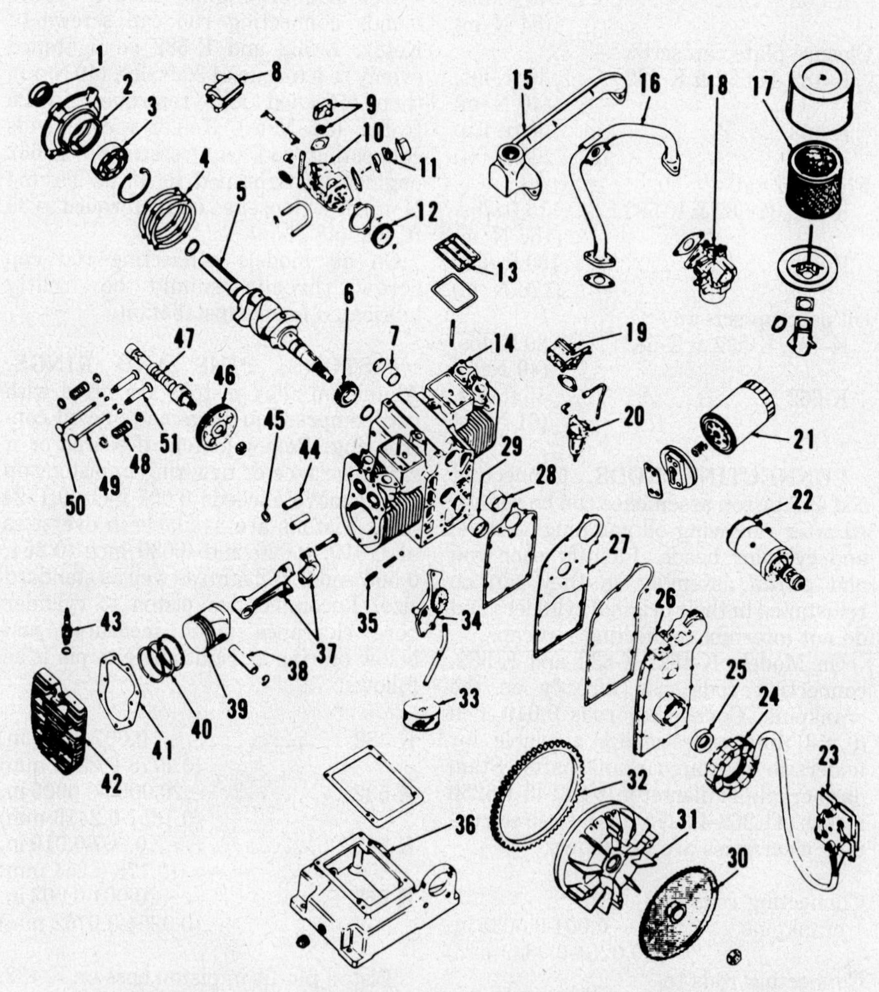

Fig. KO77—Exploded view of K-482, K-532 and K-582 model basic engine assembly.

1. Oil seal
2. Closure plate
3. Rear main (ball) bearing
4. Gaskets
5. Crankshaft
6. Crankshaft gear
7. Camshaft rear bushing
8. Ignition coil
9. Breather valve assy.
10. Governor assy.
11. Breaker points
12. Governor gear
13. Valve cover
14. Crankcase & cylinder block
15. Exhaust manifold
16. Intake manifold
17. Air cleaner element
18. Carburetor
19. Fuel pump
20. Fuel filter
21. Oil filter
22. Starting motor
23. Rectifier-regulator
24. Alternator stator
25. Oil seal
26. Gear cover
27. Gear cover plate
28. Camshaft front bushing
29. Front main (sleeve)
30. Flywheel screen
31. Flywheel
32. Starter ring gear
33. Oil strainer
34. Oil pump
35. Oil pressure adjusting screw
36. Oil pan
37. Connecting rod
38. Piston pin
39. Piston
40. Piston rings
41. Head gasket
42. Cylinder head
43. Spark plug
44. Valve guides
45. Exhaust valve seat
46. Camshaft gear
47. Camshaft
48. Valve spring
49. Intake valve
50. Exhaust valve
51. Valve tappets

with hollow driver that contacts outer edge of seals.

CRANKSHAFT AND MAIN BEARINGS (K-662). Crankshaft is supported by tapered roller bearings at each end. Crankshaft end play should be 0.0035-0.0055 inch (0.0889-0.1397 mm) and is controlled by varying thickness of shim gaskets between crankcase and closure (bearing) plate. Shim gaskets are available in a variety of thicknesses.

Oil transfer sleeves which are pressed into flywheel end of crankcase and closure convey oil to connecting rod bearings. When installing oil sleeves, make certain oil holes in sleeves are aligned with oil holes in closure plate and crankcase.

Standard crankpin diameter is 1.8745-1.8750 inches (47.61-47.63 mm). Connecting rod bearings of 0.002 inch (0.0508 mm) oversize are available for use with moderately worn crankpin. Rod bearings of 0.010 inch (0.254 mm) and 0.020 inch (0.508 mm) oversize are available for use with reground crankpins.

Renewable crankshaft gear is a light press fit on crankshaft. When installing crankshaft gear, align single punch mark on crankshaft gear with two punch marks on camshaft gear as shown in Fig. KO75.

To install rear (pto end) oil seal, carefully work seal, with lip inward, over end of shaft. Use a close fitting seal driver and tap seal into closure plate. Front oil seal (Fig. KO76) is a compression type seal which rotates with crankshaft and seals against gear cover.

CAMSHAFT. Camshaft is supported in two renewable bushings. To remove camshaft first remove flywheel, gear cover, governor assembly and fuel pump. On Model K-662, remove magneto assembly. On all models, remove valve covers, valve spring keepers, spring retainers and valve springs. Wedge a piece of wood between camshaft gear and crankshaft gear and remove gear retaining nut and washer. Attach a suitable puller and remove camshaft gear and Woodruff key. Unbolt and remove gear cover plate from crankcase. Move tappets away from camshaft and carefully withdraw camshaft.

NOTE: To prevent tappets from sliding into crankcase, tie a wire around adjusting nuts on each set of tappets.

Camshaft bushings can now be renewed. Camshaft bushings are presized and if carefully installed need no final sizing. Normal camshaft to bushing clearance is 0.0005-0.0035 inch (0.0127-0.0889 mm). Install new expansion plug behind rear camshaft bushing. Align two punch marked teeth on camshaft gear with single punch marked tooth on crankshaft gear (Fig. KO75) and reassemble engine by reversing disassembly procedure.

Camshaft end play for K-582 models is 0.004-0.010 inch (0.1016-0.254 mm) and end play for all other models is 0.017-0.038 inch (0.432-0.965 mm). End play is controlled by gear cover plate and gasket. If end play is excessive check plate for excessive wear and renew as necessary.

Tighten camshaft gear retaining nut and flywheel retaining nut to torque specified in **TIGHTENING TORQUES** paragraph for model being serviced.

Adjust valves as outlined in **VALVE SYSTEM** paragraph.

VALVE SYSTEM. Exhaust valve seats are renewable hardened valve seat inserts on all models. Intake valves in K-662 models seat on renewable inserts but intake valves in K-482, K-532 and K-582 models may seat on renewable inserts or seat may be machined directly into block casting. Seating surfaces should be 1/32-inch (0.794 mm) in width and must be reconditioned if over 1/16-inch (1.588 mm) wide. Valves and seats are ground to 45° angle and Kohler recommends lapping valves to assure proper valve to seat seal.

Maximum intake valve stem clearance in guide is 0.0045 inch (0.1143 mm) and maximum exhaust valve stem clearance in guide is 0.0065 inch (0.1651 mm) for all models. All models have renewable intake and exhaust valve guides which must be reamed to size after installation. K-662 models have a 0.010 inch (0.254 mm) oversize outside diameter guide available for crankcases with damaged guide bores.

Remove guides by driving them into valve stem chamber and breaking the end off until guide is completely removed. Use care not to damage engine block. Press new guides of K-482, K-532 and K-582 models in to a depth of 1-11/32 inch (34.131 mm) measured from cylinder head surface of block to end of guide. Press new guides of K-662 model

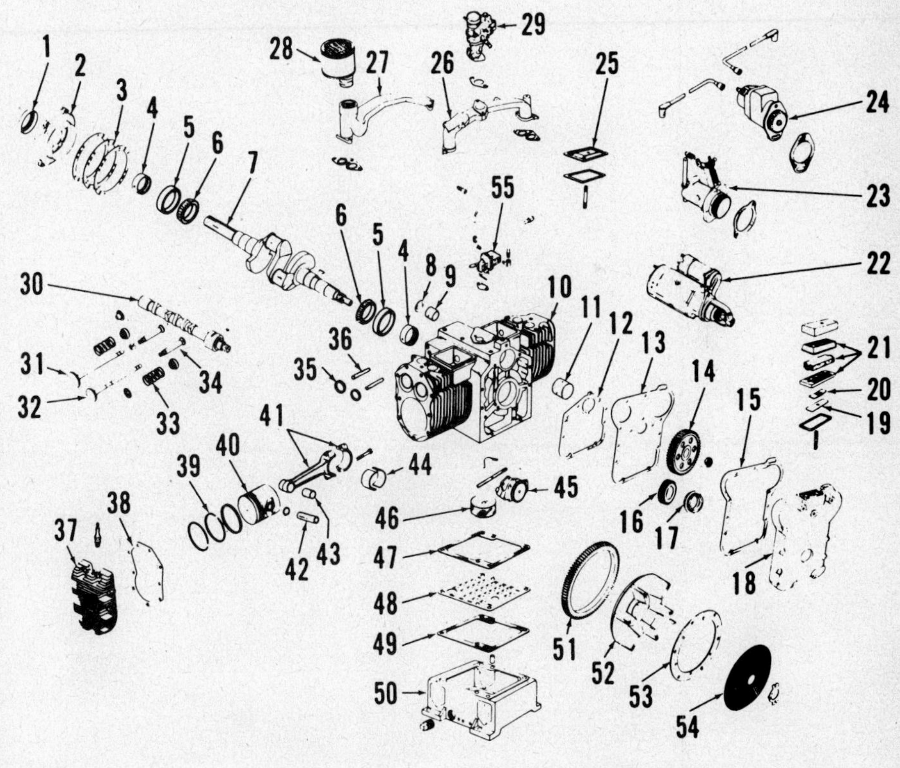

Fig. KO78—Exploded view of K-662 model basic engine assembly.

1. Rear oil seal
2. Closure plate
3. Shim gaskets
4. Oil transfer sleeve
5. Bearing cup
6. Bearing cone
7. Crankshaft
8. Expansion plug
9. Camshaft bushing (rear)
10. Crankcase and cylinder block
11. Camshaft bushing (front)
12. Gasket
13. Gear cover plate
14. Camshaft gear
15. Gasket
16. Crankshaft gear
17. Front oil seal
18. Gear cover
19. Breather reed
20. Reed stop
21. Breather filter
22. Starting motor
23. Governor assy.
24. Magneto
25. Valve cover
26. Intake manifold
27. Exhaust manifold
28. Muffler
29. Carburetor
30. Camshaft
31. Intake valve
32. Exhaust valve
33. Valve springs
34. Valve tappets
35. Valve seat inserts
36. Valve guides
37. Cylinder head
38. Head gasket
39. Piston rings
40. Piston
41. Connecting rod
42. Piston pin
43. Pin bushing
44. Connecting rod bearing
45. Oil pump
46. Oil strainer
47. Gasket
48. Baffle
49. Gasket
50. Oil pan
51. Starter ring gear
52. Flywheel
53. Flywheel plate
54. Flywheel screen
55. Fuel pump

in to a depth of 1-7/16 inch (35.513 mm) measured from cylinder head surface of block to end of guide.

Ream new guides installed in K-482, K-532 and K-582 models to 0.312-0.313 inch (7.8248-7.9502 mm) and ream new guides installed in K-662 model to 0.3430-0.3445 inch (8.7122-8.7503 mm).

Valve spring free length for K-482, K-532 and K-582 models should be 1.793 inches (45.54 mm) and for K-662 model should be 2.250 inches (57.15 mm). If rotator is used on exhaust valve, exhaust valve spring free length for K-482, K-532 and K-582 models should be 1.531 inches (38.8874 mm) and for K-662 model should be 1.812 inches (46.03 mm).

Valve tappet gap (clearance) for K-482, K-532 andK-582 models should be adjusted so intake valves have 0.008-0.010 inch (0.203-0.254 mm) clearance and exhaust valves have 0.017-0.020 inch (0.432-0.508 mm) clearance when cold. Valve tappet gap (clearance) for K-662 model should be adjusted so intake valves have 0.006-0.008 inch (0.152-0.203 mm) clearance and exhaust valves have 0.015-0.017 inch (0.381-0.432 mm) clearance when cold.

OIL PUMP. A gear type oil pump supplies oil via internal galleries to front main bearing, camshaft bearings, connecting rods and other wear areas. Model K-662 uses transfer sleeves to direct oil to the tapered roller main bearings.

Oil pressure on Models K-482, K-532 and K-582 is controlled by an adjustable pressure valve. Refer to **LUBRICATION** paragraph for operating pressures.

If faulty, oil pump must be renewed as it is serviced as an assembly only.

FUEL PUMP. A mechanically operated diaphragm type fuel pump is used. Pump is actuated by a lever which rides on an eccentric on camshaft. A priming lever is provided on the pump and repair kit is available for reconditioning pump.

KOHLER

KOHLER COMPANY
Kohler, Wisconsin 53044

Model	No. Cyls.	Bore	Stroke	Displacement	Power Rating
KT-17*	2	3.12 in. (79.3 mm)	2.75 in. (69.9 mm)	42.18 cu. in. (690.5 cc)	17 hp (12.7 kW)
KT-19*	2	3.12 in. (79.3 mm)	3.06 in. (78 mm)	46.98 cu. in. (770.5 cc)	19 hp (14.2 kW)
KT-21	2	3.31 in. (84.07 mm)	3.06 in. (78 mm)	52.76 cu. in. (866 cc)	21 hp (15.7 kW)

*Series II engines included.

Engines in this section are four-cycle, twin-cylinder opposed, split crankcase type. Model KT-17 engines with specification number 24299 or below and KT-19 engines with specification number 49199 or below have a pressure spray lubrication system. KT-17 engines with specification number 24300 or above and KT-19 engines with specification number 49200 or above are designated Series II engines and have full pressure lubrication systems and improved crankshafts and rods. All KT-21 engines have full pressure lubrication systems.

Engine identification and serial number decals are located on top of engine shrouding usually around ignition coil. Significance of each number is explained in Fig. KO79.

Series II engines have **SERIES II** logo on the front shroud from the factory. When ordering parts or service material always give model number, specification number and serial number of engine being serviced.

MAINTENANCE

SPARK PLUG. Recommended spark plug is Champion RBL15Y or equivalent. Electrode gap is 0.025 inch (0.635 mm).

CARBURETOR. A single Kohler side draft carburetor is used on all models and carburetor adjustments should be made only after engine has reached normal operating temperature.

For initial adjustment stop engine and turn main fuel and idle fuel (Fig. KO80) mixture screws in until **LIGHTLY** seated. On KT-17 and KT-19 models turn main fuel adjusting screw out 2½ turns and on KT-21 models turn main fuel adjusting screw out 3 turns.

On all models turn idle mixture screw out 1 to 1¼ turns.

For final adjustment run engine at half throttle and when operating temperature has been reached turn main fuel mixture screw in until speed decreases and note position. Now turn screw out past point where speed increases until speed again decreases. Note position of screw. Set mixture screw midway between the two points noted. Set throttle so engine is at idle and adjust idle mixture screw in the same manner.

Set idle speed screw so engine is idling at 1200 rpm.

If carburetor is to be cleaned, refer to Fig. KO80 and disassemble. Clean carburetor with suitable solvent making certain fiber and rubber seals are not damaged. Blow out all passages with compressed air and renew any worn or damaged parts.

For reassembly torque new fuel needle seat to 35 in.-lbs. (4 N·m). Install

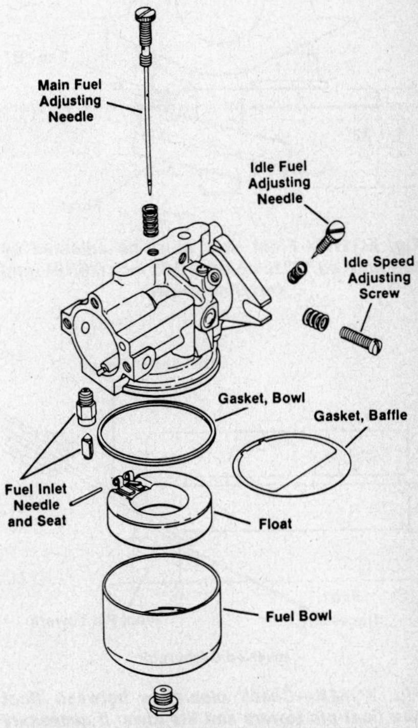

Fig. KO80—Exploded view of Kohler side draft carburetor used on KT series engines.

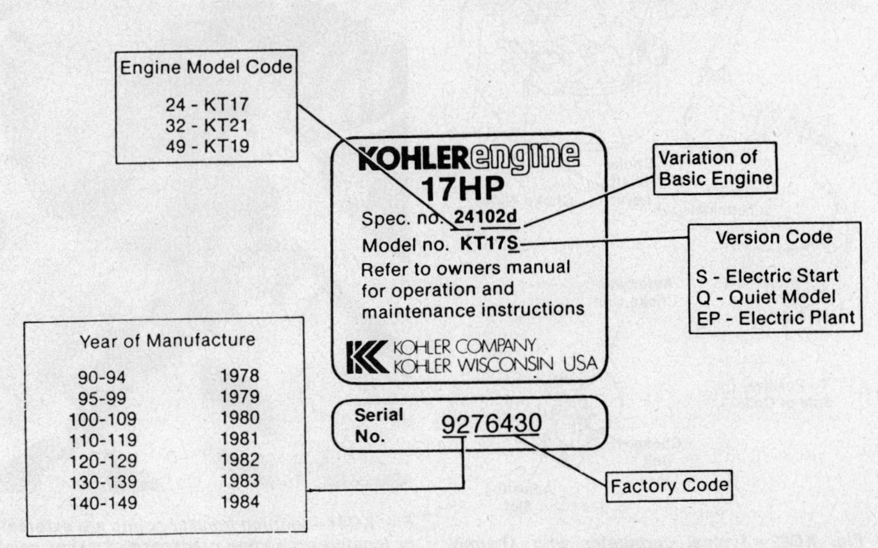

Fig. KO79—Significance of each digit of the specification, model and serial numbers are explained as shown.

needle, float and pin. Set float to 11/64-inch (4.366 mm) as shown in Fig. KO81. Late production carburetors have a tab to limit float travel (B–Fig. KO81A) and float drop must be set to 1-1/32 inch (26.194 mm) by bending tab (B). Float clearance should be checked as shown in Fig. KO81B. File float pin tower as necessary to obtain 0.010 inch (0.254 mm) clearance. Reassemble carburetor, install and adjust.

AUTOMATIC CHOKE. Some models are equipped with a "Thermo-Electric" automatic choke and fuel shutdown solenoid (Fig. KO82).

When installing or adjusting choke unit, position unit leaving mounting screw slightly loose. Hold choke plate wide open and rotate choke unit clockwise with slight pressure until it can no longer be rotated. Hold choke unit and tighten mounting screws. Choke valve should close 5°-10° at 75° F (24° C). As temperature decreases, choke will close even more. To check choke, remove spark plug lead and crank engine. Choke valve should close a minimum of 45° at 75° F (24° C).

With ignition on, shutdown solenoid plunger raises and opens float bowl vent to air horn. With ignition off, solenoid plunger closes vent and stops fuel flow. To check solenoid, remove solenoid and plunger from carburetor body. With lead wire connected, pull plunger approximately ¼-inch (6.35 mm) from solenoid and ground casing to engine surface. Turn ignition switch on. Renew solenoid unit if solenoid does not draw plunger in. After renewing solenoid, reset main fuel mixture adjusting screw.

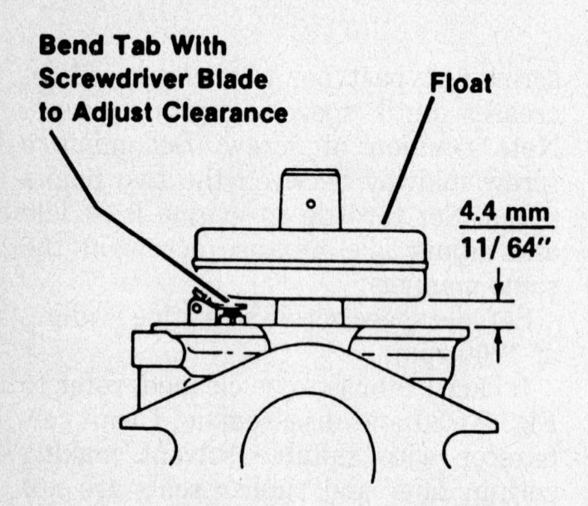

Fig. KO81—Check float setting by inverting carburetor throttle body and float assembly and measuring distance from free end of float to machined surface of casting. Correct clearance is 11/64-inch (4.4 mm).

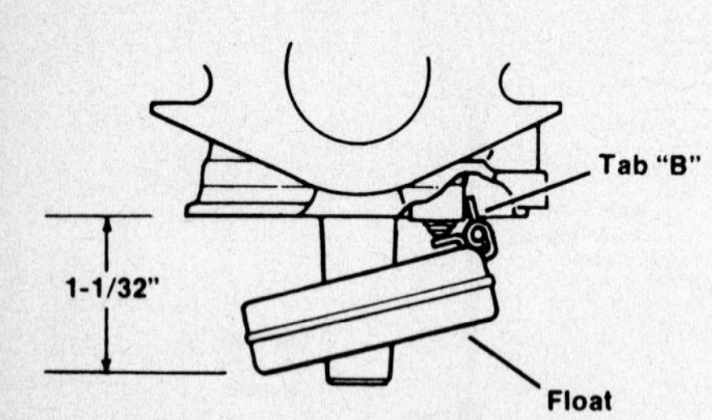

Fig. KO81A—Float drop must be adjusted by bending tab "B" to obtain 1-1/32 inch (26.194 mm) travel as shown.

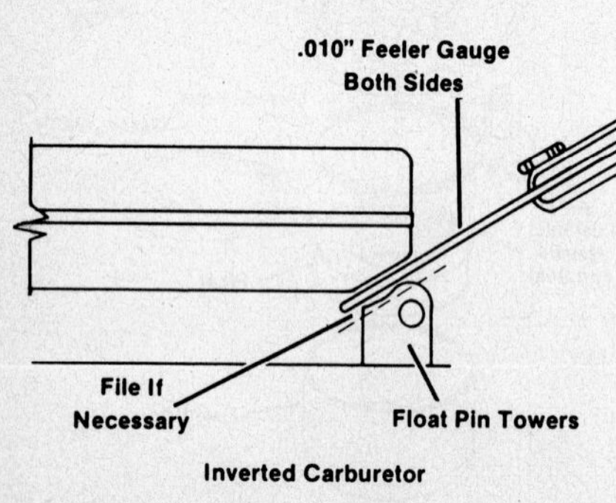

Fig. KO81B—Check clearance between float and float pin towers and file tower if necessary to obtain 0.010 inch (0.254 mm) clearance as shown.

GOVERNOR. KT-series engines are equipped with centrifugal flyweight mechanical governor with weight gear mounted within crankcase and driven by camshaft gear.

Governors are adjusted at factory and further adjustment should not be necessary unless governor arm or linkage works loose or becomes disconnected. Readjust linkage if engine speed surges or hunts with changing load or if speed drops considerably when normal load is applied.

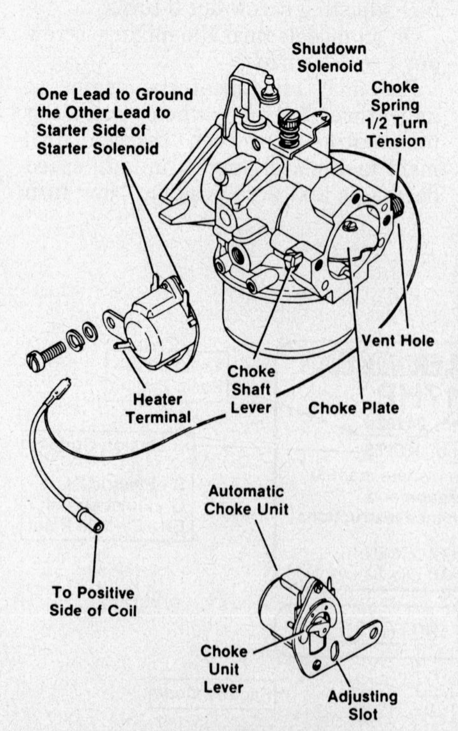

Fig. KO82—Typical carburetor with Thermo-Electric automatic choke and shutdown control. Refer to text for details.

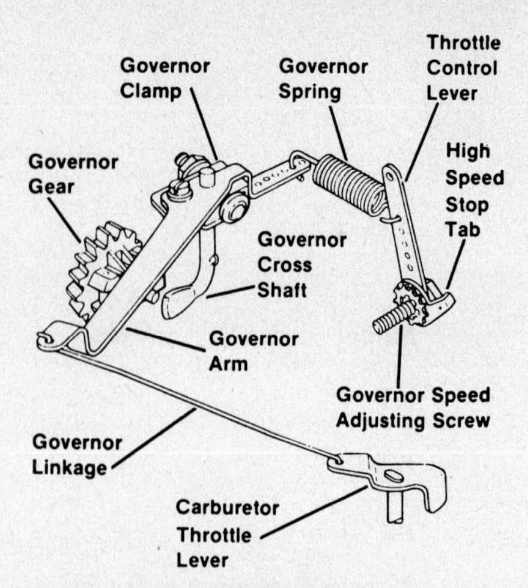

Fig. KO83—View of typical governor linkage used on KT series engines.

Maximum no load speed for all models is 3600 rpm. If adjustment is necessary, loosen governor speed adjusting screw and pivot high speed stop tab until desired speed is reached, then tighten screw. Governor sensitivity can be adjusted by repositioning governor spring. Normally, governor spring is placed in the fifth hole from pivot of governor arm and in the sixth hole from pivot on throttle control lever. To increase sensitivity, move spring end closer to center of governor arm. To allow a wider range of governor control with less sensitivity, move spring toward end of arm. Refer to Fig. KO83.

IGNITION TIMING. Breaker points are externally mounted on crankcase and are activated by a push rod which rides on camshaft. A timing sight hole is located in number "1" side of blower housing.

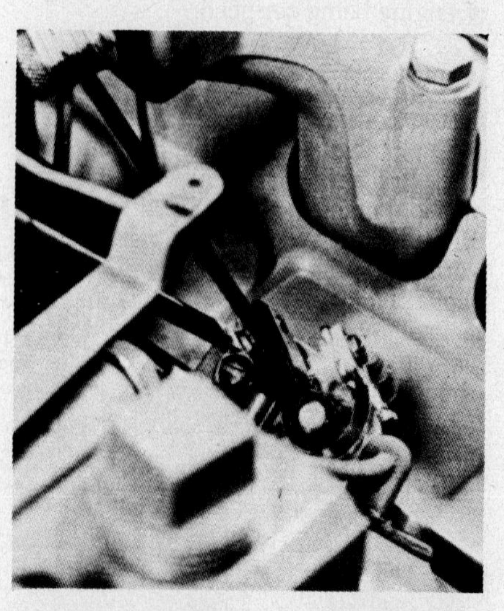

Fig. KO84—Ignition breaker points are externally located on engine crankcase. Breaker point gap will vary from 0.017-0.023 inch (0.432-0.584 mm) depending on engine performance.

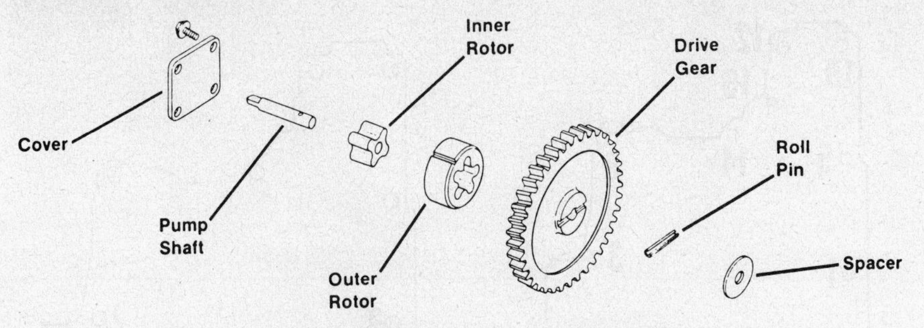

Fig. KO85—Exploded view of rotor type oil pump.

Two different timing location marks are stamped on flywheel. "T" mark locates TDC while "S" mark locates 23° BTDC, which is where timing is set.

To set timing, remove point cover and reinstall cover screws to prevent oil loss. Attach timing light to number one spark plug, start engine and run at 1200 rpm. Vary point gap between 0.017 and 0.023 inch (0.432 and 0.584 mm) until "S" mark is in line with roll pin in cylinder barrel. If double image is seen through sight hole, time engine so roll pin is exactly between the two images. Apply thread sealant to screws when reinstalling point cover. See Fig. KO84.

COMPRESSION TEST. To check compression, remove spark plugs, make certain air cleaner is clean and set throttle and choke in wide open position. Crank engine with starting motor. Compression should test approximately 115-125 psi (793-862 kPa).

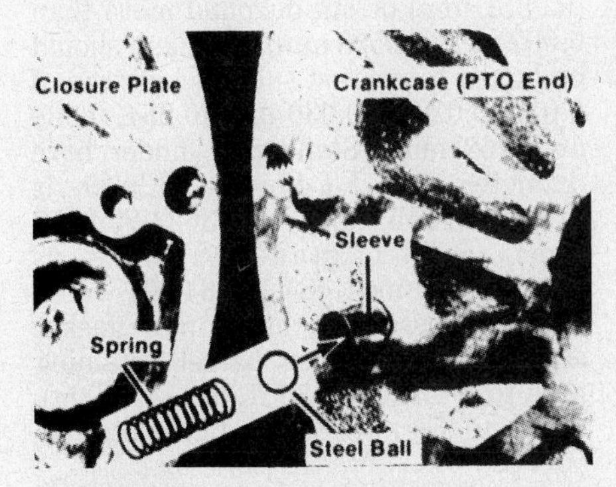

Fig. KO85A-View of relief valve spring, ball and sleeve on Series II engines.

Fig. KO85B—View of test port plug in oil gallery of Series II engines. Refer to text.

LUBRICATION. Oil pump (Fig. KO85) is driven by crankshaft and is located behind closure plate on pto side of engine on all models.

KT-17 engines with specification number 24299 and below and KT-19 engines with specification number 49199 and below use a pressure spray lubrication system. In this system oil is supplied to main bearings under pressure and travels through the hollow camshaft and sprays out two holes in camshaft to lubricate rods. Rods on these engines have oil hole drilling (Fig. KO91A) through which oil reaches bearing surface of rod.

KT-17 Series II engines, specification number 24300 and above, KT-19 series II engines, specification number 49200 and above and all KT-21 engines use a full pressure lubrication system with oil supplied to rods under pressure via drillings in crankshaft. In these engines oil pressure is regulated by a ball and spring type oil pressure relief valve located in crankcase behind closure plate (Fig. KO85A). Remove the 1/16-inch pipe plug (Fig. KO85B) in pto end of crankcase and install a 0-100 psi (0-690 kPa) pressure gage to check oil pressure. Pressure should be 20-30 psi (138-207 kPa).

Oil pump rotors and cover on all models and oil pressure relief valve and spring on models so equipped can be serviced without splitting crankcase.

Oil pump shaft to crankcase clearance should be 0.001-0.0025 inch (0.025-0.064 mm) and oil pump drive gear end play should be 0.010-0.029 inch (0.254-0.736 mm).

Free length of oil pressure relief valve spring should be 0.940 inch plus or minus 0.010 inch (23.87 mm plus or minus 0.254 mm).

High quality detergent type oil having API classification SC, CC, SD or SE should be used. Use SAE 30, 10W-30 or 10W-40 oil in temperatures of 32° F (0° C) and above and SAE 5W-30 oil in temperatures of 32° F (0° C) and below. Maintain crankcase oil level at full mark on dipstick, but do not overfill.

It is recommended oil be changed at 25 hour intervals under normal operating conditions. Crankcase capacity for all models is 3 pints (1.4 L).

REPAIRS

TIGHTENING TORQUES. Recommended tightening torques are as follows:

Spark plugs ....10-15 ft.-lbs. (14-20 N•m)
Flywheel retaining screws....40 ft.-lbs. (54 N•m)
Manifold screws ....13 ft.-lbs. (17 N•m)
Closure plate screws ....13 ft.-lbs. (17 N•m)
Cylinder barrel nuts.........22 ft.-lbs. (29 N•m)
Cylinder head bolts* ......15-20 ft.-lbs. (20-27 N•m)
Connecting rod bolts** ......17 ft.-lbs. (23 N•m)
Connecting rod nuts–
New ....12 ft.-lbs. (16 N•m)
Used.....................8 ft.-lbs. (11 N•m)
Crankcase stud nuts.........22 ft.-lbs. (29 N•m)
Crankcase slot head screws....3 ft.-lbs. (4 N•m)
Crankcase cap screws ....See Fig. KO89 for sequence & torque

*Lubricate with oil at assembly.
**Rod bolts should be overtorqued to 20 ft.-lbs. (27 N•m), loosened and retorqued to 17 ft.-lbs. (23 N•m) torque.

CYLINDER HEADS AND VALVE SYSTEM. If head is warped more than 0.003 inch (0.0762 mm) or has "hot spots", both head and head screws

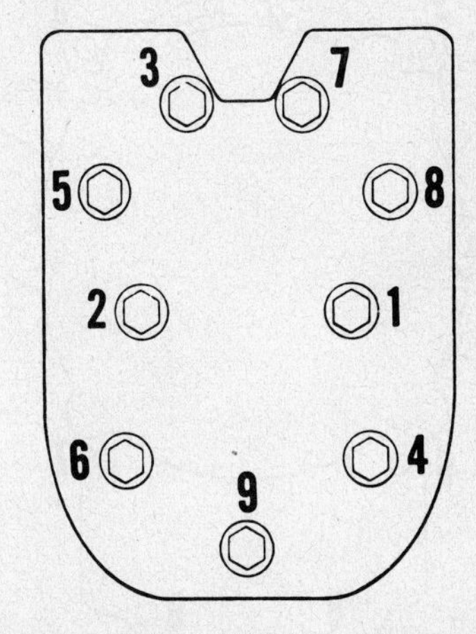

Fig. KO86—Tighten cylinder head cap screws to 15-20 ft.-lbs. (20-27 N•m) torque using sequence shown. Lubricate cap screws during assembly.

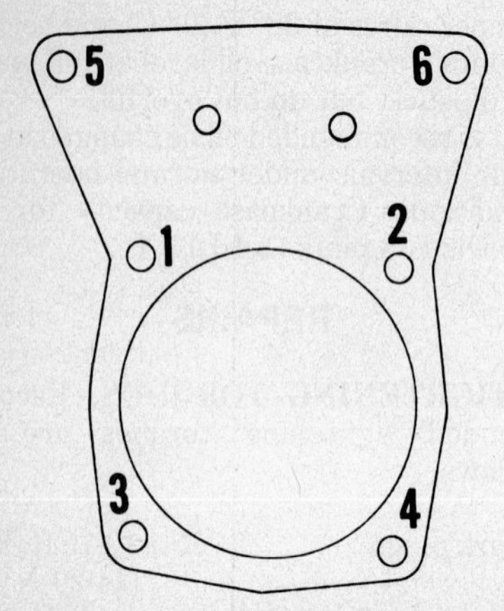

Fig. KO87 — Cylinder barrel torque tightening sequence. Tighten nuts to a torque of 22 ft.-lbs. (29 N•m).

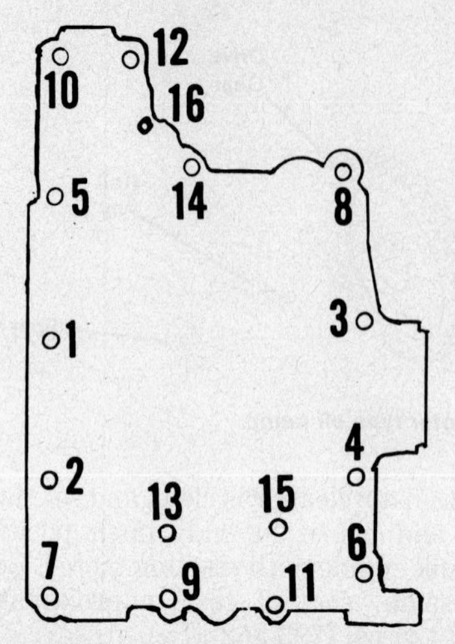

Fig. KO89 — Tighten crankcase bolts on KT-17 engines with serial numbers prior to 9755085 in the following sequence: 1 through 4, 12 and 14 to 260 in.-lbs. (29 N·m) torque. Tighten bolts 5 through 11, 13 and 15 to 150 in.-lbs. (17 N·m) torque. Tighten bolt 16 to 55 in.-lbs. (6 N·m) torque. Tighten bolts in sequence shown.

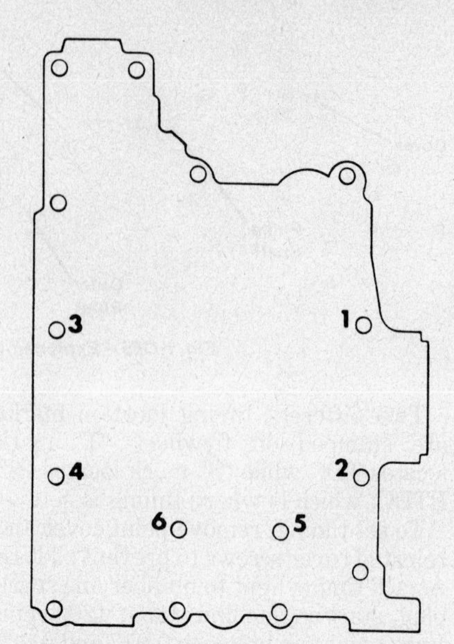

Fig. KO89A — Tighten crankcase bolts on KT-17 engines with serial number 9755085 and above, all KT-19 and KT-21 engines in the following sequence: Number 1 through 4, 12 and 14 to 260 in.-lbs. (29 N·m) torque. Tighten 5 and 6 to 200 in.-lbs. (23 N·m) torque and tighten remaining bolts to 200 in.-lbs. (23 N·m) torque.

should be renewed. Tighten cylinder head cap screws evenly to 15-20 ft.-lbs. (20-27 N·m) torque using sequence shown in Fig. KO86.

Intake valve seats are machined surfaces of cylinder casting on most KT series engines, however certain applications use a hard alloy renewable insert.

All models have renewable hardened alloy exhaust valve seats which should be measured for width before removal. Two different widths have been used due to changes in cylinder castings. Widths are 0.199 inch (0.711 mm) and 0.220 inch (0.558 mm) and are not interchangeable.

Valve face and seat angle is 45°. Desired seat width is 0.037-0.045 inch (0.9398-1.1430 mm). Valve tappet clearance (cold) is 0.003-0.006 inch (0.076-0.152 mm) for intake and 0.011-0.014 inch (0.2794-0.3556 mm) for exhaust. If adjustment is necessary, grind end of valve stem. Make certain end of valve stem is ground perfectly flat. See Fig. KO90.

Valve stem to guide clearance is 0.0025-0.0045 inch (0.0635-0.1143 mm) for intake and 0.0045-0.0065 inch (0.1143-0.1651 mm) for exhaust. To renew valve guides, press old guides into valve chamber and carefully break off protruding ends until guides are removed. New guides should have an interference fit of 0.0005-0.002 inch (0.0127-0.051 mm). Press new guides in and ream to 0.312-0.313 inch (7.9248-7.9502 mm). Depth from top of block to valve guide for KT-17, KT-17 Series II and KT-19 engines is 1.125 inch (28.58 mm) and is 1.390 inch (26.39 mm) for KT 19 Series II and KT-21 engines.

Valve spring free length (no rotators) should be 1.6876 inches (42.865 mm). If rotator is used on exhaust valve, spring free length should be 1.542 inches (39.1668 mm).

CYLINDERS, PISTONS, PINS AND RINGS. If cylinder walls are scored or bores are tapered 0.0015 inch (0.0381 mm) or out-of-round more than 0.002 inch (0.0508 mm), cylinders should be honed to nearest suitable oversize of 0.010, 0.020 or 0.030 inch (0.254, 0.508 or 0.762 mm). Standard cylinder bore diameter for KT-17 and KT-19 is 3.1245-3.1255 inch (79.3623-79.3877 mm), KT-21 standard bore is 3.312-3.315 inch (84.125-84.201 mm). During reassembly, use a new gasket and tighten cylinder barrel retaining nuts to a torque of 30 ft.-lbs. (41 N·m). Refer to Fig. KO87 for tightening sequence.

Cam ground aluminum alloy pistons are fitted with two compression rings

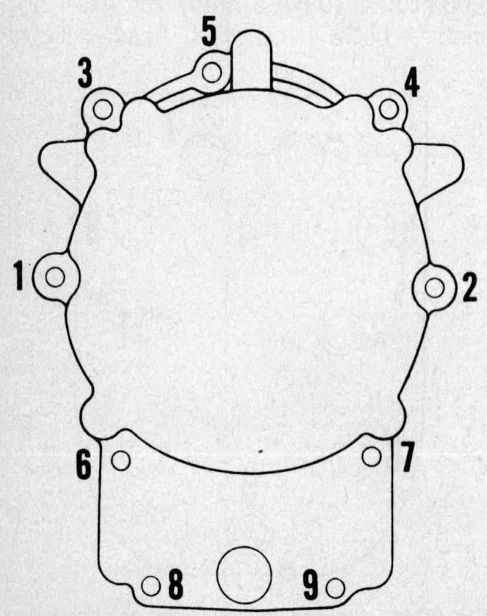

Fig. KO88 — Closure plate torque tightening sequence. Torque screws to 150 in.-lbs. (17 N·m).

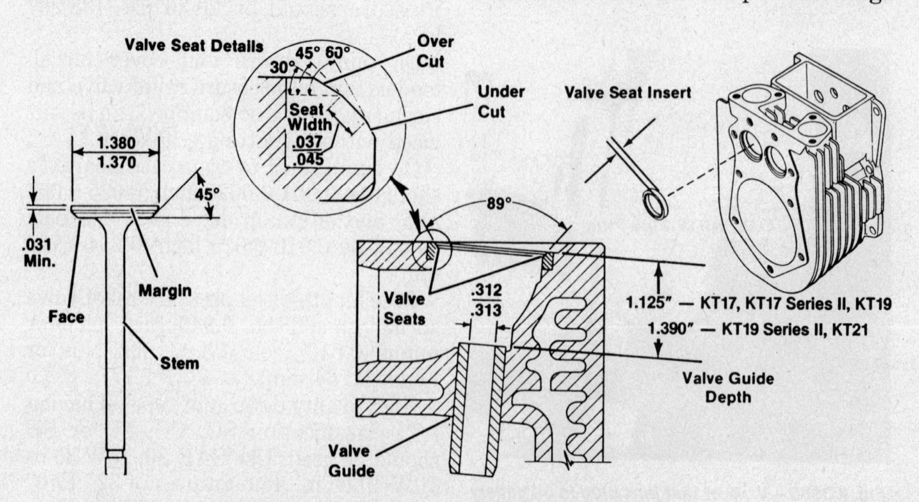

Fig. KO90 — Cross-sectional view of valve seat and valve.

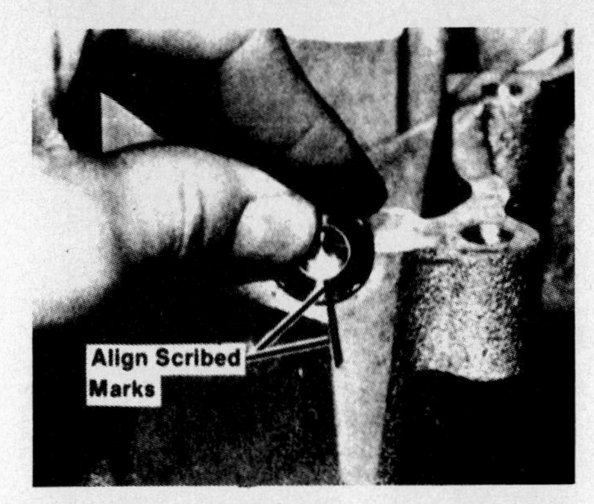

Fig. KO90A—Scribe mark across machined camshaft plug so it can be reinstalled in its original position.

and one oil control ring. Renew pistons if scored or if side clearance of new ring in top groove exceeds 0.004 inch (0.1016 mm). Pistons are available in oversizes of 0.010, 0.020 and 0.030 inch (0.254, 0.508 and 0.762 mm) as well as standard size. Recommended piston to cylinder bore clearance measured just below oil ring at right angle to pin is 0.006-0.008 inch (0.1524-0.2032 mm).

Piston pin fit in piston should be from 0.000 inch (0.000 mm) (light interference) to 0.0003 inch (0.0076 mm) (loose). Piston pin fit in connecting rod should be 0.0006-0.0011 inch (0.0152-0.0279 mm). Piston pins of 0.005 inch (0.127 mm) oversize are available. Always renew pin retaining rings.

Kohler recommends piston rings always be renewed when they are removed. Piston ring specifications are as follows:

Piston ring side clearance–
Top ring .... 0.002-0.004 in. (0.051-0.102 mm)
Middle ring .... 0.001-0.003 in. (0.025-0.076 mm)
Oil ring .... 0.001-0.003 in. (0.025-0.076 mm)

Fig. KO91—Identify one side of engine from the other before making repairs. Number one side contains governor assembly, oil pump and dipstick. To split crankcase, lay on number one side, place large flat blade screwdriver in notch and carefully pry apart.

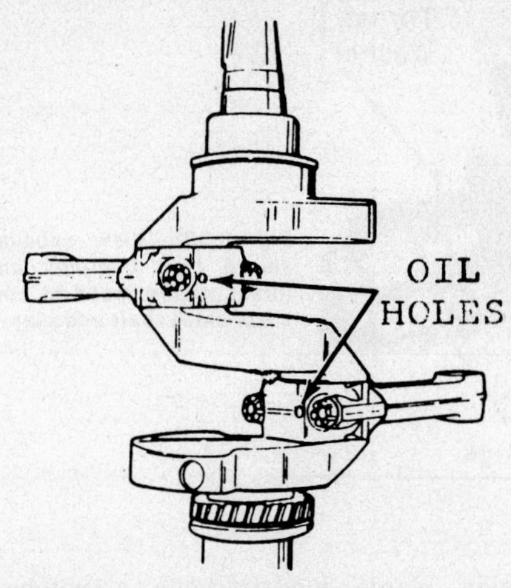

Fig. KO91A—Oil holes in rod caps of engines with pressure spray lubrication system must be facing up with crankcase in upright position when installed.

Piston ring end gap,
top and middle ring–
Used cylinder ...... 0.010-0.030 in. (0.254-0.762 mm)
New cylinder bore ... 0.010-0.020 in. (0.254-0.508 mm)

Piston ring end gap,
oil ring .... 0.060 in. (1.52 mm)

CRANKCASE, CRANKSHAFT, CONNECTING ROD. When making repairs on crankcase assembly, it will be necessary to identify one side of engine from the other. Number "1" side contains governor assembly, oil pump gear and dipstick assembly. To service internal parts, remove closure plate retaining screws and slide it off crankshaft. Lay crankcase down so number "2" side is up. Put tape around tappet stems to prevent them from falling into case when halves are split. On Series II engines with full pressure lubricating system, mark across machined camshaft plug and crankcase so plug may be reinstalled in its original position. See Fig. KO90A. Locate crankcase splitting notches. Place a large flat blade screwdriver in one notch and carefully pry halves apart (Fig. KO91).

Connecting rods must be removed with crankshaft. Identify each rod and piston unit so they can be reinstalled in correct cylinder and do not intermix connecting rod caps. Connecting rods ride directly on crankshaft. If crankpin is 0.0005 inch (0.0127 mm) out-of-round or is tapered 0.001 inch (0.0254 mm), regrind crankpin. Rods with 0.010 inch (0.254 mm) undersize crankpin bore are available for reground crankshaft. KT-17 and KT-19 standard crankpin diameter is 1.3735-1.3740 inches (34.8869-34.8996 mm). KT-21 standard crankpin bore is 1.4993-1.4998 inches (38.082-38.095 mm). Desired running clearances are as follows:

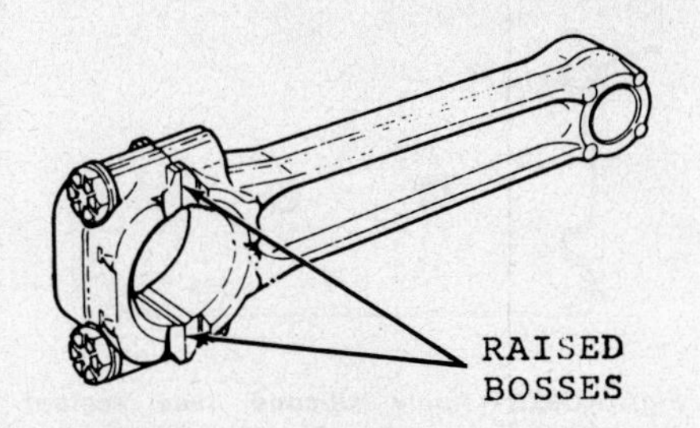

Fig. KO91B—Raised bosses on rods of Series II engines must be installed towards flywheel.

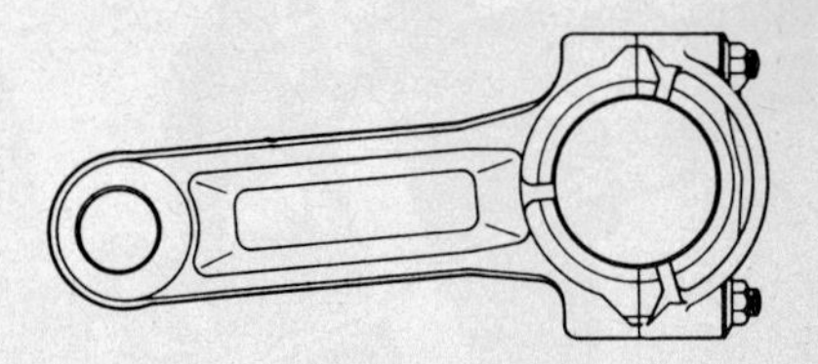

Fig. KO91C—Connecting rods in KT-19 engines are angled. Angle must be down away from camshaft when installed.

Connecting rods to
crankpins .... 0.0012-0.0024 in. (0.030-0.060 mm)
Connecting rod to
piston pin .... 0.0006-0.0011 in. (0.015-0.028 mm)
Rod side play on
crankpin .... 0.005-0.016 in. (0.127-0.406 mm)

Lubricate connecting rod journals, then assemble on crankshaft. Connecting rods and caps must be assembled so match marks on both are aligned.

Early production KT-17 engines with pressure spray lubrication system use connecting rods which have small oiling holes in rod caps and must be installed facing up with crankcase in upright position. See Fig. KO91A.

Series II engines use connecting rods with raised bosses which must be installed towards flywheel. See Fig. KO91B.

Model KT-19 connecting rods are angled as shown in Fig. KO91C. Rods must be installed on piston so they will

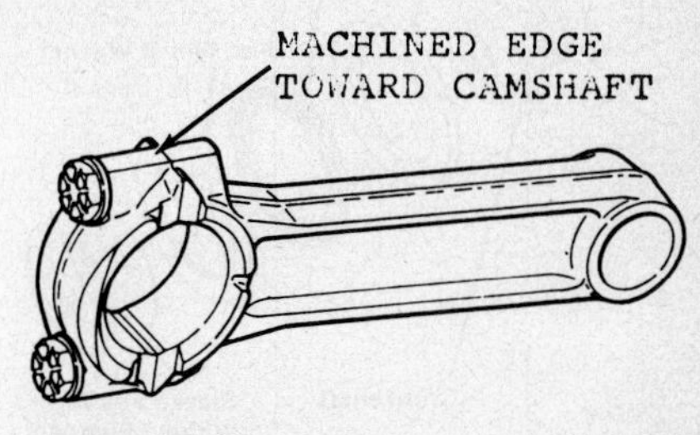

Fig. KO91D—Machined edge of KT-21 connecting rod must be toward camshaft when installed.

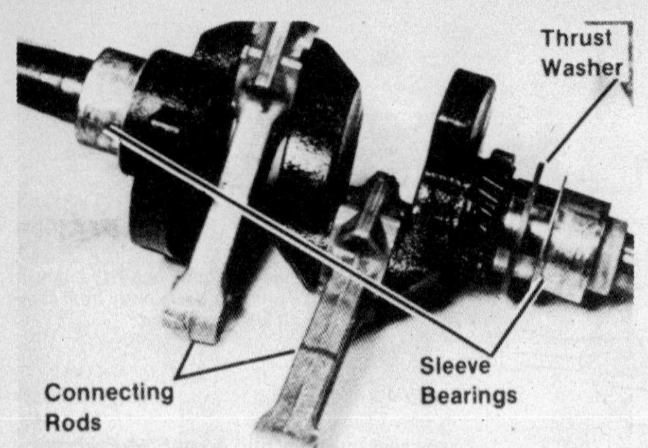

Fig. KO92—View showing sleeve type bearings and thrust washer used to control crankshaft end play.

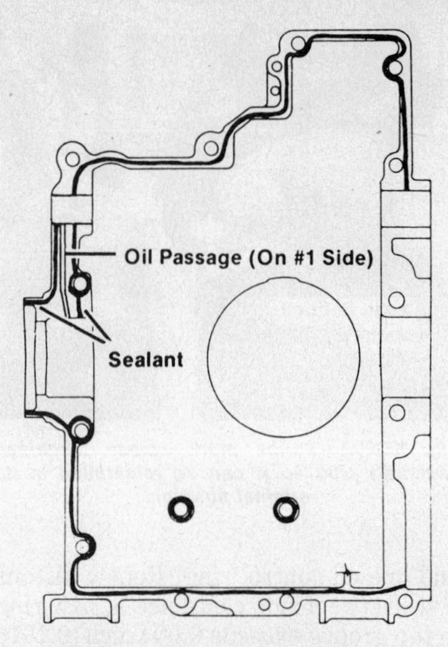

Fig. KO92C—Apply silicone base sealant around edge of crankcase, as indicated by the heavy dark line, making certain sealant does not enter oil passage on number one side of Series II engines.

be angled down away from camshaft when installed on crankshaft.

Model KT-21 connecting rods have a machined edge (Fig. KO91D) which must be toward camshaft when installed.

On all models, if connecting rod bolts are used to retain connecting rod cap, overtorque bolts to 20 ft.-lbs. (27 N•m), loosen and retorque to 17 ft.-lbs. (23 N•m). If "Posi-Lock" nuts are used to retain connecting rod caps, tighten nuts to 12 ft.-lbs. (16 N•m) for new rods and nuts and to 8 ft.-lbs. (11 N•m) for used rods and nuts.

Crankshaft may be supported by either ball bearings or sleeve type bearings. Crankshaft end play if supported by ball bearings is 0.002-0.023 inch (0.051-0.584 mm) and is adjusted by varying thickness of spacers on flywheel end of crankshaft. Spacers are available in a variety of thicknesses.

Sleeve type bearings have a running clearance of 0.0013-0.0033 inch (0.0330-0.0838 mm). On early production engines with pressure spray lubrication system and sleeve type bearings, crankshaft end play should be 0.003-0.013 inch (0.0762-0.3302 mm) and is adjusted by varying thickness of thrust washer (Fig. KO92) on pto end of crankshaft. Install washer with chamfer toward inside. Some models have a notched spacer between lip of bearing and crankcase on flywheel end. The bearing tab should fit into notch. Make certain oil hole in sleeve type bearings are aligned with hole in crankcase.

Some Series II engines use a roller thrust bearing and flat washers on flywheel end of crankshaft. Refer to Fig. KO92A for assembly sequence and note shaft locating washer is 0.1575 inch (4.0 mm) thick.

Early production engines with pressure spray lubrication system use a cupped plug to seal camshaft opening at flywheel end. Bolt crankcase halves together applying sealant to edges as shown in Fig. KO92B. Tighten crankcase bolts as shown in Fig. KO89. Apply sealant to plug and drive in until flush with crankcase. Cupped end of plug must face outward.

All Series II and KT-21 engines use a machined steel plug and "O" ring to seal camshaft opening (Fig. KO90A). Apply silicone base sealant around edge of crankcase as indicated in Fig. KO92C, install "O" ring on machined plug and hold plug in opening while bolting halves together. Tighten bolts as shown in Fig. KO89A. Stake plug around edges of crankcase at original stake marks (Fig. KO90A).

Crankshaft oil seals can be installed after crankcase and closure plate are assembled. Carefully work oil seals over crankshaft and drive seals into place with hollow driver that contacts outer edge of seals.

CAMSHAFT. Camshaft is supported directly in machined bores in crankcase halves. Camshaft bearing running clearance is 0.0010-0.0025 inch (0.0254-0.0635 mm).

Renew camshaft if gear teeth or lobes are badly worn or chipped. Check camshaft end play with a feeler gage. End

Fig. KO92A—View showing location of roller thrust bearing used on some Series II engines.

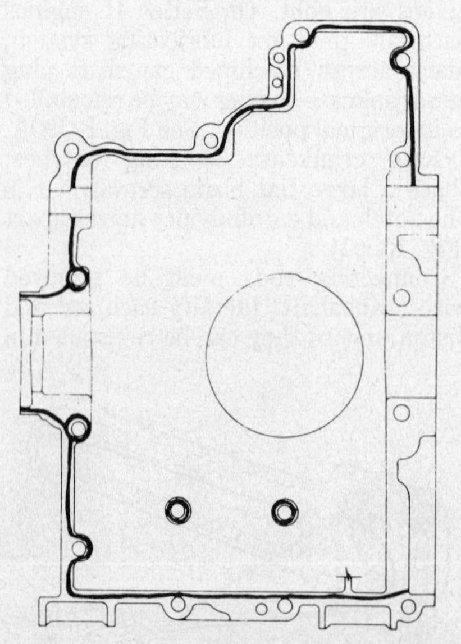

Fig. KO92B—Apply silicone base sealant around edge of crankcase, as indicated by the heavy dark line, on early production engines with pressure spray lubrication system.

Fig. KO93—Assemble crankshaft and camshaft in number one side of crankcase, making certain timing marks are aligned.

play should be 0.003-0.013 inch (0.0762-0.3302 mm). If end play exceeds 0.013 inch (0.3302 mm), renewal of camshaft or crankcase will be necessary. Assemble camshaft in number "1" side of crankcase, making sure timing mark on cam gear aligns with mark on crankshaft. See Fig. KO93.

FUEL PUMP. A mechanically operated diaphragm type fuel pump is used. Pump is actuated by a lever which rides on an eccentric on camshaft. Repair kit is available for reconditioning pump. Replacement pumps are non-metallic in construction and only thin mounting gaskets can be used. Use flat washers under mounting screws to prevent damaging flange. Never install fuel pump on crankcase before case halves are assembled as pump lever may be below camshaft. Proper position is above the camshaft. Pump breakage and possible engine damage could result if lever is positioned below camshaft.

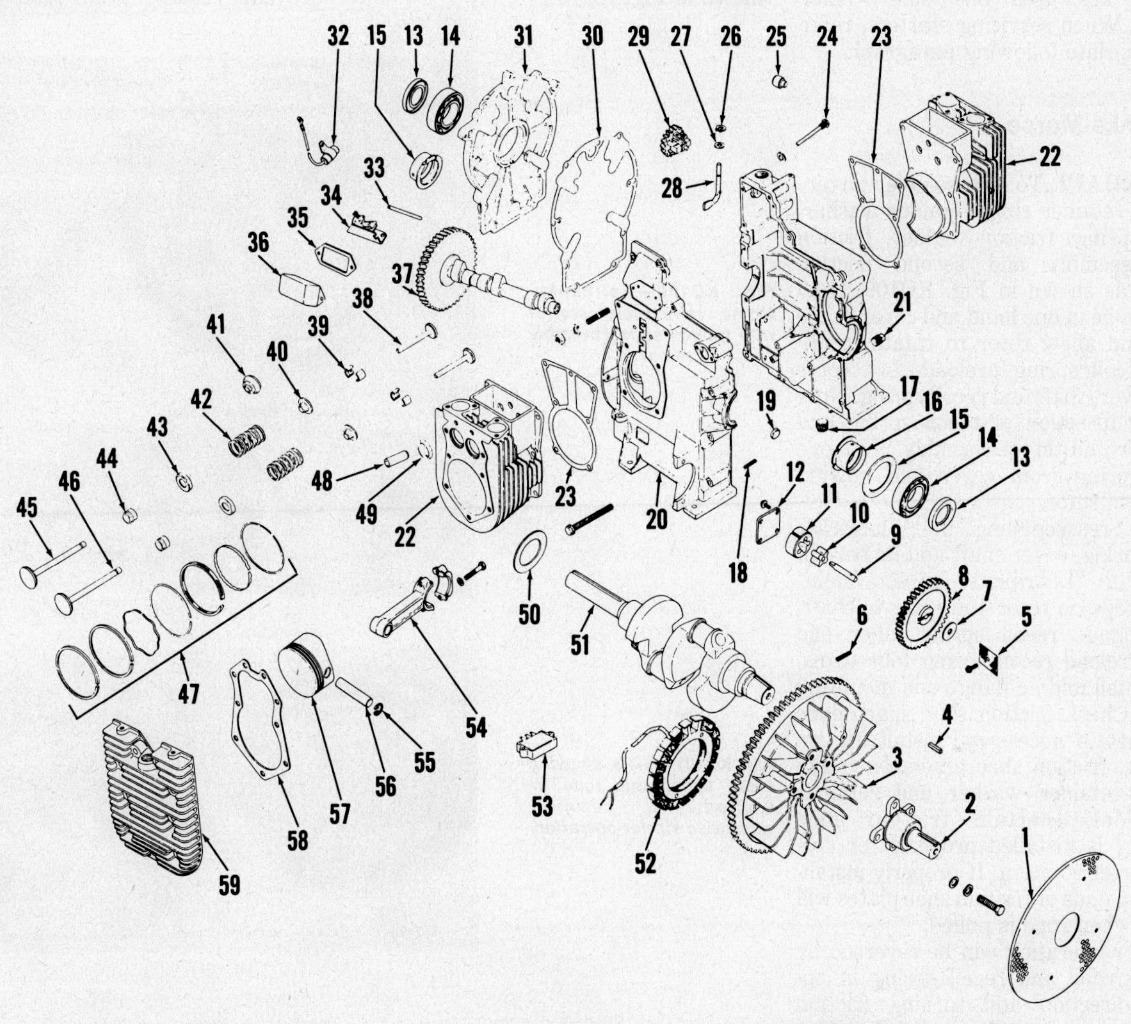

Fig. KO94—Exploded view of basic engine assembly for KT-17 series engines. Series KT-19 and KT-21 are similar.

1. Grass screen
2. Front drive shaft
3. Flywheel
4. Key
5. Screen
6. Roll pin
7. Spacer
8. Oil pump gear
9. Oil pump shaft
10. Inner rotor
11. Outer rotor
12. Pump cover
13. Oil seal
14. Ball bearing
15. Spacer (0.018-0.022 inch)
16. Sleeve bearing
17. Drain plug
18. Roll pin
19. Cap plug (3 used)
20. Crankcase halves
21. Drain plug
22. Cylinder barrel
23. Cylinder barrel gasket
24. Governor stop pin
25. Governor shaft
26. Retainer
27. Washer
28. Governor cross shaft
29. Governor gear
30. Closure plate gasket
31. Closure plate
32. Condenser
33. Breaker push rod
34. Breaker points
35. Gasket
36. Breaker cover
37. Camshaft
38. Tappet (4)
39. Retainer
40. Spring retainer
41. Valve rotator
42. Valve spring (4)
43. Spring retainer
44. Valve seal (4)
45. Exhaust valve
46. Intake valve
47. Piston rings
48. Valve guide (4)
49. Valve seat insert (4)
50. Thrust washer
51. Crankshaft
52. Stator
53. Rectifier-regulator
54. Connecting rod
55. Retainer
56. Piston pin
57. Piston
58. Head gasket
59. Cylinder head

KOHLER

SERVICING KOHLER ACCESSORIES

RETRACTABLE STARTERS

Fairbanks-Morse or Eaton retractable starters are used on some Kohler engines. When servicing starters, refer to appropriate following paragraph.

Fairbanks-Morse

OVERHAUL. To disassemble starter, remove retainer ring, retainer washer, brake spring, friction washer, friction shoe assembly and second friction washer as shown in Fig. KO100. Hold rope handle in one hand and cover in the other and allow rotor to rotate to unwind recoil spring preload. Lift rotor from cover, shaft and recoil spring. Note winding direction of recoil spring and rope for aid in reassembly. Remove recoil spring from cover and unwind rope from rotor.

When reassembling unit, lubricate recoil spring, cover shaft and its bore in rotor with "Lubriplate" or equivalent. Install rope on rotor and rotor to shaft, then engage recoil spring inner end hook. Preload recoil spring four turns, then install middle flange and mounting flange. Check friction shoe sharp ends and renew if necessary. Install friction washers, friction shoe assembly, brake spring, retainer washer and retainer ring. Make certain friction shoe assembly is installed properly for correct starter rotation. If properly installed, sharp ends of friction shoe plates will extend when rope is pulled.

Starter operation can be reversed by winding rope and recoil spring in opposite direction and turning friction shoe assembly upside down. See Fig. KO101 for counter-clockwise assembly.

Eaton

OVERHAUL. To disassemble starter, first release tension of rewind spring as follows: Hold starter assembly with pulley facing up. Pull starter rope until notch in pulley is aligned with rope hole in cover. Use thumb pressure to prevent pulley from rotating. Engage rope in notch of pulley and slowly release thumb pressure to allow spring to unwind until all tension is released.

When removing rope pulley, use extreme care to keep starter spring confined in housing. Check starter spring for breaks, cracks or distortion. If starter spring is to be renewed, carefully remove it from housing, noting direction of rotation of spring before removing. Exploded view of clockwise starter is shown in Fig. KO102.

Check pawl, brake, spring, retainer and hub for wear and renew as necessary. If starter rope is worn or frayed, remove from pulley, noting

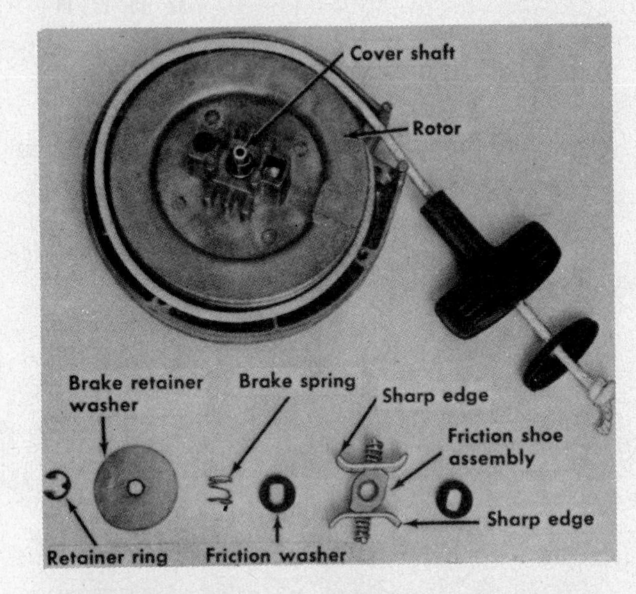

Fig. KO100 – Fairbanks-Morse retractable starter with friction shoe assembly removed.

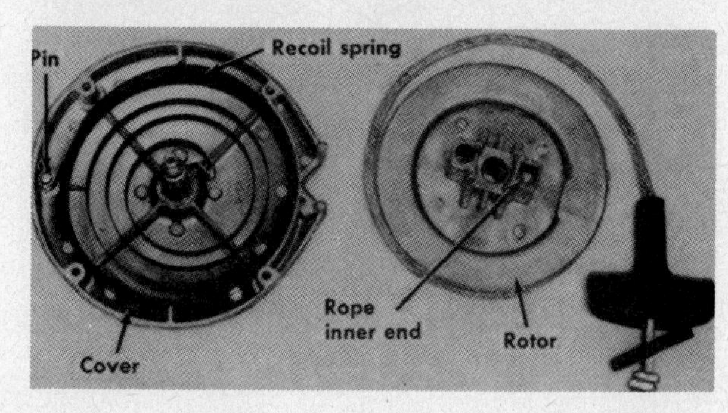

Fig. KO101 – View showing recoil spring and rope installed for counter-clockwise starter operation.

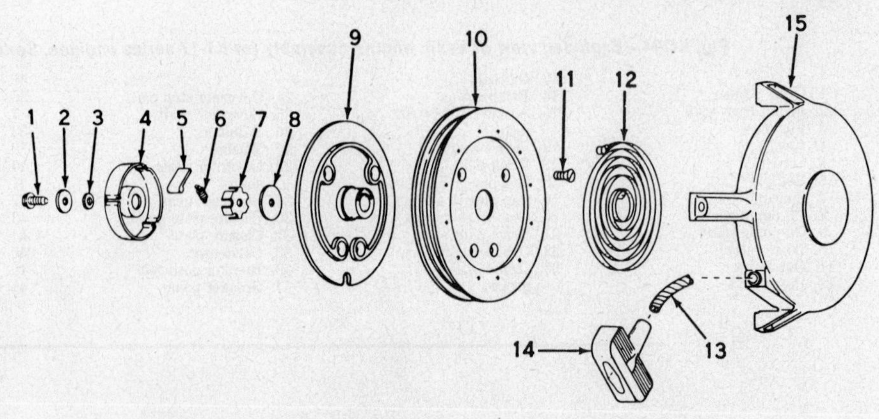

Fig. KO102 – Exploded view of Eaton retractable starter assembly.

1. Retainer screw
2. Brake washer
3. Spacer
4. Retainer
5. Pawl
6. Spring
7. Brake
8. Thrust washer
9. Pulley hub
10. Pulley
11. Screw
12. Recoil spring
13. Rope
14. Handle
15. Starter housing

direction it is wrapped on pulley. Renew rope and install pulley in housing, aligning notch in pulley assembly in housing, align notch in pulley hub with hook in end of spring. Use a wire bent to form a hook to aid in positioning spring in hub.

After securing pulley assembly in housing, engage rope in notch and rotate pulley at least two full turns in same direction it is pulled to properly preload starter spring. Pull rope to fully extended position. Release handle and if spring is properly preloaded, rope will fully rewind.

Before installing starter on engine, check teeth in starter driven hub (165-Fig. KO103) for wear and renew hub if necessary.

12-VOLT STARTER-GENERATOR

A combination 12-volt starter-generator manufactured by Delco-Remy is used on some Kohler engines. Starter-generator functions as a cranking motor when starting switch is closed. When engine is operating and with starting switch open, unit operates as a generator. Generator output and circuit voltage for battery and various operating requirements are controlled by a current-voltage regulator.

Kohler recommends starter-generator belt tension be adjusted until about 10 pounds pulling pressure (4.5 kg) applied midway between pulleys will deflect belt ½-inch (12.7 mm).

To determine cause of abnormal operation starter-generator should be given a "no-load" test or a "generator output" test. Generator output test can be performed with starter-generator on or off engine. No-load test must be made with starter-generator removed from engine. Refer to Fig. KO104 for exploded view of starter-generator assembly. Parts are available from Kohler as well as authorized Delco-Remy service stations.

Starter-generator brush spring tension for all models should be 24-32 oz. (0.68-0.91 kg).

Starter-generator and regulator service test specifications are as follows:

Starter-Generators 1101940, 1101970, 1101973 & 1101980.

Field draw–
Amperes 1.52-1.62
Volts 12
Cold output–
Amperes 12
Volts 12
Rpm 4950
No-load test–
Volts 11
Amperes (max.) 18
Rpm (min.) 2500
Rpm (max.) 2900

Starter-Generators 1101932, 1101948, 1101968, 1101972 & 1101974.

Field draw–
Amperes 1.45-1.57
Volts 12
Cold output–
Amperes 10
Volts 14
Rpm 5450
No-load test–
Volts 11
Amperes (max.) 17
Rpm (min.) 2500
Rpm (max.) 2900

Starter-Generator 1101951 & 1101967.

Field draw–
Amperes 1.52-1.62
Volts 12
Cold output–
Amperes 15
Volts 14
Rpm 3400
No-load test–
Volts 11
Amperes (max.) 14
Rpm (min.) 1650
Rpm (max.) 1950

Starter-Generator 1101996.

Field draw–
Amperes 1.52-1.62
Volts 12
Cold output–
Amperes 12
Volts 14
Rpm 4950
No-load test–
Volts 11
Amperes (max.) 18
Rpm (min.) 2500
Rpm (max.) 2900

Regulators 1118984, 1118988 &1118999.

Ground polarity . . . . Negative
Cut-out relay–
Air gap . . . . 0.020 in. (0.508 mm)
Point gap . . . . 0.020 in. (0.508 mm)
Closing voltage, range 11.8-14.0
Adjust to 12.8
Voltage regulator–
Air gap . . . . 0.075 in. (1.905 mm)
Setting voltage, range 13.6-14.5
Adjust to 14.0

Regulator 1118985.

Ground polarity . . . . Positive
Cut-out relay–
Air gap . . . . 0.020 in. (0.508 mm)
Point gap . . . . 0.020 in. (0.508 mm)
Closing voltage, range 11.8-14.0
Adjust to 12.8
Voltage regulator–
Air gap . . . . 0.075 in. (1.905 mm)
Setting voltage, range 13.6-14.5
Adjust to 14.0

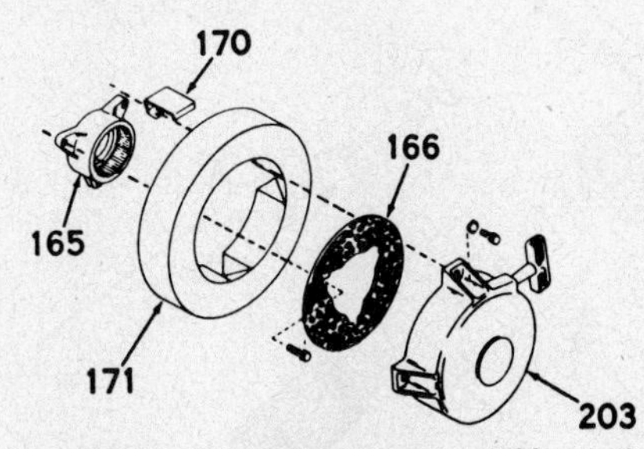

Fig. KO103—View showing retractable starter and starter hub.

165. Starter hub
166. Screen
170. Bracket
171. Air director
203. Retractable starter assy.

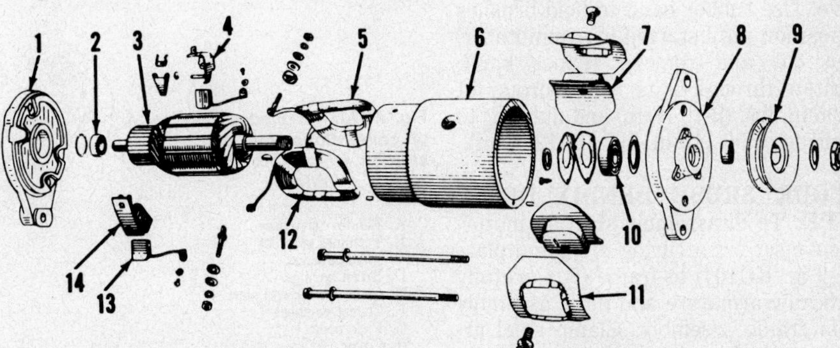

Fig. K0104—Exploded view of typical Delco-Remy starter generator assembly.

1. Commutator end frame
2. Bearing
3. Armature
4. Ground brush holder
5. Field coil (L.H.)
6. Frame
7. Pole shoe
8. Drive end frame
9. Pulley
10. Bearing
11. Field coil insulator
12. Field coil (L.H.)
13. Brush
14. Insulated brush holder

12-VOLT GEAR DRIVE STARTERS

Four types of gear drive starters are used on Kohler engines. Refer to Figs. KO105, KO106, KO107 and KO108 for exploded view of starter motors and drives.

TWO BRUSH COMPACT TYPE. To disassemble starting motor, clamp mounting bracket in a vise. Remove through-bolts (H–Fig. KO105) and slide commutator end plate (J) and frame assemble (A) off armature. Clamp steel armature core in a vise and remove Bendix drive (E), drive end plate (F), thrust washer (D) and spacer (C) from armature (B).

Renew brushes if unevenly worn or worn to a length of 5/16-inch (7.938 mm) or less. To renew ground brush (K), drill out rivet, then rivet new brush lead to end plate. Field brush (P) is soldered to field coil lead.

Reassemble by reversing disassembly procedure. Lubricate bushings with a light coat of SAE 10 oil. Inspect Bendix drive pinion and splined sleeve for damage. If Bendix is in good condition, wipe clean and install completely dry. Tighten Bendix drive retaining nut to a torque of 130-150 in.-lbs. (15-18 N•m). Tighten through-bolts (H) to a torque of 40-55 in.-lbs. (4-7 N•m).

PERMANENT MAGNET TYPE. To disassemble starting motor, clamp mounting bracket in a vise and remove through-bolts (19–Fig. KO106). Carefully slide end cap (10) and frame (11) off armature. Clamp steel armature core in a vise and remove nut (18), spacer (17), anti-drift spring (16), drive assembly (15), end plate (14) and thrust washer (13) from armature (12).

The two input brushes are part of terminal stud (6). Remaining two brushes (9) are secured with cap screws. When reassembling, lubricate bushings with American Bosch lubricant #LU3001 or equivalent. Do not lubricate starter drive. Use rubber band to hold brushes in position until started in commutator, then cut and remove rubber band. Tighten through-bolts to a torque of 80-95 in.-lbs. (8-10 N•m) and nut (18) to a torque of 90-110 in.-lbs. (11-12 N•m).

FOUR BRUSH BENDIX DRIVE TYPE. To disassemble starting motor, remove screws securing drive end plate (K–Fig. KO107) to frame (I). Carefully withdraw armature and drive assembly from frame assembly. Clamp steel armature core in a vise and remove Bendix drive retaining nut, then remove drive unit (A), end plate (K) and thrust washer from armature (J). Remove cover (H) and screws securing end plate (E) to frame. Pull field brushes (C) from brush holders and remove end plate assembly.

The two ground brush leads are secured to end plate (E) and the two field brush leads are soldered to field coils. Renew brush set if excessively worn.

Inspect bushing (L) in end plate (K) and renew bushing if necessary. When reassembling, lubricate bushings with light coat of SAE 10 oil. Do not lubricate Bendix drive assembly.

Note starter may be reinstalled with Bendix in engaged or disengaged posi-

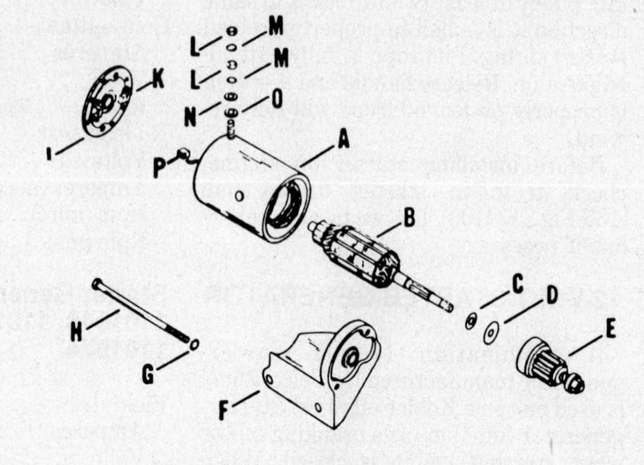

Fig. KO105—Exploded view of two brush compact gear drive starting motor.

A. Frame & field coil assy.
B. Armature
C. Spacer
D. Thrust washer
E. Bendix drive assy.
F. Drive end plate & mounting bracket
G. Lockwasher
H. Through-bolt
J. Commutator end plate
K. Ground brush
L. Terminal nuts
M. Lockwashers
N. Flat washer
O. Insulating washer
P. Field brush

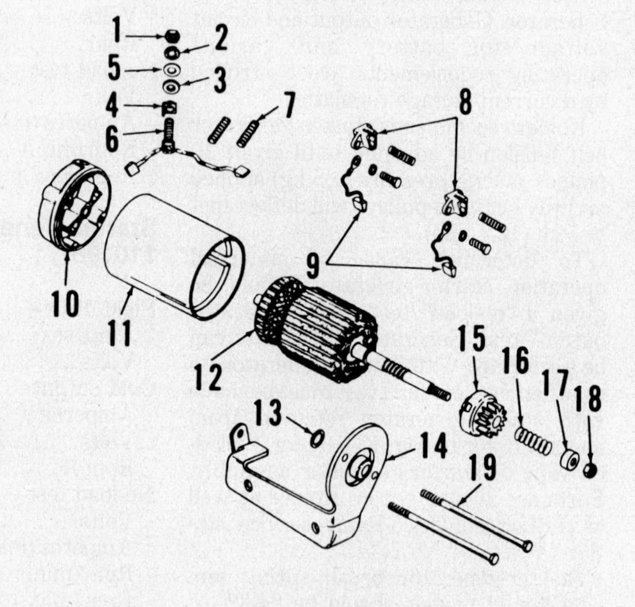

Fig. KO106—Exploded view of permanent magnet type starting motor.

1. Terminal nut
2. Lockwasher
3. Insulating washer
4. Terminal insulator
5. Flat washer
6. Terminal stud & input brushes
7. Brush springs (4 used)
8. Brush holders
9. Brushes
10. Commutator end cap
11. Frame & permanent magnets
12. Armature
13. Thrust washer
14. Drive end plate & mounting bracket
15. Drive assy.
16. Anti-drift spring
17. Spacer
18. Nut
19. Through-bolts

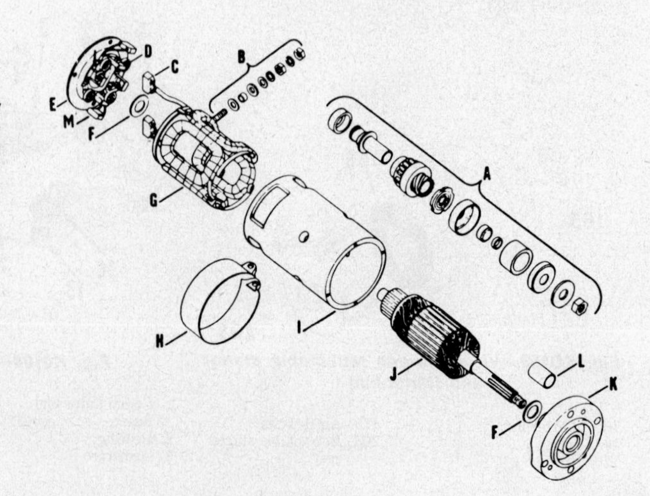

Fig. KO107—Exploded view of conventional four brush starting motor with Bendix drive.

A. Bendix drive assy.
B. Terminal stud set
C. Field brushes
D. Brush springs
E. Commutator end plate
F. Thrust washers
G. Field coils
H. Cover
I. Frame
J. Armature
K. Drive end plate
L. Bushing
M. Ground brushes

tion. Do not attempt to disengage Bendix if it is in the engaged position.

FOUR BRUSH SOLENOID SHIFT TYPE. To disassemble starting motor, refer to Fig. KO108; then unbolt and remove solenoid switch assembly (items 1 through 6). Remove through-bolts (23), end plate (24) and frame (30) with brushes (26), brush holders (27 and 29) and field coil assembly (33). Remove screws retaining center bearing (21) to drive housing (12), remove shift lever pivot bolt, raise shift lever (9) and carefully withdraw armature and drive assembly. Drive unit (16) and center bearing (21) can be removed from armature (22) after snap ring (14) and retainer (15) are removed. Drive out shift lever pin and separate plunger (7), seal (8) and shift lever (9) from drive housing. Any further disassembly is obvious after examination of unit. Refer to Fig. KO108. Renew brushes (26), center bearing (21) and bushings in end plate (24) and drive housing (12) as necessary.

Fig. KO108—Exploded view of conventional four brush starting motor with solenoid shift engagement.

1. Switch cover
2. Spring
3. Contact disc
4. Gasket
5. Coil assy.
6. Return spring
7. Plunger
8. Seal
9. Shift lever
10. Bushing
11. Lubrication wick
12. Drive housing
13. Drive end thrust washer
14. Snap ring
15. Retainer
16. Drive unit
17. Spring
18. Shift collar
19. Snap ring
20. Brake washer
21. Center bearing
22. Armature
23. Through-bolt
24. End plate
25. Thrust washer
26. Brush (4 used)
27. Insulated brush holder
28. Brush spring
29. Ground brush
30. Frame
31. Field coil insulator
32. Pole shoe
33. Field coil assy.

FLYWHEEL ALTERNATORS

3 AMP ALTERNATOR. The 3 amp alternator consists of a permanent magnet ring with five or six magnets on flywheel rim, a stator assembly attached to crankcase and a diode in charging output lead. See Fig. KO109.

To avoid possible damage to charging system, the following precautions must be observed:

1. Negative post of battery must be connected to engine ground and correct battery polarity must be observed at all times.
2. Prevent alternator leads (AC) from touching or shorting.
3. Remove battery or disconnect battery cables when recharging battery with battery charger.
4. Do not operate engine for any length of time without a battery in system.
5. Disconnect plug before electric welding is done on equipment powered by and in common ground with engine.

TROUBLESHOOTING. Defective conditions and possible causes are as follows:

1. No output. Could be caused by:
   - A. Faulty windings in stator.
   - B. Defective diode.
   - C. Broken lead wire.
2. No lighting. Could be caused by:
   - A. Shorted stator wiring.
   - B. Broken lead.

If "no output" condition is the trouble, run following tests:

1. Connect ammeter in series with charging lead. Start engine and run at 2400 rpm. Ammeter should register 2 amp charge. Run engine at 3600 rpm. Ammeter should register 3 amp charge.
2. Disconnect battery charge lead from battery, measure resistance of lead to ground with an ohmmeter. Reverse ohmmeter leads and take another reading. One reading should be about mid-scale with meter set at R x 1. If both readings are high, diode or stator is open.
3. Expose diode connections on battery charge lead. Check resistance on stator side to ground. Reading should be

Fig. KO109—Typical electrical wiring diagram for engines equipped with 3 amp alternator.

1 ohm. If 0 ohms, winding is shorted. If infinity ohms, stator winding is open or lead wire is broken.

If "no lighting" condition is the trouble, use an AC voltmeter and measure open circuit voltage from lighting lead to ground with engine running at 3000 rpm. If 15 volts, wiring may be shorted.

Check resistance of lighting lead to ground. If 0.5 ohms, stator is good, 0 ohms indicates shorted stator and a reading of infinity indicates stator is open or lead is broken.

3/6 AMP ALTERNATOR. The 3/6 amp alternator consists of a permanent magnet ring with six magnets on flywheel rim, a stator assembly attached to crankcase and two diodes located in battery charging lead and auxiliary load lead. See Fig. KO110.

To avoid possible damage to charging system, the following precautions must be observed.

1. Negative post of battery must be connected to engine ground and correct battery polarity must be observed at all times.
2. Prevent alternator leads (AC) from touching or shorting.
3. Do not operate for any length of time without a battery in system.
4. Remove battery or disconnect battery cables when recharging battery with battery charger.
5. Disconnect plug before electric welding is done on equipment powered by and in common ground with engine.

TROUBLESHOOTING. Defective conditions and possible causes are as follows:

1. No output. Could be caused by:
   A. Faulty windings in stator.
   B. Defective diode.
   C. Broken lead.
2. No lighting. Could be caused by:
   A. Shorted stator wiring.
   B. Broken lead.

If "no output" condition is the trouble, run the following tests:

1. Disconnect auxiliary load lead and measure voltage from lead to ground with engine running 3000 rpm. If 17 volts or more, stator is good.
2. Disconnect battery charging lead from battery. Measure voltage from charging lead to ground with engine running at 3000 rpm. If 17 volts or more, stator is good.
3. Disconnect battery charge lead from battery and auxiliary load lead from switch. Measure resistance of both leads to ground. Reverse ohmmeter leads and take readings again. One reading should be infinity and the other reading should be about mid-scale with meter set at R x 1. If both readings are low, diode is shorted. If both readings are high, diode or stator is open.
4. Expose diode connections on battery charging lead and auxiliary load lead. Check resistance on stator side of diodes to ground. Readings should be 0.5 ohms. If reading is 0 ohms, winding is shorted. If infinity ohms, stator winding is open or lead wire is broken.

If "no lighting" condition is the trouble, disconnect lighting lead and measure open circuit voltage with AC voltmeter from lighting lead to ground with engine running at 3000 rpm. If 22 volts or more, stator is good. If less than 22 volts, wiring may be shorted.

Check resistance of lighting lead to ground. If 0.5 ohms, stator is good, 0 ohms reading indicates shorted stator and an infinity reading indicates an open stator winding or broken lead wire.

10 AND 15 AMP ALTERNATOR. Either a 10 or 15 amp alternator is used on some engines. Alternator output is controlled by a solid state rectifier-regulator.

To avoid possible damage to charging system, the following precautions must be observed:

1. Negative post of battery must be connected to engine ground and correct battery polarity must be observed at all times.
2. Rectifier-regulator must be con-

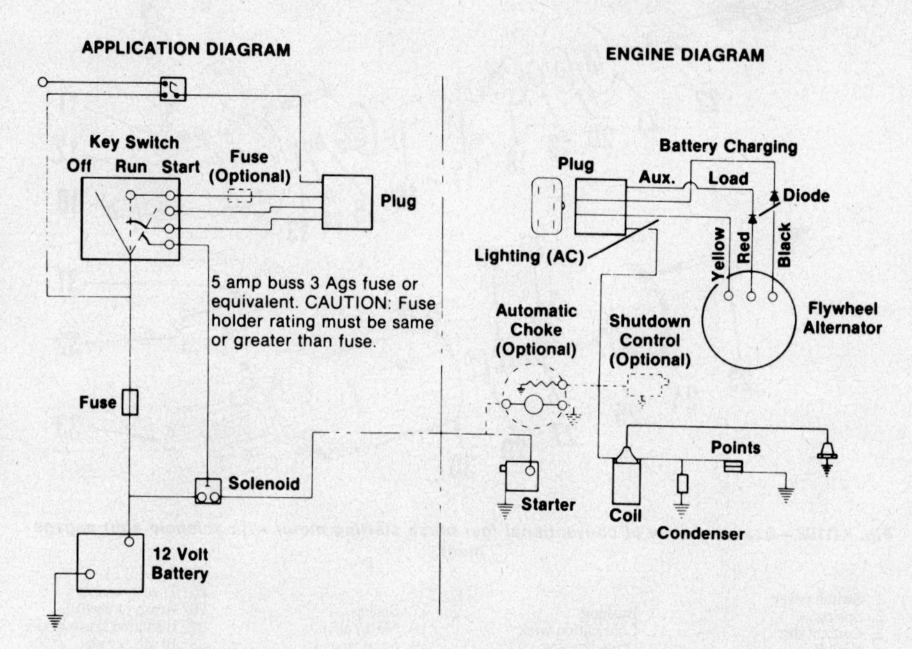

Fig. KO110—Typical electrical wiring diagram for engines equipped with 3/6 amp alternator.

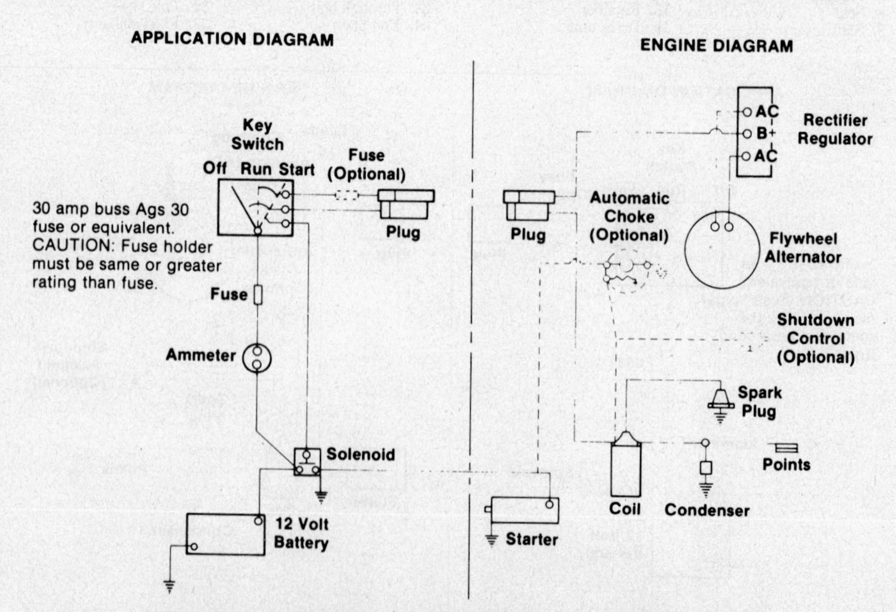

Fig. Ko111—Typical electrical wiring diagram for engines equipped with 15 amp alternator and breaker point ignition. The 10 amp alternator is similar.

nected in common ground with engine and battery.

3. Disconnect leads at rectifier-regulator if electric welding is to be done on equipment in common ground with engine.

4. Remove battery or disconnect battery cables when recharging battery with battery charger.

5. Do not operate engine with battery disconnected.

6. Make certain AC leads are prevented from being grounded at all times.

OPERATION. Alternating current (AC) produced by alternator is changed to direct current (DC) in rectifier-regulator. See Fig. KO112. Current regulation is provided by electronic devices which "sense" counter-voltage created by battery to control or limit charging rate. No adjustments are possible on alternator charging system. Faulty components must be renewed. Refer to the following troubleshooting paragraph to help locate possible defective parts.

TROUBLESHOOTING. Defective conditions and possible causes are as follows:

1. No output. Could be caused by:
   A. Faulty windings in stator.
   B. Defective diode(s) in rectifier.
   C. Rectifier-regulator not properly grounded.
2. Full charge-no regulation. Could be caused by:
   A. Defective rectifier-regulator.
   B. Defective battery.

If "no output" condition is the trouble, disconnect B+ cable from rectifier-regulator. Connect a DC voltmeter between B+ terminal on rectifier-regulator and engine ground. Start engine and operate at 3600 rpm. DC voltage should be above 14 volts. If reading is above 0 volts but less than 14 volts, check for defective rectifier-regulator. If reading is 0 volts, check for defective rectifier-regulator or defective stator by disconnecting AC leads from rectifier-regulator and connecting an AC voltmeter to the two AC leads. Check AC voltage with engine running at 3600 rpm. If reading is less than 20 volts (10 amp alternator) or 28 volts (15 amp alternator), stator is defective. If reading is more than 20 volts (10 amp alternator) or 28 volts (15 amp alternator), rectifier-regulator is defective.

If "full charge-no regulation" is the condition, use a DC voltmeter and check B+ to ground with engine operating at 3600 rpm. If reading is over 14.7 volts, rectifier-regulator is defective. If reading is under 14.7 volts but over 14.0 volts, alternator and rectifier-regulator are satisfactory and battery is probably defective (unable to hold a charge).

30 AMP ALTERNATOR. A 30 amp flywheel alternator consisting of a permanent field magnet ring (on flywheel) and an alternator stator (on bearing plate on single cylinder engines or gear cover on two cylinder engines) is used on some models. Alternator output is controlled by a solid state rectifier-regulator.

To avoid possible damage to charging system, the following precautions must be observed:

1. Negative post of battery must be connected to engine ground and correct battery polarity must be observed at all times.

2. Rectifier-regulator must be connected in common ground with engine and battery.

3. Disconnect wire from rectifier-regulator terminal marked "BATT. NEG." if electric welding is to be done on equipment in common ground with engine.

4. Remove battery or disconnect battery cables when recharging battery with battery charger.

5. Do not operate engine with battery disconnected.

6. Make certain AC leads are prevented from being grounded at all times.

OPERATION. Alternating current (AC) produced by alternator is carried by two black wires to full wave bridge rectifier where it is changed to direct current (DC). Two red stator wires serve to complete a circuit from regulator to secondary winding in stator. A zener diode is used to sense battery voltage and it controls a Silicon Controlled Rectifier (SCR). SCR functions as a switch to allow current to flow in secondary winding in stator when battery voltage gets above a specific level.

An increase in battery voltage increases current flow in secondary winding in stator. This increased current flow in secondary winding brings about a corresponding decrease in AC current in primary winding, thus controlling output.

When battery voltage decreases, zener diode shuts off SCR and no current flows to secondary winding. At this time, maximum AC current is produced by primary winding.

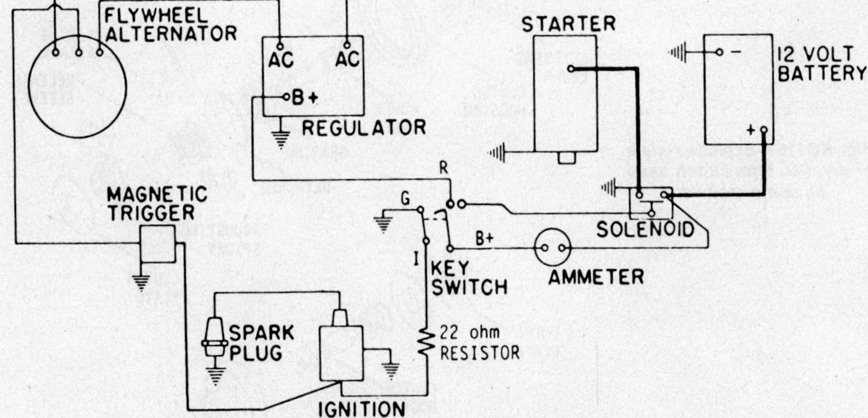

Fig. KO112—Typical electrical wiring diagram for engine equipped with 15 amp flywheel alternator and breakerless ignition system. The 10 amp alternator is similar.

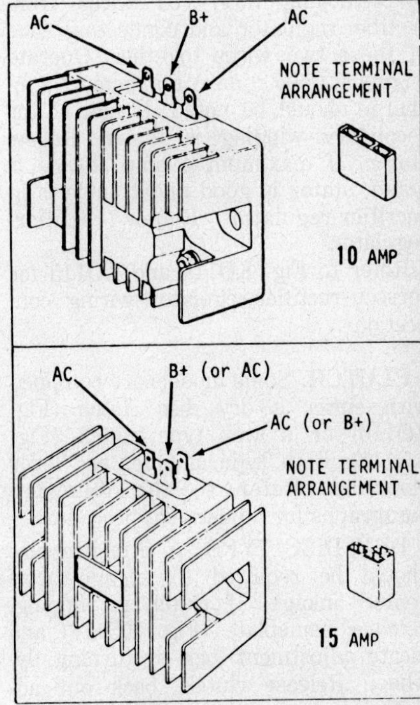

Fig. KO113—Rectifier-regulators used with 10 amp and 15 amp alternators. Although similar in appearance, units must not be interchanged.

TROUBLESHOOTING. Defective conditions and possible causes are as follows:

1. No output. Could be caused by:
   A. Faulty windings in stator.
   B. Defective diode(s) in rectifier.
2. No charge (when normal load is applied to battery). Could be caused by:
   A. Faulty secondary winding in stator.
3. Full charge-no regulation. Could be caused by:
   A. Faulty secondary winding in stator.
   B. Defective regulator.

If "no output" condition is the trouble, check stator windings by disconnecting all four stator wires from rectifier-regulator. Check resistance on R x 1 scale of ohmmeter. Connect ohmmeter leads to the two red stator wires. About 2.0 ohms should be noted. Connect ohmmeter leads to the two black stator wires. Approximately 0.1 ohm should be noted. If readings are not at test values, renew stator. If ohmmeter readings are correct, stator is good and trouble is in rectifier-regulator. Renew rectifier-regulator.

If "no charge when normal load is applied to battery" is the trouble, check stator secondary winding by disconnecting red wire from "REG" terminal on rectifier-regulator. Operate engine at 3600 rpm. Alternator should now charge at full output. If full output of at least 30 amps is not attained, renew stator.

If "full charge-no regulation" is the trouble, check stator secondary winding by removing both red wires from rectifier-regulator and connecting ends of these two wires together. Operate engine at 3600 rpm. A maximum 4 amp charge should be noted. If not, stator secondary winding is faulty. Renew stator. If maximum 4 amp charge is noted, stator is good and trouble is in rectifier-regulator. Renew rectifier-regulator.

Refer to Fig. KO114 and KO115 for correct rectifier-regulator wiring connections.

CLUTCH. Some models are equipped with either a dry disc clutch (Fig. KO116) or a wet type clutch (Fig. KO118). Both type clutches are lever operated. Refer to the following paragraphs for adjustment procedure.

DRY DISC TYPE. A firm pressure should be required to engage over-center linkage. If clutch is slipping, remove nameplate (Fig. KO116) and locate adjustment lock by turning flywheel. Release clutch, back out adjusting lock screw, then turn adjusting spider clockwise until approximately 20 pounds (9 kg) pull is required to snap clutch over-center. Tighten adjusting lock screw. Every 50 hours, lubricate clutch bearing collar through inspection cover opening.

WET TYPE CLUTCH. To adjust wet type clutch, remove nameplate and use a screwdriver to turn adjusting ring (Fig.

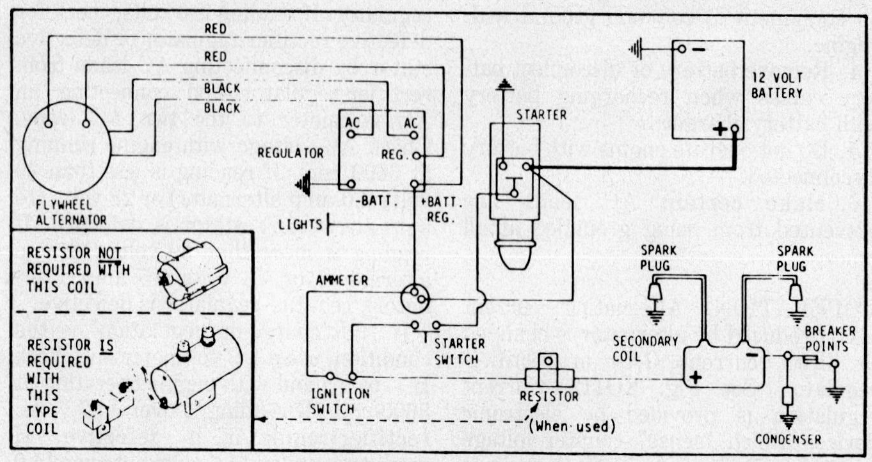

Fig. KO114—Typical electrical wiring diagram of two cylinder engine equipped with 30 amp alternator charging system. The 30 amp alternator on single cylinder engines is similar.

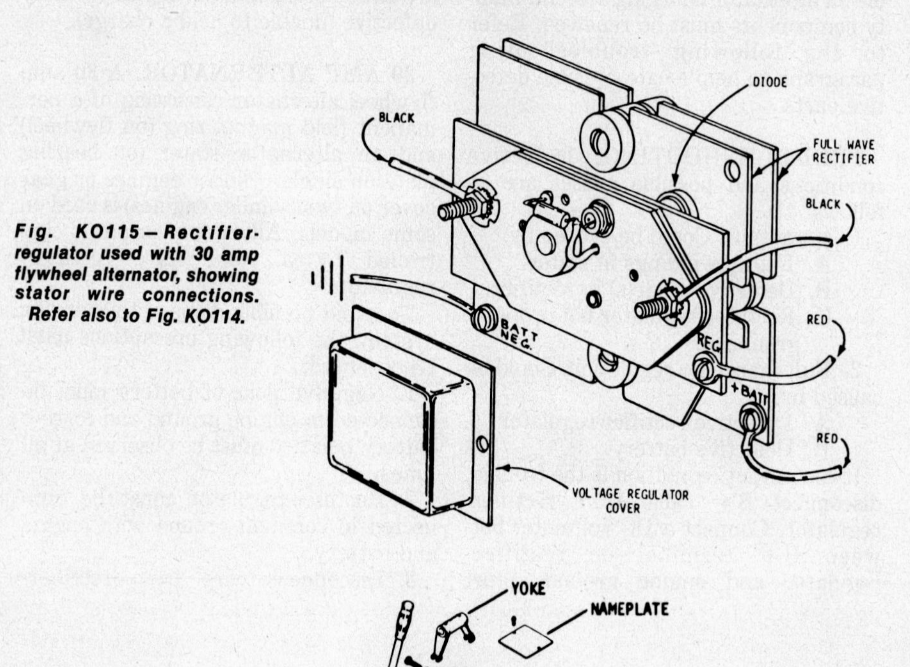

Fig. KO115—Rectifier-regulator used with 30 amp flywheel alternator, showing stator wire connections. Refer also to Fig. KO114.

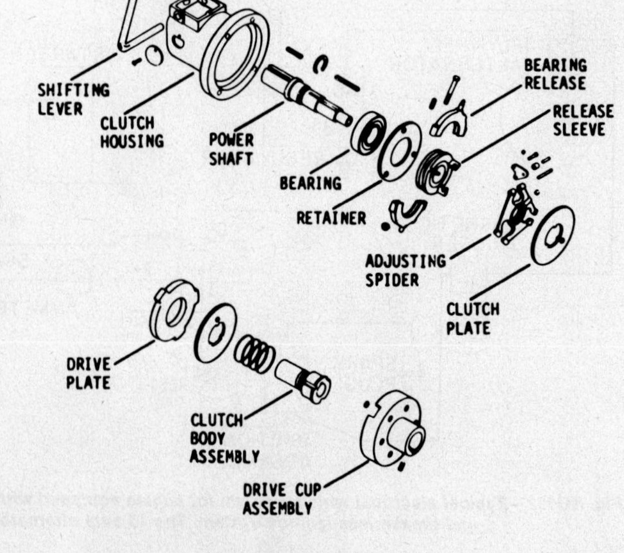

Fig. KO116—Exploded view of dry disc type clutch used on some models.

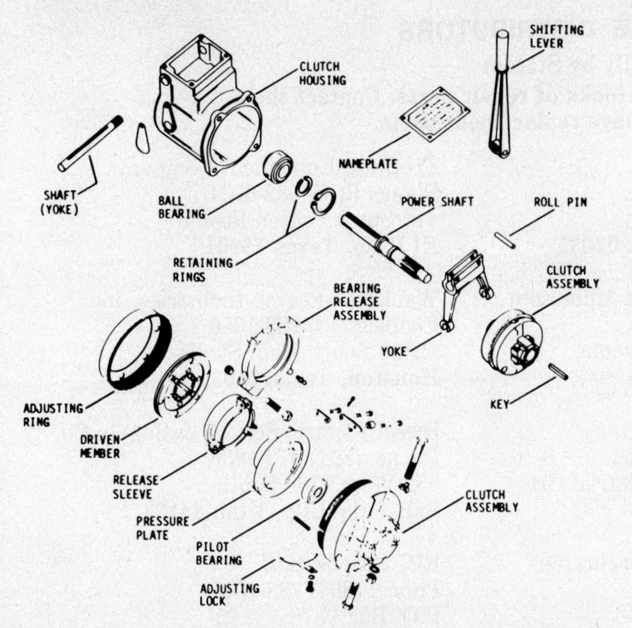

Fig. KO118—Exploded view of wet type clutch used on some models.

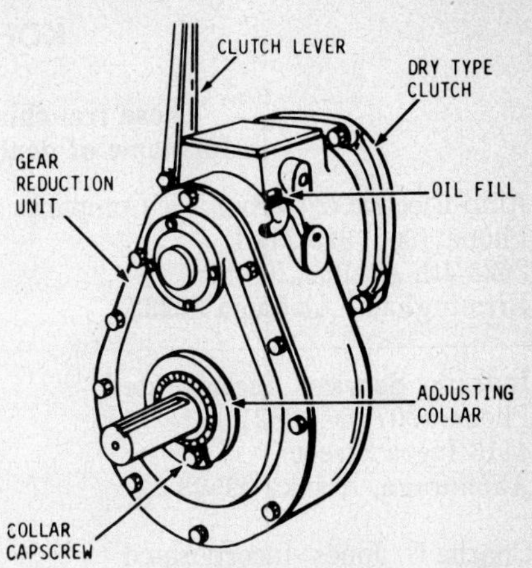

Fig. KO121—Output shaft end play on combination clutch and reduction drive must be adjusted to 0.0015-0.003 inch (0.0381-0.0762 mm). To adjust end play, loosen cap screw and rotate adjusting collar.

KO118) in clockwise direction until a pull of 40-50 pounds (18-23 kg) at hand grip lever is required to snap clutch over-center.

NOTE: Do not pry adjusting lock away from adjusting ring as spring type lock prevents adjusting ring from backing off during operation.

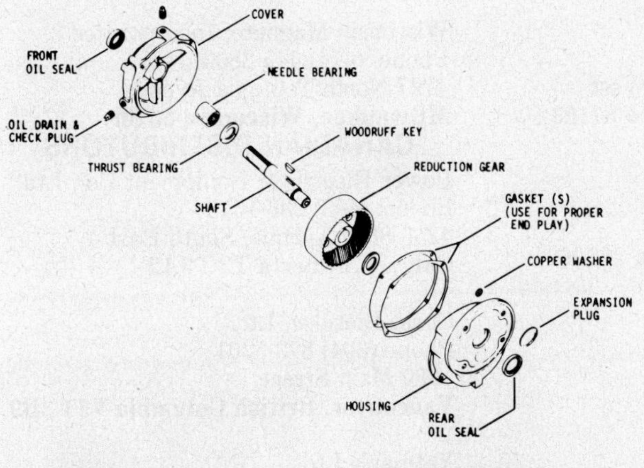

Fig. KO119—Exploded view of gear reduction drive used on some models.

Change oil after each 100 hours of normal operation. Fill housing to level plug opening with non-detergent oil. Use SAE 30 oil in temperatures above 50° F (10° C), SAE 20 oil in temperatures 50° F (10° C) to freezing and SAE 10 oil in temperatures below freezing.

REDUCTION DRIVE (GEAR TYPE). The 6:1 ratio reduction gear unit (Fig. KO119) is used on some models. To remove unit, first drain lubricating oil, then unbolt cover from housing. Remove cover and reduction gear. Unbolt and remove gear housing from engine. Separate reduction gear, shaft and thrust washer from cover. Renew oil seals and needle bearings (bronze bushings on early units) as necessary.

When reassembling, wrap tape around gear on crankshaft to protect oil seal and install gear housing. Use new copper washers on two cap screws on inside of housing. Wrap tape on shaft to prevent keyway from damaging cover oil seal and install thrust washer, shaft and reduction gear in cover. Install cover and gear assembly using new gaskets as required to provide a shaft end play of 0.001-0.006 inch (0.0254-0.1524 mm). Gaskets are available in a variety of thicknesses. Fill unit to oil check plug opening with same grade oil as used in engine.

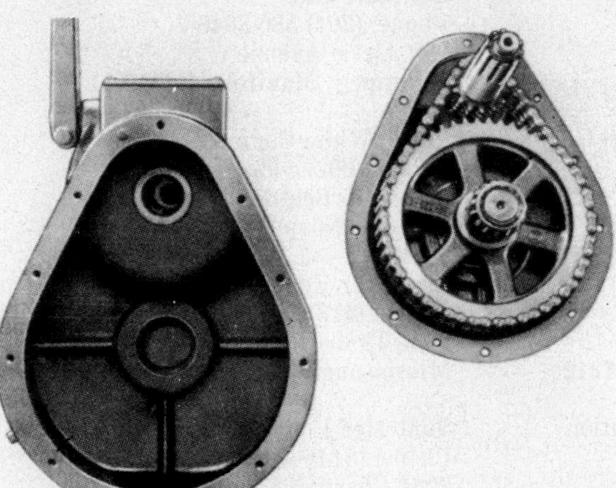

Fig. KO120—Combination clutch and chain type reduction drive used on some models.

CLUTCH AND REDUCTION DRIVE (CHAIN TYPE). Some models are equipped with a combination clutch and reduction drive unit. Clutch is a dry type and method of adjustment is the same as for clutch shown in Fig. KO116. Clutch release collar should be lubricated each 50 hours of normal operation. Remove clutch cover for access to lubrication fitting. Reduction drive unit is a chain and sprocket type. See Fig. KO120. Fill reduction housing to level hole with same grade oil as used in engine. Capacity is 3 pints (1.4 L) and should be changed each 50 hours of normal operation. The tapered roller bearings on output shaft should be adjusted to provide 0.0015-0.003 inch (0.0381-0.0762 mm) shaft end play. Adjustment is by means of a collar which is locked in position by a 5/16-inch cap screw. See Fig. KO121.

KOHLER CENTRAL ENGINE DISTRIBUTORS

(Arranged Alphabetically by States)
These franchised firms carry extensive stocks of repair parts. Contact them for name of dealer in their area who will have replacement parts.

Auto Electric & Carburetor Company
Phone: (205) 323-7113
2625 4th Avenue, South
Birmingham, Alabama 35233

Industry Services, Incorporated
Phone: (907) 562-2621
4113 Ingra Street
Anchorage, Alaska 99503

Charlie C. Jones, Incorporated
Phone: (602) 272-5621
2440 West McDowell Road
Phoenix, Arizona 85009

Generator Equipment Company
Phone: (213) 731-2401
3409 West Jefferson Boulevard
Los Angeles, California 90018

H. G. Makelim Company
Phone: (415) 873-4753
219 Shaw Road
South San Francisco, California 94080

Spitzer Industrial Products Company
Phone: (303) 287-3414
6601 North Washington Street
Denver, Colorado 80229

Spencer Engine, Incorporated
Phone: (813) 253-6035
1114 West Cass Street
Tampa, Florida 33606

Sedco, Incorporated
Phone: (404) 925-4706
1414 Red Plum Road
Norcross, Georgia 30093

Small Engine Clinic
Phone: (808) 488-0711
98019 Kam Highway, Box 98
Aiea, Hawaii 96701

Midwest Engine Warehouse
Phone: (312) 833-1200
515 Romans Road
Elmhurst, Illinois 60126

Medart Engines & Parts of Kansas
Phone: (913) 888-8828
15500 West 109th Street
Lenexa, Kansas 66215

The Grayson Company of Louisiana
Phone: (318) 222-3211
100 Fannin Street
Shreveport, Louisiana 71101

C. V. Foster Equipment Company
Phone: (301) 235-3351
2502 Harford Road
Baltimore, Maryland 21218

W. J. Connell Company
Phone: (617) 543-3600
65 Green Street
Foxboro, Massachusetts 02035

Carl A. Anderson, Inc. of Minnesota
Phone: (612) 542-2010
2737 South Lexington Avenue
Eagan, Minnesota 55121

Medart Engines & Parts
Phone: (314) 343-0505
100 Larkin Williams Industrial Crt.
Fenton, Missouri 63026

Original Equipment, Incorporated
Phone: (406) 245-3081
905 Second Avenue, North
Billings, Montana 59101

Carl A Anderson, Inc. of Nebraska
Phone: (402) 339-4944
7410 "L" Street
Omaha, Nebraska 68127

Power Distributors, Incorporated
Phone: (201) 225-5922
102 Mayfield Avenue
Edison, New Jersey 08817

Spitzer Engines & Parts
Phone: (505) 842-6472
1016 Third Street, North West
Albuquerque, New Mexico 87103

AEA, Incorporated
Phone: (704) 377-6991
700 West 28 Street
Charlotte, North Carolina 28206

Gardner, Incorporated
Phone: (614) 488-7951
1150 Chesapeake Avenue
Columbus, Ohio 43212

MICO, Incorporated
Phone: (918) 627-1448
7450 East 46th Place
Tulsa, Oklahoma 74145

Truck & Industrial Equipment Company
Phone: (503) 234-8401
7 Northeast Oregon Street
Portland, Oregon 97232

Pitt Auto Electric Company
Phone: (412) 766-9112
2900 Stayton Street
Pittsburgh, Pennsylvania 15212

Automotive Electric Corporation
Phone: (901) 345-0300
3250 Millbranch Road
Memphis, Tennessee 38116

Tri-State Equipment Company
Phone: (915) 532-6931
410 South Cotton Street
El Paso, Texas 79901

Waukesha-Pearch Industries, Inc.
Phone: (713) 723-1050
12320 South Main Street
Houston, Texas 77035

Diesel Electric Service & Supply Co.
Phone: (801) 972-1836
652 West 1700 South
Salt Lake City, Utah 84104

RBI Corporation
Phone: (804) 798-1541
P.O. Box 9378
Richmond, Virginia 23227

Northwest Motor Parts & Mfg. Co.
Phone: (206) 624-4448
2930 Sixth Avenue South
Seattle, Washington 98134

Air-Cooled Engine Supply
Phone: (509) 624-8926
127 South Walnut
Spokane, Washington 99204

Wisconsin Magneto, Incorporated
Phone: (414) 445-2800
4727 North Teutonia Avenue
Milwaukee, Wisconsin 53209

CANADIAN DISTRIBUTORS

Power Electric & Equipment Co., Ltd.
Phone: (403) 236-1515
4250 80th Avenue, South East
Calgary, Alberta T2C 3A2

Coast Dieselec, Ltd.
Phone (604) 872-5201
1920 Main Street
Vancouver, British Columbia V5T 3B9

Yetman's Ltd.
Phone: (204) 586-8046
949 Jarvis Avenue
Winnipeg, Manitoba R2X 0A1

W. N. White Company, Ltd.
Phone: (902) 443-5000
2-213-215 Bedford Highway
Halifax, Nova Scotia B3M 2J9

Suntester Equipment, Ltd.
Phone: (416) 624-6200
5466 Timberlea Boulevard
Mississauga, Ontario L4W 2T7

Suntester Equipment, Ltd.
Phone: (514) 636-1921
2081 Chartier Avenue
Dorval, P.Q. H9P 1H3

ONAN

A DIVISION OF ONAN CORPORATION
1400 73rd Avenue N.E.
Minneapolis, Minnesota 55432

Model	No. Cyls.	Bore	Stroke	Displacement	Power Rating
LK	1	3.25 in. (82.55 mm)	3.0 in. (76.2 mm)	24.9 cu. in. (407.8 cc)	5 hp. (3.7 kW)
LKB	1	3.25 in. (82.55 mm)	3.0 in. (76.2 mm)	24.9 cu. in. (407.8 cc)	8.5 hp. (6.3 kW)

Models LK and LKB engines are one cylinder, four-cycle, horizontal shaft engines. Crankshaft is supported at each end in precision type sleeve bearings.

Connecting rod in LK model is aluminum alloy, rides directly on crankpin journal and may be splash lubricated if in early model or pressure lubricated in later models.

Connecting rod in LKB model has renewable precision inserts which are pressure lubricated by a gear type oil pump driven off of crankshaft gear.

Various ignition systems utilizing externally mounted points and condenser with generating coil located under flywheel and high tension coil externally located or a combination generating and ignition coil under flywheel are used.

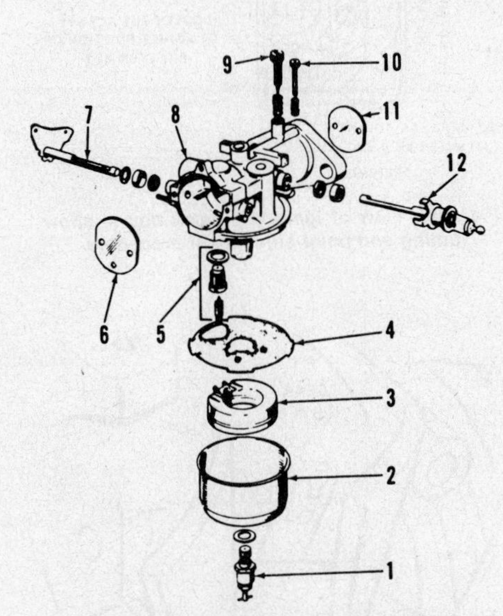

Fig. O1 – Exploded view of gasoline carburetor showing component parts.

1. Main adjusting screw
2. Fuel bowl
3. Float
4. Gasket
5. Needle valve & seat
6. Choke valve
7. Choke shaft
8. Carburetor body
9. Idle mixture needle
10. Idle speed screw
11. Throttle valve
12. Throttle shaft

A float type side draft carburetor which may be equipped with an automatic choke is used on all models.

LK engines maximum speed should be 1500 rpm when used on 50 cycle electric generating units and 1800 rpm when used on 60 cycle electric generating units.

LKB engines maximum speed should be 3000 rpm when used on 50 cycle electric generating units and 3600 rpm when used on 60 cycle electric generating units.

MAINTENANCE

SPARK PLUG. Recommended spark plug is Champion H8 or equivalent. Electrode gap is 0.025 inch (0.635 mm) for gasoline and 0.018 inch (0.4572 mm) for LP-Gas or natural gas fuel.

CARBURETOR. The same carburetor with a variety of modifications for use with gasoline, LP-Gas, natural gas or a combination of fuels is used on all models. Fig. O1 shows an exploded view of carburetor equipped for gasoline and Fig. O2 shows a carburetor for combination gas or gasoline fuel. Unnecessary parts are eliminated when unit is equipped for single fuel operation.

For initial adjustment, open both the main fuel mixture and the idle fuel mixture screws to 1 to 1¼ turns. Make final adjustment with engine at operating temperature and running. Place engine under load and adjust main fuel needle to leanest mixture that will allow satisfactory acceleration and steady governor operation while maintaining correct generating frequency required.

Run engine at idle speed, no load and adjust idle mixture screw for smoothest idle operation. As each adjustment affects the other, adjustment procedure may have to be repeated.

Throttle idle stop screw should be adjusted to clear throttle shaft stop by 1/32-inch (0.794 mm) when operating at desired speed and no-load condition. This helps prevent erratic governor operation under varying load conditions.

Carburetors used on LK models equipped for LP-Gas and natural gas operation beginning with specification letter "E" have no idle adjustment.

To check float level of gasoline fuel carburetor, invert carburetor body and float assembly. There should be 11/64-inch (4.4 mm) clearance between gasket surface of body and nearest edge of float. Adjust float by bending float lever tang that contacts inlet valve.

AUTOMATIC CHOKE. A variety of automatic choke styles have been used which vary in construction.

Some manual start models use a counterweighted choke which is closed

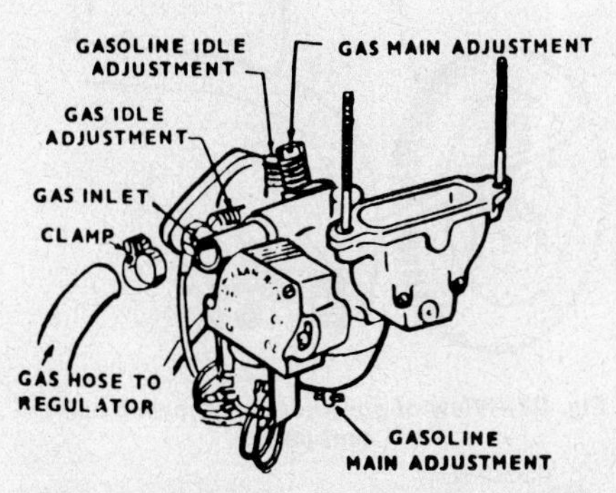

Fig. O2 – Installed view of carburetor equipped for gasoline and gaseous fuel.

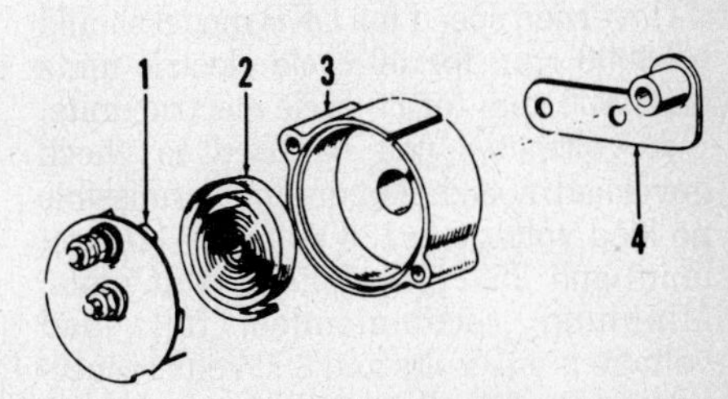

Fig. O3 – Exploded view of electric automatic choke used on some models.

Fig. O4—Schematic view showing choke adjustment.

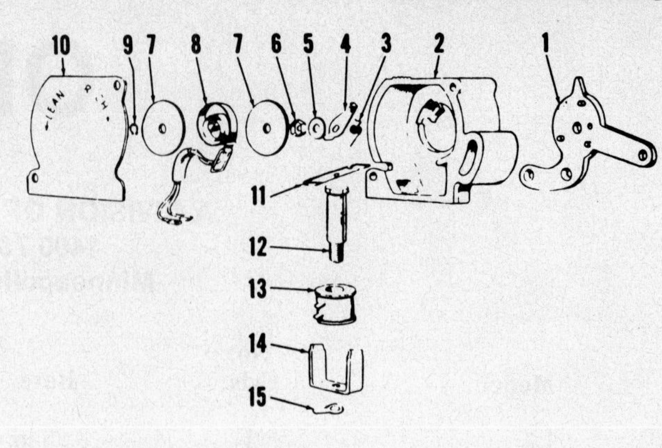

Fig. O5—Exploded view of solenoid-operated electric automatic choke used on some models.

1. Mounting plate
2. Body
3. Spring
4. Lever
5. Washer
6. Pal nut
7. Insulator
8. Heater assembly
9. Snap ring
10. Cover
11. Armature
12. Core
13. Coil
14. Frame
15. Terminal

when engine is stopped, but is opened by air stream when engine is running.

Remote control units are equipped with an electric thermal action choke (Fig. O3). When unit starts, electric current begins heating element in choke housing which causes choke to open. At temperature of 70° F (21°C) choke should be approximately 1/8-inch (3.175 mm) open before engine is started and fully open after engine reaches operating temperature and is running.

Extreme temperature conditions may require choke adjustment. Loosen screws (A–Fig. O4) and rotate housing for leaner or richer setting as shown and retighten screws.

Some models are equipped with thermomagnetic choke (Fig. O5). An electric solenoid closes choke for starting and a bi-metal heating element opens choke as engine reaches operating temperature. A continuous flow of electric current exists in heating element while engine is running. Heating element resistance should be 30.6-37.4 ohms. Choke body mounting screw (Fig. O6) can be loosened and body rotated to adjust choke for best starting performance.

GOVERNOR. Governor ball and cup assembly is located on end of camshaft. Refer to **CAMSHAFT** section for assembly information.

For linkage adjustment with engine not running, governor spring should hold carburetor throttle in full open position. Control link (2–Fig. O7) should just be able to be connected with throttle plate in full open position.

Generating unit engines are governed at a constant speed at all loads, with a minimum of change during variation of load.

Governed speed for LK model should be 1500 rpm for 50 cycle electric units and 1800 rpm for 60 cycle electric units.

Governed speed for LKB model should be 3000 rpm for 50 cycle electric units and 3600 rpm for 60 cycle electric units.

A voltmeter can be used to check governed speed. Maximum permissible no load voltage is 126 volts for 120 volt units and 252 volts for 240 volt units. Minimum recommended full load voltage is 110 volts and 220 volts respectively. Preferred voltage spread is 5 volts for 120 volt unit and 9 volts for 240 volt unit. To adjust governed speed, make certain carburetor and governor control link are correctly adjusted and engine and generator are at normal operating temperature. With all electrical load removed, turn speed adjusting nut (N) to obtain exact rated voltage of 120 or 240 volts. Apply a full electrical load and recheck voltage. Adjust sensitivity screw (S) to obtain least variation in voltage without fluctuation during changes in load. Changing sensitivity adjustment may necessitate a change in speed adjustment.

IGNITION SYSTEM. Several different types of ignition systems have been used. Early LK models with manual start were equipped with an energy transfer magneto system with a generating coil mounted underneath flywheel and a separate high tension ignition coil mounted externally on engine (Fig. O10).

LK remote start units use a battery ignition system with high tension coil mounted externally on engine.

Late LK and all LKB models use a combined generating and high tension coil mounted underneath flywheel (Fig. O11).

All models have externally mounted condenser and breaker points. Breaker point gap is 0.020 inch (0.508 mm) on all models (Fig. O8).

Timing gear cover is marked at "TC",

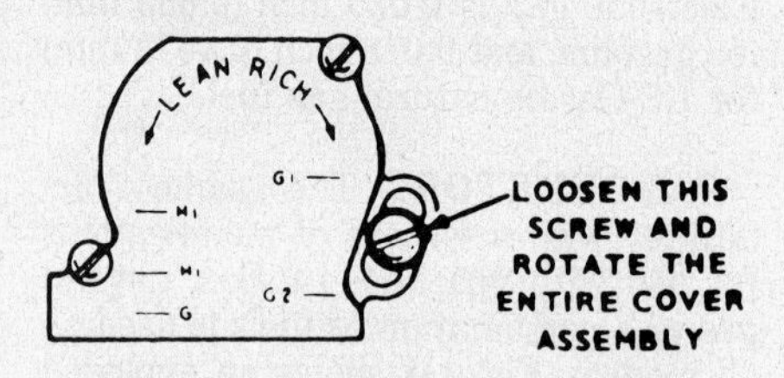

Fig. O6—Choke adjustment procedure.

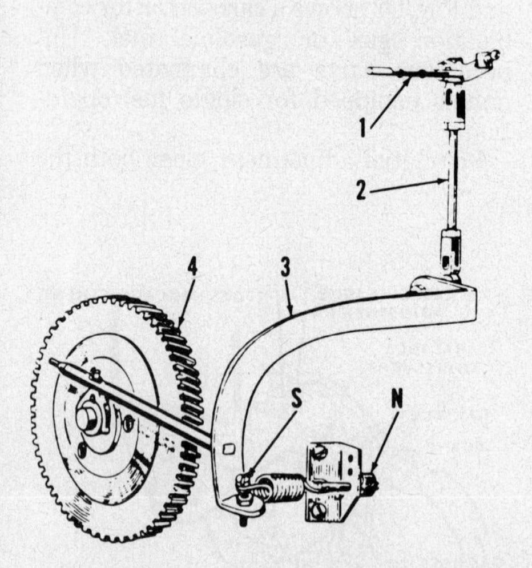

Fig. O7—View of governor unit showing component parts.

1. Throttle shaft
2. Link
3. Governor arm
4. Cam gear

N. Speed adjusting nut
S. Sensitivity adjusting screw

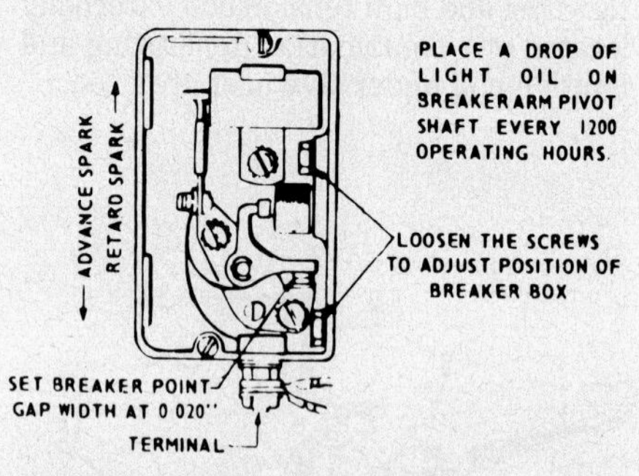

Fig. O8—View of ignition breaker box to show timing and point adjustment procedure.

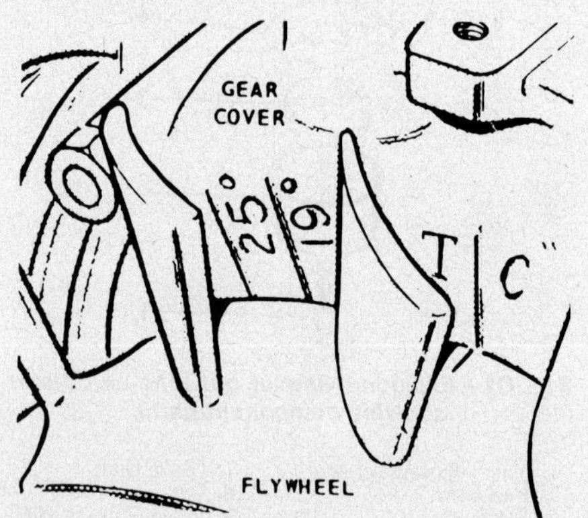

Fig. O9—Most models are furnished with flywheel timing marks as shown.

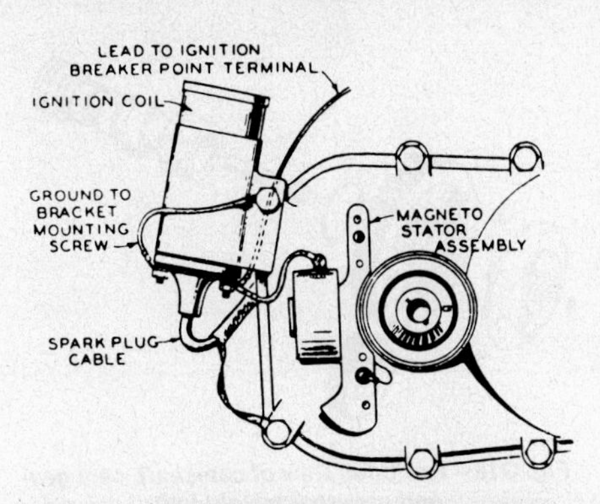

Fig. O10 – View of magneto and coil installation used on early magneto units.

19° and 25° as shown in Fig. O9 and flywheel has a timing mark. Timing is set at 19° BTDC on all LK models and is adjusted by moving breaker box up to retard timing or down to advance timing. Timing on LKB engines is set at 24° BTDC with engine running at rated speed.

Manual start LKB engines are equipped with a centrifugal advance mechanism which provides retarded timing for starting only. Static spark should occur at 5° BTDC on these models and advance should occur at 800 rpm. Fig. O14 shows an exploded view of centrifugal advance mechanism located on rear (generator) end of camshaft and is accessible for service by removing crankcase housing end plug or camshaft. If end plug is removed do not dent when reinstalling as interference with weight mechanism will occur.

LUBRICATION. Early production engines are splash lubricated by oil dipper on rod cap. When assembling, make certain oil dipper is installed to splash oil toward camshaft side of engine.

Late production engines are pressure lubricated by a gear type pump driven by crankshaft gear. Pump parts are not serviced separately so entire pump must be renewed if worn or damaged. Clearance between pump gear and crankshaft gear should be 0.005 inch (0.127 mm) and normal operating pressure should be 25 psi (173 kPa).

Oils approved by manufacturer must meet requirements of API service classification "SE" or "SE/CC".

Use SAE 30 oil in temperatures of 32° F (0° C) to 90° F (32° C), 10W-40 or 5W-30 oil in temperatures of 0° F (-18° C) to 32° F (0° C) and 5W-30 oil in temperatures below 0° F (-18° C).

Check oil level with engine stopped and maintain level at top of fill plug.

Change oil every 25 hours of normal operation. Crankcase capacity is 2 quarts (1.9 L) for all models.

REPAIRS

TIGHTENING TORQUES. Recommended tightening torques are as follows:

Connecting rod	24-26 ft.-lbs. (33-35 N·m)
Cylinder head	27-29 ft.-lbs. (37-39 N·m)
Flywheel	40-45 ft.-lbs. (54-61 N·m)
Gear cover	15-20 ft.-lbs. (20-27 N·m)
Oil base	43-48 ft.-lbs. (57-65 N·m)
Rear bearing plate	20-25 ft.-lbs. (27-34 N·m)

CONNECTING ROD. Connecting rod and piston unit is removed from above after removal of cylinder head and oil base.

LK model aluminum alloy rod rides directly on crankpin journal and should have a rod to journal clearance of 0.002-0.003 inch (0.0508-0.0762 mm) and should have side clearance of 0.013-0.038 inch (0.3302-0.9652 mm).

Models with forged rod and renewable bearing insert should have rod bearing to journal clearance of 0.0005-0.002 inch (0.0127-0.0508 mm) and should have side clearance of 0.002-0.016 inch (0.0508-0.4064 mm).

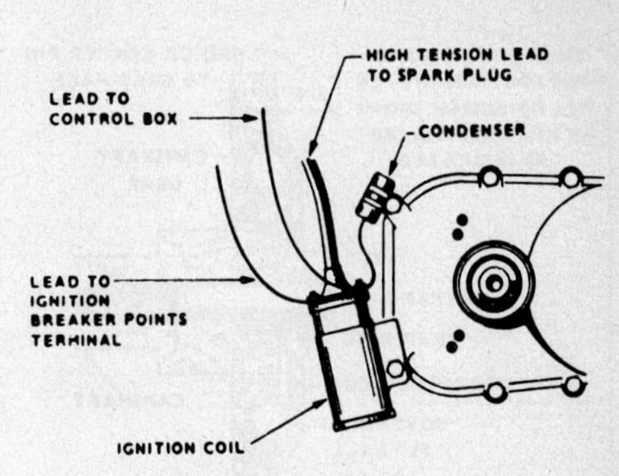

Fig. O13 – Coil mounting and wiring used on late battery ignition models.

Aluminum alloy rods or bearing inserts for forged rods are available in a variety of sizes for undersize crankshaft journals as well as standard.

When assembling engines with splash lubrication system make certain oil dipper on rod cap is installed to splash oil toward camshaft side of engine.

PISTON, PIN AND RINGS. An aluminum, cam ground piston is fitted with two compression rings and one oil control ring. Piston ring end gap should be 0.010-0.023 inch (0.254-0.5842 mm).

The floating type piston pin should be a hand push fit in piston and thumb push fit in connecting rod at 72° F (22° C) and is retained in piston by two snap rings. Pin is available in 0.002 inch (0.0508 mm) oversize as well as standard.

Standard cylinder bore is 3.249-3.250 inches (82.53-82.55 mm) and recommended piston skirt clearance is 0.0015-0.0035 inch (0.0381-0.0889 mm) when measured at bottom of skirt. Pistons and rings are available in a variety of oversizes as well as standard.

Some engines were factory equipped with 0.005 inch (0.127 mm) oversize pistons during manufacture and are identified by an "E" suffix on serial number. Standard rings are used as 0.005 inch (0.127 mm) oversize rings are not available.

CRANKSHAFT, BEARINGS AND SEALS. Crankshaft is supported at each end by precision type sleeve bearings located in cylinder block housing

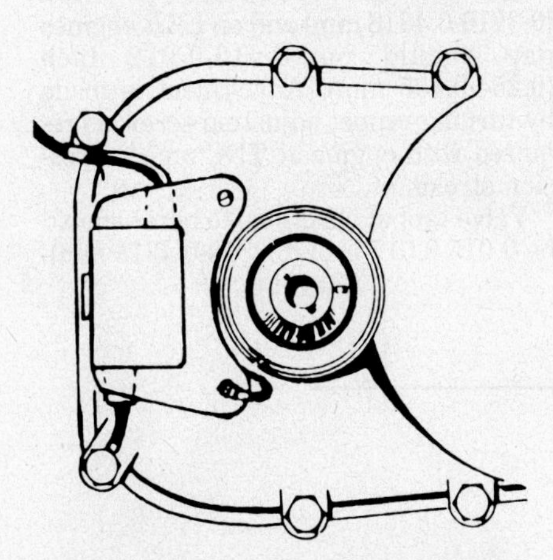

Fig. O11 – On late magneto units, high tension coil is combined with generating coil as shown.

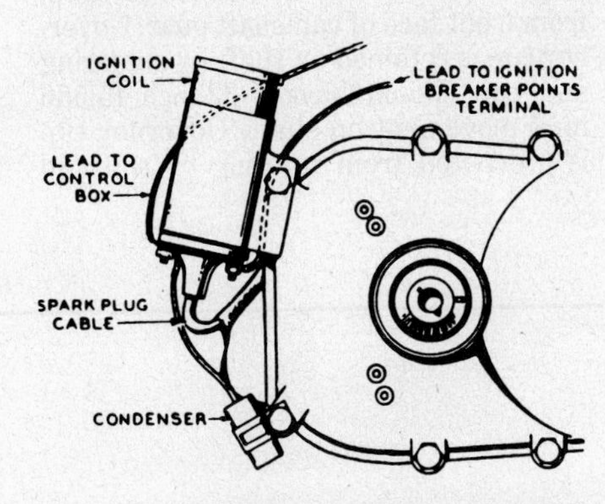

Fig. O12 – View of coil mounting used on early battery ignition units.

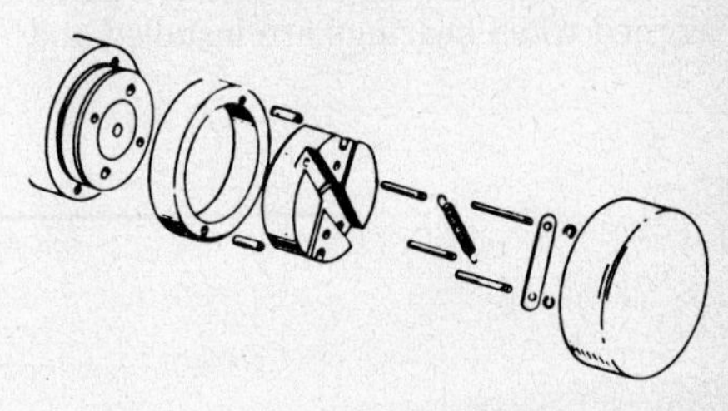

Fig. O14 – Exploded view of centrifugal advance mechanism used on Series LKB with manual start.

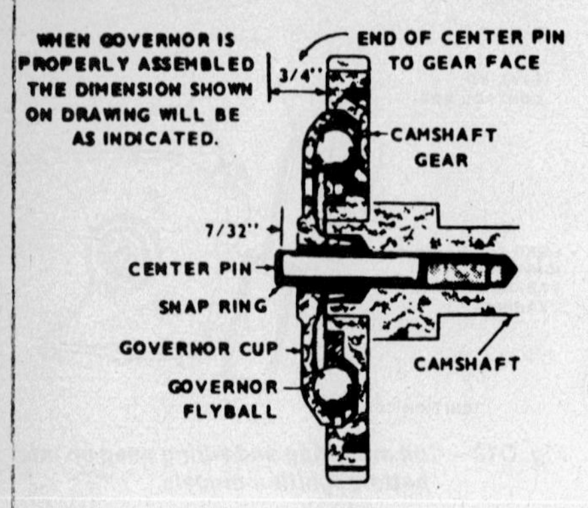

Fig. O15—Cross-sectional view of camshaft gear and governor.

Fig. O17—Exploded view of crankshaft and associated parts.

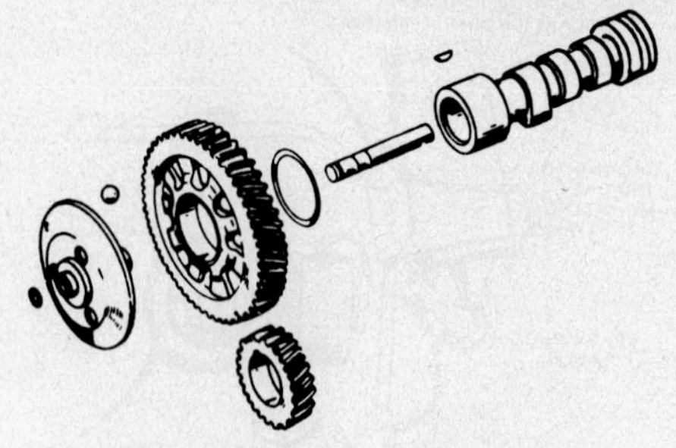

Fig. O18—Exploded view of camshaft, cam gear and governor weight unit.

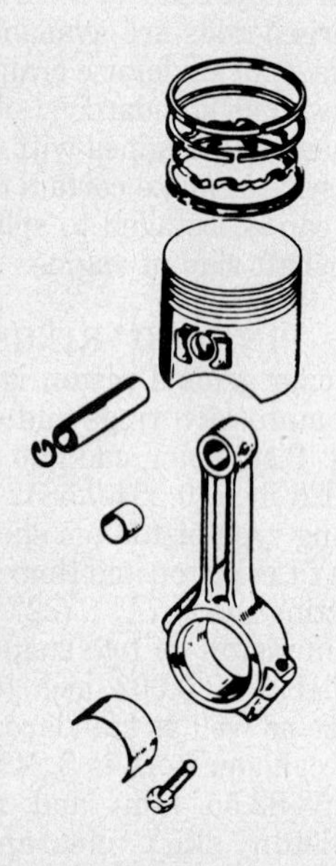

Fig. O16—Exploded view of piston and connecting rod assembly used on Series LKB. Models without pressure lubrication do not have precision bearing inserts and rod cap is equipped with oil dipper.

and rear bearing plate. Main bearing journal standard diameter should be 1.9995-2.000 inches (50.7873-50.800 mm) and bearing operating clearance should be 0.0025-0.0038 inch (0.0635-0.0965 mm). Main bearings are available in 0.002, 0.010, 0.020 and 0.030 inch (0.0508, 0.254, 0.508 and 0.762 mm) oversize as well as standard.

Oil holes in bearing and bore **MUST** be aligned when bearings are installed and bearings should be pressed into bores so inside edge of bearing is 1/16 to 3/32-inch (1.589-2.381 mm) back from inside end of bore to allow clearance for radius of crankshaft. Oil grooves in thrust washers must face crankshaft and washers two alignment notches must fit over lock pins. Thrust washers must be in good condition or excessive crankshaft end play will result.

Recommended crankshaft end play should be 0.006-0.012 inch (0.1524-0.3048 mm) and is adjusted by varying thickness of gaskets between bearing plate and cylinder block. Gaskets are available in a variety of thicknesses.

It is recommended rear bearing plate and front timing gear cover be removed for seal installation. Rear seal is pressed in until flush with seal bore and old style front seal is driven inward 31/32-inch (24.606 mm) and new thin, open face seal is driven 1-7/64 inches (28.179 mm) from mounting face of cover.

CAMSHAFT, BEARINGS AND GOVERNOR. Camshaft is supported at each end by precision sleeve type bearings pressed into bearing bores. Make certain oil holes in bearings and bores are aligned during installation. Recommended bearing to camshaft clearance is 0.0015-0.003 inch (0.0381-0.0762 mm).

Governor weight unit is mounted on camshaft gear and governor cup rides on center pin pressed into camshaft (Fig. O15) so it extends ¾-inch (19 mm) from front face of camshaft gear. Governor cup is retained on shaft by snap ring and cup should have 7/32-inch (5.556 mm) movement on shaft. Governor cup is prevented from rotating by a pin in timing gear cover which should enter metal lined hole in cup during cover installation.

Camshaft gear is a press fit on camshaft and a camshaft gear thrust washer between camshaft gear and cylinder block is used to adjust end play of camshaft to 0.003 inch (0.0762 mm). Make certain timing marks on camshaft gear and crankshaft gear are aligned after installation.

VALVE SYSTEM. Valve seats are renewable insert type and are available in a variety of oversizes as well as standard. Seats are ground to 45° angle and seat width should be 1/32 to 3/64-inch (0.794 to 1.191 mm).

Valves should be ground to 44° angle to provide an interference angle of 1°. Stellite valves and seats are available and roto-caps are standard on both intake and exhaust valves.

Recommended valve stem to guide clearance is 0.001-0.0025 inch (0.254-0.0635 mm) for intake valves and 0.0025-0.004 inch (0.0635-0.1016 mm) for exhaust valves. Renewable valve guides are shouldered and are pushed out from above. Early models use a gasket between guide and cylinder block.

Recommended valve tappet gap (cold) for intake and exhaust valves on LK engines is 0.015-0.017 inch (0.3810-0.4318 mm) and on LKB engines gap should be 0.010-0.012 inch (0.254-0.305 mm). Adjustment is made by turning tappet adjusting screw as required with engine at TDC on compression stroke.

Valve tappet clearance to bores should be 0.015-0.017 inch (0.3810-0.4318 mm).

ONAN

A DIVISION OF ONAN CORPORATION
1400 73rd Avenue, N.E.
Minneapolis, Minnesota 55432

Model	No. Cyls.	Bore	Stroke	Displacement	Power Rating
BF	2	3.125 in. (79.38 mm)	2.625 in. (66.68 mm)	40.3 cu. in. (660 cc)	16 hp. (11.9 kW)
BG	2	3.250 in. (82.55 mm)	3 in. (76.2 mm)	49.8 cu. in. (815.7 cc)	18 hp. (13.4 kW)
B43M, B43E	2	3.250 in. (82.55 mm)	2.620 in. (66.55 mm)	43.3 cu. in. (712.4 cc)	16 hp. (11.9 kW)
B43G	2	3.250 in. (82.55 mm)	2.620 in. (66.55 mm)	43.3 cu. in. (712.4 cc)	18 hp. (13.4 kW)
B48G	2	3.250 in. (82.55 mm)	2.875 in. (73.03 mm)	47.7 cu. in. (781.7 cc)	20 hp. (14.9 kW)
B48M	2	3.250 in. (82.55 mm)	2.875 in. (73.03 mm)	47.7 cu. in. (781.7 cc)	18 hp. (13.4 kW)

Engines in this section are four-cycle, twin-cylinder opposed, horizontal crankshaft type in which crankshaft is supported at each end in precision sleeve type bearings.

Connecting rods ride directly on crankshaft journals and all except early BF engines are pressure lubricated by a gear type oil pump driven by crankshaft gear. Early BF engines were splash lubricated.

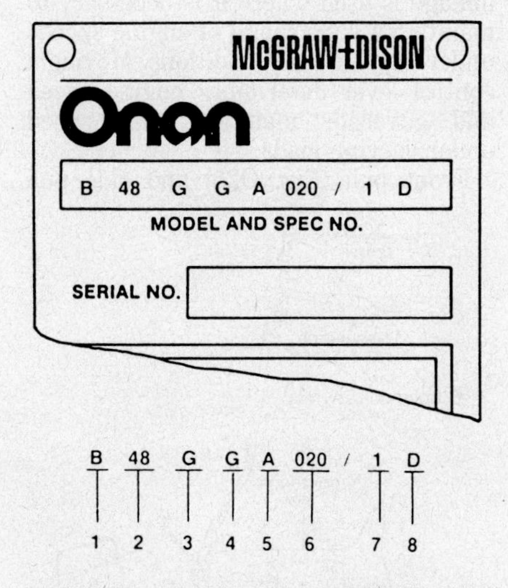

Fig. O19—Typical model, specification and serial number plate on Onan engine showing digit interpretation.

1. Identification of basic engine series.
2. Displacement (in cubic inches).
3. Engine duty cycle.
4. Fuel required.
5. Cooling system designation.
6. Power rating (in BHP)
7. Designated optional equipment.
8. Production modifications.

All models use a battery ignition system consisting of points, condenser, coil, battery and spark plug. Timing is adjustable only by varying point gap.

All models use a single down draft float type carburetor and may have a vacuum operated fuel pump attached to carburetor or mounted separately according to model and application.

Refer to model and specification number (Fig. O19) for engine model identification and interpretation.

Always give model, specification and serial numbers when ordering parts or service information.

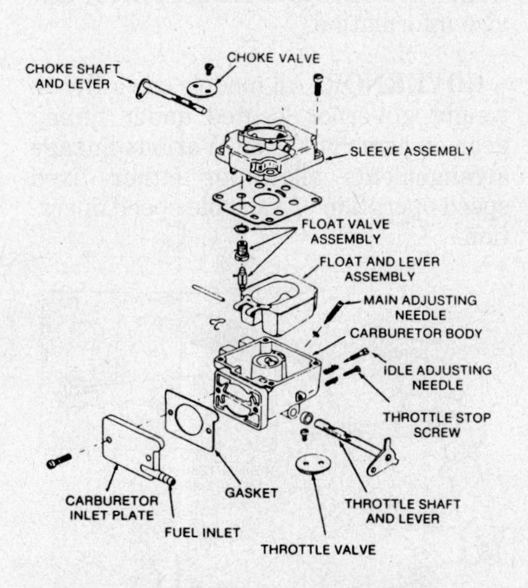

Fig. O20—Exploded view of Marvel Schebler carburetor showing component parts and their relative positions.

MAINTENANCE

SPARK PLUG. Recommended spark plug is Champion H8 or equivalent. Electrode gap for all models using gasoline fuel is 0.025 inch (0.635 mm). Electrode gap for all models using vapor (butane, LP or natural gas) fuel is 0.018 inch (0.457 mm).

CARBURETOR. According to model and application all engines use a single down draft float type carburetor manufactured by Marvel Schebler, Walbro or Nikki. Exploded view of each

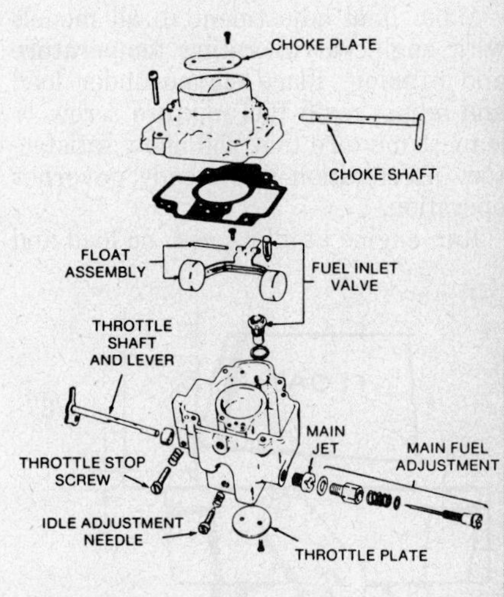

Fig. O20A—Exploded view of Walbro carburetor showing component parts and their relative positions.

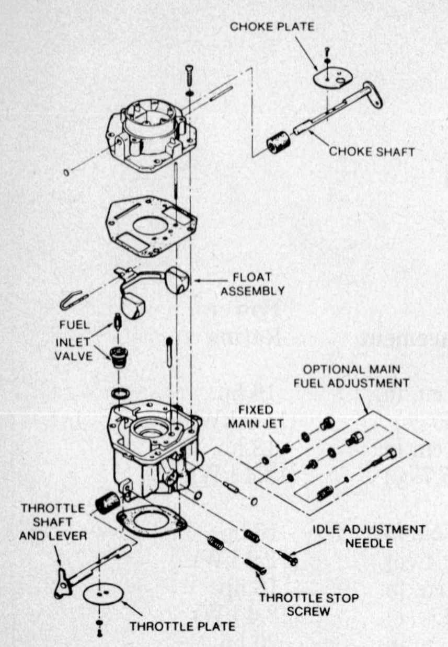

Fig. O20B—Exploded view of Nikki carburetor showing component parts and their relative positions.

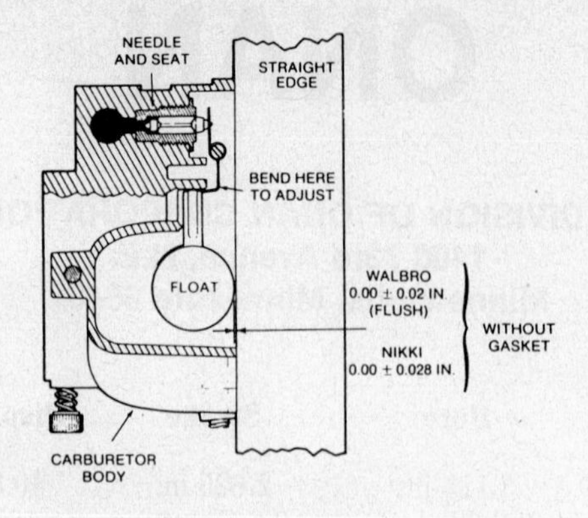

Fig. O22—To adjust float level on Walbro and Nikki carburetors, position carburetor as shown. Clearance should be measured from machined surface without gasket to float edge.

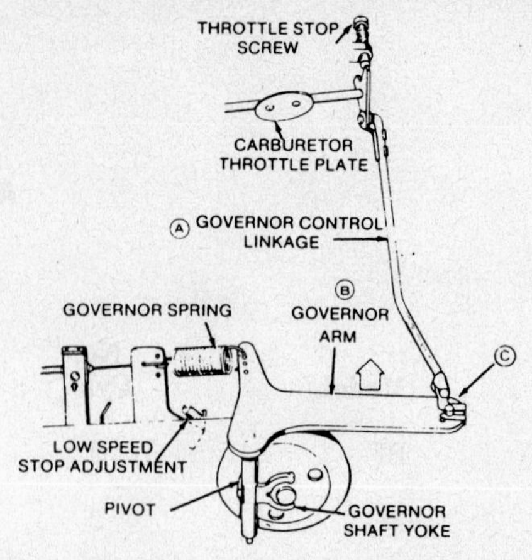

Fig. O25—View of front pull variable speed governor linkage.

carburetor is shown in either Fig. O20, Fig. O20A or Fig. O20B. Refer to appropriate Figure for model being serviced.

For initial adjustment of Marvel Schebler carburetor refer to Fig. O20. Open idle mixture screw 1 turn and main fuel mixture screw 1¼ turns.

For initial adjustment of Walbro carburetor refer to Fig. O20A. Open idle mixture screw 1-1/8 turns. If equipped with optional main fuel adjustment open main fuel mixture screw 1½ turns.

For initial adjustment of Nikki carburetor refer to Fig. O20B. Open idle mixture screw ¾-turn. If equipped with optional main fuel adjustment open main fuel mixture screw 1½ turns.

Make final adjustment to all models with engine at operating temperature and running. Place engine under load and adjust main fuel mixture screw to leanest mixture that will allow satisfactory acceleration and steady governor operation.

Run engine at idle speed, no load and adjust idle mixture screw for smoothest idle operation. As each adjustment affects the other, adjustment procedure may have to be repeated.

To check float level of Marvel Schebler Carburetor refer to Fig. O21. Invert carburetor throttle body and float assembly. There should be 1/8-inch (3.175 mm) clearance between gasket and float as shown. Adjust float by bending float lever tang that contacts inlet valve.

To check float level of Walbro or Nikki carburetor refer to Fig. O22. Position carburetor as shown. Walbro carburetor should have 0.00 inch, plus or minus 0.02 inch (0.00 mm, plus or minus 0.508 mm) clearance and Nikki carburetor should have 0.00 inch, plus or minus 0.028 inch (0.00 mm, plus or minus 0.7112 mm) clearance between machined surface without gasket, to float edge. Adjust float by bending float lever tang that contacts inlet valve.

All models use a pulsating diaphragm fuel pump. Refer to **SERVICING ONAN ACCESSORIES** section for service information.

GOVERNOR. All models use a flyball weight governor located under timing gear on camshaft gear. Various linkage arrangements allow for either fixed speed operation or variable speed operation.

FIXED SPEED. Fixed speed linkage is used on electric generators and welders or any application which requires a single constant speed under varying load conditions. Fig. O24 shows typical linkage arrangement.

Governor link length should be adjusted so stop on throttle shaft almost touches stop on side of carburetor with engine stopped and governor spring under tension.

Sensitivity is adjusted by varying tension of governor spring and changing location of governor spring on governor arm.

Engine speed should be set to manufacturers specifications and output of generators should be checked with a frequency meter.

VARIABLE SPEED. Variable speed linkage is used where it is necessary to maintain a wide range of engine speeds under varying load conditions. Moving a control lever determines engine speed and governor maintains this speed under varying loads.

Front pull (Fig. O25) and side pull

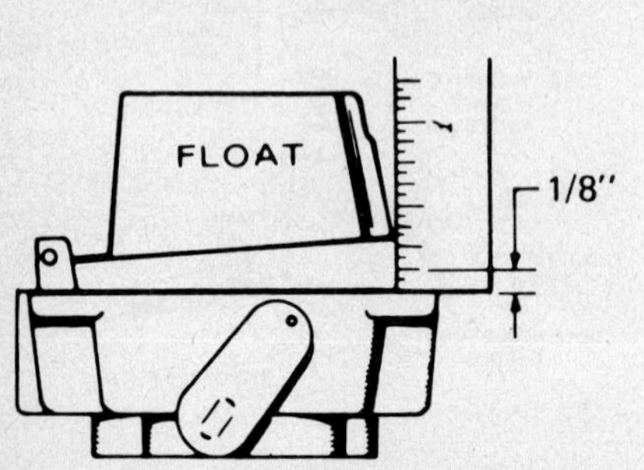

Fig. O21—Float valve setting for Marvel Schebler carburetor. Measure between inverted float and surface of gasket.

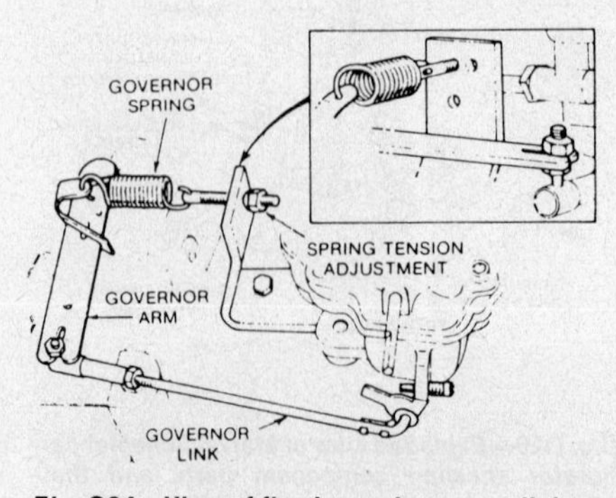

Fig. O24—View of fixed speed governor linkage.

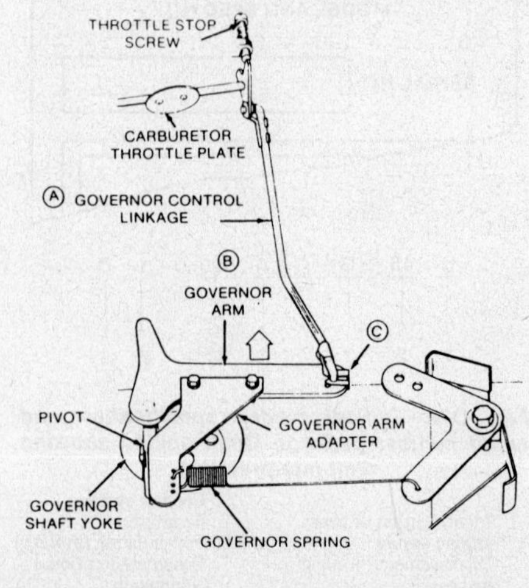

Fig. O26—View of side pull variable speed governor linkage.

(Fig. O26) linkage arrangements are used and service procedures are similar.

For correct governor link installation on either front or side pull system, disconnect linkage (A) from hole (C) and push linkage and governor arm (B) toward carburetor as far as they will go. While held in this position insert end of linkage into nearest hole in governor arm. If between two holes insert in next hole out.

Normal factory setting is in third hole from pivot for side pull linkage and second hole from pivot for front pull linkage.

Sensitivity is increased by connecting governor spring closer to pivot and decreased by connecting governor spring further from pivot.

Engine speed should be set to manufacturers specifications and speed should be checked after linkage adjustments.

IGNITION. A battery type ignition system is used on all models. Breaker point box is non-movable and only means of changing timing is to vary point gap slightly. Static timing should be checked if point gap is changed or new points are installed. Use a continuity test light across breaker points to check timing and remove air intake hose from blower housing for access to timing marks. Refer to ignition specifications as follows:

Model	Initial Point Gap	Static Timing BTDC
BF	0.025 in. (0.635 mm)	21°
BF*	0.025 in. (0.635 mm)	21°
BF**	0.025 in. (0.635 mm)	26°
BG, B43M, B48M	0.021 in. (0.533 mm)	21°
B48G	0.020 in. (0.508 mm)	20°
B48G*	0.016 in. (0.406 mm)	16°

*Specification letter "C" and after.
**"Power Drawer" models.

LUBRICATION. Early Model BF is splash lubricated and all other models are pressure lubricated by a gear type pump driven by crankshaft gear. Internal pump parts are not serviced separately so entire pump must be renewed if worn or damaged. Clearance between pump gear and crankshaft gear should be 0.002-0.005 inch (0.0508-0.127 mm) and normal operating pressure should be 30 psi (207 kPa). Oil pressure is regulated by an oil pressure relief valve (Fig. O27) located in engine block near timing gear cover. Spring (3) free length should be 1.00 inch (25.4 mm) and valve (4) diameter should be 0.3365-0.3380 inch (8.55-8.59 mm).

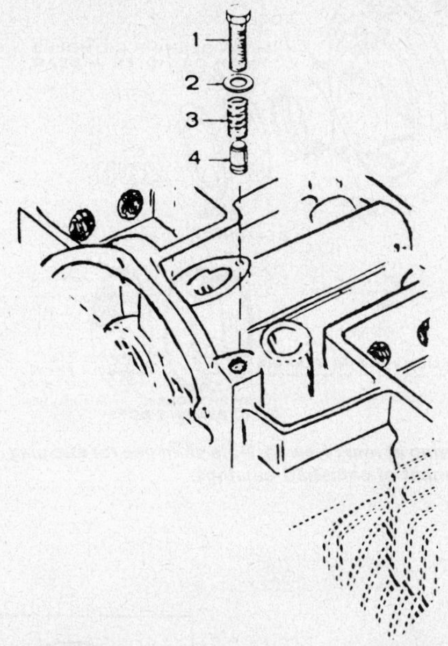

Fig. O27 — View showing location of oil pressure relief valve and spring.

1. Cap screw
2. Sealing washer
3. Spring
4. Valve

Oils approved by manufacturer must meet requirements of API service classification "SE" or "SE/CC".

Use SAE 5W-30 oil in below freezing temperatures and SAE 20W-40 oil in above freezing temperatures.

Check oil level with engine stopped and maintain at "FULL" mark on dipstick. **DO NOT** overfill.

Recommended oil change intervals for models without oil filter is every 25 hours of normal operation. If equipped with a filter, change oil at 50 hour intervals and filter every 100 hours.

Crankcase capacity without filter is 3.5 pints (1.66 L) for BG model, 4 pints (1.86 L) for BF model, 1.5 quarts (1.4 L) for B43E, B43G and B48G models. If filter is changed add an additional 0.5 pint (0.24 L) for BF and BG models and an additional 0.5 quart (0.47 L) for B43E, B43G and B48G models. Oil filter is screw on type and gasket should be lightly lubricated before installation. Screw filter on until gasket contacts, then turn an additional ½-turn. Do not overtighten.

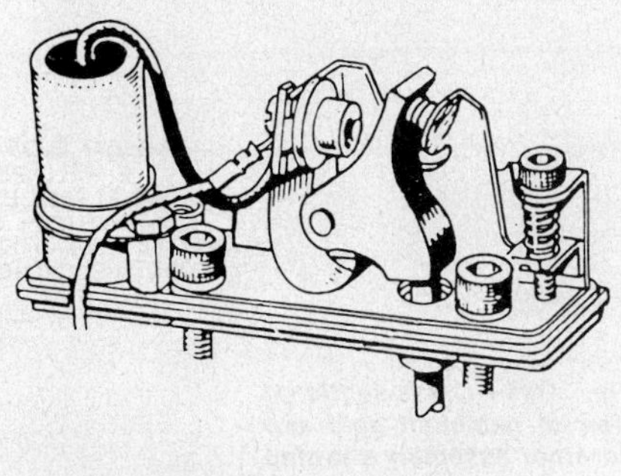

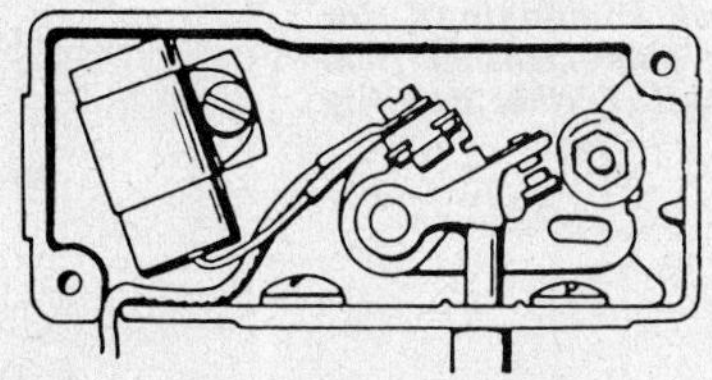

Fig. O28 — Two different type point boxes are used according to model and application. Upper illustration shows top adjust point models while lower illustration shows side adjust point model.

REPAIRS

TIGHTENING TORQUES. Recommended tightening torques are as follows:

Cylinder heads –
- BF, BG . . . . . 14-16 ft.-lbs. (19-22 N·m)
- B43E,* B43G,* B48G,* B43M,* B48M* . . . . . 16-18 ft.-lbs. (22-24 N·m)

*If graphoil gasket is used, tighten to 14-16 ft.-lbs. (19-22 N·m) torque.

Rear bearing plate . . . . . 25-27 ft.-lbs. (34-37 N·m)

Connecting rod –
- BG, BF . . . . . 14-16 ft.-lbs. (19-22 N·m)
- All others . . . . . 12-14 ft.-lbs. (16-19 N·m)

Flywheel . . . . . 35-40 ft.-lbs. (48-54 N·m)

Oil base . . . . . 18-23 ft.-lbs. (24-31 N·m)

Timing cover . . . . . 8-10 ft.-lbs. (11-13 N·m)

Oil pump . . . . . 7-9 ft.-lbs. (10-12 N·m)

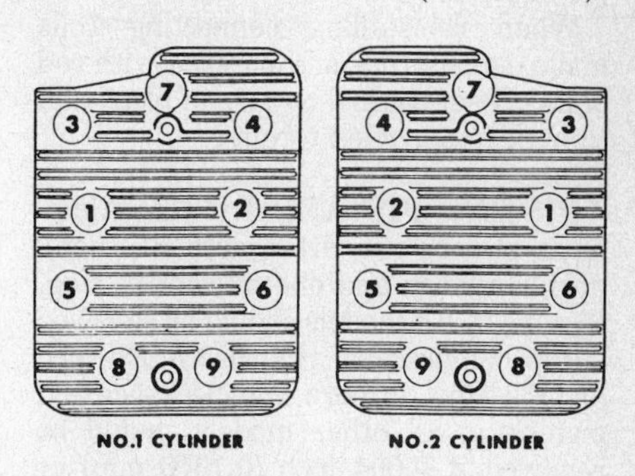

Fig. O29 — Torque sequence shown is for all models even though spark plug may be in different location in head.

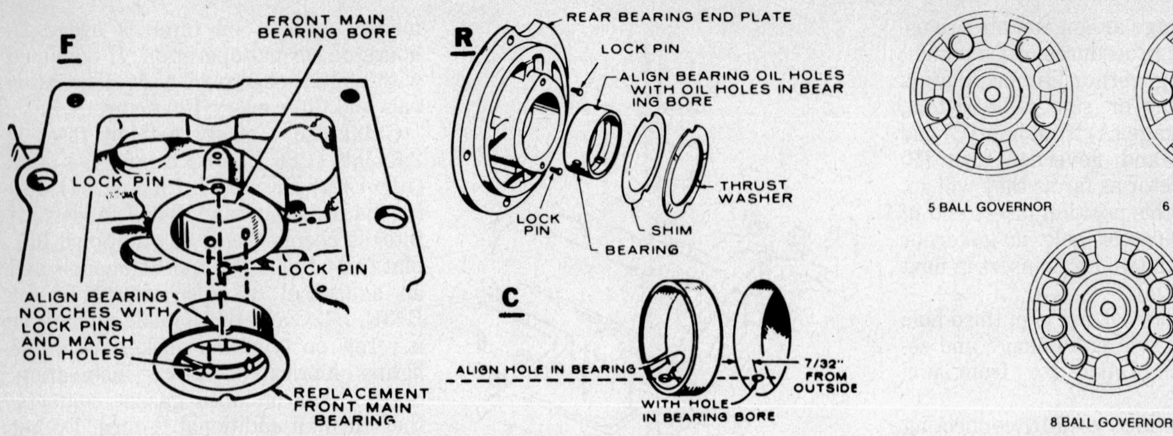

Fig. O31 – Alignment of precision main camshaft bearing at rear (View R). Note shim use for end play adjustment. View C shows placement of camshaft bearings.

Fig. O32 – Flyballs must be arranged in pockets as shown, according to total number used to obtain governor sensitivity desired. Refer to text.

CYLINDER HEADS. It is recommended cylinder heads be removed and carbon cleaned at 200 hour intervals.

CAUTION: Cylinder heads should not be unbolted from block when hot due to danger of head warpage.

When installing cylinder heads torque bolts in sequence shown in Fig. O29 in gradual, even steps until correct torque for model being serviced is obtained. Note spark plug location on some models varies but torque sequence is the same.

CONNECTING RODS. Connecting rod and piston are removed from cylinder head surface end of block after removing cylinder head and oil base. Connecting rods ride directly on crankshaft journal on all models and standard journal diameter should be 1.6252-1.6260 inches (41.28-41.30 mm). Connecting rods are available for 0.005, 0.010, 0.020, 0.030 and 0.040 inch (0.127, 0.254, 0.508, 0.762 and 1.016 mm) undersize crankshafts as well as standard.

Connecting rod to crankshaft journal running clearance should be 0.002-0.0033 inch (0.0508-0.0838 mm) and side play should be 0.002-0.016 inch (0.0508-0.406 mm).

When reinstalling connecting rods make certain rods are installed with rod bolts off-set toward outside of block and tighten to specified torque.

PISTON, PIN AND RINGS. Aluminum pistons are fitted with two compression rings and one oil control ring. Pistons in BF models should be renewed if top compression ring has 0.008 inch (0.2032 mm) or more side clearance and pistons in all other models should be renewed if 0.004 inch (0.1016 mm) or more side clearance is present.

Pistons and rings are available in 0.005, 0.010, 0.020, 0.030 and 0.040 inch (0.127, 0.254, 0.508, 0.762 and 1.016 mm) oversize as well as standard.

Recommended piston to cylinder wall clearance for BF engines when measured 0.10 inch (2.54 mm) below oil control ring 90° from pin should be

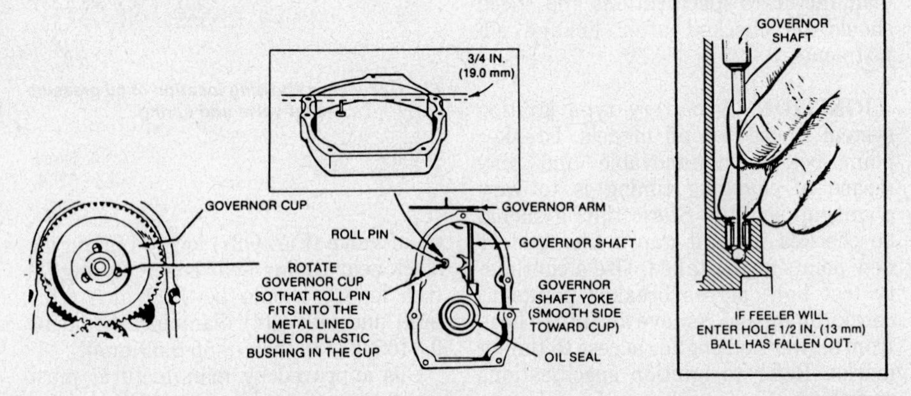

Fig. O33 – View of governor shaft, timing cover and governor cup showing assembly details. Refer to text.

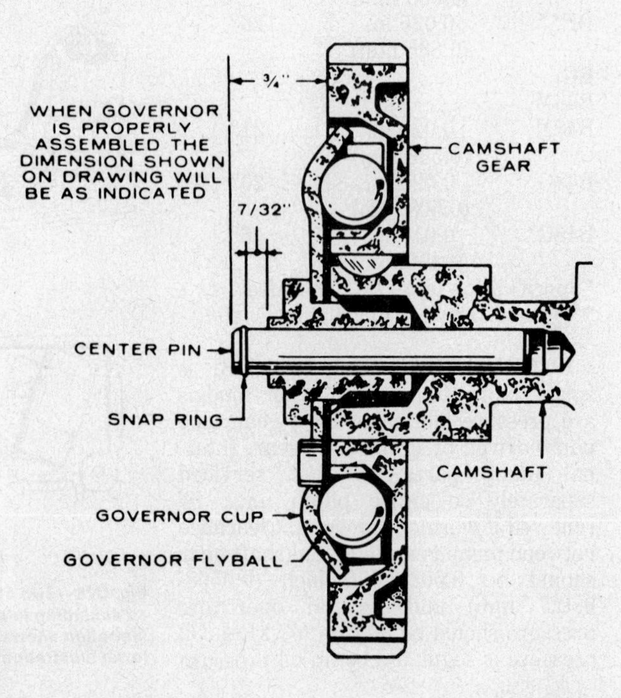

Fig. O34 – Cross-sectional view of camshaft gear and governor assembly showing correct dimensions for center pin extension from camshaft. Further detail in text.

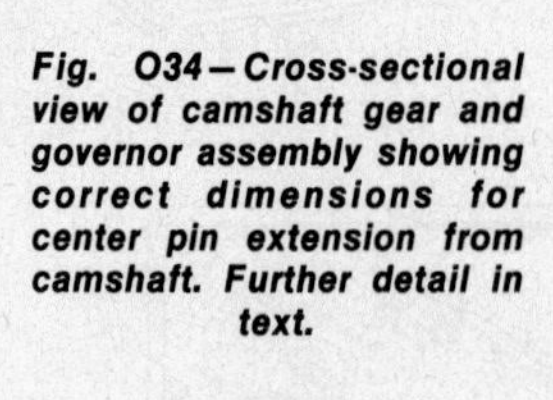

0.001-0.003 inch (0.0254-0.0762 mm).

Recommended piston to cylinder wall clearance for B43M and B48M engines when measured 0.35 inch (8.89 mm) below oil control ring 90° from pin should be 0.0033-0.0053 inch (0.0838-0.1346 mm).

Recommended piston to cylinder wall clearance for B43E, B43G and B48G engines when measured 0.35 inch (8.89 mm) below oil control ring 90° from pin should be 0.0044-0.0064 inch (0.1118-0.1626 mm).

Ring end gap should be 0.010-0.020 inch (0.254-0.508 mm) clearance in standard bore.

Piston pin to piston bore clearance should be 0.0002-0.0004 inch (0.0051-0.0102 mm) and pin to rod clearance should be 0.0002-0.0007 inch (0.0051-0.0178 mm) for all models.

Standard cylinder bore of BF engine is 3.1245-3.1255 inches (79.36-79.39 mm) and standard cylinder bore for all other models is 3.249-3.250 inches (82.53-82.55 mm).

Engines should be rebored if taper exceeds 0.005 inch (0.127 mm) or if 0.003 inch (0.0762 mm) out-of-round.

CRANKSHAFT, BEARINGS AND SEALS. Crankshaft is supported at each end by precision type sleeve bearings located in cylinder block housing and rear bearing plate. Main bearing journal standard diameter should be 1.9992-2.000 inches (50.7797-50.800 mm) and bearing operating clearance should be 0.0025-0.0038 inch (0.0635-0.0965 mm). Main bearings are available for a variety of undersize crankshafts as well as standard.

Oil holes in bearing and bore **MUST** be aligned when bearings are installed and bearings should be pressed into bores so inside edge of bearing is 1/16 to 3/32-inch (1.588 to 2.381 mm) back from inside end of bore to allow clearance for radius of crankshaft. Oil grooves in thrust washers must face crankshaft. The two alignment notches on thrust washers must fit over lock pins. Thrust washers must be in good condition or excessive crankshaft end play will result.

NOTE: Replacement front bearing and thrust washer is a one piece assembly, do not install a separate thrust washer.

Recommended crankshaft end play should be 0.006-0.012 inch (0.1524-0.3048 mm) and is adjusted by varying thrust washer thickness or placing shim between thrust washer and rear bearing plate. Shims are available in a variety of thicknesses. See Fig. O31.

It is recommended rear bearing plate and front timing gear cover be removed for seal renewal. Rear seal is pressed in until flush with seal bore. Timing cover seal should be driven in until it is 1-1/32 inch (26.19 mm) from mounting face to cover.

CAMSHAFT, BEARINGS AND GOVERNOR. Camshaft is supported at each end in precision type sleeve bearings. Make certain oil holes in bearings are aligned with oil holes in block, press front bearing in flush with outer surface of block and press rear bearing in until flush with bottom of counterbore. Camshaft to bearing clearance should be 0.0015-0.0030 inch (0.0381-0.0762 mm).

Camshaft end play should be 0.003 inch (0.0762 mm) and is adjusted by varying thickness of shim located between camshaft timing gear and engine block.

Camshaft timing gear is a press fit on end of camshaft and is designed to accommodate 5, 8 or 10 flyballs arranged as shown in Fig. O32. Number of flyballs is varied to alter governor sensitivity. Fewer flyballs are used on engines with variable speed applications and greater number of flyballs are used for continuous speed application. Flyballs must be arranged in "pockets" as shown according to total number used. Make certain timing mark on cam gear aligns with timing mark on crankshaft gear when reinstalling.

Center pin (Fig. O34) should extend ¾-inch (19 mm) from end of camshaft to allow 7/32-inch (5.6 mm) in-and-out movement of governor cup. If distance is incorrect remove pin and press new pin in to correct depth.

Always make certain governor shaft pivot ball is in timing cover by measuring as shown in Fig. O33 inset and check length of roll pin which engages bushed hole in governor cup. This pin must extend 25/32-inch (19.844 mm) from timing cover mating surface.

VALVE SYSTEM. Valve seats are renewable insert type and are available in a variety of oversizes as well as standard. Seats are ground at a 45° angle and seat width should be 1/32 to 3/64-inch (0.794 to 1.191 mm).

Valves should be ground at a 44° angle to provide an interference angle of 1°. Stellite valves and seats are

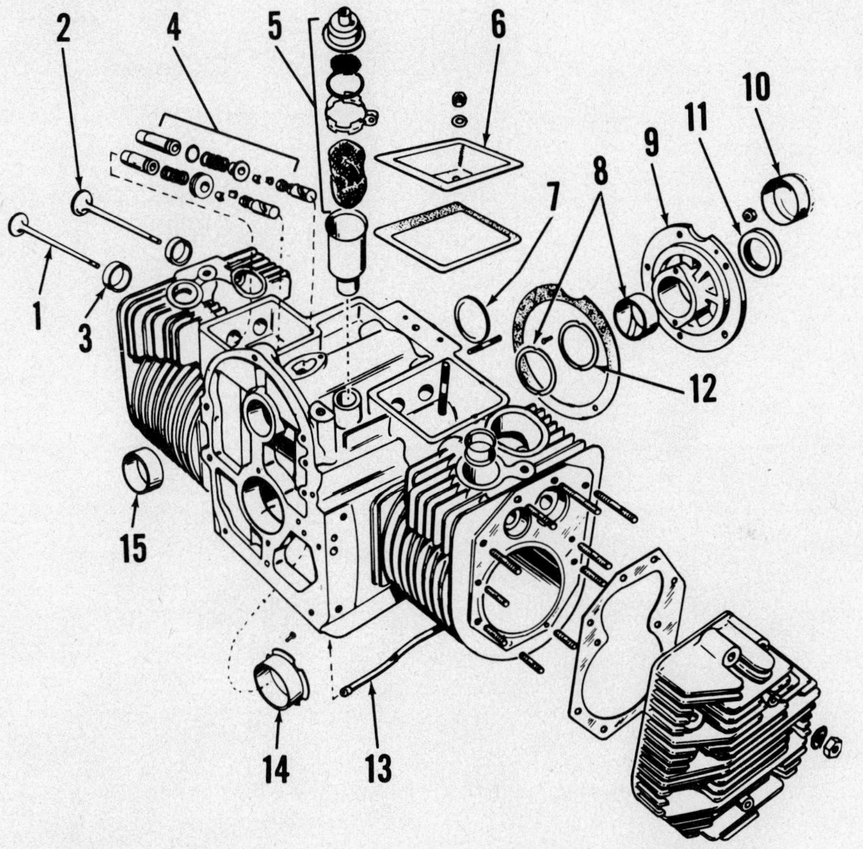

Fig. O35 — Exploded view of cylinder and crankcase assembly typical of all models.

1. Exhaust valve
2. Intake valve
3. Seat insert (2)
4. Guide, spring, tappet group
5. Crankcase breather group
6. Valve compartment cover
*7. Camshaft expansion plug
8. Two-piece main bearing
9. Rear main bearing plate
*10. Timing control cover
11. Crankshaft seal
12. Crankshaft thrust shim
13. Oil tube
14. One-piece main bearing
15. Camshaft bearing (2)

*Plug is installed on engine without timing control.

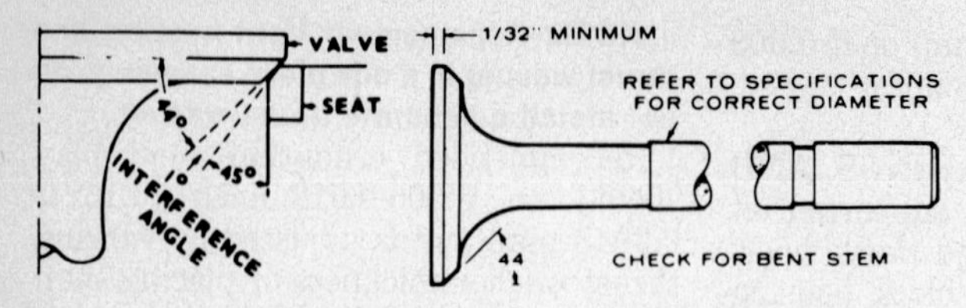

Fig. O36—Principal valve service specifications illustrated. Refer to text for further detail.

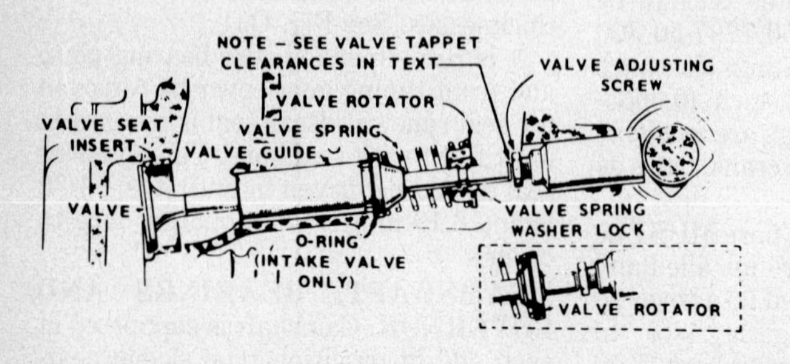

Fig. O37—Typical valve train on all models. Refer to text.

available and rotocaps are available for exhaust valves.

Recommended valve stem to guide clearance is 0.001-0.0025 inch (0.0254-0.0635 mm) for intake valves and 0.0025-0.004 inch (0.0635-0.1016 mm) for exhaust valve. Renewable valve guides are shouldered and are pushed out from above. Early models use a gasket between guide and cylinder block and some models may be equipped with valve stem seal on intake valve which must be renewed if valve is removed.

Recommended valve tappet gap (cold) for B43E model is 0.003 inch (0.0762 mm) for intake valves and 0.010 inch (0.254 mm) for exhaust valves. Recommended valve tappet gap (cold) for all remaining models is 0.008 inch (0.2032 mm) for intake valves and 0.013 inch (0.3302 mm) for exhaust valves.

Adjustment is made by turning tappet adjusting screw as required and valves of each cylinder must be adjusted with cylinder at "top dead center" on compression stroke. At this position both valves will be fully closed.

ONAN

A DIVISION OF ONAN CORPORATION
1400 73rd Avenue N.E.
Minneapolis, Minnesota 55432

Model	No. Cyls.	Bore	Stroke	Displacement	Power Rating
NH	2	3.563 in. (90.50 mm)	3 in. (76.2 mm)	60 cu. in. (980.3 cc)	25 hp. (18.6 kW)
NHA	2	3.563 in. (90.50 mm)	3 in. (76.2 mm)	60 cu. in. (980.3 cc)	18 hp. (13.4 kW)
NHB	2	3.563 in. (90.50 mm)	3 in. (76.2 mm)	60 cu. in. (980.3 cc)	20 hp. (14.9 kW)
NHC	2	3.563 in. (90.50 mm)	3 in. (76.2 mm)	60 cu. in. (980.3 cc)	25 hp. (18.6 kW)
T-260G	2	3.563 in. (90.50 mm)	3 in. (76.2 mm)	60 cu. in. (980.3 cc)	24 hp. (17.9 kW)

Engines in this section are four-cycle, twin cylinder opposed, horizontal crankshaft engines. Crankshaft is supported at each end in precision sleeve type bearings.

Connecting rods in T-260G engines ride directly on crankshaft journal and all remaining models use renewable precision insert rod bearings. All models are pressure lubricated by a gear type oil pump and engines are equipped with "spin-on" oil filter with a by-pass.

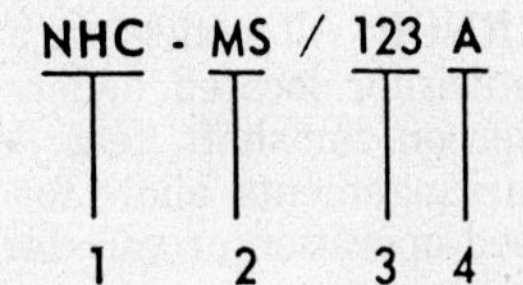

Fig. O38—Interpretation of engine model and specification number as an aid to identification of various engines.

1. General engine model identification
2. Specific type: S–Manual starting MS–Electric starting
3. Optional equipment identification
4. Specification letter which advances with factory production modifications

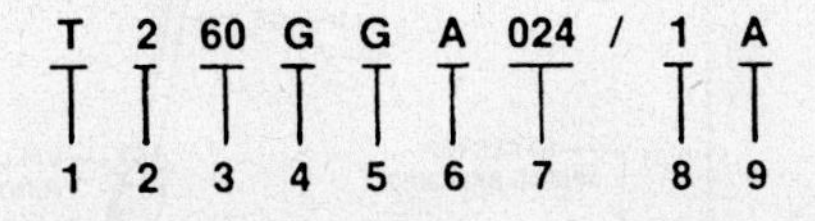

Fig. O38A—Interpretation of engine model and specification number as an aid to identification of various engines.

1. General engine model identification
2. Number of cylinders
3. Cubic inch displacement
4. Engine duty cycle (M=medium duty)
5. Fuel required (G=gasoline)
6. Cooling system description (A=air cooling pressure)
7. BHP rating
8. Optional equipment identification
9. Specification letter which advances with factory production modifications

Most models are equipped with a battery ignition system, however a magneto system is available for manual start models.

Side draft or down draft carburetor may be used according to model or application and fuel is supplied by either a mechanically operated or vacuum operated fuel pump.

Refer to Fig. O38 or O38A for interpretation of engine specification and model numbers. Always give specification, model and serial numbers when ordering parts or service material.

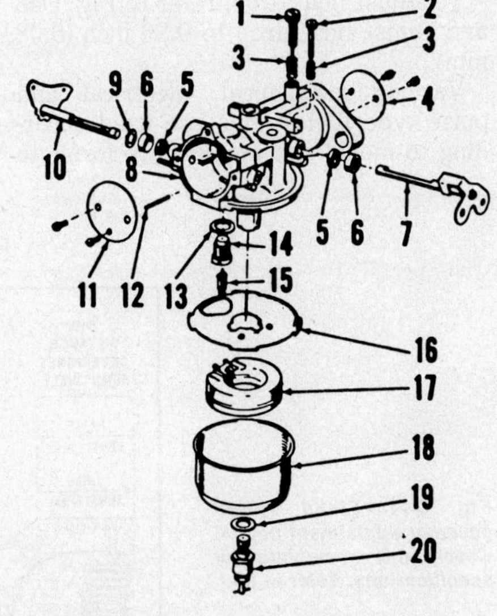

Fig. O39—Exploded view of Onan side draft carburetor used on NH and NHC model engines.

1. Idle mixture needle
2. Throttle stop screw
3. Springs (2)
4. Throttle plate
5. Shaft seal (2)
6. Seal retainer (2)
7. Throttle shaft
8. Body
9. Washer
10. Choke shaft
11. Choke plate
12. Float pin
13. Washer
14. Inlet valve seat
15. Inlet valve needle
16. Body gasket
17. Float
18. Float bowl
19. Washer
20. Main fuel valve

MAINTENANCE

SPARK PLUG. Recommended spark plug for NHC and NHCV models with specification letter A through C is Onan part number 167-0240 for non-resistor plug and 167-0247 for a resistor type plug or their equivalent. Recommended spark plug for all remaining models is Onan part number 167-0291 or equivalent. Electrode gap for all models using vapor (butane, LP or natural gas) fuel is 0.018 inch (0.4572 mm) and gap

Fig. O40—Exploded view of downdraft carburetor used on NHA, NHB and NHC models.

1. Pump cover
2. Diaphragm plunger & spring
3. Diaphragm
4. Gasket
5. Pump body
6. Gasket
7. Air horn
8. Choke shaft
9. Choke plate
10. Inlet valve
11. Float
12. Main fuel needle
13. Idle mixture neelde
14. Throttle stop screw
15. Throttle plate
16. Throttle shaft
17. Carburetor body
18. Gasket

for models using gasoline for fuel is 0.025 inch (0.635 mm).

CARBURETOR. According to model and application, engine may be equipped with either a side draft or a downdraft carburetor. Refer to appropriate paragraph for model being serviced.

SIDE DRAFT CARBURETOR. Side draft carburetor is used on some NH and NHC model engines. Refer to Fig. O39 for exploded view of carburetor and location of mixture adjustment screws.

For initial carburetor adjustment, open idle mixture screw and main fuel mixture screw to 1 to 1½ turns. Make final adjustments with engine at normal operating temperature and running. Place engine under load and adjust main fuel mixture screw for leanest setting that will allow satisfactory acceleration and steady governor operation. Set engine at idle speed, no load and adjust idle mixture screw for smoothest idle operation.

As each adjustment affects the other, adjustment procedure may have to be repeated.

To check float level refer to Fig. O41. Invert carburetor throttle body and float assembly. There should be 1/8-inch (3.2 mm), plus or minus 1/16-inch (1.6 mm), clearance between gasket and free end of float. See Fig. O41.

Adjust float by bending float lever tang that contacts inlet valve.

DOWNDRAFT CARBURETOR. Downdraft carburetor is used on T-260G model and some NH and NHC models. Refer to Fig. O40 for exploded view of carburetor and location of mixture adjustment screws.

For initial carburetor adjustment, open idle mixture screw 1-3/8 to 1-5/8 turns and open main fuel mixture screw 1¼ to 1½ turns. Make final adjustments with engine at normal operating temperature and running. Place engine under load and adjust main fuel mixture screw for leanest setting that will allow satisfactory acceleration and steady governor operation. Set engine at idle speed, no load and adjust idle mixture screw for smoothest idle operation.

As each adjustment affects the other, adjustment procedure may have to be repeated.

To check float level refer to Fig. O42. Position carburetor as shown. There should be 0.00-0.04 inch (0.00-1.02 mm) clearance between float and straightedge.

Adjust float by bending float lever tang that contacts inlet valve.

CAUTION: Remove float assembly to adjust float level. Failure to do so could result in deformation of inlet needle and seat.

To adjust float drop, refer to Fig. O43 and adjust float drop to 0.20 inch (5.08 mm).

Various mechanical, electrical and pulse type fuel pumps are used according to model and application. Refer to SERVICING ONAN ACCESSORIES section for service information.

GOVERNOR. All models use a flyball weight governor located under timing gear cover on camshaft gear. Various linkage arrangements allow for either fixed speed operation or variable speed operation.

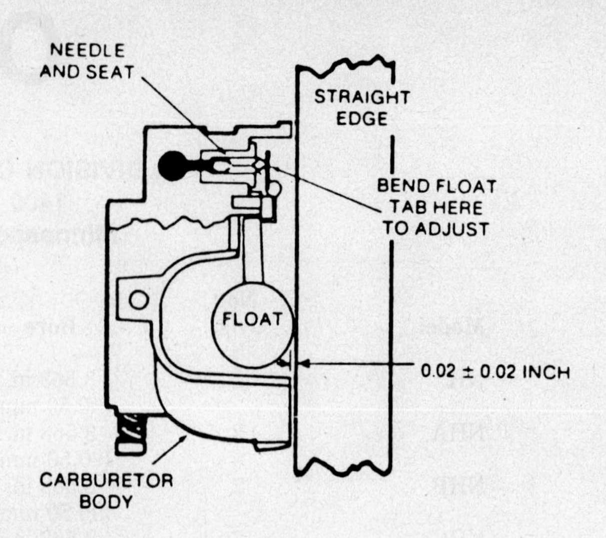

Fig. O42 – View showing use of straightedge to check float level on downdraft type carburetor. Float level clearance should be 0.00-0.04 inch (0.00-1.02 mm).

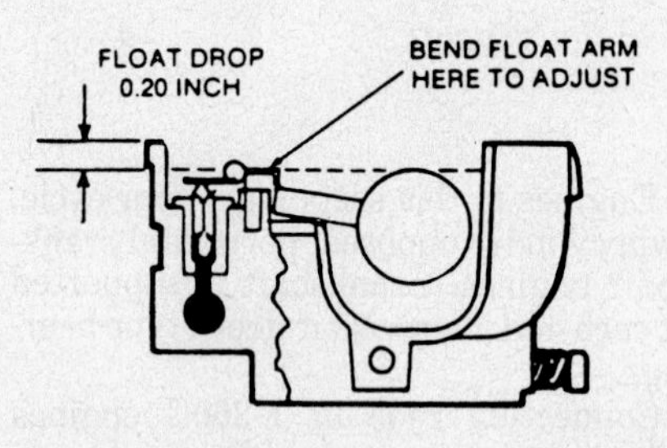

Fig. O43 – View showing proper procedure for checking and adjusting float drop on downdraft style carburetor.

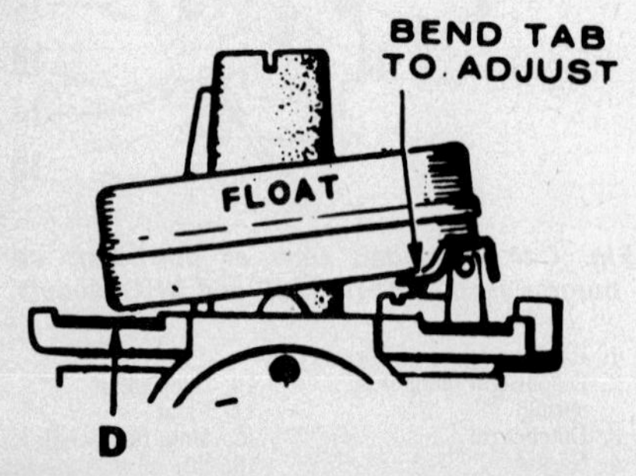

Fig. O41 – Side draft carburetor inlet valve setting. Dimension (D), measured between inverted float and gasket surface is ⅛-inch (3.2 mm), plus or minus 1/16-inch (1.6 mm). Refer to text.

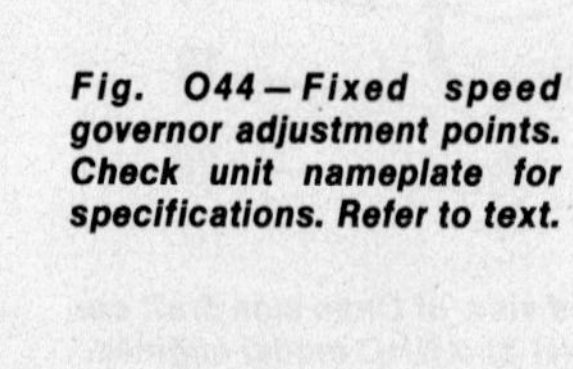

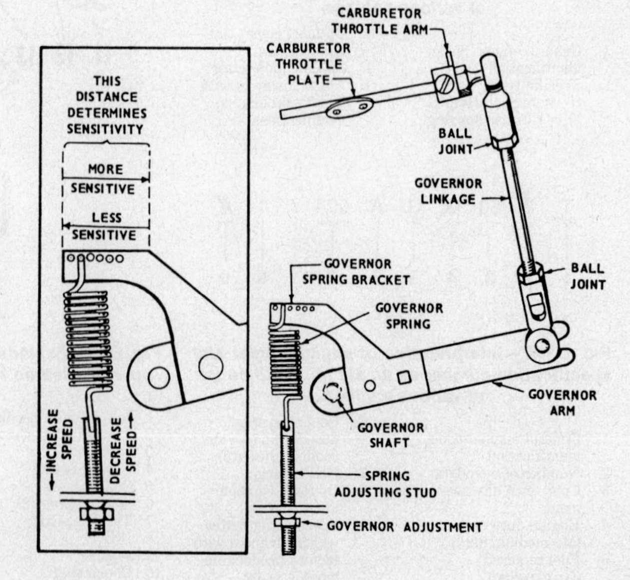

Fig. O44 – Fixed speed governor adjustment points. Check unit nameplate for specifications. Refer to text.

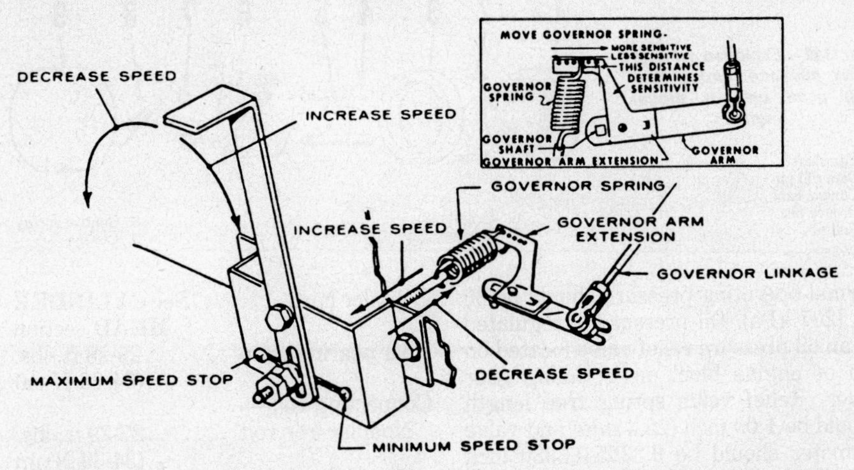

Fig. O45—Variable speed governor linkage adjustment points and procedure. Refer to text.

FIXED SPEED. Fixed speed linkage is used on electric generators and welders or any application which requires a single constant speed under varying load conditions. Fig. O44 shows typical linkage arrangement.

Engines with fixed speed governors start at wide open throttle and engine speed is pre-set at factory to 2400 rpm unless special ordered with different specified speed.

With engine stopped, adjust length of linkage connecting throttle arm to governor arm so stop on carburetor throttle lever is 1/32-inch (0.794 mm) from stop boss. This allows immediate governor control at engine start and synchronizes travel of governor arm and throttle shaft.

Engine speed is determined by governor spring tension. Increasing spring tension results in higher engine speeds and decreasing spring tension results in lower engine speeds. Adjust spring tension as necessary by turning nut on spring adjusting stud. See Fig. O44.

Governor sensitivity is determined by location of governor spring at governor spring bracket. Refer to Fig. O44 for adjustment.

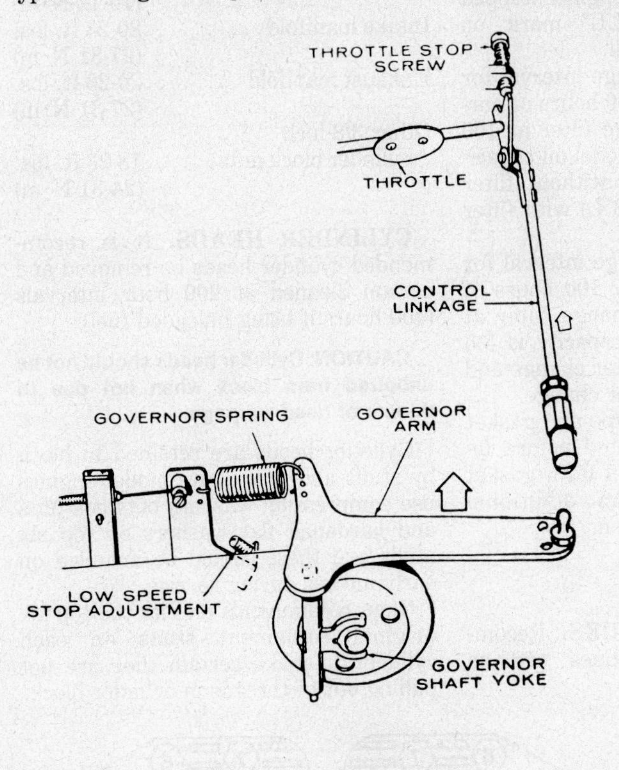

Fig. O46—View of variable speed governor linkage when "Bowden" cable is used. Refer to text for adjustment procedure.

VARIABLE SPEED. Variable speed linkage is used where it is necessary to maintain a wide range of different engine speeds under varying load conditions. Moving a control lever determines engine speed and governor maintains this speed under varying loads.

To adjust variable speed governor linkage adjust throttle stop screw on carburetor so engine idles at 1100 rpm, then adjust governor spring tension so engine idles at 1500 rpm when manual control lever is at minimum speed position (Fig. O45) or "Bowden" cable control knob (Fig. O46) is at first notch (low speed) position.

Adjust sensitivity with engine running at minimum speed to obtain smoothest engine operation by moving governor spring outward on extension arm to decrease sensitivity or inward to increase sensitivity. Refer to Fig. O45 or O46.

Maximum full load speed should not exceed 3000 rpm for continuous operation. To adjust, apply full load to engine and move control lever or knob to maximum speed position and adjust set screw in bracket slot to stop lever travel or turn knob until desired speed is attained.

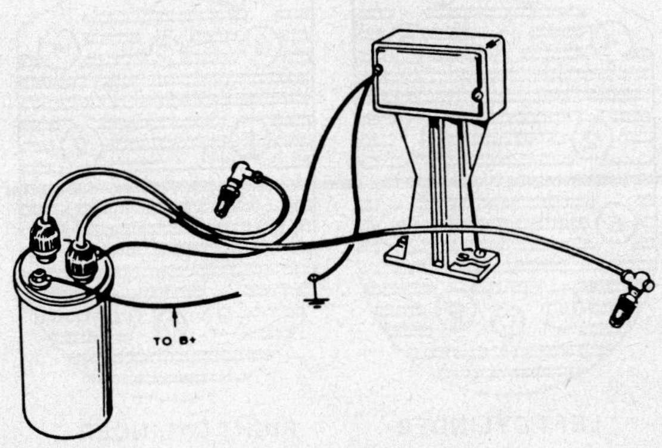

Fig. O47—View of external wiring of engines with battery ignition. Breaker box is non-adjustable and timing is set by varying point gap slightly.

IGNITION SYSTEM. Most models are equipped with battery ignition system, however a magneto system is available for manual start models. Refer to appropriate paragraph for model being serviced.

BATTERY IGNITION. Breaker points are located in a non-movable breaker box (Fig. O47) and timing is adjusted by varying point gap slightly. Static timing should be checked if point gap is changed or new points are installed.

Remove air intake hose from blower housing on pressure cooled engines or remove sheet metal plug in air shroud over right cylinder of "Vacu-Flo" engines to gain access to timing marks.

Initial point gap is 0.016 inch (0.41 mm) and static timing should be checked by connecting a test light across ignition points and rotating engine in direction of normal rotation (clockwise) until light comes on and then just goes out. This should be 20° BTDC and is obtained by varying point gap.

MAGNETO IGNITION. Magneto ignition system is used on manual start models and a stop button wired across breaker points is used to ground primary circuit to stop engine (Fig. O48).

To prevent engine recoil starter damage during starting, spark is retarded to 3° ATDC and advance mechanism shown in Fig. O49 automatically advances timing to 22° BTDC for normal operation.

Initial point gap is 0.020 inch (0.508 mm) and running timing should be checked using a timing light. Timing should be 22° BTDC at 1500 rpm and is adjusted by varying point gap.

If spark advance does not respond when engine is over 1500 rpm, or if advance is sluggish, remove cup-shaped cover (9–Fig. O49) at rear of camshaft on engine block and check advance mechanism condition. Clean assembly thoroughly and renew worn parts as necessary.

LUBRICATION. All models are pressure lubricated by a gear type pump driven by crankshaft gear. Internal pump parts are not serviced separately so entire pump must be renewed if worn or damaged. Clearance between pump gear and crankshaft gear should be 0.002-0.005 inch (0.0508-0.127 mm) and normal operating pressure should be 30 psi (207 kPa). Oil pressure is regulated by an oil pressure relief valve located on top of engine block near timing gear cover. Relief valve spring free length should be 1.00 inch (25.4 mm) and valve diameter should be 0.3365-0.3380 inch (8.55-9.59 mm).

Oils approved by manufacturer must meet requirements of API service classification "SE" or "SE/CC".

Use SAE 5W-30 oil in below freezing temperatures and SAE 20W-40 oil in above freezing temperatures.

Check oil level with engine stopped and maintain at "FULL" mark on dipstick. **DO NOT** overfill.

Recommended oil change interval for T-260G models is every 50 hours of normal operation and change filter at 100 hour intervals. T-260G model oil capacity is 2.5 quarts (2.4 L) without filter change and 3 quarts (2.8 L) with filter change.

Recommended oil change interval for all other models is every 100 hours of normal operation and change filter at 200 hour intervals. Oil capacity is 3.5 quarts (3.3 L) without filter change and 4 quarts (3.8 L) with filter change.

Oil filter is screw on type and gasket should be lightly lubricated before installation. Screw filter on until gasket contacts, then turn an additional ½-turn. Do not overtighten.

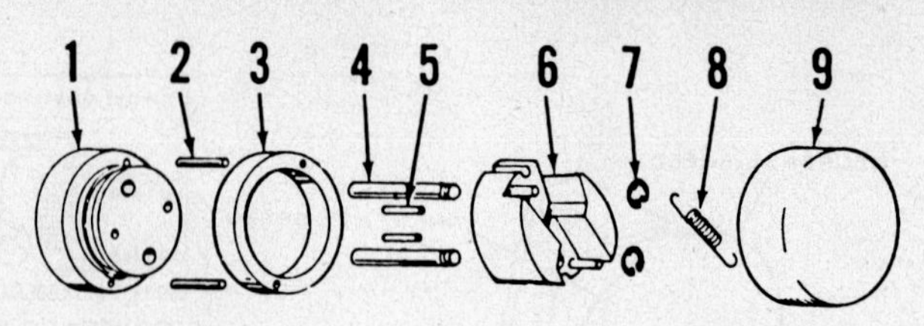

Fig. O49—Exploded view of spark advance (timing control) used on NH model engine.

1. Camshaft
2. Cam roll pin
3. Control cam
4. Groove pin
5. Roll pin
6. Weights (2)
7. Retainers (2)
8. Control spring
9. Cover

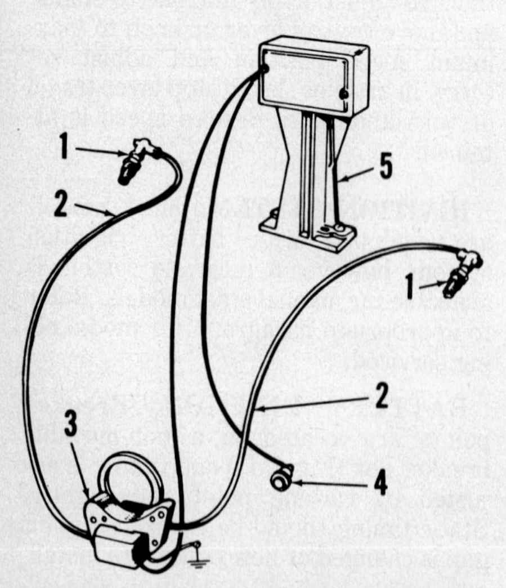

Fig. O48—Magneto ignition used on NH model. Stop button (4) grounds breaker points to halt engine.

1. Spark plugs
2. Plug leads
3. Magneto coil
4. Stop button
5. Breaker box & stand

REPAIRS

TIGHTENING TORQUES. Recommended tightening torques are as follows:

Cylinder head	See **CYLINDER HEAD** section
Rear bearing plate	25-28 ft.-lbs. (34-38 N·m)
Connecting rod–	
Nodular iron rod	27-29 ft.-lbs. (34-39 N·m)
Aluminum rod	14-16 ft.-lbs. (19-22 N·m)
Flywheel cap screws	35-40 ft.-lbs. (48-54 N·m)
Gear case cover	8-10 ft.-lbs. (11-13 N·m)
Oil pump	7-9 ft.-lbs. (10-12 N·m)
Intake manifold	20-23 ft.-lbs. (27-31 N·m)
Exhaust manifold	20-23 ft.-lbs. (27-31 N·m)
Other 3/8-inch cylinder block nuts	18-23 ft.-lbs. (24-31 N·m)

CYLINDER HEADS. It is recommended cylinder heads be removed and carbon cleaned at 200 hour intervals (400 hours if using unleaded fuel).

CAUTION: Cylinder heads should not be unbolted from block when hot due to danger of head warpage.

Cylinder heads are retained to block by studs and nuts. Late model engines use compression washers between nuts and hardened flat washers on top six studs and these should be installed on early models during service.

Onan recommends testing the top six original equipment studs on each cylinder to make certain they are not pulling out of threads in cylinder block.

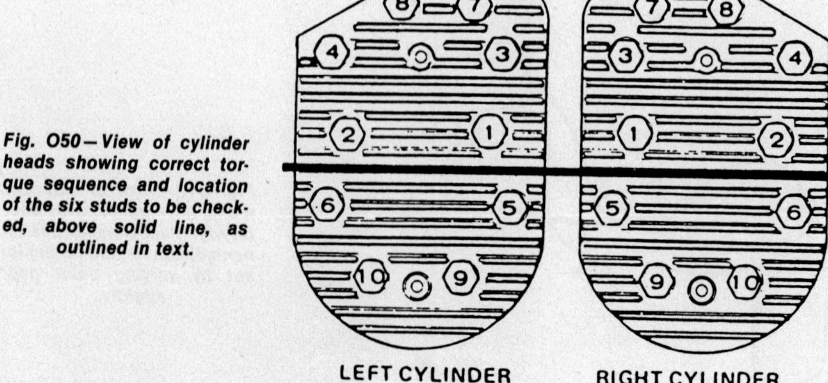

Fig. O50—View of cylinder heads showing correct torque sequence and location of the six studs to be checked, above solid line, as outlined in text.

To test studs, remove nuts and compression washers from top six studs (Fig. O50) leaving hardened flat washers in place. Reinstall nuts and tighten to 30 ft.-lbs. (40 N·m) torque. Studs with weak threads will pull out of head before 30 ft.-lbs. (40 N·m) torque is reached.

A special stepped renewal stud, part number 520-0912 and a drilling fixture, part number 420-0398, are available from Onan to service damaged stud hole threads.

To install stepped renewal stud, remove cylinder head and gasket. Examine head and block surface. If head or block is warped or has a depression of more than 0.005 inch (0.127 mm) it may be resurfaced with a maximum total of 0.010 inch (0.254 mm) of material removed. Remove studs from holes with damaged threads and install drilling fixture (Fig. O52) securing it in position with nuts and flat washers on studs indicated.

NOTE: Some engines may require addition of flat washers between block and fixture plate to clear sheet metal scroll backing plate. If so, be certain drilling fixture remains parallel with head surface of block.

Place bushing with small hole in it which is furnished with drilling plate, in plate hole over stud hole with damaged threads. Use a 27/64-inch drill bit to drill damaged threads out of block. Holes at side of block should be drilled through to the fourth cooling fin and holes at top of block should penetrate into corresponding intake or exhaust valve port. Manifolds should be removed during drilling procedure. Remove small bushing and install bushing with larger hole and lock in place. Use a ½-13 tap to thread hole for new stepped stud. Repeat process for any other hole with damaged threads. Remove any ridge around holes using a flat file or a 45° chamfer tool. When using chamfer tool chamfer depth should be 1/32-1/16 inch (0.794-1.588 mm). Apply "Loctite 242" to stepped stud threads and install in holes making certain entire stepped portion is below gasket surface of cylinder block.

NOTE: Stepped studs installed in holes corresponding with exhaust ports must have 3/16-1/4 inch (4.763-6.350 mm) cut off large end of stud so it does not extend into exhaust port.

Remove any metal particles from engine ports and place new graphoil type gasket on cylinder head and install gasket and head on studs at the same time to avoid damaging gasket.

NOTE: Graph-oil type gaskets become soft and gummy at temperatures above 100° F (38° C). Avoid installation or removal if engine temperature exceeds this.

Install hardened flat washers on all studs, two compression washers on each of the top six long studs (Fig. O51) so outside edges of compression washers are in contact with each other. Install nuts and tighten in several even steps until top six studs reach 12 ft.-lbs. (16 N·m) torque and the bottom studs reach 15 ft.-lbs. (20 N·m) torque.

Tighten in sequence shown in Fig. O50 and recheck all nuts after initial torque has been reached.

CAUTION: Too much torque will flatten compression washers and could result in engine damage.

Recheck torque before engine has a total of 50 hours operation.

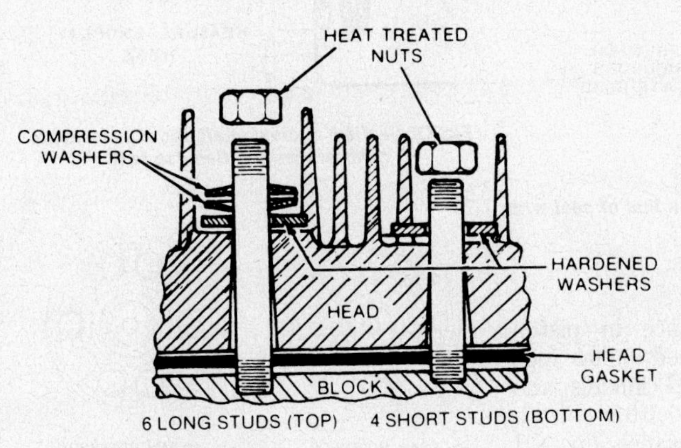

Fig. O51—View showing correctly installed compression washers. Compression washers are installed on top six (long) studs only. Hardened flat washers are installed on all studs.

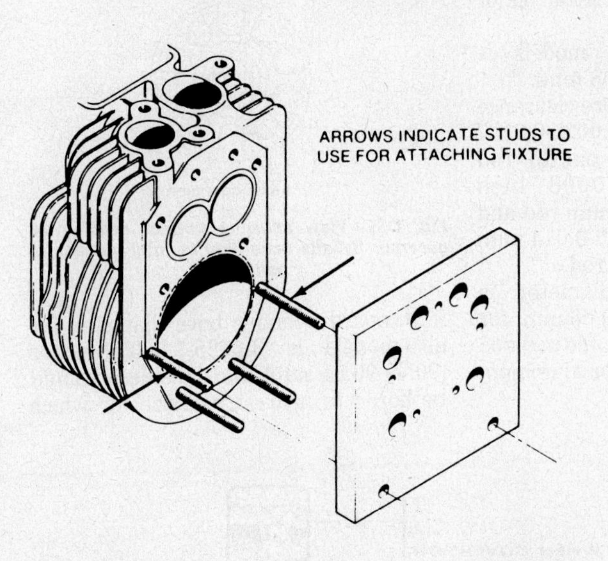

Fig. O52—Special drilling fixture is required to drill out damaged threads in stud holes. Fixture part number is 420-398.

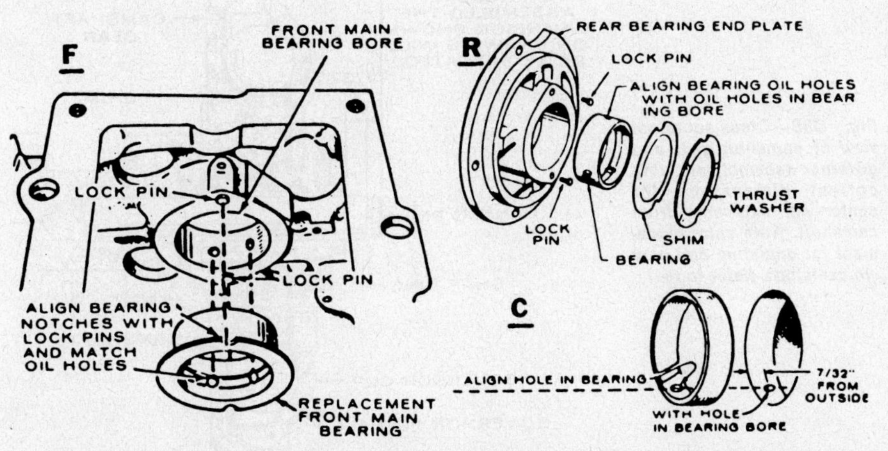

Fig. O53—Alignment of precision main and camshaft bearings. One-piece bearing is used at front (view F), two-piece bearing at rear (view R). Note shim use for end play adjustment. View C shows placement of camshaft bearings.

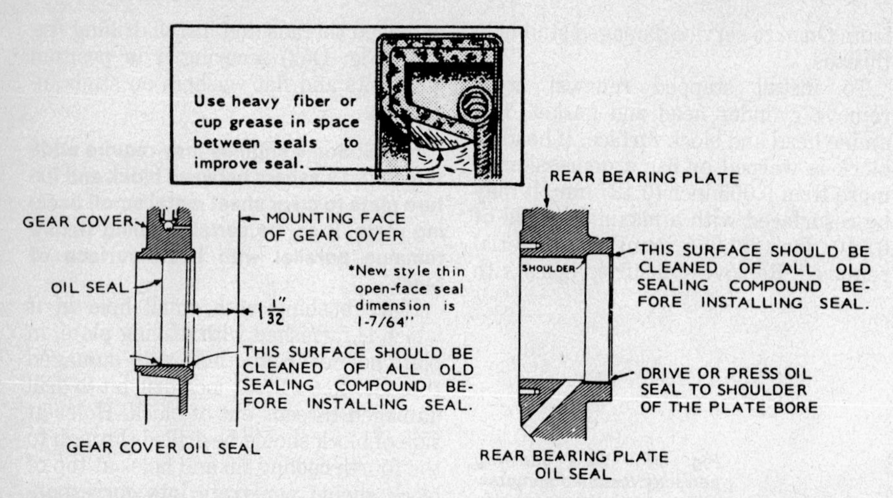

Fig. O53A – View showing correct seal installation procedure. Fill cavity between lips of seal with heavy grease to improve sealing efficiency.

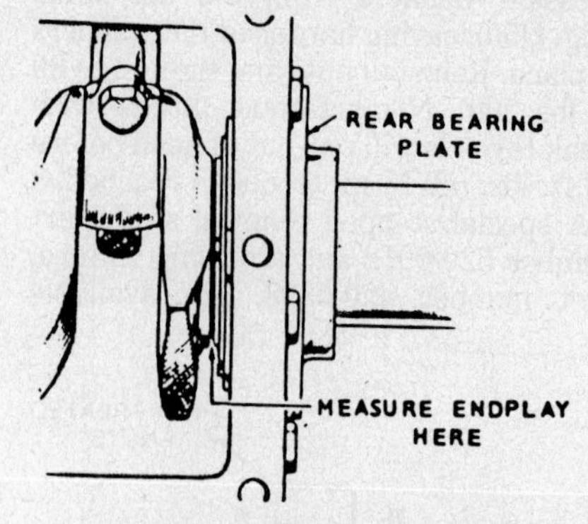

Fig. O54 – View of crankshaft end play measurement procedure. Refer to text.

CONNECTING ROD. Connecting rod and piston are removed from cylinder head surface end of block after removing cylinder head and oil base.

T-260G model connecting rod rides directly on crankpin journal and all remaining models have renewable insert type rod bearings and a renewable piston pin bushing.

Standard crankpin journal diameter is 1.6252-1.6260 inches (41.28-41.30 mm) for all models and connecting rod or bearing inserts are available in a variety of sizes for undersize crankshafts as well as standard.

Connecting rod to crankshaft journal running clearance for T-260G engine or engines with aluminum rod is 0.002-0.003 inch (0.051-0.076 mm) and running clearance for bearings in nodular iron rod should be 0.0005-0.0028 inch (0.013-0.071 mm).

Piston and rod assembly should be installed so rod bolts are offset toward outside of cylinder block. If rod has an oil hole, hole must be toward camshaft. Tighten connecting rod cap screws to specified torque.

PISTON, PIN AND RINGS. NH model engines with specification letters "A" through "C" use a three ring strut type piston which should have 0.0015-0.0035 inch (0.0381-0.0889 mm) clearance between cylinder and piston when measured 0.10 inch (2.54 mm) below oil control ring 90° from piston pin. If side clearance of top ring in piston groove exceeds 0.006 inch (0.1524 mm) piston should be renewed.

NH models with specification letter D or after and all T-260G engines use a three ring piston. Piston to cylinder clearance should be 0.007-0.009 inch (0.178-0.229 mm) when measured 0.10 inch (2.54 mm) below oil control ring 90° from piston pin.

Top ring side clearance in piston groove should not exceed 0.004 inch (0.1016 mm) for T-260G models and should be 0.002-0.008 inch (0.0508-0.2032 mm) clearance for NH models with specification letter D or after.

Ring end gap for all models is 0.010-0.020 inch (0.254-0.508 mm).

Piston pin to piston pin bore clearance is 0.0001-0.0005 inch (0.0025-0.0127 mm) for all models and pin to rod clearance is 0.0002-0.0008 inch (0.005-0.020 mm) for aluminum rod and 0.00005-0.00055 inch (0.0013-0.014 mm) for bushing in nodular iron rod.

Standard piston pin diameter is 0.7500-0.7502 inch (19.05-19.06 mm) for all models and pins are available in 0.002 inch (0.0508 mm) oversize for aluminum rod.

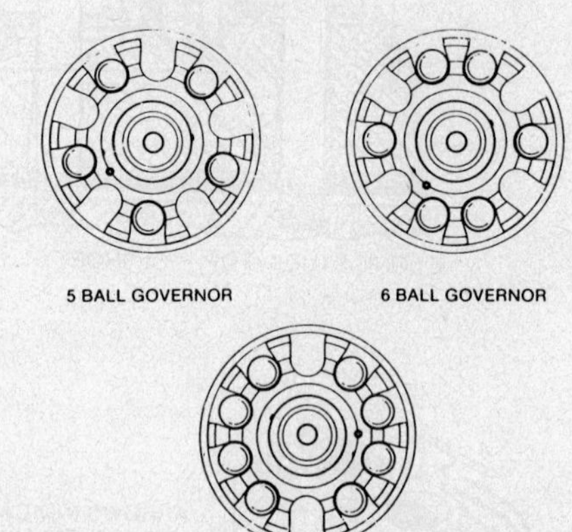

Fig. O55 – View showing correct spacing of governor flyballs according to total number of balls used.

Standard cylinder bore diameter for all models is 3.5625-3.5635 inches (90.49-90.51 mm) and cylinders should be bored to nearest oversize for which

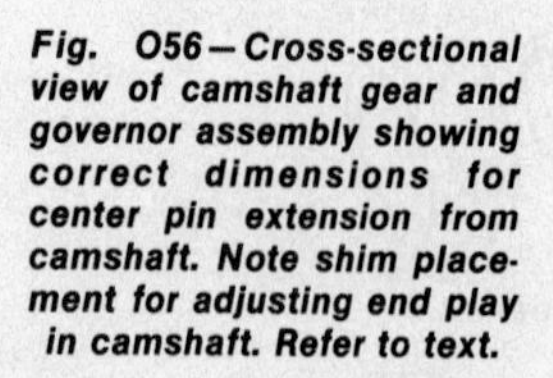
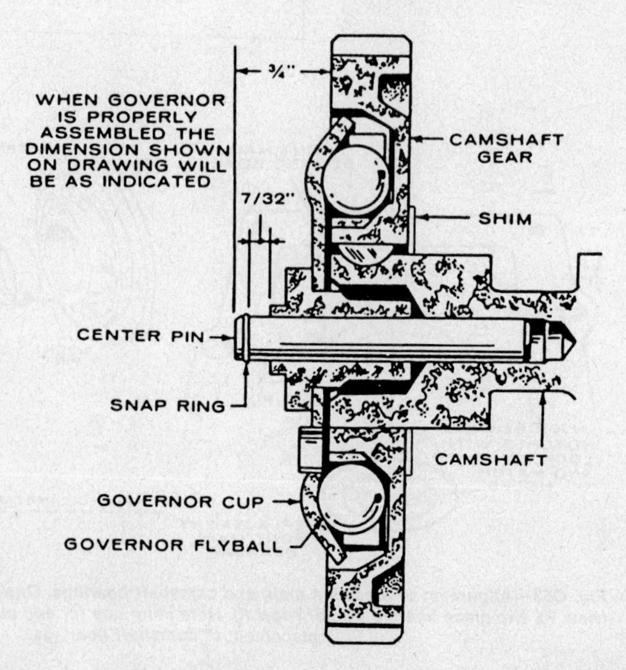

Fig. O56 – Cross-sectional view of camshaft gear and governor assembly showing correct dimensions for center pin extension from camshaft. Note shim placement for adjusting end play in camshaft. Refer to text.

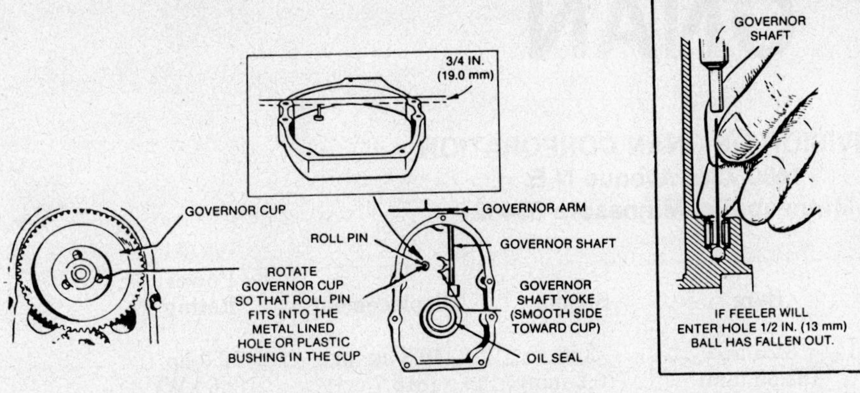

Fig. O57—View of governor shaft, timing gear cover and governor cup showing assembly details. Refer to text.

piston and rings are available if cylinder is scored or out-of-round more than 0.003 inch (0.0762 mm) or if taper exceeds 0.005 inch (0.127 mm). Pistons and rings are available in a variety of oversizes as well as standard.

CRANKSHAFT, BEARINGS AND SEALS. Crankshaft is supported at each end by precision type sleeve bearings located in cylinder block housing and rear bearing plate. Main bearing journal standard diameter should be 1.9992-2.000 inches (50.7797-50.800 mm) and bearing operating clearance should be 0.0025-0.0038 inch (0.0635-0.0965 mm) for T-260G model and 0.0015-0.0043 inch (0.0381-0.1092 mm) for all other models. Main bearings are available for a variety of undersize crankshafts as well as standard.

Oil holes in bearing and bore **MUST** be aligned when bearings are installed and rear bearing should be pressed into bearing plate until flush, or recessed into bearing plate 1/64-inch (0.406 mm). Make certain bearing notches are aligned with lock pins (Fig. O53) during installation.

Apply "Loctite Bearing Mount" to front bearing and press bearing in until flush with block. Make certain bearing notches are aligned with lock pins (Fig. O53) during installation.

NOTE: Replacement front bearing and thrust washer is a one piece assembly, do not install a separate thrust washer.

Rear thrust washer is installed with oil grooves toward crankshaft. Measure crankshaft end play as shown in Fig. O54 and add or remove shims or renew thrust washer (Fig. O53) as necessary to obtain 0.005-0.009 inch (0.13-0.23 mm) end play.

It is recommended rear bearing plate and front timing gear cover be removed for seal installation. Rear seal is pressed in until flush with seal bore. Timing cover seal should be driven in until it is 1-1/32 inch (26.19 mm) from mounting face of cover for old style seal and 1-7/64 inch (28.18 mm) from mounting face of cover for new style, thin, open faced seal. See Fig. O53A.

CAMSHAFT, BEARINGS AND GOVERNOR. Camshaft is supported at each end in precision type sleeve bearings. Make certain oil holes in bearings are aligned with oil holes in block (Fig. O53), press front bearing in flush with outer surface of block and press rear bearing in until flush with bottom of counterbore. Camshaft to bearing clearance should be 0.0015-0.0030 inch (0.0381-0.0762 mm).

Camshaft end play should be 0.003 inch (0.0762 mm) and is adjusted by varying thickness of shim located between camshaft timing gear and engine block. See Fig. O56.

Camshaft timing gear is a press fit on end of camshaft and is designed to accomodate 5, 8 or 10 flyballs arranged as shown in Fig. O55. Number of flyballs is varied to alter governor sensitivity. Fewer flyballs are used on engines with variable speed applications and greater number of flyballs are used for continuous (fixed) speed application. Flyballs must be arranged in "pockets" as shown according to total number used.

Center pin (Fig. O56) should extend ¾-inch (19 mm) from end of camshaft to allow 7/32-inch (5.6 mm) in-and-out movement of governor cup. If distance is incorrect, remove pin and press new pin in to correct depth.

Always make certain governor shaft pivot ball is in timing cover by measuring as shown in Fig. O57 inset and check length of roll pin which engages bushed hole in governor cup. This pin must extend to within ¾-inch (19 mm) of timing gear cover mating surface.

Make certain timing mark on camshaft gear is aligned with timing mark on crankshaft gear during installation.

VALVE SYSTEM. Valve seats are renewable insert type and are available in a variety of oversizes as well as standard. Seats are ground at a 45° angle and seat width should be 1/32 to 3/64-inch (0.794 to 1.191 mm).

Valves should be ground at a 44° angle to provide an interference angle of 1°.

Recommended valve stem to guide clearance is 0.001-0.0025 inch (0.0254-0.0635 mm) for intake valves and 0.0025-0.004 inch (0.0635-0.1016 mm) for exhaust valves. Renewable shouldered valve guides are pushed out from head surface end of block. An "O" ring is installed on intake valve guide of some models and a valve stem seal is also available for intake valve. See Fig. O58.

Recommended valve tappet gap (cold) is 0.003 inch (0.0762 mm) for intake valves and 0.010 inch (0.254 mm) for exhaust valves.

Adjustment is made by turning tappet adjusting screw as required and valves of each cylinder must be adjusted with cylinder at "top dead center" on compression stroke. At this position both valves will be fully closed.

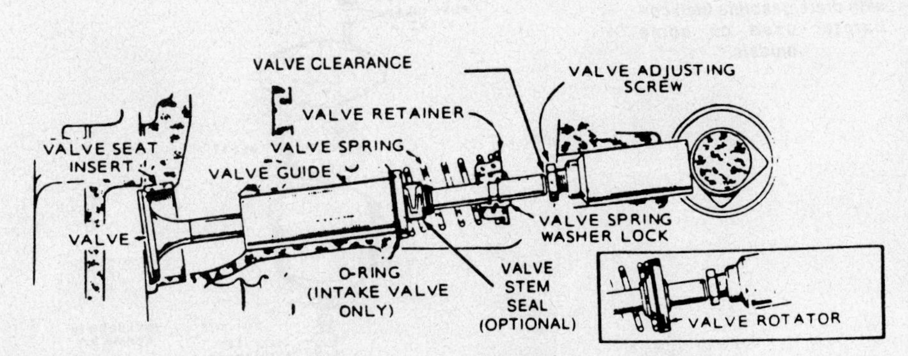

Fig. O58—Valve train typical of NH and T-260G engines. Refer to text.

ONAN

A DIVISION OF ONAN CORPORATION
1400 73rd Avenue N.E.
Minneapolis, Minnesota 55432

Model	No. Cyls.	Bore	Stroke	Displacement	Power Rating
CCK	2	3.25 in. (82.55 mm)	3 in. (76.2 mm)	49.8 cu. in. (815.7 cc)	12.9 hp. (9.6 kW)
CCKA	2	3.25 in. (82.55 mm)	3 in. (76.2 mm)	49.8 cu. in. (815.7 cc)	16.5 hp. (12.3 kW)
CCKB	2	3.25 in. (82.55 mm)	3 in. (76.2 mm)	49.8 cu. in. (815.7 cc)	20 hp. (14.9 kW)

Engines in this section are four-cycle, twin cylinder opposed, horizontal crankshaft engines. Crankshaft is supported at each end in precision sleeve type bearings.

CCK and CCKA engines prior to specification letter "D" have aluminum connecting rods which ride directly on crankshaft journal.

CCK and CCKA engines with specification letter "D" or after and all other models have forged steel rods equipped with renewable bearing inserts.

All models are pressure lubricated by a gear type oil pump. A "spin-on" oil filter is available.

Most models are equipped with a battery ignition system, however a magneto system is available for manual start models.

Engines may be equipped with a side draft or downdraft carburetor which may be used with gasoline or vapor (butane, LP or natural gas) fuels according to model and application. Mechanical, pulsating diaphragm or electric fuel pumps are available.

Refer to Fig. O59 for interpretation of engine specification and model numbers.

CCKA engines may be visually distinguished by the cup-shaped advance mechanism cover, rather than the flat-shaped expansion plug, in rear camshaft opening just below ignition point breaker box.

High compression heads are identified by a 3/32-inch radius boss located on corner of head nearest spark plug.

Always give specification, model and serial numbers when ordering parts or service material.

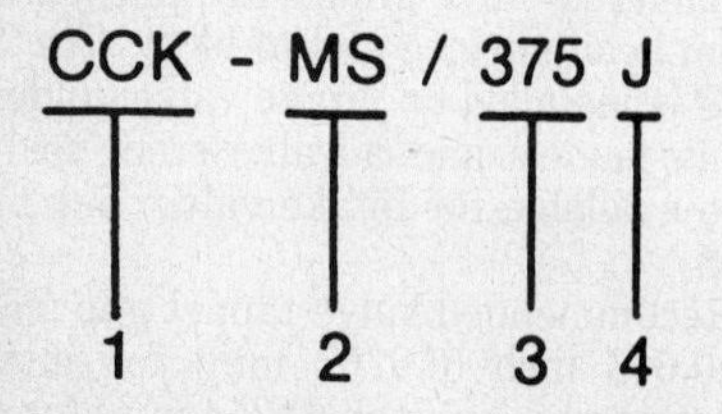

Fig. O59—Interpretation of engine model and specification number as an aid to identification number of various engines.

1. General engine model identification
2. Specific type
   S–Manual starting
   MS–Electric starting
3. Optional equipment identification
4. Specification letter which advances with factory production modifications.

MAINTENANCE

SPARK PLUG. Recommended spark plug is Onan part number 167-0241 for non-resistor plug and 167-0237 for a resistor type plug or their equivalent. Electrode gap for all models using vapor (butane, LP or natural gas) fuel is 0.018 inch (0.4572 mm) and gap for models using gasoline for fuel is 0.025 inch (0.635 mm).

CARBURETOR. According to model and application, engine may be equipped with either a side draft or a downdraft carburetor which uses either gasoline or vapor (butane, LP or natural gas) as fuel. Refer to appropriate paragraph for model being serviced.

SIDE DRAFT CARBURETOR (GASOLINE). Side draft carburetor is used on some CCKB models. Refer to Fig. O60 for exploded view of carburetor and location of mixture adjustment screws.

For initial carburetor adjustment, open idle mixture screw and main fuel

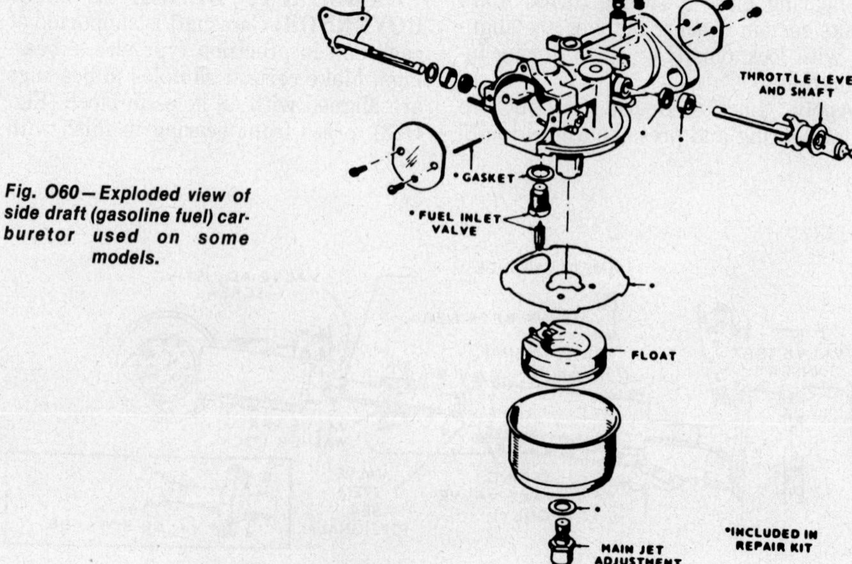

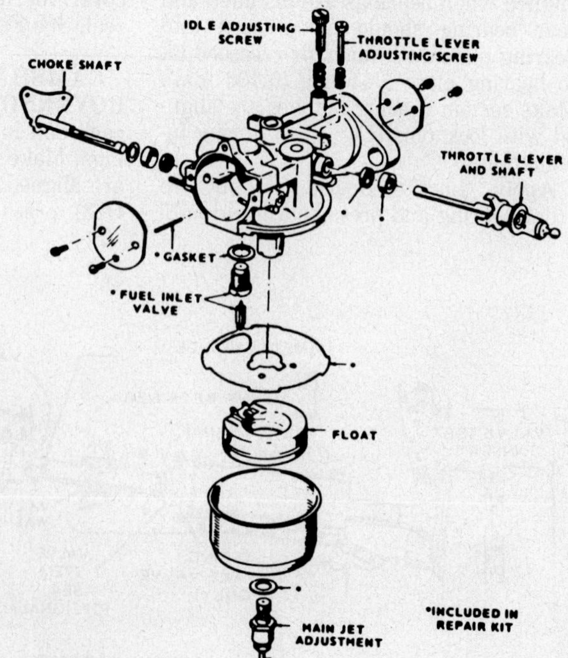

Fig. O60—Exploded view of side draft (gasoline fuel) carburetor used on some models.

mixture screw 1 to 1½ turns. Make final adjustments with engine at normal operating temperature and running. Place engine under load and adjust main fuel mixture screw for leanest setting that will allow satisfactory acceleration and steady governor operation. Set engine at idle speed, no load and adjust idle mixture screw for smoothest idle operation.

As each adjustment affects the other, adjustment procedure may have to be repeated.

To check float level, refer to Fig. O61. Invert carburetor throttle body and float assembly. Float clearance should be 11/64-inch (4.366 mm) for models prior to specification letter "H" and 1/8 to 3/16-inch (3.2 to 4.8 mm) for models with specification letter "H" or after.

Adjust float by bending float lever tang that contacts inlet valve.

SIDE DRAFT CARBURETOR (VAPOR FUEL/GASOLINE). A side draft carburetor designed to use either gasoline or vapor (butane, LP or natural gas) fuel is available for some models.

For initial adjustment, refer to Fig. O62 for location of appropriate adjustment screws and make initial and final adjustments as outlined in **SIDE DRAFT CARBURETOR (GASOLINE)** section.

To change system from gasoline fuel operation to vapor fuel operation, follow the procedure below.

Shut off gasoline supply valve.
Install lock wire on choke.
Set spark plug gap at 0.018 inch (0.4572 mm).
Lock float in position.*
Open supply valve for vapor fuel.

*It may be necessary to remove float assembly and inlet valve on models which are not equipped with a float locking mechanism for extended vapor fuel use.

To change system from vapor fuel operation to gasoline fuel operation, follow the procedure below.

Close vapor fuel supply valve.
Remove lock wire from choke.
Set spark plug gap at 0.025 inch (0.635 mm).
Unlock carburetor float.*
Open gasoline fuel supply valve.

*Replace float assembly and inlet valve if they were removed because carburetor was not equipped with float lock.

To check float level, refer to procedure outlined in **SIDE DRAFT CARBURETOR (GASOLINE)** section.

DOWNDRAFT CARBURETOR (GASOLINE). A downdraft carburetor is used on most CCK and CCKA models. Refer to Fig. O63 for exploded view of carburetor and location of mixture adjustment screws.

For initial carburetor adjustment, open idle mixture screw 1 turn. If equipped with main fuel adjustment open main fuel mixture screw about 2 turns.

Make final adjustments with engine at normal operating temperature and running. Place engine under load and adjust main fuel mixture screw (if equipped) for leanest setting that will allow satisfactory acceleration and steady governor operation. Onan special tool number 420-0169 (Fig. O64) aids main fuel mixture screw adjustment. Set engine at idle speed, no load and adjust idle mixture screw for smoothest idle operation.

As each adjustment affects the other, adjustment procedure may have to be repeated.

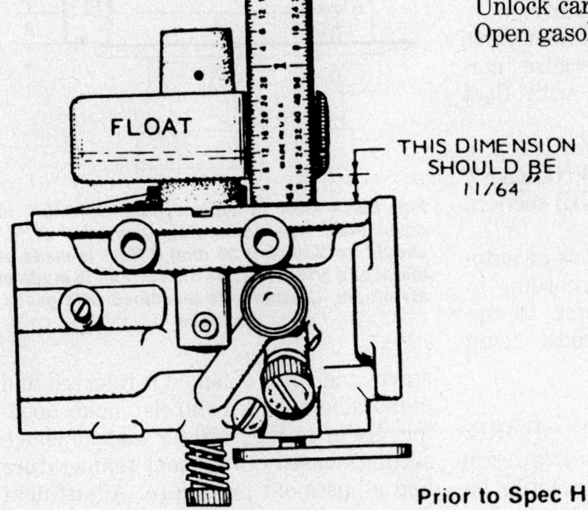

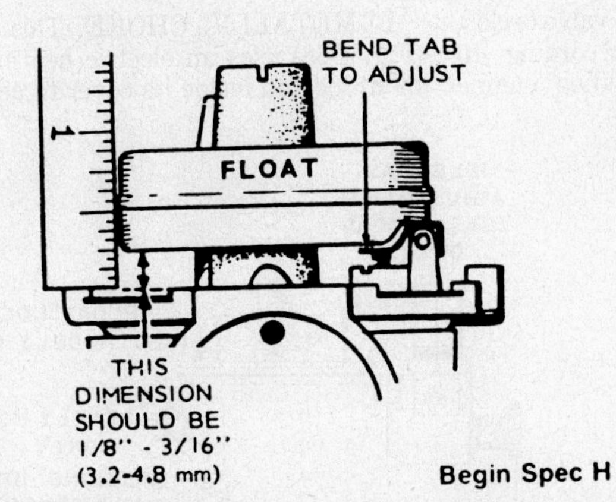

Fig. O61 — Measure and set float level as shown noting upper view is for engines with specification letter prior to "H" and lower view is for engines with specification letter "H" and after.

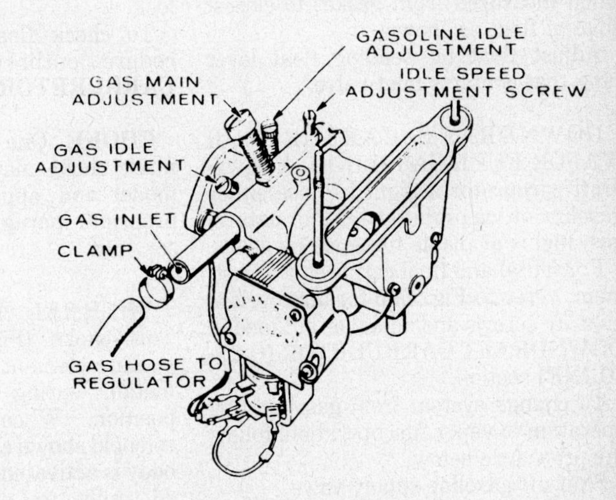

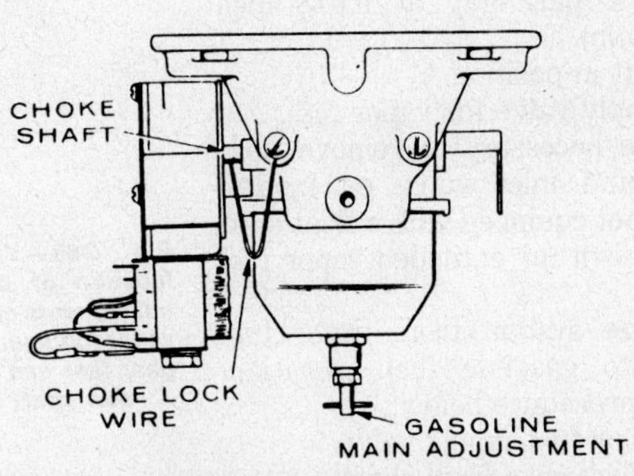

Fig. O62 — View showing location of mixture adjustment screws on carburetor designed for either vapor (butane, LP or natural gas) fuel or gasoline fuel. Refer to text.

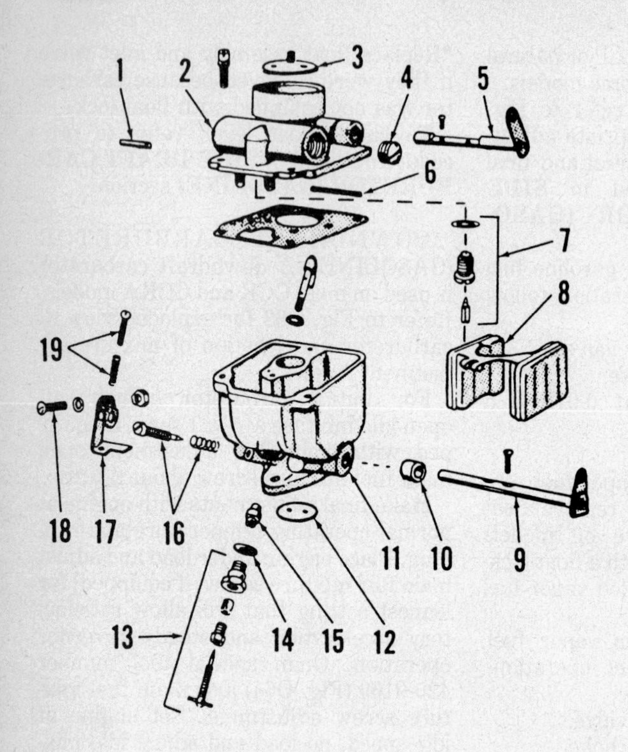

Fig. O63—Exploded view of downdraft carburetor used on CCK and CCKA models.

1. Choke stop pin
2. Cover assy.
3. Choke valve
4. Gas inlet plug
5. Choke shaft
6. Float pin
7. Inlet valve assy.
8. Float
9. Throttle shaft
10. Shaft bushing
11. Body assy.
12. Throttle plate
13. Main fuel valve
14. Gasket
*15. Main jet
16. Idle mixture adjustment needle
17. Idle stop lever
18. Clamp screw
19. Stop screw & spring

*Either 13 or 15 is used.

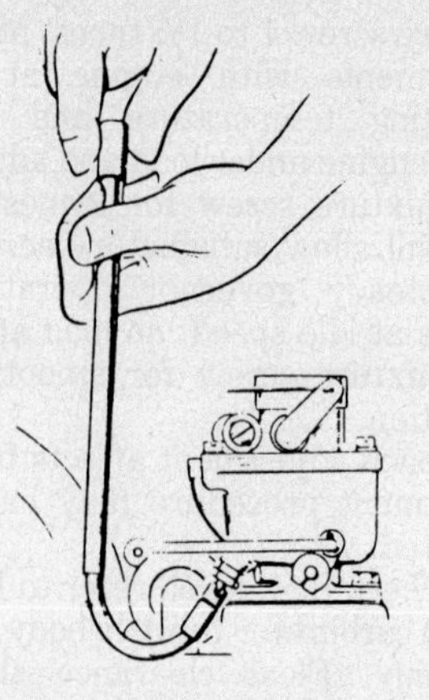

Fig. O64—View showing use of Onan main fuel adjusting tool, part number 420-0169, used to adjust main fuel mixture needle. Refer to text.

To check float level, refer to Fig. O65. Invert carburetor throttle body and float assembly. Float clearance should be ¼-inch (6.35 mm) for metal float or 5/16-inch (7.94 mm) for styrofoam float when measured from gasket to closest edge of float as shown.

Adjust float by bending float lever tang that contacts inlet valve.

DOWNDRAFT CARBURETOR (VAPOR FUEL/GASOLINE). A downdraft carburetor designed to use either gasoline or vapor (butane, LP or natural gas) fuel is available for some models.

For initial and final carburetor adjustment, refer to Fig. O66 for appropriate mixture screws and adjust as outlined in **DOWNDRAFT CARBURETOR (GASOLINE)** section.

To change system from gasoline fuel operation to vapor fuel operation, follow the procedure below.

Shut off gasoline supply valve.

Install lock wire on choke.

Set spark plug gap at 0.018 inch (0.4572 mm).

Lock float in position.*

Open supply valve for vapor fuel.

*It may be necessary to remove float assembly and inlet valve on models which are not equipped with a float locking mechanism for extended vapor fuel use.

To change system from vapor fuel operation to gasoline fuel operation, follow the procedure below.

Close vapor fuel supply valve.

Remove lock wire from choke.

Set spark plug gap at 0.025 inch (0.635 mm).

Unlock carburetor float.*

Open gasoline fuel supply valve.

*Replace float assembly and inlet valve if they were removed because carburetor was not equipped with float lock.

To check float level, refer to procedure outlined in **DOWNDRAFT CARBURETOR (GASOLINE)** section.

CHOKE. One of three types of automatic choke may be used according to model and application. Refer to appropriate paragraph for model being serviced.

THERMAL MAGNETIC CHOKE. This choke (Fig. O67) uses a strip heating element and a heat-reactive bimetallic spring to control choke valve position. When engine is cranked, solenoid shown at lower portion of choke body is activated and choke valve is closed, fully or partially, according to temperature conditions. When engine starts and runs, solenoid is released and bi-metallic spring controls choke opening. Refer to Fig. O70 for table of choke settings based on ambient temperature and adjustment procedure. Adjustment is made by rotating entire cover as shown.

BI-METALLIC CHOKE. This choke (Fig. O68) uses an electric heating element located inside its cover to activate

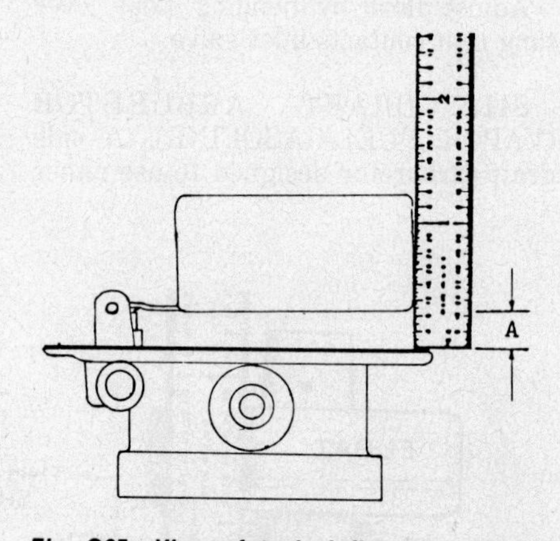

Fig. O65—View of typical float assembly of down draft type carburetor. Clearance at "A" should be ¼-inch (6.35 mm) if float is made of metal and 5/16-inch (7.94 mm) if float is made of styrofoam. Clearance is measured with gasket in place.

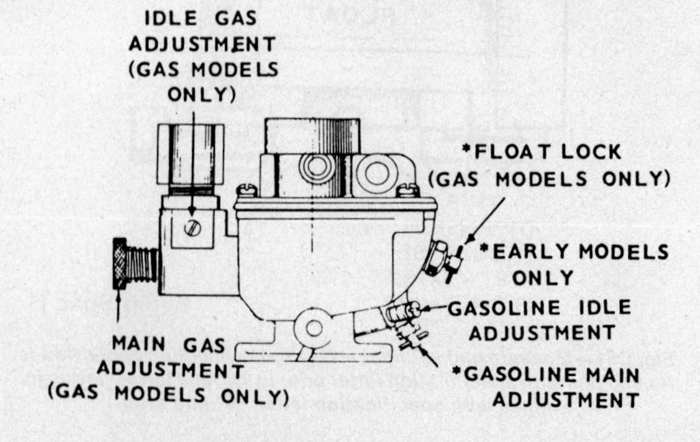

Fig. O66—View showing location of mixture screw adjustments on combination vapor (butane, LP or natural gas) fuel and gasoline fuel downdraft carburetor.

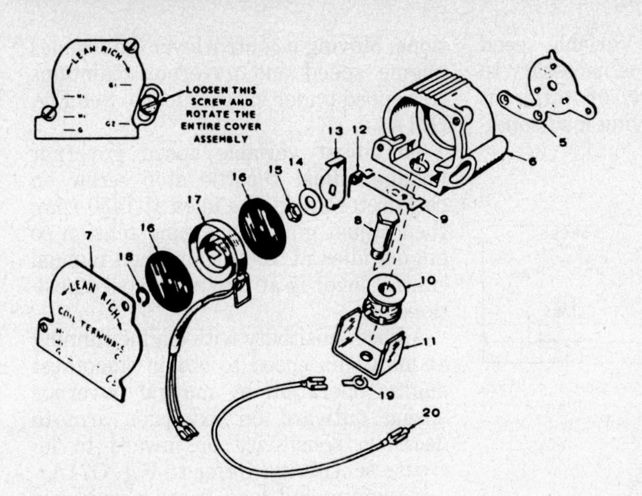

Fig. O67—Exploded view of thermal magnetic choke used on some engines.

5. Mounting plate
6. Body housing
7. Cover
8. Solenoid core
9. Armature
10. Solenoid coil
11. Frame
12. Lever spring
13. Lever
14. Washer
15. Pal nut
16. Insulator disc (2)
17. Heater & spring
18. Retainer (to shaft)
19. Ground terminal
20. Choke lead wire

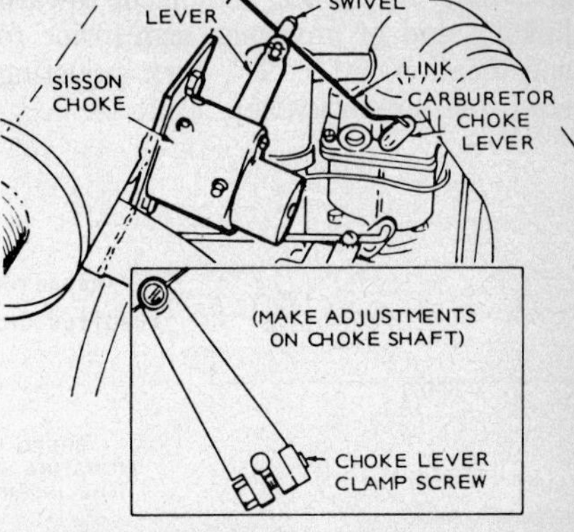

Fig. O69—View showing various locations of component parts for Sisson choke. Refer to text for adjustment procedure.

a bi-metallic spring which opens or closes choke valve according to temperature. Electric current is supplied to heating element by leads from generator exciter. Refer to Fig. O70 for choke settings according to ambient temperature and adjustment procedure. Adjustment is made by rotating cover to open or close choke valve.

SISSON CHOKE. This choke (Fig. O69) uses a bi-metallic strip which reacts to manifold temperatures and a magnetic solenoid which is wired in series with starter switch circuit.

To adjust, pull choke lever up and insert a 1/16-inch diameter rod through shaft hole to engage notch in mounting flange and lock shaft against rotation. Loosen choke lever clamp screw and adjust choke lever so choke plate is completely closed, or not more than 5/16-inch (7.94 mm) open. Tighten clamp screw and remove 1/16-inch rod.

FUEL PUMP. Various mechanical, electric and pulse type fuel pumps are used according to model and application. Refer to **SERVICING ONAN ACCESSORIES.**

GOVERNOR. All models use a flyball weight governor located under timing gear cover on camshaft gear. Various linkage arrangements allow for either fixed speed operation or variable speed operation.

Some engines may be equipped with vacuum operated speed booster, automatic idle control or two-speed electric solenoid controls. Refer to **SERVICING ONAN ACCESSORIES** for service.

Before any governor linkage adjustment is made, make certain all worn or binding linkage is repaired and carburetor is properly adjusted. Clean plastic ball joints, but do not lubricate. Beginning with specification letter "J", metal ball joints are used and these should be cleaned and lubricated with graphite.

FIXED SPEED. Fixed speed linkage is used on electric generators and welders or any application which requires a single constant speed under varying load conditions. Fig. O71 shows typical linkage arrangement.

Engines with fixed speed governors start at wide open throttle and engine speed is pre-set at factory to 2400 rpm unless special ordered with different specified speed.

With engine stopped, adjust length of linkage connecting throttle arm to governor arm so stop on carburetor throttle lever is 1/32-inch (0.794 mm) from stop boss. This allows immediate governor control at engine start and synchronizes travel of governor arm and throttle shaft.

Engine speed is determined by governor spring tension. Increasing spring tension results in higher engine speeds and decreasing spring tension results in lower engine speeds. Adjust spring tension as necessary by turning nut on spring adjusting stud. See Fig. O71.

Governor sensitivity is determined by location of governor spring at governor spring bracket. Refer to Fig. O71. Governor sensitivity is increased by shifting sliding clip toward governor shaft (prior to specification letter "D", turn adjusting stud clockwise). Decrease

AMBIENT TEMP.	CHOKE SETTING
60° F (16° C)	1/8-in. (3.2 mm)
65° F (18° C)	9/64-in. (3.6 mm)
70° F (21° C)	5/32-in. (4 mm)
75° F (24° C)	11/64-in. (4.4 mm)
80° F (27° C)	3/16-in. (4.8 mm)
85° F (29° C)	13/64-in. (5.2 mm)
90° F (32° C)	7/32-in. (5.6 mm)
95° F (35° C)	15/64-in. (6 mm)
100° F (38° C)	1/4-in. (6.4 mm)

Fig. O70—Let engine cool at least one hour before adjusting choke setting. Drill bits may be used as gages to set choke opening according to ambient temperature.

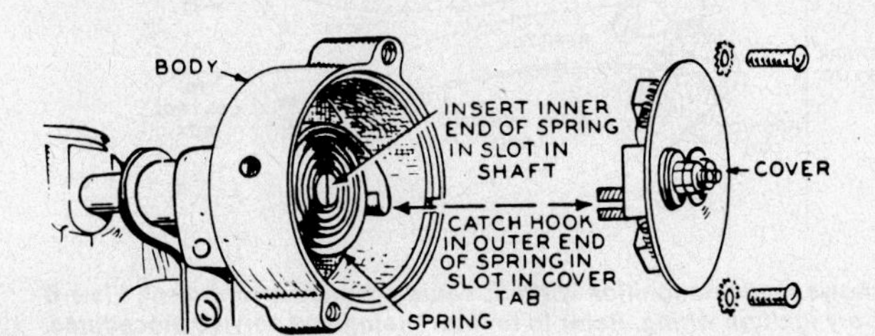

Fig. O68—Bi-metallic choke used on some models. Refer to text.

sensitivity by shifting sliding clip toward linkage end of governor arm (prior to specification letter "D", turn adjusting stud counter-clockwise).

VARIABLE SPEED. Variable speed linkage is used where it is necessary to maintain a wide range of different engine speeds under varying load conditions. Moving a control lever determines engine speed and governor maintains this speed under varying loads. See Fig. O71A.

To adjust variable speed governor linkage adjust throttle stop screw on carburetor so engine idles at 1450 rpm, then adjust governor spring tension so engine idles at 1500 rpm when manual control lever is at minimum speed position.

Adjust sensitivity with engine running at minimum speed to obtain smoothest engine operation by moving governor spring outward on extension arm to decrease sensitivity or inward to increase sensitivity. Refer to Fig. O71A.

Maximum full load speed should not exceed 3000 rpm for continuous operation. To adjust, apply full load to engine and move control lever to maximum speed position. Adjust set screw in bracket slot to stop lever travel at desired speed.

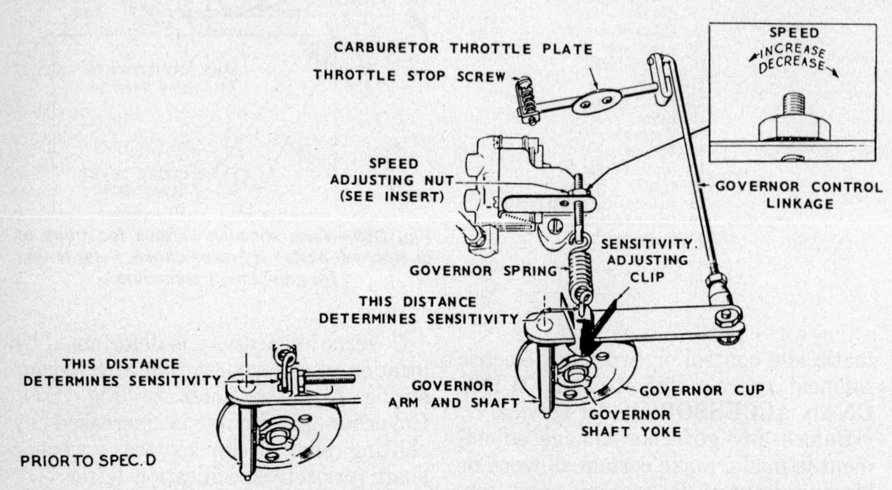

Fig. O71 – View of typical governor linkage arrangement used on engines with fixed speed application.

IGNITION SYSTEM. Most models are equipped with battery ignition

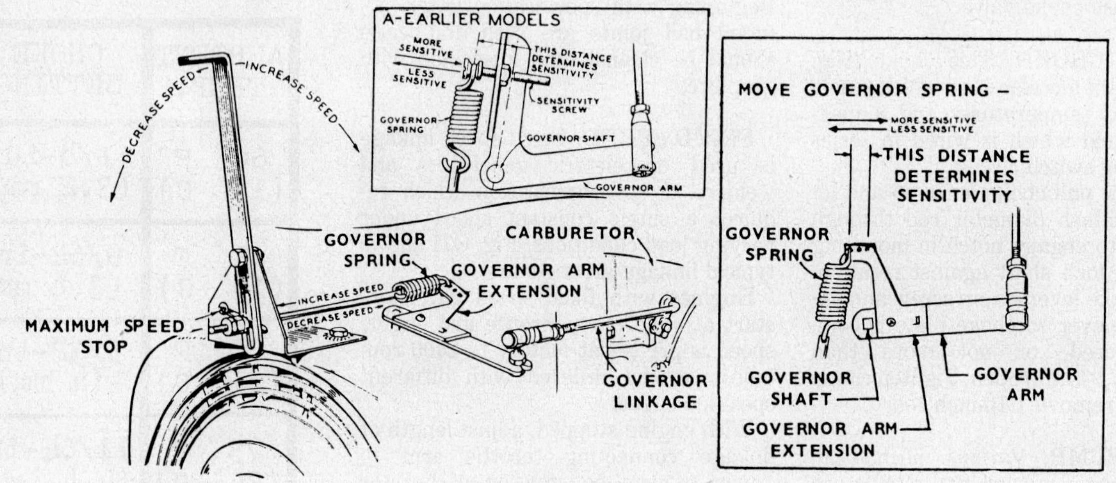

Fig. O71A – View of typical governor linkage arragnement used on engines with variable speed application.

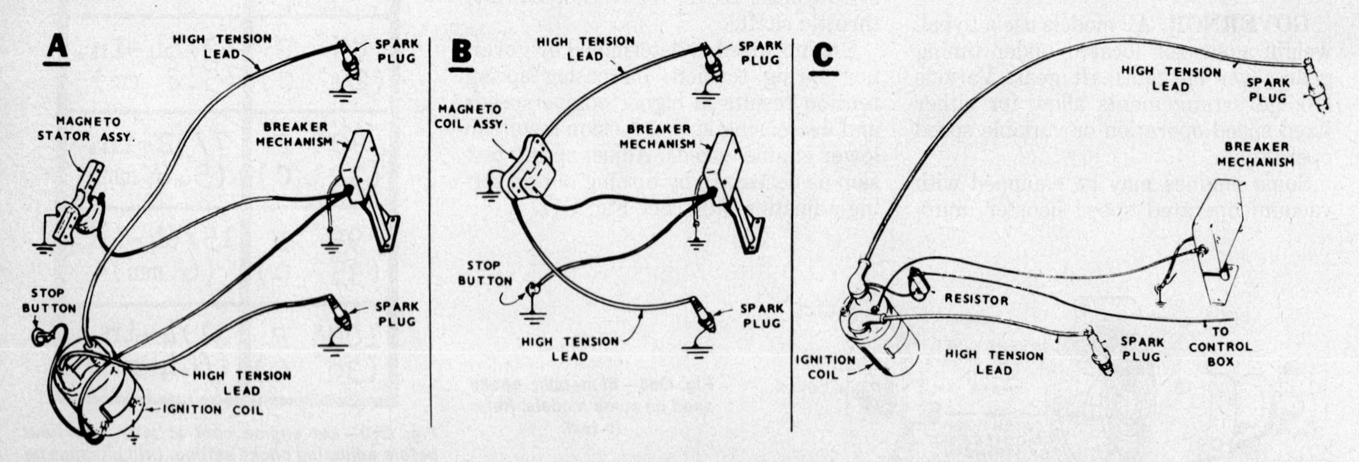

Fig. O72 – Layout of ignition wiring arrangements for CCK series engines. View A shows magneto ignition without spark advance mechanism. View B shows flywheel magneto system with spark advance mechanism. View C shows battery ignition wiring. Refer to text for timing and service procedures.

system, however a magneto system is available for manual start models. Ignition timing specifications are stamped on crankcase adjacent to breaker box.

Refer to appropriate paragraph for model being serviced.

BATTERY IGNITION. Ignition points and condenser are located in a breaker box at center, rear of engine. Refer to Fig. O72 for view of ignition wiring layout and Fig. O72A for view of points box and timing information.

To check timing, remove air intake hose from blower housing on pressure cooled engines or remove sheet metal plug in air shroud over right cylinder of "Vacu-Flo" engines to gain access to timing marks. Make certain points are set at 0.020 inch (0.51 mm) for all models.

Static timing is checked by connecting a test light across ignition points and rotating engine in direction of normal rotation (clockwise) until light comes on, then just goes out. This should be at 19° BTDC for all CCK models, 20° BTDC for CCKA and CCKB models without automatic spark advance and 1° ATDC for CCKA model with automatic spark advance (Fig. O73). Adjust by moving breaker box as required (Fig. O72A).

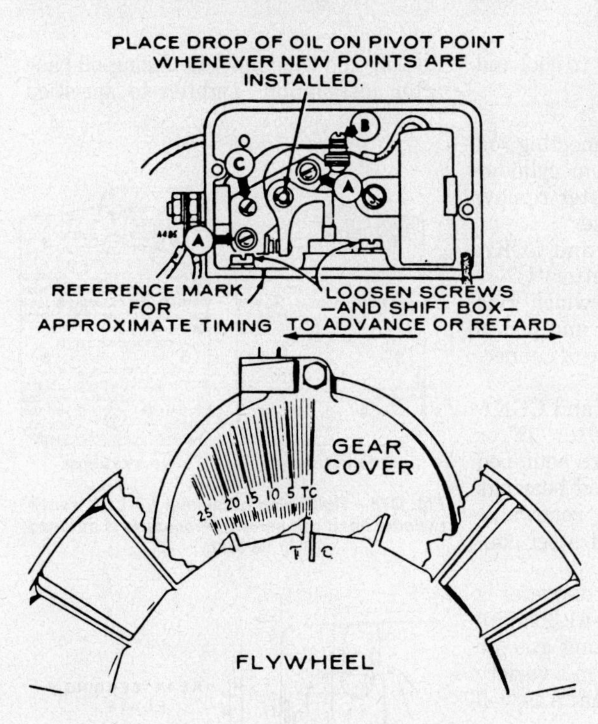

Fig. O72A—View showing location of points in breaker box of all models and location of timing marks on pressure cooled engines.

To check running timing, connect timing light to either spark plug and start and run engine. Operate engine at 1500 rpm or over. Light should flash when flywheel mark is aligned with 19° BTDC mark for all CCK models, 20° BTDC mark for CCKA model without automatic spark advance, 24° BTDC for CCKA model with automatic spark advance and 24° BTDC mark for CCKB models using vapor (butane, LP or natural gas) fuel. Adjust by moving breaker box as required (Fig. O72A).

MAGNETO IGNITION. If engine is equipped with automatic spark advance mechanism (Fig. O73), magneto coil, mounted on gear cover (B–Fig. O72), contains both primary and secondary windings.

If engine is not equipped with automatic spark advance, stator contains primary (low voltage) windings only and a separate coil is used to develop secondary voltage (A–Fig. O72).

Ignition points and condenser are located in a breaker box at center, rear of engine. Refer to Fig. O72 for view of ignition wiring layout and Fig. O72A for view of points box and timing information.

1 2 3 4 5 6 7 8 9

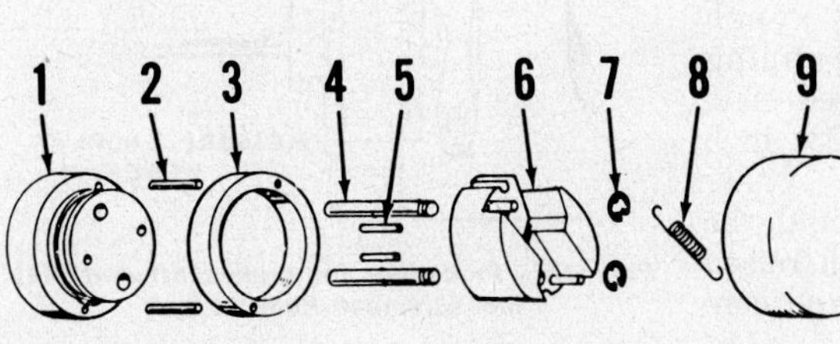

Fig. O73—Exploded view of spark advance (timing control) used on some series engines.

1. Camshaft
2. Cam roll pin
3. Control cam
4. Groove pin
5. Roll pin
6. Weights (2)
7. Retainers (2)
8. Control spring
9. Cover

To check timing, remove air intake hose from blower housing on pressure cooled engines or remove sheet metal plug in air shroud over right cylinder of "Vacu-Flo" engines to gain access to timing marks. Make certain points are set at 0.020 inch (0.51 mm) for all models.

Static timing is checked by connecting a test light across ignition points and rotating engine in direction of normal rotation (clockwise) until light comes on, then just goes out. This should be at 19° BTDC for CCK model, 1° ATDC for CCKA model with automatic spark advance, 20° BTDC for CCKA model without automatic spark advance, 5° BTDC for CCKB model with electric start and 5° BTDC to 1° ATDC for CCKB model with manual start. Adjust by moving breaker box as required (Fig. O72A).

To check running timing, connect timing light to either spark plug and start and run engine. Operate engine at 1500 rpm or over. Light should flash when flywheel mark is aligned with 19° BTDC mark for all CCK models, 20° BTDC for CCKA model without automatic spark advance, 24° BTDC for CCKA model with automatic spark advance and 24° BTDC mark for CCKB models. Adjust by moving breaker box as required (Fig. O72A).

LUBRICATION. All models are pressure lubricated by a gear type pump driven by crankshaft gear. Internal pump parts are not serviced separately so entire pump must be renewed if worn or damaged. Clearance between pump gear and crankshaft gear should be 0.002-0.005 inch (0.0508-0.127 mm) and normal operating pressure should be 30 psi (207 kPa). Oil pressure is regulated by an oil pressure relief valve located on top of engine block near timing gear cover. Relief valve spring free length should be 2-5/16 inch (58.74 mm) and valve diameter should be 0.3365-0.3380 inch (8.55-8.59 mm).

Oils approved by manufacturer must meet requirements of API service classification "SE" or "SE/CC".

Use SAE 5W-30 oil in below freezing temperatures and SAE 20W-40 oil in above freezing temperatures.

Check oil level with engine stopped and maintain at "FULL" mark on dipstick. **DO NOT** overfill.

Recommended oil change interval for all models is every 100 hours of normal operation and change filter at 200 hour intervals. Oil capacity for all models with electric start is 3.5 quarts (3.3 L) without filter change and 4 quarts (3.8 L) with filter change. Oil capacity for all manual start models is 3 quarts (2.8 L) without filter change and 3.5 quarts (3.3 L) with filter change.

Oil filter is a screw on type and gasket should be lightly lubricated before installation. Screw filter on until gasket contacts, then turn an additional ½-turn. Do not overtighten.

CRANKCASE BREATHER. Engines are equipped with a crankcase breather (Fig. O73A) which helps maintain a partial vacuum in crankcase during engine operation to help control oil loss and to ventilate crankcase.

Clean at 200 hour intervals of normal engine operation. Wash valve in suitable solvent and pull baffle out of breather tube to clean. Reinstall valve with perforated disk toward engine.

REPAIRS

TIGHTENING TORQUES. Recommended tightening torques are as follows:

Cylinder head	29-31 ft.-lbs. (39-42 N·m)
Connecting rod–	
Aluminum	24-26 ft.-lbs. (33-35 N·m)
Forged steel	27-29 ft.-lbs. (37-39 N·m)
Rear bearing plate	20-25 ft.-lbs. (27-34 N·m)
Flywheel	35-40 ft.-lbs. (48-54 N·m)
Oil base	43-48 ft.-lbs. (58-65 N·m)
Blower housing screws	8-10 ft.-lbs. (11-14 N·m)
Exhaust manifold	15-20 ft.-lbs. (20-27 N·m)
Intake manifold	15-20 ft.-lbs. (20-27 N·m)
Timing gear cover	10-13 ft.-lbs. (14-18 N·m)
Valve cover nut	4-8 ft.-lbs. (6-11 N·m)
Starter bolts	25-28 ft.-lbs. (34-38 N·m)
Magneto stator screws	15-20 ft.-lbs. (20-27 N·m)
Spark plug	25-30 ft.-lbs. (34-41 N·m)

CYLINDER HEAD. Cylinder heads should be removed and carbon and lead deposits cleaned at 200 hour intervals (400 hours if using unleaded fuel) or as a reduction in engine power or excessive pre-ignition occurs.

CAUTION: Cylinder heads should not be unbolted from block when hot due to danger of head warpage.

Always install new head gaskets and torque retaining cap screws in 5 ft.-lbs. (7 N·m) steps in sequence shown (Fig. O74) until 29-31 ft.-lbs. (39-42 N·m) torque is obtained. Operate engine at normal operating temperature and light load for a short period, allow to cool and re-torque head bolts.

CONNECTING ROD. Connecting rod and piston are removed from cylinder head surface end of block after removing cylinder head and oil base.

Connecting rods for CCK and CCKA engines with specification letter "C" or before are aluminum rods which ride directly on crankpin journal and piston pin operates in unbushed bore of connecting rod.

Connecting rods for CCK and CCKA engines with specification letter "D" or after and all other models are equipped with precision type insert rod bearings and piston pin operates in renewable bushing pressed into forged steel connecting rod.

Standard crankpin journal diameter is 1.6252-1.6260 inches (41.28-41.30 mm) for all models and connecting rod or bearing inserts are available in a variety of sizes for undersize crankshafts as well as standard.

Connecting rod to crankshaft journal running clearance for aluminum rod should be 0.002-0.0033 inch (0.0508-0.0838 mm) and running clearance for bearings in forged steel rod should be 0.0005-0.0023 inch (0.0127-0.0584 mm).

Connecting rod caps should be reinstalled on original rod with raised lines (witness marks) aligned and connecting rod caps must be facing oil base after installation. Tighten to specified torque.

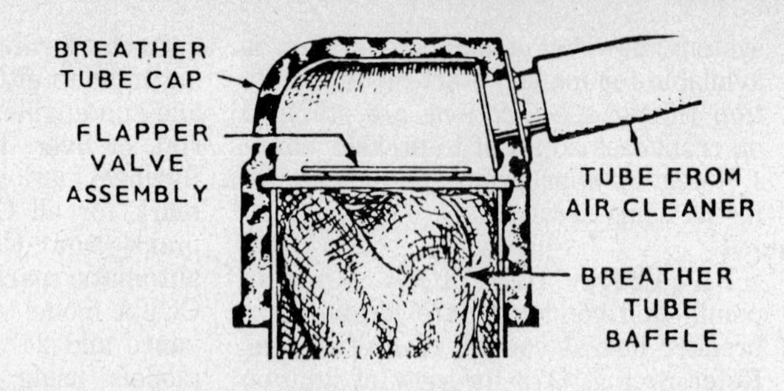

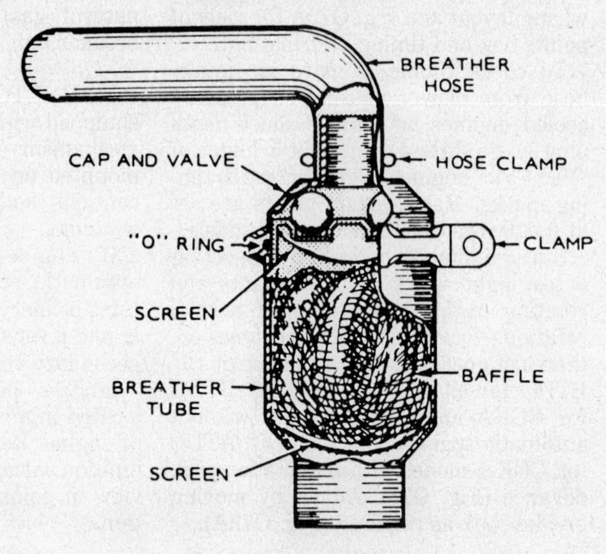

Fig. O73A – Crankcase breather assembly should be cleaned at 200 hour intervals of normal engine operation. Upper breather is used on engines before specification letter "H" and lower breather is used on engines after specification letter "H".

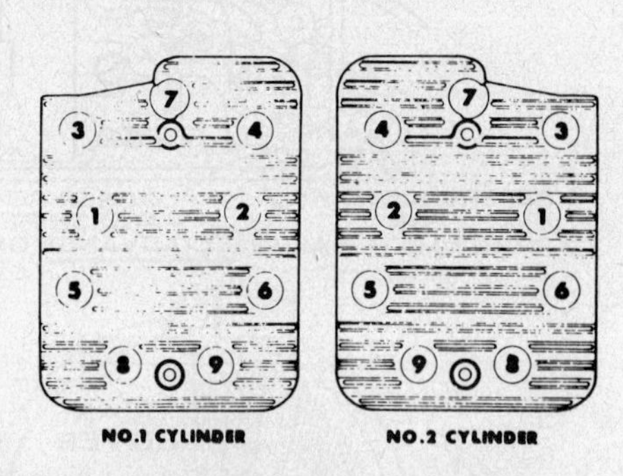

Fig. O74 – Tightening sequence for CCK series cylinder head cap screws. Procedure is outlined in text.

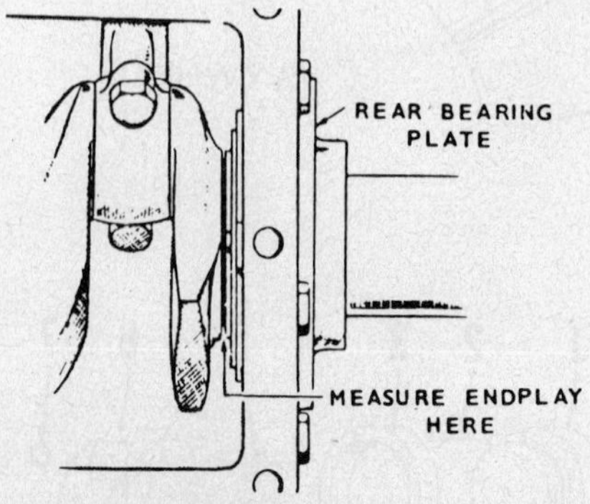

Fig. O75 – Procedure for crankshaft end play measurement. Refer to text.

PISTON, PIN AND RINGS. Engines may be equipped with one of three different type three ring pistons. Do not intermix different type pistons in engines during service.

STRUT TYPE PISTON. Strut type piston may be visually identified by struts cast in underside of piston which run parallel to pin bosses. Piston to cylinder clearance when measured just below oil ring, 90° from pin, should be 0.0025-0.0045 inch (0.0635-0.1143 mm).

CONFORMATIC PISTON. Conformatic piston may be visually identified by smooth contours in underside of piston and no struts are visible. Top ring groove is 5/32-inch (3.969 mm) from top of piston and slots on opposite sides of piston, behind oil control ring allow oil return and expansion. Piston to cylinder clearance when measured just below oil ring, 90° from pin, should be 0.0015-0.0035 inch (0.0381-0.0889 mm).

VANASIL PISTON. Vanasil type piston may be visually identified by smooth contours in underside of piston and no struts are visible. Top ring groove is 9/32-inch (7.144 mm) from top of piston and round holes behind oil control ring allow oil return and expansion. Piston to cylinder clearance when measured just below oil ring, 90° from pin, should be 0.006-0.008 inch (0.1524-0.2032 mm).

Side clearance of top ring in piston groove for all models should be 0.002-0.008 inch (0.051-0.203 mm).

Ring end gap for all models should be 0.010-0.023 inch (0.254-0.584 mm) and end gaps should be staggered around piston circumference.

Piston pin to piston pin bore clearance is 0.0001-0.0005 inch (0.0025-0.0127 mm) for all models and pin to rod clearance is 0.0002-0.0008 inch (0.005-0.020 mm) for aluminum rod and 0.00005-0.00055 inch (0.001-0.014 mm) for bushing in forged steel rod.

Standard piston pin diameter is 0.7500-0.7502 inch (19.05-19.06 mm) for all models and pins are available in 0.002 inch (0.0508 mm) oversize for aluminum rod.

Standard cylinder bore diameter for all models is 3.249-3.250 inches (82.53-82.55 mm) and cylinders should be bored to nearest oversize for which piston and rings are available if cylinder is scored or out-of-round more than 0.003 inch (0.0762 mm) or taper exceeds 0.005 inch (0.127 mm). Pistons and rings are available in a variety of oversizes as well as standard.

CRANKSHAFT, BEARINGS AND SEALS. Crankshaft is supported at each end by precision type sleeve bearings located in cylinder block housing and rear bearing plate. Main bearing journal standard diameter should be 1.9992-2.000 inches (50.7797-50.800 mm) and bearing operating (running) clearance for early model flanged type bearing (before specification letter "F")

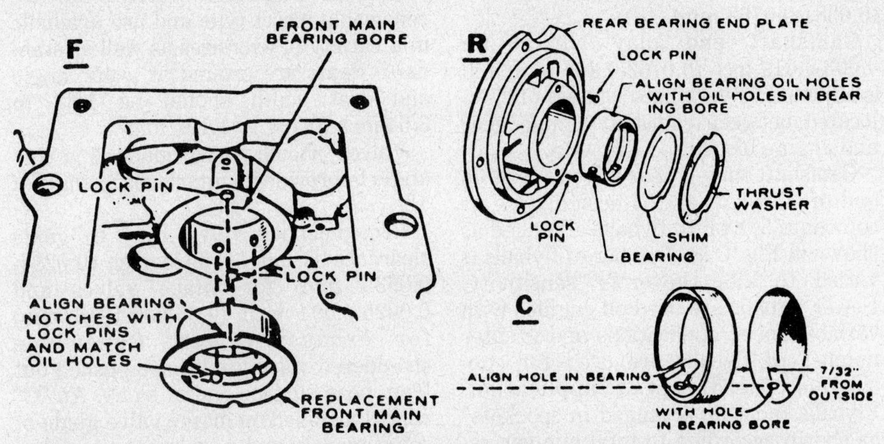

Fig. O76—Alignment of precision main and camshaft bearings. One-piece bearing is used at front (view F), two-piece bearing at rear (view R). Note shim use for end play adjustment. View C shows placement of camshaft bearing.

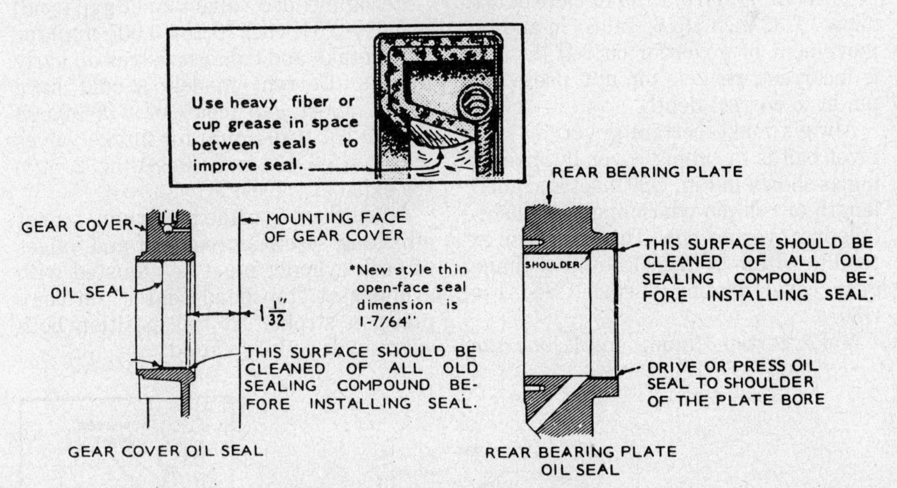

Fig. O77—View showing correct seal installation procedure. Fill cavity between lips of seal with heavy grease to improve sealing efficiency.

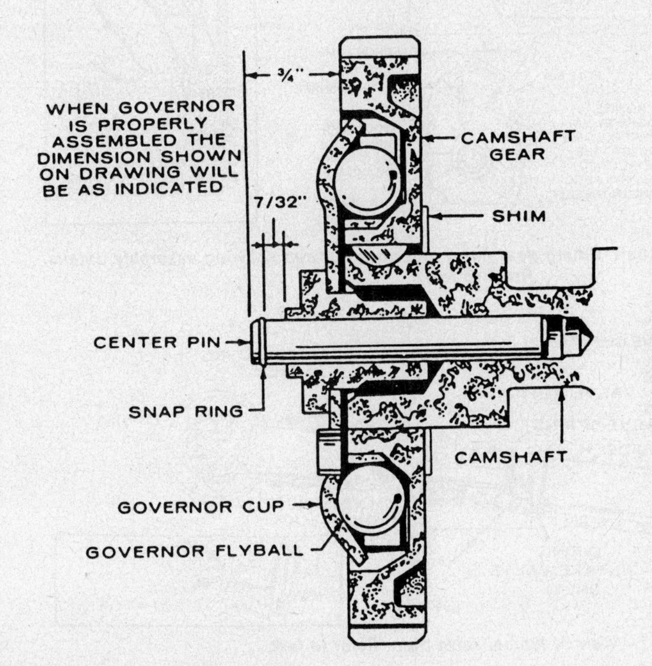

Fig. O78—Cross-sectional view of camshaft gear and governor assembly showing correct dimensions for center pin extension from camshaft. Note shim placement for adjusting end play of camshaft. Refer to text.

should be 0.002-0.003 inch (0.0508-0.0762 mm) and for all remaining models clearance should be 0.0024-0.0042 inch (0.061-0.107 mm). Main bearings are available for a variety of sizes for undersize crankshafts as well as standard.

Oil holes in bearing and bore **MUST** be aligned when bearings are installed and rear bearing should be pressed into bearing plate until flush, or recessed into bearing plate 1/64-inch (0.40 mm). Make certain bearing notches are aligned with lock pins (Fig. O76) during installation.

Apply "Loctite Bearing Mount" to front bearing and press bearing in until flush with block. Make certain bearing notches are aligned with lock pins (Fig. O76) during installation.

NOTE: Replacement front bearing and thrust washer is a one piece assembly, do not install a separate thrust washer.

Rear thrust washer is installed with oil grooves toward crankshaft. Measure crankshaft end play as shown in Fig. O75 and add or remove shims or renew thrust washer (Fig. O76) as necessary to obtain 0.006-0.012 inch (0.15-0.30 mm) end play.

It is recommended rear bearing plate and front timing gear cover be removed for seal installation. Rear seal is pressed in until flush with seal bore. Timing cover seal should be driven in until it is 1-1/32 inch (26.19 mm) from mounting face of cover for old style seal and 1-7/64 inch (28.18 mm) from mounting face of cover for new style, thin, open faced seal. See Fig. O77.

CAMSHAFT, BEARINGS AND GOVERNOR. Camshaft is supported at each end in precision type sleeve bearings. Make certain oil holes in bearings are aligned with oil holes in block (Fig. O76), press front bearing in flush with outer surface of block and press rear bearing in until flush with bottom of counterbore. Camshaft to bearing clearance should be 0.0015-0.0030 inch (0.0381-0.0762 mm).

Camshaft end play should be 0.003-0.012 inch (0.0762-0.305 mm) and is adjusted by varying thickness of shim located between camshaft timing gear and engine block. See Fig. O78.

Camshaft timing gear is a press fit on end of camshaft and is designed to accomodate 5, 8 or 10 flyballs arranged as shown in Fig. O79. Number of flyballs is varied to alter governor sensitivity. Fewer flyballs are used on engines with variable speed applications and greater number of flyballs are used for continuous (fixed) speed application. Flyballs must be arranged in "pockets" as shown according to total number used.

Center pin (Fig. O78) should extend ¾-inch (19 mm) from end of camshaft to allow 7/32-inch (5.6 mm) in-and-out movement of governor cup. If distance is incorrect, remove pin and press new pin in to correct depth.

Always make certain governor shaft pivot ball is in timing cover by measuring as shown in Fig. O80 inset and check length of roll pin which engages bushed hole in governor cup. This pin must extend to within ¾-inch (19 mm) of timing gear cover mating surface. See Fig. O80.

Make certain timing mark on camshaft gear is aligned with timing mark on crankshaft gear during installation.

VALVE SYSTEM. Valve seats are renewable insert type and are available in a variety of oversizes as well as standard. Seats are ground at a 45° angle and seat width should be 1/32 to 3/64-inch (0.794 to 1.191 mm).

Valves should be ground at a 44° angle to provide an interference angle of 1°.

Recommended valve stem to guide clearance is 0.001-0.0025 inch (0.0254-0.0635 mm) for intake valves and 0.0025-0.004 inch (0.0635-0.1016 mm) for exhaust valves. Renewable shouldered valve guides are pushed out from head surface end of block. An "O" ring is installed on intake valve guide of some models and a valve stem seal is also available for intake valve. See Fig. O81.

Recommended valve tappet gap (cold) is 0.010-0.012 inch (0.254-0.305 mm) for both intake and exhaust valves on early models. Current models should have valve tappet gap (cold) of 0.006-0.008 inch (0.152-0.203 mm) for intake valves and 0.015-0.017 inch (0.381-0.432 mm) for exhaust valves.

Adjustment is made by turning tappet adjusting screw as required and valves of each cylinder must be adjusted with cylinder at "top dead center" on compression stroke. At this position both valves will be fully closed.

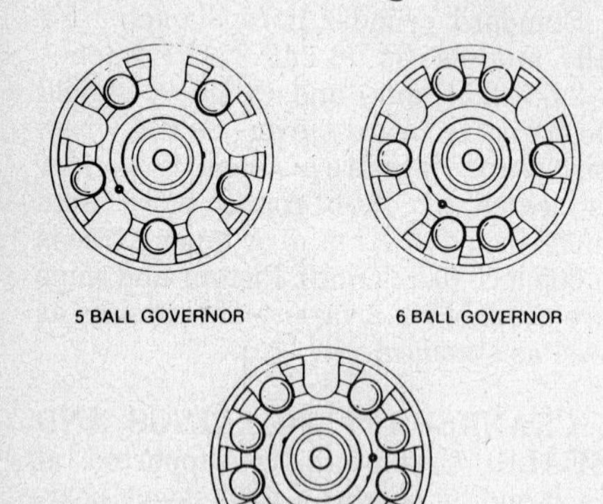

Fig. O79—View showing correct spacing of governor flyballs according to total number of balls used.

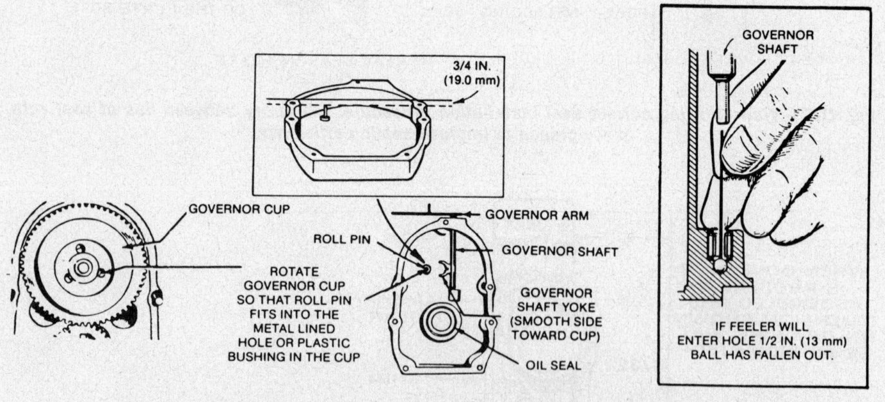

Fig. O80—View of governor shaft, timing gear cover and governor cup showing assembly details. Refer to text.

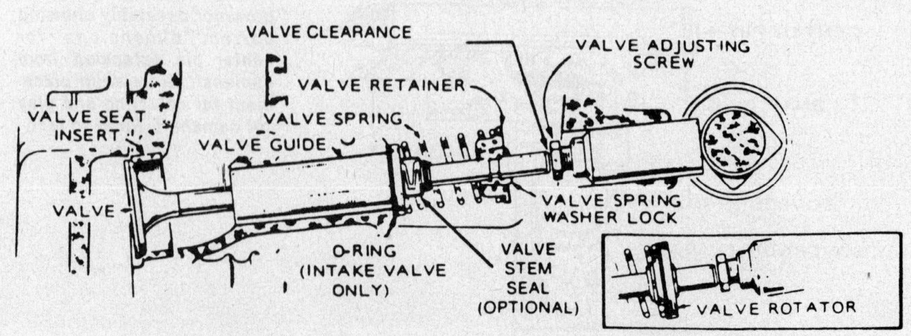

Fig. O81—View of typical valve train. Refer to text.

ONAN

A DIVISION OF ONAN CORPORATION
1400 73rd Avenue N.E.
Minneapolis, Minnesota 55432

Model	No. Cyls.	Bore	Stroke	Displacement	Power Rating
JB	2	3.25 in. (82.55 mm)	3.63 in. (92.08 mm)	60 cu. in. (985.7 cc)	21.6 hp. (16.1 kW)
JC	4	3.25 in. (82.55 mm)	3.36 in. (92.08 mm)	120 cu. in. (1971.4 cc)	42.5 hp. (31.7 kW)

Engines in this section are four-cycle, vertical in-line two or four cylinder horizontal crankshaft engines. Valves are located in cylinder heads and crankshaft is supported at each end in precision sleeve type bearings and four cylinder models have a two-piece precision type center main bearing.

Connecting rods are equipped with renewable bearing inserts and all models are pressure lubricated by a gear type oil pump. A "spin-on" oil filter is available.

Most models are equipped with a battery ignition system, however a magneto system is available for manual start models.

All models are equipped with a side draft carburetor which may be used with gasoline or vapor (butane, LP or natural gas) fuels according to model and application.

Refer to engine nameplate for engine model numbers and specification letter. Always give engine serial, model numbers and specification letter when ordering parts or service material.

MAINTENANCE

SPARK PLUG. Recommended spark plug for all models is Champion H8 or equivalent. Electrode gap for JB model and all models using vapor (butane, LP or natural gas) fuel is 0.025 inch (0.635 mm) and gap for JC model using gasoline for fuel is 0.035 inch (0.889 mm).

CARBURETOR. All models are equipped with a side draft carburetor which may use gasoline or vapor (butane, LP or natural gas) fuel.

SIDE DRAFT CARBURETOR (GASOLINE). Refer to Fig. O87 for exploded view of side draft carburetor used after specification letter "R" and location of mixture adjustment screws. Note Fig. O87 is similar to carburetor used prior to specification letter "R" except for location of adjustment screws. Refer to Fig. O88.

For initial carburetor adjustment, open idle mixture screw and main fuel mixture screw 1 to 1½ turns. Make final adjustments with engine at normal operating temperature and running. Place engine under load and adjust main fuel mixture screw for leanest setting that will allow satisfactory acceleration and steady governor operation. Set engine at idle speed, no load and adjust idle mixture screw for smoothest idle operation.

As each adjustment affects the other, adjustment procedure may have to be repeated.

To check float level, refer to Fig. O89. Invert carburetor throttle body and float assembly. Float clearance should be 11/64-inch (4.366 mm) for models prior to specification letter "R" and 1/8-inch (3.2 mm) for models with specification letter "R" or after.

Adjust float by bending float lever tang that contacts inlet valve.

SIDE DRAFT CARBURETOR (VAPOR FUEL/GASOLINE). A side draft carburetor designed to use either

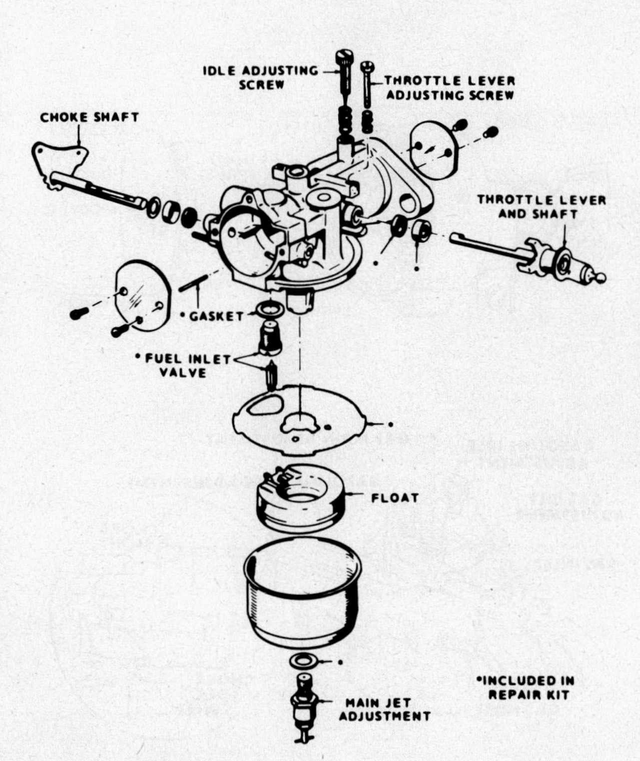

Fig. O87—Exploded view of carburetor used on late production JB and JC engines. Earlier models used carburetors which were similar in construction but all mixture adjustments were in upper casting with main jet enclosed in a fuel well at bottom of float bowl. Refer to Fig. O88 for location of mixture adjustment screws in early carburetor.

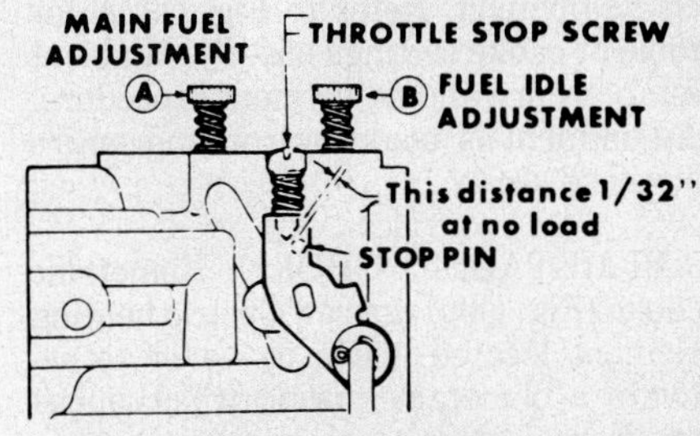

Fig. O88—Location of fuel mixture adjustments of early JB and JC carburetors.

gasoline or vapor (butane, LP or natural gas) fuel is available for some models.

For initial carburetor adjustment, refer to Fig. O91 for location of appropriate adjustment screws and make initial and final adjustments as outlined in **SIDE DRAFT CARBURETOR (GASOLINE)** section.

To change system from gasoline fuel operation to vapor fuel operation, use the following procedure:

Shut off gasoline supply valve.
Install lock wire on choke.
Set spark plug gap at 0.025 inch (0.635 mm).
Lock float in position.*
Open supply valve for vapor fuel.

*It may be necessary to remove float assembly and inlet valve on models which are not equipped with a float locking mechanism for extended vapor fuel use.

To change system from vapor fuel operation to gasoline operation, use the following procedure:

Close vapor fuel supply valve.
Remove lock wire from choke.
Set spark plug gap for model JC at 0.035 inch (0.889 mm).
Unlock carburetor float.*
Open gasoline fuel supply valve.

*Replace float assembly and inlet valve if they were removed because carburetor was not equipped with float lock.

To check float level, refer to procedure outlined in **SIDE DRAFT CARBURETOR (GASOLINE)** section.

CHOKE. One of two types of automatic choke may be used according to model and application. Refer to appropriate paragraph for model being serviced.

THERMAL MAGNETIC CHOKE. Thermal magnetic choke (Fig. O92) uses a strip heating element and a heat-reactive bi-metallic spring to control choke valve position. When engine is cranked, solenoid shown at lower portion of choke body is activated and choke valve is closed, fully or partially, according to temperature conditions. When engine starts and runs, solenoid is released and bi-metallic spring controls choke opening. Refer to Fig. O92A for table of choke settings based on ambient temperature and adjustment procedure. Adjustment is made by rotating entire cover as shown.

BI-METALLIC CHOKE. Bi-metallic choke (Fig. O90) uses an electric heating element located inside its cover to activate a bi-metallic spring which opens or closes choke valve according to temperature.

Refer to Fig. O90 for adjustment procedure.

Heating element may be checked with an ohmmeter. When element is at room temperature, resistance should be 5-6 ohms for 12 volt models, approximately

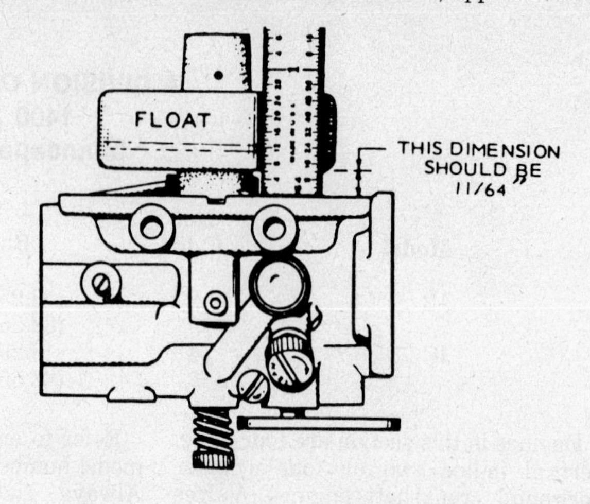

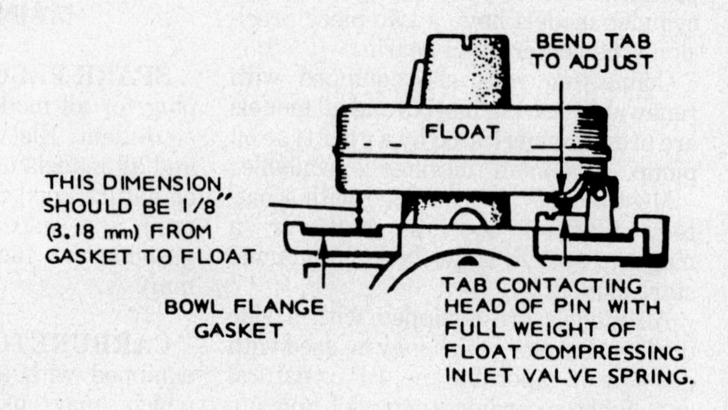

Fig. O89—Measure float level as shown noting gasket must be in place for measurement.

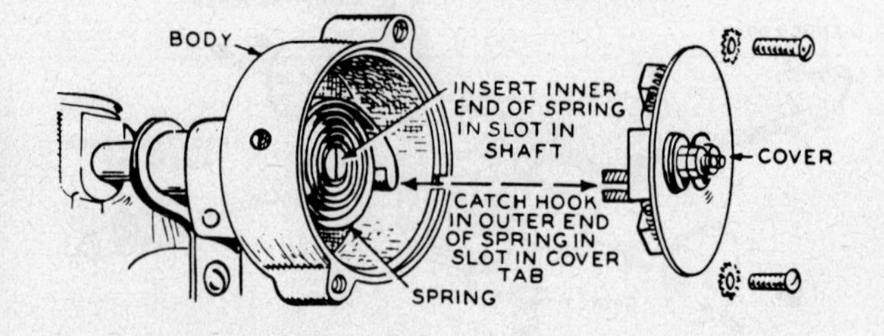

Fig. O90—Bi-metallic choke used on some models. Refer to text. If ambient temperature is 72° F (22° C) choke valve should be open 7/32-inch (5.56 mm) for JB models and 1/64-inch (0.40 mm) for JC models. Loosen cover screws and rotate cover to adjust.

GASOLINE IDLE ADJUSTMENT
GAS MAIN ADJUSTMENT
GASOLINE MAIN ADJUSTMENT
GAS IDLE ADJUSTMENT
GAS INLET
CHOKE SHAFT
CHOKE LOCK WIRE
CLAMP
GAS HOSE TO REGULATOR

Fig. O91—View of adjustments for typical gas (vapor) gasoline carburetor. Refer to text.

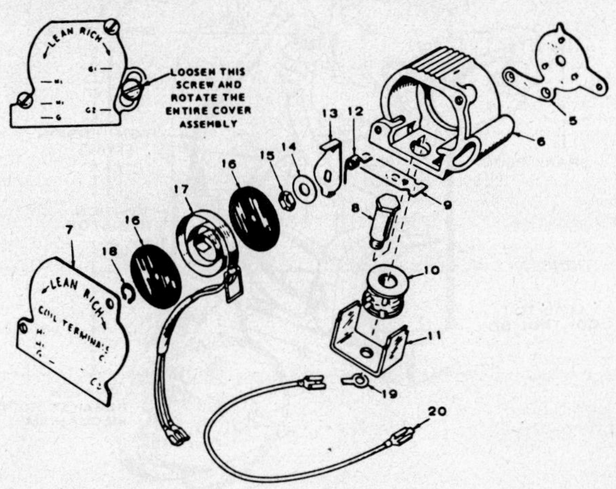

Fig. O92—Exploded view of thermal magnetic choke used on some engines.

5. Mounting plate
6. Body housing
7. Cover
8. Solenoid core
9. Armature
10. Solenoid coil
11. Frame
12. Lever spring
13. Lever
14. Washer
15. Pal nut
16. Insulator disc (2)
17. Heater & spring
18. Retainer (to shaft)
19. Ground terminal
20. Choke lead wire

16 ohms for 24 volt models, and 25 ohms for 32 volt models. Renew cover if element is defective.

FUEL PUMP. All models are equipped with a mechanical type fuel pump.

AMBIENT TEMP.	CHOKE SETTING
60° F (16° C)	1/8-in. (3.2 mm)
65° F (18° C)	9/64-in. (3.6 mm)
70° F (21° C)	5/32-in. (4 mm)
75° F (24° C)	11/64-in. (4.4 mm)
80° F (27° C)	3/16-in. (4.8 mm)
85° F (29° C)	13/64-in. (5.2 mm)
90° F (32° C)	7/32-in. (5.6 mm)
95° F (35° C)	15/64-in. (6 mm)
100° F (38° C)	1/4-in. (6.4 mm)

Fig. O92A—Let engine cool at least one hour before adjusting choke setting. Drill bits may be used as gages to set choke opening according to ambient temperature.

Refer to **SERVICING ONAN ACCESSORIES** section.

GOVERNOR. All models use a flyball weight governor located on camshaft gear.

Before any governor linkage adjustment is made, make certain all worn or binding linkage is repaired and carburetor is properly adjusted.

With engine stopped, adjust length of linkage connecting throttle arm to governor arm so stop on carburetor throttle lever is 1/32-inch (0.794 mm) from stop boss. This allows immediate governor control at engine start and synchronizes travel of governor arm and throttle shaft.

Engine speed is determined by governor spring tension. Increasing spring tension results in higher engine speeds and decreasing spring tension results in lower engine speeds. Adjust spring tension as necessary until rated speed is obtained by turning nut on spring adjusting stud. See Fig. O93.

Sensitivity adjustment on models prior to specification letter "R" is made by turning spring stud as necessary for maximum sensitivity without engine "hunting".

Sensitivity adjustment on models beginning with specification letter "R" is made by adjusting ratchet nut (Fig. O93) which is accessible through hole in side of blower housing for maximum sensitivity without engine "hunting".

If sensitivity cannot be correctly adjusted with stud or ratchet nut, spring may have to be repositioned on governor arm and normal adjustment procedure repeated.

IGNITION SYSTEM. Most models are equipped with battery ignition system, however a magneto system is available for manual start models. Refer to appropriate paragraph for model being serviced.

BATTERY IGNITION (JB). A single coil is used to fire both spark plugs simultaneously. Fig. O95 shows complete layout of battery ignition system.

Breaker point gap is 0.020 inch (0.508 mm) at maximum opening.

Static timing is checked by connecting a test light across ignition points and rotating engine in direction of normal rotation (clockwise) until light comes on and then just goes out. This should be 5° ATDC and is adjusted by shifting breaker plate to advance or retard as shown in Fig. O97.

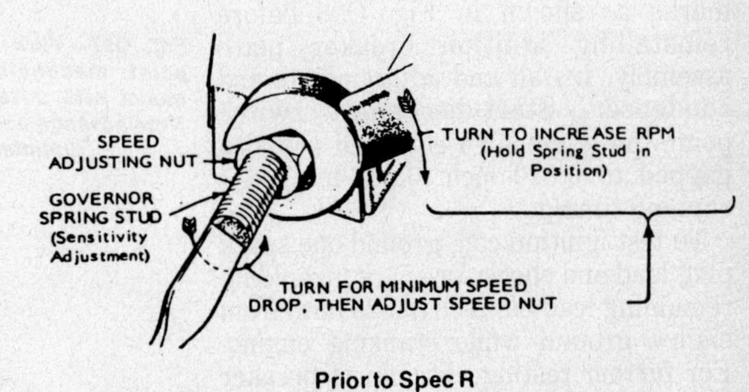

Prior to Spec R

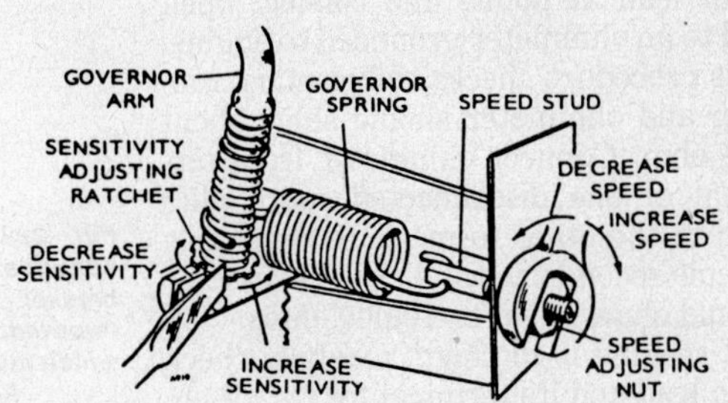

Begin Spec R

Fig. O93—View showing locations of adjustment points for governor used on JB and JC models. Refer to text.

To check running timing, connect timing light to either spark plug and start and run engine at rated speed. Light should flash when flywheel mark is aligned with 25° BTDC mark for gasoline fuel models and 35° BTDC for vapor (butane, LP or natural gas) or combination fuel models. Adjust by shifting breaker plate to advance or retard as shown in Fig. O97.

CAUTION: Engine should not be run for more than two or three minutes with access door open or removed. Engine cooling is affected and engine will overheat rapidly. Run without load.

If breaker points do not maintain adjustment or cannot be adjusted as specified, it is likely breaker point cam drive gear is not correctly installed or centrifugal advance mechanism is defective. Refer to Fig. O98 for correct timing and alignment of breaker cam drive gear.

To service advance mechanism, remove primary lead, breaker plate retaining screws and breaker plate. Remove cam and weight assembly being careful not to lose spacer mounted on gear shaft. Remove points and condenser from breaker plate and pull out plunger and plunger diaphram.

Clean and inspect all parts and renew any that are worn or damaged. If cam is loose on gear or binding, renew entire assembly. Weights should snap outward between 1000 and 1075 rpm. If necessary, adjust by lightly bending one or both spring anchor pins toward center of gear. Make certain oil trickle holes (Fig. O97) are open.

Reassemble cam gear, weights and springs making certain gears and weights are aligned on sets of "O" index marks as shown in Fig. O98 before reinstalling ignition breaker plate assembly. Install and adjust points and condenser. Start-disconnect switch points on engines so equipped are also gapped to 0.020 inch (0.508 mm). Set running timing.

To test ignition coil, ground one spark plug lead and check for spark by holding remaining lead 3/8-inch (9.525 mm) from engine ground while cranking engine. For further testing, disconnect breaker point lead at points and connect open end to an ohmmeter grounded to engine. This procedure checks coil primary winding and ohmmeter should show about one ohm. Connect ohmmeter from terminal of one disconnected spark plug lead to the other to measure resistance of coil secondary windings. Ohmmeter should show 7,000 to 10,000 ohms.

If reading is too high, carefully check each lead and its terminal for corrosion, broken wire strands or bad connections. Renew or correct as necessary.

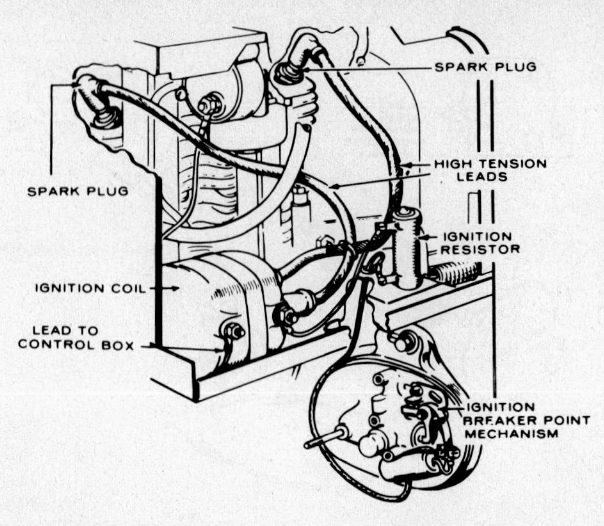

Fig. O95—Battery ignition system wiring for JB models.

If reading to too low, indication is that coil is internally shorted and should be renewed.

If any continuity appears between end of primary lead and a spark plug (secondary) lead, whether checked by test lamp or ohmmeter, a short circuit is indicated.

CAUTION: If commercial tester is used, test voltage should not exceed six volts.

BATTERY IGNITION (JC). JC model battery ignition system components consist of an ignition coil, spark plugs, points, condenser and distributor.

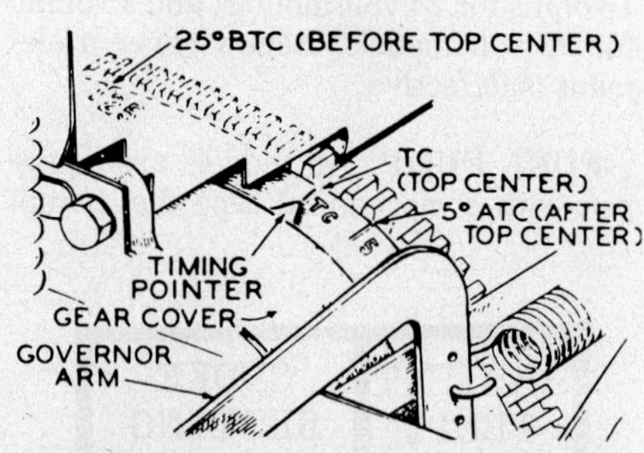

Fig. O96—Alignment of TC mark with timing pointer of JB model. Timing marks of JC model are similar.

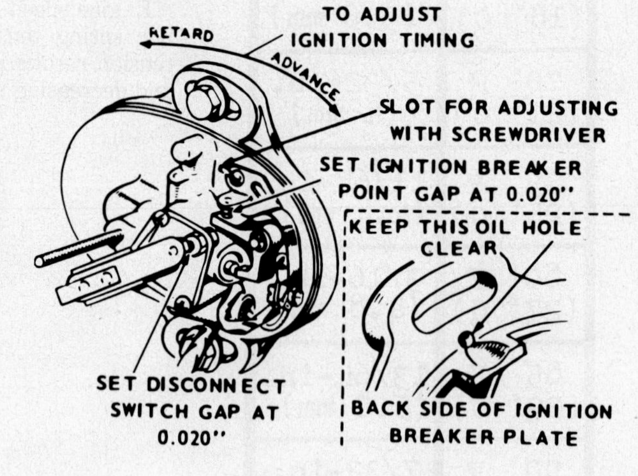

Fig. O97—View of breaker point mechanism of JB model with cover removed. Note advance and retard adjustment.

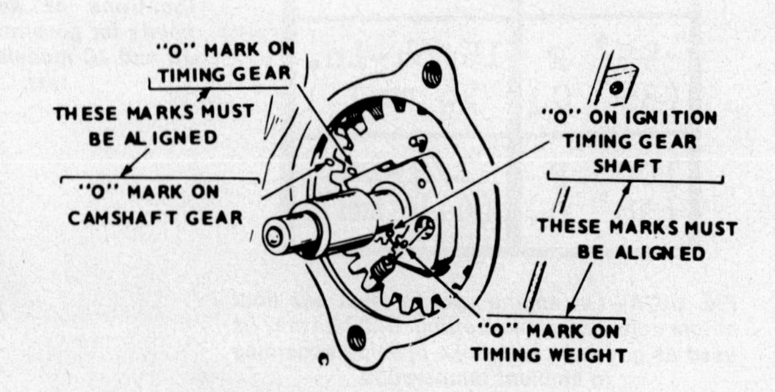

Fig. O98—View of breaker cam-gear assembly with breaker plate and points removed. Note index marks which must be in alignment. Refer to text.

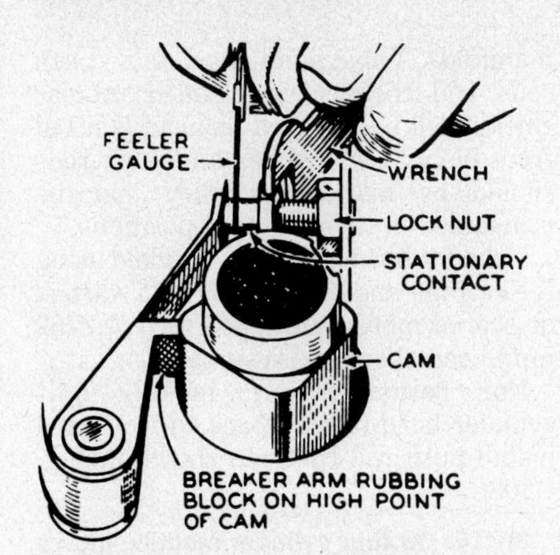

Fig. O99—Procedure for setting breaker point gap on four cylinder JC model.

Points are located inside distributor which has centrifugal spark advance mechanism.

Breaker point gap is 0.020 inch (0.508 mm) at maximum opening. Engine firing order is 1-2-4-3 and distributor turns in a clockwise direction when viewed from the top.

Static timing is checked by connecting a test light across ignition points and rotating engine in direction of normal rotation (clockwise) until number one piston is coming up on compression stroke. As piston comes up on compression stroke, test light should be on. When test light just goes out when rotating engine, TC mark on flywheel should align with 25° BTDC mark for gasoline operation and 35° BTDC mark for vapor (butane, LP or natural gas) operation.

Adjust by rotating distributor as necessary.

To check running timing, connect timing light to number one spark plug and start and run engine at rated speed. Light should flash when flywheel mark is aligned with 25° BTDC mark for gasoline fuel models and 35° BTDC mark for vapor (butane, LP or natural gas) or combination fuel models. Adjust by rotating distributor as necessary.

CAUTION: Engine should not be run for more than two or three minutes with access door open or removed. Engine cooling is affected and engine will overheat rapidly. Run without load.

If ignition timing appears to be erratic after installation of new points, a worn or loose breaker plate or worn distributor shaft and bushing should be suspected. Refer to Fig. O101 for exploded view of distributor. For reassembly note distributor shaft end play when measured between gear and thrust washer, should be 0.003-0.010 inch (0.0762-0.254 mm). If end play is below specifications, tap lower end of distributor drive shaft lightly with soft hammer to increase end play. If it is too great, check thrust washer installation or reinstall gear. See Fig. O101A.

To install distributor on JC model, refer to Fig. O100. Make certain number one cylinder is at "top dead center" on compression stroke and install distributor so rotor rotates into position shown after engaging drive gear. Set timing to appropriate specification.

To test coil, ohmmeter should show resistance of 7,000 to 10,000 ohms when connected at high tension terminal and ground (–) terminal. Resistance between primary terminals should be about 1 ohm.

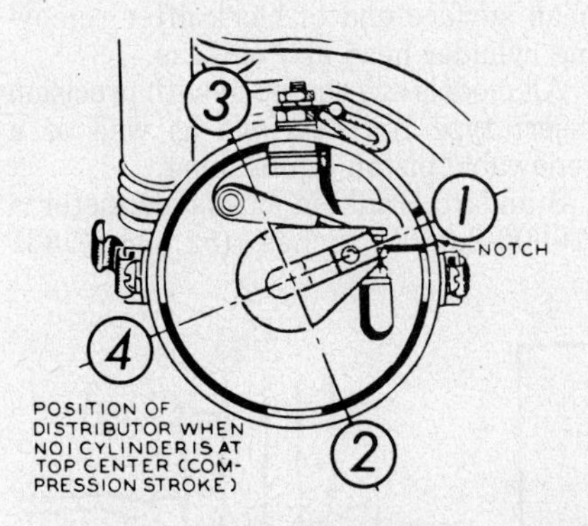

Fig. O100—To install distributor on JC model, make certain number one cylinder is at "top dead center" on compression stroke and install distributor so rotor rotates into position shown after engaging drive gear.

MAGNETO IGNITION (JB). JB model magneto ignition system consists of permanent magnets mounted on inner perimeter of flywheel which energizes secondary windings and stator of magneto which is located behind flywheel.

Maintenance, breaker point setting and timing procedures are the same as for battery ignition system with the exception that running timing is set at 25° BTDC regardless of type fuel used. Refer to **BATTERY IGNITION (JB)** section.

To test magneto coil, disconnect ignition point lead wire and connect ohmmeter between lead wire and engine ground. Ohmmeter should show 0.6 ohm reading. Remove spark plug leads and check resistance between leads. Ohmmeter should show approximately 11,000 ohms. Greater reading indicates high tension coils or leads with high resistance or open circuit. Check primary and secondary circuits for short by measuring resistance between ignition point lead and spark plug lead. Any continuity indicates a defective coil.

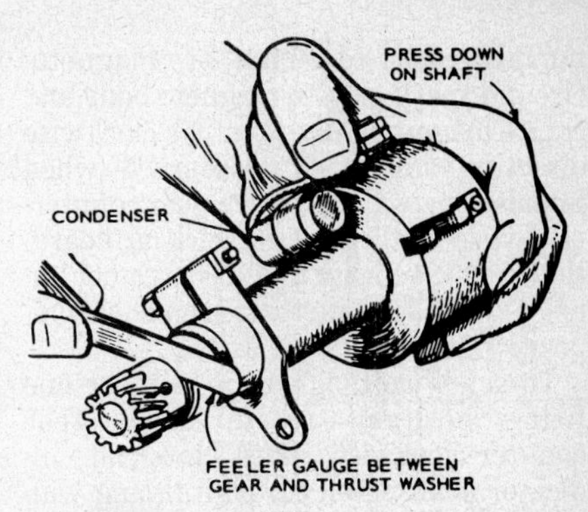

Fig. O101A—Distributor shaft end play is 0.003-0.010 inch (0.0762-0.254 mm) when measured between gear and thrust washer. Refer to text.

MAGNETO IGNITION (JC). Some JC models may be equipped with either a Wico or a Fairbanks-Morse magneto.

Breaker point gap is 0.015 inch (0.381 mm) for either magneto and firing order is 1-2-4-3. Magneto turns in a clockwise direction when viewed from terminal end.

To install either magneto, make certain number one piston is on compression stroke and align flywheel 19½ teeth before TDC (top dead center) mark with pointer on front cover.

To set Fairbanks-Morse magneto in number one firing position, install spark plug wire with spark plug attached, in

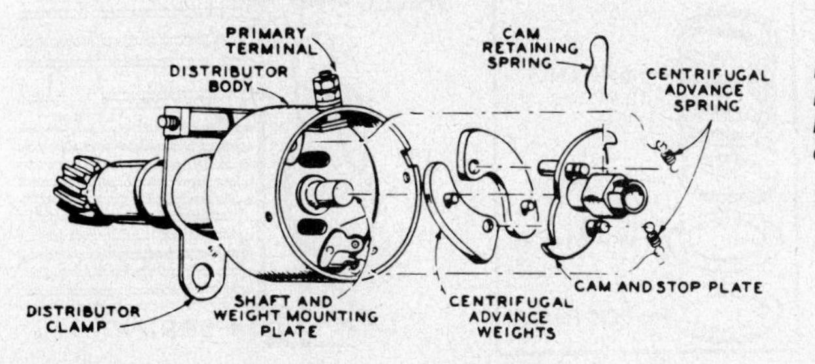

Fig. O101—View of JC model distributor with breaker plate removed and centrifugal advance mechanism disassembled.

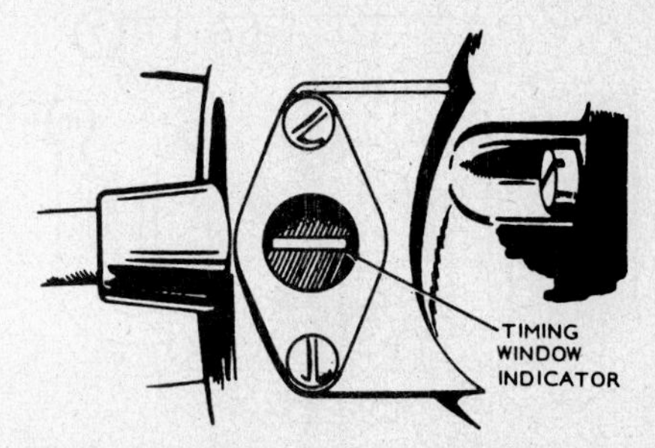

Fig. O101B—View of timing window and internal timing indicator of Wico magneto. Refer to text.

number one terminal of magneto. Ground spark plug to magneto body and rotate magneto drive gear in a clockwise direction until spark is observed (when impulse snaps). Turn drive gear counter-clockwise until a slight click is heard, then turn clockwise and hold against impulse. Install magneto on engine at mid-position of adjustment slots.

To set Wico magneto in number one firing position, rotate drive gear counter-clockwise until internal indicator is observed through timing window (Fig. O101B). Rotate drive gear a little more until a slight click is heard and hold drive gear against impulse. Install magneto on engine at mid-position of adjustment slots.

Set running timing on all magneto equipped models as outlined in **BATTERY IGNITION (JC)** section.

LUBRICATION. All models are pressure lubricated by a gear type pump driven by crankshaft gear. Internal pump parts are not serviced separately so entire pump must be renewed if worn or damaged. Clearance between pump gear and crankshaft gear should be 0.002-0.005 inch (0.0508-0.127 mm) and normal operating pressure should be 25 psi (172 kPa). Oil pressure is regulated by an oil pressure relief valve located in rear bearing plate. Relief valve spring free length should be 2-5/16 inch (58.74 mm) and valve diameter is 0.3365-0.3380 inch (8.55-8.59 mm).

Oils approved by manufacturer must meet requirements of API service classification "SE" or "SE/CC".

Use SAE 30 oil for temperatures above 30° F (-1° C), SAE 10W oil between 0° F (-18° C) and 30° F (-1° C) and SAE 5W (5W-30 if 5W is unavailable) for below 0° F (−18° C) temperatures.

Check oil level with engine stopped and maintain at "FULL" mark on dipstick. **DO NOT** overfill.

Recommended oil change interval for all models is every 100 hours of normal operation and change filter at 200 hour intervals. Oil capacity for all JB models is 3 quarts (2.8 L) without filter change and 3.5 quarts (3.3 L) with filter change. Oil capacity for all JC models is 6 quarts (5.68 L) without filter change and 6.5 quarts (6.15 L) with filter change.

Oil filter is screw on type and gasket should be lightly lubricated before installation. Screw filter on until gasket contacts, then turn an additional ½-turn. Do not overtighten.

REPAIRS

TIGHTENING TORQUES. Recommended tightening torques are as follows:

Cylinder head	28-30 ft.-lbs. (38-40 N·m)
Connecting rods	27-29 ft.-lbs. (37-39 N·m)
Rear bearing plate	40-45 ft.-lbs. (54-61 N·m)
Flywheel	65-70 ft.-lbs. (88-95 N·m)
Oil base	45-50 ft.-lbs. (61-68 N·m)
Exhaust manifold	13-15 ft.-lbs. (18-20 N·m)
Intake manifold	13-15 ft.-lbs. (18-20 N·m)
Timing gear cover	18-20 ft.-lbs. (24-27 N·m)
Rocker arm cover	8-10 ft.-lbs. (11-13 N·m)
Rocker arm stud	25-30 ft.-lbs. (34-40 N·m)
Center main bearing (JC model)	97-102 ft.-lbs. (132-138 N·m)
Spark plug	25-30 ft.-lbs. (34-40 N·m)

CYLINDER HEAD. To remove JB or JC models cylinder head, or heads, remove rocker arm cover, oil lines to cylinder heads, intake and exhaust manifolds. Valve stem caps, valve push rods and their spring loaded tubular shields will be released as head is lifted from block. Caps, tubes and push rods should be marked so they may be reinstalled in their original positions.

Cylinder head should be checked using feeler gage and straightedge. If warped or scored more than 0.003 inch (0.0762 mm) head should be resurfaced.

For reinstallation, loosely bolt cylinder head (heads) back to block and install push rod tubes as shown in Fig. O103.

NOTE: On four cylinder models, intake manifold must be installed on heads and bolts tightened to 13-15 ft.-lbs. (18-20 N·m) while heads are loosely bolted to block.

On all models, tighten cylinder head cap screws in several even steps in sequence shown (Fig. O102) until 28-30 ft.-lbs. (38-40 N·m) torque is obtained. Install exhaust manifold and oil lines. Install valve stem caps, push rods, rocker arms and rocker arm nuts.

NOTE: Rocker arm nuts are self locking nuts which are used to adjust valve gap. A torque of 4-10 ft.-lbs. (5-13 N·m) should be required to turn nuts on studs.

Adjust valves as outlined in **VALVE SYSTEM** section and install rocker arm cover (covers).

Cylinder head bolt torque should be checked after first 50 hours of operation.

CONNECTING ROD. Connecting rod and piston are removed from cylinder head surface end of block after removing cylinder head and oil base.

All models are equipped with precision insert type rod bearings as well as a renewable piston pin bushing.

Standard crankpin journal diameter is 2.0600-2.0605 inches (52.324-52.337

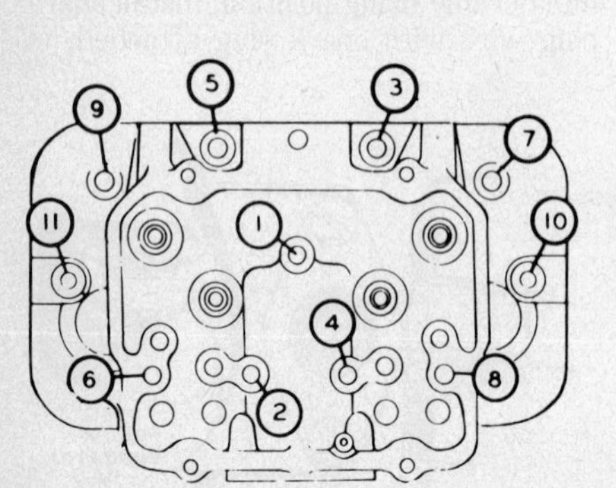

Fig. O102—Cylinder head tightening sequence for JB and JC models. Refer to text.

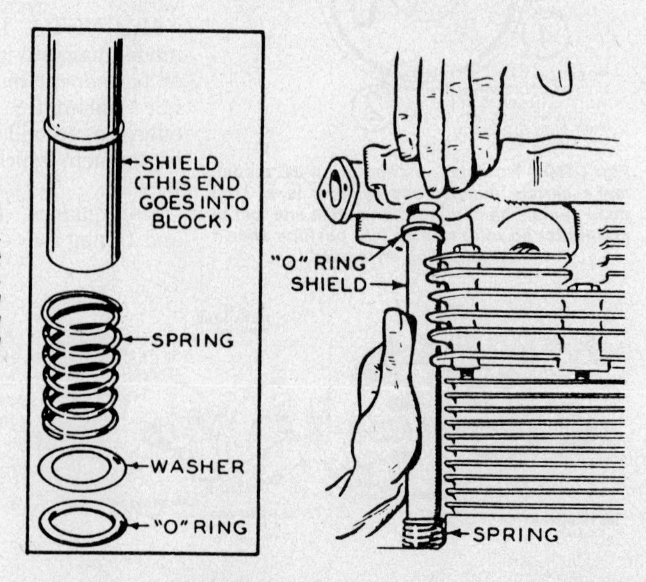

Fig. O103—Before tightening cylinder head bolts, lift head slightly while depressing valve push rod shield and spring. Insert upper end of shield into hole provided in head.

mm) for all models and connecting rod bearings are available in a variety of sizes for undersize crankshafts as well as standard.

Connecting rod bearing to crankshaft journal running clearance should be 0.0010-0.0033 inch (0.0254-0.0838 mm) for all models.

Piston pin clearance in bushing should be 0.0002-0.0007 inch (0.0051-0.0178 mm) and bushing inside diameter is 1.044-1.045 inches (26.518-26.543 mm). Install bushing as shown in Fig. O104 noting difference between early (A) style single bushing and late style double bushings (B).

Connecting rod must be assembled with piston so oil spray hole at rod bearing end of connecting rod is on the same side as "V" notch in top of piston.

PISTON, PIN AND RINGS. All models use three ring, cam ground aluminum pistons which are tapered. A full floating piston pin is held in position in piston bore by a snap ring at each end.

Engines which have an "E" following engine serial number are equipped at factory with 0.005 inch (0.127 mm) oversize pistons. Standard size rings are used on these models.

Clearance between piston and cylinder when measured 3/8-inch (9.525 mm) below oil control ring, 90° from pin, should be 0.0012-0.0032 inch (0.0305-0.0813 mm) on all models.

Side clearance of top ring in piston groove should be 0.002-0.008 inch (0.0508-0.2032 mm) and piston ring end gap should be 0.010-0.020 inch (0.254-0.508 mm).

Stagger ring end gaps around diameter of piston during installation and note JB models do not use rails above or below oil control ring (Fig. O104A).

Piston pin diameter is 0.9899-0.9901 inch (25.144-25.149 mm) and pin should be a light push fit in piston bore.

Standard cylinder bore diameter for

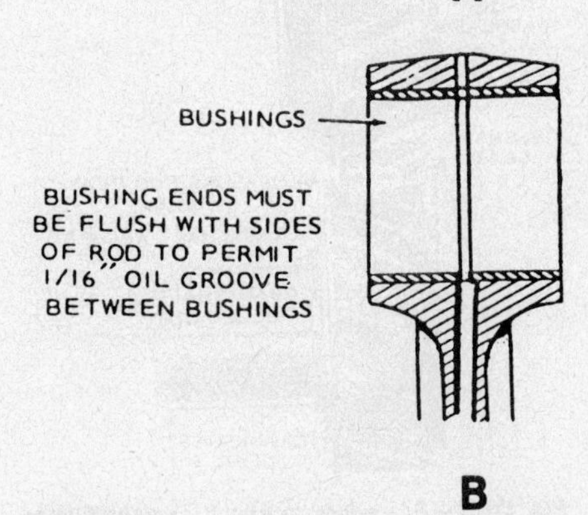

Fig. O104—View "A" shows early style bushing installation in connecting rod. If bushing has a "split" or "seam", install so "seam" is at right angle to oil hole. View "B" shows late style bushings. Press bushings in until flush with each side. A 1/16-inch (1.588 mm) space must be left between bushings as shown.

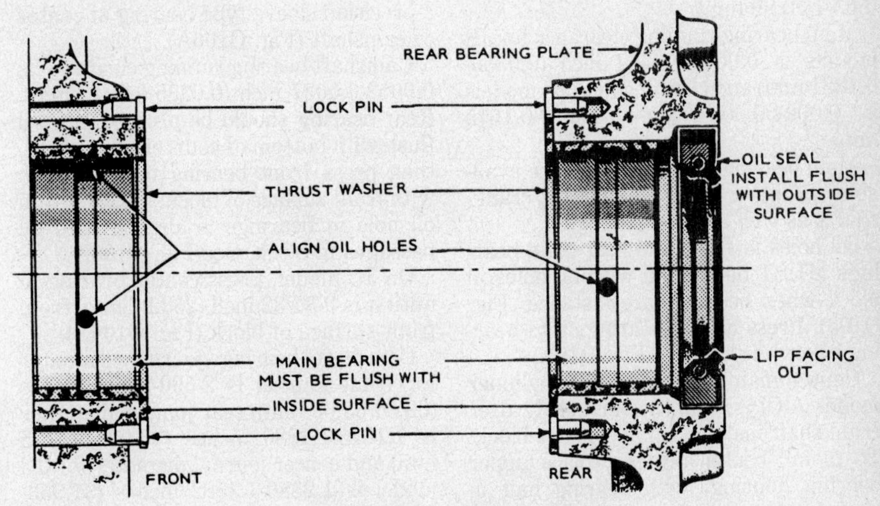

Fig. O105—Install front and rear main bearings as shown. Bearings should be chilled so they will press into bores with less resistance.

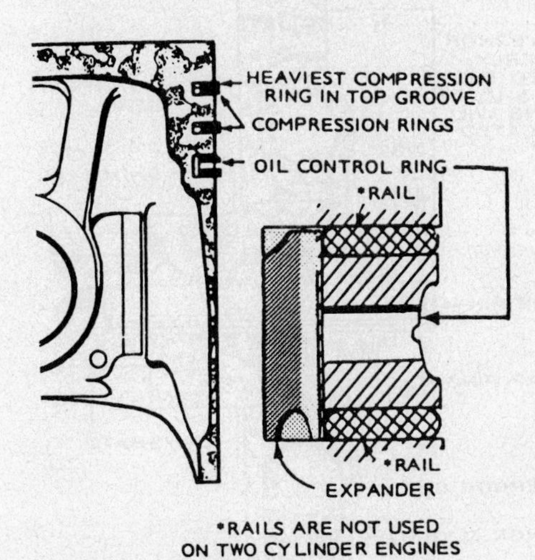

Fig. O104A—Install rings on pistons as shown staggering end gaps around diameter of piston. JB models do not use rails above or below oil ring.

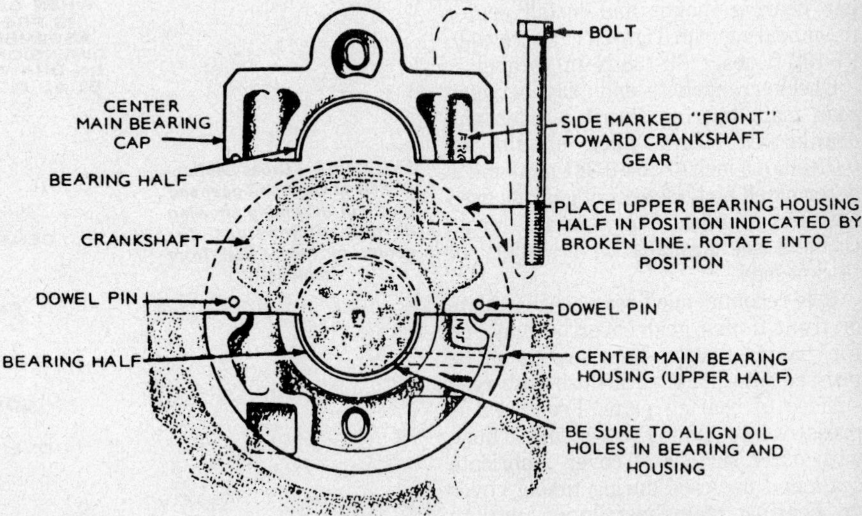

Fig. O106—Model JC has a center main bearing and housing which is installed after crankshaft installation. Refer to text.

all models is 3.2495-3.2505 inches (82.537-82.563 mm) and cylinders should be bored to nearest oversize for which piston and rings are available if cylinder is scored or out-of-round more than 0.003 inch (0.0762 mm) or if taper exceeds 0.005 inch (0.127 mm). Pistons and rings are available in a variety of oversizes as well as standard and pistons should be installed in block with "V" shaped notch in top of piston toward timing gear cover end of engine.

CRANKSHAFT, BEARINGS AND SEALS. Crankshaft is supported at each end in precision sleeve type bearings and four cylinder models have a two-piece precision type center main bearing.

Standard main bearing journal diameter for JB models is 2.2437-2.2445 inches (56.99-57.01 mm) and for JC models is 2.2327-2.2435 inches (56.71-56.99 mm).

Main bearing running clearance for JB models is 0.0014-0.0052 inch (0.0356-0.1321 mm) and clearance for JC models is 0.0024-0.0062 inch (0.0610-0.1575 mm).

Main bearings for all models are available for a variety of undersize crankshafts as well as standard.

Oil holes in front and rear main bearings **MUST** be aligned with oil holes in block when bearings are installed (Fig. O105). Press chilled bearings into bearing bores as shown in Fig. O105.

Center main bearing of four cylinder model (JC) should be installed after crankshaft installation in cylinder block. To install center bearing, place upper bearing housing, with bearing half in place, on crankshaft journal making certain side marked FRONT is facing crankshaft gear and rotate into place (Fig. O106). Place remaining lower bearing half in center main bearing cap, install the two positioning dowels on upper bearing mount and install center main bearing cap. Tighten cap screws to 97-102 ft.-lbs. (132-138 N·m) torque.

Check crankshaft end play between rear main bearing thrust washer and crankshaft. End play should be 0.010-0.015 inch (0.254-0.381 mm) and is determined by thickness of gaskets used between rear bearing plate and block. Gaskets are available in a variety of thicknesses.

It is recommended rear bearing plate or front timing gear cover be removed for seal installation. Rear seal should be pressed into place until flush with rear surface of bearing plate. Front seal is pressed into timing cover until flush with outer surface of cover. Lubricate seals and use care during timing cover or bearing plate installation not to damage seals.

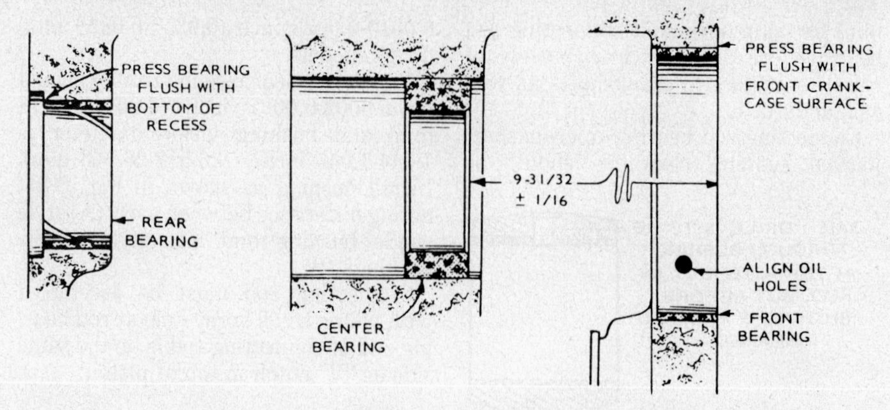

Fig. O106A—Camshaft bearings should be pressed into block as shown. Center main bearing is used on JC models only.

CAMSHAFT, BEARINGS AND GOVERNOR. Camshaft is supported at each end in precision sleeve type bearings and four cylinder JC model also has a precision sleeve type bearing at center of camshaft (Fig. O106A).

Camshaft bearing running clearance is 0.0012-0.0037 inch (0.0305-0.0940 mm). Rear bearing should be pressed in until flush with bottom of counterbore recess, then press front bearing in until flush with front surface of block. Make certain oil hole in bearings is aligned with oil passages in block. See Fig. O106A.

On JC model, press center bearing in until it is 9-31/32 inch (253.21 mm) from front surface of block (Fig. O106A).

Camshaft bearing journal diameter for front journal is 2.500-2.505 inches (63.50-63.63 mm), rear journal diameter is 1.1875-1.1880 inches (30.073-30.175 mm) and center journal diameter for JC model is 1.2580-1.2582 inches (31.953-31.958 mm).

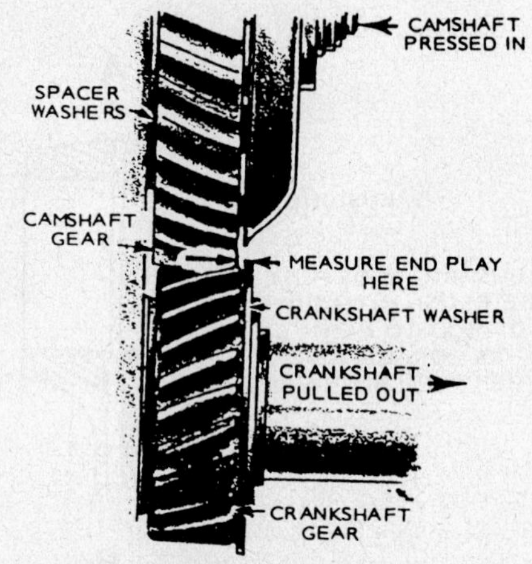

Fig. O107—Procedure to check camshaft gear end play. Refer to text.

When reinstalling camshaft, align timing marks on camshaft gear with timing marks on crankshaft gear. Install crank-

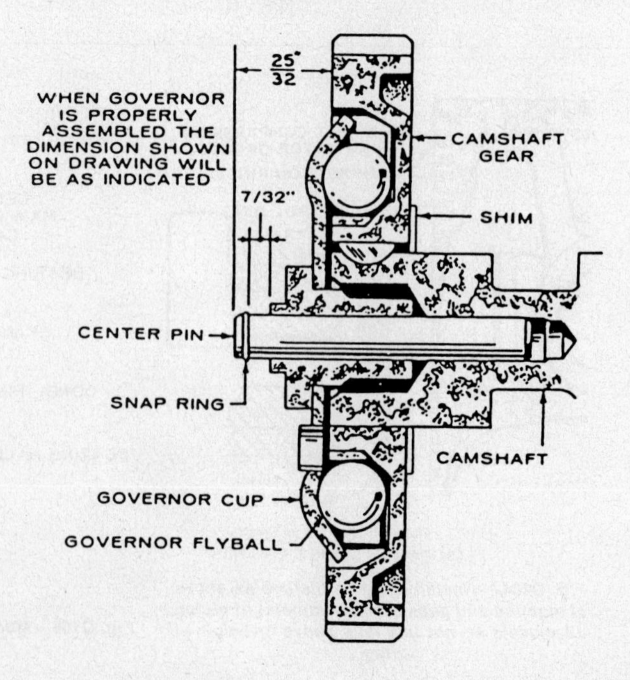

Fig. O107A—Cross-sectional view of camshaft gear and governor assembly showing correct dimensions for center pin extension from camshaft.

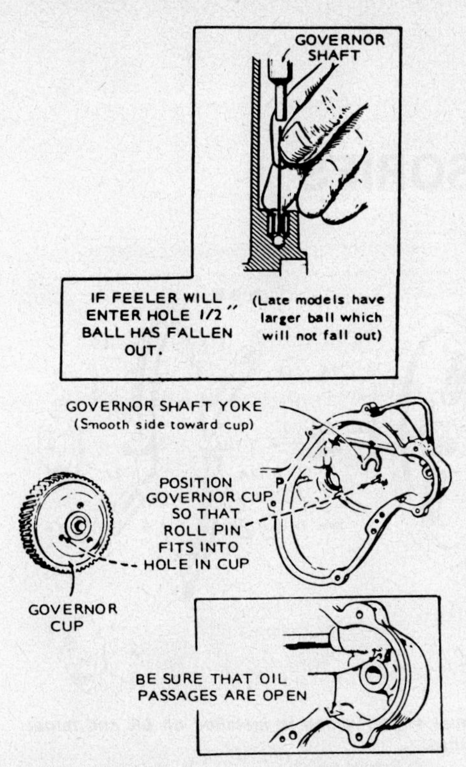

Fig. O108 — View of governor shaft, timing gear cover and governor cup showing assembly details. Refer to text.

shaft washer and snap ring and check camshaft end play as shown in Fig. O107. End play should be 0.007-0.039 inch (0.1778-0.9906 mm). If clearance is excessive, renew spacer washer or crankshaft washer.

Camshaft timing gear is a press fit on end of camshaft and governor weight assembly is mounted on camshaft gear and governor cup rides on a center pin pressed into camshaft as shown in Fig. O107A.

All models except JC model with specification letter "A" use 10 steel balls for governor weights. JC model with specification letter "A" uses 12 steel balls for governor weights.

Center pin (Fig. O107A) should extend 25/32-inch (19.844 mm) from end of camshaft to allow 7/32-inch (5.6 mm) in-and-out movement of governor cup. If distance is incorrect, remove pin and press new pin in to correct depth.

Make certain governor shaft pivot ball is in timing cover by measuring as shown in Fig. O108 and note roll pin must enter hole in governor cup during timing cover installation.

VALVE SYSTEM. Valve seats are renewable insert type and are available in a variety of oversizes as well as standard. Onan special tool number 420-0272 is used in drill press to cut out old seat. Set tool to cut all but a thin outside shell of seat and do not bottom in seat counterbore. Chill new inserts to shrink and install. Use a hammer and punch to peen head material against seat in the three areas between machine roll marks (Fig. O108A).

Valve seats are ground at a 45° angle for both intake and exhaust valves and seat width should be 3/64 to 1/16-inch (1.191 to 1.588 mm).

Intake valve face is ground at a 42° angle to provide a 3° interference angle and exhaust valve face is ground at a 45° angle.

Recommended valve stem to guide clearance is 0.001-0.003 inch (0.0254-0.0762 mm) for intake valves and 0.003-0.005 inch (0.0762-0.127 mm) for exhaust valves.

Drive old guides into valve chamber to remove and press new guides in until they extend 11/32-inch (8.731 mm) above machined valve spring seat in head. Ream new intake valve guide to 0.3425-03435 inch (8.700-8.725 mm) and new exhaust valve guide to 0.3445-0.3455 inch (8.750-8.776 mm).

Engines built after June, 1962 have a valve seal on intake valve guide. To install, push seal onto intake valve guide and clamp in place. Lubricate inside surface of seal and carefully insert valve through guide and seal. When compressing spring to install keepers, do not let spring or retainer contact seal.

Recommended valve tappet gap (cold) for models with specification letter prior to "D" and using gasoline for fuel is 0.010 inch (0.254 mm) for intake valves and 0.013 inch (0.3302 mm) for exhaust valves.

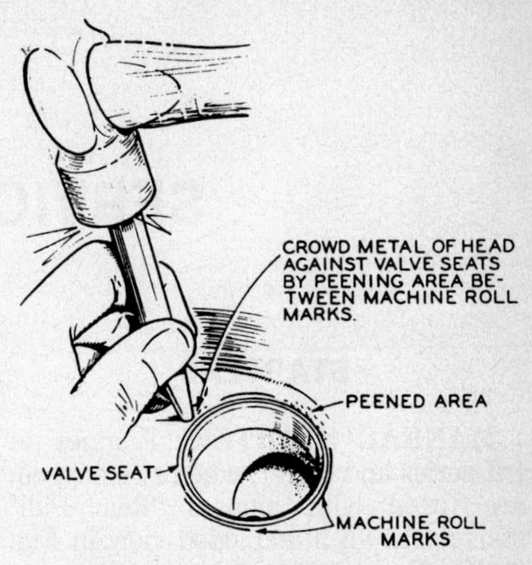

Fig. O108A — View showing proper installation of valve seat insert. Refer to text.

Recommended valve tappet gap (cold) for models with specification letter "D" or after, using gasoline for fuel, is 0.012 inch (0.3048 mm) for intake valves and 0.015 inch (0.3810 mm) for exhaust valves.

Recommended valve tappet gap (cold) for all models using vapor (butane, LP or natural gas) fuel or combination vapor/gasoline fuel is 0.013 inch (0.3302 mm) for intake valves and 0.020 inch (0.508 mm) for exhaust valves.

Adjust valves on JB model by rotating engine until number one piston is 10° to 45° past TDC on compression stroke. At this position adjust number one cylinder valves. Rotate engine one full revolution and adjust number two cylinder valves.

Adjust JC model valves according to firing order sequence by rotating engine until number one piston is 10° to 45° past TDC on compression stroke. At this position adjust number one cylinder valves. Rotate engine ½-turn clockwise and adjust number two cylinder valves. Rotate engine ½-turn clockwise and adjust number four cylinder valves. Rotate engine another ½-turn and adjust number three cylinder valves.

All valves are adjusted by turning lock-nut on rocker arm stud to gain specified clearance.

ONAN

SERVICING ONAN ACCESSORIES

STARTERS

MANUAL STARTER. Engines of LK series and larger, when so equipped, are fitted with manual "Readi-Pull" starter shown in exploded view in Fig. O140. For convenience, direction of starter rope pull may be adjusted to suit special cases by loosening clamps which hold starter cover (5) in position on its mounting ring (20), then turning cover (5) so rope (2) exits in desired direction.

Mounting ring (20) must be firmly attached to engine blower housing which must be as rigid as possible. If blower housing is damaged or if mounting holes for starter are misshaped or worn it may be necessary to renew entire blower housing. See Fig. O141 for cross-section detail of starter mounting ring attachment to blower housing.

To attach starter to earlier production engines, refer to Fig. O142, and use a pair of 10-penny (3 inch) common nails passed through holes in cover to insert into recesses in heads of special screws which retain ratchet wheel to engine flywheel. In later production (after specification D), spirol pin (12A–Fig. O140) is centered upon and engages drilled head screws (17) which secure ratchet wheel (22) and rope sheave hub bearing (16) for alignment during assembly.

Common repairs to manual starter can be made with minimum disassembly. Mounting ring (20) is left in place and only cover assembly (5) need be removed after its four clamps (19) are released. To renew starter rope (2), remove cover (5) from mounting ring (20), release clamp (15) to remove old rope from sheave (10). Then rotate sheave (10) in normal direction of crankshaft rotation so as to tighten spring (8) all the way. Align rope hole in sheave with rope slot in cover, secure new rope by its clamp (15), and when sheave (10) is released, spring (8) will recoil and wind rope on sheave. If renewal of recoil spring (8) is required, sheave (10) must be lifted out of cover (5). Remove starter rope (2) from sheave. Starting at outer end, wrap new spring (8) into a coil small enough to fit into recess in cover with loop at inner end of spring engaging roll pin (7) in cover. It may be necessary to secure wound-up spring with a temporary restraint such as a wire while doing so. Then reinstall rope sheave (10) so tab on sheave fits into loop at outer end of recoil spring. Install thrust washer (9) and spring washer (6A) if starter is so equipped.

Whenever starter is disassembled for any service, it is advisable to add a small amount of grease to factory-packed sheave hub bearing (16) and to clean and lubricate pawls (11) and ratchet arms (13) at pivot and contact points. If ratchet arms (13) require renewal due to wear, pawls (11) must first be removed. If kept clean, securely mounted and lightly lubricated, this starter will give long term reliable service.

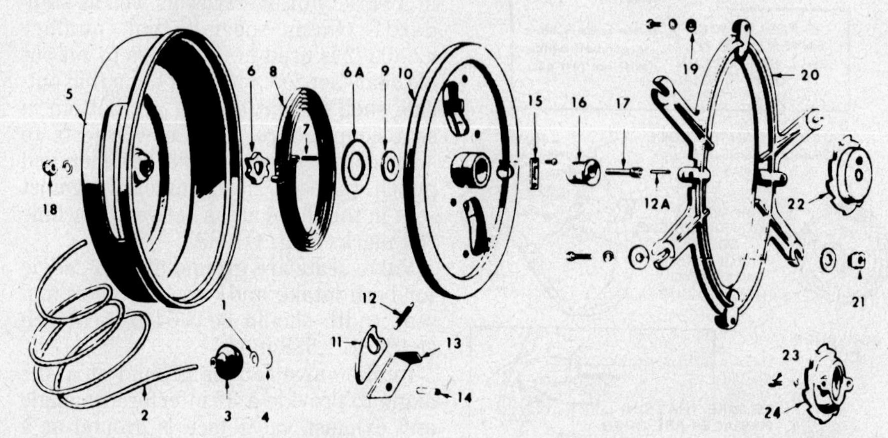

Fig. O140—Exploded view of "Ready-Pull" manual starter which is installed on LK and larger engines.

1. Rope retainer
2. Starter rope
3. Starter grip
4. Grip plug
5. Cover
6. Anti-backlash cog
6A. Spring washer
7. Roll pin
8. Recoil spring
9. Thrust washer
10. Rope sheave
11. Pawl (2)
12. Pawl spring
12A. Spiral pin
13. Ratchet arm (2)
14. Pivot roll pin (2)
15. Rope clamp
16. Hub bearing
17. Recessed screw
18. Flexlock nut
19. Washer
20. Mounting ring
21. Speed grip nut
22. Ratchet wheel (late)
23. Special cap screw
24. Ratchet wheel (early)

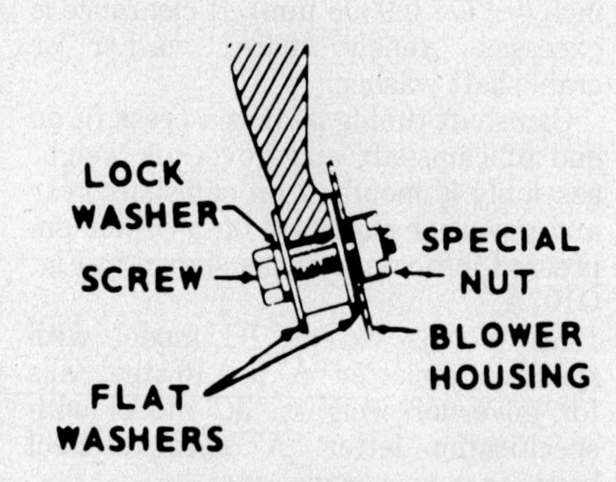

Fig. O141—Cross-section of mounting ring bolt to show housing attachment detail.

AUTOMOTIVE TYPE ELECTRIC STARTERS. Two styles of battery-driven electric starter motors are used on ONAN engines. Bendix-drive starter shown at A–Fig. O143 is designed to engage teeth of flywheel ring gear when starter switch is depressed to close cir-

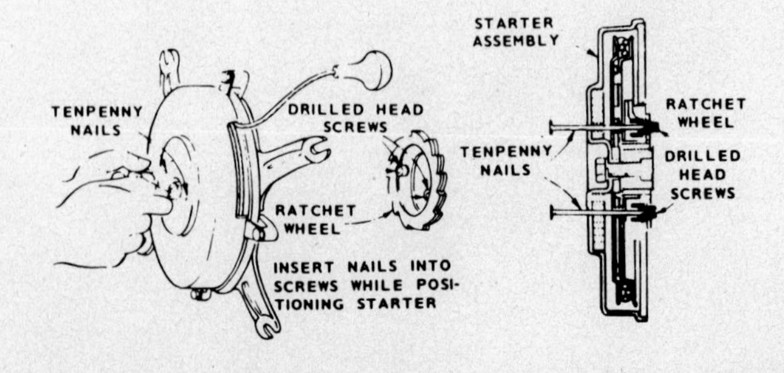

Fig. O142—Technique for mounting older style starter to ratchet wheel on engine flywheel. Refer to text.

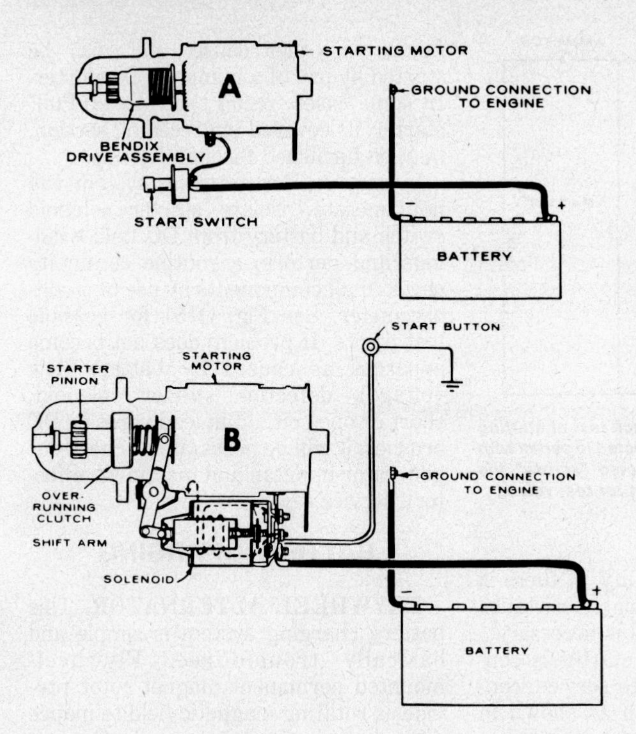

Fig. O143—Views of Bendix-drive starter and solenoid-shift starter in basic electric circuits. Refer to text.

cuit causing starter motor to turn. Engagement of starter pinion with ring gear by means of spiral shaft screw within Bendix pinion is cushioned by action of its coiled drive spring so starting motor can absorb sudden loading shock of engagement. Engine manufacturer recommends complete Bendix drive unit be renewed in case of failure, however, if a decision is made to overhaul starter drive by obtaining parts from manufacturer of starter (Prestolite), be sure correct drive spring is used. Length of spring is critical to mesh and engagement of starter pinion to flywheel ring gear. There are no procedures for adjustment of this starter.

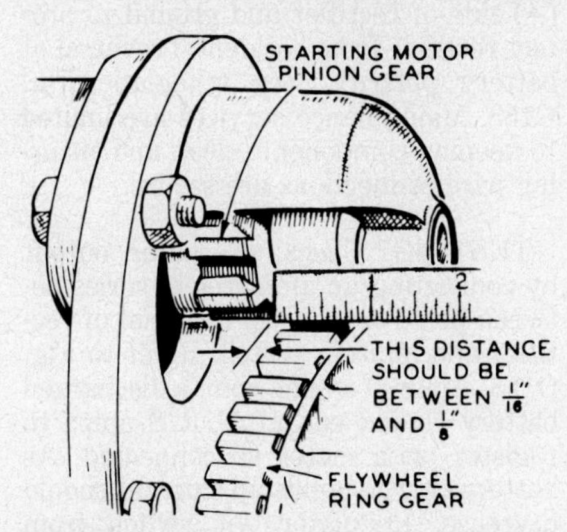

Fig. O144—Check starter pinion to ring gear clearance as shown. Adjustment details in text.

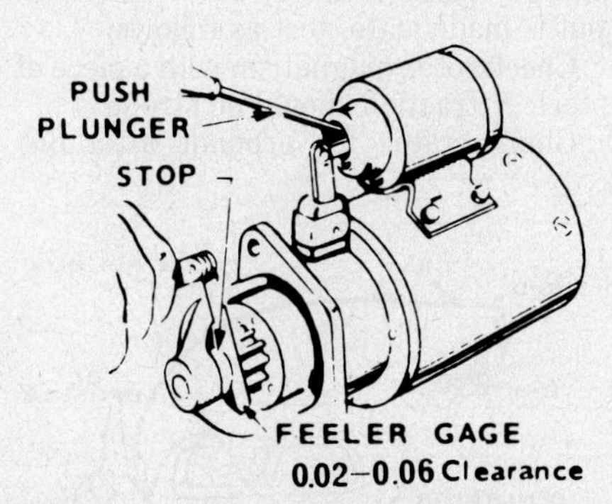

Fig. O145—Measure clearance between starter pinion and pinion stop with feeler gage as shown.

Service is generally limited to cleaning and careful lubrication, renewal of starting motor brushes (4 used), brush tension springs and starter motor and drive housing bearings. See STARTER MOTOR TESTS for electrical check-out procedures for starter armature and field windings.

Solenoid-shift style starter (B–Fig. O143) uses a coil solenoid to shift starter pinion into mesh with flywheel ring gear and an over-running (one-way) clutch to ease disengagement of pinion from flywheel as engine starts and runs.

NOTE: Starter clutch will burn out if held in contact with flywheel for long periods and starter switch must be released quickly as engine starts.

All parts of starter are available for service. Solenoid unit and starter clutch are renewed as complete assemblies.

Refer to Figs. O144 and O145 for adjustment check points to measure flywheel ring gear to starter pinion clearance and gap between pinion and pinion stop on starter shaft. Starter pinion to ring gear clearance (Fig. O144) is adjusted during starter assembly by proper selection of spacer washers fitted to armature shaft as installed in housing. To adjust pinion clearance against pinion stop (Fig. O145), remove mounting screws which attach solenoid magnetic coil to front bracket and pinion housing assembly and select a proper thickness of fiber packing gaskets to set required clearance. Be sure plunger is pressed inward as shown when measuring.

Starting motor brushes require renewal when worn away by 0.3 inch (7.62 mm). Original brush length is 0.55 inches (13.97 mm).

Commutator must be clean and free from oil. Use No. 00 sandpaper to clean and lightly polish segments of commutator; never use emery cloth or any abrasive which may have a metallic content. Starter motor commutators do not need to have mica separators between segments undercut. Mica may be flush with surface.

STARTER MOTOR TESTS

Armature Short Circuit. Place armature in growler as shown in Fig. O146 and hold a hack saw blade or similar piece of thin steel stock above and parallel to core. Turn growler "ON". A short circuit is indicated by vibration of blade and attraction to core. If this condition appears, renew armature.

Grounded Armature. Check each segment of commutator for grounding to shaft (or core) using ohmmeter setup as shown in Fig. O147. A low (R×1 scale) continuity reading indicates armature is grounded and renewal is necessary.

It is good procedure to mount armature on a test bench or between lathe centers to check for runout of commutator shaft. If shaft is worn badly,

Fig. O146—Use of growler to check armature for short circuit. Follow procedure in text.

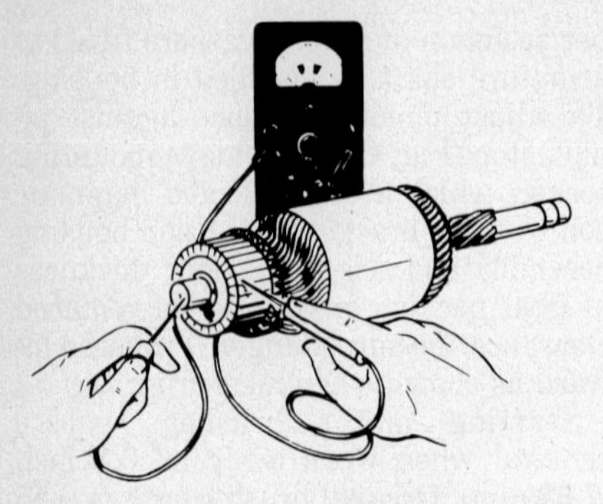

Fig. O147—Test for grounded armature commutator by placing ohmmeter test probes as shown. Probe shown touching shaft may also be held against core. See text.

renewal is recommended. If commutator runout exceeds 0.004 inch (0.1016 mm), reface by turning.

Grounded Field Coils. Refer to Fig. O148 and touch one ohmmeter probe to a clean, unpainted spot on frame and the other to connector as shown, after unsoldering field coil shunt wire. A low range reading (R×1 scale) indicates grounded coil winding. Be sure to check for possible grounding at connector lead which can be corrected, while grounded field coil cannot be repaired and calls for renewal.

Open in Field Coils. Use procedure shown in Fig. O149 and check all four

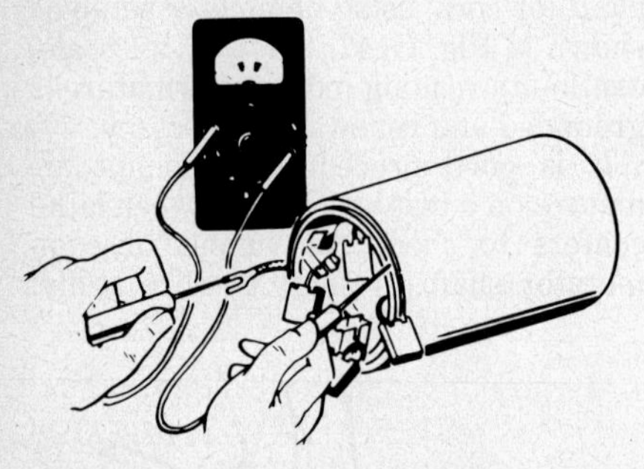

Fig. O148—Use ohmmeter as shown to check field coil for suspected internal grounding.

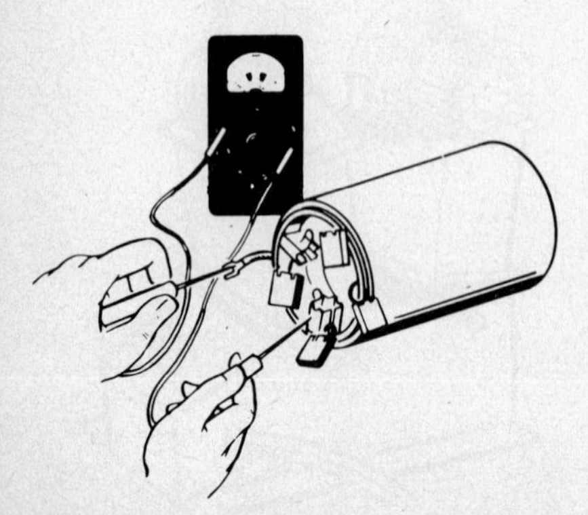

Fig. O149—Ohmmeter used to check for breaks or opens in field coil windings. Be sure to check lead wires.

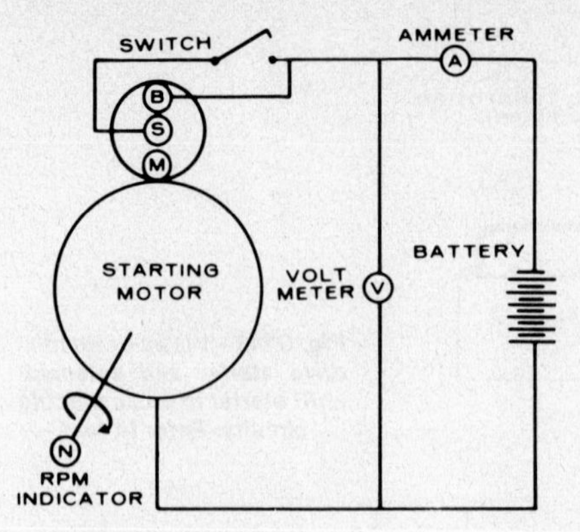

Fig. O150—Schematic for bench test of starting motor. Note ammeter is connected in series with load and voltmeter is connected "across" the load in parallel. Refer to text for test values.

brush holders for continuity. If there is no continuity or if a high resistance reading appears, renewal is necessary.

No-Load Test. When starter is considered ready to return to service, connect motor on bench top as shown in Fig. O150. Acceptable test readings are:

Minimum speed ....3700 rpm
Voltmeter reading ....11.5 Volts
Maximum current draw ...60 Amperes

If starter motor does not check out as satisfactory on this test, make further checks for:

Weak brush springs

Brushes not squarely seated

Dirty commutator

Poor electrical connections. May be caused by "cold" or corroded solder joints.

Tight armature. Not sufficient end play. End play should be 0.004-0.020 inch (0.1016-0.508 mm).

Open or ground in field coil

Short circuit, open or ground in armature.

EXCITER CRANKING. Exciter cranking, with cranking torque furnished by switching battery current through a separate series winding of generator field coils and DC brushes, using DC portion of generator armature is wired as shown in Fig. O151.

This starting procedure may be standard for LK or CCK series, or for NH or JB models. In cases where exciter cranking is inoperative, due to battery failure or other cause, unit may be started by use of a manual rope starter. In some cases, recoil type "Readi-Pull" starter, as covered in preceding section, may be furnished for standby use.

In case exciter cranking system will not operate, isolate starter solenoid switch and battery from DC field windings and perform a routine continuity check of all components by use of a voltohmmeter. See Fig. O151 for possible test points. If problem does not become apparent as caused by battery (low voltage), defective starter solenoid, short or open circuit in lead wires or DC brushes, it will be necessary to check out generator in detail and may involve factory service.

BATTERY CHARGING

FLYWHEEL ALTERNATOR. This battery charging system is simple and basically trouble-free. Flywheel-mounted permanent magnet rotor provides a rotating magnetic field to induce AC voltage in fixed stator coils. Current is then routed through a two-step mechanical regulator to a full-wave rectifier which converts this regulated alternating current to direct current for battery charging. Later models are equipped with a fuse between negative (−) side of rectifier and ground to protect rectifier from accidental reversal of battery polarity. See schematic Fig. O153. Maintenance services are limited to keeping components clean and ensuring wire connections are secure.

TESTING. Check alternator output by connecting an ammeter in series between positive (+), red terminal of rectifier and ignition switch. Refer to Fig. O153. At 1800 engine rpm, a discharged battery should cause about 8 amps to register on a meter so connected. As battery charge builds up, current should decrease. Regulator will switch from high charge to low charge at about 14½ volts with low charge current of about 2 amps. Switch from low charge to high charge occurs at about 13 volts. If output is inadequate, test as follows:

Check rotor magnetism with a piece of steel. Attraction should be strong.

Check stator for grounds after dis-

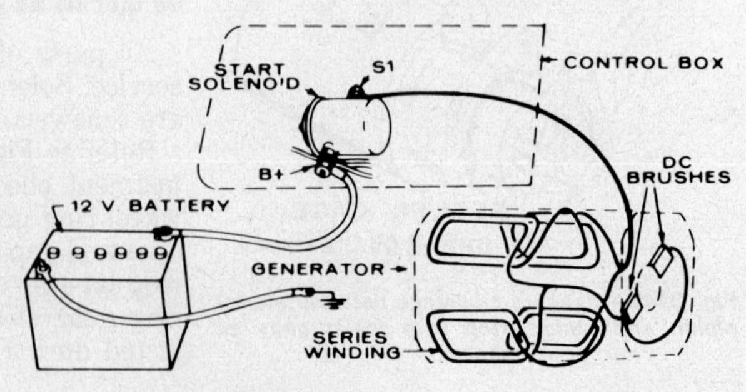

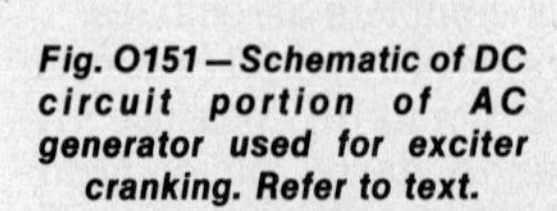

Fig. O151—Schematic of DC circuit portion of AC generator used for exciter cranking. Refer to text.

connecting by grounding each of the three leads through a 12-V test lamp. If grounding is indicated by lighted test lamp, renew stator assembly.

To check stator for shorts or open circuits, use an ohmmeter of proper scale connected across open leads to check for correct resistance values. Identify leads by reference to schematic.

From lead 7 to lead 8 . . . . 0.25 ohms
From lead 8 to lead 9 . . . . 0.95 ohms
From lead 9 to lead 7 . . . . 1.10 ohms

Variance by over 25% from these values calls for renewal of stator.

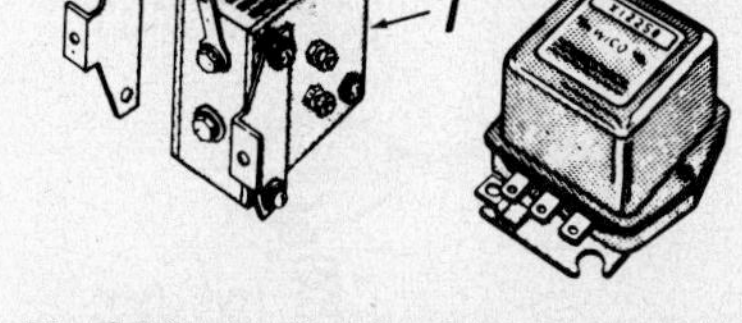

Fig. O152—Typical flywheel alternator shown in exploded view. In some models, regulator (6) and rectifier (7) are combined in a single unit.

1. Flywheel
2. Rotor
3. Fuse holder
4. Fuse
5. Stator & leads
6. Regulator
7. Rectifier assy

RECTIFIER TESTS. Use an ohmmeter connected across a pair of terminals as shown in Fig. O154. All rectifier leads should be disconnected when testing. Check directional resistance through each of the four diodes by comparing resistance reading when test leads are reversed. One reading should be much higher than the other.

NOTE: Forward-backward ratio of a diode is on the order of several hundred ohms.

If a 12-V test lamp is used instead of an ohmmeter, bulb should light, but dimly. Full bright or no light at all indicates diode being tested is defective.

Voltage regulator may be checked for high charge rate by installing a jumper lead across regulator terminals (B and C–Fig. O153). With engine running, battery charge rate should be about 8 amperes. If charge rate is low, alternator or its wiring is defective.

If charge rate is correct (near 8 amps), defective regulator or its power circuit is indicated. To check, use a 12-V test lamp to check input at regulator terminal (A). If lamp lights, showing adequate input, regulator is defective and should be renewed.

NOTE: Regulator, being mechanical, is sensitive to vibration. Be sure to mount it on bulkhead or firewall separate from engine for protection from shock and pulsating motion.

Engine should not be run with battery disconnected, however, this alternator system will not be damaged if battery terminal should be accidentally separated from binding post.

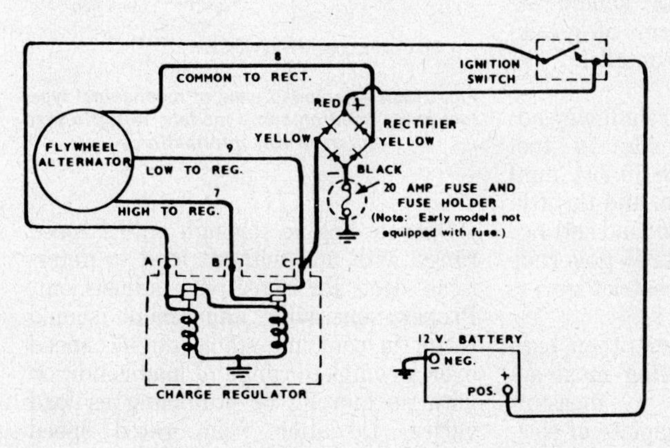

Fig. O153—Schematic of flywheel alternator circuits for location of test and check points. Refer to text for procedures.

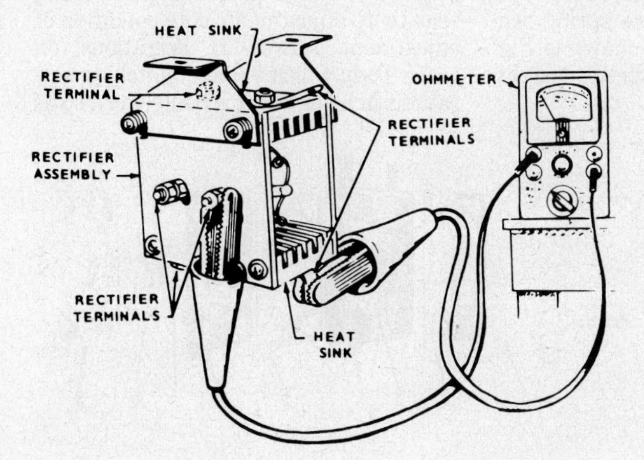

Fig. O154—Test each of four diodes in rectifier using Volt-Ohmmeter hook-up as shown. See text for procedure.

FUEL SYSTEMS

ELECTRIC FUEL PUMP. Some engines may be furnished with Bendix Electric Fuel Pump, code R-8 or after.

Maintenance service to these electric pumps is generally limited to simple disassembly and cleaning of removable components, not electrical overhaul. Stored gasoline is prone to deterioration and gum residues formed can foul internal precision units of a fuel system so as to cause sticking and sluggish operation of functional parts.

Refer to Fig. O155 for sequence of disassembly, beginning with cover (1) by turning 5/8-inch hex to release cover from bayonet lugs after fuel lines have been disconnected. Wash parts in solvent and blow dry using air pressure. Renew damaged or deteriorated parts, especially gasket (2) or filter element (4). Use needle-nose pliers to remove retainer (5) and withdraw remainder of parts (6 through 10) from plunger tube (11). Clean and dry each item and inspect carefully for wear or damage. Plunger (10) calls for special attention. Clean rough spots very gently using crocus cloth if necessary. Clean bore of plunger tube (11) thoroughly and blow dry. For best results, use a swab on a stick to remove stubborn deposits from inside tube.

During reassembly, check fit of plunger (10) in tube (11). Full, free, in-and-out motion with no binding or sticking is required. If movement of plunger does not produce an audible click, interrupter assembly within housing (12) is defective. Renew entire pump.

If all parts appear serviceable, reassemble pump parts in order shown.

Pump output pressure can be raised or lowered by selection of a different plunger return spring (9). Consult authorized parts counter for special purpose spring or other renewable electric fuel pump parts.

NOTE: Seal at center of pump case mounting bracket retains a dry gas in pump electrical system. Be sure this seal is not damaged during disassembly for servicing.

MECHANICAL FUEL PUMP. A mechanical fuel pump (Fig. O155A) is used on some models. Pump operation may be checked by disconnecting fuel line at carburetor and slowly cranking engine by hand. Fuel should discharge from line.

CAUTION: Make certain engine is cool and there is nothing present to ignite discharged fuel.

To recondition pump, scribe a locating mark across upper and lower pump bodies and remove retaining screws. Noting location for reassembly, remove upper pump body, valve plate screw and washer, valve retainer, valves, valve springs and valve gasket. To remove lower diaphragm, hold mounting bracket and press down on diaphragm to compress spring. Turn bracket 90° to unhook diaphram. Clean all parts thoroughly.

To reassemble, hold pump cover with diaphragm surface up. Place valve gasket and assembled valve springs and valves in cavity. Assemble valve retainer and lock in position by inserting and tightening valve retainer screw. To reassemble lower diaphragm section hold mounting bracket and press down on diaphragm to compress spring. Turn bracket 90° to hook diaphragm. Assemble pump upper and lower bodies, but do not tighten screws. Push pump lever to its limit of travel, hold in this position and tighten the four screws. This prevents stretching diaphragm. Reinstall pump.

PULSATING DIAPHRAGM FUEL PUMP. A pulsating diaphragm type fuel pump (Fig. O155B) is used on some models. Pump may be mounted directly to side of carburetor or at a remote mounting location.

Pump relies on a combination of crankcase pressure and spring pressure for correct operation.

Refer to Fig. O155B for disassembly and reassembly noting air bleed hole (10) must be open for correct pump operation.

HOOF GOVERNOR. This governor is flyweight-operated, camgear driven to control engine rpm at all points of engine speed range.

Pressure generated by centrifugal force of revolving flyweights is exerted against thrust sleeve and bearing shown in Fig. O156 so as to impart rotary motion to governor rocker shaft through rocker shaft lever. Governor mechanism is liberally lubricated from engine crankcase and severe wear to parts is unusual. If erratic governor performance is traced to its speed sensor (flyweight mechanism), refer to Fig. O156 and proceed as follows:

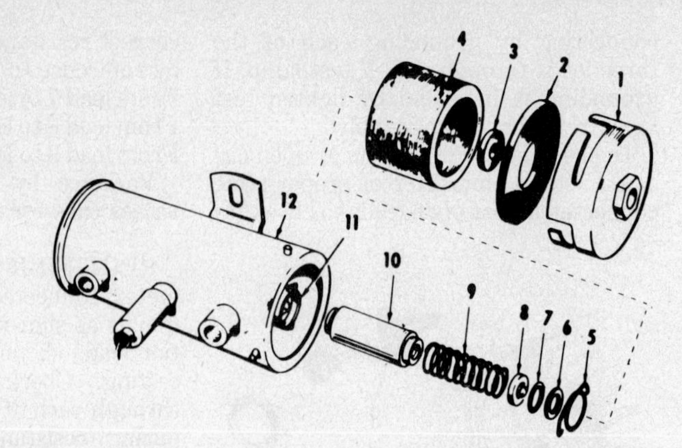

Fig. O155—Exploded view of electric fuel pump used on some engine models.

1. Cover
2. Cover gasket
3. Magnet
4. Filter element
5. Retainer spring
6. Washer
7. "O" ring
8. Cup valve
9. Plunger spring
10. Plunger
11. Plunger tube
12. Pump housing

Disconnect linkages.

Unbolt governor body flange and remove governor from engine.

Entire shaft assembly with gear and front cover can be removed after backing out the one cover to body screw.

At this point, any abnormal condition should be apparent when all sludge and contaminated oil is flushed away. Further disassembly is not likely to be needed. Flyweights, pivots and limiting stops should be checked. If necessary, remove retainers and flyweight pins to disassemble governor completely. Thrust bearing and washers can be removed from thrust sleeve after removal of lock ring. After flyweight assembly and thrust sleeve are removed from shaft and shaft is withdrawn, shaft ball bearing can be pressed from cover. Necessary renewal parts should be ordered from manufacturer of governor. Check nameplate for details.

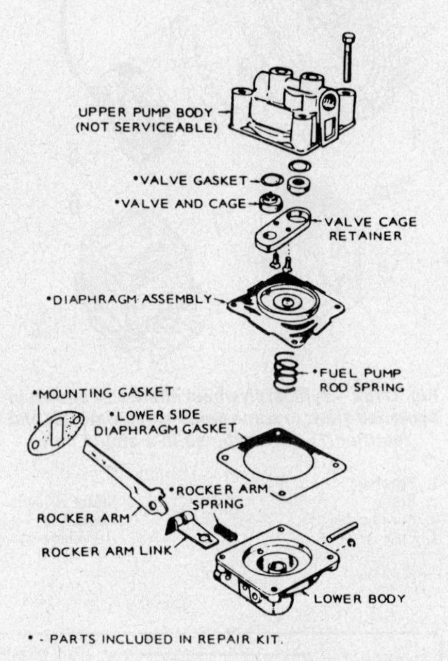

Fig. O155A—Exploded view of mechanical type fuel pump used on some models. Refer to text for assembly information.

ADJUSTMENTS. All preliminary adjustments should be made to fuel system. Be sure 1/32-inch (0.794 mm) gap is set between stop pin and throttle stop screw as low speed-no load setting. See Fig. O157. Be sure governor bumper screw does not restrict governor action.

Set low speed limits first, then high speed to specifications using most accurate means available to measure engine rpm. Refer to nameplate of electric generator sets for correct speed settings. Increase or decrease spring tension at adjustment points shown in Fig. O157 to set low and high speeds correctly.

Operate engine through entire speed range with and without load to determine need for sensitivity adjustment. Proper sensitivity adjustment should result in constant stable engine speed over a complete range of load condition with no hunting or stumbling as load varies. Deviation from rated speed should not exceed 50 rpm. When making sensitivity adjustment, note condition of adjustment screw. If serrations on screw body which engage matching serrations in lever slot are badly worn so as

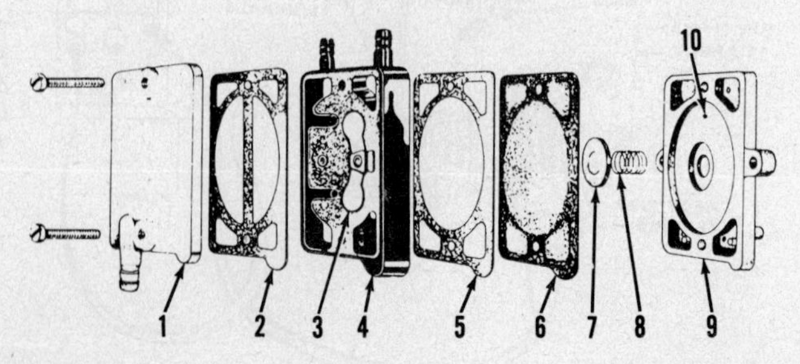

Fig. O155B—Exploded view of pulsating diaphragm fuel pump used on some models. Hole (10) must be open before reassembly.

1. Pump cover
2. Gasket
3. Reed valve
4. Valve body
5. Gasket
6. Diaphragm
7. Pump plate
8. Spring
9. Pump base
10. Air bleed hole

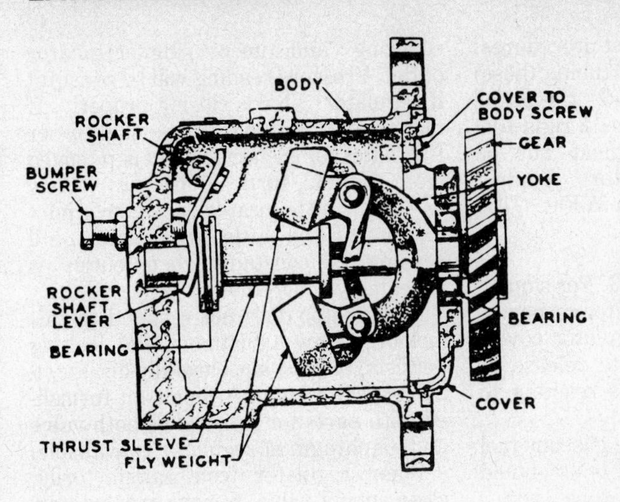

Fig. O156—Cross-section view of Hoof governor used on older four-cylinder gasoline models (early JC). Refer to text for description of parts functions.

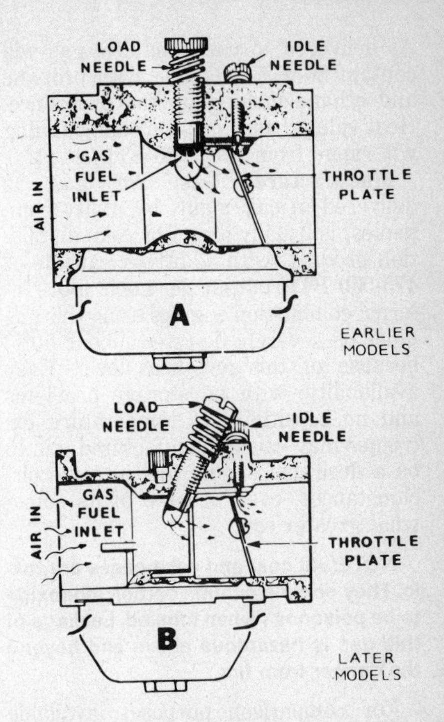

Fig. O158—Cross-section view of carburetor designed for gaseous fuel only. A-Early style. B-Late style.

not to hold adjustment, renewal of parts is in order. As in all governor adjustments, movement of sensitivity adjuster away from its pivot control point will decrease sensitivity.

GASEOUS FUEL OPERATION. This section is concerned only with exclusive operation on gaseous (vapor-type) fuels such as natural gas, methane, butane, propane or mixtures such as LPG. See CCK engine section for dual-fuel (gas-gasoline) operations.

Only one style carburetor is used for constant gaseous fuel operation. Fig. O158 shows a functional cross-section of this carburetor in both early and late models. Note adjustment points. Adjustment of vapor fuel carburetors is essentially the same as for gasoline-fueled models. It will be noted idle adjustment has only a limited effect on performance of these engines as they normally operate at rated rpm. If carburetor is entirely out of adjustment, to such extent engine will not start or run, proceed as follows:

Turn idle adjustment and main adjustment needles inward until lightly seated. Then, open idle needle from one to two turns and crank engine while opening main adjustment needle a little at a time until engine starts. As with gasoline carburetors, final adjustment is made to provide a smooth running engine at both idle speed and rated rpm.

GASEOUS FUELS. It should be noted operation on gaseous fuels sometimes involves changes in types of fuel available for use and not all gaseous fuels have the same power potential. Rating of these fuels is based on their heat output measured in BTU's per cubic foot of volume.

Butane. Butane develops about 3200 BTU/cu. ft. and is rated on a par with gasoline as a fuel. It is generally compounded with other gases for regular use.

Propane. This gas, rated at about 2500 BTU/cu. ft., will also perform near the level of gasoline. It is also frequently mixed with other gases to suit special circumstances of its use.

LPG. Liquified Petroleum Gas, LPG, sometimes called "bottled gas" is specially compounded to meet requirements for energy (heat) output and for reliability of performance in different climates. LPG is a variable proportion of butane, propane and other hydrocarbons generally "tailor-made" for a particular requirement. As with butane and propane gases, no de-rating of engine is required for its use.

Natural Gas. The principal component of natural gas is methane (sometimes called marsh gas) which has a heat output of about 500 BTU/cu. ft.

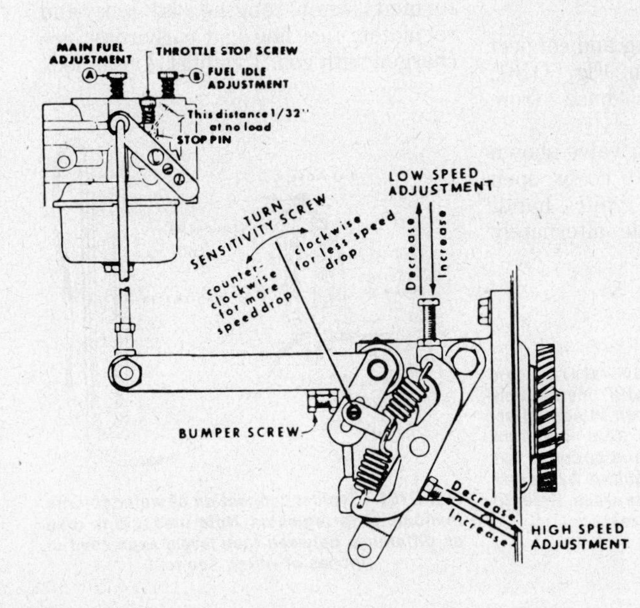

Fig. O157—View of external linkage and adjustments for Hoof governor. Refer to text.

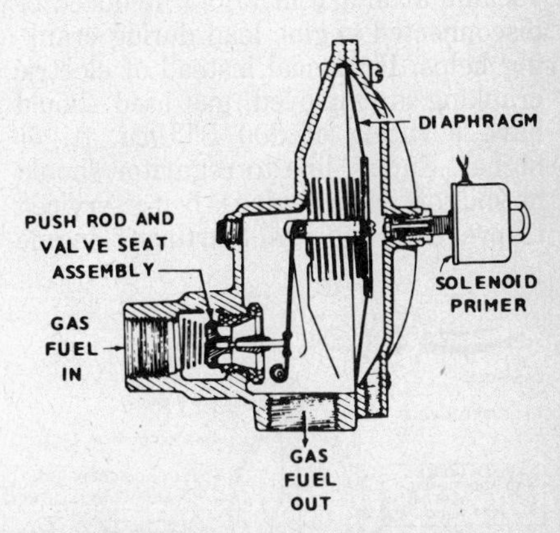

Fig. O159—Cross-section of Algas regulator used on early vapor (butane, LP or natural gas) fuel systems. Note installation of optional solenoid controlled primer. Refer to text for primer adjustment.

As delivered to users, natural gas will contain 80-95% methane with propane and ethane making up the difference. Heat values based on such compounding will range from 900-1200 BTU/cu. ft.

Manufactured Gas. This gas, as delivered in city mains by utility companies, is usually a coal or coke distillation product, with additives, capable of 475-550 BTU output per cubic foot. Internal combustion engines using this gas must be severely de-rated (about 50%) because of this low heat level. Easy availability, with no storage problems and no sensitivity to temperature extremes may cause manufactured gas to be a desirable engine fuel in some circumstances, even in spite of its somewhat greater cost.

NOTE: All coal and coke gases are toxic. They contain enough carbon monoxide to be poisonous when inhaled. Leakage of this gas is hazardous above and beyond the danger from fire.

For comparison purposes, available gaseous fuels are rated by a percentage figure against pump grade regular gasoline:

FUEL COMPARISON TABLE

GASEOUS FUEL	GASOLINE
LPG (Butane, Propane)	98-100%
1100 BTU gas*	80-95%
850 BTU gas*	80-85%
600 BTU gas*	70-75%
450 BTU gas*	50-60%

*Check supplier for BTU rating of fuel gas used.

GAS REGULATOR. Engines which operate on gaseous fuels require a pressure regulator be installed in supply line to carburetor. Regulators used are highly sensitive demand types, opening only when fuel is called for by vacuum at carburetor inlet. Ease of starting is dependent upon rapid cranking for high vacuum at intake manifold. Reduced or disconnected engine load during cranking helps. If manual instead of electric cranking is employed, gas used should have a rating of 800 BTU/cu. ft. or higher. Supply line to regulator should be shut off when engine is being serviced to prevent accidental starting if engine is turned over as part of test procedures. Factory furnished flexible tubing (hose) between carburetor and regulator should never be replaced by a rigid fuel line. Engine vibration must not be transmitted to regulator. Typical regulators used are shown in Fig. O159 and Fig. O160.

REGULATOR TESTING. For a quick operational check of regulator response, blow into vent hole in regulator cover. Hissing sound will indicate release of gas and that diaphragm is reacting to open regulator valve.

For proper operation of a gaseous fuel system, required pressure tests should be made by use of a "U-tube" water manometer having a minimum working range of 14 inches. A commercial model manometer as shown in Fig. O161 is recommended, however, a length of clear plastic tubing of about 3/8-inch inside diameter, formed into a "U" shape comparable to that shown in the figure may be attached to a board and will serve to make pressure measurements. Pour water (with coloring added if hard to see) in open end of tube so it rises into each leg of "U" for 7 or 8 inches or to ZERO level if commercial (scaled) manometer is used. It should be kept in mind pressure is read as the **difference** in fluid level in tube legs. Space between levels, high and low, measured in inches, is pressure of gas in inches of water. For conversion purposes, 1.73 inches of water is equal to one ounce of gas pressure. One inch of water lift or displacement is equal to 0.58 ounces of pressure. When testing or adjusting systems, be sure to identify unit of measure used on gas bottle gages or in utility gas mains and convert where necessary.

To test regulator for leaking, proceed as follows:

Refer to Fig. O160. Turn off gas supply valve.

Remove 1/8-inch pipe plug and connect manometer as shown in Fig. O161. Detach carburetor gas hose from regulator outlet.

Open supply valve (plug valve shown in Fig. O160) and quickly cover open regulator outlet with your hand. Observe manometer while alternately covering and uncovering regulator outlet. Pressure reading will be constant if regulator valve is closing properly. If manometer reading drops slightly or fluctuates for each time hand is removed from outlet, turn adjustment screw (G–Fig. O161) located just above inlet line inward, a little at a time, until reading is constant when outlet is repeatedly covered and uncovered.

If regulator does not respond to this lock-off screw adjustment and leaking persists, remove and carefully disassemble regulator body. A repair kit furnished with parts for renewal of both valve and diaphragm of regulator is available.

When regulator tests satisfactorily, close supply valve, remove manometer, bleed air from supply line and replace test-hole pipe plug. Reconnect carburetor and test run engine.

IMPORTANT NOTE: Many test and leak detection procedures in general use for servicing gaseous fueled equipment specify a "soap bubble" test. Do not use to test closing of these highly sensitive demand type regulators. Soap bubble tension alone over regulator outlet is enough to block and shut off this regulator and such a test will prove inaccurate.

REGULATOR ADJUSTMENT. Algas regulator (Fig. O159) is non-adjustable. It operates with an inlet pressure ranging from 6 ounces to 5 psi. It is no longer furnished on production engines. Because a number of units may still require service in the field, parts and overhaul kits are still obtainable. The Algas regulator features an optional solenoid primer as shown in Fig. O159 to assist in starting. It function is to hold regulator open during cranking. One electrical lead connects to starter solenoid switch on starter side; the other lead is grounded. Adjustment, for rapid start of a cold engine, should be performed when engine is hot and regulator, fuel line and carburetor are charged with gas. To adjust, loosen lock

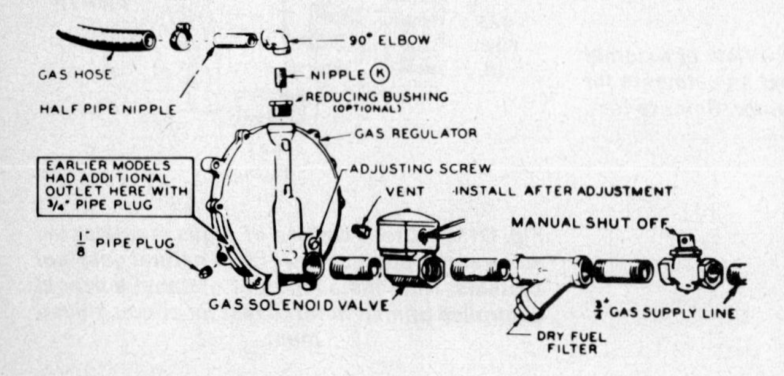

Fig. O160 – Garretson regulator with installation fittings shown in normal arrangement. Gas solenoid valve shown is optional, but may be required by safety code in some areas. Refer to text.

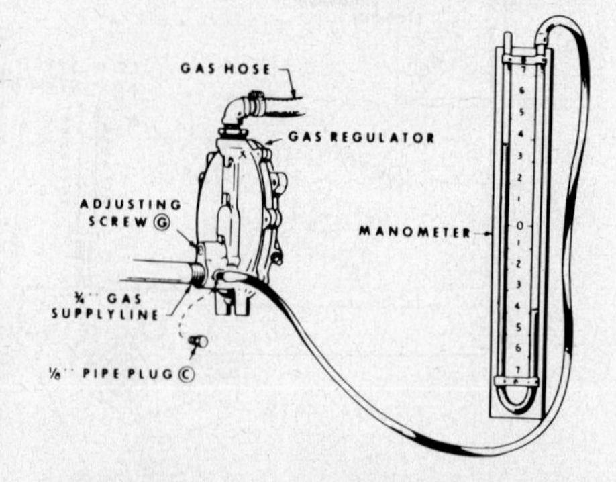

Fig. O161 – Typical connection of water column manometer to regulator. Note pressure is read as difference between tube levels expressed in inches of water. See text.

nut which secures primer stem to regulator cover and turn primer body inward (clockwise) for richer mixture. Primer is correctly set when slightly rough running and somewhat dark exhaust occur briefly when restarting the hot engine.

When a cold engine must be started and a solenoid primer is known to be out of adjustment, make coarse preliminary adjustment as follows:

Remove carburetor hose from regulator outlet and apply temporary battery voltage across solenoid primer leads.

Then, slowly turn primer body inward (clockwise) until valve opens and gas flow can be heard.

Reconnect primer leads in normal fashion and connect gas hose to regulator. If engine starts in 3 seconds or less adjustment is correct. If engine is slow to start, repeat procedure until cold engine with no gas in hose or carburetor can be rapidly started.

If engine cannot be started properly, test primer. To do so, remove primer from regulator cover and connect battery across solenoid leads while observing if primer plunger extends when power is switched on. If plunger is inoperative, check for mechanical interference or sticking in primer body and perform a continuity check to determine if there is an open circuit in windings or solenoid. Renew if defective.

Garretson regulator is shown in Fig. O160. This is the regulator furnished with current production engines. As a "demand type" regulator, it performs no metering function in regard to gas flow but is only open or closed. Maximum allowable inlet pressure is 8 ounces and minimum is 2 ounces. If supply line pressure exceeds 8 ounces (13.8 inches) a primary regulator should be installed in supply line to reduce working pressure. It is generally recognized that two-stage regulation is advisable in most gas systems. An appliance type primary regulator installed ahead of secondary (final) regulator is excellent insurance against unsafe increase of pressure, regardless of fluctuations in pressure from gas supply or source. Such primary regulators should be set to deliver controlled pressure at the same level as operating pressure of secondary regulator. Secondary regulator should be adjusted so it will close off gas flow to carburetor at supply line pressure when there is no demand. This prevents gas leaks (seepage) when engine is not running and provides maximum regulator sensitivity. Factory setting is for operation between 2 and 4 ounces (3½ to 7 inches) of pressure. If pressure in gas supply line is from 4 to 8 ounces, readjust at lockout screw as in Fig. O162. Test of regulator shut-off by use of manometer is as previously covered under REGULATOR TESTING.

Fig. O162 – Adjustment of regulator shut-off of Garretson regulator. Regulator should be set to close gas valve when there is no demand (vacuum) from carburetor. Refer to text.

PRIMARY REGULATORS. When gas source pressures range from 6 ounces to 4 pounds, a pressure-reducing primary regulator is available. Engine model LK uses part number 148P33 which has a maximum gas flow capacity of 190 cubic feet per hour (CFH). CCK series and models JB and NH use No. 148P23 which delivers 330 CFH. For model JC use No. 148P34 which has a capacity of 680 CFH. These regulators all deliver an outlet pressure of 11 inches (water column). These engine models equipped with Garretson regulators in gas line to carburetor accept an inlet pressure of from 2 to 8 ounces (3.5-13.8 inch water column).

LPG VAPORIZER. A vaporizer, sometimes referred to as a converter (converts liquid to vapor) must be used in a system designed for LPG fuel. Its function is to offset cooling by expansion of liquid petroleum gas as it vaporizes from its liquid state. Vaporizer is normally installed in the path of the warm air flow off an air-cooled engine's cooling system to provide warming effect needed.

Fig. O163 shows a cross-section of a typical vaporizer with vaporization process illustrated. Note unit also contains an adjustable pressure regulator for control of outlet gas pressure.

Adjustment procedure proposed by manufacturer recommends vaporizer-regulator be removed from LPG line and fitted to a compressed air source at its inlet port. Seventy five pounds air pressure is adequate. Fit a pressure gage of sufficient range to regulator outlet port and back knurled pressure adjusting screw out nearly all the way. Turn on air pressure and slowly screw adjuster down until test gage reads 7 psi. If pressure rises after being set to specification, regulator valve or diaphragm is leaking and overhaul will be necessary.

NOTES ON GASEOUS FUEL OPERATION. Vibration and dirt contamination are major problems in vapor fuel systems. Use of flexible lines which will not transmit engine vibration or shock to regulators or controls is essential. Fuel filters and strainers as installed or recommended by manufacturer must be used and maintained.

Local safety codes which pertain to dispersal of exhaust fumes and engine heat, venting of regulators and fuel storage areas and location of fuel tanks and service lines in relation to structures must be observed.

Many local codes require an electric shut-off valve in fuel line (see Fig. O160) which is usually a solenoid-operated gate valve so connected as to shut off gas supply when engine is halted. Some safety codes will accept the final regulator as an adequate shut-off valve.

Ensign regulator, though no longer furnished by manufacturer, may still be encountered on older production series engines. If a problem develops, contact local factory branch for advice.

Fire safety and electrical installation codes, NEC and local, must be meticulously observed.

Technical Bulletin, T-015, entitled "Use of Gaseous Fuels with Onan Elec-

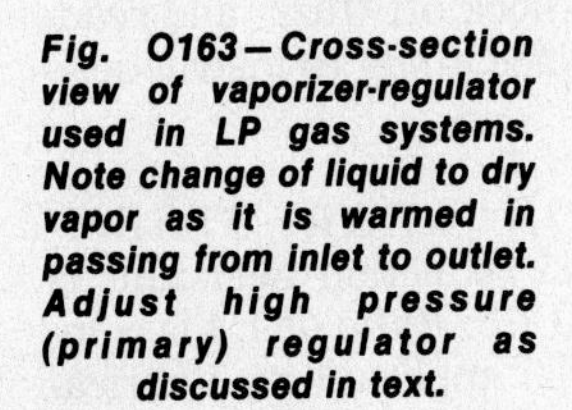

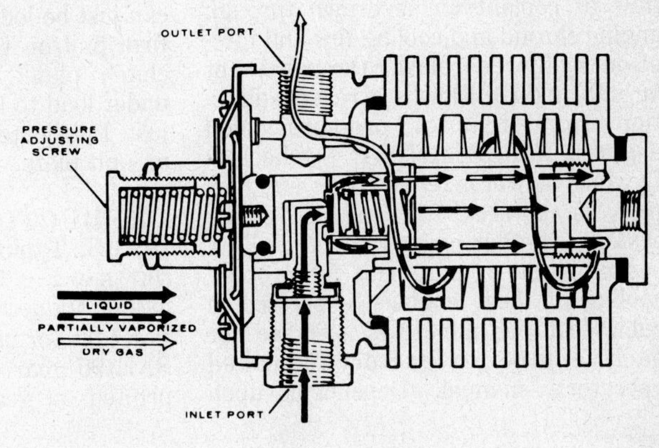

Fig. O163 – Cross-section view of vaporizer-regulator used in LP gas systems. Note change of liquid to dry vapor as it is warmed in passing from inlet to outlet. Adjust high pressure (primary) regulator as discussed in text.

tric Generating Sets" was published by ONAN in August, 1972.

It should be kept in mind when fitting gaseous system fuel line pipe assemblies, strainers or regulators that high grade non-drying pipe joint compound be used on threads. Gaseous fuels have a severe drying effect and use of oil-base paint or other unsatisfactory substitute as a thread-sealant must be avoided or dangerous leaks may result.

AIR COOLING SYSTEMS

PRESSURE COOLING. Most widely used conventional cooling system for these 1 to 4 cylinder models is referred to as pressure cooling. In this system, free air is drawn by flywheel rotation into engine sheet metal housing through flywheel grille opening and is forced through cylinder cooling fins and out through a rear or side aperture.

Larger, J-series engines may be equipped with a thermostat controlled shutter (Vernatherm) which allows engine compartment air to reach 120° F (49° C) before opening and becomes wide open at 140° F (60° C) for full ventilation of enclosure. Opening temperature of sensing element is not adjustable. To determine if this operating element is in working order, remove two screws which retain it to mounting bracket (note slotted holes for adjusting position) and test it by application of heat. Opening should begin at 120° F (49° C) and plunger should be fully extended at 140° F (60° C). Total movement should be at least 13/64-inch (5.16 mm). Reinstall so plunger when fully withdrawn into element body just touches roll pin as in Fig. O164 with shutter completely closed at ambient (free air) temperature.

If shutter operation is unsatisfactory, check for a weak shutter return (closing) spring and examine nylon shutter bearings for dirt or damage. Clean and renew as necessary.

VACU-FLO COOLING. This system is designed for cooling industrial power plants which are installed in a closed compartment. Note in Fig. O165 that flow of coolant air is drawn through engine shroud and cooling fins and forced out by flywheel blower through a vent or outside duct. Flow is in reverse direction from that of pressure-cooled engines. IMPORTANT: If flywheel or flywheel blower is renewed, be sure new part is correct for engine cooling system, whether pressure or Vacu-Flo.

Air volume requirement for proper cooling of these engines, expressed in cubic feet per minute is specified for each engine in factory-furnished operator's manual. Dependent upon engine size, this may range from 300 to 1600 cfm. Duct and vent sizes are detailed for each model and type of cooling system.

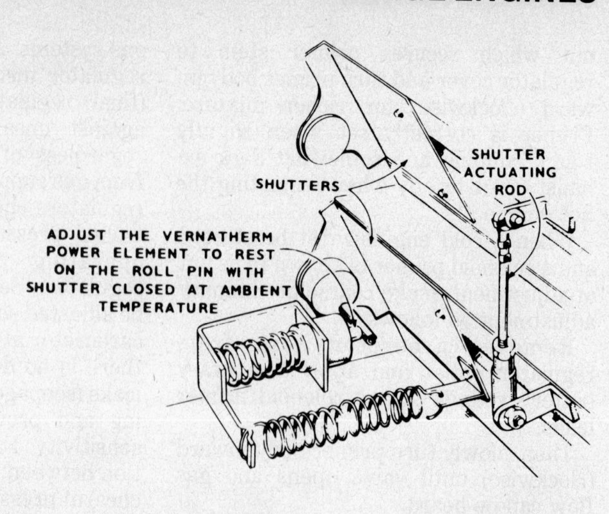

Fig. O164—View of thermostat controlled power shutter for pressure cooled engines. Note adjustment. Refer to text for details.

HIGH TEMPERATURE CUT-OFF. Some larger engine models were equipped with a high temperature safety switch for protection from overheating. Switch is normally closed, but opens to halt engine if compartment air temperature rises to 240° F (116° C) due to problem in cooling system caused by blockage or shutter failure to open. When engine compartment temperature drops to about 190° F (88° C) switch will automatically close and engine can be restarted.

CLUTCH. When optional Rockford clutches are furnished with these engines, an adaptor flange is fitted to engine output shaft for mounting clutch unit and a variety of housings are used dependent upon application and model of engine or clutch used. Refer to Fig. O172 for guidance in adjustment and proceed as follows:

Remove plate from top of housing and rotate engine manually until lock screw (1–Fig. O172) is at top of ring (2) as shown. Loosen lock screw and turn adjusting ring clockwise (as facing through clutch toward engine) until toggles cannot be locked over center. Then, turn ring in reverse direction until toggles can just be locked over center by a very firm pull on operating lever. If a new clutch plate has been installed, slip under load to knock off "fuzz" and readjust. Lubricate according to instructions on unit plate.

REDUCTION GEAR ASSEMBLIES. Typical reduction gear unit is shown in Fig. O173. Ratio of 1:4 is common in industrial applications. Lubrication calls for use of SAE 50 motor oil or SAE 90 gear oil. Refer to instructions printed on gear case for guidance. In most cases, a total of six plugs are fitted into case for lubricant fill or level check. Plug openings to be used are determined by positioning of gear box in relation to horizontal or vertical. It is recommended that square plug heads be cut off those plugs not to be used to fill, check

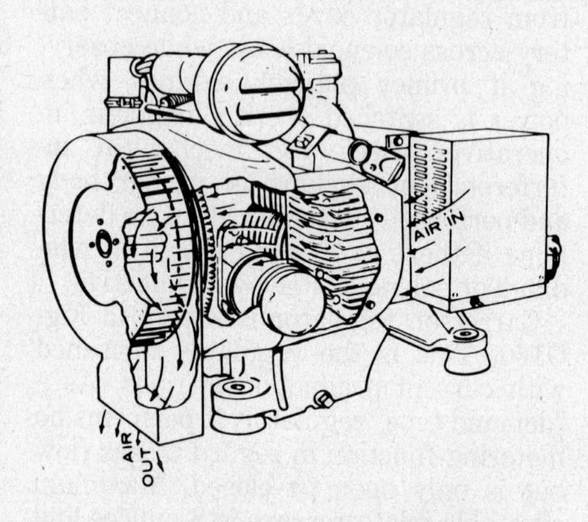

Fig. O165—Typical cooling air flow in Vacu-Flo system used for closed compartment installation.

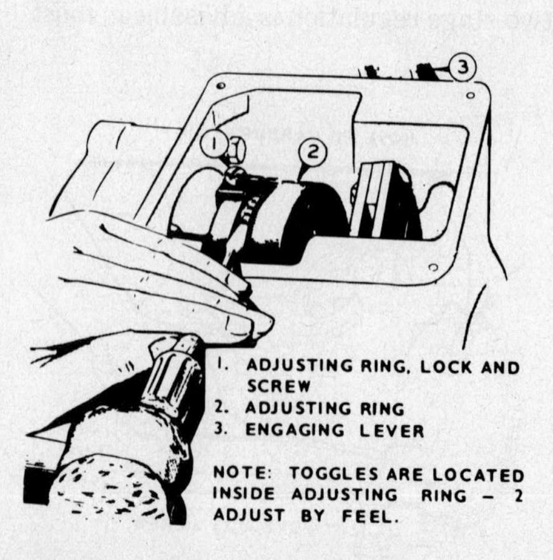

Fig. O172—Procedure for adjustment of Rockford clutch.

1. Ring lock & screw
2. Adjuster ring
3. Clutch lever

or drain so as to eliminate chance of error by overfill or underfill. All parts shown are available for renewal if needed in overhaul.

NOTE: In some installations, no shaft seal is fitted between engine crankcase and reduction gear housing. In these cases, with a common oil supply, engine oil lubricates gears and bearings of reduction gear unit and gear oil is not used. Be sure to check nameplate or operator's manual.

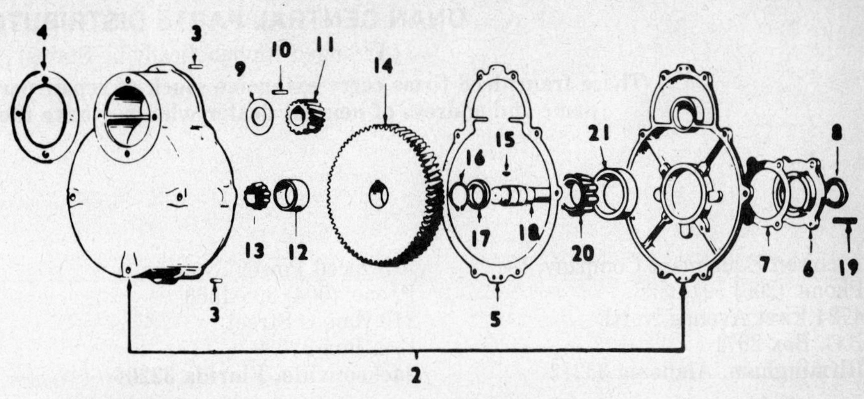

Fig. O173—Exploded view of typical reduction gear set. See text for service details.

2. Housing & cover
3. Dowel pins (2)
4. Gasket (engine)
5. Cover gasket
6. Bearing retainer
7. Shims
8. Oil seal
9. Pinion washer
10. Pinion key
11. Pinion gear
12. Bearing cup
13. Bearing cone
14. Driven gear
15. Gear key
16. Snap ring
17. Bearing spacer
18. Shaft
19. Key
20. Bearing cone
21. Bearing cup

ONAN SERVICE TOOLS

Following special tool list indicates ONAN part number, tool name, use and application.

Refer to list of **ONAN CENTRAL WAREHOUSE DISTRIBUTORS** as a source of these tools.

Valve tools,

420-0311-VALVE SEAT REMOVER ....J Series, NH, NHA, NHB, NHC, NB, N52, T-260G

420-0274-CUTTER BLADE..............For 420-0311 Seat Remover

420-0349-SEAT CUTTER ....All Series

420-0351-CUTTER BLADE..............For 420-0349 Seat Cutter

420-0071-VALVE SEAT DRIVER........ACK, CCK, CCKB, CK, LK, LKB, BF, BG, B43, B48

420-0270-VALVE SEAT DRIVER ....J Series

420-0308-VALVE SEAT DRIVER ....NH, NHA, NHB, NB, N52, T-260G

420-0310-VALVE SEAT STAKER.............(For Intake Valve) NB, NHA, NH, NHB, NHC, N52, T-260G

420-0309-VALVE SEAT STAKER ....(For Exhaust Valve) NB, NHA, NH, NHB, NHC, N52, T-260G

420-0305-VALVE GUIDE HONE...................All Series

420-0363-REPLACEMENT STONES (5/16 and 11/32-inch) ...For 420-0305 Valve Guide Hone

420-0364-REPLACEMENT STONES (3/8, 13/32 and 7/16-inch ....For 420-0305 Valve Guide Hone

420-0300-VALVE GUIDE DRIVER ....CCK, CCKB, LK, LKB, NH, NHA, NHB, NHC, N52, T-260G, NB, BF, BG, B43, B48, J Series

Oil Seal Guides And Drivers,

420-0387-BEARING PLATE ....CK, ACK, CCK, CCKB, LK, AJ, AK, NH, NHA, NHB, B48, T-260G

420-0250-BEARING PLATE ....J Series

420-0389-GEAR COVER......J Series

420-0313-GEAR COVER ....NH, NHA, NHB, NHC, NB, N52, T-260G, CCK, CCKB, BF, BG, B43, B48

Pullers,

420-0100-FLYWHEEL PULLER..................J Series BH, CK, LK, LKB, CCK, CCKB, NH, NHA, NHB, NHC, N52, NB, BF, BG, B43, B48, T-260G

Wrenches,

420-0169-CARBURETOR ADJUSTING WRENCH ....BH, CK, CCK

420-0294-CARBURETOR ADJUSTING WRENCH ....J Series, CCKB, NH, NHA, NHB, NHC, NB, CCK, T-260G

ONAN CENTRAL PARTS DISTRIBUTORS

(Arranged Alphabetically by States)

These franchised firms carry extensive stock of repair parts. Contact them for name and address of nearest dealer who may have the parts you need.

Atchison Equipment Company, Inc.
Phone: (205) 591-2328
4724 First Avenue North
P.O. Box 2971
Birmingham, Alabama 35212

Delhomme Industries
Phone: (205) 473-6626
3422 Georgia Pacific Avenue
Mobile, Alabama 36607

Fremont Electric Company
Phone: (907) 277-9558
140 East Dowling Road
Anchorage, Alaska 99502

Harrison Industries of Arizona Inc.
Phone: (602) 243-7222
3502 East Broadway Road
Phoenix, Arizona 85040

Mecelec of Arkansas
Phone: (501) 490-1801
1701 Dixon Road
P.O. Box 9297
Little Rock, Arkansas 72219

Equipment Service Company
Phone: (213) 426-0311
3431 Cherry Avenue
P.O. Box 1307
Long Beach, California 90801

Cal-West Electric, Inc.
Phone: (916) 372-5522
3939 West Capitol Avenue
West Sacramento, California 95691

Cal-West Electric, Inc.
Phone: (415) 873-7710
1341 San Mateo Avenue
P.O. Box 2364
San Francisco (South), California 94080

Equipment Service Company
Phone: (714) 562-2804
1954 Friendship Drive
El Cajon, California 92020

C.W. Silver Company, Inc.
Phone: (303) 399-7440
3945 East 50th Avenue
Denver, Colorado 80216

GLT Industries, Inc.
Phone: (203) 528-9944
29 Mascolo Road
P.O. Box 307
South Windsor, Connecticut 06074

Advanced Power Systems
Phone: (904) 355-4563
710 Haines Street
P.O. Box 38039
Jacksonville, Florida 32206

R. B. Grove, Inc.
Phone: (305) 854-5420
261 South West 6th Street
Miami, Florida 33130

Tampa Armature Works, Inc.
Phone: (305) 843-8250
3400 Bartlett Boulevard
Orlando, Florida 32805

Tampa Armature Works, Inc.
Phone: (813) 621-5661
440 South 78th Street
P.O. Box 3381
Tampa, Florida 33601

Blalock Machinery & Equipment Company
Phone: (912) 436-1507
700 South Westover Road
Albany, Georgia 31702

Blalock Machinery & Equipment Company
Phone: (404) 766-2632
5112 Blalock Industrial Boulevard
College Park, Georgia 30349

Atlas Electric Company, Inc.
Phone: (808) 524-5866
1151 Mapunapuna Street
Honolulu, Hawaii 96819

Power Systems, Division of E. C. Distributing
Phone: (208) 342-6541
4499 Market Street
Boise, Idaho 83705

Power Systems, Division of E. C. Distributing
Phone: (208) 234-2442
1060 South Main
Pocatello, Idaho 83204

Stannard Power Equipment Company
Phone: (312) 597-5500
4901 West 128th Place
Alsip, Illinois 60658

Service Automotive Warehouse, Inc.
Phone: (309) 794-0400
111 4th Avenue
Rock Island, Illinois 61201

Meco-Indiana, Inc.
Phone: (219) 262-4611
23900 County Road 6
Elkhart, Indiana 46514

Evansville Auto Parts, Inc.
Phone: (812) 425-8264
5 East Riverside Drive
Evansville, Indiana 47713

Meco-Indiana, Inc.
Phone: (317) 873-5005
5005 West 106th Street
Zionsville, Indiana 46077

Midwestern Power Products Company
Phone: (515) 278-5521
10100 Dennis Drive
Des Moines, Iowa 50322

Anderson Equipment Company, Inc.
Phone: (712) 255-8033
300 South Virginia Street
Sioux City, Iowa 51102

Mecelec of Kansas
Phone: (316) 522-4767
4631 Palisade Street
Wichita, Kansas 67217

Southern Power Systems
Phone: (502) 459-5060
2025 Old Shepherdsville Road
Louisville, Kentucky 40218

Delhomme Industries, Inc.
Phone: (318) 234-9837
337 Mecca Drive
Lafayette, Louisiana 70505

Delhomme Industries, Inc.
Phone: (318) 439-9700
2506 Elaine Street
Lake Charles, Louisiana 70601

Delhomme Industries, Inc.
Phone: (318) 365-5476
Northside Road
P.O. Box 266
New Iberia, Louisiana 70560

Menge Pump and Machinery Company, Inc.
Phone: (504) 888-8830
2740 North Arnoult
P.O. Box 8210
New-Orleans-Metairie, Louisiana 70011

Menge Pump & Machinery Company, Inc.
Phone: (318) 222-5781
1510 Grimmet Drive
P.O. Box 7323
Shreveport, Louisiana 71107

The Leen Company
Phone: (207) 989-7363
54 Wilson Street
Brewer, Maine 04412

The Leen Company
Phone: (207) 774-6266
366 West Commercial Street
Portland, Maine 04101

Curtis Engine & Equipment Company
Phone: (301) 633-5161
6120 Holabird Avenue
Baltimore, Maryland 21224

New England Engine Corporation
Phone: (617) 948-7331
RR 1
Rowley, Massachusetts 01962

Standby Power Inc.
Phone: (616) 949-7990
2745 South East 29th Street
Grand Rapids, Michigan 49508

Stanby Power Inc.
Phone: (313) 348-6400
4700 12 Mile Road
Novi, Michigan 48050

Flaherty Equipment Corporation
Phone: (612) 338-8796
2525 East Franklin Avenue
Minneapolis, Minnesota 55406

Interstate Detroit Diesel Allison Inc.
Phone: (612) 854-5511
2501 East 80th Street
Minneapolis, Minnesota 55420

Northstar Detroit Diesel Allison Inc.
Phone: (218) 749-4484
1921 West 16th Avenue
Virginia, Minnesota 55792

Menge Pump & Power Systems Division
Phone: (601) 969-9333
1327 South Gallatin
Jackson, Mississippi 39202

National Industrial Supply Company
Phone: (314) 621-0350
1100 Martin Luther King Drive
St. Louis, Missouri 62201

Comet Industries, Inc.
Phone: (816) 245-9400
4800 Deramus Avenue
Kansas City, Missouri 64120

Anderson Equipment Company, Inc.
Phone: (402) 558-1200
5532 Center Street
Omaha, Nebraska 68106

Anderson Industrial Engines
Phone: (402) 558-8700
2123 South 56th Street
Omaha, Nebraska 68106

Equipment Service Company of Nevada
Phone: (702) 382-3852
1916 Highland Avenue
Las Vegas, Nevada 89102

R. C. Equipment Company, Inc.
Phone: (609) 742-0220
522 South Broadway
Gloucester City, New Jersey 08030

GLT Industries, Inc.
Phone: (201) 767-9751
411 Clinton Avenue
Northvale, New Jersey 07647

C. W. Silver Company, Inc.
Phone: (505) 881-2454
4812 North East Jefferson
Albuquerque, New Mexico 87109

Power Plant Equipment Corporation
Phone: (518) 783-1991
6 Northway Lane
Latham, New York 12110

Ronco Communications & Electronics
Phone: (716) 424-3890
230 Metro Park
Rochester, New York 14623

Power Plant Equipment Corporation
Phone: (315) 475-7251
929 South Salina Street
Syracuse, New York 13202

Ronco Communications & Electronics, Inc.
Phone: (716) 873-0760
595 Sheridan Drive
Tonawanda-Buffalo, New York 14150

Owsley & Sons
Phone: (919) 668-2454
Interstate 40
P.O. Box 8627
Greensboro, North Carolina 27410

Interstate Detroit Diesel Allison, Inc.
Phone: (701) 258-2303
3801 Miriam Gateway
Bismarck, North Dakota 58501

Interstate Detroit Diesel Allison, Inc.
Phone: (701) 282-6556
3902 North 12th Avenue
Fargo, North Dakota 58102

Southern Ohio Power Systems, Inc.
Phone: (513) 821-6305
8148 Vine Street
Cincinnati, Ohio 45216

McDonald Equipment Company
Phone: (216) 951-8222
37200 Vine Street-Willoughby
Cleveland, Ohio 44094

Tuller Corporation
Phone: (614) 224-8246
947 West Goodale Boulevard
Columbus, Ohio 43212

G & R Equipment Company
Phone: (405) 685-5534
3826 Newcastle Road
Oklahoma City, Oklahoma 73119

Mechanical & Electrical Equipment Company
Phone: (918) 582-7777
712 Wheeling
P.O. Box 50323
Whittier Station
Tulsa, Oklahoma 74150

EC Distributing Division Electrical Construction
Phone: (503) 224-3623
2122 North West Thurman Street
Portland, Oregon 97208

A. F. Shane Company
Phone: (412) 781-8000
654 Alpha Drive RIDC Industrial Park
Pittsburgh, Pennsylvania 15238

Winter Engine Generator Service
Phone: (717) 848-3777
1600 Pennsylvania Avenue
York, Pennsylvania 17404

Owsley & Sons Inc.
Phone: (803) 548-3636
I-77 & SC Exit 72
Gold Hill Road
Fort Mill, South Carolina 29715

Hobbs Equipment Company
Phone: (615) 894-8400
6203 Provence Street
Chattanooga, Tennessee 37421

Hobbs Equipment Company
Phone: (615) 966-7550
10625 Lexington Drive
Knoxville, Tennessee 37922

Maritime & Industrial, Inc.
Phone: (901) 775-1204
P.O. Box 9397
292 East Mallory
Memphis, Tennessee 38109

Hobbs Equipment
Phone: (615) 244-4933
1327 Foster Avenue
Nashville, Tennessee 37211

Lightbourn Equipment Company
Phone: (214) 233-5151
P.O. Box 401870
13649 Beta Road
Dallas, Texas 75240

Harrison Equipment Company
Phone: (713) 879-2600
1100 West Airport Boulevard
Houston, Texas 77001

Power Support Systems
Phone: (915) 332-1429
P.O. Box 4417
3208 Kermit Highway
Odessa, Texas 79760

Lightbourn Equipment Company
Phone: (512) 333-7542
4260 Dividend Drive
San Antonio, Texas 78219

C. W. Silver Company, Inc.
Phone: (801) 355-5373
550 West 7th South
Salt Lake City, Utah 84101

T & L Electric Company
Phone: (802) 295-3114
Sykes Avenue
White River Junction, Vermont 05001

Curtis Engine & Equipment
Phone: (804) 627-9470
1114 Ballentine Boulevard
Norfolk, Virginia 23504

Owsley & Sons
Phone: (804) 275-2603
10300 Jefferson
Davis Highway
P.O. Box 34508
Richmond, Virginia 23234

Fremont Electric Company
Phone: (206) 633-2323
P.O. Box 31640
744 North 34th Street
Seattle, Washington 98103

Lay & Nord
Phone: (509) 453-5591
P.O. Box 472
511 South 3rd Street
Yakima, Washington 98901

Call Detroit Diesel Allison, Inc.
Phone: (304) 744-1511
P.O. Box 8245
Charleston Ordinance Center
Charleston, West Virginia 25303

Inland Diesel, Inc.
Phone: (414) 781-7100
13015 Custer Avenue
Butler, Wisconsin 53007

Morley-Murphy
Phone: (414) 499-3171
Box 3640
700 Morley Road
Green Bay, Wisconsin 54303

Clymar, Inc.
Phone: (414) 781-0700
N55 W13787 Oak Lane
Menomenee Falls
Milwaukee, Wisconsin 53051

Power Service Company
Phone: (307) 237-3773
P.O. Box 2880
5201 West Yellowstone Highway
Casper, Wyoming 82606

CANADIAN DISTRIBUTORS

Calgary, Alberta T2H 1J5
Simson-Maxwell
6447-2nd Street S.E.

South Edmonton, Alberta
T6H 1E7
Simson-Maxwell
Box 4446
10375 59th Avenue

Vancouver V6H 1A7, B.C.
Simson-Maxwell
1380 W. 6th Avenue

Kenora, Ontario
L.O.W.E. Power Systems, Ltd.
P.O. Box 81

Oakville, Ontario L6K 2H2
Hawker Siddeley Diesels & Electric
355 Wyecroft Road

Ottawa Ontario K1B 3V7
J.A. Faguy & Sons, Ltd.
2544 Sheffield Road

Thunder Bay, Ontario, P7B 5M5
Lake of the Woods Electric Ltd.
1177 Roland Street
Postal Station F

Montreal, Quebec H4T 1P3
J.A. Faguy & Sons, Ltd.
750 Montee de Liesse

Saskatoon, Saskatchewan S7K 1L2
Pryme Power & Diesels Ltd.
5-2949 3rd Avenue North
Phone: (306) 665-8044

Prince George, B.C.
Simson-Maxwell
729 4th Avenue

Winnipeg, Manitoba R3B 0A6
Kipp Kelly, Ltd.
68 Higgins Avenue

Fredericton, New Brunswick
Sansom Equipment, Ltd.
Woodstock Rd., P.O. Box 1263

Truro, Nova Scotia
Sansom Equipment Ltd.
P.O. Box 152
80 Glenwood Drive

Sault Ste. Marie, Ontario
Algonma Truck & Tractor Sales
815 Great Northern Road

Scarborough, Ontario
Total Power, Ltd.
330 Nantucket Boulevard

TECUMSEH

TECUMSEH PRODUCTS COMPANY
Grafton, Wisconsin 53024

TECUMSEH ENGINE IDENTIFICATION INFORMATION

In order to obtain correct service parts or service information it is necessary to correctly identify engine model and locate model, specification and serial number.

Model and serial number are stamped in blower housing at cylinder head end or on a nameplate or tag located on side of blower housing.

Letters at beginning of model number indicate basic engine type.

V-Indicates vertical shaft
VM-Indicates vertical shaft, medium frame
LAV-Indicates lightweight aluminum, vertical shaft
VH-Indicates vertical shaft, heavy duty (cast iron)
TVS-Indicates Tecumseh vertical styled
TNT-Indicates Toro N' Tecumseh
ECV-Indicates exlusive Craftsman vertical
H-Indicates horizontal shaft
HS-Indicates horizontal shaft, small frame
HM-Indicates horizontal shaft, medium frame
HHM-Indicates horizontal shaft heavy duty, medium frame
HH-Indicates horizontal heavy duty (cast iron)
ECH-Indicates exclusive Craftsman horizontal

The number following the letters indicate horsepower or cubic inch displacement.

The number following the model number is the specification number and the last three numbers of specification number indicate a variation to basic engine specification.

The serial number indicates production date.

Using model TVS90-43056A, serial number 831OC as an example, interpretation is as follows:

TVS-Indicates Tecumseh vertical styled
90-Indicates 9 cubic inch displacement
43056A-Indicates specification number used for correctly identifying engine parts

8310C is the serial number.

8-Indicates year of manufacture (1978)
310-Indicates calendar day of year of manufacture (310th day or November 6, 1978)
C-Indicates line and shift on which engine was built at factory

Medium frame engine have aluminum blocks with cast iron sleeves. Heavy frame engines have cast iron cylinder and block assemblies.

Early VH70 and HH70 model engines were identified as V70 and H70 models.

It is necessary to have correct model, specification and serial numbers to obtain parts or service material.

TECUMSEH

TECUMSEH PRODUCTS COMPANY
Grafton, Wisconsin 53024

Model	No. Cyls.	Bore	Stroke	Displacement	Power Rating
VM70*	1	3.062 in. (77.8 mm)	2.531 in. (64.3 mm)	18.65 cu. in. (305.7 cc)	7 hp. (5.2 kW)
VM70**	1	2.94 in. (74.61 mm)	2.531 in. (64.3 mm)	17.16 cu. in. (281 cc)	7 hp. (5.2 kW)
HH70	1	2.75 in. (69.9 mm)	2.531 in. (64.3 mm)	15 cu. in. (246.8 cc)	7 hp. (5.2 kW)
VH70	1	2.75 in. (69.9 mm)	2.531 in. (64.3 mm)	15 cu. in. (246.8 cc)	7 hp. (5.2 kW)
HM70*	1	3.062 in. (77.8 mm)	2.531 in. (64.3 mm)	18.65 cu. in. (305.7 cc)	7 hp. (5.2 kW)
HM70**	1	2.94 in. (74.61 mm)	2.531 in. (64.3 mm)	17.16 cu. in. (281 cc)	7 hp. (5.2 kW)
TVM170	1	2.94 in. (74.61 mm)	2.531 in. (64.3 mm)	17.16 cu. in. (281 cc)	7 hp. (5.2 kW)
V80	1	3.062 in. (77.8 mm)	2.531 in. (64.3 mm)	18.65 cu. in. (305.7 cc)	8 hp. (6 kW)
VM80***	1	3.062 in. (77.8 mm)	2.531 in. (64.3 mm)	18.65 cu. in. (305.7 cc)	8 hp. (6 kW)
VM80****	1	3.125 in. (79.38 mm)	2.531 in. (64.3 mm)	19.4 cu. in. (318 cc)	8 hp. (6 kW)
HM80***	1	3.062 in. (77.8 mm)	2.531 in. (64.3 mm)	18.65 cu. in. (305.7 cc)	8 hp. (6 kW)
HM80****	1	3.125 in. (79.38 mm)	2.531 in. (64.3 mm)	19.4 cu. in. (318 cc)	8 hp. (6 kW)
HH80†	1	3.00 in. (76.2 mm)	2.75 in. (69.9 mm)	19.4 cu. in. (318.7 cc)	8 hp. (6 kW)
HH80††	1	3.313 in. (84.15 mm)	2.75 in. (69.9 mm)	23.75 cu. in. (388.6 cc)	8 hp. (6 kW)
VH80	1	3.313 in. (84.15 mm)	2.75 in. (69.9 mm)	23.75 cu. in. (388.6 cc)	8 hp. (6 kW)
TVM195	1	3.125 in. (79.38 mm)	2.531 in. (64.3 mm)	19.4 cu. in. (318 cc)	8 hp. (6 kW)
HM100†††	1	3.187 in. (80.95 mm)	2.531 in. (64.3 mm)	20.2 cu. in. (330.9 cc)	10 hp. (7.5 kW)
HM100††††	1	3.313 in. (84.15 mm)	2.531 in. (64.3 mm)	21.82 cu. in. (357.6 cc)	10 hp. (7.5 kW)
VM100	1	3.187 in. (80.95 mm)	2.531 in. (64.3 mm)	20.2 cu. in. (330.9 cc)	10 hp. (7.5 kW)
VH100	1	3.313 in. (84.15 mm)	2.75 in. (69.9 mm)	23.75 cu. in. (388.6 cc)	10 hp. (7.5 kW)
HH100	1	3.313 in. (84.15 mm)	2.75 in. (69.9 mm)	23.75 cu. in. (388.6 cc)	10 hp. (7.5 kW)
TVM220	1	3.313 in. (84.15 mm)	2.531 in. (64.3 mm)	21.82 cu. in. (357.6 cc)	10 hp. (7.5 kW)
HH120	1	3.5 in. (88.9 mm)	2.875 in. (72.9 mm)	27.66 cu. in. (452.5 cc)	12 hp. (9 kW)

* VM70 or HM70 models prior to type letter A
** VM70 or HM70 models, type letter A and after
*** VM80 or HM80 models prior to type letter E
**** VM80 or HM80 models, type letter E and after
† Prior to type letter B
†† Type letter B and after
††† Early HM100 models
†††† Late HM100 models

Medium and heavy frame, horizontal or vertical crankshaft, four cycle, one cylinder engines are covered in this section.

Care must be taken to correctly identify engine model as outlined in **TECUMSEH ENGINE IDENTIFICATION INFORMATION** section.

Connecting rod for all models rides directly on crankshaft crankpin journal. Lubrication is provided by splash lubrication for some models or a barrel and plunger type oil pump is used on some models.

A variety of ignition systems, magneto, battery or solid state, have been used. Refer to appropriate paragraph for model being serviced.

Tecumseh or Walbro float type carburetor is used according to model and application.

Refer to **TECUMSEH ENGINE IDENTIFICATION INFORMATION** section for correct engine identification and always give complete engine model, specification and serial numbers when ordering parts or service material.

MAINTENANCE

SPARK PLUG. Recommended spark plug is Champion J8 or equivalent. Electrode gap is 0.030 inch (0.762 mm) for all models.

CARBURETOR. Either a Tecumseh or Walbro float type carburetor is used. Refer to appropriate paragraph for model being serviced.

TECUMSEH CARBURETOR. Refer to Fig. T1 for exploded view of Tecumseh float type carburetor and location of fuel mixture adjustment screws.

For initial carburetor adjustment for VM70, VM80, VM100, HM70, HM80 and HM100 models, open idle mixture and main fuel mixture screws 1½ turns each.

For initial carburetor adjustment for VH70 model, open idle mixture screw 1 turn and main fuel mixture screw 1¼ turns.

For initial carburetor adjustment for HH80, HH100, HH120 and VH100 models, open idle mixture screw 1¼ turns and main fuel mixture screw 1¾ turns.

Make final adjustment on all models with engine at normal operating temperature and running. Place engine under load and adjust main fuel mixture screw for leanest mixture that will allow satisfactory acceleration and steady governor operation. Set engine at idle speed, no load and adjust idle mixture screw to obtain smoothest idle operation.

As each adjustment affects the other, adjustment procedure may have to be repeated.

To check float level, invert carburetor throttle body and float assembly. Float setting should be 7/32-inch (5.556 mm) measured between free end of float and rim on carburetor body.

Adjust float by carefully bending float lever tang that contacts inlet valve.

Refer to Fig. T1 for exploded view of carburetor during disassembly. When reinstalling Viton inlet valve seat, grooved side of seat must be installed in bore first so inlet valve will seat against smooth side (Fig. T2). Some later models have a Viton tipped inlet needle (Fig. T3) and a brass seat.

Install throttle plate (2–Fig. T1) with the two stamped lines facing out and at 12 and 3 o'clock position. The 12 o'clock line should be parallel to throttle shaft and to top of carburetor.

Install choke plate (10) with flat side towards bottom of carburetor.

Fuel fitting (8) is pressed into body. When installing fuel inlet fitting, start fitting into bore; then, apply a light coat of "Loctite" (grade A) to shank and press fitting into place.

WALBRO CARBURETOR. Refer to Fig. T4 for exploded view of Walbro float type carburetor and location of fuel mixture adjustment screws.

For initial carburetor adjustment for VM80 and HM80 models, open idle mixture screw 1¾ turns and main fuel mixture screw 2 turns.

For initial carburetor adjustment for all remaining models, open idle mixture and main fuel mixture adjustment screws 1 turn each.

Make final adjustment on all models with engine at normal operating temperature and running. Place engine under load and adjust main fuel mixture screw for leanest mixture that will allow satisfactory acceleration and steady governor operation. Set engine at idle speed, no load and adjust idle mixture screw to obtain smoothest idle operation.

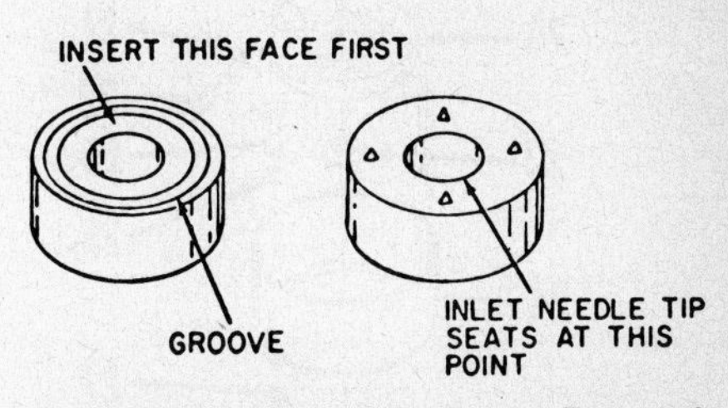

Fig. T2—Viton seat used on some Tecumseh carburetors must be installed correctly to operate properly. All metal needle is used with seat shown.

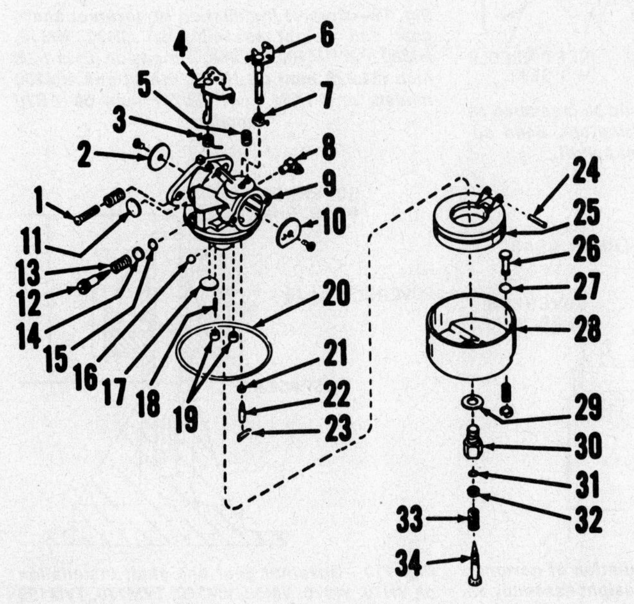

Fig. T1—Exploded view of Tecumseh carburetor.

1. Idle speed screw
2. Throttle plate
3. Return spring
4. Throttle shaft
5. Choke stop spring
6. Choke shaft
7. Return spring
8. Fuel inlet fitting
9. Carburetor body
10. Choke plate
11. Welch plug
12. Idle mixture needle
13. Spring
14. Washer
15. "O" ring
16. Ball plug
17. Welch plug
18. Pin
19. Cup plugs
20. Bowl gasket
21. Inlet needle seat
22. Inlet needle
23. Clip
24. Float shaft
25. Float
26. Drain stem
27. Gasket
28. Bowl
29. Gasket
30. Bowl retainer
31. "O" ring
32. Washer
33. Spring
34. Main fuel needle

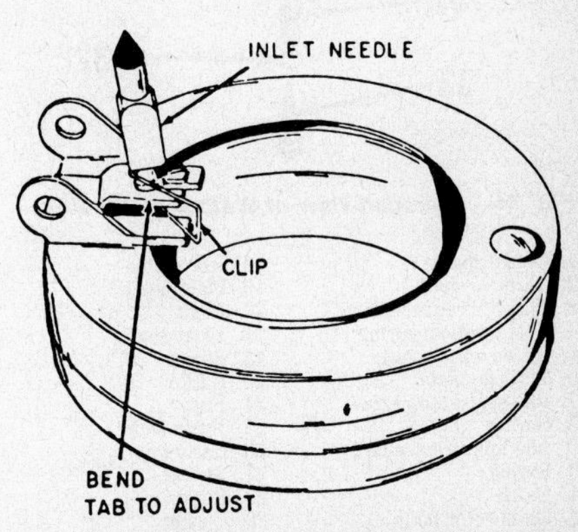

Fig. T3—View of float and fuel inlet valve needle. Valve needle shown is equipped with resilient tip and a clip. Bend tab shown to adjust float level.

As each adjustment affects the other, adjustment procedure may have to be repeated.

To check float level, invert carburetor throttle body and float assembly (Fig. T5). Float setting, dimension (H) should be 1/8-inch (3.175 mm) for horizontal crankshaft engines and 3/32-inch (2.381 mm) for vertical crankshaft engines.

Adjust float by carefully bending float lever tang that contacts inlet valve.

Float drop (travel) for all models should be 9/16-inch (14.288 mm) and is adjusted by carefully bending limiting tab on float.

Fig. T4 — Exploded view of Walbro carburetor.

1. Choke shaft
2. Throttle shaft
3. Throttle return spring
4. Choke return spring
5. Choke stop spring
6. Throttle plate
7. Idle speed stop screw
8. Spring
9. Idle mixture needle
10. Spring
11. Baffle
12. Carburetor body
13. Choke plate
14. Bowl gasket
15. Gasket
16. Inlet valve seat
17. Spring
18. Inlet valve
19. Main nozzle
20. Float
21. Float shaft
22. Spring
23. Gasket
24. Bowl
25. Drain stem
26. Gasket
27. Spring
28. Retainer
29. Gasket
30. Bowl retainer
31. Spring
32. "O" ring
33. Main fuel adjusting needle

GOVERNOR. A mechanical flyweight type governor is used on all models. Governor weight and gear assembly is driven by camshaft gear and rides on a renewable shaft which is pressed into engine crankcase or crankcase cover.

If renewal of governor shaft is necessary, press governor shaft in until shaft end is located as shown in Fig. T7, T8, T9 or T10.

To adjust governor lever position on vertical crankshaft models, refer to Fig. T11. Loosen clamp screw on governor lever. Rotate governor lever shaft counter-clockwise as far as possible. Move governor lever to left until throttle is fully open, then tighten clamp screw.

On horizontal crankshaft models, loosen clamp screw on lever, rotate governor lever shaft clockwise as far as possible. See Fig. T12. Move governor lever clockwise until throttle is wide open, then tighten clamp screw.

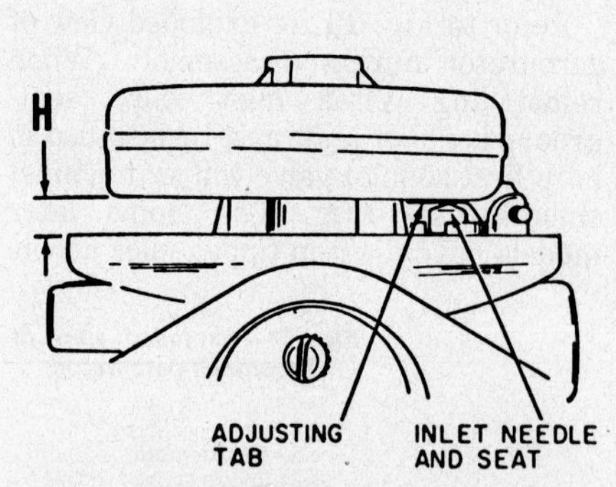

Fig. T5 — Float height (H) should be measured as shown on Walbro float carburetors. Bend adjusting tab to adjust height.

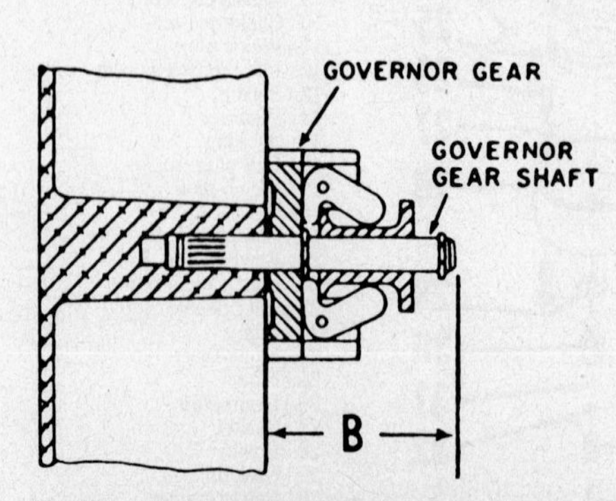

Fig. T7 — View showing installation of governor shaft and governor gear and weight assembly on HH80, HH100 and HH120 models. Dimension (B) is 1 inch (25.4 mm).

For external linkage adjustments, refer to Figs. T13 and T14. Loosen screw (A), turn plate (B) counter-clockwise as far as possible and move lever (C) to left until throttle is fully open. Tighten screw (A). Governor spring must be hooked in hole (D) as shown. Adjusting screws on bracket shown in Figs. T13 and T14 are used to adjust fixed or variable speed settings.

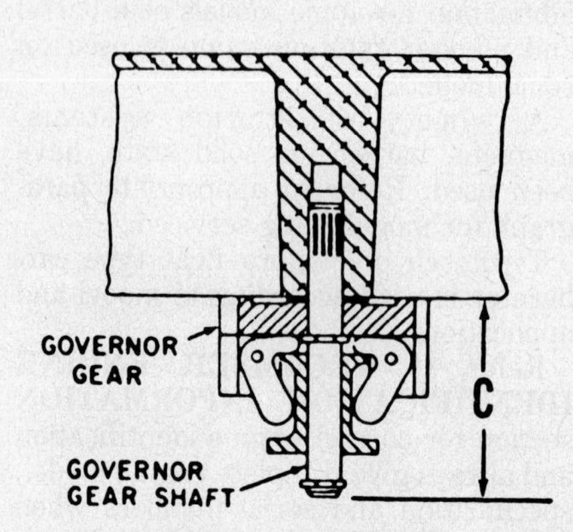

Fig. T8 — Governor gear and shaft installation on VH80 and VH100 models. Dimension (C) is 1 inch (25.4 mm).

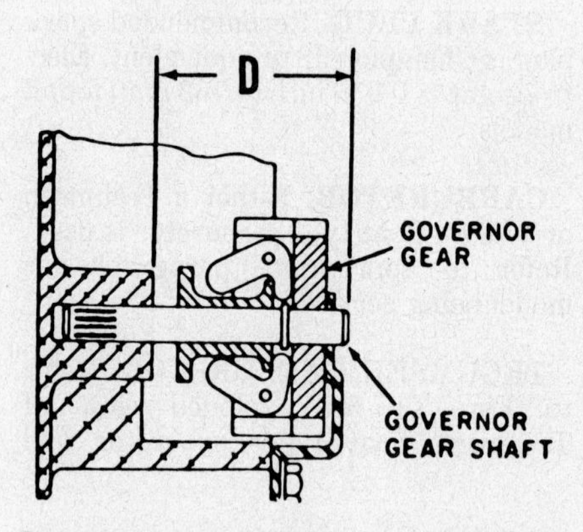

Fig. T9 — Correct installation of governor shaft, gear and weight assembly on HH70, HM70, HM80 and HM100 models. Dimension (D) is 1-3/8 inch (34.925 mm) on HM70, HM80 and HM100 models or 1-17/64 inch (32.147 mm) on HH70 model.

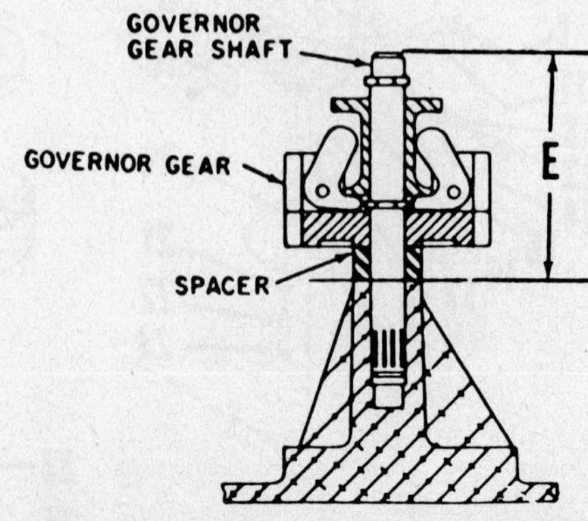

Fig. T10 — Governor gear and shaft installation on VH70, VM70, VM80, VM100, TVM170, TVM195 and TVM220 models. Dimension (E) is 1-19/32 inch (40.481 mm).

IGNITION SYSTEM. A variety of ignition systems, magneto, battery or solid state, have been used. Refer to appropriate paragraph for model being serviced.

MAGNETO IGNITION. Tecumseh flywheel type magnetos are used on some models. On VM70, HM70, VM80, HM80, VM100, HM100, HH70 and VH70 models, breaker points are enclosed by the flywheel. Breaker point gap must be adjusted to 0.020 inch (0.508 mm). Timing is correct when timing mark on stator plate is in line with mark on bearing plate as shown in Fig. T15. If timing marks are defaced, points should start to open when piston is 0.085-0.095 inch (2.159-2.413 mm) BTDC.

Breaker points on HH80, VH80, HH100, VH100 and HH120 models are located in crankcase cover as shown in Fig. T16. Timing should be correct when points are adjusted to 0.020 inch (0.508 mm). To check timing with a continuity light, refer to Fig. T17. Remove "pop" rivets securing identification plates to blower housing. Remove plate to expose a 1¼ inch hole. Connect continuity light to terminal screw (78–Fig. T16) and suitable engine ground. Rotate engine clockwise until piston is on compression stroke and timing mark is just below stator laminations as shown in Fig. T17. At this time, points should be ready to open and continuity light should be on.

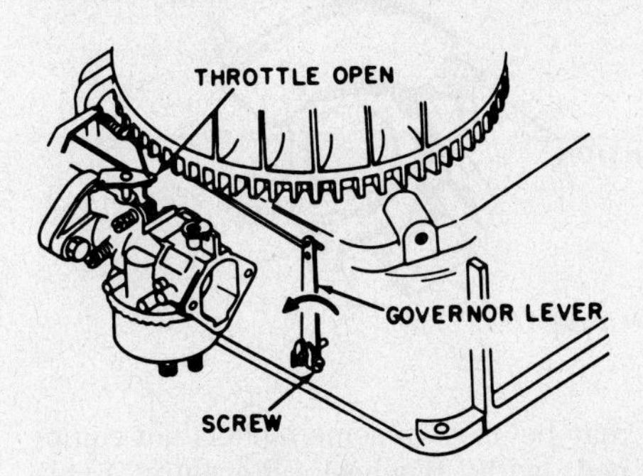

Fig. T11 – When adjusting governor linkage on VH70, VM70, VM80, VM100, TVM170, TVM195 or TVM220 models, loosen clamp screw and rotate governor lever shaft and lever counter-clockwise as far as possible.

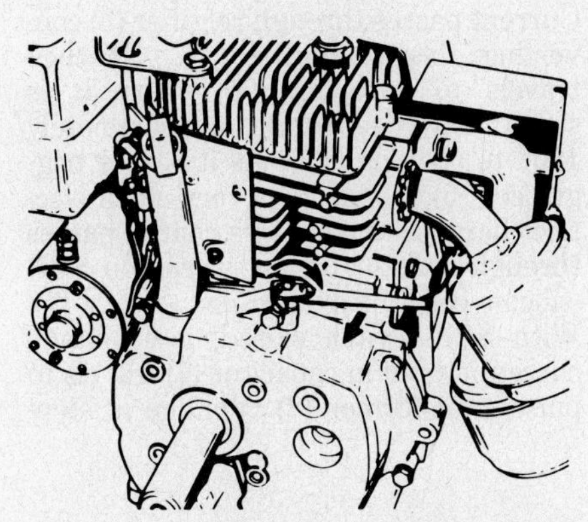

Fig. T12 – On HH70, HM70, HM80 and HM100 models, rotate governor lever shaft and lever clockwise when adjusting linkage.

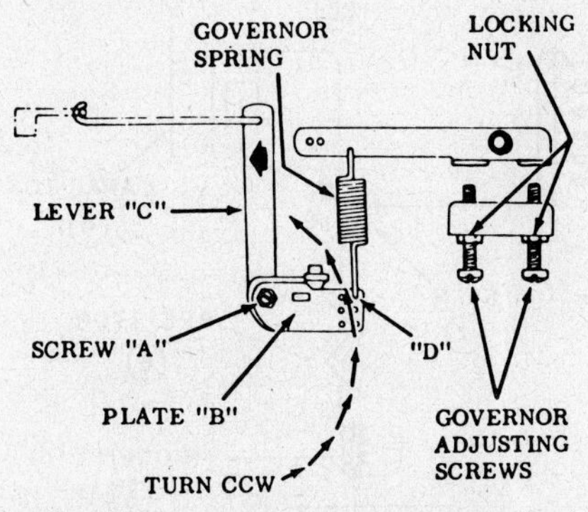

Fig. T13 – External governor linkage on VH80 and VH100 models. Refer to text for adjustment procedure.

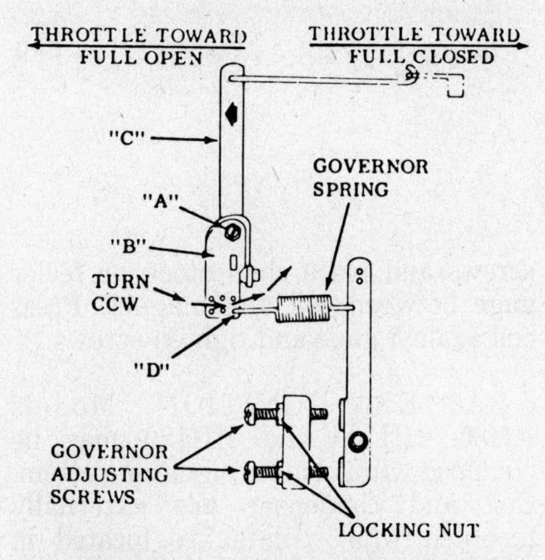

Fig. T14 – External governor linkage on HH80, HH100 and HH120 models. Refer to text for adjustment procedure.

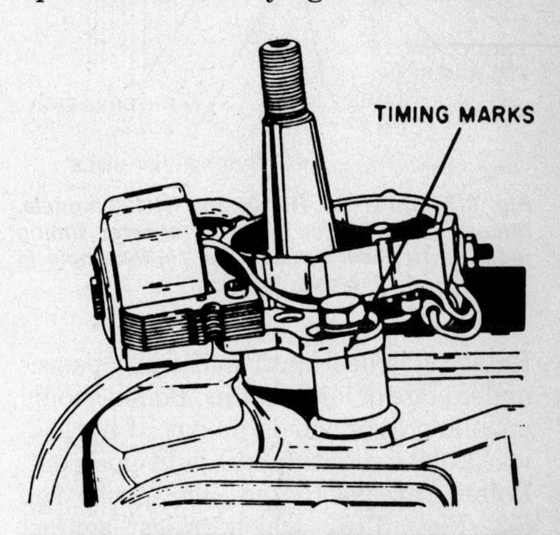

Fig. T15 – On VM70, VH70, HM70, HH70, VM80, HM80, VM100 and HM100 equipped with magneto ignition, adjust breaker point gap to 0.020 inch (0.508 mm) and align timing marks as shown.

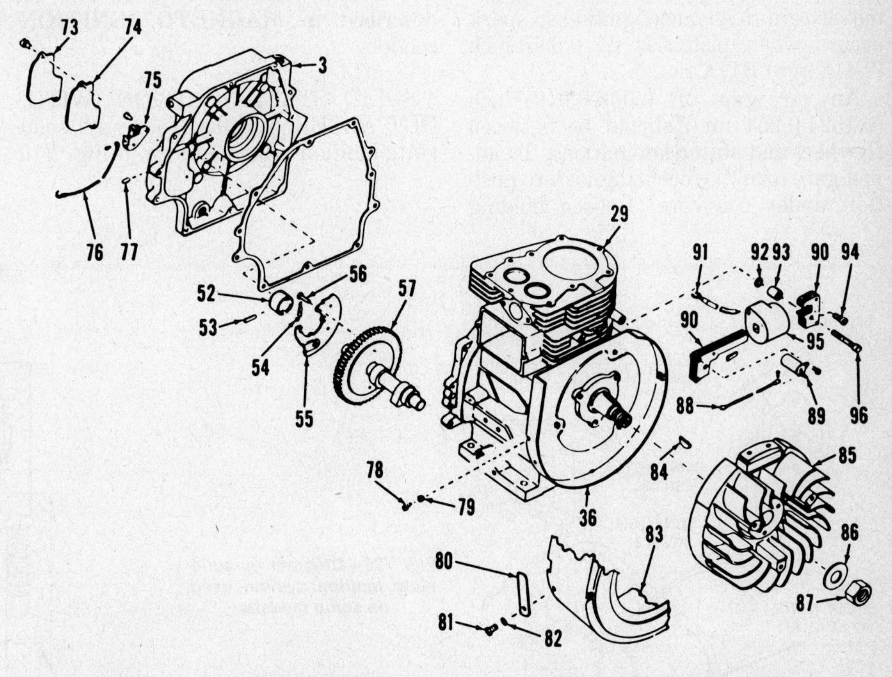

Fig. T16 – Exploded view of magneto ignition components used on HH80, HH100 and HH120 models. Timing advance and breaker points used on engines equipped with battery ignition are identical.

3. Crankcase cover
29. Cylinder block
36. Blower air baffle
52. Breaker cam
53. Push rod
54. Spring
55. Timing advance weight
56. Rivet
57. Camshaft assy.
73. Breaker box cover
74. Gasket
75. Breaker points
76. Ignition wire
77. Pin
78. Screw
79. Clip
80. Ground switch
81. Screw
82. Washer
83. Blower housing
84. Flywheel key
85. Flywheel
86. Washer
87. Nut
88. Condenser wire
89. Condenser
90. Armature core
91. High tension lead
92. Washer
93. Spacer
94. Screw
95. Coil
96. Screw

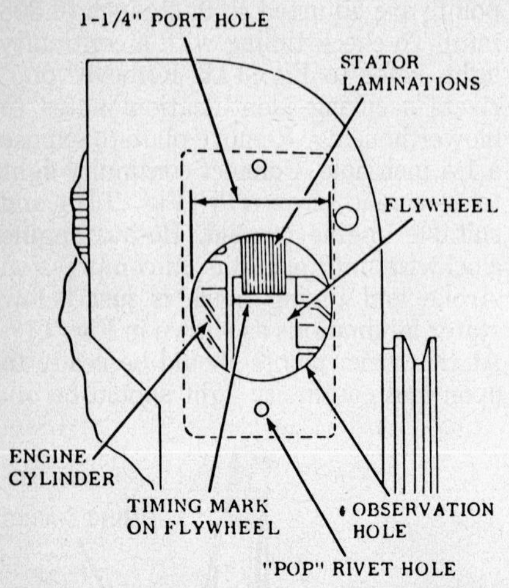

Fig. T17—On HH80, HH100 and HH120 models, remove identification plate to observe timing mark on flywheel through the 1¼-inch hole in blower housing.

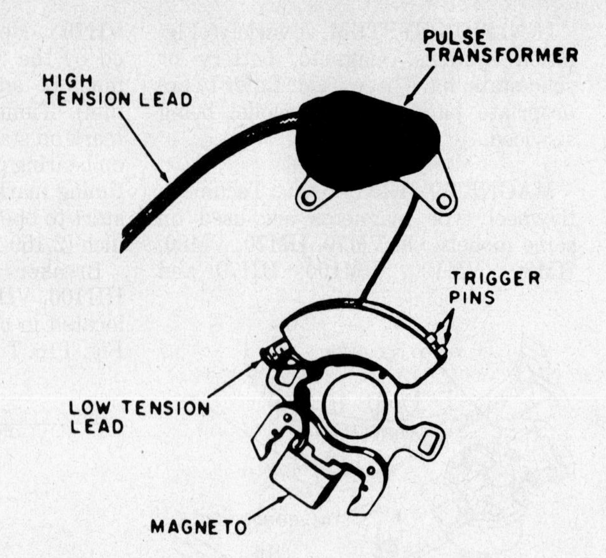

Fig. T19—View of solid state ignition system used on some models not equipped with flywheel alternator.

Rotate flywheel until mark just passes under edge of laminations. Points should open and light should be out. If not, adjust points slightly until light goes out. Points are actuated by push rod (53–Fig. T16) which rides against breaker cam (52). Breaker cam is driven by a tang on advance weight (55). When cranking, spring (54) holds advance weight in retarded position (TDC). At operating speeds, centrifugal force overcomes spring pressure and weight moves cam to advance ignition so spark occurs when piston is at 0.095 inch (2.413 mm) BTDC.

An air gap of 0.006-0.010 inch (0.1524-0.254 mm) should be between flywheel and stator laminations. To adjust gap, turn flywheel magnet into position under coil core. Loosen holding screws and place shim stock or feeler gage between coil and magnet. Press coil against gage and tighten screws.

BATTERY IGNITION. Models HH80, HH100 and HH120 may be equipped with a battery ignition system. Coil and condenser are externally mounted while points are located in crankcase cover. See Fig. T18. Points should be adjusted to 0.020 inch (0.508 mm). To check timing, disconnect primary wire between coil and points and follow the same procedure as described in **MAGNETO IGNITION** section.

SOLID STATE IGNITION (WITHOUT ALTERNATOR). Tecumseh solid state ignition system shown in Fig. T19 may be used on some models not equipped with flywheel alternator. This system does not use ignition breaker points. The only moving part of system is rotating flywheel with charging magnets. As flywheel magnet passes position (1A–Fig. T20), a low voltage AC current is induced into input coil (2). Current passes through rectifier (3) converting this current to DC. It then travels to capacitor (4) where it is stored. Flywheel rotates approximately 180° to position (1B). As it passes trigger coil (5), it induces a very small electric charge into coil. This charge passes through resistor (6) and turns on SCR (silicon controlled rectifier) switch (7). With SCR switch closed, low voltage current stored in capacitor (4) travels to pulse transformer (8). Voltage is step-

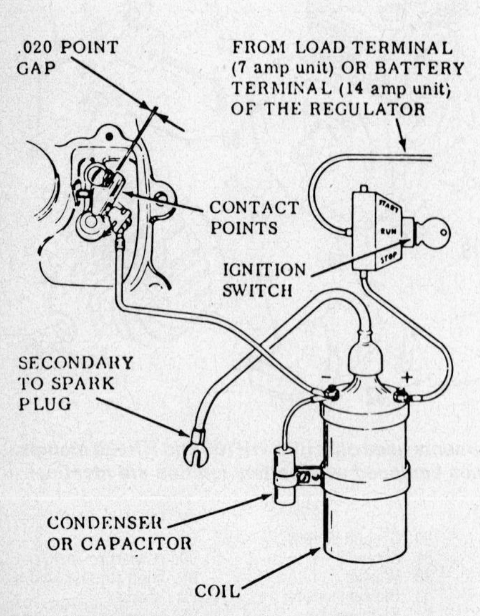

Fig. T18—Typical battery ignition wiring diagram used on some HH80, HH100 and HH120 engines.

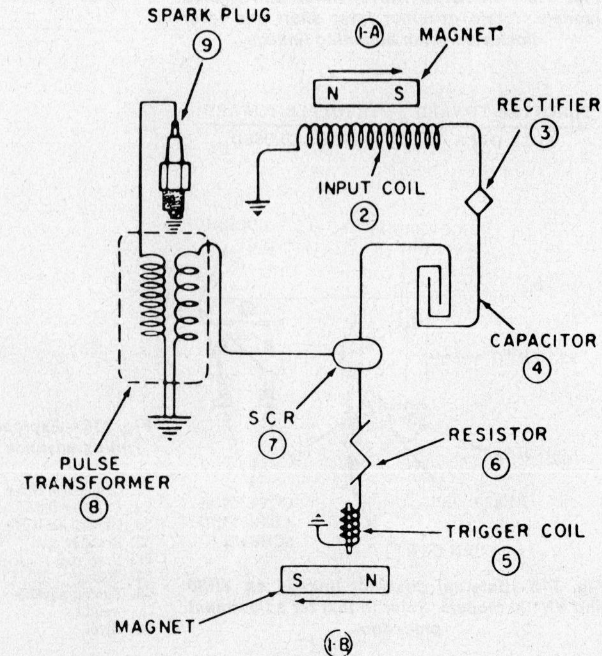

Fig. T20—Diagram of solid state ignition system used on some models.

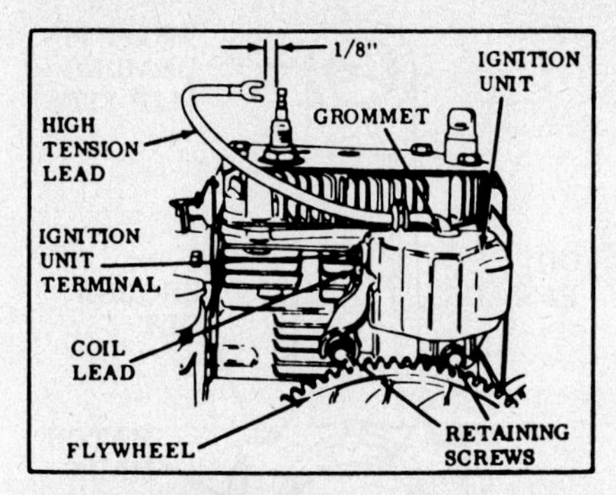

Fig. T21 — View of solid state ignition unit used on some models equipped with flywheel alternator. System should produce a good blue spark 1/8-inch (3.175 mm) long at cranking speed.

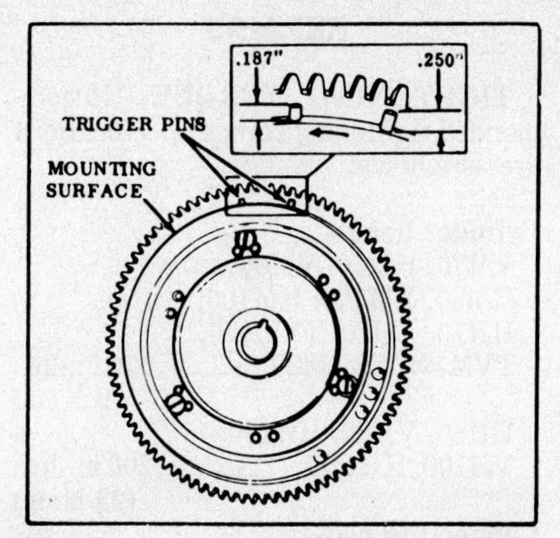

Fig. T23 — Remove flywheel and drive trigger pins in or out as necessary until long pin is extended 0.250 inch (6.35 mm) and short pin is extended 0.187 inch (4.763 mm) above mounting surface.

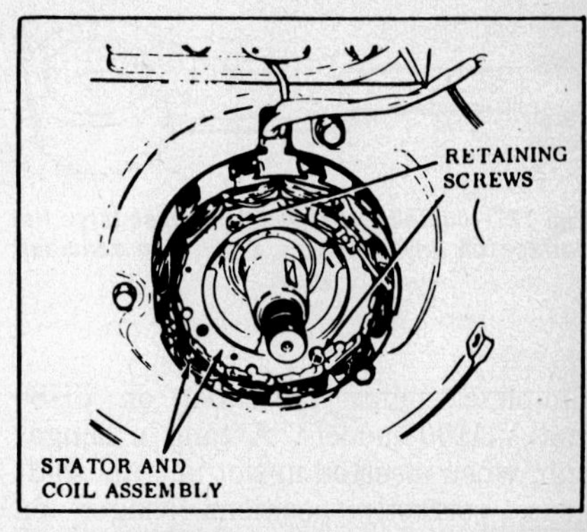

Fig. T25 — Ignition generator coil and stator serviced only as an assembly.

ped up instantaneously and current is discharged across electrodes of spark plug (9), producing a spark before TDC.

Some units are equipped with a second trigger coil and resistor set to turn SCR switch on at a lower rpm. This second trigger pin is closer to flywheel and produces a spark at TDC for easier starting. As engine rpm increases, first (shorter) trigger pin picks up small electric charge and turns SCR switch on, firing spark plug BTDC.

If system fails to produce a spark to spark plug, first check high tension lead (Fig. T19). If condition of high tension lead is questionable, renew pulse transformer and high tension lead assembly. Check low tension lead and renew if insulation is faulty. Magneto charging coil, electronic triggering system and mounting plate are available only as an assembly. If necessary to renew this assembly, place unit in position on engine. Start retaining screws, turn mounting plate counter-clockwise as far as possible, then tighten retaining screw to a torque of 5-7 ft.-lbs. (7-10 N·m).

SOLID STATE IGNITION (WITH ALTERNATOR). Tecumseh solid state ignition system used on some models equipped with flywheel alternator does not use ignition breaker points. The only moving part of system is rotating flywheel with charging magnets and trigger pins. Other components of system are ignition generator coil and stator assembly, spark plug and ignition unit.

The long trigger pin induces a small charge of current to close SCR (silicon controlled rectifier) switch at engine cranking speed and produces a spark at TDC for starting. As engine rpm increases, first (shorter) trigger pin induces current which produces a spark when piston is 0.095 inch (2.413 mm) BTDC.

Test ignition system by holding high tension lead 1/8-inch (3.175 mm) from spark plug (Fig. T21), crank engine and check for a good blue spark. If no spark is present, check high tension lead and coil lead for loose connections or faulty insulation. Check air gap between trigger pin and ignition unit as shown in Fig. T22. Air gap should be 0.006-0.010 inch (0.1524-0.254 mm). To adjust air gap, loosen the two retaining screws and move ignition unit as necessary, then tighten retaining screws.

NOTE: Long trigger pin should extend 0.250 inch (6.35 mm) and short trigger pin should extend 0.187 inch (4.763 mm), measured as shown in Fig. T23. If not, remove flywheel and drive pins in or out as required.

Remove coil lead from ignition terminal and connect an ohmmeter as shown in Fig. T24. If series resistance test of ignition generator coil is below 400 ohms, renew stator and coil assembly (Fig. T25). If resistance is above 400 ohms, renew ignition unit.

LUBRICATION. On VH70, VM70, VM100, TVM170, TVM195 and TVM220 models, a barrel and plunger type oil pump (Fig. T26 or T27) driven by an eccentric on camshaft, pressure lubricates upper main bearing and connecting rod journal. When installing early type pump (Fig. T26), chamfered side of drive collar must be against thrust bearing surface on camshaft gear. When installing late type pump, place side of drive collar with large flat surface shown in Fig. T27 away from camshaft gear.

An oil slinger (59 – Fig. T28), installed on crankshaft between gear and lower bearing is used to direct oil upward for

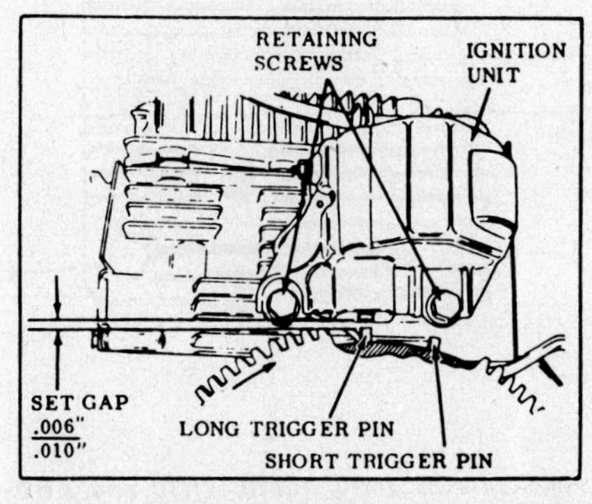

Fig. T22 — Adjust air gap between long trigger pin and ignition unit to 0.006-0.010 inch (0.1524-0.254 mm).

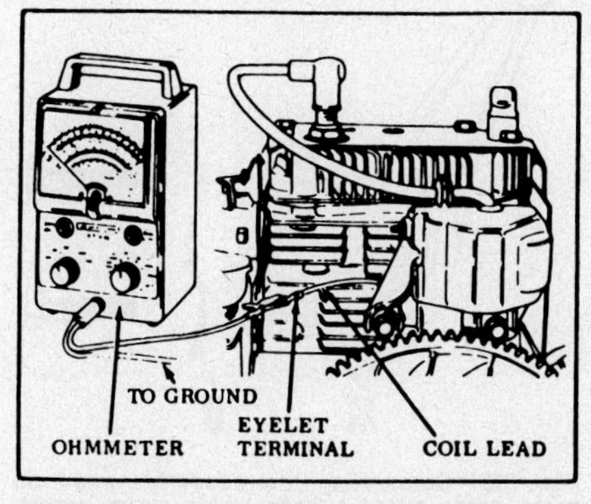

Fig. T24 — View showing an ohmmeter connected for resistance test of ignition generator coil.

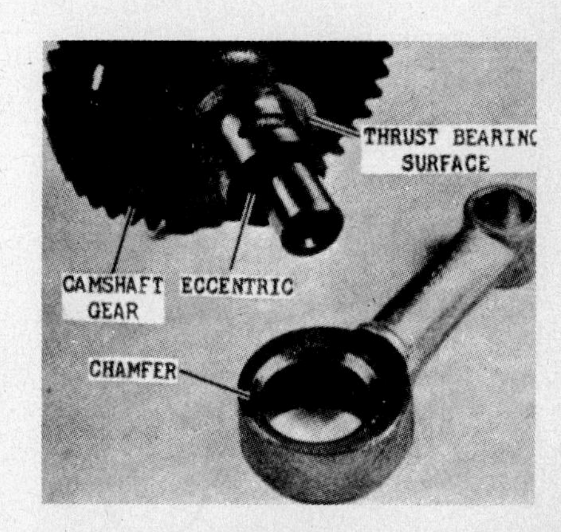

Fig. T26 — View of early type oil pump used on VH70, VM70 and VM80 models. Chamfered face of drive collar should be toward camshaft gear.

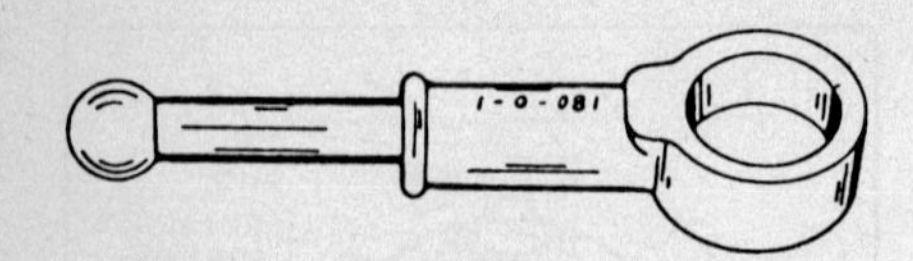

Fig. T27—Install late type oil pump so large flat surface on drive collar is away from camshaft gear.

complete engine lubrication on VH80 and VH100 models. A tang n slinger hub, when inserted in slot in crankshaft gear, correctly positions slinger on crankshaft as shown in Fig. T28.

Splash lubrication system on all other models is provided by use of an oil dipper on connecting rod. See Figs. T30 and T31.

Oils approved by manufacturer must meet requirements of API service classification SC, SD, SE or SF.

Use an SAE 30 oil for temperatures above 32° F (0° C) and SAE 5W20 or 5W30 oil for temperatures below 32° F (0° C).

Check oil level at 5 hour intervals or before initial start-up of engine. Maintain oil level at edge of filler hole or at "FULL" mark on dipstick, as equipped.

Recommended oil change interval for all models is every 25 hours of normal operation.

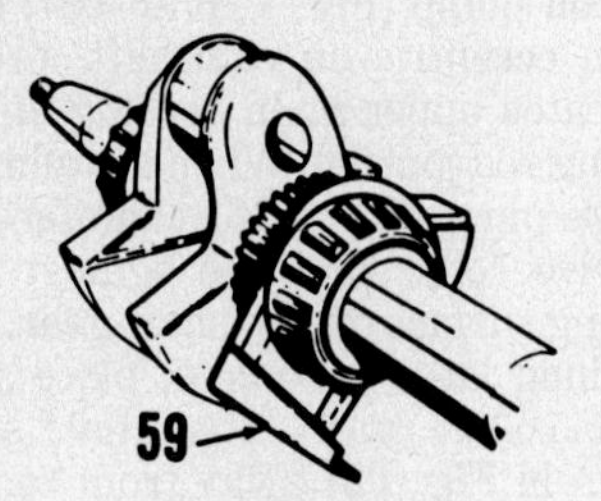

Fig. T28—Oil slinger (59) on VH80 and VH100 models must be installed on crankshaft as shown.

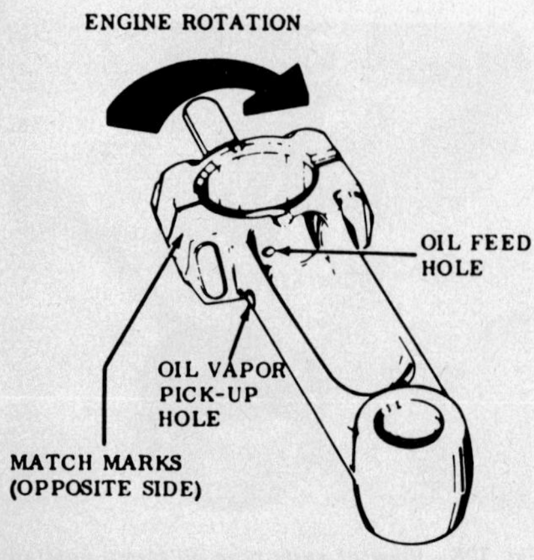

Fig. T29—Connecting rods used on VH80 and VH100 have two oil holes.

REPAIRS

TIGHTENING TORQUE. Recommended tightening torque specifications are as follows:

Cylinder head—
VM70, HM70, VM80, HM80, VM100, HM100, HH70, VH70, TVM170, TVM195, TVM220 .... 170 in.-lbs. (19 N·m)
HH80, VH80, HH100, VH100, HH120 .... 200 in.-lbs. (23 N·m)

Connecting rod—
VM70, HM70, VM80, HM80, VM100, HM100, HH70, VH70, TVM170, TVM195, TVM220 .... 120 in.-lbs. (13 N·m)
HH80, VH80, HH100, VH100, HH120 .... 110 in.-lbs. (12 N·m)

Crankcase cover—
All models .... 65-110 in.-lbs. (5-12 N·m)

Bearing retainer—
HH80, VH80, HH100, VH100, HH120 .... 65-110 in.-lbs. (5-12 N·m)

Ball bearing retainer nut—
All models so equipped .... 15-22 in.-lbs. (1-3 N·m)

Flywheel nut—
Light frame .... 400-440 in.-lbs. (45-50 N·m)
Medium frame .... 430-500 in.-lbs. (49-57 N·m)
Heavy frame .... 600-660 in.-lbs. (68-75 N·m)

CYLINDER HEAD. When removing cylinder head, note location of different length cap screws for aid in correct reassembly. Always install new head gasket and tighten cap screws evenly in sequence shown in Figs. T32, T33, T34 or T35. Refer to **TIGHTENING TORQUE** section for correct torque specifications for model being serviced.

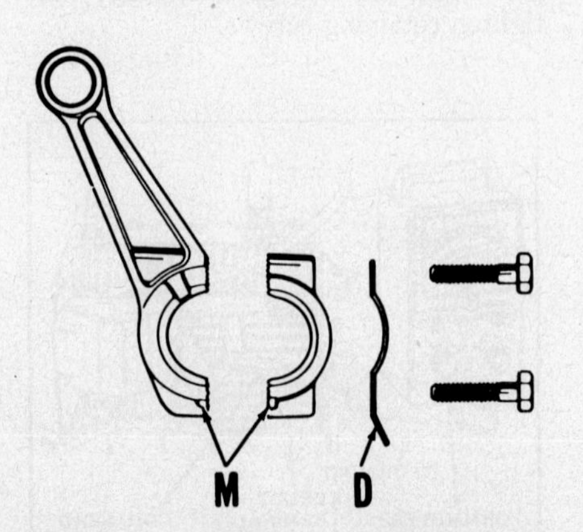

Fig. T30—Connecting rod assemble used on VH70, VM70, VM80, VM100, HH70, HM70, HM80, HM100, TVM170, TVM195 and TVM220 models. Note position oil dipper (D) and match marks (M).

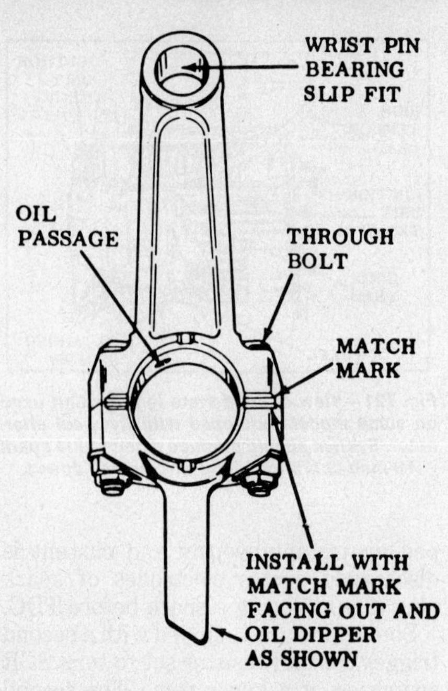

Fig. T31—Connecting rod assembly used on HH80, HH100 and HH120 models.

CONNECTING ROD. Piston and connecting rod assembly is removed from cylinder head end of engine. Connecting rod rides directly on crankpin journal of crankshaft.

Standard crankpin journal diameter is 1.1865-1.1870 inches (30.14-30.15 mm) for VM70, HM70, VM80, HM80, VM100, HM100, HH70, VH70, TVM170, TVM195 and TVM220 models and 1.3750-1.3755 inches (34.93-34.94 mm) for all remaining models.

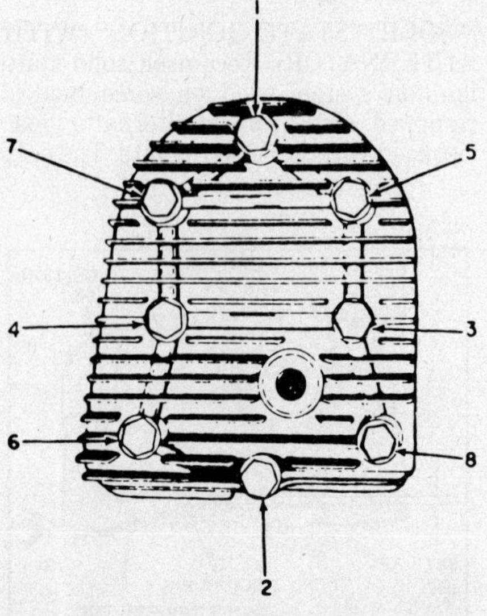

Fig. T32—On VM70, HM70, VH70 and HH70 models, tighten cylinder head cap screws evenly to a torque of 170 in.-lbs. (19 N·m) using tightening sequence shown.

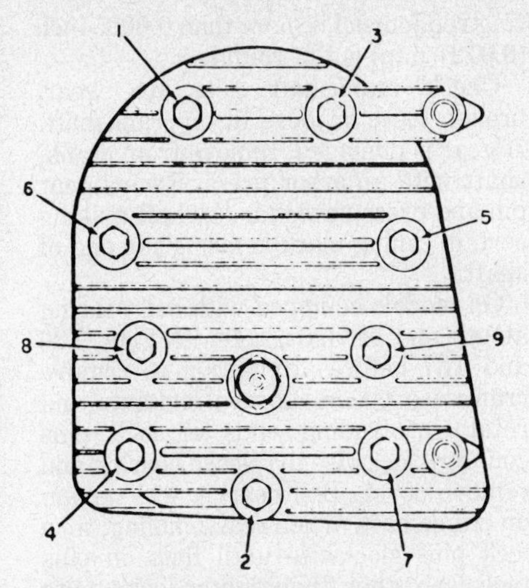

Fig. T33—Tighten cylinder head cap screws on HM80, VM80, HM100, VM100, TVM170, TVM195 and TVM220 models in sequence shown to 170 in.-lbs. (19N·m) torque.

Connecting rod to crankpin journal running clearance should be 0.002 inch (0.0508 mm) for all models.

Connecting rods are equipped with match marks which must be aligned and face pto end of crankshaft after installation. See Figs. T29, T30 or T31.

Connecting rod is available in a variety of sizes for undersize crankshafts, as well as standard.

PISTON, PIN AND RINGS. Aluminum alloy piston is fitted with two compression rings and one oil control ring.

Piston skirt diameter and piston skirt to cylinder wall clearance is measured at bottom edge of skirt at a right angle to piston pin for all models.

Piston skirt diameter for HH70 and VH70 models is 2.7450-2.7455 inches (69.72-69.74 mm) and skirt to cylinder bore clearance is 0.0045-0.0060 inch (0.1143-0.1524 mm).

Piston skirt diameter for HH80 model prior to type letter B, VM70 and HM70 models with type letter A and after and TVM170 models is 2.9325-2.9335 inches (74.49-74.51 mm) and skirt to cylinder bore clearance is 0.004-0.006 inch (0.1016-0.1524 mm).

Piston skirt diameter for VM70 and HM70 models prior to type letter A, V80 model, VM80 and HM80 models prior to type letter E is 3.0575-3.0585 inches (77.66-77.69 mm) and skirt to cylinder bore clearance is 0.0035-0.0055 inch (0.0889-0.1397 mm).

Piston skirt diameter for VM80 and HM80 models with type letter E or after and TVM195 models is 3.1195-3.1205 inches (79.24-79.26 mm) and skirt to cylinder bore clearance is 0.004-0.006 inch (0.1016-0.1524 mm).

Piston skirt diameter for HM100 (early production) and VM100 models is 3.1817-3.1842 inches (80.815-80.823 mm) and skirt to cylinder bore clearance is 0.0028-0.0063 inch (0.0711-0.1600 mm).

Piston skirt diameter for HH80 models with type letter B and after, VH80, VH100, HM100 (late production), HH100 and TVM220 models is 3.308-3.310 inches (84.02-84.07 mm) and skirt to cylinder bore clearance is 0.002-0.005 inch (0.0508-0.1270 mm).

Piston skirt diameter for HH120 model is 3.4950-3.4970 inches (88.77-88.82 mm) and skirt to cylinder bore clearance is 0.003-0.006 inch (0.0762-0.1524 mm).

Side clearance of top ring in piston groove for VM70 (prior to type letter A), V80 model and VM80 model (prior to type letter E) is 0.003-0.004 inch (0.0762-0.1016 mm), for HM and TVM170 models is 0.0028-0.0051 inch (0.0711-0.1296 mm), for VM80 (type letter E and after), HM80, TVM195, HM100, VM100, TVM220, H80, VH100, HH100 and VH80 models is 0.002-0.005 inch (0.0508-0.1270 mm).

Standard piston pin diameter is 0.625-0.6254 inch (15.88-15.89 mm) for VM70, HM70, VM80, HM80, HH70, VH70, TVM170 and TVM195 models and 0.6873-0.6875 inch (17.457-17.462 mm) for all other models.

Piston pin clearance should be 0.0001-0.0008 inch (0.0025-0.0203 mm) in connecting rod and 0.0002-0.0005 inch (0.0051-0.0127 mm) in piston. If excessive clearance exists, both piston and pin must be renewed as pin is not available separately.

Pistons should be assembled on connecting rod so arrow on top of piston is pointing toward carburetor side of engine after installation.

Pistons and rings are available in a variety of oversizes as well as standard.

CYLINDER. If cylinder is scored or excessively worn, or if taper or out-of-round exceeds 0.004 inch (0.1016 mm), cylinder should be rebored to nearest

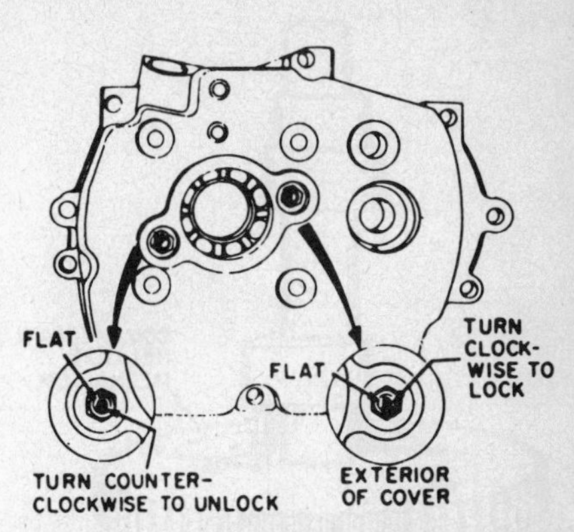

Fig. T36—View showing bearing locks on HM70, HH70, HM80 and HM100 models equipped with ball bearing main bearings. Locks must be released before removing crankcase cover. Refer to Fig. T37 for interior view of cover and locks.

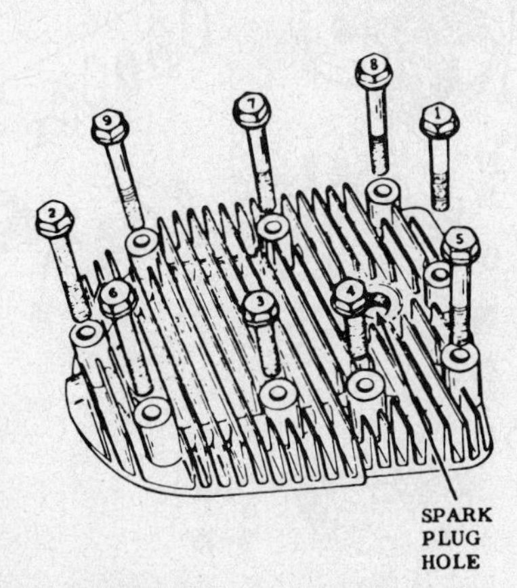

Fig. T34—View showing cylinder head cap screw tightening sequence used on early HH80, HH100 and HH120 engines. Tighten cap screw to 200 in.-lbs. (23 N·m) torque. Note type and length of cap screws.

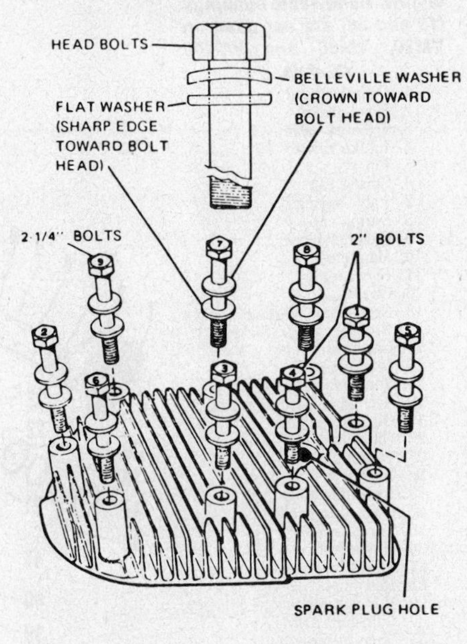

Fig. T35—Flat washers and Belleville Washers are used on cylinder head cap screws on late HH80, HH100 and HH120 and all VH80 and VH100 engines. Tighten cap screws in sequence shown to a torque of 200 in.-lbs. (23 N·m).

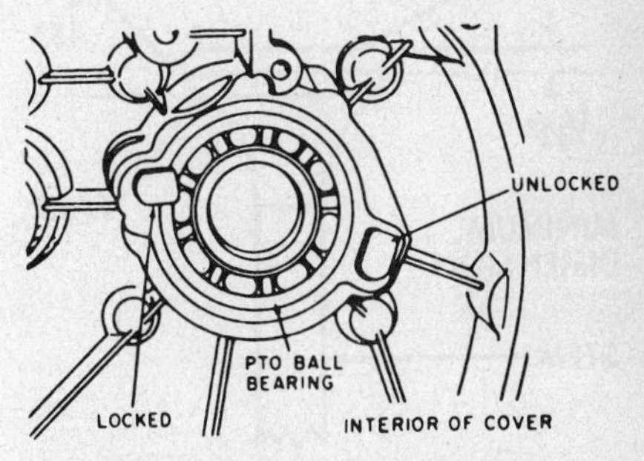

Fig. T37—Interior view of crankcase cover and ball bearing locks used on HM70, HH70, HM80 and HM100 models.

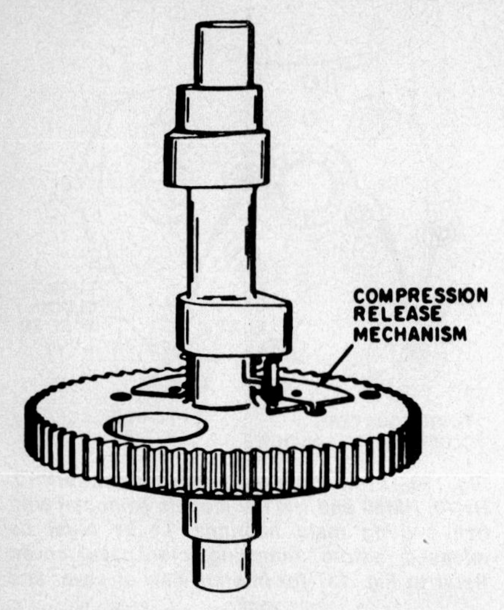

Fig. T38—View of Insta-matic Ezee-Start compression release camshaft assembly used on all models except HH80, HH100 and HH120.

oversize for which piston and rings are available.

Standard cylinder bore for HH70 and VH70 models is 2.750-2.751 inches (69.850-69.854 mm).

Standard cylinder bore for HH80 (prior to type letter B), VM70 and HH70 models with type letter A and after and TVM170 models is 2.9375-2.9385 inches (74.61-74.64 mm).

Standard cylinder bore for VM70 and HM70 models prior to type letter A, VM80 and HM80 models prior to type letter E and V80 models is 3.062-3.063 inches (77.78-77.80 mm).

Standard cylinder bore for VM80 and HM80 models with type letter E or after and TVM195 models is 3.125-3.126 inches (79.38-79.40 mm).

Standard cylinder bore for HM100 (early production) and VM100 models is 3.187-3.188 inches (80.95-80.98 mm).

Standard cylinder bore for HH80 (type letter B and after), VH80, HM100 (late production), VH100, HH100 and TVM220 models is 3.312-3.313 inches (84.13-84.15 mm).

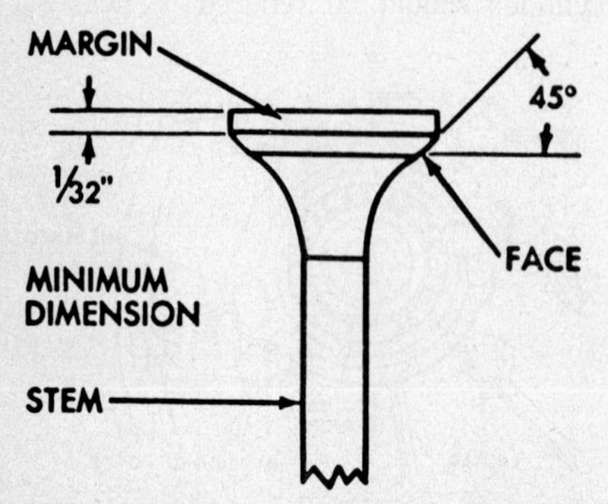

Fig. T39—Valve face angle should be 45°. Minimum valve head margin is 1/32-inch (0.794 mm).

Standard cylinder bore for HH120 model is 3.500-3.501 inches (88.90-88.93 mm).

CRANKSHAFT AND MAIN BEARINGS. Crankshaft main journals ride directly in aluminum alloy bearings in crankcase and mounting flange (engine base) on vertical crankshaft engines or in two renewable steel backed bronze bushings. On some horizontal crankshaft engines, crankshaft rides in a renewable sleeve bushing at flywheel end and a ball bearing or bushing at pto end. Models HH80, VH80, HH100, VH100 and HH120 are equipped with taper roller bearings at both ends of crankshafts.

Standard main bearing bore diameter for bushing type main bearings for VM70, VH70 and HH70 models should be 1.0005-1.0010 inches (25.41-25.43 mm) for either bearing.

Standard main bearing bore diameter for bushing type main bearings for all remaining models should be 1.0005-1.0010 inches (25.41-25.43 mm) for main bearing in cylinder block section and 1.1890-1.1895 inches (30.20-30.21 mm) for main bearing in cover section.

Normal running clearance of crankshaft journals in aluminum bearings or bronze bushings is 0.0015-0.0025 inch (0.038-0.064 mm). Renew crankshaft if main journals are more than 0.001 inch (0.025 mm) out-of-round or renew or regrind crankshaft if connecting rod journal is more than 0.0005 inch (0.0127 mm) out-of-round.

Check crankshaft gear for wear, broken teeth or loose fit on crankshaft. If gear is damaged, remove from crankshaft with an arbor press. Renew gear pin and press new gear on shaft making certain timing mark is facing pto end of shaft.

On models equipped with ball bearing at pto end of shaft, refer to Figs. T36 and T37 before attempting to remove crankcase cover. Loosen locknuts and rotate protruding ends of lock pins counter-clockwise to release bearing and remove cover. Ball bearing will remain on crankshaft. When reassembling, turn lock pins clockwise until flats on pins face each other, then tighten locknuts to 15-22 in-lbs. (2-3 N·m) torque.

Crankshaft end play for all models except those with taper roller bearing main bearings is 0.005-0.027 inch (0.1270-0.6858 mm) and is controlled by varying thickness of thrust washers between crankshaft and cylinder block or crankcase cover.

Crankshaft main bearing preload for all models with taper roller bearing main bearings is 0.001-0.007 inch (0.0254-0.1778 mm) and is controlled by varying thickness of shim gasket (32–Fig. T42) on all models.

Taper roller bearings are a press fit on crankshaft and must be renewed if removed. Heat bearings in hot oil to aid installation.

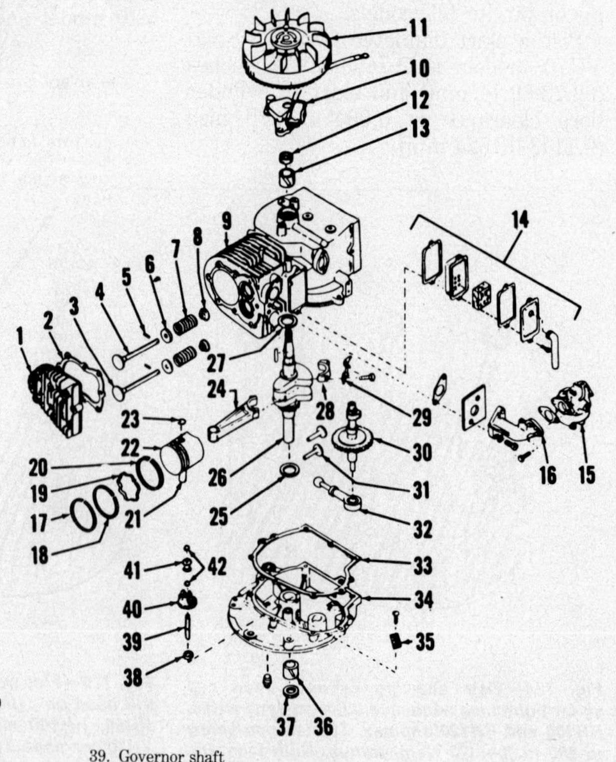

Fig. T40—Exploded view of typical vertical crankshaft engine. Renewable bushings (13 and 36) are not used on VM70, VM80 and VM100 models.

1. Cylinder head
2. Head gasket
3. Exhaust valve
4. Intake valve
5. Pin
6. Spring cap
7. Valve spring
8. Spring cap
9. Cylinder block
10. Magneto
11. Flywheel
12. Oil seal
13. Crankshaft bushing
14. Breather assy.
15. Carburetor
16. Intake pipe
17. Top compression ring
18. Second compression ring
19. Oil ring expander
20. Oil control ring
21. Piston pin
22. Piston
23. Retaining ring
24. Connecting rod
25. Thrust washer
26. Crankshaft
27. Thrust washer
28. Rod cap
29. Rod bolt lock
30. Camshaft assy.
31. Valve lifters
32. Oil pump
33. Gasket
34. Mounting flange (engine base)
35. Oil screen
36. Crankshaft bushing
37. Oil seal
38. Spacer
39. Governor shaft
40. Governor gear assy.

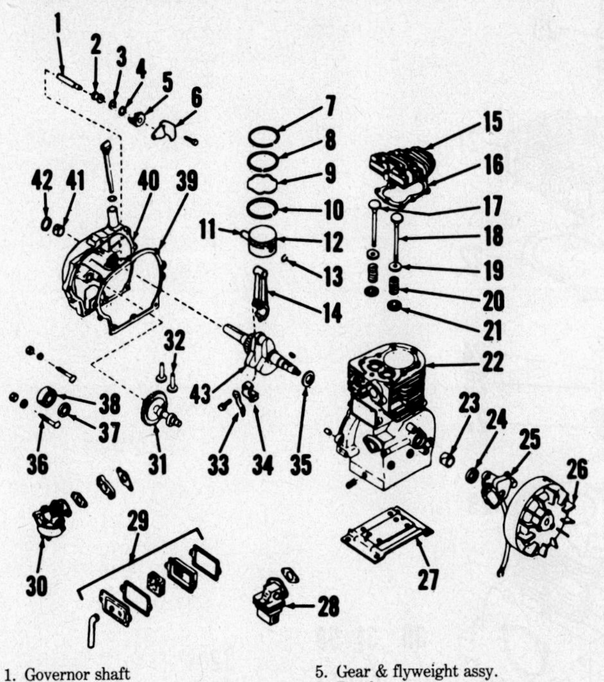

Fig. T41 – Exploded view of typical horizontal crankshaft engine. Engines may be equipped with crankshaft bushing (41) or ball bearing (38) at pto end of shaft.

1. Governor shaft
2. Spool
3. Washer
4. Retaining ring
5. Gear & flyweight assy.
6. Bracket
7. Top compression ring
8. Second compression ring
9. Oil ring expander
10. Oil control ring
11. Piston pin
12. Piston
13. Retaining ring
14. Connecting rod
15. Cylinder head
16. Head gasket
17. Exhaust valve
18. Intake valve
19. Spring cap
20. Valve spring
21. Spring retainer
22. Cylinder block
23. Crankshaft bushing
24. Oil seal
25. Magneto
26. Flywheel
27. Mounting plate
28. Fuel pump
29. Breather assy.
30. Carburetor
31. Camshaft assy.
32. Valve lifter
33. Rod bolt lock
34. Rod cap
35. Thrust washer
36. Bearing lock pin
37. Thrust washer
38. Ball bearing
39. Gasket
40. Crankcase cover
41. Bushing
42. Oil seal
43. Crankshaft

Bronze or aluminum bushing type main bearings are renewable and finish reamers are available from Tecumseh.

Standard crankshaft journal diameters are as follows:

Main journal diameter –

VM70, HM70,
HH70, VH70,
(Both journals) . . . . 0.9985-0.9990 inch (25.36-25.38 mm)

HH80, VH80, HH100,
VH100, HH120,
(Both journals) . . 1.1865-1.1870 inches (30.14-30.15 mm)

VM80, HM80, HM100, VM100,
TVM170, TVM195, TVM220
Flywheel end . . . . . 0.9985-0.9990 inch (25.36-25.38 mm)
Pto end . . . . 1.1870-1.1875 inches (30.15-30.16 mm)

Crankpin journal diameter –

HH80, VH80
HH100, HH120 . 1.3750-1.3755 inches (34.93-34.94 mm)

All other
models . . . . 1.1865-1.1870 inches (30.14-30.15 mm)

Connecting rods are available in a variety of sizes for reground crankshaft crankpin journals.

When reinstalling crankshaft, make certain timing marks on camshaft and crankshaft are properly aligned.

Fig. T42 – Exploded view of VH80 or VH100 model vertical crankshaft engine.

1. Governor arm bushing
2. Oil seal
3. Mounting flange (engine base)
7. Governor arm
8. Thrust spool
9. Snap ring
10. Governor gear & weight assy.
11. Governor shaft
12. Bearing cup
13. Gasket
14. Piston & pin assy.
15. Top compression ring
16. Second compression ring
17. Ring expander
18. Oil control ring
19. Retaining ring
20. Spark plug
21. Cylinder head
22. Head gasket
23. Exhaust valve
24. Intake valve
25. Pin
26. Exhaust valve spring
27. Intake valve spring
28. Spring cap
29. Cylinder block
30. Bearing cone
31. Bearing cup
32. Shim gasket
33. Steel washer (0.010 inch)
34. Oil seal
35. Bearing retainer cap
36. Blower air baffle
38. Gasket
39. Breather
40. Breather tube
42. Rod cap
43. Self locking nut
44. Washer
45. Crankshaft gear pin
46. Crankshaft
47. Connecting rod
48. Rod bolt
49. Crankshaft gear
50. Valve lifters
51. Bearing cone
57. Camshaft assy.
58. "O" ring
59. Oil slinger

CAMSHAFT. Camshaft and camshaft gear are an integral part which rides on journals at each end of camshaft. Camshaft journal diameter is 0.6235-0.6240 inch (15.84-15.85 mm) and journal to bearing running clearance should be 0.003 inch (0.0762 mm).

Renew camshaft if gear teeth are worn or if journal or lobe surfaces are worn or damaged.

Camshaft equipped with compression release mechanism (Fig. T38) is easier to remove and install if crankshaft is rotated three teeth past aligned position which allows compression release mechanism to clear exhaust valve lifter. Compression release mechanism parts should work freely with no binding or sticking. Parts are not serviced separate from camshaft assembly.

Some models are equipped with an automatic timing advance mechanism (52 through 56 – Fig. T43) which must work freely with no binding or sticking. Renew damaged parts as necessary.

Make certain timing marks on camshaft gear and crankshaft gear are aligned after installation.

VALVE SYSTEM. Valve seats are machined directly into cylinder block assembly. Seats are ground at a 45° angle and should not exceed 3/64-inch (1.191 mm) in width.

Valve face is ground at a 45° angle and margin should not be less than 1/32-inch (0.794 mm). See Fig. T39. Valves are available with 1/32-inch oversize stem for use with oversize valve guide bore.

Valve guides are not renewable. If guides are excessively worn, ream to 0.3432-0.3442 inch (8.72-8.74 mm) and install valve with 1/32-inch (0.794 mm) oversize valve stem. Drill upper and

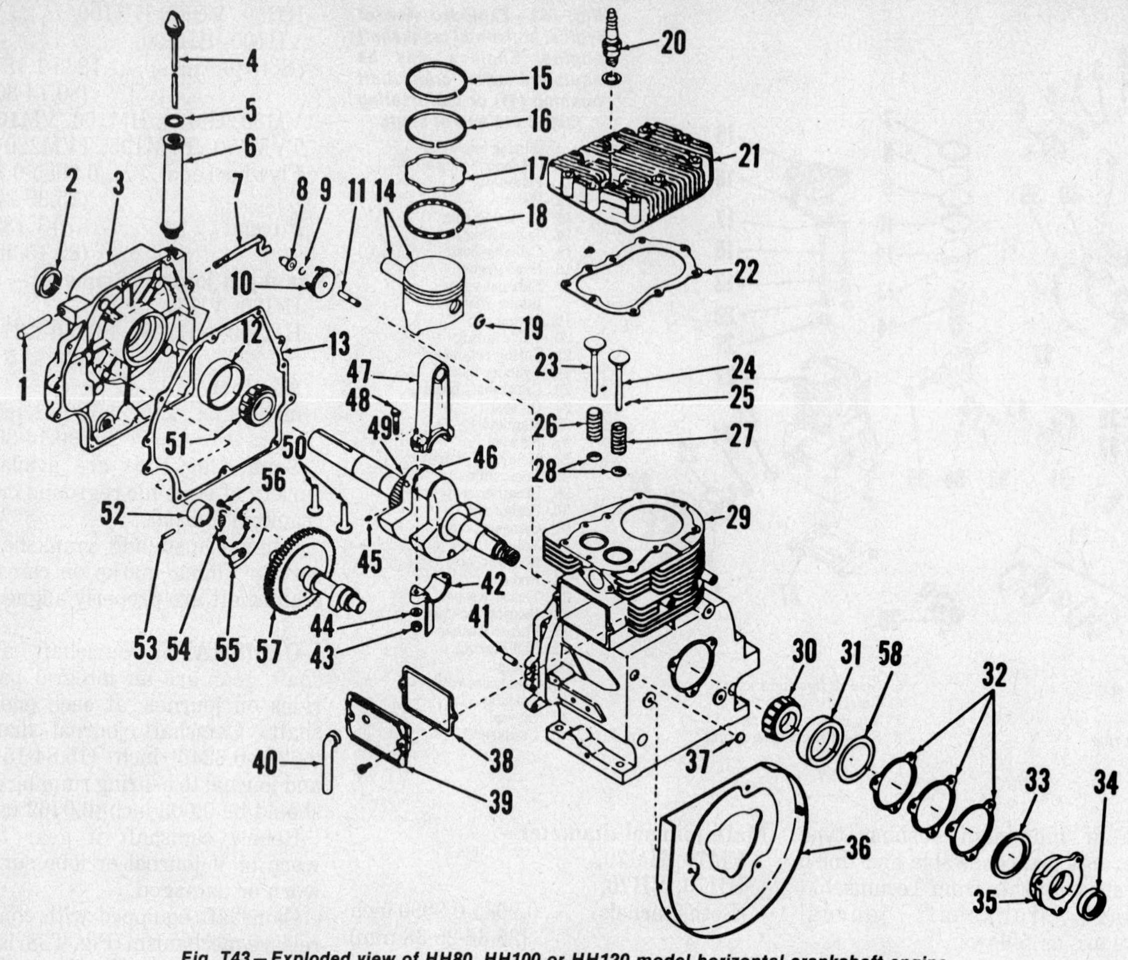

Fig. T43—Exploded view of HH80, HH100 or HH120 model horizontal crankshaft engine.

1. Governor arm bushing
2. Oil Seal
3. Crankcase cover
4. Dipstick
5. Gasket
6. Oil filler tube
7. Governor arm
8. Thrust spool
9. Snap ring
10. Governor gear & weight assy.
11. Governor shaft
12. Bearing cup
13. Gasket
14. Piston & pin assy.
15. Top compression ring
16. Second compression ring
17. Oil ring expander
18. Oil control ring
19. Retaining ring
20. Spark plug
21. Cylinder head
22. Head gasket
23. Exhaust valve
24. Intake valve
25. Pin
26. Exhaust valve spring
27. Intake valve spring
28. Spring cap
29. Cylinder block
30. Bearing cone
31. Bearing cup
32. Shim gaskets
33. Steel washer (0.010 inch)
34. Oil seal
35. Bearing retainer cap
36. Blower air bafflė
37. Plug
38. Gasket
39. Breather assy.
40. Breather tube
41. Dowel pin
42. Rod cap
43. Self locking nut
44. Washer
45. Crankshaft gear pin
46. Crankshaft
47. Connecting rod
48. Rod bolt
49. Crankshaft gear
50. Valve lifters
51. Bearing cone
52. Breaker cam
53. Push rod
54. Spring
55. Timing advance weight
56. Rivet
57. Camshaft assy.
58. "O" ring

lower valve spring caps as necessary for oversize valve stem.

Valve lifters should be identified before removal so they may be reinstalled in their original positions. Some models use lifters of different length. Short lifter is installed at intake position and longer lifter is installed at exhaust position.

Valve tappet gap (cold) is measured with piston at TDC on compression stroke. Correct clearance for all models is 0.010 inch (0.254 mm).

Adjust clearance by squarely grinding valve stem to increase clearance or grinding seat deeper to decrease clearance.

DYNA-STATIC BALANCER. Dyna-Static engine balancer used on some models consists of counterweighted

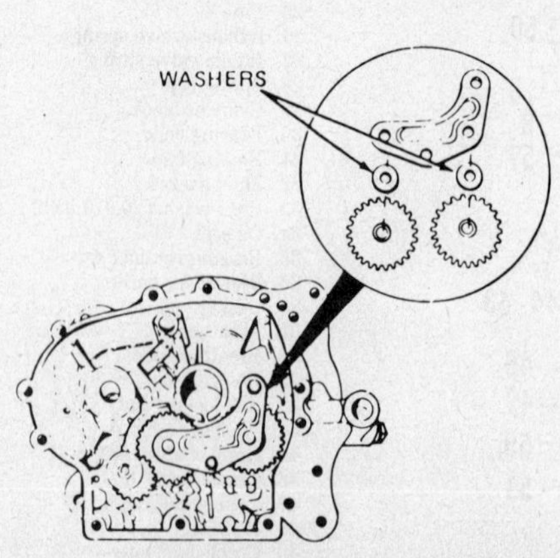

Fig. T44—View showing Dyna-Static balancer gears installed in models so equipped. Note location of washers between gears retaining bracket.

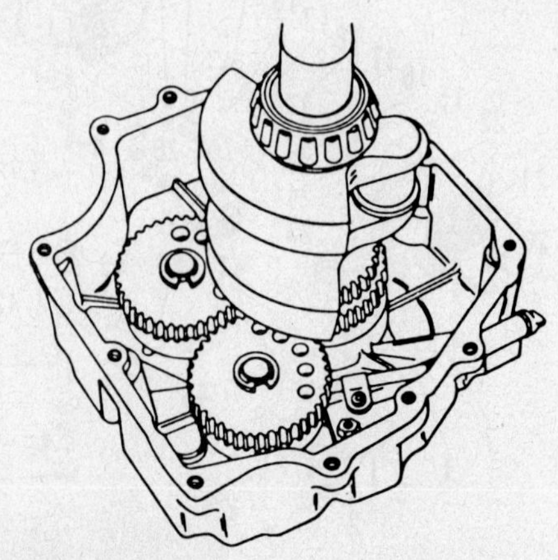

Fig. T45—View showing Dyna-Static balancer gears installed in HH80, HH100 or HH120 models. Note gear retaining snap rings. Refer to Fig. T44 also.

Fig. T46 – On HM80, VM80, HM100 and VM100 models, balancer gear shafts must be pressed into cover or engine base so a distance of 1.757-1.763 inches (44.52-44.78 mm) exists between shaft bore boss and edge of step cut as shown.

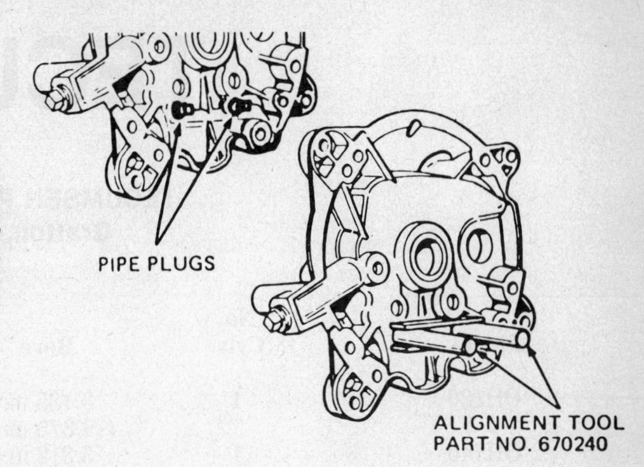

Fig. T50 – To time engine balancer gears, remove pipe plugs and insert alignment tool number 670240 through crankcase cover (HM80 and HM100) or engine base (VM80 and VM100) and into slots in balancer gears. Refer also to Fig. T52.

Fig. T47 – On HH80, HH100 and HH120 models, press balancer gear shafts into cover until 1.7135-1.7185 inches (43.52-43.64 mm) exists between cover boss and outer edge of snap ring groove as shown.

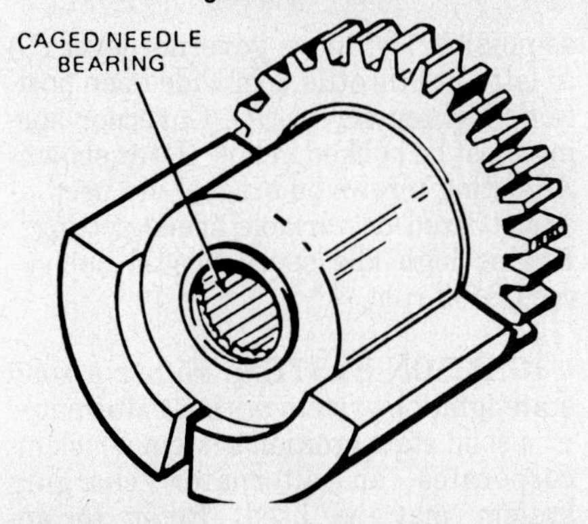

Fig. T48 – Using tool number 670210, press new needle bearing into HM80, VM80, HM100 or VM100 models balancer gears until bearing cage is flush to 0.015 inch (0.381 mm) below edge of bore.

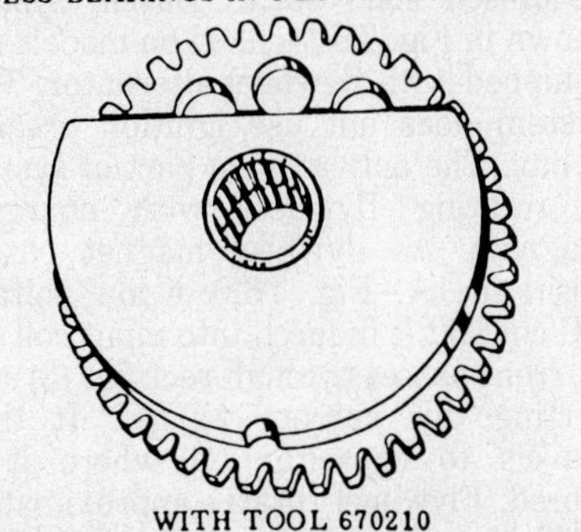

Fig. T49 – Using tool number 670210, press new needle bearings into HH80, HH100 and HH120 models balancer gears until bearing cage is flush to 0.015 inch (0.381 mm) below edge of bore.

gears driven by crankshaft gear to counteract unbalance caused by counterweights on crankshaft.

Counterweight gears on medium frame models are held in position on shafts by a bracket bolted to crankcase or engine base (Fig. T44). Snap rings are used on heavy frame models to retain counterweight gears on shafts which are pressed into crankcase cover (Fig. T47).

Renewable balancer gear shafts are pressed into crankcase cover or engine base. On medium frame models, press shafts into cover or engine base until a distance of 1.757-1.763 inches (44.52-44.78 mm) exists between shaft bore boss and edge of step cut on shafts as shown in Fig. T46. Heavy frame model shafts should be pressed into cover until a distance of 1.7135-1.7185 inches (43.52-43.64 mm) exists between cover boss and outer edge of snap ring groove as shown in Fig. T47.

All balancer gears are equipped with renewable cage needle bearings. Using tool number 670210, press new bearings into gears until cage is flush to 0.015 inch (0.381 mm) below edge of bore.

When reassembling engine, balancer gears must be timed with crankshaft for correct operation. Refer to Figs. T50 and T51 and remove pipe plugs. Insert alignment tool number 670240 through crankcase cover or engine base of medium frame models and into timing slots in balancer gears. On heavy frame models use timing tool number 670239. On all models, rotate engine until piston is at TDC on compression stroke and install cover and gear assembly while tools retain gears in correct position. See Figs. T52 and T53.

When correctly assembled, piston should be on TDC and weights on balancer gears should be in directly opposite position. See Figs. T52 and T53.

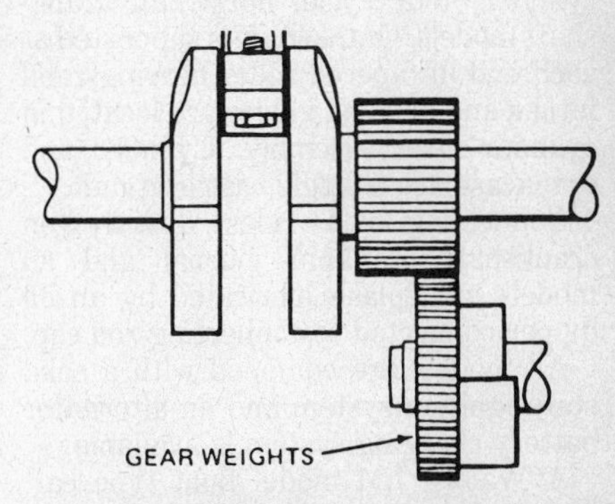

Fig. T52 – View showing correct balancer gear timing to crankshaft gear on HM80, VM80, HM100 and VM100 models. With piston at TDC, weights should be directly opposite.

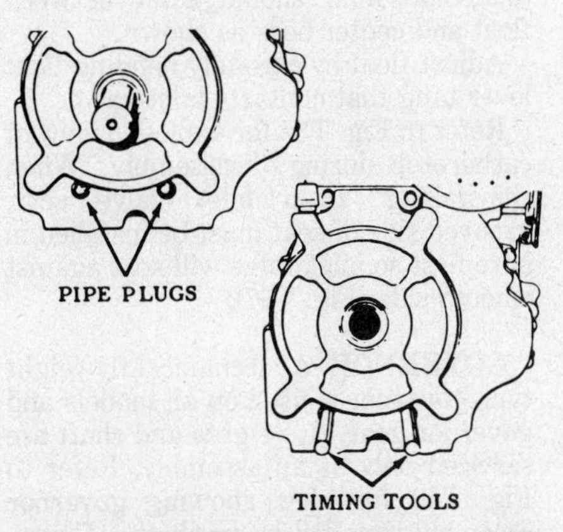

Fig. T51 – To time balancer gears on HH80, HH100 and HH120 models, remove pipe plugs and insert timing tools number 67039 through crankcase cover and into timing slots in balancer gears. Refer also to Fig. T53.

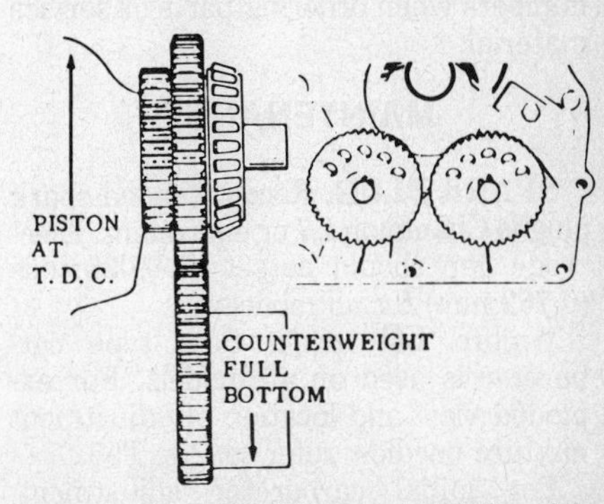

Fig. T53 – On HH80, HH100 and HH120 models, balancer gears are correctly timed to crankshaft when piston is at TDC and weights are at full bottom position.

TECUMSEH

TECUMSEH PRODUCTS COMPANY
Grafton, Wisconsin 53024

Model	No. Cyls.	Bore	Stroke	Displacement	Power Rating
OH120	1	3.125 in. (79.375 mm)	2.75 in. (69.85 mm)	21.1 cu. in. (155.8 cc)	12 hp. (8.9 kW)
OH140	1	3.312 in. (84.125 mm)	2.75 in. (69.85 mm)	23.7 cu. in. (175 cc)	14 hp. (10.4 kW)
OH150	1	3.500 in. (88.90 mm)	2.875 in. (72.625 mm)	27.66 cu. in. (203.2 cc)	15 hp. (11.2 kW)
OH160	1	3.500 in. (88.90 mm)	2.875 in. (72.625 mm)	27.66 cu. in. (203.2 cc)	16 hp. (11.9 kW)
OH180	1	3.625 in. (92.075 mm)	2.875 in. (72.625 mm)	30 cu. in. (218 cc)	18 hp. 13.4 kW)
HH140	1	3.312 in. (84.125 mm)	2.75 in. (69.85 mm)	23.7 cu. in. (175 cc)	14 hp. (10.4 kW)
HH150	1	3.500 in. (88.90 mm)	2.875 in. (72.625 mm)	27.66 cu. in. (203.2 cc)	15 hp. (11.2 kW)
HH160	1	3.500 in. (88.90 mm)	2.875 in. (72.625 mm)	27.66 cu. in. (203.2 cc)	16 hp. (11.9 kW)

Engines in this section are one cylinder, four cycle, horizontal crankshaft models. Crankshaft is supported at each end in tapered roller bearings and intake and exhaust valves are located in cylinder head assembly. Cylinder and crankcase are a single cast iron unit.

Connecting rod rides directly on crankshaft crankpin journal and all models are splash lubricated by an oil dipper connected to connecting rod cap.

All models are equipped with a solid state ignition system and an alternator battery charging system is available.

A Walbro LM model float type carburetor is used on all models.

Refer to **TECUMSEH ENGINE IDENTIFICATION INFORMATION** section for engine identification. Always give model, serial and specification numbers when ordering parts or service material.

MAINTENANCE

SPARK PLUG. Recommended spark plug is Champion L7 or equivalent. Electrode gap should be set at 0.030 inch (0.762 mm) for all models.

Walbro LM model float type carburetor is used on all models. For exploded view and location of adjustment mixture needles, refer to Fig. T55.

For initial carburetor adjustment, open idle mixture screw and main fuel mixture screw one turn. Make final adjustments with engine at normal operating temperature and running. Place engine under load at rated engine speed and adjust main fuel mixture screw for leanest setting that will allow satisfactory acceleration and steady governor operation. Set engine at idle speed, no load and adjust idle mixture screw for smoothest idle operation. Adjust idle speed stop screw (8 – Fig. T55) so engine idles at 1200 rpm.

As each adjustment affects the other, adjustment procedures may have to be repeated.

To check float level, refer to Fig. T56. Invert carburetor throttle body and float assembly. A distance of 0.275-0.315 inch (6.99-8.00 mm) should exist between float and center boss as shown.

Adjust float by carefully bending float lever tang that contacts inlet valve.

Refer to Fig. T55 for exploded view of carburetor during disassembly. When reinstalling Viton inlet valve seat, grooved side of seat must be installed in bore first so inlet valve will seat against smooth side (Fig. T57).

GOVERNOR. A mechanical flyweight type governor is used on all models and governor gear, flyweights and shaft are serviced only as an assembly. Refer to Fig. T59 for view showing governor assembly installed in crankcase. Governor gear is driven by camshaft gear.

To adjust external governor linkage, refer to Fig. T60. Loosen screw (A), turn plate (B) counter-clockwise as far as possible and move governor lever (C) to left until throttle is in wide open position. Tighten screw (A). Governor spring must be hooked in hole (D) as shown. Adjusting screws on bracket are used to adjust fixed or variable speed settings. Engine high idle speed should not exceed 3600 rpm.

IGNITION SYSTEM. Either a solid state ignition system without alternator or a solid state ignition system which incorporates an alternator charging system may be used. Refer to appropriate paragraph for model being serviced.

SOLID STATE IGNITION (WITHOUT ALTERNATOR). Tecumseh solid state ignition system shown in Fig. T61 is used on models not equipped with flywheel alternator. This system does not use ignition breaker points. The only moving part of system is rotating flywheel with charging magnets. As flywheel magnet passes position (1A – Fig. T62), a low voltage AC current is induced into input coil (2). Current passes through rectifier (3) converting this current to DC. It then travels to capacitor (4) where it is stored. Flywheel rotates approximately 180° to position (1B). As it passes trigger coil (5), it induces a very small electric charge into coil. This charge passes through resistor (6) and turns on SCR (silicon controlled rectifier) switch (7).

With SCR switch closed, low voltage current stored in capacitor (4) travels to pulse transformer (8). Voltage is stepped up instantaneously and current is discharged across electrodes of spark plug (9), producing a spark before "top dead center".

Units may be equipped with a second trigger coil and resistor set to turn SCR switch on at a lower rpm. This second trigger pin is closer to flywheel and produces a spark at TDC for easier starting. As engine rpm increases, first (shorter) trigger pin turns SCR switch on, firing spark plug before "top dead center".

If system fails to produce a spark at spark plug, first check high tension lead (Fig. T61). If condition of high tension lead is questionable, renew pulse transformer and high tension lead assembly. Check low tension lead and renew if insulation is faulty. Magneto charging coil, electronic triggering system and mounting plate are available only as an assembly. If necessary to renew this assembly, place unit in position on engine. Start retaining screws, turn mounting plate counter-clockwise as far as possible, then tighten retaining screws to 5-7 ft.-lbs. (7-10 N·m) torque.

SOLID STATE IGNITION (WITH ALTERNATOR). Tecumseh solid state ignition system used on models equipped with flywheel alternator does not use ignition points. The only moving part of system is rotating flywheel with charging magnets and trigger pins. Other components of system are ignition generator coil and stator assembly, spark plug and ignition unit.

Long trigger pin induces a small charge of current to close SCR (silicon controlled rectifier) switch at engine

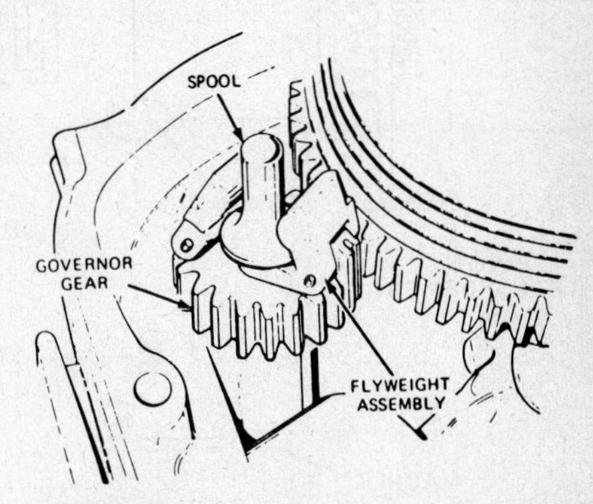

Fig. T59—View showing governor assembly installed in crankcase. Governor gear is driven by camshaft gear.

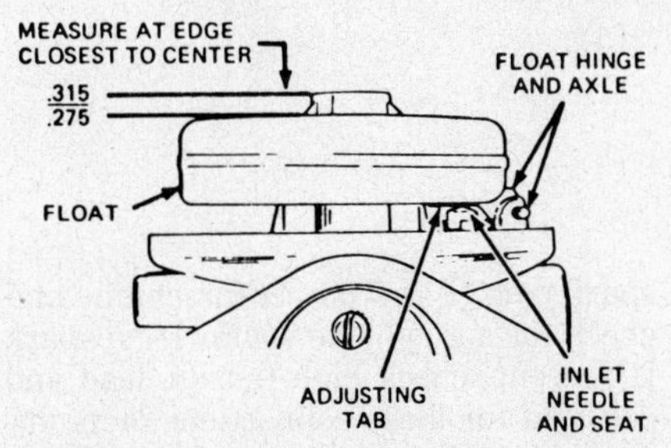

Fig. T56—Float setting should be measured as shown. Bend adjusting tab to adjust float setting

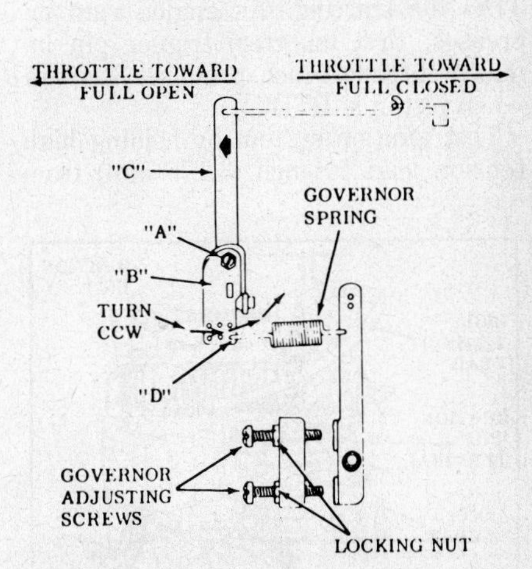

Fig. T60—Typical external governor linkage. Refer to text for adjustment procedures.

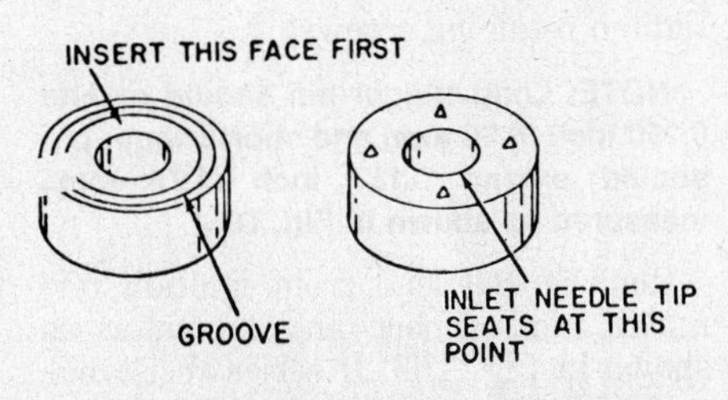

Fig. T57—Viton inlet fuel valve seat must be installed grooved side first.

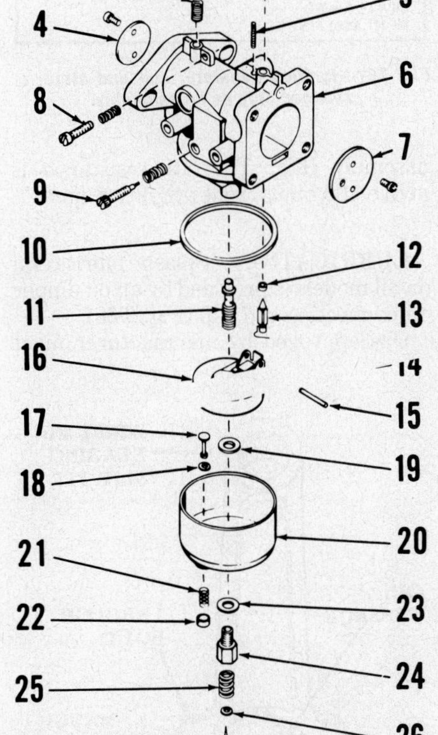

Fig. T55—Exploded view of typical Walbro carburetor used on all models.

1. Choke shaft
2. Throttle shaft
3. Throttle return spring
4. Throttle plate
5. Choke stop spring
6. Carburetor body
7. Choke plate
8. Idle speed stop screw
9. Idle mixture needle
10. Bowl gasket
11. Main nozzle
12. Inlet valve seat
13. Inlet valve
14. Float spring
15. Float shaft
16. Float
17. Drain stem
18. Gasket
19. Gasket
20. Bowl
21. Spring
22. Retainer
23. Gasket
24. Bowl retainer
25. Spring
26. "O" ring
27. Main fuel adjusting needle

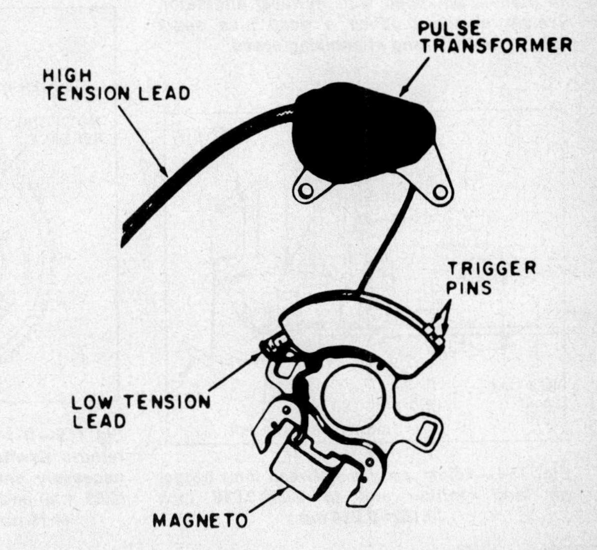

Fig. T61—View of solid state ignition system used on models not equipped with flywheel alternator.

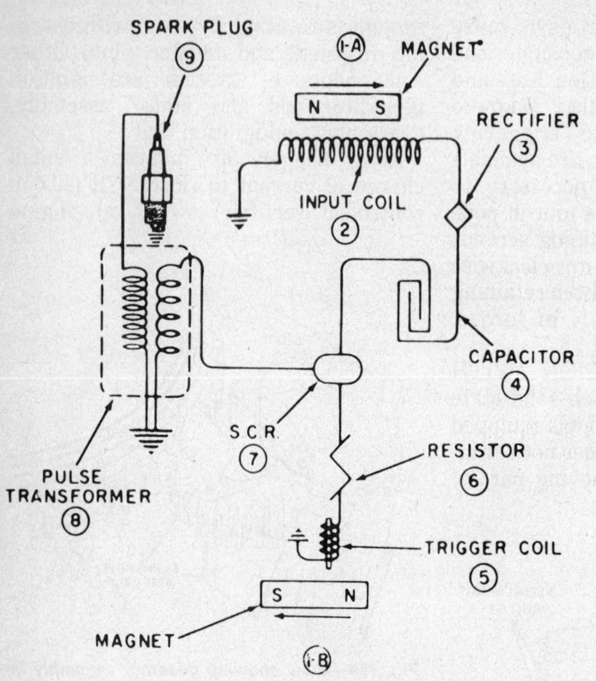

Fig. T62 — Operational diagram of solid state ignition system used on some models.

cranking speed and produces a spark at TDC for starting. As engine rpm increases, first (shorter) trigger pin induces current which produces a spark when piston is BTDC.

Test ignition system by holding high tension lead 1/8-inch (3.175 mm) from spark plug (Fig. T63), crank engine and check for a good blue spark. If no spark is present, check high tension lead and coil lead for loose connections or faulty insulation. Check air gap between long trigger pin and ignition unit as shown in Fig. T64. Air gap should be 0.006-0.010 inch (0.1524-0.254 mm). To adjust air gap, loosen two retaining screws and move ignition unit as necessary, then tighten retaining screws.

NOTE: Long trigger pin should extend 0.250 inch (6.35 mm) and short trigger pin should extend 0.187 inch (4.75 mm), measured as shown in Fig. T65.

Remove coil lead from ignition terminal and connect an ohmmeter as shown in Fig. T66. If series resistance test of ignition generator coil is below 400 ohms, renew stator and coil assembly (Fig. T67). If resistance is above 400 ohms, renew ignition unit.

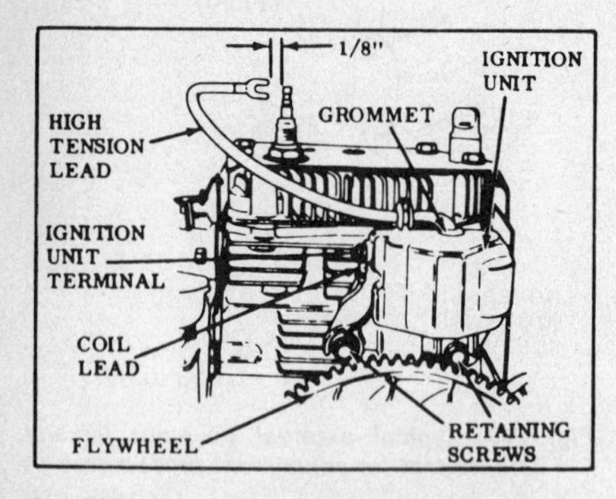

Fig. T63 — View of solid state ignition unit used on models equipped with flywheel alternator. System should produce a good blue spark 1/8-inch long at cranking speed.

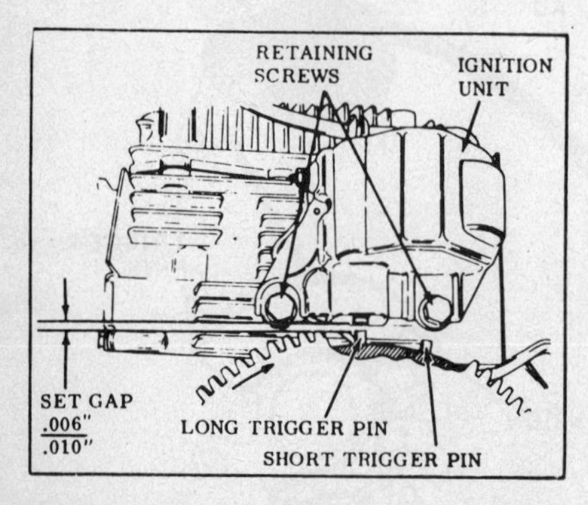

Fig. T64 — Adjust air gap between long trigger pin and ignition unit to 0.006-0.010 inch (0.1524-0.254 mm).

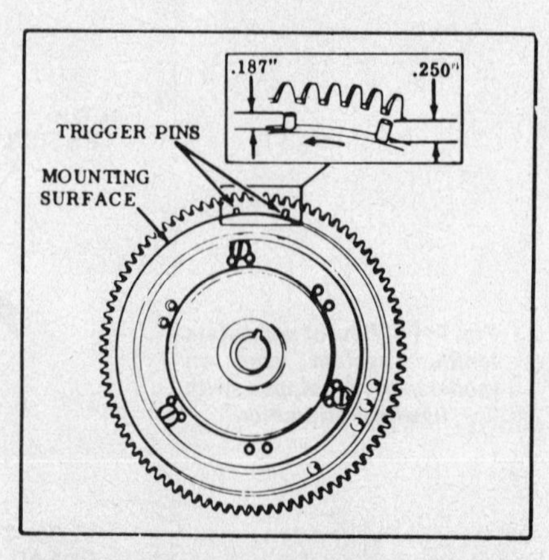

Fig T65 — If trigger pin extension is incorrect, remove flywheel and drive pins in or out, as necessary, until long pin is extended 0.250 inch (6.35 mm) and short pin is extended 0.187 inch (4.75 mm) above mounting surface.

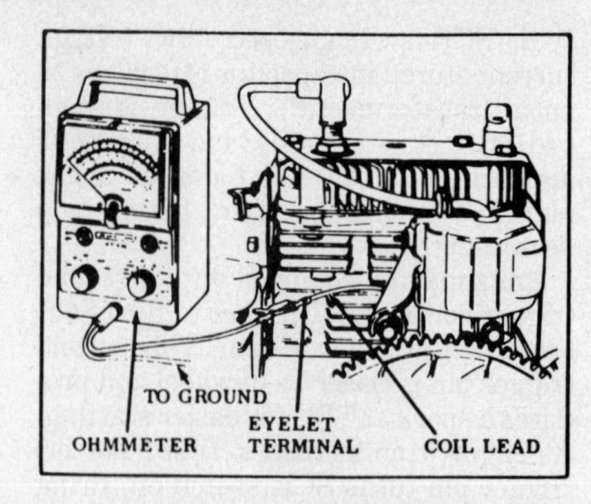

Fig. T66 — View showing ohmmeter connected for resistance test of ignition generator coil.

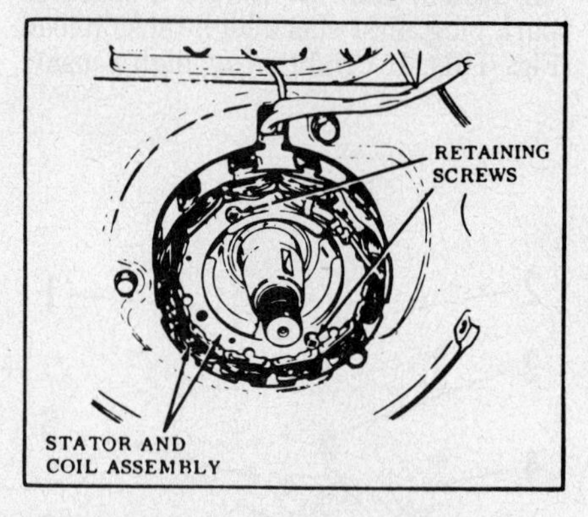

Fig. T67 — Ignition generator coil and stator is serviced only as an assembly.

LUBRICATION. Splash lubrication on all models is provided by an oil dipper on connecting rod cap (Fig. T68).

Oils approved by manufacturer must

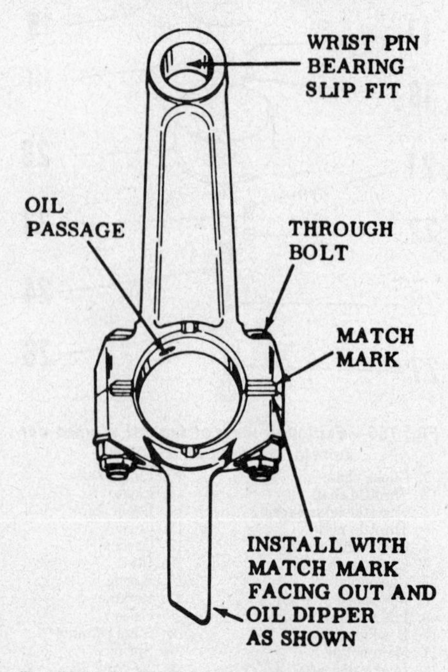

Fig. T68 — Connecting rod assembly used on all models. Note oil dipper on rod cap.

meet requirements of API service classifications MS, SC, SD or SE.

Use SAE 30 oil for operating temperatures above 32° F (0° C) and SAE 10W oil for temperatures below 32° F (0° C).

REPAIRS

TIGHTENING TORQUES. Recommended tightening torques are as follows:

Cylinder head bolts .... 15-20 ft.-lbs. (20-27 N·m)
Connecting rod .... 7-9 ft.-lbs. (10-12 N·m)
Crankcase cover .... 5-9 ft.-lbs. (7-12 N·m)
Bearing retainer .... 5-9 ft.-lbs. (7-12 N·m)
Flywheel nut .... 50-55 ft.-lbs. (68-75 N·m)
Spark plug .... 18-23 ft.-lbs. (24-31 N·m)
Stator mounting .... 5-7 ft.-lbs. (7-10 N·m)
Carburetor to inlet pipe ..... 4-5 ft.-lbs. (5-7 N·m)
Inlet pipe to head .... 5-8 ft.-lbs. (7-11 N·m)
Rocker arm housing to head .. 7-8 ft.-lbs. (10-11 N·m)
Rocker arm shaft screw ... 15-18 ft.-lbs. (20-24 N·m)
Rocker arm cover .... 1-2 ft.-lbs. (1-3 N·m)

CYLINDER HEAD. To remove cylinder head, unbolt and remove blower housing, valve cover and breather assembly. Turn crankshaft until piston is at "top dead center" on compression stroke (Fig. T69), loosen locknuts on rocker arms and back off adjusting screws. Remove snap rings from rocker shaft and remove rocker arms. Using valve spring compressor tool (special tool number 670237) as shown in Fig. T70, remove valve retainers. Remove upper spring cap, valve spring, lower spring cap and "O" ring from each valve. Remove the three cap screws, washers and "O" rings from inside rocker arm housing and carefully lift off housing. Push rods and push rod tubes can now be withdrawn. Unbolt and remove carburetor and inlet pipe assembly from cylinder head. Remove cylinder head cap screws and lift off cylinder head, taking care not to drop intake and exhaust valves.

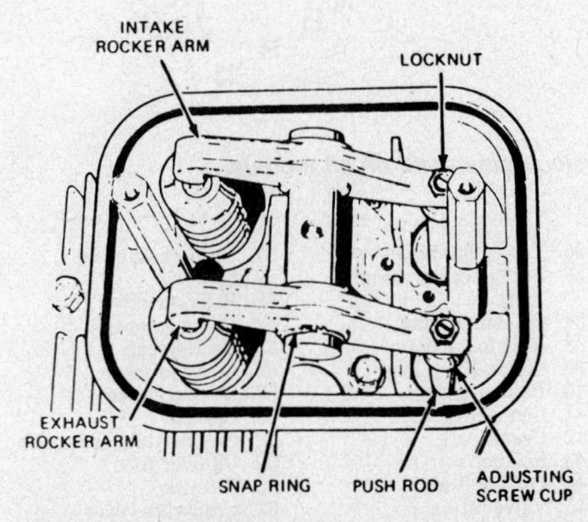

Fig. T69 – View showing rocker arms used on all models. Slotted adjusting screws were used on early production engines. Later engines have adjusting nut on screw below rocker arm.

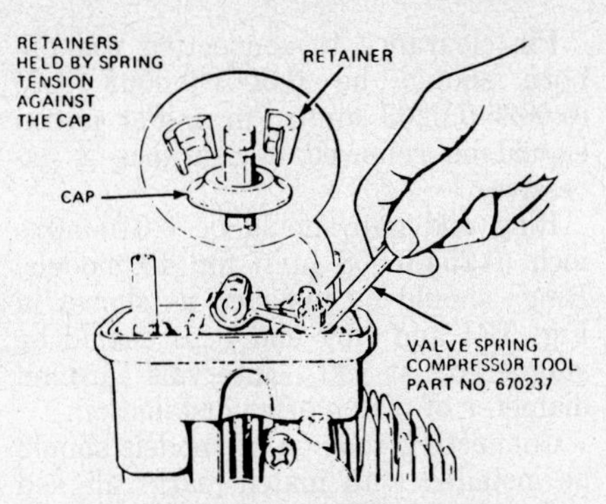

Fig. T70 – Use tool No. 670237 to compress valve springs while removing retainers.

Always use new head gasket when reinstalling cylinder head and make certain Belleville washers and flat washers are properly installed (Fig. T71). Note location of the two short cap screws (Fig. T72) for correct installation. Tighten cylinder head cap screws evenly to 15-20 ft.-lbs. (20-27 N·m) torque using sequence shown in Fig. T72. Place new "O" rings on push rod tubes and install push rods and tubes. Install rocker arm housing and using new "O" rings on the three mounting cap screws, tighten cap screws to 7-8 ft.-lbs. (10-11 N·m) torque. Install new "O" ring, lower spring cap, valve spring and upper spring cap on each valve. Use valve spring compressor (special tool number 670237) to compress valve springs and install retainers. Install rocker arms and secure them with snap rings. Refer to **VALVE SYSTEM** paragraph for valve adjustment procedure and after correctly adjusting valves, install rocker arm cover, breather assembly, carburetor and inlet pipe assembly and blower housing.

CONNECTING ROD. Aluminum alloy connecting rod rides directly on crankshaft crankpin journal for all models.

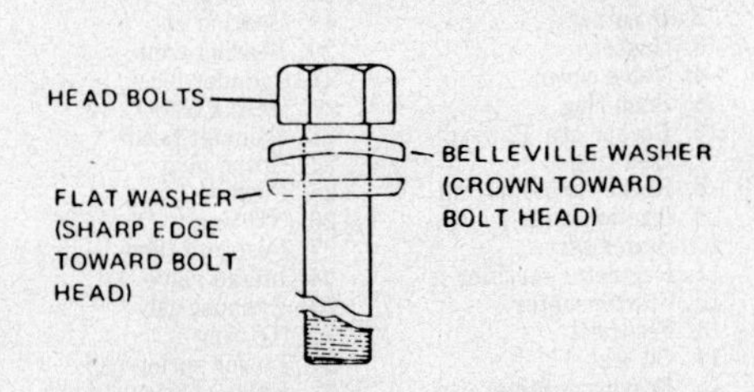

Fig. T71 – Install Belleville washer and flat washer on cylinder head cap screws as shown.

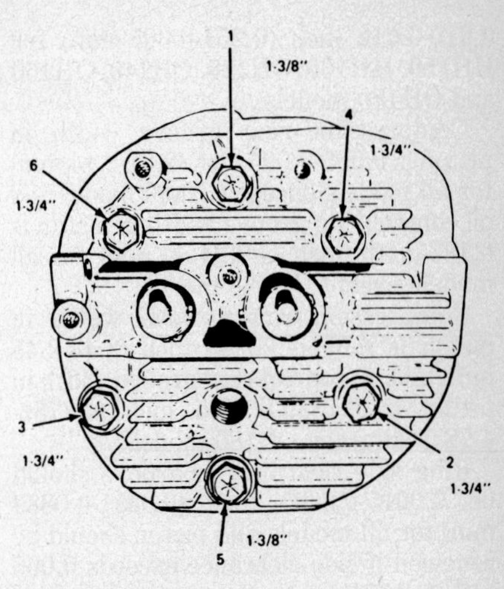

Fig. T72 – Tighten cylinder head cap screws evenly to 15-20 ft.-lbs. (20-27 N·m) torque using tightening sequence shown. Note location of different length cap screws.

Piston and connecting rod assembly is removed from above after removing rocker arm housing, cylinder head, crankcase cover and connecting rod cap.

Standard crankpin journal diameter is 1.3750-1.3755 inches (34.925-34.938 mm) and connecting rods are available in a variety of sizes for undersize crankshafts as well as standard.

When installing piston and connecting rod assembly, make certain match marks on connecting rod and rod cap (Fig. T68) are aligned and marks are facing pto end of shaft. Always renew self-locking nuts on connecting rod bolts and tighten nuts to 7-9 ft.-lbs. (10-12 N·m) torque.

PISTON, PIN AND RINGS. Aluminum alloy piston for all models is fitted with two compression rings and one oil control ring. Piston should be renewed if visibly scored or damaged.

Piston skirt clearance in cylinder, measured at thrust side of piston just below oil control ring should be 0.006-0.008 inch (0.1524-0.2032 mm) for OH140 and HH140 models and

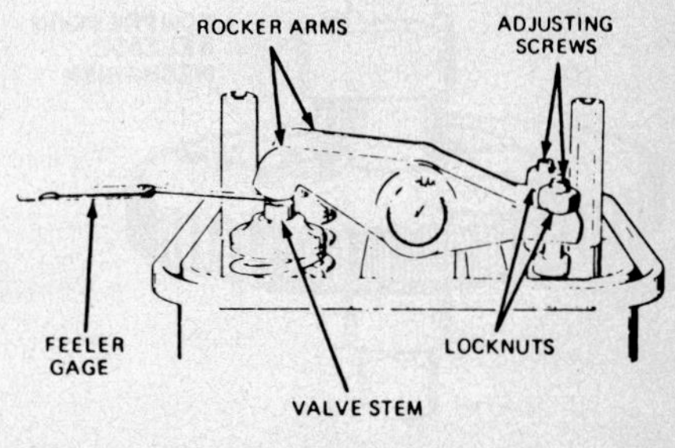

Fig. T73 – Use a feeler gage when adjusting valve tappet gap. Refer to text for adjustment procedure.

0.010-0.012 inch (0.254-0.305 mm) for HH150, HH160, OH120, OH140, OH160 and OH180 models.

Compression ring groove width in piston is 0.095-0.096 inch (2.41-2.44 mm) for all models except OH180 model and oil control ring groove width in piston is 0.188-0.189 inch (4.78-4.80 mm) for all models except OH180 model.

Compression ring groove width in piston is 0.0955-0.0965 inch (2.43-2.45 mm) and oil control ring groove width in piston is 0.1880-0.1885 inch (4.7752-4.7879 mm) for OH180 model.

Ring side clearance in groove should be 0.0015-0.0035 inch (0.0381-0.0889 mm) for all models and piston should be renewed if side clearance exceeds 0.006 inch (0.1524 mm).

Standard piston pin diameter is 0.6876-0.6880 inch (17.465-17.475 mm) for all models except OH180.

Standard piston pin diameter is 0.7810-0.7812 inch (20.07-20.12 mm) for OH180 model.

Pin clearance to connecting rod pin bore should be 0.0001-0.0008 inch (0.0025-0.0203 mm). Pin and/or piston should be renewed if clearance is excessive.

Ring end gap should be 0.010-0.020 inch (0.254-0.508 mm) for all models. Rings should be installed as shown in Fig. T74 and ring end gaps should be staggered at 90° intervals around diameter of piston after installation.

Connecting rods on all models should be installed with match marks aligned and facing pto end of crankshaft.

Piston should be assembled to connecting rod so arrow on piston is pointing to carburetor side of engine after installation.

A variety of oversize pistons and rings are available for oversize bores for all models.

CYLINDER. Standard cylinder bore is 3.125-3.126 inches (79.375-79.400 mm) for OH120 model, 3.312-3.313 inches (84.125-84.150 mm) for HH140 and OH140 models, 3.500-3.501 inches 88.90-88.93 mm) for HH150, OH150, HH160 and OH160 models and 3.625-3.626 inches (92.075-92.100 mm) for OH180 model.

If cylinder is scored or if taper or out-of-round exceeds 0.004 inch (0.1016 mm), cylinder should be rebored to nearest oversize for which piston and rings are available.

CRANKSHAFT. Crankshaft for all models is supported at each end in tapered roller bearings and crankshaft bearings should have 0.001-0.007 inch (0.0254-0.1778 mm) preload. Preload is controlled by varying thickness of shim (17–Fig. T76) on all models.

Bearings are a press fit on crankshaft and should be renewed if removed or if rough or damaged.

Standard crankpin journal diameter

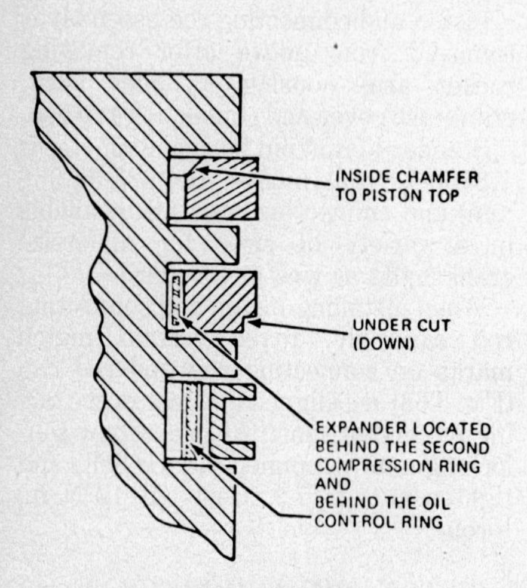

Fig. T74—Cross-sectional view showing correct installation of piston rings.

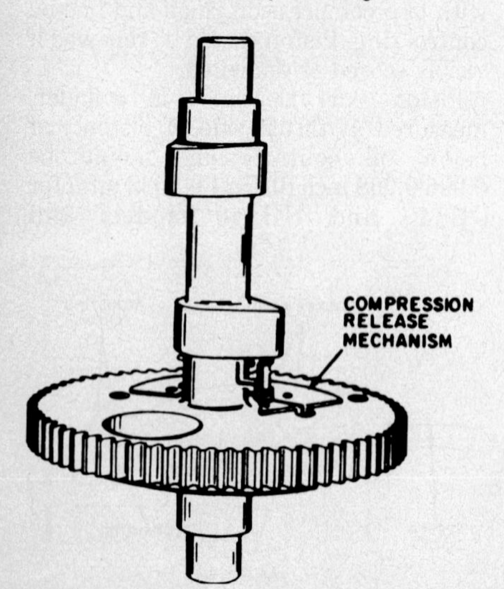

Fig. T75—View of Insta-matic Ezee-Start compression release camshaft assembly used on all models.

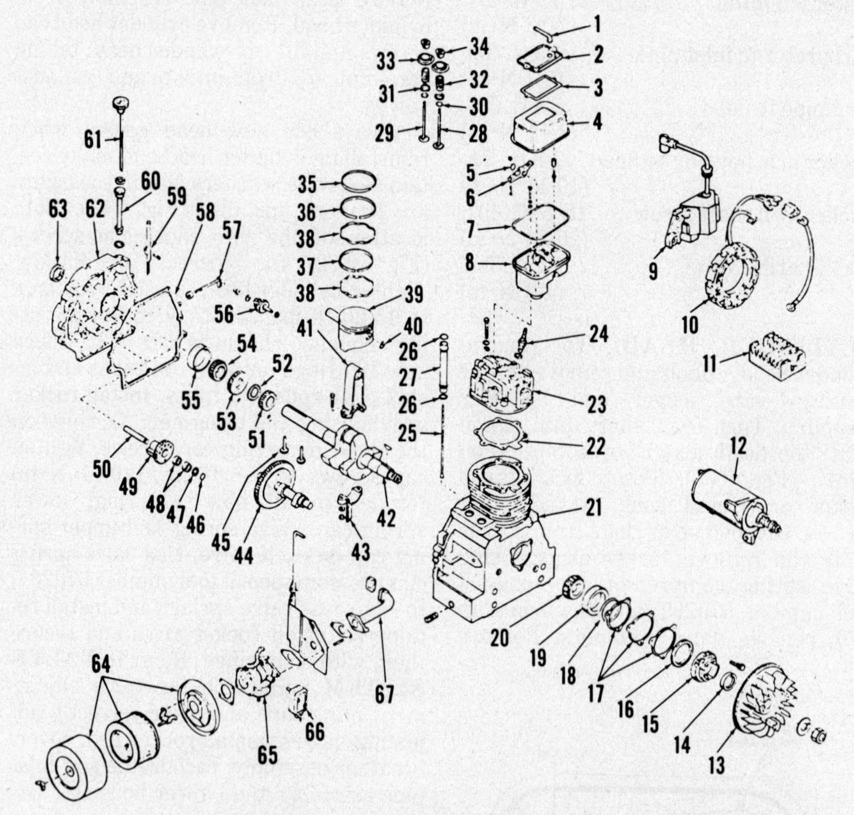

Fig. T76—Exploded view of basic engine used on all models.

1. Breather tube
2. Breather
3. Gasket
4. Valve cover
5. Snap ring
6. Rocker arm (2 used)
7. Seal ring
8. Rocker arm housing
9. Ignition unit
10. Stator assy.
11. Regulator rectifier
12. Starter motor
13. Flywheel
14. Oil seal
15. Bearing retainer cap
16. Steel washer
17. Shim gaskets
18. "O" ring
19. Bearing cup
20. Bearing cone
21. Cylinder block
22. Head gasket
23. Cylinder head
24. Spark plug
25. Push rod
26. "O" ring
27. Push rod tube
28. Intake valve
29. Exhaust valve
30. "O" ring
31. Lower spring cap
32. Valve spring
33. Upper valve cap
34. Valve retainers
35. Top compression ring
36. Second compression ring
37. Oil control ring
38. Ring expanders
39. Piston & pin assy.
40. Retaining ring
41. Connecting rod
42. Crankshaft
43. Rod cap
44. Camshaft assy.
45. Valve lifters
46. Snap ring
47. Thrust washer
48. Needle bearing
49. Dyna-Static balancer
50. Balancer shaft
51. Crankshaft gear
52. Spacer
53. Balancer drive gear
54. Bearing cone
55. Bearing cup
56. Governor assy.
57. Governor arm
58. Gasket
59. Oil filler tube extension
60. Oil filler tube
61. Dipstick
62. Crankcase cover
63. Oil seal
64. Air cleaner assy.
65. Carburetor
66. Fuel pump
67. Inlet pipe

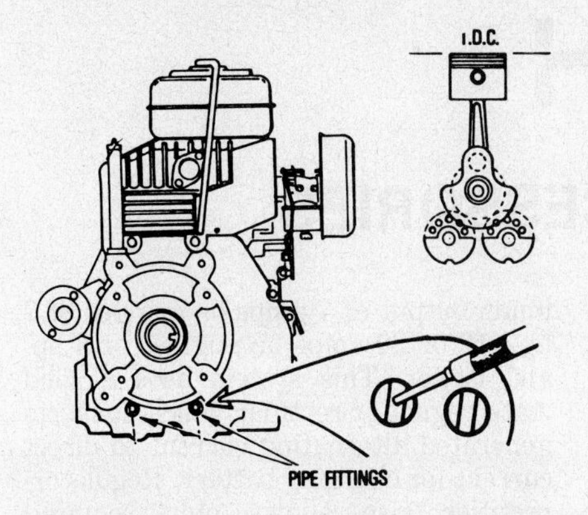

Fig. T77 — View showing location of pipe plugs covering timing alignment holes on models with side-by-side Dyna-Static balancer gears. Note location of notches in balancer gears when properly installed.

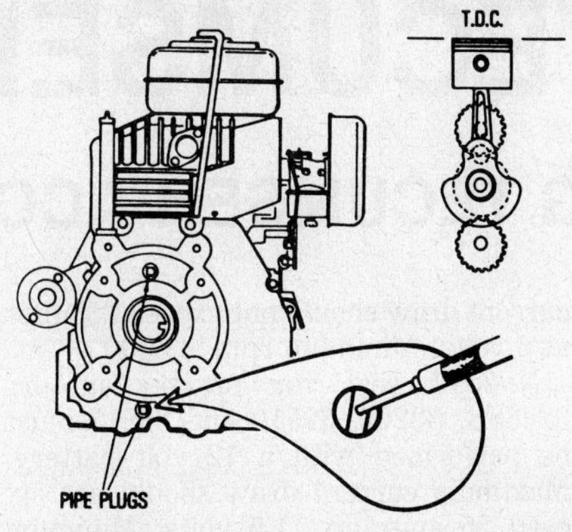

Fig. T78 — View showing location of pipe plugs covering timing alignment holes on models with above and below crankshaft gear balance gears. Note location of notches in balancer gears when properly installed.

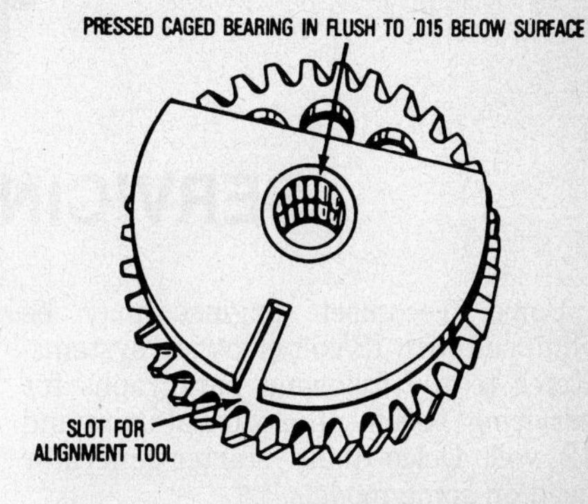

Fig. T79 — Using tool number 670210, press new needle bearing into balancer gears until bearing cage is flush to 0.015 inch (0.381 mm) below edge of bore. Note alignment notch at lower side of balancer.

for all models, is 1.3750-1.3755 inches (34.925-34.938 mm) and connecting rods are available in a variety of sizes for undersize crankshafts as well as standard.

Crankshaft should be renewed or reground if crankpin journal is tapered or worn over 0.002 inch (0.0508 mm) or is out-of-round more than 0.005 inch (0.0127 mm).

When installing crankshaft, align timing mark on crankshaft gear (chamfered tooth) with timing mark on camshaft gear.

Crankshaft oil seals should be installed flush to 0.025 inch (0.635 mm) below surface, with lips on seals facing inward.

CAMSHAFT. Camshaft and camshaft gear are an integral part which ride on journals at each end of camshaft. Camshaft is equipped with a compression release mechanism (Fig. T75). Check compression release parts for binding, excessive wear or other damage. If any parts are excessively worn or damaged, renew complete camshaft assembly. Parts are not serviced separately for compression release mechanism.

Renew camshaft if gear teeth are excessively worn or bearing surfaces or lobes are worn or scored. Camshaft lobe nose-to-heel diameter should be 1.3117-1.3167 inches (33.317-33.444 mm) for all models.

Diameter of camshaft journals is 0.6235-0.6240 inch (15.837-15.850 mm) for all models.

Maximum allowable clearance between camshaft journal and bearing bore is 0.003 inch (0.0762 mm).

When installing camshaft, align timing mark on camshaft gear with timing mark (chamfered tooth) on crankshaft gear.

DYNA-STATIC BALANCER. Dyna-Static engine balancer consists of counterweighted gears, mounted side-by-side (Fig. T77) or above and below crankshaft gear (Fig. T78) and driven by crankshaft to counteract unbalance caused by counterweights on crankshaft. Balancer gears are held in position on balancer shafts by snap rings. Renewable balancer shafts are pressed into cover until a distance of 1.7135-1.7185 inches (43.52-43.65 mm) exists between boss on cover and outer edge of snap ring groove on shafts.

Balancer gears are equipped with renewable caged needle bearings (Fig. T79). Using tool number 670210, press new bearings into balancer gears until bearing cage is flush to 0.015 inch (0.381 mm) below edge of bore.

When reassembling engine, balancer gears must be timed with crankshaft for correct operation. To time balancer gears, refer to Fig. T77 or T78 and remove pipe plugs. Rotate crankshaft so piston is at TDC and install cover and weight assembly so slots in weights are visible through pipe plug openings. See Fig. T77 and T78. When correctly assembled, piston should be exactly at TDC and weights should be at full bottom position.

VALVE SYSTEM. Seats are machined directly into cylinder head surface. Seats should be ground at a 46° angle and seat width should be 0.042-0.052 inch (1.067-1.321 mm).

Valves should be ground at a 45° angle and valve should be renewed if margin is 0.060 inch (1.524 mm) or less.

Valves are available with 1/32-inch (0.794 mm) oversize stems for installation in guides which are badly worn. Guides must be reamed to proper dimension.

Standard valve guide inside diameter is 0.312-0.313 inch (7.925-7.950 mm) for all models. Guides may be reamed to 0.343-0.344 inch (8.71-8.74 mm) for use with 1/32-inch (0.794 mm) oversize valve stems.

To renew valve guides, remove and submerge head in large pan of oil. Heat on a hot plate until oil begins to smoke, about 15-20 minutes. Remove head from pan and place head on arbor press with valve seats facing up. Use a drift punch ½-inch in diameter to press guides out.

CAUTION: Be sure to center punch. DO NOT allow punch to contact head when pressing guides out.

To install new guides, place guides in freezer or on ice for 30 minutes prior to installation. Submerge head in pan of oil. Heat on hot plate until oil begins to smoke, about 15-20 minutes. Remove head and place, gasket surface down, on a 6 x 12 inch piece of wood. Using snap rings to locate both guides, insert silver color guide in intake side and brass colored guide in exhaust side. It may be necessary to use a rubber or rawhide mallet to fully seat snap rings. **DO NOT** use metal hammer or guide damage will result. Allow head to cool and reface both valve seats.

Recommended valve tappet gap (cold) for all models is 0.005 inch (0.127 mm) for intake valves and 0.010 inch (0.254 mm) for exhaust valves.

Valves must be adjusted with engine at TDC on compression stroke. Refer to Fig. T69 for view of rocker arm adjusting screw.

TECUMSEH

SERVICING TECUMSEH ACCESSORIES

Some Tecumseh engines may be equipped with 12 volt electrical systems. Refer to the following paragraphs for servicing Tecumseh electrical units and 12 volt Delco-Remy starter-generator used on some models.

12 VOLT STARTER MOTOR (BENDIX DRIVE TYPE). Refer to Fig. T82 for exploded view of 12 volt starter motor and Bendix drive unit used on some engines. To identify starter, refer to service number stamped on end cap. When assembling starter motor use spacers (15) of varying thicknesses to obtain an armature end play of 0.005-0.015 inch (0.127-0.381 mm). Tighten armature nut (1) to 100 in.-lbs. (11 N·m) torque on motor numbers 29965, 32468, 32468A, 32468B and 33202, to 130-150 in.-lbs. (15-17 N·m) torque on motor number 32817. Tighten through-bolts to 30-35 in.-lbs. (3-4 N·m) torque on motor numbers 29965, 32468, 32468A, 32468B and 33202, to 35-44 in.-lbs. (4-5 N·m) torque on motor number 32817 and to 45-50 in.-lbs. (5-6 N·m) torque on motor number 32510.

To perform no-load test for starter motors 29965, 32468 and 32468A, use a fully charged 6 volt battery. Maximum current draw should not exceed 25 amps at 6 volts. Minimum rpm is 6500.

No-load test for starter motors 32468B, 33202, 32510 and 32817 must be performed with a 12 volt battery. Maximum current draw should not exceed 25 amps at 11.5 volts. Minimum rpm is 8000.

ALTERNATOR CHARGING SYSTEMS. Flywheel alternators are used on some engines for the charging system. Generated alternating current is converted to direct current by two rectifiers on rectifier panel (Fig. T83 and T84) or regulator-rectifier (Fig. T85).

System shown in Fig. T83 has a maximum charging output of about 3 amps at 3600 rpm. No current regulator is used on this low output system. Rectifier panel includes two diodes (rectifiers) and a 6 amp fuse for overload protection.

System shown in Fig. T84 has a maximum output of 7 amps. To prevent overcharging battery, a double pole switch is used in low output position to reduce output to 3 amps for charging battery. Move switch to high output position (7 amps) when using accessories.

System shown in Fig. T85 has a maximum output of 7 amps on engines of 7 hp.; 10 or 20 amps on engines of 8 hp. and larger. This system uses a solid state regulator-rectifier which converts generated alternating current to direct current for charging battery. Regulator-rectifier also allows only required amount of current flow for existing battery conditions. When battery is fully charged, current output is decreased to prevent overcharging battery.

TESTING. On models equipped with rectifier panel (Figs. T83 or T84), remove rectifiers and test them with

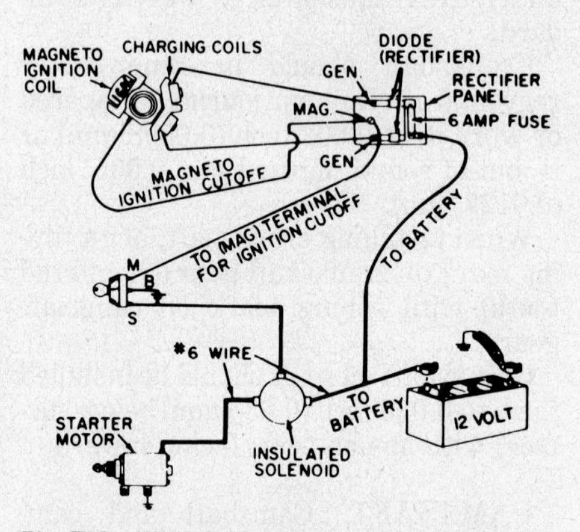

Fig. T83 — Wiring diagram of typical 3 amp alternator and rectifier panel charging system.

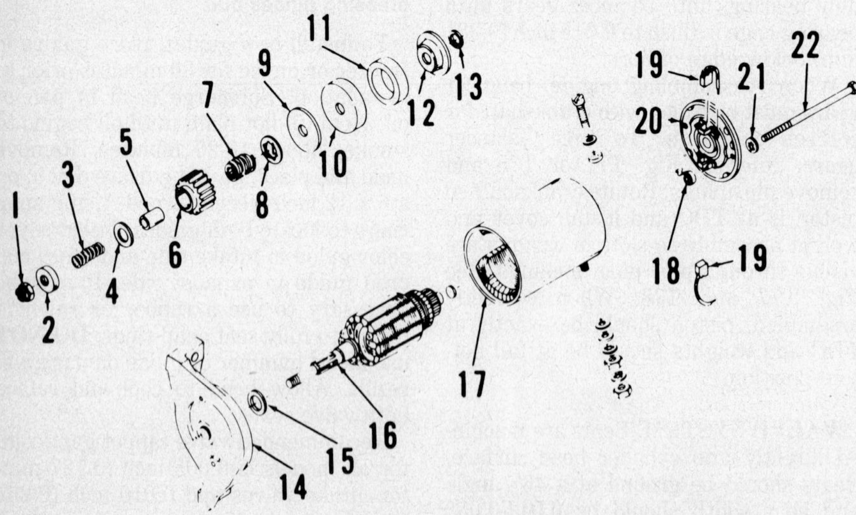

Fig. T82 — Exploded view of typical 12 volt starter motor assembly. Spacer (15) is available in different thicknesses to adjust armature end play.

1. Nut
2. Pinion stop
3. Anti-drift spring
4. Washer
5. Anti-drift sleeve
6. Pinion gear
7. Screw shaft
8. Stop washer
9. Thrust washer
10. Cushion cup
11. Rubber cushion
12. Thrust washer
13. Thrust bushing
14. Drive end cap
15. Spacer washer
16. Armature
17. Frame & field coil assy.
18. Brush spring
19. Brushes
20. End cap
21. Washer
22. Bolt

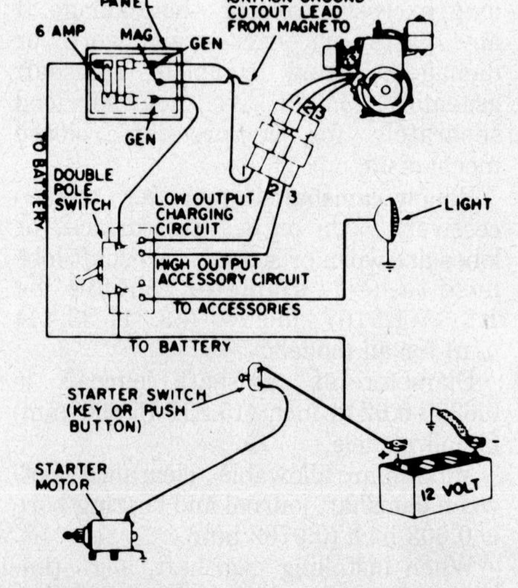

Fig. T84 — Wiring diagram of typical 7 amp alternator and rectifier panel charging system. The double pole switch in one position reduces output to 3 amp for charging or increases output to 7 amp in other position to operate accessories.

either a continuity light or an ohmmeter. Rectifiers should show current flow in one direction only. Alternator output can be checked using an induction ammeter over positive lead wire to battery.

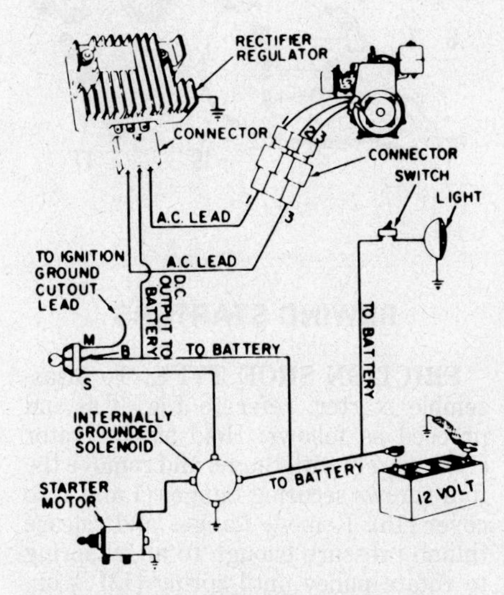

Fig. T85—Wiring diagram of typical 7, 10 or 20 amp alternator and regulator-rectifier charging systems.

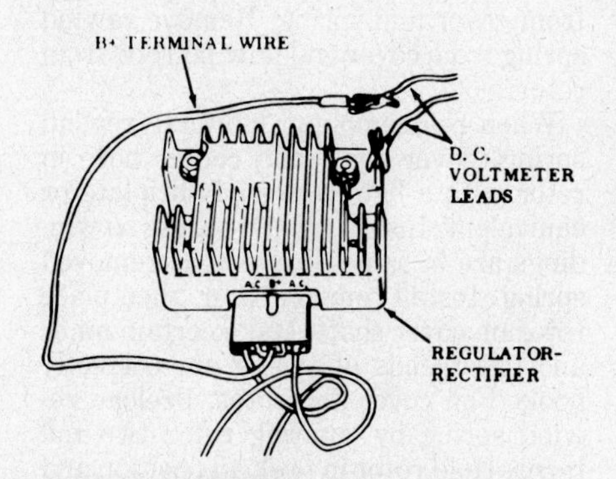

Fig. T86—Connect DC voltmeter as shown when checking regulator-rectifier.

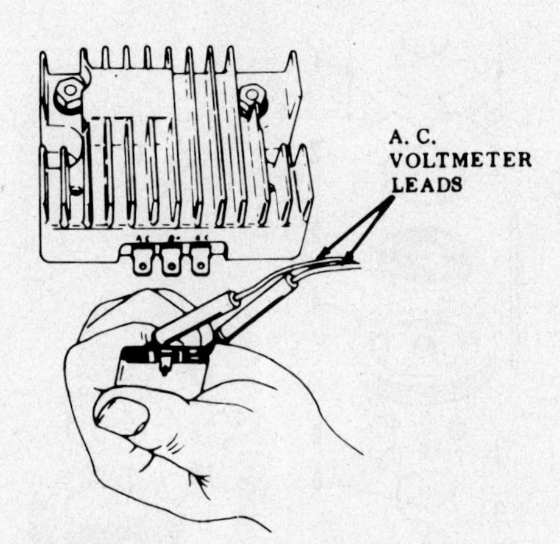

Fig. T87—Connect AC voltmeter to AC leads as shown when checking alternator coils.

On models equipped with regulator-rectifier (Fig. T85), check system as follows: Disconnect B+ lead and connect a DC voltmeter as shown in Fig. T86. With engine running near full throttle, voltage should be 14.0-14.7. If voltage is above 14.7 or below 14.0 but above 0, regulator-rectifier is defective. If voltmeter reading is 0, regulator-rectifier or alternator coils may be defective. To test alternator coils, connect an AC voltmeter to AC leads as shown in Fig. T87. With engine running at near full throttle, check AC voltage. If voltage is less than 20.0 volts on 10 amp or 32.0 volts on 20 amp system, alternator is defective.

MOTOR-GENERATOR. Combination motor-generator (Fig. T88) functions as a cranking motor when starting switch is closed. When engine is operating and starting switch is open, unit operates as a generator. Generator output and circuit voltage for battery and various accessories are controlled by current-voltage regulator.

To determine cause of abnormal operation, motor-generator should be given a "no-load" test or a "generator output" test. Generator output test can be performed with a motor-generator on or off engine. No-load test must be made with motor-generator removed from engine.

Motor-generator test specifications are as follows:

Motor-Generator Delco-Remy No. 1101980.

Brush spring tension, oz. 24-32
Field draw,
  Amperes 1.52-1.62
  Volts 12
Cold output,
  Amperes 12
  Volts 14
  Rpm 4950
No-load test,
  Amperes (max.) 18
  Volts 11
  Rpm (min.) 2500
  Rpm (max.) 2900

CURRENT-VOLTAGE REGULATORS. Two types of current-voltage regulators are used with motor-generator system. One is a low output unit which delivers a maximum of 7 amps. High output unit delivers a maximum of 14 amps.

Low output (7 amp) unit is identified by its four connecting terminals (three on one side of unit and one on underside of regulator). Battery ignition coil has a 3 amp draw. This leaves a maximum load of 4 amps which may be used on accessory lead.

High output (14 amp) unit has only three connecting terminals (all on side of unit). So with a 3 amp draw for battery ignition coil, a maximum of 11 amps can be used for accessories.

Regulator service test specifications are as follows:

Regulator Delco-Remy No. 1118988 (7 amp)

Ground polarity . . . . Negative
Cut-out relay,
  Air gap . . . . 0.020 inch
  (0.508 mm)

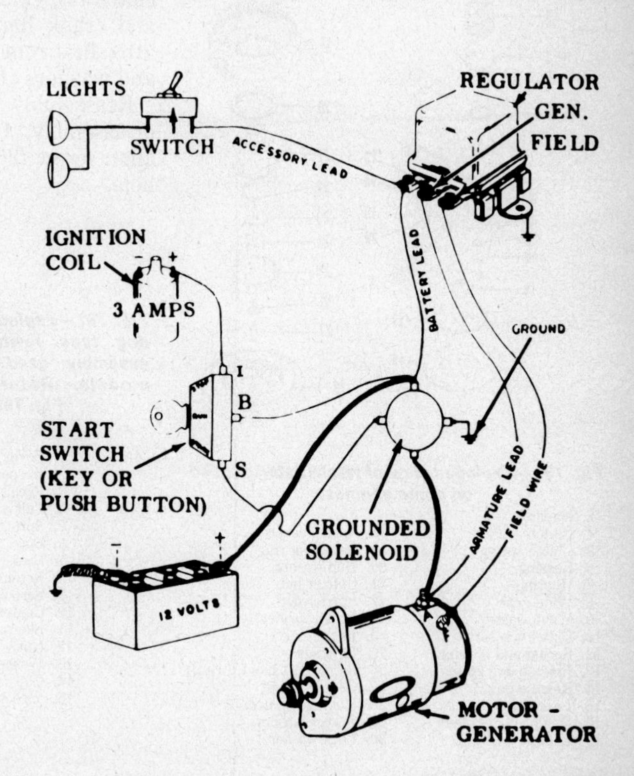

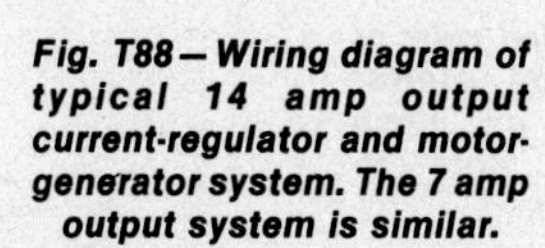

Fig. T88—Wiring diagram of typical 14 amp output current-regulator and motor-generator system. The 7 amp output system is similar.

Point gap . . . . 0.020 inch (0.508 mm)
Closing voltage, range 11.8-14.0
Adjust to 12.8
Voltage regulator,
Air gap . . . . 0.075 inch (1.9 mm)
Setting volts, range 13.6-14.5
Adjust to 14.0

Regulator Delco-Remy No. 1119207 (14 amp)
Ground polarity . . . . Negative
Cut-out relay,
Air gap . . . . 0.020 inch (0.508 mm)
Point gap . . . . 0.020 inch (0.508 mm)
Closing voltage, range 11.8-13.5
Adjust to 12.8
Voltage regulator,
Air gap . . . . 0.075 inch (1.9 mm)
Voltage setting:
14.4-15.5 @ 65° (18° C)
14.2-15.2 @ 85° (29° C)
14.0-14.9 @ 105° (41° C)
13.8-14.8 @ 125° (52° C)
13.5-14.3 @ 145° (63° C)
13.1-13.9 @ 165° (74° C)
Current regulator,
Air gap . . . . 0.075 inch (1.9 mm)
Current setting 13-15

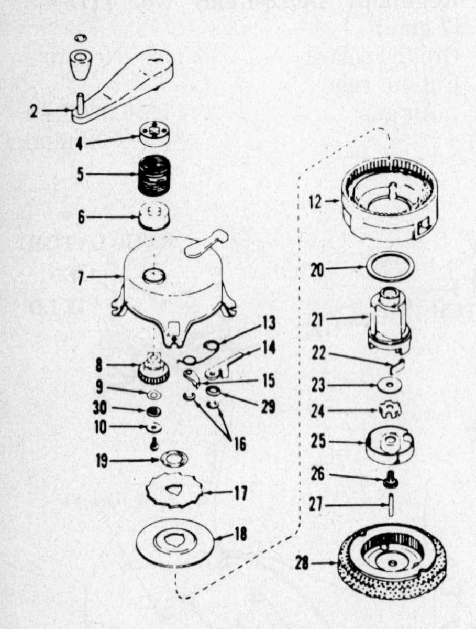

Fig. T89—Exploded view of ratchet starter used on some engines.

2. Handle
4. Clutch
5. Clutch spring
6. Bearing
7. Housing
8. Wind gear
9. Wave washer
10. Clutch washer
12. Spring and housing
13. Release dog spring
14. Release dog
15. Lock dog
16. Dog pivot retainers
17. Release gear
18. Spring cover
19. Retaining ring
20. Hub washer
21. Starter hub
22. Starter dog
23. Brake washer
24. Brake
25. Retainer
26. Screw (left hand thread)
27. Centering pin
28. Hub and screen
29. Spacer washers
30. Lock washer

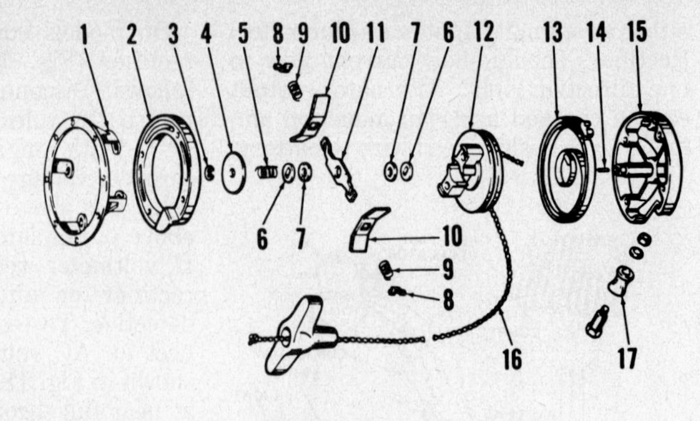

Fig. T90—Exploded view of typical friction shoe rewind starter assembly.

1. Mounting flange
2. Flange
3. Retaining ring.
4. Washer
5. Spring
6. Slotted washer
7. Fibre washer
8. Spring retainer
9. Spring
10. Friction shoe
11. Actuating lever
12. Rotor
13. Rewind spring
14. Centering pin
15. Cover
16. Rope
17. Roller

WIND-UP STARTER

RATCHET STARTER. On models equipped with ratchet starter, refer to Fig. T89 and proceed as follows: Move release lever to "RELEASE" position to remove tension from main spring. Remove starter assembly from engine. Remove left hand thread screw (26), retainer hub (25), brake (24), washer (23) and six starter dogs (22). Note position of starter dogs in hub (21). Remove hub (21), washer (20), spring and housing (12), spring cover (18), release gear (17) and retaining ring (19) as an assembly. Remove retaining ring, then carefully separate these parts.

CAUTION: Do not remove main spring from housing (12). The spring and housing are serviced only as an assembly.

Remove snap rings (16), spacer washers (29), release dog (14), lock dog (15) and spring (13). Winding gear (8), clutch (4), clutch spring (5), bearing (6) and crank handle (2) can be removed after first removing the retaining screw and washers (10, 30 and 9).

Reassembly procedure is reverse of disassembly. Centering pin (27) must align screw (26) with crankshaft center hole.

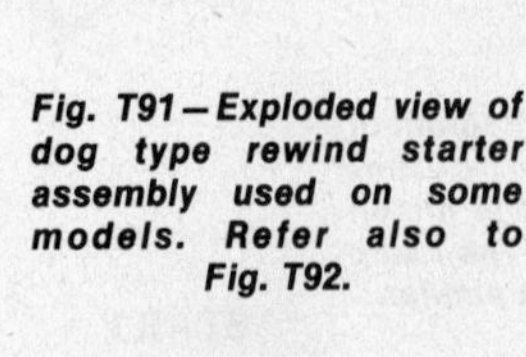

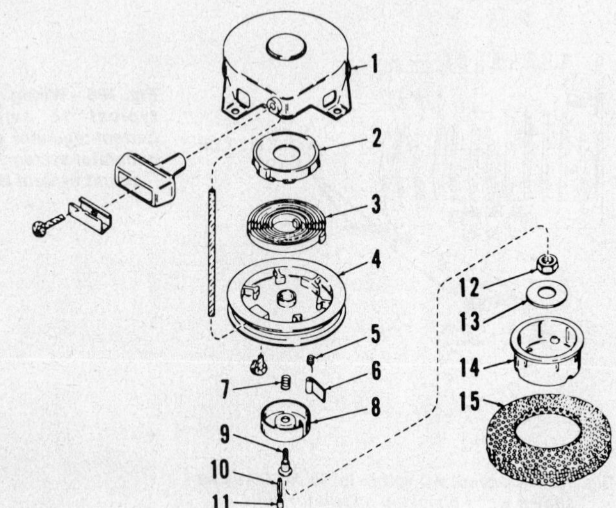

Fig. T91—Exploded view of dog type rewind starter assembly used on some models. Refer also to Fig. T92.

1. Cover
2. Keeper
3. Recoil spring
4. Pulley
5. Spring
6. Dog
7. Brake spring
8. Retainer
9. Screw
10. Centering pin
11. Sleeve
12. Nut
13. Washer
14. Cup
15. Screen

REWIND STARTERS

FRICTION SHOE TYPE. To disassemble starter, refer to Fig. T90 and proceed as follows: Hold starter rotor (12) securely with thumb and remove the four screws securing flanges (1 and 2) to cover (15). Remove flanges and release thumb pressure enough to allow spring to rotate pulley until spring (13) is unwound. Remove retaining ring (3), washer (4), spring (5), slotted washer (6) and fibre washer (7). Lift out friction shoe assembly (8, 9, 10 and 11), then remove second fibre washer and slotted washer. Withdraw rotor (12) with rope from cover and spring. Remove rewind spring from cover and unwind rope from rotor.

When reassembling, lubricate rewind spring, cover shaft and center bore in rotor with a light coat of Lubriplate or equivalent. Install rewind spring so windings are in same direction as removed spring. Install rope on rotor, then place rotor on cover shaft. Make certain inner and outer ends of spring are correctly hooked on cover and rotor. Preload rewind spring by rotating rotor two full turns. Hold rotor in preload position and install flanges (1 and 2). Check sharp end of friction shoes (10) and sharpen or renew as necessary. Install washers (6 and 7), friction shoe assembly, spring (5), washer (4) and retaining ring (3). Make certain friction shoe assembly is installed properly for correct starter rotation. If properly installed, sharp ends of friction shoes will extend when rope is pulled.

Remove brass centering pin (14) from cover shaft, straighten pin if necessary, then reinsert pin 1/3 of its length into cover shaft. When installing starter on engine, centering pin will align starter with center hole in end of crankshaft.

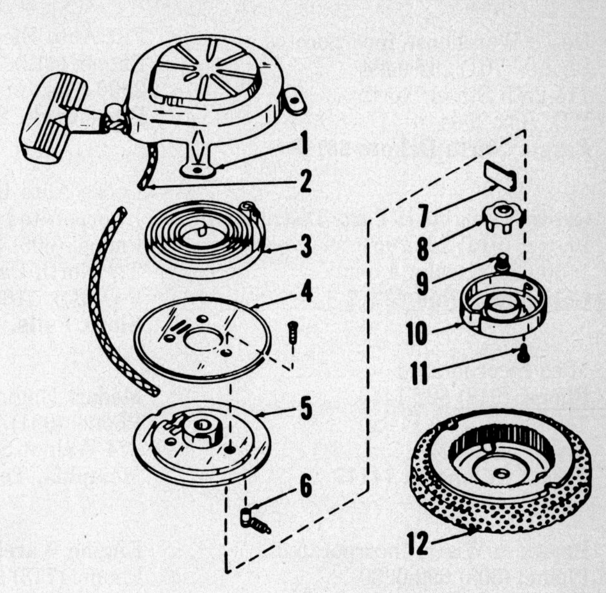

Fig. T92—Exploded view of dog type rewind starter assemby used on some models. Some units use three starter dogs (7).

1. Cover
2. Rope
3. Rewind spring
4. Pulley half
5. Pulley half and hub
6. Retainer spring
7. Starter dog
8. Brake
9. Brake screw
10. Retainer
11. Retainer screw
12. Hub & screen assy.

DOG TYPE. Two dog type starters may be used as shown in Fig. T91 and Fig. T92. Disassembly and assembly is similar. To disassemble starter shown in Fig. T91, remove starter from engine and while holding pulley remove rope handle. Allow recoil spring to unwind. Remove starter components in order shown in Fig. T91 noting position of dog (6) and direction spring (3) is wound. Be careful when removing recoil spring (3). Reassemble by reversing disassembly procedure. Turn pulley six turns before passing rope through cover so spring (3), is preloaded. Tighten retainer screw (9) to 45-55 in.-lbs. (5-6 N•m) torque.

To disassemble starter shown in Fig. T92, pull starter rope until notch in pulley half (5) is aligned with rope hole in cover (1). Hold pulley and prevent from rotating. Engage rope in notch and allow pulley to slowly rotate so recoil spring will unwind. Remove components as shown in Fig. T92. Note direction recoil spring is wound being careful when removing spring from cover. Reassemble by reversing disassembly procedure. Preload recoil spring by turning pulley two turns with rope.

TECUMSEH CENTRAL WAREHOUSE DISTRIBUTORS

(Arranged Alphabetically by States)

These franchised firms carry extensive stocks of repair parts. Contact them for name of dealer in their area who will have replacement parts.

Charlie C. Jones Battery & Electric Co., Inc.
Phone: (602) 272-5621
2440 West McDowell Road
P.O. Box 6654
Phoenix, Arizona 85005

Pacific Power Equipment Company
Phone: (415) 692-1094
1565 Adrian Road
Burlingame, California 94010

Pacific Power Equipment Company
Phone: (303) 371-4081
500 Oakland Street
Denver, Colorado 80239

Spencer Engine Incorporated
Phone: (813) 253-6035
1114 West Cass Street
P.O. Box 2579
Tampa, Florida 33601

Sedco Incorporated
Phone: (404) 925-4706
1414 Red Plum Road NW
Norcross, Georgia 30093

Small Engine Clinic
Phone: (808) 488-0711
98019 Kam Highway
Honolulu, Hawaii 96701

Industrial Engine & Parts
Phone: (312) 927-4100
1133 West Pershing Road
Chicago, Illinois 60609

Medart Engines & Parts of Kansas
Phone: (913) 888-8828
15500 West 109th Street
Lenexa, Kansas 66219

Grayson Company of Louisiana
Phone: (318) 222-3211
100 Fannin
P.O. Box 206
Shreveport, Louisiana 71102

W. J. Connell Company
Phone: (617) 543-3600
65 Green Street
Route 106
Foxboro, Massachusetts 02035

Carl A. Anderson Inc. of Minnesota
Phone: (612) 452-2010
2737 South Lexington
Eagan, Minnesota 55121

Medart Engines & Parts
Phone: (314) 343-0505
100 Larkin William Industrial Ct.
Fenton, Missouri 63026

Original Equipment Incorporated
Phone: (406) 245-3081
905 Second Avenue North
Box 2135
Billings, Montana 59103

Carl A. Anderson Incorporated
Phone: (404) 339-4944
7410 "L" Street
P.O. Box 27139
Omaha, Nebraska 68127

E. J. Smith & Sons Company
Phone: (704) 394-3361
4250 Golf Acres Drive
P.O. Box 668887
Charlotte, North Carolina 28266

Uesco Warehouse Incorporated
Phone: (701) 237-0424
715 25th Street North
P.O. Box 2904
Fargo, North Dakota 58108

Gardner Engine & Parts Distribution
Phone: (614) 488-7951
1150 Chesapeake Avenue
Columbus, Ohio 43212

Mico Incorporated
Phone: (918) 627-1448
7450 East 46th Place
P.O. Box 470324
Tulsa, Oklahoma 74147

Brown & Wiser, Incorporated
Phone: (503) 692-0330
9991 South West Avery Street
Tualatin, Oregon 97062

Sullivan Brothers Incorporated
Phone: (215) 942-3686
Creek Road & Langoma Avenue
P.O. Box 140
Elverson, Pennsylvania 19520

Pitt Auto Electric Company
Phone: (412) 766-9112
2900 Stayton Street
Pittsburgh, Pennsylvania 15212

Locke Auto Electric Service Incorporated
Phone: (605) 336-2780
231 North Dakota Avenue
P.O. Box 1165
Sioux Falls, South Dakota 57101

Medart Engines & Parts of Memphis
Phone: (901) 774-6371
674 Walnut Street
Memphis, Tennessee 38126

Engine Warehouse Incorporated
Phone: (713) 937-4000
7415 Empire Central Drive
Houston, Texas 77040

Frank Edwards Company
Phone: (801) 363-8851
100 South 300 West
P.O. Box 2158
Salt Lake City, Utah 84110

R B I Corporation
Phone: (804) 798-1541
101 Cedar Run Drive
Lake-Ridge Park
Ashland, Virginia 23005

BITCO Western Incorporated
Phone: (206) 682-4677
4030 1st Avenue South
P.O. Box 24707
Seattle, Washington 98124

Wisconsin Magneto Incorporated
Phone: (414) 445-2800
4727 North Teutonia Avenue
P.O. Box 09218
Milwaukee, Wisconsin 53209

CANADIAN DISTRIBUTORS

Suntester Equipment (Central) Ltd.
Phone: (403) 453-5791
13315 146th Street
Edmonton, Alberta, Canada T5L 4S8

Suntester Equipment (Central) Ltd.
Phone: (416) 624-6200
5466 Timberlea Boulevard
Mississauga, Ontario, Canada L4W 2T7

WISCONSIN

TELEDYNE WISCONSIN MOTOR
Milwaukee, Wisconsin 53246

Model	No. Cyls.	Bore	Stroke	Displacement	Power Rating
ADH	1	2.75 in. (69.8 mm)	3.25 in. (82.55 mm)	19.3 cu. in. (308.7 cc)	5.1 hp. (3.8 kW)
AE, AEH, AEHS, AEN, AENL, AENS	1	3.0 in. (76.2 mm)	3.25 in. (82.55 mm)	23.0 cu. in. (376.5 cc)	9.2 hp. (6.9 kW)
AFH	1	3.25 in. (82.55 mm)	4.0 in. (101.6 mm)	33.2 cu. in. (543.8 cc)	7.2 hp. (5.4 kW)
AGH, AGND	1	3.5 in. (88.90 mm)	4.0 in. (101.6 mm)	38.5 cu. in. (630.7 cc.)	12.5 hp. (9.3 kW)
AHH	1	3.625 in. (92.08 mm)	4.0 in. (101.6 mm)	41.3 cu. in. (676.6 cc)	9.2 hp. (6.9 kW)
AK	1	2.875 in. (73.025 mm)	2.75 in. (69.8 mm)	17.8 cu. in. (292.3cc)	4.1 hp. (3.1 kW)
AKS	1	2.875 in. (73.025 mm)	2.75 in. (69.8 mm)	17.8 cu. in. (292.3 cc)	4.7 hp. 3.5 kW)
AKN	1	2.875 in. (73.025 mm)	2.75 in. (69.8 mm)	17.8 cu. in. (292.3 cc)	6.2 hp. (4.6 kW)

Engines in this section are four-cycle, one cylinder, horizontal crankshaft engines. Crankshaft is supported at each end by taper roller bearings.

Connecting rod rides directly on crankpin journal and shims between rod cap and rod provide running clearance adjustment. Pressure spray or splash lubrication is provided by a plunger type oil pump or a dipper located on end of connecting rod cap. Pump is driven off of camshaft lobe.

Ignition system consists of either a battery type system or one of a variety of magneto systems.

Float type carburetors from a variety of manufacturers are available. A mechanical fuel pump is available for most models.

Engine identification information is located on engine instruction plate and engine model, specification and serial number are required when ordering parts or service material.

MAINTENANCE

SPARK PLUG. Recommended spark plug is Champion D16J or equivalent. Electrode gap should be 0.030 inch (0.762 mm).

CARBURETOR. A variety of float type carburetors are used as listed. Marvel Schebler VH53 and VH70 models are replacements for older VH12 and VH14 models.

Marvel Schebler VH53 model is used on AK, AKS and AKN engines, TSX147 model is used on AHH engines and TSX676 model is used on AGND engines.

Stromberg OH 5/8 model is used on AK, AKS and AKN engines, UC 3/4 model is used on ADH, AE, AEH, AEHS and AFH engines, UC 7/8 model is used on AGH and AHH engines and UR 3/4 model is used on ADH, AE, AEH, AEHS and AENL engines.

Zenith 87B5 and 87BY6 models were used on AKN, AK and AKS engines, 161-7 model is used on AE, AEH, AEHS, AFH, AGH, AHH, AEN and AENS engines and 68-7 model is used on some AENL and AGND engines.

For initial carburetor adjustment open idle mixture screw 1-1/4 turns and main fuel mixture screw, if so equipped, 1-1/4 turns. Make final adjustments with engine at normal operating temperature and running. Place engine under load and adjust main fuel mixture screw for leanest setting that will allow satisfactory acceleration and steady governor operation. Set engine at idle speed, no load and adjust idle mixture screw for smoothest idle operation.

As each adjustment affects the other, adjustment procedure may have to be repeated.

Recommended float level for Marvel Schebler VH12 and VH14 models is 1/2-inch (12.7 mm) and 1/4-inch (6.35 mm) for VH53, VH70 and TSX models.

Recommended float level for Zenith 87B5 and 87BY6 models is 31/32-inch (24.606 mm) and 1-5/32 inch (29.369 mm) for 161-7 and 68-7 models. Recommended float level for Stromberg models is 1/2 to 9/16-inch (12.7 to 14.3 mm).

GOVERNOR. Flyweight type governors are used on all models.

On Models AK, AKN and AKS flyweights are attached to camshaft and operate a sleeve and shaft arrangement (Fig. W1) to control engine speed.

Governor flyweights on remaining models are attached to a gear assembly which is driven by camshaft gear and actuates an internal and external lever arrangement to control engine speed.

Speed changes on all models are obtained by using springs of varying tension and locating spring or governor rod in alternate holes of governor control lever.

Governor-to-carburetor rod should be adjusted so it just will enter hole in governor lever when governor weights are "in" and carburetor throttle is wide open.

IGNITION SYSTEM. Either a battery type ignition system with breaker points located in a distributor or a magneto type ignition system are used according to model and application.

BATTERY IGNITION. Breaker points on battery system are located inside a distributor which is driven by camshaft gear. Condenser is mounted externally on distributor body. Point gap is 0.020 inch (0.508 mm) for both type Auto-Lite distributors. See Fig. W5.

To correctly time distributor to engine during installation, remove screen over flywheel air intake, make certain piston is at "top dead center" on compression stroke (Fig. W3) and install distributor so points are just beginning to open. Connect timing light to engine, start and run engine at rated speed and rotate distributor as necessary to set timing to specifications as follows:

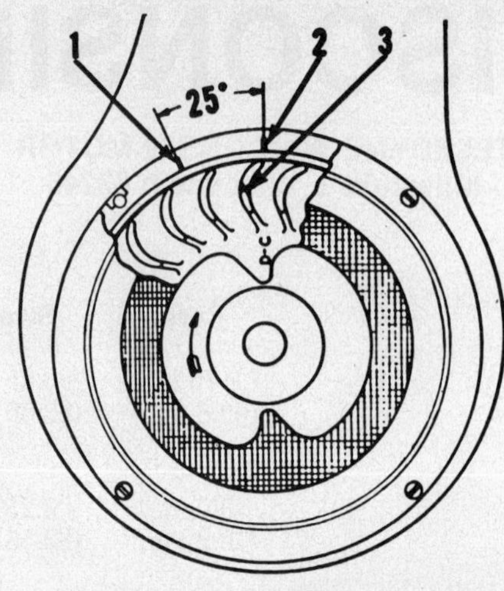

Fig. W3—Ignition timing marks for a typical "A" series engine. Piston is at top dead center position when leading edge of flywheel vane (3), marked "X" and "DC" is in register with static ignition timing mark (2).

ADH, AE, AEH, AEHS, AEN, AENS, AFH, AGH, AHH .... 25° BTDC

AK, AKN, AKS .... 28° BTDC

AENL, AGND .... 20° BTDC

MAGNETO IGNITION. Eisemann and Edison magnetos were obsoleted and replaced by Fairbanks-Morse or Wico magnetos.

Breaker point gap for all models is 0.015 inch (0.381 mm).

To correctly time magneto to engine during installation, remove screen over flywheel air intake opening and timing hole plug from crankcase. Make certain piston is at "top dead center" on compression stroke and leading edge of flywheel vane (3-Fig. W3) marked "X" and "DC" is in register with timing mark (2) on air shroud. Install magneto so marked tooth (Figs. W4 or W6) on magneto drive gear is visible through port in timing gear cover as shown. If timing is correct, impulse coupling will trip when crankshaft keyway is up as engine is being cranked.

Connect timing light and set timing to specifications listed for battery ignition system.

LUBRICATION. A plunger type oil pump driven off of a camshaft lobe may either pressure spray lubricate (Fig. W7) connecting rod and internal engine components or maintain proper oil level in an internal oil trough below connecting rod which enables dipper on rod cap to splash lubricate (Fig. W8) internal engine components.

Oil pump plunger to bore clearance should be 0.003-0.007 inch (0.0762-0.1778 mm) and plunger should be renewed if clearance exceeds 0.008 inch (0.203 mm).

Maintain oil level at full mark on dipstick or at top of filler plug but do not overfill. High quality detergent oil having API classification "SD" or "SE" is

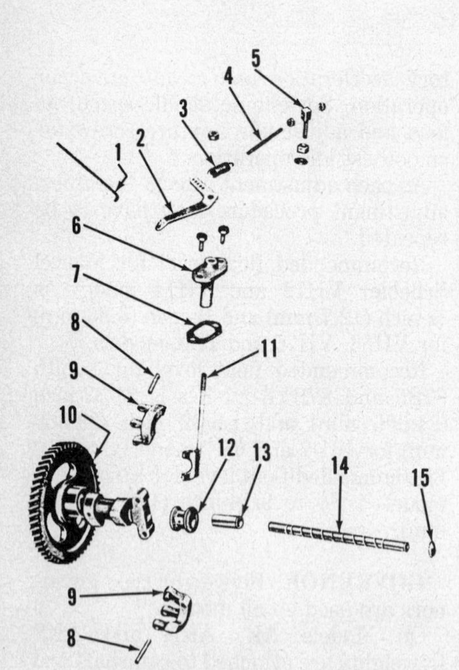

Fig. W1—Exploded view of governor assembly showing component parts and their relative positions.

1. Governor control rod
2. Governor lever
3. Governor spring
4. Adjusting screw
5. Pin
6. Support bracket
7. Gasket
8. Roll pin
9. Flyweight
10. Camshaft & gear assy.
11. Governor yoke
12. Thrust sleeve
13. Spacer
14. Support pin
15. Core plug

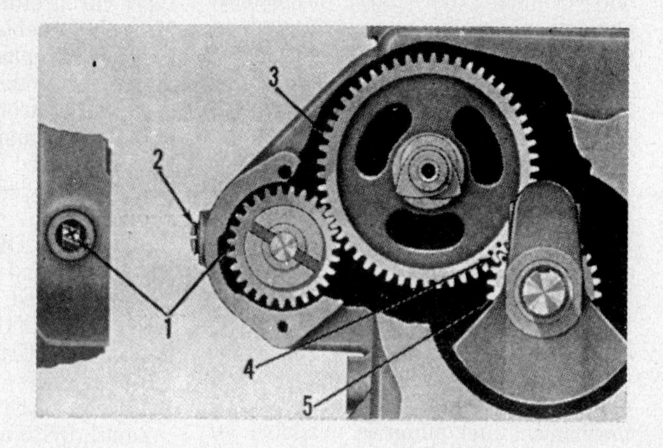

Fig. W4—Timing gears and marks of AEN, AENL and AENS models shown in register. With marks on gears (3 and 5) aligned as shown at (4), "X" marked magneto gear tooth must show through port (2) when piston is at TDC.

1. Magneto gear
2. Timing port
3. Camshaft gear
4. Timing marks
5. Crankshaft gear

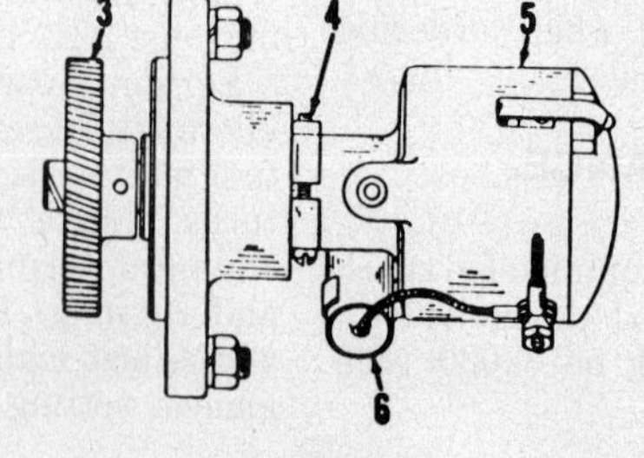

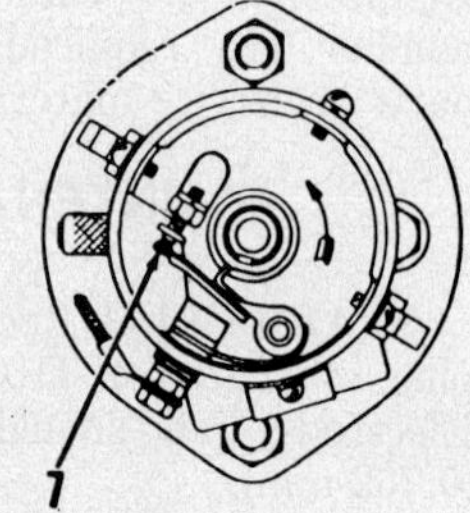

Fig. W5—Distributor used on Models AEN, AENL and AENS with battery ignition is driven from camshaft gear.

3. Drive gear
4. Screw clamp
5. Distributor housing
6. Condenser
7. Breaker points

recommended. Recommended oil change interval is every 50 hours of normal operation.

Use SAE 30 oil when operating temperatures are above 40° F (4° C), SAE 20 oil for temperatures between 40° F (4° C) and 5° F (-15° C) and SAE 10 oil if temperature is below 5° F (-15° C).

REPAIRS

TIGHTENING TORQUES. Recommended torque specifications are as follows:

Cylinder head–
AK, AKS & AKN . . . 14-18 ft.-lbs. (19-24 N•m)
All other models . . . . 26-32 ft.-lbs. (35-43 N•m)

Connecting rod–
AFH, AGH, AHH, AGND . . . . 32 ft.-lbs. (43 N•m)
All other models . . . . 18 ft.-lbs. (24 N•m)

Cylinder to block (where applicable)
AGND . . . . 40-50 ft.-lbs. (54-68 N•m)
All other models . . . . 62-78 ft.-lbs. (84-106 N•m)

Main bearing plate–
AK, AKS & AKN . . . 14-18 ft.-lbs. (19-24N•m)
All other models . . . . 26-32 ft.-lbs. (35-43 N•m)

CONNECTING ROD. Connecting rod and piston assemblies can be removed from above after removing cylinder head and engine base.

Connecting rod rides directly on crankpin journal and rods are available for 0.010, 0.020 and 0.030 inch (0.254, 0.508 and 0.762 mm) undersize crankshafts.

Rod to journal running clearance should be 0.001-0.0018 inch (0.0254-0.0457 mm) for AEN, AENS and AENL models and 0.0007-0.002 inch (0.0178-0.0508 mm) for all other models. Clearance is controlled to some degree by varying thickness of shims between connecting rod and cap.

Connecting rod side clearance should be 0.004-0.013 inch (0.1016-0.3302 mm) for AEN, AENS and AENL models and 0.004-0.011 inch (0.1016-0.2794 mm) for all other models.

Connecting rod on all models must be installed with oil hole in rod cap facing oil pump side of engine for proper lubrication.

Crankpin standard journal diameter is 1.00 inch (25.4 mm) for AK, AKS and AKN models, 1.1255 inches (28.59 mm) for AE, AEH, AEN, AEHS, AENS, ADH and AENL models, 1.375 inches (34.925 mm) for AFH, AGH and AHH models and 1.750 inches (44.45 mm) for AGND models.

PISTON, PIN AND RINGS. Piston may be of three ring or four ring design according to model and application. Pistons and rings are available in a variety of oversizes as well as standard.

Piston ring end gap for all models should be 0.012-0.022 inch (0.3048-0.5588 mm) and ring side clearance should be 0.002-0.003 inch (0.0508-0.0762 mm).

Early models were equipped with a split skirt piston and split should be toward valve side of engine after installation. Later model cam ground piston has a wide and a narrow thrust face on piston skirt and wide side must be toward thrust side (side opposite valve) of engine after installation.

Split skirt piston should have 0.0045-0.005 inch (0.1143-0.1270 mm) clearance in cylinder measured 90° from piston pin at bottom edge of piston skirt and cam ground piston should have 0.003-0.0035 inch (0.0762-0.0889 mm) clearance in cylinder measured along thrust surface of skirt.

Floating type piston pin is retained by snap rings in bore of piston and should have 0.0002-0.0008 inch (0.0051-0.0203 mm) clearance in connecting rod.

Fig. W7—View of pressure spray lubricating system used on some models.

CYLINDER. Cylinder bore section on ADH, AE, AEH, AEHS, AFH, AGH, AHH and AGND is separate from crankcase and may be renewed as an individual unit. Cylinder bore and crankcase for all other models are cast as a unit.

Cylinders for all models may be bored to oversize if worn or out-of-round more than 0.005 inch (0.127 mm).

Standard cylinder bore for ADH, AK, AKS and AKN is 2.750 inches (69.85 mm), for AEN, AENL and AENS bore is 3.0 inches (76.2 mm), for AE, AEHS and AFH bore is 3.250 inches (82.55 mm), for AGH and AGND bore is 3.5 inches (88.90 mm) and for AHH bore is 3.625 inches (92.08 mm).

Pistons are available in 0.005, 0.010, 0.020 and 0.030 inch (0.1270, 0.254, 0.508 and 0.762 mm) oversizes as well as standard.

CRANKSHAFT AND MAIN BEARINGS. Crankshaft for all models is supported at each end by taper roller bearings. Crankshaft end play should be 0.002-0.004 inch (0.0508-0.1016 mm)

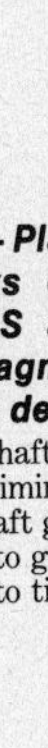
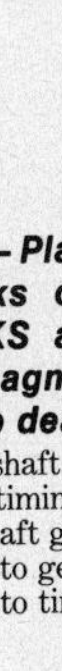

Fig. W6—Placement of timing marks on Models AK, AKN, AKS and BKN when timing magneto to piston at top dead center.

2. Crankshaft gear
3. Valve timing marks
4. Camshaft gear
5. Magneto gear
6. Magneto timing marks

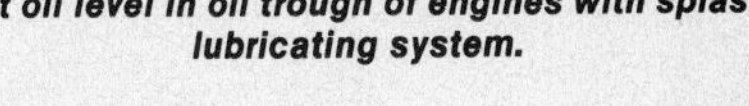

Fig. W8—View of oil pump used to maintain correct oil level in oil trough of engines with splash lubricating system.

and is controlled by varying thickness of gaskets between crankcase and bearing plate. Gaskets are available in a variety of thicknesses.

Main bearings are press fit on crankshaft. Refer to Fig. W4 or W6 for correct timing mark position for installation.

CAMSHAFT. The hollow camshaft turns on a stationary support pin (14-Fig. W1) and should have a running clearance of 0.001-0.0025 inch (0.0254-0.0635 mm). Support pin is pressed into case from flywheel end while camshaft and lifters are held in place. Camshaft plug (15) should be renewed.

Governor flyweights are attached to camshaft on some models and care should be taken to avoid damaging when camshaft is being installed.

MAGNETO DRIVE GEAR. On Models ADH, AE, AEH, AEHS, AFH, AGH and AHH magneto drive shaft should have a running clearance of 0.002-0.0035 inch (0.0508-0.0889 mm) in shaft bushing. Drive shaft end play should be 0.004-0.005 inch (0.1016-0.1270 mm) and is adjusted by varying thickness of gasket (2). Magneto drive gear is pressed on shaft so distance from coupling face of shaft to centerline of magneto mounting hole is 2.584 inches (65.6336 mm) as shown.

VALVE SYSTEM. Valve seats are renewable type inserts in cylinder block and seats should be ground to a 45° angle. Width should be 3/32-inch (2.381 mm).

Valve face is ground at 45° angle and stellite valves and seats and valve rotocaps are available for some models.

Valve stem clearance in block or guide should be 0.003-0.005 inch (0.0762-0.127 mm). AGND and AENL models have renewable guides. Valves with oversize stems are available for models which do not have renewable guides if stem clearance exceeds 0.006 inch (0.1524 mm).

Valve tappet gap (cold) for intake valve is 0.008 inch (0.2032 mm) and for exhaust valve is 0.016 inch (0.4064 mm). Gap is adjusted by turning adjusting screws on tappets if equipped or grinding end of valve stem if non-adjustable tappets are installed.

WISCONSIN

TELEDYNE WISCONSIN MOTOR
Milwaukee, Wisconsin 53246

Model	No. Cyls.	Bore	Stroke	Displacement	Power Rating
HBKN	1	2.875 in. (73.03 mm)	2.75 in. (69.85 mm)	17.8 cu. in. (292.6 cc)	7 hp. (5.2 kW)
HAENL	1	3.0 in. (76.2 mm)	3.25 in. (82.55 mm)	23.0 cu. in. (376.5 cc)	9.2 hp. (6.9 kW)

Engines in this section are four-cycle, one cylinder, vertical crankshaft engines. Crankshaft is supported at each end by taper roller bearings.

Connecting rod rides directly on crankpin journal and shims between connecting rod cap and rod provide running clearance adjustment. Pressure spray lubrication is provided by a vane type oil pump mounted to and driven by crankshaft.

All models are equipped with a magneto ignition system.

Zenith float type carburetors are standard equipment for all models and a mechanical fuel pump is available.

Engine identification information is located on engine instruction plate and engine model, specification and serial number are required when ordering parts or service material.

MAINTENANCE

SPARK PLUG. Recommended spark plug is Champion D16 or equivalent. Electrode gap should be 0.030 inch (0.762 mm).

CARBURETOR. Zenith 68-7 model (Fig. W9) carburetor is used on HAENL engines and Zenith 87B5 model (Fig. W10) carburetor is standard equipment for HBKN engines with a variety of other carburetors used for special applications. Refer to carburetor identification number for proper carburetor identification.

For initial adjustment on all models, open idle mixture screw and main fuel mixture screw (as equipped) 1-¼ turns each. Make final adjustments with engine at normal operating temperature and running. Place engine under load and adjust main fuel mixture screw (as equipped) for leanest setting that will allow satisfactory acceleration and steady governor operation. Set engine at idle speed no load and adjust idle mixture screw for smoothest idle operation.

As each adjustment affects the other, adjustment procedure may have to be repeated.

To check float level, invert carburetor throttle body and float assembly (Fig. W13) and measure distance (A) from machined surface of throttle body to highest point on float. Model 87B5 carburetor float level should be 31/32-inch (24.606 mm) and 68-7 model carburetor float level should be 1-5/32 inch (29.369 mm). Adjust as necessary by bending float lever tang that contacts inlet valve.

GOVERNOR. Flyweight type governors are used on all models and flyweights are attached to camshaft and operate a sleeve and shaft arrangement (Fig. W14) to control engine speed.

Speed changes on all models is obtained by using springs (3) of varying tension, flyweights (9) of varying weights or locating spring or governor rod in alternate holes of governor control lever (2).

Governor-to-carburetor rod should be adjusted so it just will enter hole in governor lever when governor weights are "in" and carburetor throttle is wide open.

Camshaft must be removed from engine to service governor flyweights.

IGNITION SYSTEM. Fairbanks-Morse magnetos are used for all models and magneto is mounted in suspension below timing gear train. An oil drain

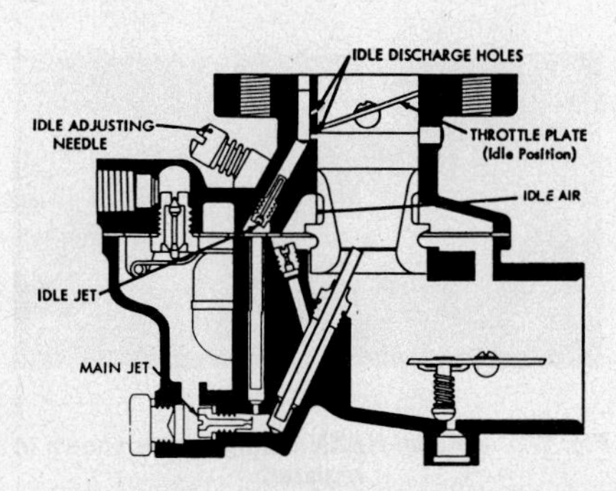

Fig. W9—Sectional view of Zenith 68-7 model carburetor showing location of idle adjusting screw and main fuel jet.

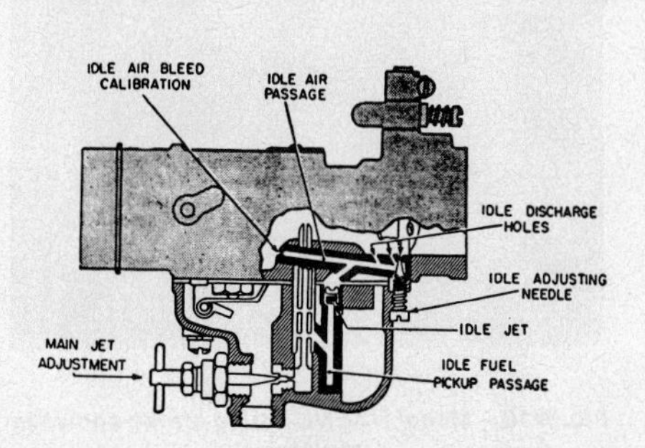

Fig. W10—Sectional view of Zenith 87B5 model carburetor showing location of idle adjusting screw and main fuel adjusting screw.

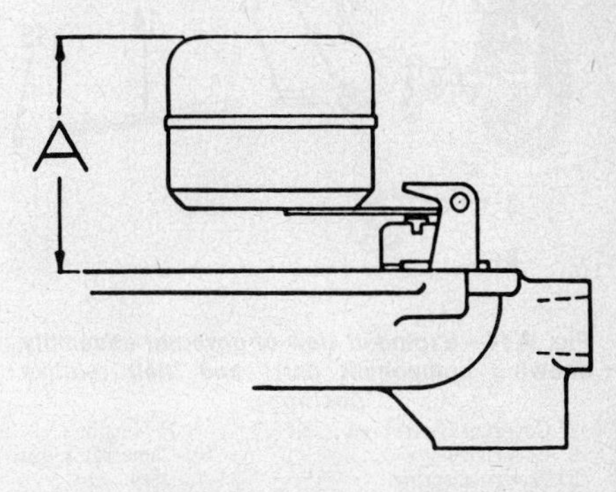

Fig. W13—Float level measurement for Zenith carburetors. On 87B5 model dimension "A" is 31/32-inch (24.6 mm). On 68-7 model, it is 1-5/32 inch (29.36 mm).

tube is connected to magneto body to allow excess oil to drain back into engine crankcase.

Magneto breaker points and condenser are accessible by removing magneto end cover. Breaker point gap should be adjusted to 0.015 inch (0.381 mm) clearance.

To correctly time magneto to HBKN engine, remove screen over flywheel air intake opening and timing hole plug from crankcase. Make certain piston is at "top dead center" on compression stroke and leading edge of flywheel vane (Fig. W17) marked "X" and "DC" is in register with timing mark (2) on air shroud. Install magneto so marked tooth (1) on magneto drive gear is visible through port in timing gear as shown. If timing is correct, impulse coupling will trip when crankshaft keyway is up as engine is being cranked.

To correctly time magneto to HAENL engine, remove screen over flywheel air intake opening and timing hole plug from crankcase. Make certain piston is at "top dead center" on compression stroke and leading edge of flywheel vane (Fig. W19) is in register with timing mark (3) on air shroud. Install magneto so marked tooth (1-Fig. W20) on magneto drive gear is visible through port in timing gear as shown.

To set running timing on either model,

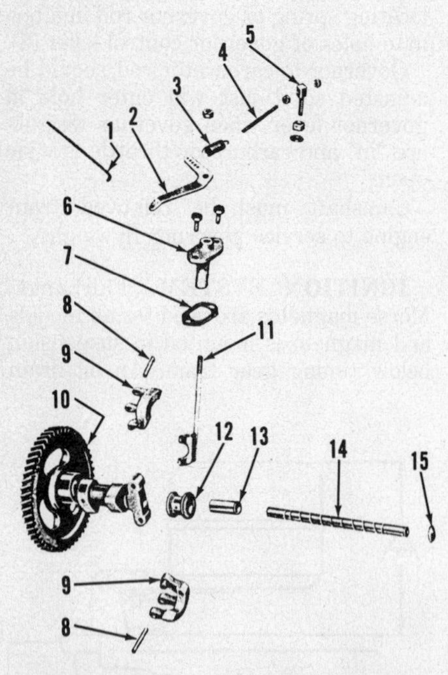

Fig. W14—Exploded view of governor assembly showing component parts and their relative positions.

1. Governor control rod
2. Governor lever
3. Governor spring
4. Adjusting screw
5. Pin
6. Support bracket
7. Gasket
8. Roll pin
9. Flyweight
10. Camshaft & gear assy.
11. Governor yoke
12. Thrust sleeve
13. Spacer
14. Support pin
15. Core plug

Fig. W17—Magneto timing for HBKN engines.

1. Magneto timing marks
2. Centerline mark
3. Running timing mark
4. Flywheel keyway

connect timing light, start and run engine at 1800 rpm and set timing on HBKN engines at 17° BTDC and set timing on HAENL engines 20° BTDC.

LUBRICATION. A vane type oil pump mounted to and driven by crankshaft is used to provide pressure spray lubrication on all models. Pump is located in engine adapter base (Fig. W21).

Pump vane retainer is held in position on crankshaft by a set screw which also holds pump body in place. A strap attached to engine bearing plate keeps oil pump body stationary.

Oil pressure relief valve reed should fit firmly against oil pressure relief hole. Renew reed if it is bent or out of shape.

Maintain oil level at filler plug but do not overfill. High quality detergent oil having API classification "SD" or "SE" is recommended. Recommended oil change interval is every 50 hours of normal operation.

Use SAE 30 oil when operating temperatures are above 40°F (4°C), SAE 20 oil for temperatures between 40°F (4°C) and 5°F (-15°C) and SAE

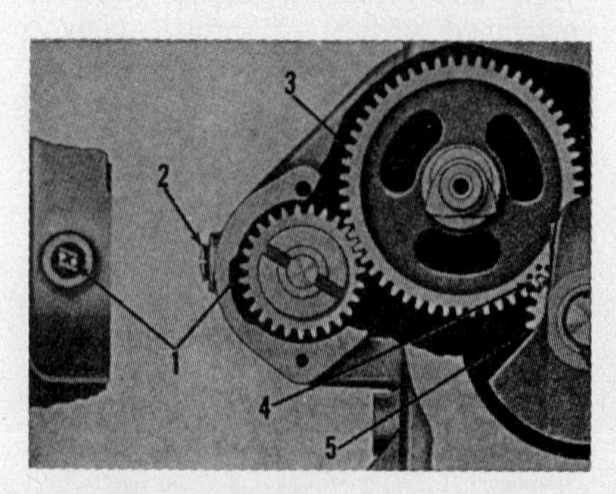

Fig. W18—Model HAENL timing marks shown in register.

1. Magneto timing mark
2. Timing port plug
3. Camshaft gear
4. Cam gear timing marks
5. Crankshaft gear timing mark

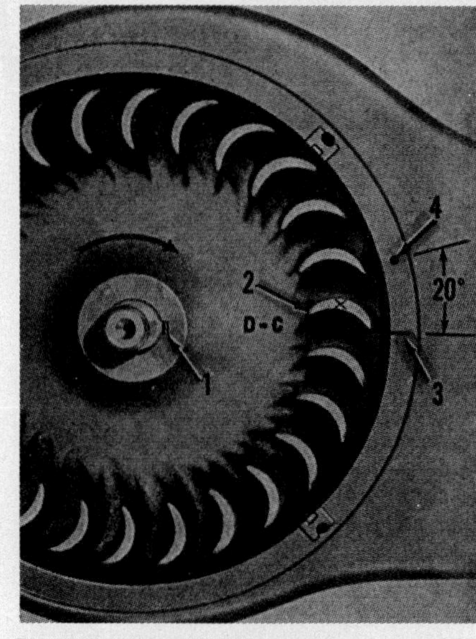

Fig. W19—View of HAENL model ignition timing marks lined up.

1. Keyway
2. Marked air vane
3. Centerline mark
4. Running timing mark

10 oil if temperature is below 5°F (-15°C).

REPAIRS

TIGHTENING TORQUES. Recommended torque specifications are as follows:

Cylinder head–
- HBKN .... 14-18 ft.-lbs. (19-24 N·m)
- HAENL .... 26-32 ft.-lbs. (35-43 N·m)

Connecting rod–
- HBKN .... 14-18 ft.-lbs. (19-24 N·m)
- HAENL .... 18 ft.-lbs. (24 N·m)

Main bearing plate–
- HBKN .... 14-18 ft.-lbs. (19-24N·m)
- HAENL .... 26-32 ft.-lbs. (35-43 N·m)

Fig. W20—Model HBKN timing marks shown in register.

1. Magneto timing marks
2. Crankshaft gear
3. Camshaft timing marks
4. Camshaft gear
5. Magneto gear

Crankcase cover plate –
HBKN . . . . 6-8 ft.-lbs. (8-11 N•m)
HAENL . . . . 7-9 ft.-lbs. (10-12 N•m)

Engine Adapter base –
All models . . . . 24-26 ft.-lbs. (33-35 N•m)

Spark plug –
All models . . . . 25-30 ft.-lbs. (34-40 N•m)

CONNECTING ROD. Connecting rod and piston assemblies can be removed from cylinder head surface end of block after removing cylinder head and side cover.

Connecting rod rides directly on crankpin journal and rods are available for 0.010, 0.020 and 0.030 inch (0.254, 0.508 and 0.762 mm) undersize crankshafts.

Rod to journal running clearance should be 0.0007-0.002 inch (0.0178-0.0508 mm) for all models and is controlled to some degree by varying thickness of shims between connecting rod and cap.

Connecting rod side clearance should be 0.004-0.010 inch (0.1016-0.254 mm) for HBKN engines and 0.006-0.013 inch (0.1524-0.3302 mm) for HAENL engines.

Connecting rod on all models must be installed with oil hole in rod cap facing oil pump side of engine for proper lubrication.

Crankpin standard diameter for HBKN engines is 1.000 inch (25.4 mm) and is 1.125 inches (31.75 mm) for HAENL engines.

PISTON, PIN AND RINGS. Pistons are equipped with two compression rings, one scraper ring and one oil control ring. Install rings as shown in Fig. W22. View "A" shows ring cross-section on HBKN engines and view "B" shows ring cross-section on HAENL engines.

HBKN engines operated at 3000 rpm or below require a different piston than HBKN engines operating at speed above 3000 rpm.

Ring end gap for all models should be 0.012-0.022 inch (0.3048-0.5588 mm) and side clearance of top ring in groove should be 0.002-0.0035 inch (0.0508-0.0889 mm). Side clearance of second or third ring should be 0.001-0.0025 inch (0.0254-0.0635 mm) and side clearance of oil control ring should be 0.0025-0.004 inch (0.0635-0.1016 mm). Ring gaps should be installed so they are at 90° intervals around piston.

Piston skirt clearance for HBKN engines operating at 3000 rpm or below should be 0.0055-0.006 inch (0.1397-0.1524 mm) and for HBKN engines operating above 3000 rpm clearance should be 0.006-0.0065 inch (0.1524-0.1651 mm).

Piston skirt clearance for all HAENL engines should be 0.003-0.0035 inch (0.0762-0.0889 mm).

Pistons and rings are available in 0.005, 0.010, 0.020 and 0.030 inch (0.1270, 0.254, 0.508 and 0.762 mm) oversizes as well as standard.

Floating type piston pin is retained by snap rings in each end of piston and pin clearance in connecting rod for HBKN engines should be 0.0002-0.0008 inch (0.0051-0.0203 mm) and for HAENL engines clearance should be 0.0005-0.001 inch (0.0127-0.0254 mm). Pins are available in a variety of oversizes as well as standard.

CYLINDER. Cylinder bore is an integral part of crankcase casting and if worn or out-of-round more than 0.005 inch (0.127 mm), cylinder should be bored to nearest size for which piston and rings are available.

If crankcase and cylinder renewal is required make certain number (Fig. W24) stamped as shown is added to basic part number shown in Wisconsin parts catalog when ordering new cylinder and crankcase.

CRANKSHAFT AND MAIN BEARINGS. Crankshaft on all models is supported at each end by taper roller bearings which are press fit on crankshaft. Races are driven into bearing plate or engine block. Crankshaft end play for HBKN engines should be 0.002-0.004 inch (0.0508-0.1016 mm), should be 0.001-0.003 inch (0.0254-0.0762 mm) for HAENL engines and is controlled on all

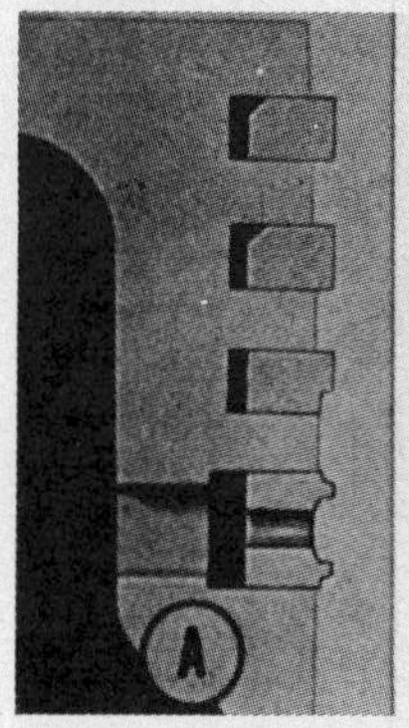

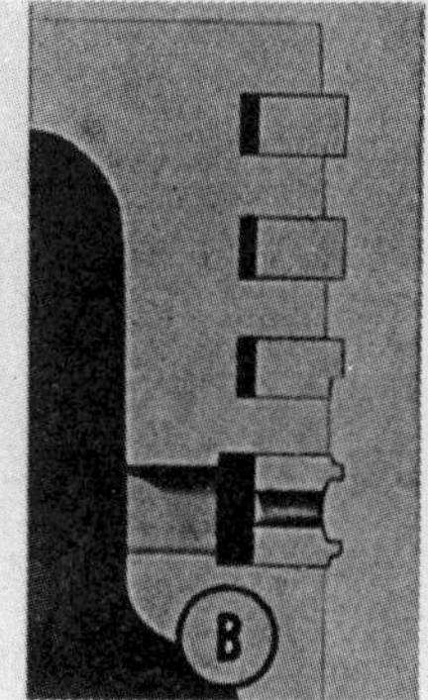

Fig. W22 – Cross-section of piston rings as installed in vertical crankshaft engines. View "A" shows HBKN models and view "B" shows HAENL models.

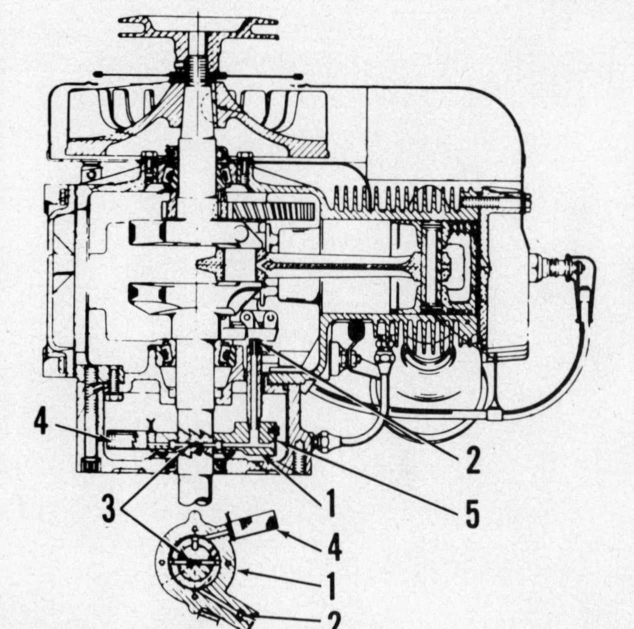

Fig. W21 – Sectional view of engine oil pump as installed on vertical shaft models.

1. Pump body
2. Vertical spray jet
3. Vane assembly
4. Intake oil screen
5. Relief valve

Fig. W23 – Crankshaft end play is controlled by number of gaskets (G) used under main bearing plate at pto end of crankcase.

models by varying number of gaskets between bearing plate and crankcase (Fig. W23).

HAENL engine is also equipped with a ball bearing at lower end of crankshaft which is mounted in engine adapter base below oil pump.

Install crankshaft so timing marks on crankshaft gear and camshaft gear are in register as shown in Figs. W18 or W20.

If necessary to renew crankshaft, refer to Fig. W25 for location of crankshaft part number.

CAMSHAFT. The hollow camshaft turns on a stationary support pin (14-Fig. 14) and should have a running clearance of 0.001-0.0025 inch (0.0254-0.0635 mm). Support pin is pressed into case from flywheel end while camshaft and lifters are held in place. Camshaft plug (15) should be renewed.

Governor flyweights are attached to camshaft on all models and care should be taken to avoid damaging when camshaft is being installed.

VALVE SYSTEM. Valve seats are renewable type inserts in cylinder block

Fig. W24—Location of specification number (A) for special type crankcase and cylinder. If number is present, add to basic part number when renewing crankcase and cylinder.

and seats should be ground to 45° angle. Width should be 3/32-inch (2.381 mm).

Valve face is ground at 45° angle and stellite valves and seats and valve rotocaps are available for some models.

Valve stem clearance in guide should be 0.003-0.005 inch (0.0762-0.127 mm).

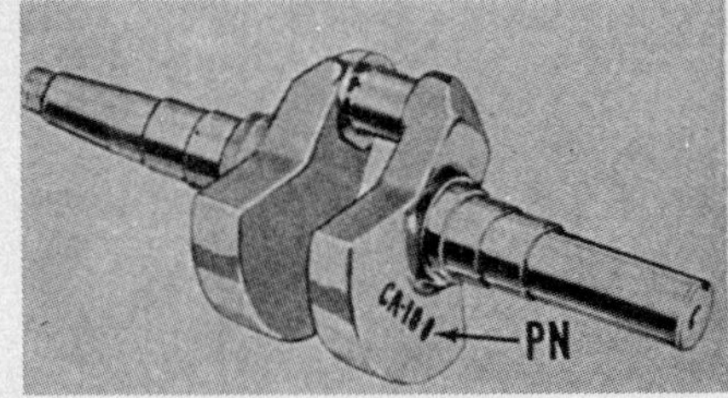

Fig. W25—When renewing crankshaft, verify part number from crankshaft being removed from engine. Number appears on counterweight as shown at PN.

All models have renewable guides and valves with oversize stems are available for some models.

Valve tappet gap (cold) for HBKN engines should be 0.008 inch (0.2032 mm) for intake valve and 0.014 inch (0.3556 mm) for exhaust valve.

Valve tappet gap (cold) for HAENL engines should be 0.008 inch (0.2032 mm) for intake valve and 0.016 inch (0.4064 mm) for exhaust valve.

Valve gap is adjusted by turning adjusting screws on tappets if equipped or grinding end of valve stem if non-adjustable tappets are installed.

WISCONSIN

TELEDYNE WISCONSIN MOTOR
Milwaukee, Wisconsin 53246

Model	No. Cyls.	Bore	Stroke	Displacement	Power Rating
S-7D, HS-7D	1	3 in. (76.2 mm)	2.625 in. (66.68 mm)	18.6 cu. in. (304.1 cc)	7.25 hp. (5.4 kW)
S-8D, HS-8D	1	3.125 in. (79.375 mm)	2.625 in. (66.68 mm)	20.2 cu. in. (330 cc)	8.25 hp. (6.2 kW)
TR-10D	1	3.125 in. (79.375 mm)	2.625 in. (66.68 mm)	20.2 cu. in. (330 cc)	10 hp. (7.5 kW)
TRA-10D	1	3.125 in. (79.375 mm)	2.875 in. (73.03 mm)	22.1 cu. in. (361.4 cc)	10.1 hp. (7.8 kW)
S-10D	1	3.25 in. (82.55 mm)	3 in. (76.2 mm)	24.9 cu. in. (407.8 cc)	10.1 hp. (7.8 kW)
TRA-12D	1	3.5 in. (88.9 mm)	2.875 in. (73.03 mm)	27.7 cu. in. (453 cc.)	12 hp. (9.0 kW)
S-12D	1	3.5 in. (88.9 mm)	3 in. (76.2 mm)	28.9 cu. in. (473 cc)	12.5 hp. (9.3 kW)
S-14D	1	3.75 in. (95.3 mm)	3 in. (76.2 mm)	33.1 cu. in. (543.5cc)	14.1 hp. (10.5 kW)

Engines in this section are four-cycle, one cylinder engines and all models except HS-7D and HS-8D are horizontal crankshaft engines. HS-7D and HS-8D engines have an adapter-base for vertical mounting. Crankshaft for all models is supported at each end in taper roller bearings.

Connecting rod used in S-10D, S-12D and S-14D engines have renewable insert type bearings. Connecting rods on all other models ride directly on crankpin journal. All horizontal crankshaft models are splash lubricated by an oil dipper located on end of connecting rod and all vertical crankshaft models are pressure spray lubricated and have a plunger type oil pump located on lower end of crankshaft.

Various magneto, battery or solid-state electronic ignition systems are used according to model and application.

Zenith or Walbro float type carburetor is used and a mechanical fuel pump is available.

Engine identification information is located on engine instruction plate and engine model, specification and serial number are required when ordering parts or service material.

MAINTENANCE

SPARK PLUG. Recommended spark plug is Champion D16J or equivalent. Electrode gap should be 0.030 inch (0.762 mm).

CARBURETOR. A variety of float type carburetors are used as listed.

Zenith 72Y6 carburetor is standard on S-7D, S-8D, HS-7D and HS-8D engines. See Fig. W30.

Zenith 68-7 carburetor is standard on TR-10D and TRA-10D engines. See Fig. W31.

Walbro LME-35 carburetor is standard on TRA-12D engines. See Fig. W34.

Zenith 1408 carburetor is used on S-10D, S-12D and S-14D engines. See Fig. W35.

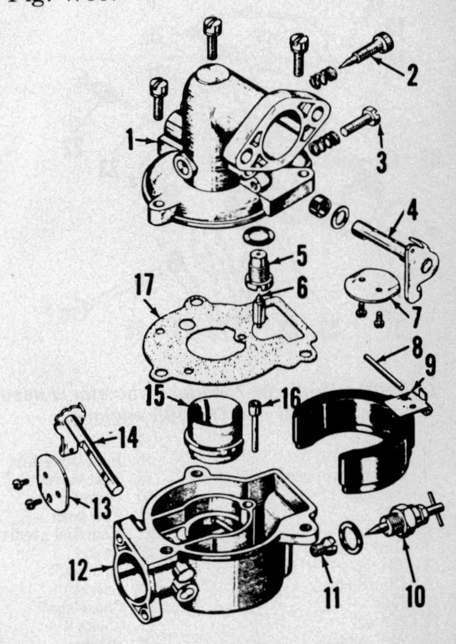

1. Throttle body
2. Idle mixture needle
3. Idle speed adjusting screw
4. Throttle shaft
5. Needle valve seat
6. Needle valve
7. Throttle plate
8. Float pin
9. Float
10. Main fuel adjusting needle
11. Main jet
12. Fuel bowl
13. Choke plate
14. Choke shaft
15. Venturi
16. Idle tube
17. Gasket

Fig. W30—Exploded view of Zenith 72Y6 carburetor used on S-7D and S-8D engines and on HS-7D and HS-8D vertical shaft engines.

For initial adjustment of Zenith 72Y6 carburetor (Fig. W30) open idle mixture screw (2) ½-turn and main fuel mixture screw (10) 2 turns.

For initial adjustment of Zenith 68-7 or Walbro LME-35 carburetor open idle mixture screw and main fuel mixture screw 1¼ turns. Refer to Fig. W31 or W34 for location of mixture screw of model being serviced.

For initial adjustment of Zenith 1408 carburetor (Fig. W35) open idle mixture screw (5) 1½ turns and main fuel mixture screw (13) 2¼ turns.

On all models make final adjustments with engine at normal operating temperature and running. Place engine under load and adjust main fuel mixture screw for leanest setting that will allow satisfactory acceleration and steady governor operation. Set engine at idle speed, no load and adjust idle mixture screw for smoothest idle operation.

As each adjustment affects the other, adjustment procedure may have to be repeated.

To check float level on all models, invert carburetor body and float assembly and refer to appropriate Figure for model being serviced. Adjust as necessary by bending float tang that contacts inlet valve.

GOVERNOR. Flyweight type governors driven by camshaft gear are used on all models.

Refer to Fig. W42 for S-7D, S-8D, HS-7D, HS-8D, TR-10D, TRA-10D and TRA-12D engine governor component parts and to Fig. W43 for S-10D, S-12D and S-14D engine governor component parts.

Major speed changes on all models is

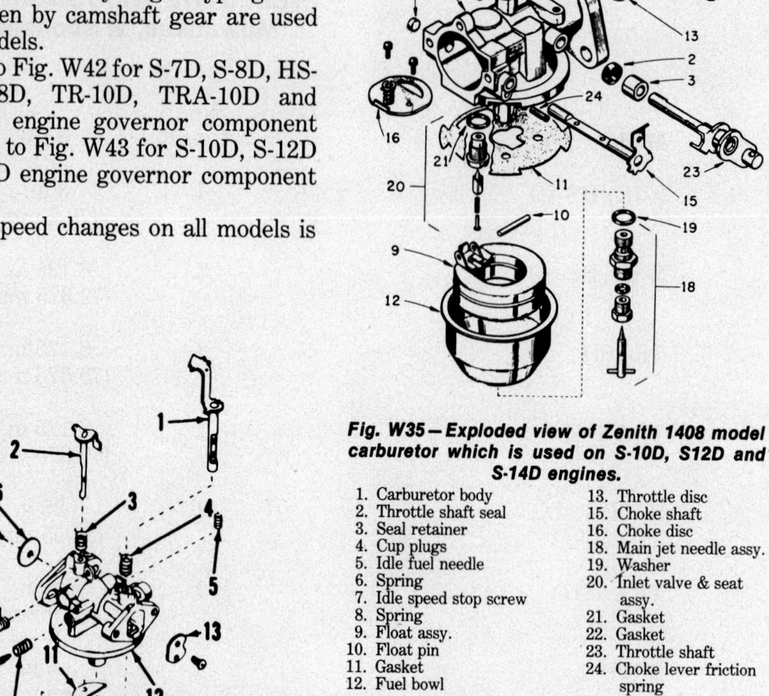

Fig. W35—Exploded view of Zenith 1408 model carburetor which is used on S-10D, S12D and S-14D engines.

1. Carburetor body
2. Throttle shaft seal
3. Seal retainer
4. Cup plugs
5. Idle fuel needle
6. Spring
7. Idle speed stop screw
8. Spring
9. Float assy.
10. Float pin
11. Gasket
12. Fuel bowl
13. Throttle disc
15. Choke shaft
16. Choke disc
18. Main jet needle assy.
19. Washer
20. Inlet valve & seat assy.
21. Gasket
22. Gasket
23. Throttle shaft
24. Choke lever friction spring

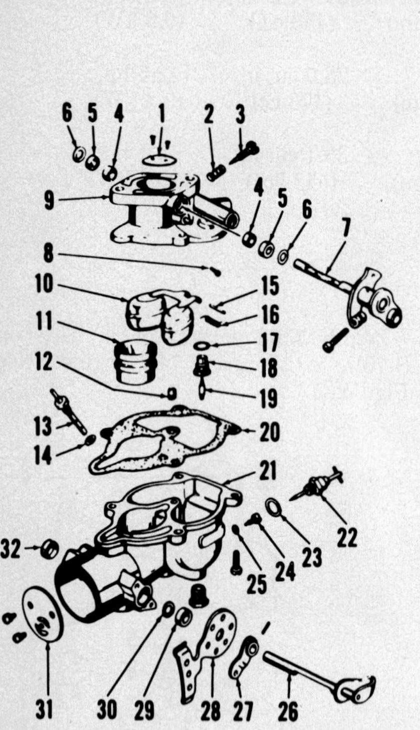

Fig. W31—Zenith 68-7 model carburetor is used on TR-10D and TRA-10D engines.

1. Throttle plate
2. Spring
3. Idle mixture needle
4. Bushing
5. Seal
6. Retainer
7. Throttle shaft
8. Idle jet
9. Throttle body
10. Float
11. Venturi
12. Well vent
13. Discharge nozzle
14. Gasket
15. Float shaft
16. Float spring
17. Gasket
18. Inlet valve seat
19. Inlet valve
20. Gasket
21. Fuel bowl
22. Main fuel needle
23. Gasket
24. Main jet
25. Gasket
26. Choke shaft
27. Choke lever
28. Bracket
29. Retainer
30. Seal
31. Choke plate
32. Plug

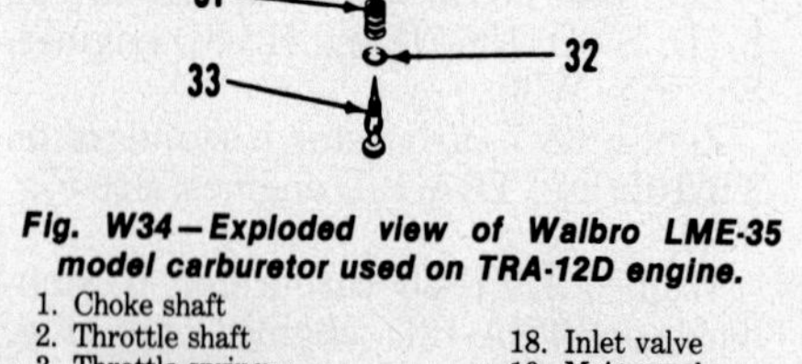

Fig. W34—Exploded view of Walbro LME-35 model carburetor used on TRA-12D engine.

1. Choke shaft
2. Throttle shaft
3. Throttle spring
4. Choke spring
5. Choke stop spring
6. Throttle plate
7. Idle speed screw
8. Spring
9. Idle mixture needle
10. Spring
11. Baffle
12. Carburetor body
13. Choke plate
14. Bowl gasket
15. Gasket
16. Inlet valve seat
17. Spring
18. Inlet valve
19. Main nozzle
20. Float
21. Float shaft
22. Spring
23. Gasket
24. Bowl
25. Drain stem
26. Gasket
27. Spring
28. Retainer
29. Gasket
30. Bowl retaier
31. Spring
32. "O" ring
33. Main fuel needle

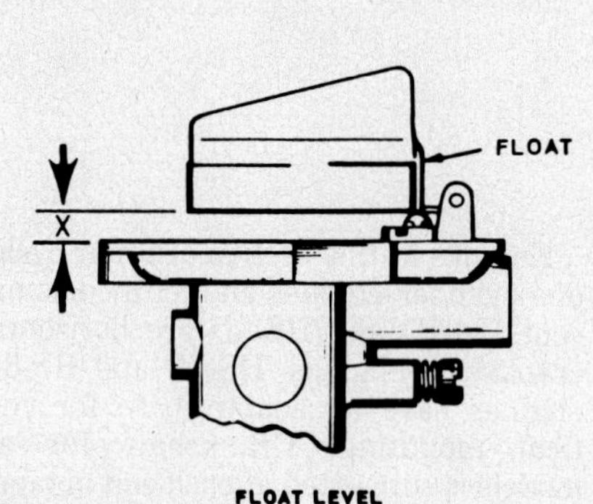

Fig. W36—Zenith 72Y6 carburetor float must be parallel to casting. Dimension "X" should be the same, measured near hinge pin and at outer end of float.

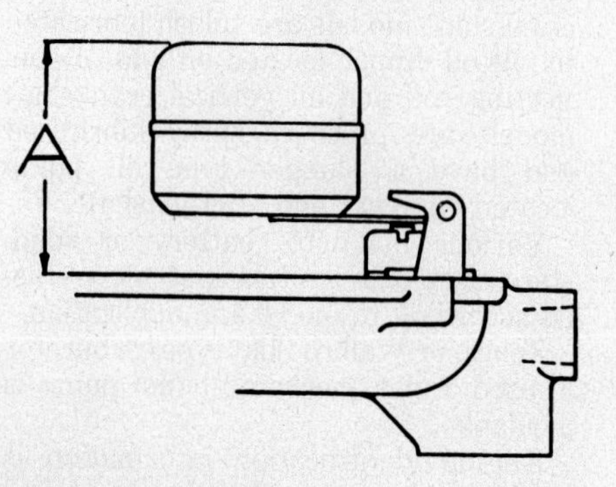

Fig. W37—Zenith 68-7 carburetor float should be adjusted so dimension "A" is 1-5/32 inch (29.37 mm).

obtained by using springs of varying tension, flyweights of varying weights or locating spring or governor rod in alternate holes of governor control lever.

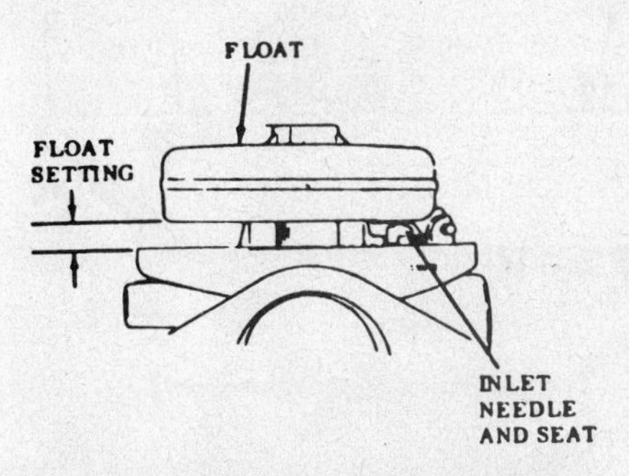

Fig. W38—Walbro LME-35 carburetor float should have 5/32-inch (3.97 mm) space between free end of float and gasket surface as shown.

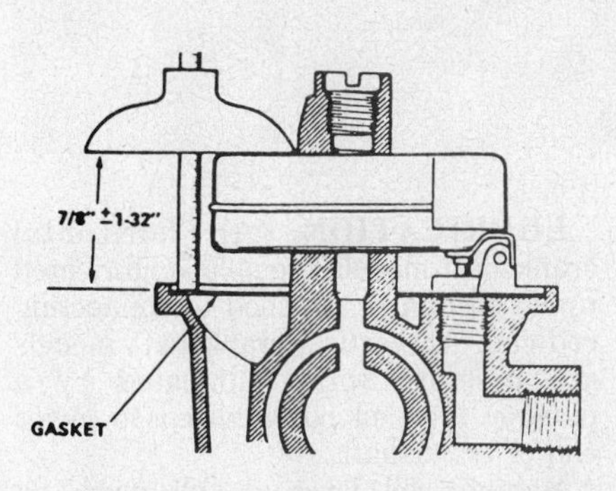

Fig. W39—Zenith 1408 carburetor float measurement should be 7/8-inch (22.23 mm) plus or minus 1/32-inch (0.794 mm) when measured as shown with gasket in place.

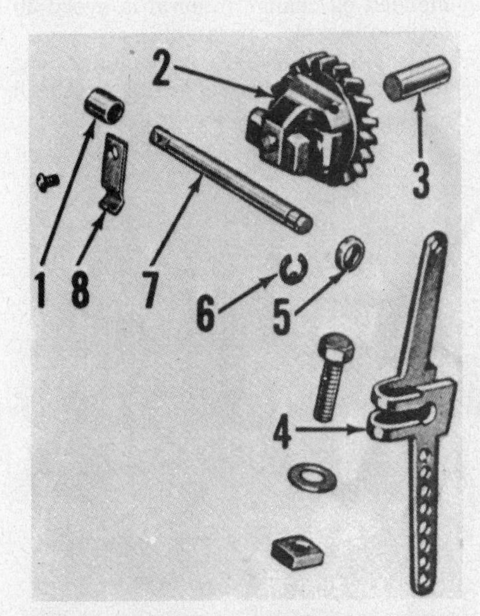

Fig. W42—Exploded view of governor asembly typical of that used on Models S-7D, HS-7D, S-8D, HS-8D, TR-10D, TRA-10D and TRA-12D.

1. Spacer
2. Gear/flyweight assy.
3. Governor shaft
4. Governor lever
5. Oil seal
6. Retaining ring
7. Fulcrum shaft
8. Vane

To correctly set governed speeds, use a tachometer to accurately record crankshaft rpm. Refer to appropriate table (I through IV) corresponding to engine model being serviced to determine proper governor lever hole for attaching governor spring to set required speed.

Table I—Models S-7D, S-8D Engines to and including Serial Number 3909151

Desired Rpm Under Load	Hole Number	Adjust No-Load Rpm To:
1600	3	1880
1700	3	1940
1800	3	1990
1900	3	2080
2000	4	2260
2100	4	2360
2200	4	2410
2300	4	2510
2400	4	2590
2500	4	2680
2600	5	2830
2700	5	2920
2800	5	2970
*2900	5	3040
*3000	6	3230
*3100	6	3330
*3200	6	3420
*3300	6	3510
*3400	6	3590
*3500	7	3750
*3600	7	3840

1600-2800 rpm, use 5-1/4 inch adjusting screw.
2900 rpm (*) and higher, use 5 inch adjusting screws.

Engines beginning with Serial Number 3909152

Desired Rpm Under Load	Hole Number	Adjust No-Load Rpm To:
1800	2	2030
1900	2	2125
2000	2	2220
2100	2	2320
2200	3	2430
2300	3	2520
2400	4	2690
2500	4	2720
2600	4	2845
2700	4	2930
2800	4	3010
2900	5	3150
*3000	5	3230
*3100	5	3300
*3200	5	3350
*3300	6	3575
*3400	6	3650
*3500	6	3750
*3600	6	3800

1800-2900 rpm, use 5-5/8 inch adjusting screw.
3000 rpm (*) and higher, use 5-1/4 inch adjusting screw.

Table II—Models HS-7D, HS-8D

Desired Rpm Under Load	Hole Number	Adjust No-Load Rpm To:
1800	4	2200
1900	4	2290
2000	4	2380
2100	4	2465
2200	4	2550
2300	5	2690
2400	5	2770
2500	5	2850
2600	6	3000
2700	6	3060
2800	6	3120
2900	6	3200
*3000	6	3280
*3100	6	3340
*3200	6	3400
*3300	7	3560
*3400	7	3620
*3500	7	3685
*3600	7	3750

1800-2900 rpm, use 5-5/8 inch adjusting screw.
3000 rpm (*) and higher, use 5-1/4 inch adjusting screw.

Table III—Models TRA-10D, TRA-12D, TR-10D Use Lever Spring Hole Number and Set No Load Rpm:

Desired Rpm Under Load	TRA-10D**		TRA-12D	
2000	1	2520	3	2230
2100	1	1580	4	2430
2200	1	2610	4	2515
2300	1	2690	4	2590
2400	1	2740	4	2660
2500	1	2800	4	2750
2600	1	2890	4	2810
2700	1	2935	5	3020
2800	2	3065	5	3100
2900	2	3160	5	3180
*3000	3	3230	5	3260
*3100	3	3300	5	3325
*3200	3	3380	6	3535
*3300	3	3460	6	3620
*3400	4	3615	6	3700
*3500	4	3690	6	3790
*3600	5	3850	6	3860

**Applies to TR-10D engines Serial No. 3909152 and after.
2000-2900 rpm, TRA-10D uses 3-5/8 inch adjusting screw; TRA-12D uses 5-5/8 adjusting screw.
3000 rpm (*) and higher, TRA-10D uses 5 inch adjusting screw; TRA-12D uses 5-1/4 inch adjusting screw.

Table IV—Models S-10D, S-12D, S-14D

Desired Rpm Under Load	Hole Number	Adjust No-Load Rpm To:
1600	1	1760
1800	2	1975
1900	2	2040
2000	2	2120
2100	3	2260
2200	3	2340
2300	3	2400
2400	4	2580
2500	4	2650
2600	4	2720
2700	4	2810
2800	5	2910
2900	5	3010
*3000	6	3150
*3100	6	3230
*3200	7	3360
*3300	7	3455
*3400	7	3520
*3500	7	3590
*3600	7	3680

1600-2900 rpm, use 3-15/16 inch adjusting screw.
3000 rpm (*) and higher, use 3-5/8 inch adjusting screws.

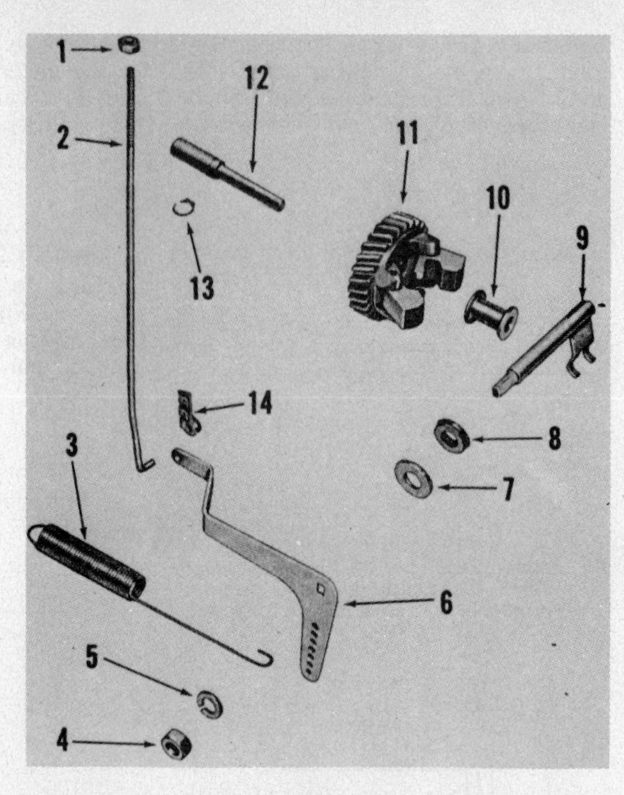

Fig. W43—Governor mechanism as used on Models S-10D, S-12D and S-14D. Refer to text for service information.

1. Locknut
2. Throttle rod
3. Governor spring
4. Nut
5. Lockwasher
6. Governor lever
7. Flat washer
8. Oil seal
9. Cross (fulcrum) shaft
10. Thrust sleeve
11. Gear weight assy.
12. Shaft
13. Snap ring
14. Clip

IGNITION SYSTEM. Various magneto, battery or solid-state electronic ignition systems are used according to model and application. Refer to appropriate paragraph for model being serviced.

MAGNETO. Fairbanks-Morse or Wico magneto assembly located under flywheel is used (Fig. W47) and points and condenser are located in an external breaker box (Fig. W49 or W50) and are actuated by camshaft lobe via a short push rod.

Initial point gap is 0.018-0.020 inch (0.457-0.508 mm) and point gap is varied slightly to obtain correct timing as outlined in **BATTERY IGNITION** section.

BATTERY IGNITION. Battery ignition system uses a conventional ignition coil and points and condenser are the same as used for magneto system. Points and condenser location remains the same (Fig. W49 or W50) and initial point gap is 0.023 inch (0.5842 mm) and gap is varied slightly to obtain correct engine timing.

To set timing on either magneto system or battery system, position engine flywheel so timing marks appear in hole in flywheel shroud (Fig. W49) or align with timing pointer (Fig. W50), connect test light across points and adjust point gap so light just goes out.

Timing should be set at 15° BTDC on S-7D and HS-7D engines and 18° BTDC on all remaining models.

SOLID-STATE ELECTRONIC IGNITION. Breakerless capacitive-discharge (CD) ignition system is available for S-10D, S-12D and S-14D engines and is standard ignition system for TRA-12D engines.

No adjustments are possible and system has only three components which are magnet (part of flywheel), stator (containing trigger coil, rectifier diode and a silicone-controlled rectifier) located on bearing plate at flywheel side of engine and a special ignition coil.

If visual inspection of component parts fail to find possible cause of an ignition failure, a continuity test of ignition switch and coil should be made. If switch and coil are in working order but ignition system still fails, renew stator.

LUBRICATION. All horizontal crankshaft models are splash lubricated by an oil dipper attached to connecting rod cap. All vertical crankshaft models are pressure spray lubricated by a plunger type oil pump fitted to lower end of crankshaft.

Maintain oil level at full mark on dipstick or level with filler plug as equipped, but do not overfill. High quality detergent oil having API classification "SD" or "SE" is recommended. Recommended oil change interval is every 50 hours of normal operation.

Use SAE 30 oil when operating temperature is between 120°F (49°C)

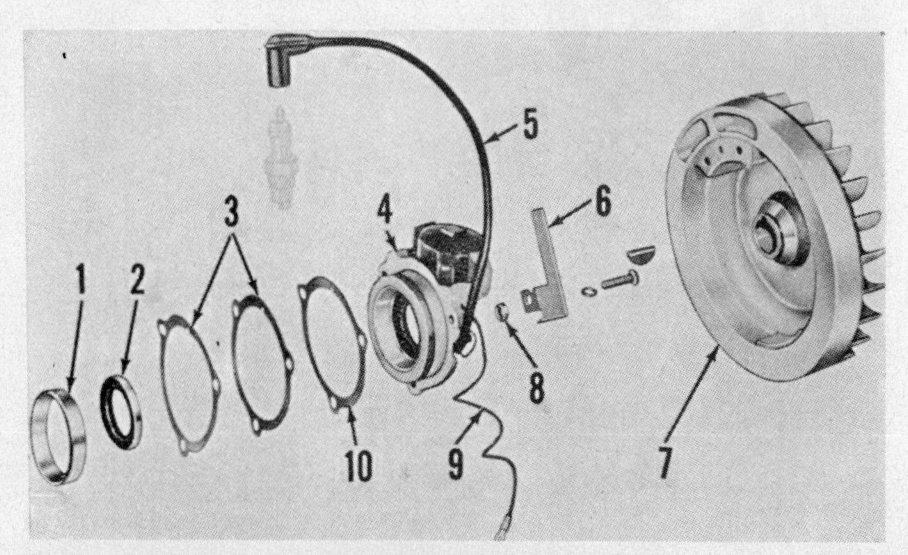

Fig. W47—Typical flywheel magneto assembly. Note crankshaft end play is adjusted by shims (3) which are offered in a variety of thicknesses. Flywheel must be removed for access to magneto stator plate.

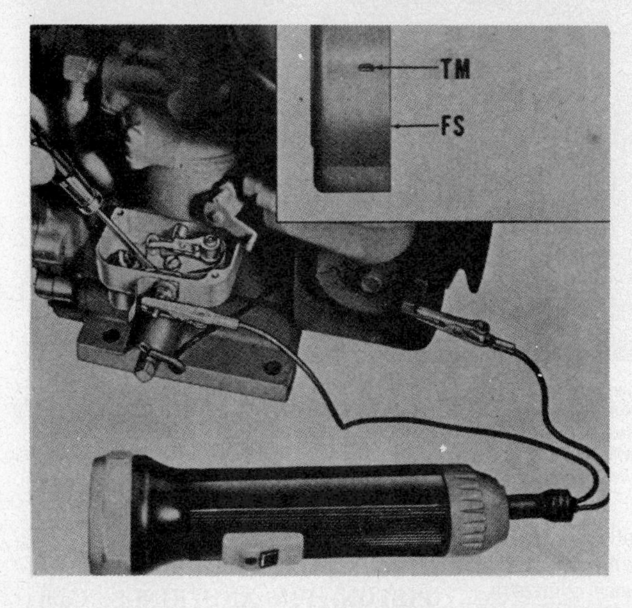

Fig. W49—Timing mark (TM) should appear in hole in flywheel shroud (FS) on S-7D and HS-7D engines as timing light goes out.

and 40°F (4°C), SAE 20 oil between 40°F (4°C) and 15°F (−9°C), SAE 10 oil between 15°F (−9°C) and 0°F (−18°C) and SAE 5W-20 oil for below 0°F (−18°C) temperature.

REPAIRS

TIGHTENING TORQUES. Recommended tightening torques are as follows:

Models S-7D, S-8D, HS-7D and HS8D

Gear cover . . . . 8 ft.-lbs. (11N•m)
Stator plate . . . . 8 ft.-lbs. (11 N•m)
Connecting rod . . . . 18 ft.-lbs. (24 N•m)
Spark plug . . . . 29 ft.-lbs. (39 N•m)
Cylinder head . . . . 18 ft.-lbs. (24N•m)

Models TR-10D, TRA-10D and TRA-12D

Gear cover . . . . 8 ft.-lbs. (11N•m)
Stator Plate . . . . 8 ft.-lbs. (11N•m)
Connecting rod . . . . 22 ft.-lbs. (30N•m)
Spark plug . . . . 29 ft.-lbs. (39 N•m)
Flywheel nut . . . . 55 ft.-lbs. (75 N•m)
Cylinder head . . . . 18 ft.-lbs. (24 N•m)

Models S-10D, S-12D and S-14D

Gear cover . . . . 18 ft.-lbs. (24 N−m)
Stator plate . . . . 18 ft.-lbs. (24 N•m)
Connecting rod . . . . 32 ft.-lbs. (43 N•m)
Spark plug . . . . 29 ft.-lbs. (39 N•m)
Flywheel nut . . . . 55 ft.-lbs. (75 N•m)
Cylinder block nut . . . . 50 ft.-lbs. (68 N•m)
Cylinder head . . . . 32 ft.-lbs. (43 N•m)

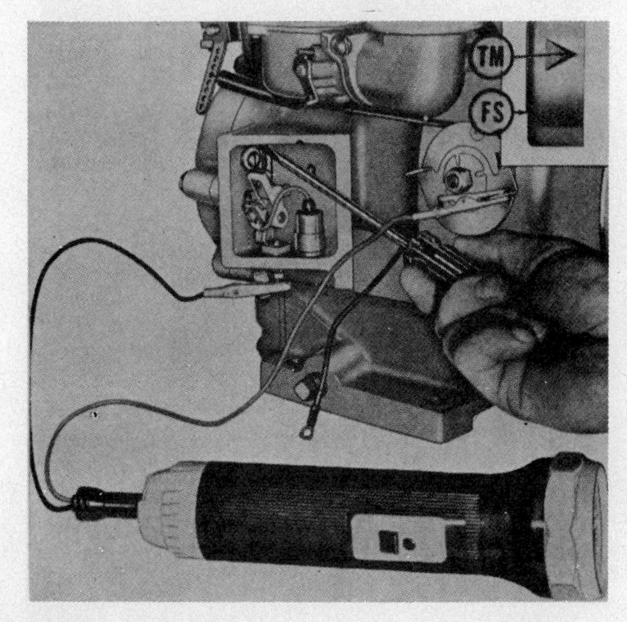

Fig. W50-Timing mark (TM) should be aligned with timing pointer on flywheel shroud (FS) on all models except S-7D and HS-7D engines as timing light just goes out. Refer to Fig. W49 for S-7D and HS-7D engines.

CYLINDER HEAD. Always install a new head gasket when installing cylinder head. Note different lengths and styles of studs and cap screws for correct reinstallation and lightly lubricate threads.

Cylinder heads should be tightened in three equal steps to recommended torque using a criss-cross pattern working from the center out.

CONNECTING ROD. Connecting rod and piston are removed from cylinder head end of block after removing head and crankcase side cover. See Fig. W54 or W55.

Connecting rod for all models except S-10D, S-12D or S-14D ride directly on crankpin journal. S-10D, S-12D and S-14D models connecting rods have renewable insert type bearings.

Running clearance for insert type bearings should be 0.0005-0.0015 inch (0.0127-0.0381 mm). Running clearance for S-7D and HS-7D models rod should be 0.0012-0.002 inch (0.0305-0.0508 mm), for S-8D and HS-8D models clearance should be 0.0007-0.0015 inch (0.0178-0.0381 mm) and clearance for all TR and TRA models should be 0.0005-0.0015 inch (0.0127-0.0381 mm).

Side clearance for connecting rods

Fig. W54—Piston (P) is removed from top. Oil dipper (D) provides lubrication for horizontal crankshaft engines. Note placement of connecting rod index arrow at (A), location of governor shaft (GS) and that camshaft gear is fitted with a compression release (CR), typical of TR-TRA and larger models.

Fig. W55—Open view of typical S-10D, S-12D or S-14D engine. Compare to Fig. W54, note difference in placement of governor shaft, style of oil dipper and that cylinder and crankcase are separate castings. Flywheel should be left in place to balance crankshaft when gear cover is removed.

with insert type bearing should be 0.004-0.013 inch (0.1016-0.3302 mm). Side clearance for S-7D, HS-7D, S-8D and HS-8D models should be 0.006-0.013 inch (0.1524-0.3302 mm) and side clearance for all TR and TRA models should be 0.009-0.016 inch (0.2286-0.4064 mm).

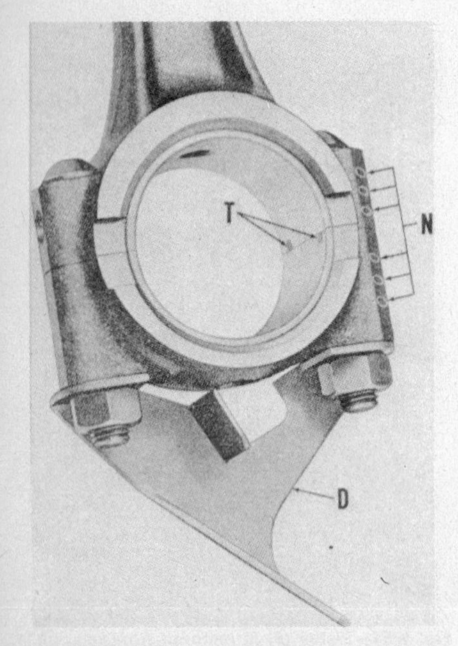

Fig. W56—Install connecting rod cap (S-10D, S-12D and S-14D) so tangs (T) of bearing inserts are on same side. Numbers (N) on rod end and cap should be aligned and installed toward gear cover side of crankcase. Oil dipper (D) open side faces outward.

Fig. W57—Sectional view showing proper arrangement of piston rings. In this typical view, top ring may not be chamfered inside as shown, however, all rings are marked with "TOP" or pit mark for correct installation.

When installing piston and connecting rod assembly on models which are not equipped with renewable bearing inserts align index arrows (A-Fig. W54) on rod end and cap. Arrows must face toward open end of crankcase. Horizontal shaft engines which are equipped with oil dipper (D) should have dipper installed so connecting rod cap screws are accessible from open end of crankcase. Refer to Fig. W56 when assembling cap to connecting rod on S-10D, S-12D and S-14D models and make certain fitting tangs (T) are on same side as shown. Stamped numbers (N) and oil dipper (D) should face open side of crankcase.

Standard connecting rod journal diameter for S-7D and HS-7D models should be 1.3750-1.3755 inches (34.93-34.94 mm), standard crankshaft diameter for S-8D, HS-8D, TR-10D, TRA-10D and TRA-12D should be 1.3755-1.3760 inches (34.94-34.95 mm) and standard diameter for S-10D, S-12D and S-14D models should be 1.4984-1.4990 inches (38.06-38.08 mm).

Connecting rods or bearing inserts are available in a variety of sizes to fit reground crankshafts.

Fig. W58—Location of specification number (A) on engines with cylinder cast as an integral part of crankcase.

Fig. W60—Locate timing mark (A) on camshaft gear between two marked teeth (B) of crankshaft gear. View is typical of S-7D, S-8D and all TR and TRA engines.

PISTON, PIN AND RINGS. Cam ground piston for all models is equipped with one chrome faced compression ring, one scraper ring and an oil control ring (Fig. W57). Top side of rings are marked for correct installation. Ring end gap for all models should be 0.010-0.020 inch (0.254-0.508 mm) and ring end gaps should be spaced at 120° intervals around piston. A variety of oversize pistons and rings are available.

Ring side clearance in groove for S-10D, S-12D and S-14D should be: Top and second ring, 0.002-0.004 inch (0.0508-0.1016 mm) and for oil control ring, 0.0015-0.0035 inch (0.0381-0.0889 mm).

Ring side clearance in groove for all remaining models should be: Top ring, 0.002-0.0035 inch (0.0508-0.0889 mm), second ring, 0.001-0.0025 inch (0.0254-0.0635 mm) and oil control ring, 0.002-0.0035 inch (0.0508-0.0889 mm).

Fig. W61—View of timing marks lined up in S-10D, S-12D and S-14D. In current prouction, camshaft gear will support a compression release as shown in Fig. W64. Camshaft thrust spring (S) and governor thrust sleeve (10) must be in place before installation of gear cover. Use heavy grease to hold camshaft thrust ball in cover hole during installation.

Piston to cylinder clearance should be measured at 90° angle to piston pin at lower edge of skirt. Clearance for S-7D, S-8D, HS-7D, HS-8D, TR-10D and TRA-10D models should be 0.004-0.0045 inch (0.1016-0.1143 mm), clearance for TRA-12D, S-10D and S-12D models should be 0.0025-0.003 inch (0.0635-0.0762 mm) and clearance for S-14D model should be 0.0025-0.004 inch (0.0635-0.1016 mm).

Floating type piston pin is retained by snap rings in bore of piston and should have 0.0002-0.0008 inch (0.0051-0.0203 mm) clearance in connecting rod of all models except S-10D, S-12D and S-14D. S-10D, S-12D and S-14D models use a bushing in connecting rod end and clearance in bushing should be 0.0005-0.0011 inch (0.0127-0.0279 mm).

Pins are available in a variety of oversizes as well as standard. Connecting rod end or bushing and piston pin bore must be reamed to fit available oversize pin.

CYLINDER. Cylinder is an integral part of crankcase casting for S-7D, S-8D, HS-7D, HS-8D, TR-10D, TRA-10D and TRA-12D models (Fig. W58) but is a separate casting bolted to crankcase for S-10D, S-12D and S-14D models (Fig. W55).

Standard cylinder bore diameter for S-7D and HS-7D models should be 3 inches (76.2 mm), standard diameter for S-8D, HS-8D, TR-10D and TRA-10D models should be 3.125 inches (79.38 mm), standard diameter for S-10D model should be 3.25 inches (82.55 mm), standard diameter for TRA-12D and S-12D models should be 3.5 inches (88.9 mm) and standard diameter for S-14D model should be 3.75 inches (95.2 mm).

Cylinder should be bored to nearest oversize for which piston and rings are available if worn or out-of-round more than 0.005 inch (0.1270 mm).

Cylinder section of two piece assembly is bolted to crankcase section and one cap screw is concealed within valve spring compartment (Fig. W67). Tighten concealed cap screw to 32 ft.-lbs. (43 N•m) torque and all remaining stud nuts to 42-50 ft.-lbs. (57-68 N•m) torque.

CRANKSHAFT, MAIN BEARINGS AND SEALS. Crankshaft on all models is supported at each end by taper roller bearings. Crankshaft end play should be 0.001-0.004 inch (0.0254-0.1016 mm) and is controlled by varying number of shims between crankcase and stator plate (main bearing support). See Fig. W47.

Main bearings are a press fit on crankshaft and bearing cups are pressed into stator plate and crankcase. Bearings and cups must be fully seated to correctly set crankshaft end play.

To properly time crankshaft to related engine components during installation, refer to Fig. W60 or W61 according to model being serviced.

Seals should be installed in stator plate and crankcase prior to crankshaft installation and seal protectors (Fig. W62 or W63) should be used to prevent seal damage during crankshaft installation.

GOVERNOR GEAR AND WEIGHT ASSEMBLY. Governor gear and weight assemblies rotate on a shaft which is a press fit in a bore of crankcase. Exploded views are shown in Figs. W42 and W43. On S-7D, S-8D, HS-7D, HS-8D and all TR and TRA models shaft (3-Fig. W42) has had its depth in block held by a snap ring beginning with production serial number 3090152. Models S-10D, S-12D and S-14D require end play of 0.003-0.005 inch (0.0762-0.1270 mm) be maintained on governor gear shaft between gear and its snap ring retainer. See Fig. W65 for measurement technique to be used on these models. Press-fit shaft is driven in or out of bore to make adjustment. On models with straight governor lever and governor gear mounted above cam gear,

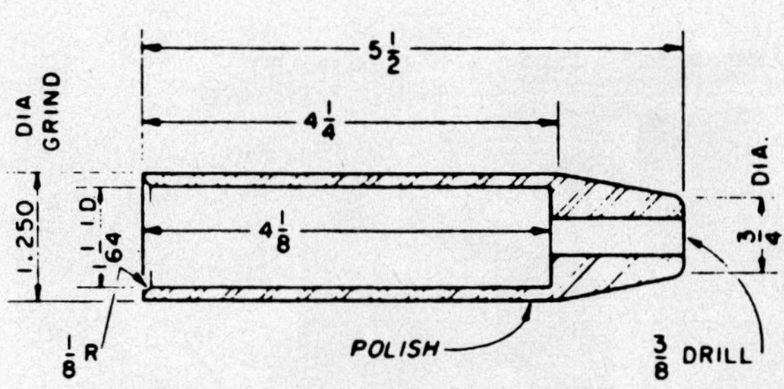

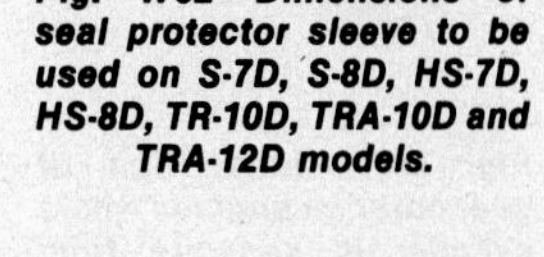
Fig. W62—Dimensions of seal protector sleeve to be used on S-7D, S-8D, HS-7D, HS-8D, TR-10D, TRA-10D and TRA-12D models.

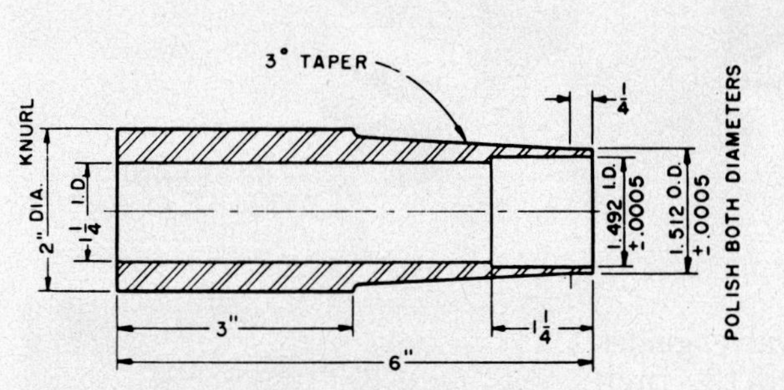

Fig. W63—Seal protector sleeve to be used on S-10D, S-12D and S-14D models is fabricated to dimensions shown.

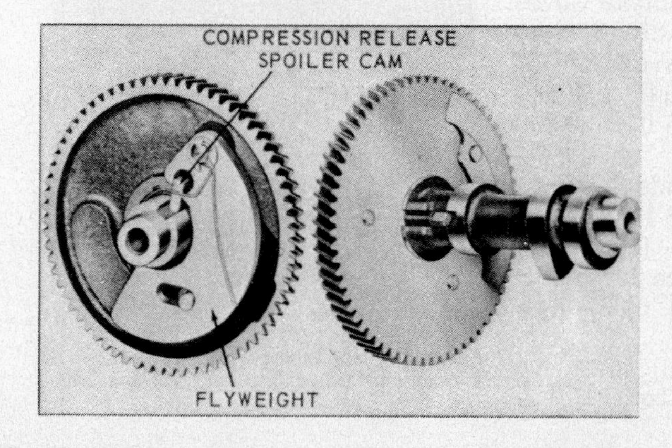

Fig. W64—View of both sides of camshaft gear to show compression release assembly installed.

Fig. W65—Make certain snap ring is correctly seated and use feeler gage to measure end play of governor gear on shaft.

note in Fig. W66 upper end of governor lever must tilt toward engine so governor vane will not be fouled or interfere with flyweights as gear cover is installed. On S-10D, S-12D and S-14D models, governor thrust sleeve must be in place as shown in Fig. W67 when gear cover is placed in position.

Camshafts on all models are supported at each end in unbushed bores in crankcase and gear cover. Camshaft end play is controlled by a thrust spring (Fig. W66 and W67) fitted into shaft hub which centers upon steel ball in a socket in gear cover. During assembly, ball is held in place by a coating of heavy grease. To remove or reinstall camshaft place block on its side as shown if Fig. W68 to prevent tappets from falling out.

On models equipped with a compression release type camshaft (Fig. W64), a spoiler cam holds exhaust valve slightly open during part of compression stroke while cranking. Reduced compression pressure allows for faster cranking speed with lower effort. When crankshaft reaches 650 rpm during cranking, centrifugal force swings flyweight on front of cam gear so as to turn spoiler cam to inoperative position allowing exhaust valve to seat and restore full compression. Whenever camshaft is removed, compression release mechanism should be checked for damage to spring or excessive wear on spoiler cam. Flyweight and spoiler cam must move easily with no binding.

See Figs. W60 and W61 to set timing marks in register during reassembly.

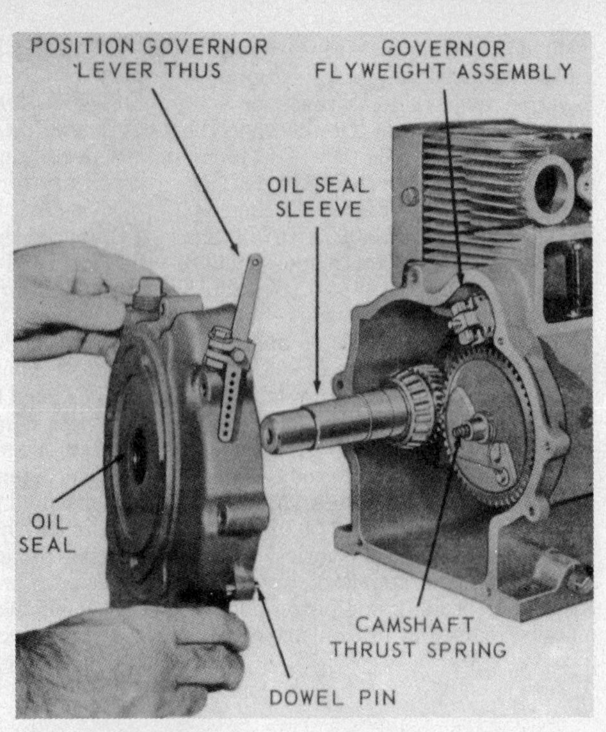

Fig. W66—To install gear cover on all models where cylinder is an integral part of crankcase, governor assembly, camshaft thrust spring and oil seal protector must be in place. Make certain governor lever is tilted as shown and governor thrust ball is held in cover with heavy grease.

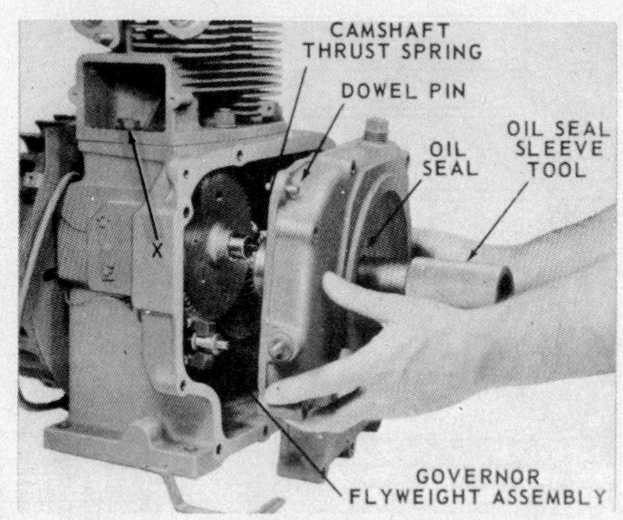

Fig. W67—Installation of gear cover on engines where cylinder is separate from crankcase. Note cap screw (X) in valve compartment, referred to in text. Protect cover oil seal with sleeve tool as shown and make certain thrust sleeve is in place on governor shaft.

VALVE SYSTEM. Exhaust valve seats on all models are renewable insert type and intake seat may be machined directly into block or be of renewable insert type according to model and application. Seats should be ground to a 45° angle and width should not exceed 3/32-inch (2.381 mm).

Valve faces are ground at 45° angles and stellite exhaust valves and rotocaps are standard on all models.

Renewable guides are pressed in or out from top side of block. Tool, DF-72 guide driver is available from Wisconsin. Internal chamfered end of guide is installed toward tappet end of valve.

Inside diameter of valve guides should be 0.312-0.313 inch (7.93-7.95 mm) and valve stem diameter should be 0.310-0.311 inch (7.87-7.90 mm) for all intake valves and 0.309-0.310 inch (7.85-7.87 mm) for all exhaust valves except for S-10D, S-12D and S-14D exhaust valves which should be 0.308-0.309 inch (7.82-7.85 mm).

Maximum stem to guide clearance for all models except S-10D, S-12D and S-14D should be 0.006 inch (0.1524 mm). Clearance for S-10D and S-14D should be 0.007 inch (0.1778 mm).

Valve tappet gap (cold) for S-10D, S-12D and S-14D should be 0.007 inch

Fig. W68—Place engine on its side as shown to prevent tappets dropping from block bores when camshaft is removed.

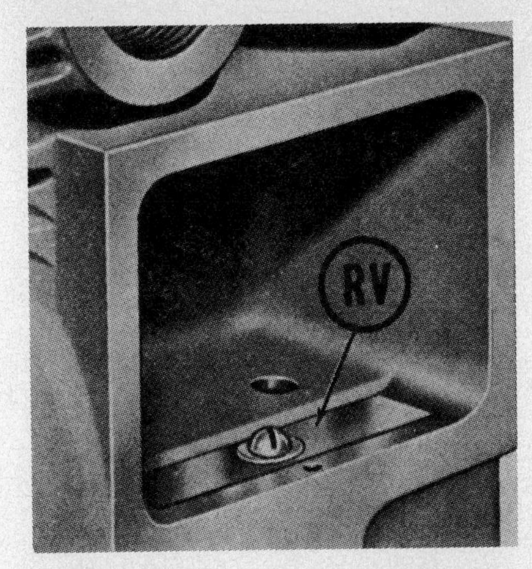

Fig. W69—Reed type breather valve located in valve spring compartment of S-7D, S-8D, HS-7D, HS-8D, TR-10D, TRA-10D and TRA-12D models.

(0.1778 mm) for intake valves and 0.016 inch (0.4064 mm) for exhaust valves. Adjust by turning self-locking cap screw on tappet as requred.

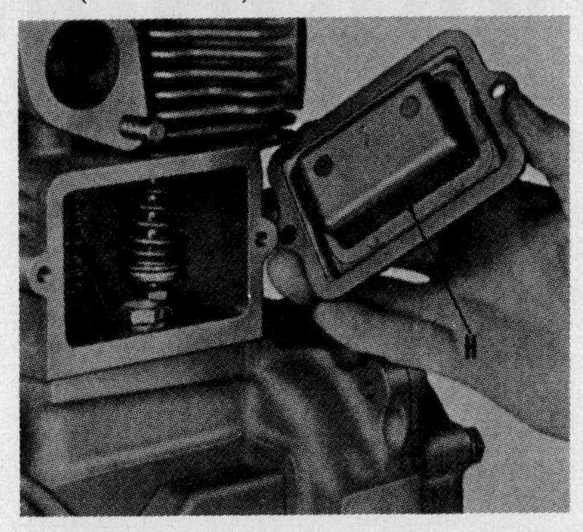

Fig. W70—Breather valve used on S-10D, S-12D and S-14D models is located in valve spring compartment cover. Drain hole (H) must be kept open.

Valve tappet gap (cold) for all remaining models should be 0.006 inch (0.1524 mm) for intake and 0.012 inch (0.3048 mm) for all exhaust valves except ones used on TRA-12D which should be 0.015 inch (0.3810 mm). Adjustment is made by carefully grinding valve stem end until required clearance is reached.

BREATHER A reed type breather valve (Fig. W69 or W70) is located in valve spring compartment. These reed valve assemblies should be kept clean and renewed whenever found to be inoperable.

If oil fouling occurs in ignition breaker box, condition of breather valve should be checked.

WISCONSIN

TELEDYNE WISCONSIN MOTOR
Milwaukee, Wisconsin 53246

Model	No. Cyls.	Bore	Stroke	Displacement	Power Rating
TE	2	3 in. (76.2 mm)	3.25 in. (82.55 mm)	45.9 cu. in. (753 cc)	11.2 hp. (8.4 kW)
TF	2	3.25 in. (82.55 mm)	3.25 in. (82.55 mm)	53.9 cu. in. (883.6 cc)	14.6 hp. (10.9 kW)
TH, THD	2	3.25 in. (82.55 mm)	3.25 in. (82.55 mm)	53.9 cu. in. (883.6 cc)	18 hp. (13.4 kW)
TJD	2	3.25 in. (82.55 mm)	3.25 in. (82.55 mm)	53.9 cu. in. (883.6 cc.)	18.2 hp. (13.6 kW)

Engines in this section are four-cycle, two cylinder upright, horizontal crankshaft engines. Crankshaft for all models is supported at each end in taper roller bearings.

On some early engines connecting rod rides directly on crankpin journal while later models have insert type renewable connecting rod bearings. All models are pressure spray lubricated by a plunger type oil pump driven off of a camshaft lobe.

Various magneto or battery type ignition systems are used according to model and application.

Zenith or Marvel-Schebler float type carburetor is used and a mechanically operated fuel pump is available.

Engine identification information is located on engine instruction plate and engine model, specification and serial number are required when ordering parts or service material.

MAINTENACE

SPARK PLUG. Recommended spark plug is Champion D16J or equivalent. Electrode gap should be 0.030 inch (0.762 mm).

CARBURETOR. A variety of float type carburetors are used as listed.

Zenith 161-7 carburetor is used on TE and TF engines.

Zenith 68-7 carburetor is used on TH, THD and TJD engines, however some TJD engines were equipped with Marvel-Schebler TSX-954 carburetor.

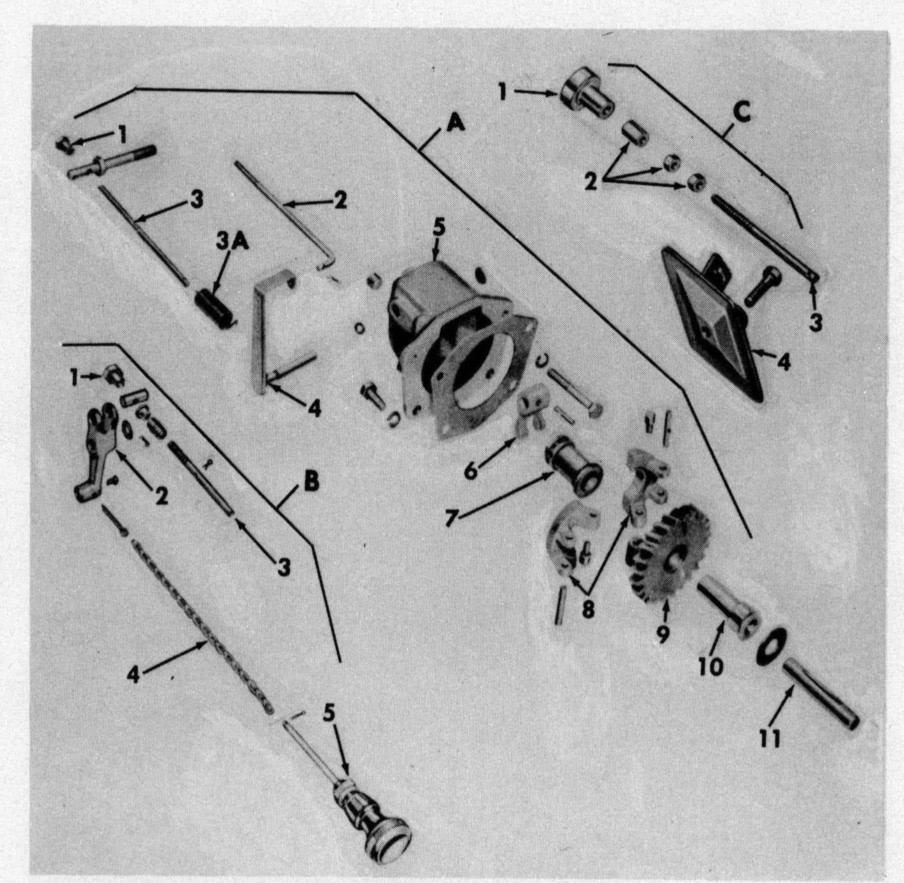

Fig. W72—Exploded view of governor assembly (A) to show general arrangement of working parts. View B is variable speed control and View C shows idle control assembly. Refer to text. Item 3, (A-B-C) connects to 3A.

A. Governor assy.
1. Adjustment nut
2. Throttle rod to governor
3. Adjustment screw
3A. Governor spring
4. Governor lever
5. Governor housing
6. Governor yoke
7. Thrust sleeve & bearing
8. Flyweights
9. Governor gear
10. Bushing
11. Shaft

B. Variable speed control
1. Locknut
2. Variable speed lever
3. Adjusting screw
4. Control chain
5. Control knob

C. Idle control
1. Control knob
2. Locknuts
3. Control rod
4. Tappet cover rod support

For initial adjustment of all models note main fuel jet, except in possible special application, are fixed and nonadjustable.

For initial idle mixture screw adjustment of Zenith 161-7 or Marvel-Schebler TSX-954 carburetor, open idle mixture screw 1 turn.

For initial idle mixture screw adjustment of Zenith 68-7 carburetor, open idle mixture screw 1¼ turn.

On all models make final adjustments with engine at normal operating temperature and running. Set engine at idle speed and adjust idle mixture screw for smoothest idle operation.

To check float level on Zenith carburetor, invert carburetor throttle body and float assembly and measure from machined surface of cover (no gasket) to high point of float furthest from pivot. For 68-7 carburetor, this distance should be 1-5/32 inch (29.37 mm) plus or minus 1/16-inch (1.59 mm). For 161-7 carburetor this distance should be 1-5/32 inch (29.37 mm) plus or minus 3/64-inch (1.19 mm).

To check float level on Marvel-Schebler carburetor, invert carburetor throttle body and float assembly and measure from gasket on gasket surface to near side of float furthest away from pivot. Measurement should be ¼-inch (6.35 mm).

Adjust float level on all models by bending float tang that contacts inlet valve.

GOVERNOR. Flyweight type governor driven by camshaft gear is used on all models. Refer to Fig. W72 for exploded view of typical governor component parts.

Major speed changes on all models is obtained by using springs of varying tension, flyweights of varying weights or locating spring or governor rod in alternate holes of governor control lever.

To correctly set governed speeds, use a tachometer to accurately record crankshaft rpm. Refer to appropriate table (I through III) corresponding to engine model being serviced to determine proper governor lever hole for attaching governor spring to set required speed.

Table I—Model TJD

Desired Rpm Under Load	Hole Number	Adjust No-Load Rpm To:
1600	3	1725
1700	3	1800
1800	4	1925
1900	4	2000
2000	5	2140
2100	5	2210
2200	6	2365
2300	6	2420
2400	7	2540
2500	8	2675
2600	6	2775
2700	6	2870
2800	6	2935
2900	7	3090
3000	7	3160
3100	7	3230
3200	8	3390
3300	8	3430
3400	9	3590
3500	9	3640
3600	10	3775

1600-2500 rpm, use 20 coil spring PM-75
2600-3600 rpm, use 24 coil spring PM-76

Table II—Models TH, THD

Desired Rpm Under Load	Hole Number	Adjust No-Load Rpm To:
1600	3	1775
1700	3	1825
1800	4	1975
1900	4	2025
2000	5	2175
2100	5	2250
2200	6	2350
2300	6	2425
2400	7	2550
2500	8	2675
*2600	8	2750
2700	9	2875
2800	9	2950
2900	9	3025
3000	10	3160
3100	10	3225
3200	11	3350
3300	12	3450
3400	12	3525
3500	12	3600
3600	12	3675

*Rpm under load, TH model must not exceed 2600 rpm.

Table III—Models TE, TF

Desired Rpm Under Load	Hole Number	Adjust No-Load Rpm To:
1400	2	1550
1500	3	1650
1600	3	1725
1700	4	1850
1800	4	1925
1900	5	2025
2000	6	2150
2100	6	2225
2200	7	2350
2300	7	2425
2400	8	2550
2500	9	2650
2600	9	2725

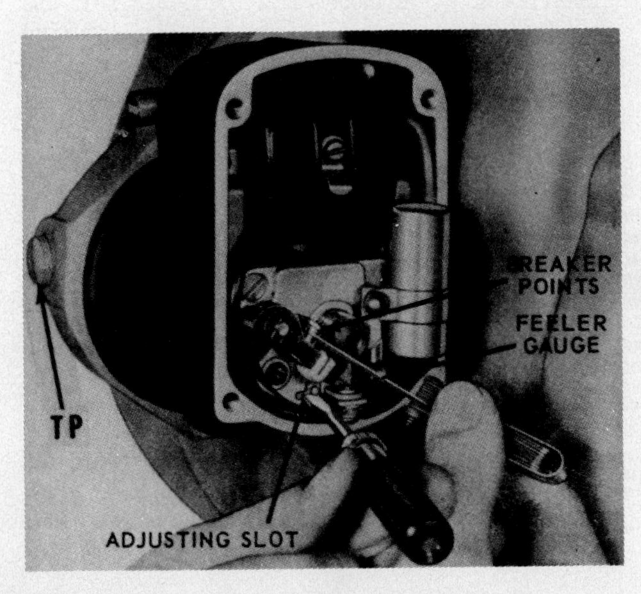

Fig. W73—Adjustment procedure for magneto breaker points. Magneto shown is typical for TJD model. Timing plug (TP) must be removed so marked tooth is visible. See text.

Governor-to-carburetor rod should be adjusted so it just will enter hole in governor lever when governor weights are "in" and carburetor throttle is wide open.

IGNITION SYSTEM. Either a battery type ignition system with breaker points located in a distributor or a magneto type ignition system are used according to model and application.

BATTERY IGNITION. Breaker points for battery ignition system are located inside a distributor which may be mounted to a gear driven generator (Fig. W77) or on an adapter attached to gear cover at magneto mounting location. Points should be set at 0.020 inch (0.508 mm) on all models.

Distributor rotates in a counterclockwise direction at one-half crankshaft speed. Automatic advance flyweights pivot on distributor shaft below breaker point mounting plate to provide automatic timing advance.

To set timing, connect timing light and with engine running slightly above 2000 rpm, timing should be set at 27° BTDC on TE and TF engines and 20° BTDC on TH, THD and TJD engines. Refer to Fig. W78.

To install distributor, make certain number "1" cylinder is at "top dead center" on compression stroke and install distributor as shown in Fig. W77. Rotate distributor counter-clockwise until points are completely closed, then turn distributor clockwise until points just open. Set final timing with timing light as previously outlined.

MAGNETO IGNITION. Either a Fairbanks-Morse or Wico magneto may be used, however magnetos may not be interchangeable between models due to different firing intervals.

Breaker point gap for all models should be 0.015 inch (0.381 mm).

To install magneto to engine, make certain number "1" cylinder (cylinder nearest flywheel side of engine) is at "top dead center" on compression stroke, remove air intake screen at flywheel and timing hole plug (7-Fig. W74). Leading edge of flywheel vane marked "X" and "DC" (2-Fig. W74) should be in register with static timing mark on air shroud. Before mounting magneto in place, turn drive gear clockwise until impulse coupling snaps, align on number one firing position. Install magneto so "X" marked tooth appears in timing port as shown (8-Fig. W74).

Set ignition timing as outlined in **BATTERY IGNITION** section.

LUBRICATION. Pressure spray lubrication is provided by a plunger type oil pump (Fig. W76) which is driven off of a camshaft lobe. Oil is also pumped through an external oil line and restricter to governor mechanism and timing gears.

Maintain oil level at full mark on dipstick or at top of filler plug but do not overfill. High quality detergent oil having API classification 'SD" or "SE" is recommended. Recommended oil change interval is every 50 hours of normal operation.

Use SAE 30 oil when operating temperature is between 120°F (49°C) and 40°F (4°C), SAE 20 oil between 40°F (4°C) and 15°F (−9°C) and SAE 10 oil between 15°F (−9°C) and 0°F (−18°C) and SAE 5W-20 oil for below 0°F (−18°C) temperature.

REPAIRS

TIGHTENING TORQUES. Recommended torque specifications are as follows:

Cylinder head–
All models . . . . 16-18 ft.-lbs. (22-24 N·m)

Cylinder block–
All models . . . . 32-34 ft.-lbs. (43-46 N·m)

Connecting rod–
All models . . . . 22-28 ft.-lbs. (30-38 N·m)

Main bearing plate–
All models . . . . 24-26 ft.-lbs. (33-35N·m)

Engine base–
All models . . . . 22-24 ft.-lbs. (30-33·m)

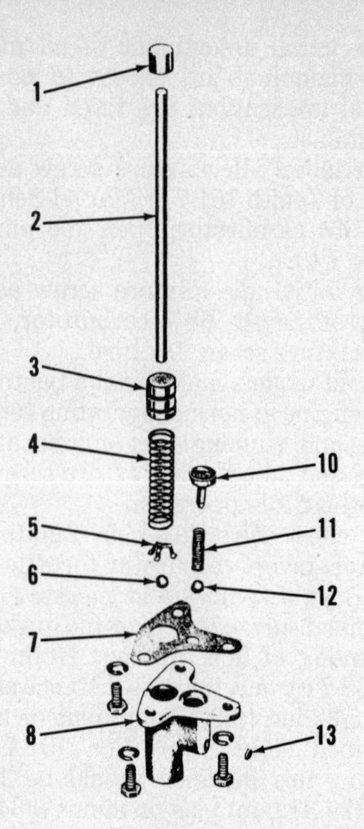

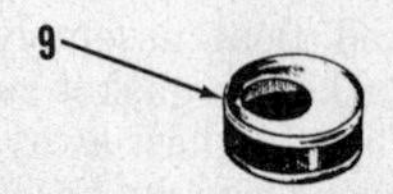

Fig. W76—Exploded view of plunger type oil pump which is driven off of a camshaft lobe. Pump body (8) may vary according to model but basic construction is similar.

1. Cap
2. Push rod
3. Plunger
4. Pump spring
5. Retainer
6. Check ball
7. Gasket
8. Pump body
9. Sump screen
10. Retainer
11. Spring
12. Check ball
13. Core plug

Gear cover–
All models . . . . 16-18 ft.-lbs. (22-24 N·m)

Manifold mounting nuts–
All models . . . . 26 ft.-lbs. (35 N·m)

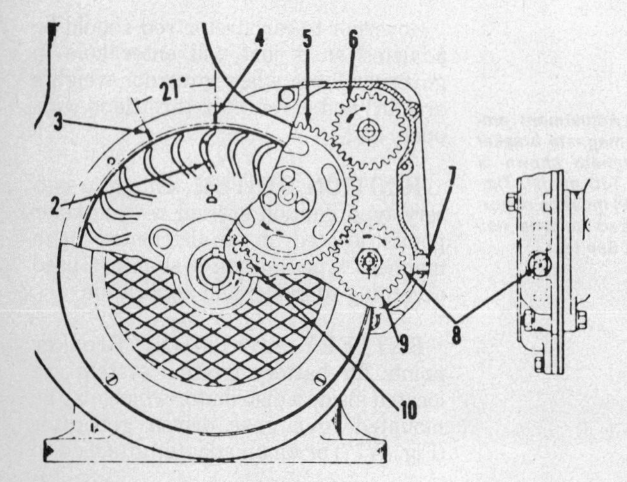

Fig. W74—Typical magneto ignition timing mark alignment. Timing should be set at 27° BTDC for TE and TF models with engine running at 2000 rpm. Timing should be set at 20° BTDC for all other models with engine running at 2000 rpm.

2. Marked flywheel vane
3. Running advance mark
4. Static timing index
5. Camshaft gear
6. Governor gear
7. Timing plug
8. Marked magneto gear tooth
9. Magneto gear
10. Crankshaft gear

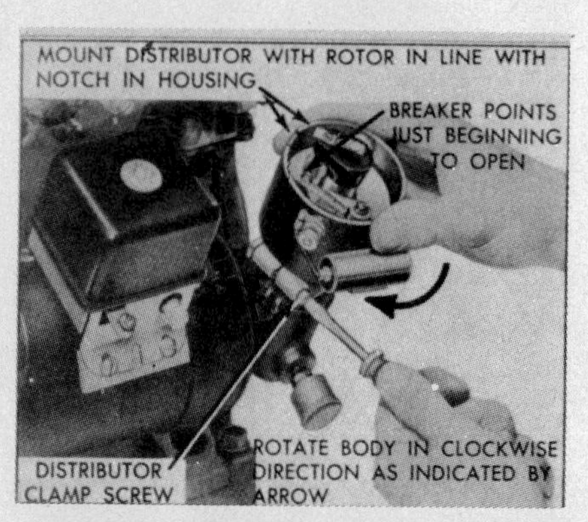

Fig. W77—View to show distributor alignment in static timing procedure. See text. Note distributor is driven through generator shaft of model shown.

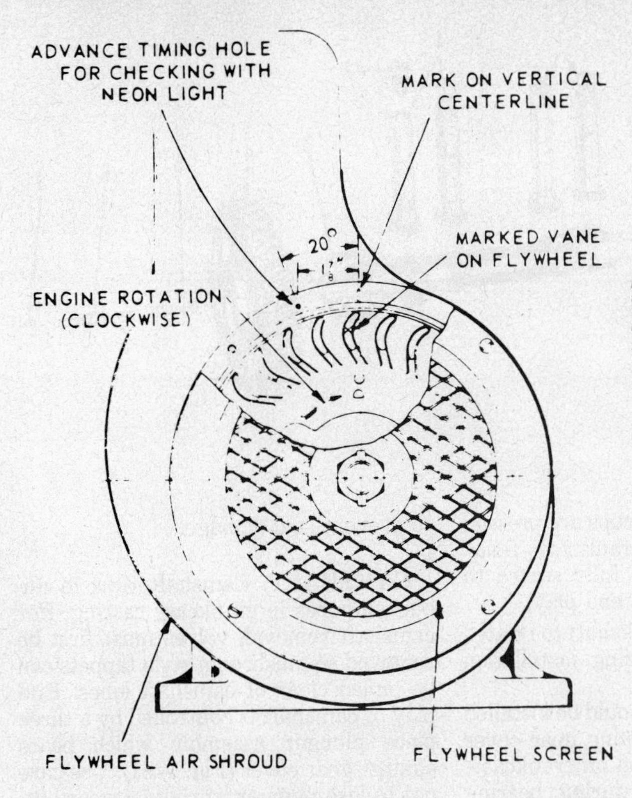

Fig. W78—View of flywheel timing marks for static timing of battery ignition system. Refer to text for correct running timing adjustment procedure.

CONNECTING ROD. Connecting rod and piston assemblies can be removed from above after removing cylinder head and engine base.

TE, TF and early TH and THD engines have connecting rods which ride directly on crankpin journal and running clearance is controlled to some degree by varying thickness of shims found between rod cap and rod. TH and THD early model rods can be renewed with current production renewable insert type rod.

Rod to journal running clearance should be 0.0007-0.002 inch (0.0178-0.0508 mm) for TE, TF and early TH and THD engines where rod rides directly on crankpin journal, 0.0012-0.0034 inch (0.305-0.0864 mm) for TH and THD insert type rod and 0.0012-0.0033 inch (0.0305-0.0838 mm) for TJD model.

Connecting rod side clearance should be 0.004-0.010 inch (0.1016-0.254 mm) for TE and TF models, 0.009-0.018 inch (0.2286-0.4572 mm) for TH and THD models and 0.009-0.014 inch (0.2286-0.3556 mm) for TJD model.

Connecting rods which ride directly on crankshaft must be installed with oil hole in rod cap pointing away from camshaft side of engine and rod caps should be assembled on their respective rods so numbers stamped on side of rod and side of cap are on the same side when installed.

Oil hole in rod cap of TH, THD and TJD engines with renewable insert type bearing has no significance but rod caps should be assembled on their respective rods so numbers stamped on side of rod and side of cap are on the same side when installed.

Crankpin standard journal diameter is 1.750-1.751 inches (44.45-44.48 mm) for TE, TF, TH and THD models and 1.8756-1.8764 inches (47.64-47.66 mm) for TJD model. Connecting rods or bearing inserts are available for a variety of undersize crankshafts.

PISTON, PIN AND RINGS. Pistons are equipped with two compression rings, one scraper ring and one oil control ring for all models except TJD model which changed to a three ring piston effective with serial number 5219324. Pistons and rings are available in a variety of oversizes as well as standard.

Early engines used a split skirt piston which has been replaced with a cam ground piston similar to ones used for TJD models.

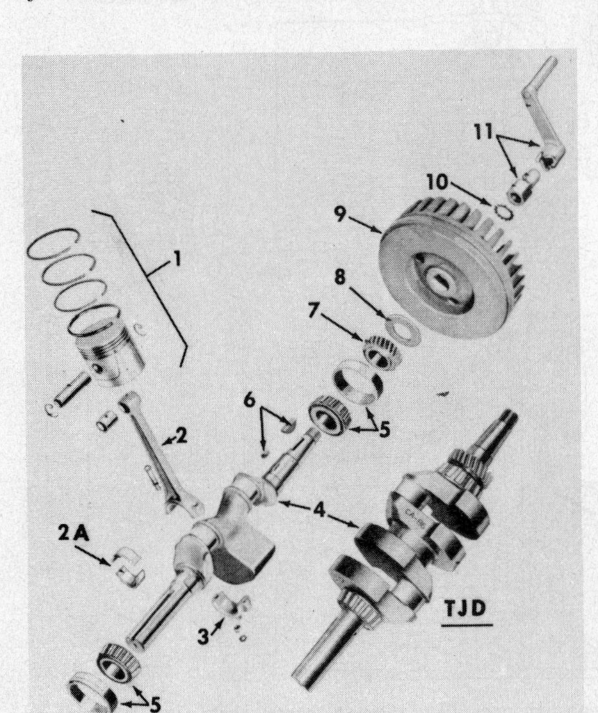

Fig. W79—Exploded view of crankshaft, piston, connecting rod parts groups. Note differences in TJD crankshaft (4) compared to earlier style.

1. Piston & ring assy.
2. Connecting rod assy.
2A. Bearing inserts (late models)
3. Connecting rod cap
4. Crankshaft (2 types)
5. Main bearing cup & cone assy.
6. Woodruff keys
7. Crankshaft gear
8. Oil slinger
9. Flywheel
10. Lockwasher
11. Crank nut & crank

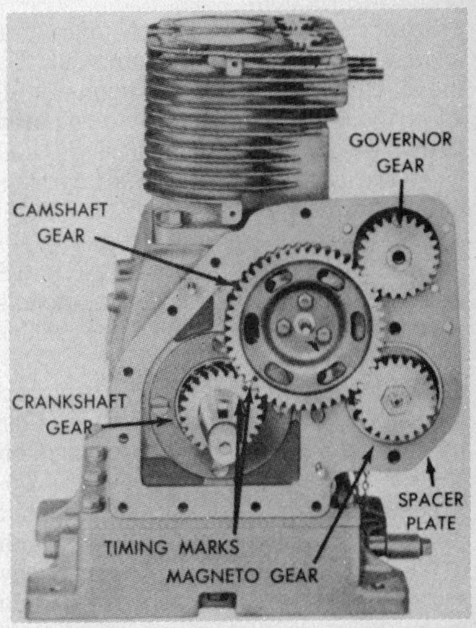

Fig. W80—View of timing gear train with gear cover removed. Note position of timing index marks on camshaft and crankshaft gears. Refer to text.

Piston to cylinder clearance should be measured at 90° angle to piston pin at lower edge of skirt. Clearance for all split skirt type pistons should be 0.004-0.0045 inch (0.1016-0.1143 mm). Clearance for cam ground pistons used in TH and THD engines should be 0.0032-0.0037 inch (0.0813-0.0940 mm) and clearance should be 0.0025-0.003 inch (0.0635-0.0762 mm) for TJD engines.

Ring end gap for early split skirt piston should be 0.004-0.0045 inch (0.1016-0.1143 mm) for top ring, 0.001-0.0025 inch (0.0254-0.0635 mm) for second and third ring and 0.0025-0.004 inch (0.0635-0.1016 mm) for oil control ring.

Ring end gap for cam ground pistons should be 0.010-0.020 inch (0.254-0.508 mm).

On all models rings should be installed with end gaps located at 90° intervals around piston.

Ring side clearance in piston grooves for all pistons should be 0.002-0.003 inch (0.0508-0.0762 mm).

Wide thrust face on skirt of cam ground piston must face thrust side of engine when installed, see Fig. W82.

Split skirt type pistons must be installed with split toward manifold side of engine, see Fig. W82.

Piston pin should be 0.000-0.0008 inch (0.000-0.0203 mm) tight fit in piston and is retained at each end by a snap ring in bore of piston. Pin should have 0.0005-0.0011 inch (0.0127-0.0279 mm) clearance in bushing end of connecting rod. Pins are available in a variety of oversizes as well as standard.

CYLINDER. Cylinder bore section of all models is a separate section bolted to crankcase section.

Standard cylinder bore diameter should be 3 inches (76.2 mm) for TE model and 3.25 inches (82.55 mm) for all other models. Cylinders should be bored to nearest oversize for which pistons and rings are available if excessively worn or out-of-round.

CRANKSHAFT, MAIN BEARINGS AND SEALS. Crankshaft for all models is supported at each end by taper roller bearings.

TJD model crankshaft connecting rod journals are 180° apart and all other models have connecting rod journals on the same side with a counterweight opposite. See Fig. W79. Crankshaft end play should be 0.001-0.004 inch (0.0254-0.1016 mm) and is adjusted by varying thickness of gaskets between main bearing plate and crankcase. Gaskets are available in a variety of thicknesses.

Main bearings are a press fit on crankshaft and bearing cups are pressed into stator plate and crankcase. Bearings and cups must be fully seated to correctly set crankshaft end play.

To properly time crankshaft to related engine components during installation refer to Fig. W80.

Crankshaft oil seals should be installed in bearing plate and timing gear cover prior to their installation on crankcase. Care should be used during bearing plate or timing gear cover installation to prevent seal damage.

CAMSHAFT. Camshaft runs in unbushed bores in crankcase casting. For camshaft removal, valves must first be removed so mushroom style tappets can be pulled clear of camshaft lobes. End play of camshaft is controlled by a three piece plunger assembly which bears against gear cover (Fig. W81). Use care not to lose plunger, spring or wear button during reassembly.

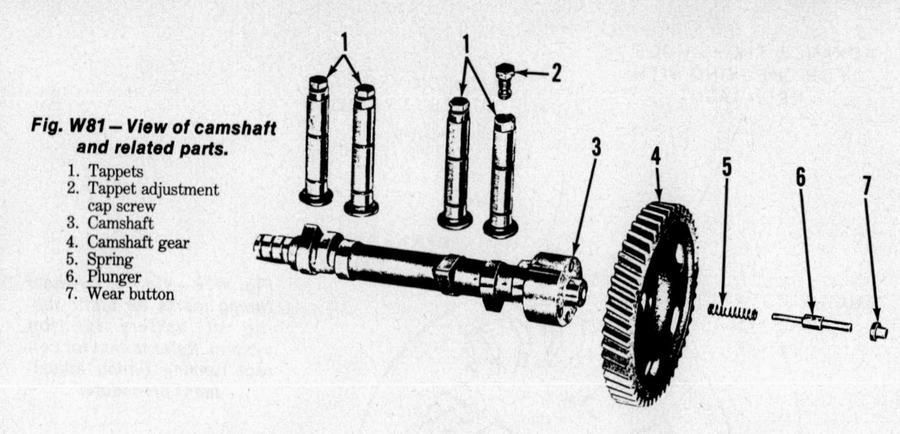

Fig. W81 – View of camshaft and related parts.
1. Tappets
2. Tappet adjustment cap screw
3. Camshaft
4. Camshaft gear
5. Spring
6. Plunger
7. Wear button

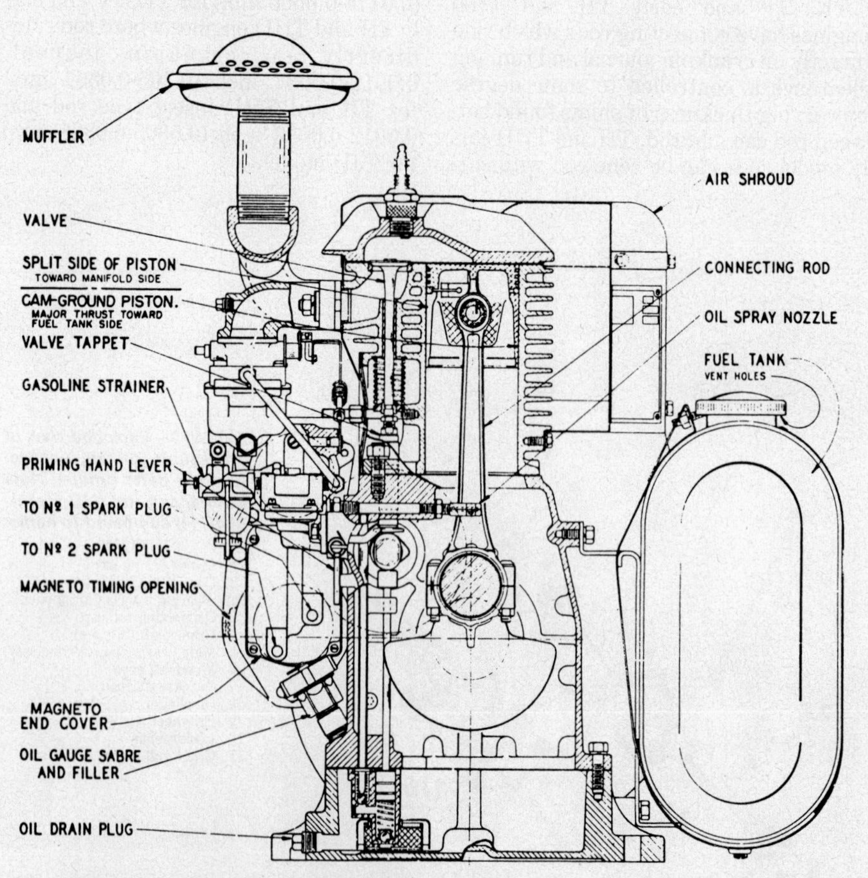

Fig. W82 – Sectional view of engine component parts showing general location and relationship.

OIL SPRAY NOZZLES. Oil spray nozzle is installed so both metered holes can be seen when looking directly into bottom of crankcase and when installed correctly, flats on hex body of nozzle will be parallel with top and bottom machined surfaces of crankcase. End of spray nozzle should extend 1-½ inches (38.1 mm) from boss it is screwed into or be in line with the centerline of crankshaft when it is installed. See Fig. W82.

VALVE SYSTEM. Valve seats on all models are renewable insert type and should be ground at 45° angle. Seat width on all models should not exceed 3/32-inch (2.381 mm).

Valve face is ground at 45° angle and stellite exhaust valves and seats and rotocaps are standard on most late models.

Early TE, TF, TH and THD models had valve guides made directly in cylinder block and valves with oversize stems are available if clearance between valve stem and guide exceeds 0.007 inch (0.1778 mm).

Late production TE, TF, TH, THD and all TJD models have renewable valve guides. Guides should be renewed if clearance between valve stem and guide exceeds 0.007 inch (0.1778 mm).

Normal clearance between valve stem and guide is 0.003-0.005 inch (0.0762-0.1270 mm) for all models.

Wisconsin Motor DF-72 guide driver is available to aid valve guide removal and installation.

Valve tappet gap (cold) for TE and TF models, both intake and exhaust, is 0.011-0.013 inch (0.2794-0.3302 mm).

Valve tappet gap (cold) for all remaining models is 0.008 inch (0.2032 mm) for intake valve and 0.016 inch (0.4064 mm) for exhaust valve.

Valve adjustment on all models is made by turning self-locking cap screw on tappet as required.

WISCONSIN

TELEDYNE WISCONSIN MOTOR
Milwaukee, Wisconsin 53246

Model	No. Cyls.	Bore	Stroke	Displacement	Power Rating
VE4, VE4D	4	3 in. (76.2 mm)	3.25 in. (82.55 mm)	91.9 cu. in. (1505.8 cc)	21.5 hp. (16 kW)
VF4, VF4D	4	3.25 in. (82.55 mm)	3.25 in. (82.55 mm)	107.7 cu. in. (1767.3 cc)	25 hp. (18.6 kW)
VP4, VP4D	4	3.5 in. (88.9 mm)	4 in. (101.6 mm)	154 cu. in. (2522.6 cc)	31 hp. (23.1 kW)

Engines in this section are four-cycle, four cylinder "V" type with a horizontal crankshaft. Crankshaft for all models is supported at each end in taper roller bearings.

Connecting rods in early models ride directly on crankpin journal but are replaced by renewable insert type rods in later models and during service. Lubrication is provided by pressure spray system and gear type oil pump is driven by an idler gear.

Magneto ignition is used except on electric start engines which are equipped with a battery type ignition system with points located in a distributor.

Various float type carburetors are used and a mechanical type fuel pump with priming lever is available.

Engine identification information is located on engine instruction plate and engine model, specification and serial number are required when ordering parts or service material. Suffix "D" engines are equipped with stellite exhaust valves and seats.

MAINTENANCE

SPARK PLUG. Recommended spark plug is Champion D16 or equivalent. Electrode gap should be 0.030 inch (0.762 mm).

CARBURETOR. A variety of float type carburetors are used as listed.

Zenith 161-7, Stromberg UC-7/8 or Marvel-Schebler TSX-148 carburetor is used for VE4 and VE4D engines.

Zenith 161-7 or Stromberg UC-7/8 carburetor is used for VF4 and VF4D engines.

Marvel-Schebler VH-69 carburetor is used for VP4 and VP4D engines.

For initial adjustment of all models note main fuel jet, except in possible special application, is fixed and nonadjustable.

For initial idle mixture screw adjustment for Zenith 161-7, Stromberg UC-7/8 and Marvel-Schebler TSX-148 carburetors, open idle mixture screw 1-1/4 turns.

For initial idle mixture adjustment of Marvel-Schebler VH-69 carburetor, open idle mixture screw 1 to 1-1/4 turns.

On all models make final adjustments with engine at normal operating temperature and running. Set engine at idle speed and adjust idle mixture screw for smoothest idle operation.

To check float level on Zenith 161-7 or Stromberg UC-7/8 carburetor, invert carburetor throttle body and float assembly and measure from furthest face of float to machined gasket surface of throttle body. Measurement should be 1-5/32 inch (29.37 mm) for Zenith 161-7 model and 1-1/4 inch (31.75 mm) for Stromberg UC-7/8 model.

To check float level on Marvel-Schebler TSX-148 or VH-69 carburetor, invert carburetor throttle body and float assembly and measure from nearest face of float to machined gasket surface of throttle body. Measurement should be 1/4-inch (6.35 mm) for TSX-148 model and 37/64-inch (14.68 mm) for VH-69 model.

Adjust float level on all models by bending float tang that contacts inlet valve.

GOVERNOR. Flyweight type governor driven by camshaft gear is used on all models. Refer to Fig. W85 for exploded view of typical governor component parts.

Speed changes on all models is obtained by varying spring (12-Fig. W85) tension by adjusting screw (13) or locating spring in alternate holes of governor control lever (11).

To correctly set governed speeds, use a tachometer to accurately record crankshaft rpm. Refer to Fig. W86 or W87 to determine correct hole in which

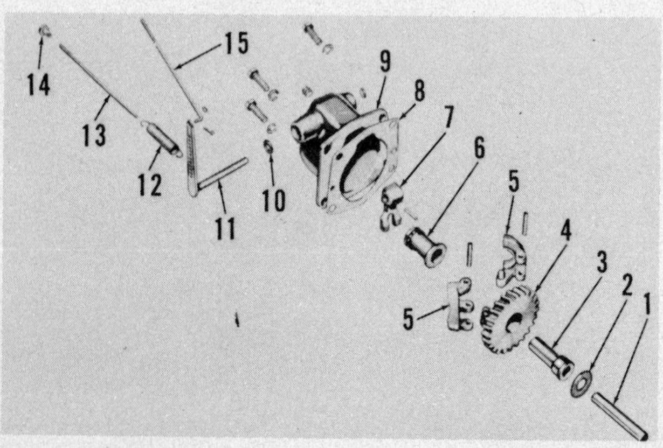

Fig. W85—Exploded view of typical governor assembly used on four cylinder Wisconsin engines.

1. Shaft
2. Thrust washer
3. Bushing
4. Governor gear
5. Flyweights
6. Thrust sleeve & bearing
7. Yoke
8. Gasket
9. Housing
10. Oil seal
11. Governor lever & shaft
12. Governor spring
13. Control rod
14. Adjusting screw
15. Throttle rod

to locate spring for desired rpm.

Governor-to-carburetor rod should be adjusted so it just will enter hole in governor lever when governor weights are "in" and carburetor throttle is wide open.

IGNITION SYSTEM. Either a battery type ignition system with breaker points located in a distributor or a magneto type ignition system are used according to model and application.

BATTERY IGNITION. Autolite IGW-4159A or IAD-4036A battery ignition distributor is used when engine is equipped with electric starting motor and generator. Early IGW-4159A distributor rotates at engine speed and later IAD-4036A distributor rotates at half engine speed. Breaker point gap on both distributors should be 0.018-0.022 inch (0.457-0.559 mm). Firing order is 1-3-4-2 on all models. Cylinders are numbered from flywheel end of engine, and (facing flywheel) are as follows: Left bank, Nos. 1 and 3 and right bank Nos. 2 and 4.

To install and time early IGW-4159A engine speed distributor, remove screen over flywheel air intake opening and make certain number "1" cylinder is at "top dead center" on compression stroke and leading edge of flywheel vane marked "X" and "DC" is in register with distributor static timing mark (1-Fig. W89) on air shroud as shown. Install distributor so rotor is pointing towards lug (4–Fig. W90) on distributor body. Loosen clamp screw (2) and turn distributor counter-clockwise until breaker points are closed. Turn distributor clockwise until breaker points just begin to open and tighten clamp screw. Positioning distributor as outlined will give a static timing of 12° BTDC for VE4, VE4D, VF4 and VF4D models or 10° BTDC for VP4 and VP4D models.

Running timing should be 27° BTDC for VE4, VE4D, VF4 and VF4D models and 25° BTDC for VP4 and VP4D models. With engine operating at 1800 rpm, timing light should flash when leading edge of flywheel vane (3-Fig. W89 marked with "X" is aligned with running timing mark (4) on air shroud. If not, loosen clamp screw (2-Fig. W90) and turn distributor body until correct timing is obtained. Tighten clamp screw.

To install and time later IAD-4036A half engine speed distributor, remove screen over flywheel air intake opening and proceed as follows: Crank engine until No. 1 piston is coming up on compression stroke and leading edge of "X" marked flywheel vane is aligned with distributor static timing mark (T.C. 1 & 3) as shown in Fig. W91. Adjust breaker point gap to 0.018-0.022 inch (0.457-0.559 mm), then install distributor so rotor arm is centered on right hand edge of notch in distributor housing as shown in Fig. W92. Loosen distributor body clamp screw and turn distributor body counter-clockwise until breaker points are closed. Then, turn distributor in clockwise direction until breaker points are just beginning to open and tighten clamp screw. Running timing is 27° BTDC for VE4, VE4D, VF4 and VF4D models and 25° BTDC

LOAD R.P.M.	NO LOAD R.P.M.	HOLE NO.
1400	1525	4
1500	1650	5
1600	1725	5
1700	1850	6
1800	1950	7
1900	2025	7
2000	2150	8
2100	2225	8
2200	2350	9
2300	2425	9
2400	2550	10

GOVERNOR LEVER — HOLE NO. 12, 11, 10, 9, 8, 7, 6, 5, 4, 3, 2, 1

Fig. W86 — Governor speed table for VE4, VE4D, VF4 and VF4D models.

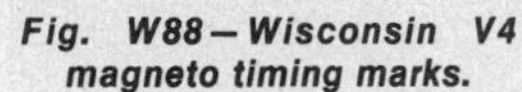

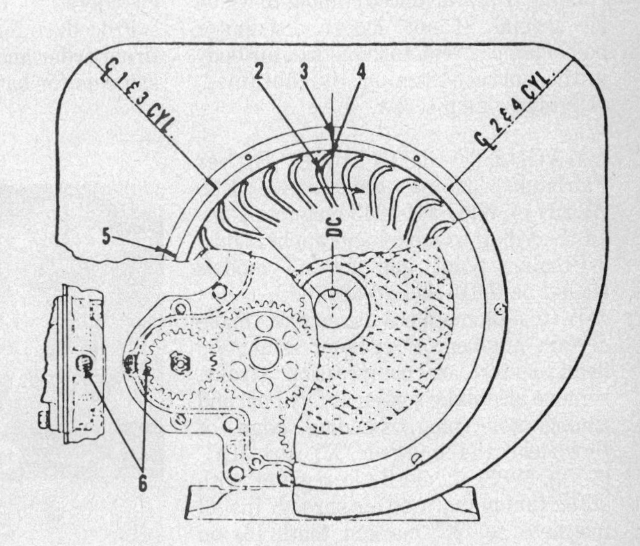

Fig. W88 — Wisconsin V4 magneto timing marks.
2. "X" marked flywheel vane
3. Magneto installation timing mark
4. Leading edge of marked flywheel vane
5. Running timing mark
6. Marked tooth on magneto gear

LOAD R.P.M.	NO LOAD R.P.M.	HOLE NO.
1400	1550	4
1500	1650	5
1600	1725	5
1700	1850	6
1800	1950	7
1900	2025	8
2000	2125	8
2100	2250	9
2200	2350	10

GOVERNOR LEVER — HOLE NO. 12, 11, 10, 9, 8, 7, 6, 5, 4, 3, 2, 1

Fig. W87 — Governor speed table for VP4 and VP4D models.

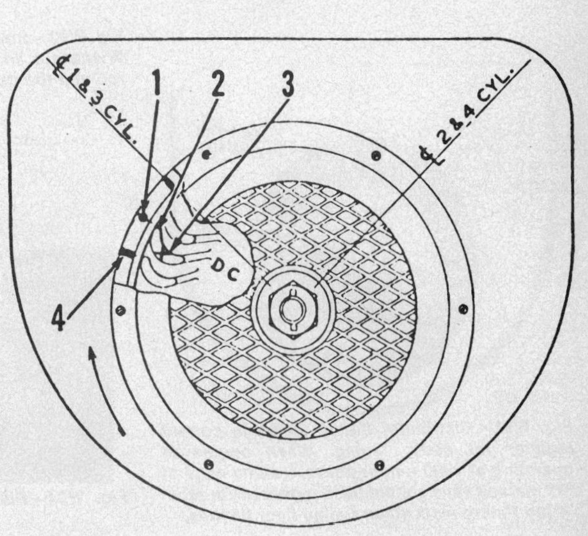

Fig. W89 — Distributor timing marks used on engines equipped with early IGW-4159A engine speed distributor.
1. Distributor static timing mark
2. Leading edge of marked flywheel vane
3. "X" marked on flywheel vane
4. Running timing mark

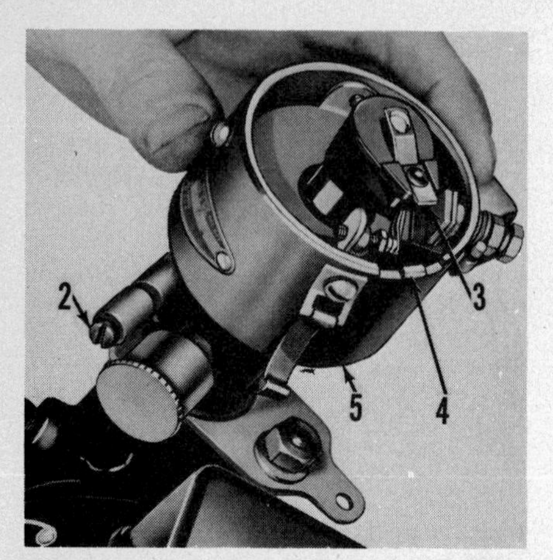

Fig. W90—Autolite IGW-4159A ignition distributor rotates at engine speed. Refer to text for timing procedure.

2. Clamp screw
3. Rotor arm
4. Lug
5. Distributor body

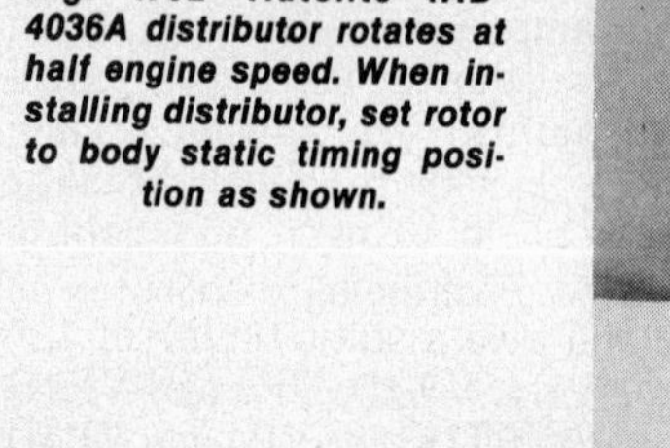

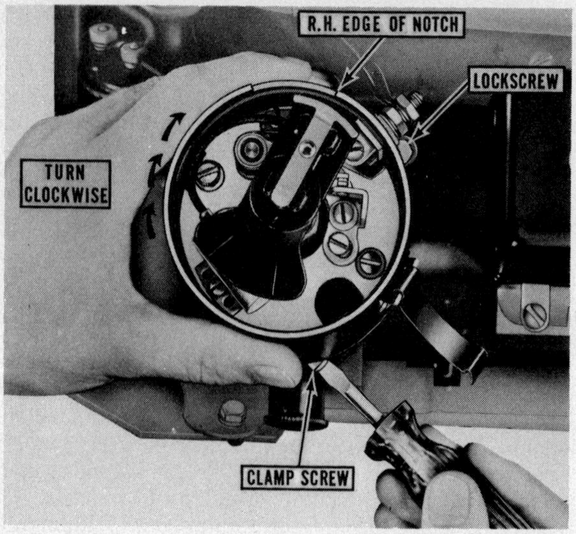

Fig. W92—Autolite IAD-4036A distributor rotates at half engine speed. When installing distributor, set rotor to body static timing position as shown.

for VP4 and VP4D models. Timing marks indicating these positions are stamped on air shrouds. With engine operating at 1800 rpm, timing light should flash when leading edge of "X" marked flywheel vane is in register with running (fully advanced) timing mark on air shroud. If not, loosen distributor body clamp screw, turn distributor body until correct timing is obtained, retighten clamp screw.

MAGNETO IGNITION. Either Fairbanks- Morse FM-XV4B7, FM-ZV4B7 or Wico XH-1343 magneto is used according to model and application.

Breaker point gap for all models should be 0.015 inch (0.381 mm).

To install magneto to engine, make certain number "1" cylinder is at "top dead center" on compression stroke, remove air intake screen at flywheel and timing hole plug. Leading edge of flywheel vane marked "X" and "DC" (3-Fig. W88) should be in register with static timing mark on air shroud. Install magneto so "X" marked tooth (6) on magneto drive gear is visible through port in timing gear housing as shown. To check for correct installation, slowly crank engine. Impulse coupling should snap when leading edge of marked vane on flywheel is aligned with center line of cylinders.

Procedure for setting running timing, firing order and cylinder sequence is the same as for battery ignition system.

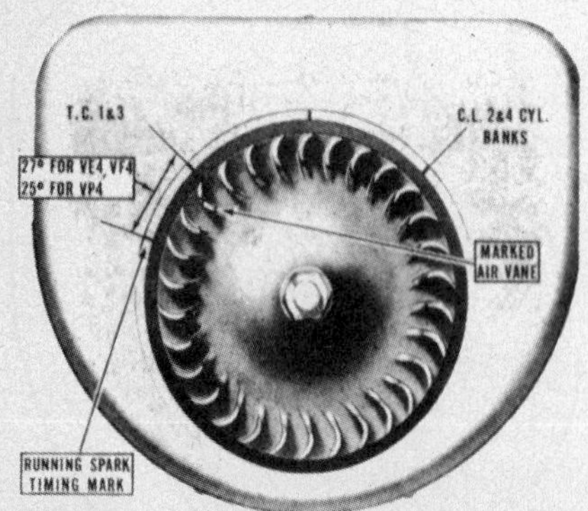

Fig. W91—Distributor timing marks in correct register for static timing. When engine is operating at 1800 rpm or above, leading edge of "X" marked vane should be in register with running timing mark when timing light flashes.

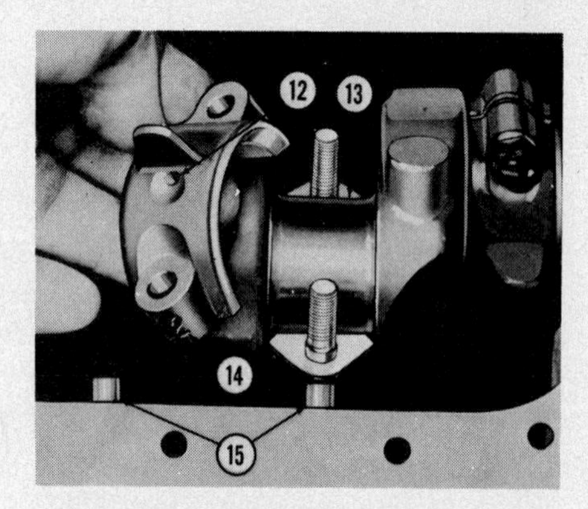

Fig. W93—Installing connecting rod and cap on Wisconsin V4 engine. Numbers on connecting rod and rod cap must face oil spray nozzle side of engine.

12. Oil hole
13. Crankshaft
14. Number on rod cap
15. Oil spray nozzles

UPPER END OF PISTON
COMPRESSION RINGS
SCRAPER RING
OIL RING

Fig. W94—Piston rings must be installed on piston as shown.

LUBRICATION. Pressure spray lubrication is provided by a gear type oil pump which is driven by crankshaft gear via an idler gear. Oil is pumped through an external oil line at 4-5 psi (28-35 kPa) to lubricate governor and nozzles spray connecting rods and surrounding engine components. Maximum system pressure should be 15 psi (103 kPa) and is controlled by a relief valve in pump. See Fig. W97.

To remove oil pump, remove crankcase oil pan and timing gear cover. Remove pump retaining screw (Fig. W96). Unscrew pump drive gear nut (1-Fig. W97) and use soft brass punch to drive shaft out of gear. Oil pump assembly may be withdrawn from crankcase. Refer to Fig. W97 for disassembly or reassembly.

Fig. W95—Wisconsin V4 engine with gear cover removed showing gear train and location of timing marks.

2. Magneto gear
3. Camshaft gear
4. Governor gear
5. Valve timing marks
6. Crankshaft gear
7. Idler gear
8. Oil pump drive gear

Maintain oil level at full mark on dipstick but do not overfill. High quality detergent oil having API classification "SD" or "SE" is recommended. Recommended oil change interval is every 50 hours of normal operation and filter should be changed every other oil change.

Crankcase capacity for VE4, VE4D, VF4 and VF4D is 4 quarts (3.8 L) and capacity for VP4 and VP4D is 5 quarts (4.7 L).

Use SAE 30 oil when operating temperature is between 120°F (49°C) and 40°F (4°C), SAE 20 oil between 40°F (4°C and 15°F (-9°C), SAE 10 oil between 15°F (-9°C) and 0°F (-18°C) and SAE 5W-20 oil for below 0°F (-18°C)temperature.

REPAIRS

TIGHTENING TORQUES. Recommended torque specifications are as follows:

Cylinder head–
- VE4, VE4D, VF4, VF4D 22-24 ft.-lbs. (30-33 N·m)
- VP4, VP4D............25-32 ft.-lbs. (34-43 N·m)

Cylinder block–
- VE4, VE4D, VF4, VF4D 40-50 ft.-lbs. (54-68 N·m)
- VP4, VP4D............62-78 ft.-lbs. (84-106 N·m)

Connecting rod–
- VE4, VE4D, VF4, VF4D 22-24 ft.-lbs. (30-33N·m)
- VP4, VP4D............28-32 ft.-lbs. (38-43 N·m)

Gear cover–
- All models.............14-18 ft.-lbs. (19-24 N·m)

Bearing plate–
- All models.............25-30 ft.-lbs. (34-40N·m)

Manifold–
- VE4, VE4D, VF4, VF4D 14-18 ft.-lbs. (19-24 N·m)
- VP4, VP4D............40-50 ft.-lbs. (54-68 N·m)

Oil pan–
- All models...............6-9 ft.-lbs. (8-12·m)

Spark plug–
- All models.............25-30 ft.-lbs. (34-40 N·m)

CONNECTING RODS. Connecting rods and piston assemblies can be removed from above after removing cylinder head and crankcase oil pan.

Fig. W96–Removing oil pump retaining screw on Wisconsin V4 engine. Allen head screw is accessible after removing slotted plug (94).

Early model engines connecting rods ride directly on crankpin journal and running clearance for these rods should be 0.0007-0.002 inch (0.0178-0.0508 mm).

Late model engines have connecting rods with renewable insert type bearings and running clearance for these bearings should be 0.0012-0.0034 inch (0.0305-0.0864 mm).

Old style rods are renewed with insert bearing type rods during service.

Connecting rod side clearance should be 0.009-0.018 inch (0.2286-0.4571 mm) for all models.

Number stamped on side of rod should correspond with number stamped on side of rod cap and oil hole in rod cap must be toward oil spray nozzles when rods are reinstalled (Fig. W93).

Crankpin standard journal diameter for VP4 and VP4D engines with serial numbers prior to 771501 and all VE4, VE4D, VF4 and VF4D should be 1.750-1.751 inches (44.45-44.48 mm). Crankpin standard diameter for VP4 and VP4D engines after serial number 771501 should be 1.875-1.876 inches (47.63-47.65 mm). Connecting rods or bearing inserts are available for a variety of undersize crankshafts.

PISTON, PIN AND RINGS. Pistons are equipped with two compression rings, one scraper ring and one oil control ring for all models. Pistons and rings are available in a variety of oversizes as well as standard.

Early engines used a split skirt piston which has been replaced with a cam ground piston in later models.

Piston to cylinder bore clearance when measured 90° from pin at lower edge of skirt should be 0.004-0.0045 inch (0.1016-0.1143 mm) for split skirt piston and 0.0035-0.004 inch (0.0089-0.1016 mm) for cam ground piston.

Ring end gap for all models should be 0.010-0.020 inch (0.254-0.508 mm).

Ring side clearance in grooves for all models should be 0.002-0.0035 inch (0.0508-0.0889 mm) for top ring, 0.001-0.0025 inch (0.0254-0.0635 mm) for second and third ring and 0.0025-0.004 inch (0.0635-0.1016 mm) for oil control ring.

Wide thrust face on skirt of cam ground piston must face thrust side of engine when installed, see Fig. W97A.

Split skirt piston should be assembled on rod so split is on oppostie side of connecting rod oil hole in connecting rod cap, see Fig. W97A.

Piston pin should be 0.000-0.0008 inch (0.000-0.0203 mm) tight fit in piston and is retained at each end by a snap ring in bore of piston. Pin should have 0.0005-0.0011 inch (0.0127-0.0279 mm) clearance in bushing in end of connecting rod. Pins are available in a variety of oversizes as well as standard.

CYLINDER HEADS. Always use new head gaskets when installing cylinder heads and make certain head bolts of varying length are in correct locations. Tighten cap screws evenly.

CYLINDER. Two cylinders are cast as a unit and bolt to crankcase. Cylinders should be bored to nearest oversize for which pistons and rings are available if worn or out-of-round more than 0.005 inch (0.127 mm).

Standard cylinder bore is 2.998-2.999 inches (76.1492-76.1746 mm) for VE4 and VE4D models, 3.248-3.249 inches (82.4992-82.5246 mm) for VF4 and VF4D models and 3.498-3.499 inches (88.8492-88.8746 mm) for VP4 and VP4D models.

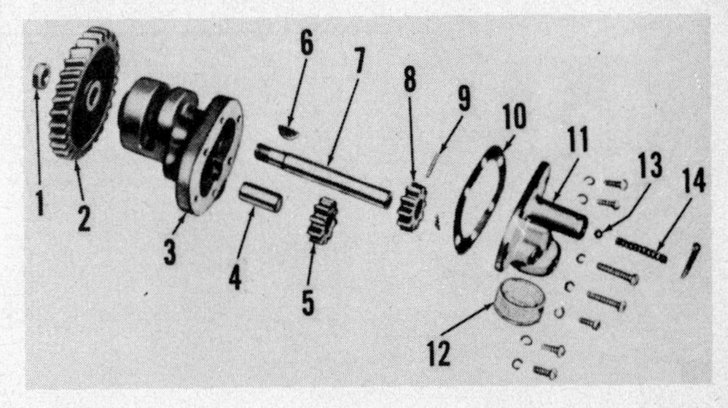

Fig. W97–Exploded view of oil pump used on all models. Items 13 and 14 are the 15 psi (103 kPa) maximum pressure relief valve.

1. Nut
2. Drive gear
3. Pump body
4. Idler shaft
5. Idler gear
6. Key
7. Pump shaft
8. Pump gear
9. Pin
10. Gasket
11. Cover
12. Screen
13. Relief ball
14. Relief spring

CRANKSHAFT, MAIN BEARINGS AND OIL SEALS. Crankshaft is supported at each end by taper roller bearings. Crankshaft end play should be 0.002-0.004 inch (0.0508-0.1016 mm) for all models and is controlled by varying thickness of gaskets between main bearing plate and crankcase. Gaskets are available in a variety of thicknesses.

Main bearings are a press fit on crankshaft and rear bearing cup is pressed into bearing plate and front bearing cup is retained in crankcase by a retainer plate and cap screws.

Timing marks should be aligned as shown in Fig. W95 during crankshaft installation.

Oil seals should be installed in timing gear cover and rear bearing plate before they are bolted in place and care should be taken to avoid seal damage during installation.

If scored or out-of-round, crankpin journals can be reground and a variety of oversize connecting rods or bearing inserts are available. See **CONNECTING ROD** section for standard crankshaft rod journal diameters.

IDLER GEAR AND SHAFT. An idler gear driven by crankshaft gear is used to drive magneto and oil pump on all models. See Fig. W95.

To remove idler gear and shaft, remove timing gear cover, magneto or generator and distributor. Remove Allen head set screws from side of crankcase which retain idler shaft in position and use a sleeve type puller over end of idler shaft and against idler gear, thread a 3/8-16 cap screw into end of shaft. Pull shaft from crankcase and idler gear.

When reassembling, make certain oil slot in idler shaft is facing upward. Press or drive idler shaft into crankcase until a clearance of 0.003-0.004 inch (0.0762-0.1016 mm) exists between gear hub and shoulder on end of idler shaft. Lock shaft in position with Allen set screws.

CAMSHAFT. Camshaft runs in unbushed bores in crankcase casting. Camshaft gear is held on end of camshaft by three cap screws and end play is controlled by a spring loaded plunger assembly and wear button which bears against timing gear cover. Use care not to lose plunger, spring or wear button during reassembly.

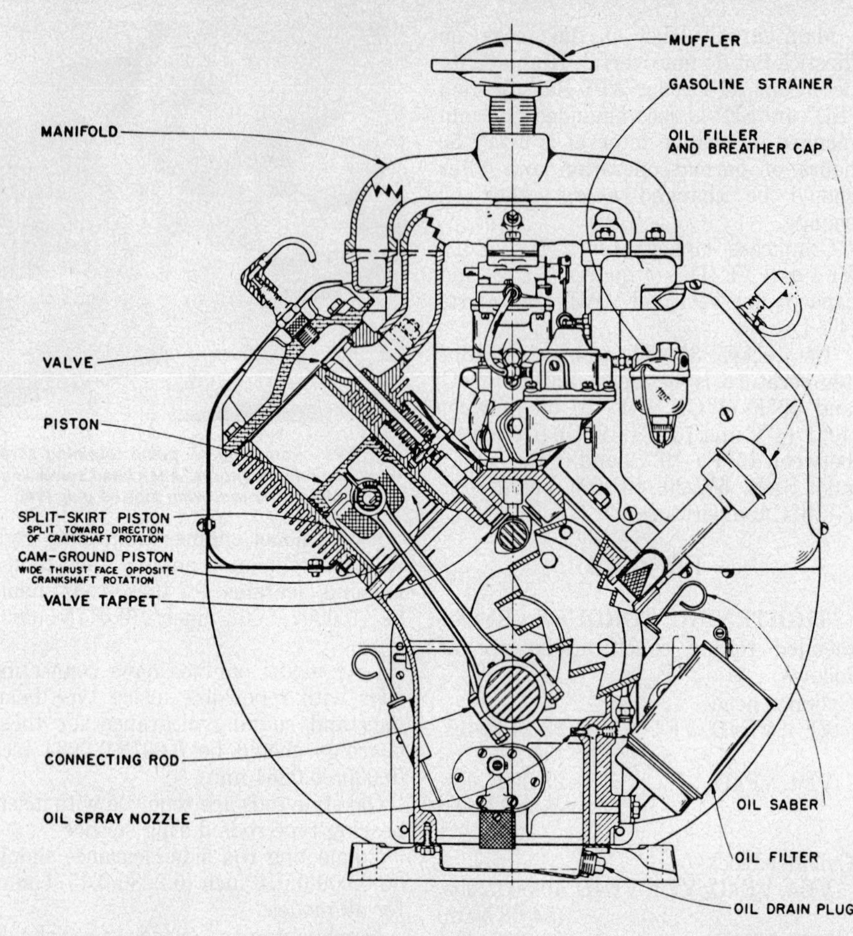

Fig. W97A – Sectional view of engine showing general location and relationship of component parts.

VALVE SYSTEM. Suffix "D" of engine model number indicates stellite exhaust valve and seat. All models have renewable valve seats which should be ground at 45° angle. Seat width should not exceed 3/32-inch (2.381 mm).

Valve faces are ground at 45°angle and valves with oversize stems are available for some models.

Early models had valve guides machined directly into cylinder casting and late models have renewable guides.

Valve stem to guide clearance for all models should be 0.003-0.005 inch (0.0762-0.127 mm) and should be reconditioned, valves with oversize stems installed or guides renewed if clearance exceeds 0.007 inch (0.1778 mm).

Valve tappet gap (cold) for VE4, VE4D, VF4 and VF4D models should be 0.008 inch (0.2032 mm) for intake valves and 0.015 inch (0.4064 mm) for exhaust valves. Valve tappet gap (cold) for VP4 and VP4D models should be 0.012 inch (0.3048 mm) for intake and 0.017 inch (0.4318 mm) for exhaust valves.

Valve adjustment on all models is made by turning self-locking cap screw on tappet as required.

WISCONSIN

TELEDYNE WISCONSIN MOTOR
Milwaukee, Wisconsin 53246

Model	No. Cyls.	Bore	Stroke	Displacement	Power Rating
VH4, VH4D	4	3.25 in. (82.55 mm)	3.25 in. (82.55 mm)	107.7 cu. in. (1767.3 cc)	30 hp. (22.4 kW)
VG4D	4	3.5 in. (88.9 mm)	4 in. (101.6 mm)	154 cu. in. (2522.6 cc)	37 hp. (28 kW)
VR4D	4	4.25 in. (107.95 mm)	4.5 in. (114.3 mm)	255 cu. in. (4184.5 cc)	56.5 hp. (42.1 kW)

Engines in this section are four-cycle, four cylinder "V" type with a horizontal crankshaft. Crankshaft for all models is supported at each end in taper roller bearings.

Connecting rods in early models ride directly on crankpin journal but are replaced by renewable insert type rods in later models and during service. Lubrication is provided by pressure spray system and gear type oil pump is driven by an idler gear.

Magneto ignition is used except on electric start engines which are equipped with a battery type ignition system with points located in a distributor.

Various float type carburetors are used and a mechanical type fuel pump with priming lever is available.

Engine identification information is located on engine instruction plate and engine model, specification and serial number are required when ordering parts or service material. Suffix "D" engines are equipped with stellite exhaust valves and seats.

MAINTENANCE

SPARK PLUG. Recommended spark plug is Champion D16 or equivalent. Electrode gap should be 0.030 inch (0.762 mm).

CARBURETOR. A variety of float type carburetors are used as listed.

Zenith 68-7 or Marvel-Schebler TSX-690 carburetor is used on VH4 and VH4D engines.

Marvel-Schebler VH-69 carburetor is used on VG4D engines.

Zenith 12A10 carburetor is used on VR4D engines.

For initial adjustment of all models note main fuel jet, except in possible special application, is fixed and non-adjustable.

For initial idle mixture screw adjustment for Zenith 68-7 or 12A10 carburetor, open idle mixture screw 1-1/4 turn.

For initial idle mixture screw adjustment for Marvel-Schebler TSX-690 or VH-69 carburetor, open idle mixture screw 7/8-turn.

On all models make final adjustments with engine at normal operating temperature and running. Set engine at idle speed and adjust idle mixture screw for smoothest idle operation.

To check float level of Zenith carburetors, invert carburetor throttle body and float assembly and measure from furthest face of float to machined gasket surface of throttle body. Measurement should be 1-5/32 inches (29.37 mm) for 68-7 model and 31/32-inch (24.606 mm) for 12A10 model.

To check float level of Marvel-Schebler carburetors, invert carburetor throttle body and float assembly and measure from nearest face of float to machined gasket surface of throttle body. Measurement should be 1/4-inch (6.35 mm) for TSX-690 model or 37/64-inch (14.68 mm) for VH-69 model.

Adjust float level on all models by bending float tang that contacts inlet valve.

GOVERNOR. Flyweight type governor driven by camshaft gear is used on all models. Refer to Fig. W98 for exploded view of typical governor component parts.

Speed changes on all models are obtained by varying spring (12 – Fig. W98) tension by adjusting screw (14) or locating spring in alternate holes of governor control lever (11).

To correctly set governed speeds, use a tachometer to accurately record crankshaft rpm. Refer to Fig. W99, W100 or W101 according to model being

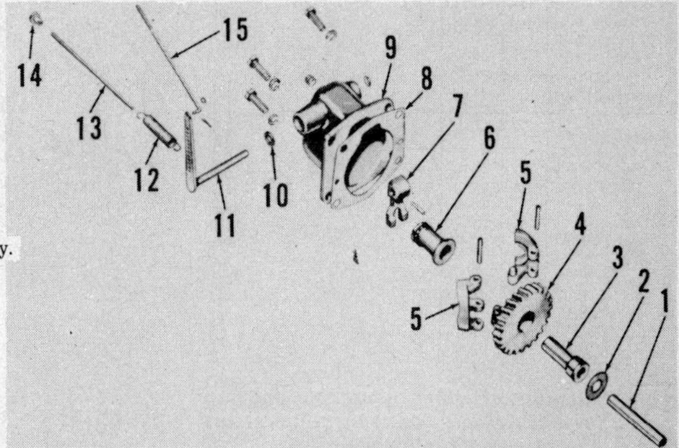

Fig. W98 – Exploded view of typical governor assembly used on Wisconsin V4 engines.

1. Shaft
2. Thrust washer
3. Bushing
4. Governor gear
5. Flyweights
6. Thrust sleeve & bearing assy.
7. Yoke
8. Gasket
9. Housing
10. Oil seal
11. Governor lever & shaft
12. Governor spring
13. Control rod
14. Adjusting screw
15. Throttle rod

serviced to determine correct hole in which to locate spring for desired rpm.

Governor-to-carburetor rod should be adjusted so it just will enter hole in governor lever when governor weights are "in" and carburetor throttle is wide open.

IGNITION SYSTEM. Either a battery type ignition system with breaker points located in a distributor or a magneto type ignition system are used according to model and application.

LOAD R.P.M.	NO LOAD R.P.M.	HOLE NO.
1400	1525	4
1500	1650	5
1600	1725	5
1700	1850	6
1800	1950	7
1900	2025	7
2000	2150	8
2100	2225	8
2200	2350	9
2300	2425	9
2400	2550	10
2500	2625	10
2600	2750	11
2700	2850	12
2800	2925	12

GOVERNOR LEVER — HOLE NO. 12, 11, 10, 9, 8, 7, 6, 5, 4, 3, 2, 1

Fig. W99—Governor speed table for VH4 and VH4D models.

LOAD R.P.M.	NO LOAD R.P.M.	HOLE NO.
1400	1550	4
1500	1650	5
1600	1725	5
1700	1850	6
1800	1950	7
1900	2025	7
2000	2125	8
2100	2250	9
2200	2350	10

GOVERNOR LEVER — HOLE NO. 12, 11, 10, 9, 8, 7, 6, 5, 4, 3, 2, 1

Fig. W100—Governor speed table for VG4D model.

	VARIABLE SPEED GOVERNOR		TWO SPEED GOVERNOR	
Load R.P.M.	No Load R.P.M.	Hole No.	No Load R.P.M.	Hole No.
1400	1590	2	1580	2
1500	1650	2	1640	2
1600	1750	4	1730	3
1700	1860	5	1810	4
1800	1920	5	1930	5
1900	2000	6	2000	5
2000	2100	6	2100	6
2100	2230	8	2200	7
2200	2300	8	2300	8

GOVERNOR LEVER — Hole No. 12, 11, 10, 9, 8, 7, 6, 5, 4, 3, 2, 1

Fig. W101—Speed table for variable and two speed (over-center idle control) type governors used on VR4D models.

BATTERY IGNITION. Autolite IAD-6004-2F distributor is used when engine is equipped for electric starting motor and generator.

Distributor is mounted on generator (Fig. W107) on VH4, VH4D and VG4D models and on accessory drive shaft housing on VR4D models.

Distributor rotates at half engine speed in a counter-clockwise direction as viewed from above. Breaker point gap should be 0.018-0.022 inch (0.457-0.558 mm) and firing order is 1-3-4-2. Cylinders are numbered from flywheel side of engine and (facing flywheel) are as follows: Left bank, Nos. 1 and 3 and right bank, Nos. 2 and 4.

To install and time distributor, remove screen over flywheel air intake opening and make certain number "1" cylinder is at "top dead center" on compression stroke and leading edge of flywheel vane marked "X" and "DC" is in register with distributor static timing mark (Fig. W106) on air shroud as shown. Install distributor so rotor arm is centered on notch in distributor body as shown in Fig. W107 or W108. Rotate distributor counter-clockwise until breaker points are closed. Turn distributor clockwise until breaker points just begin to open and tighten clamp screw.

Running timing should be 23° BTDC on VH4, VH4D and VG4D models and 20° BTDC on VR4D models. Advance timing marks indicating these positions are 1/8-inch (3.165 mm) holes in air shroud. See Fig. W104. With engine operating at 1800 rpm, timing light should flash when leading edge of "X" marked flywheel vane is in register with running (fully advanced) timing mark on air shroud. If not, loosen distributor body clamp screw, turn distributor until correct timing is obtained, retighten clamp screw.

MAGNETO IGNITION. Either Fairbanks-Morse FM-X4B7A or Wico XHG-4 magneto is used on VH4, VH4D and VG4D engines and Fairbanks-Morse FM-X4A7B magneto is used on VR4D engine.

Breaker point gap for all models

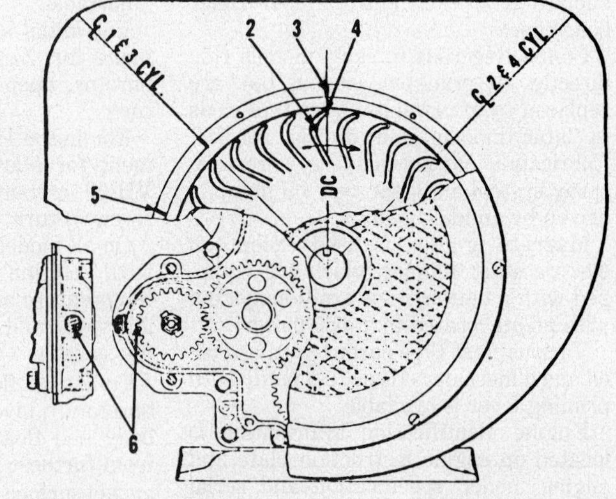

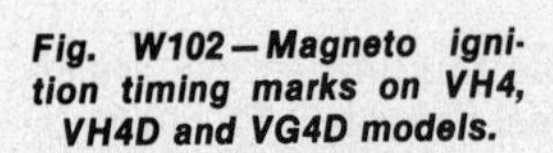

Fig. W102—Magneto ignition timing marks on VH4, VH4D and VG4D models.
2. "X" marked flywheel vane
3. Magneto installation timing mark
4. Leading edge of marked flywheel vane
5. Running timing mark
6. Marked tooth on magneto gear

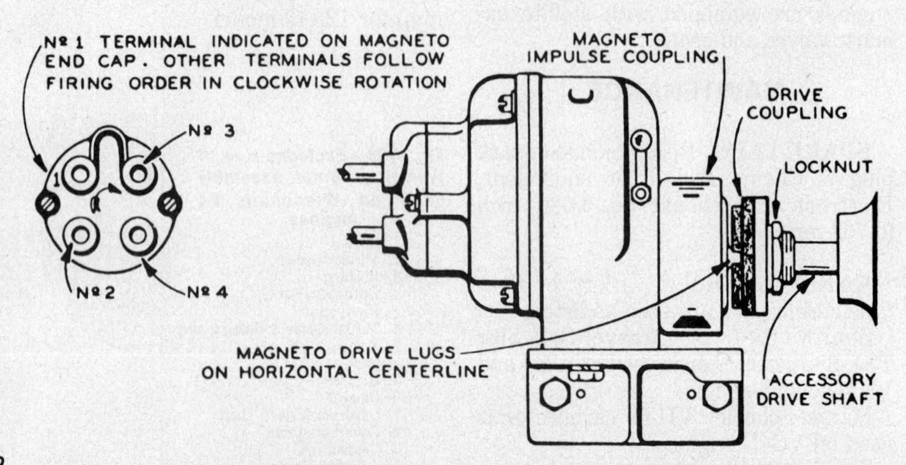

Fig. W103—On VR4D model, magneto coupling is adjustable for timing variations by means of locknut.

should be 0.015 inch (0.381 mm).

To install magneto to all models except VR4D model, remove air intake screen at flywheel and timing hole plug. Make certain number "1" cylinder is at "top dead cneter" on compression stroke. Leading edge of flywheel vane marked "X" and "DC" (Fig. W102) should be in register with static timing mark on air shroud. Install magneto so "X" marked tooth (6) on magneto drive gear is visible through port in timing gear housing as shown. To check for correct installation, slowly crank engine. Impulse coupling should snap when leading edge of marked vane on flywheel is aligned with center line of cylinders.

Procedure for setting running timing, firing order and cylinder sequence is the same as for battery ignition system.

To install magneto to VR4D engine, make certain number "1" cylinder is at "top dead center" on compression stroke, remove air intake screen at flywheel and timing hole plug. Leading edge of flywheel vane marked "X" and "DC" (Fig. W104) should be in register with static timing mark on air shroud. Remove magneto end cap and turn magneto shaft in opposite direction of normal rotation until rotor is in exact alignment with timing boss on end cap as shown in Fig. W105. Lugs on magneto coupling will be horizontal at this time as shown in Fig. W103. Holding magneto in this position, install it on engine. If coupling on engine accessory shaft is not in correct position, loosen locknut and rotate coupling (Fig. W103) until slot engages lugs of impulse coupling of magneto. Tighten locknut.

Running timing should be 20° BTDC and is adjusted by resetting coupling on engine accessory drive shaft.

Procedure for setting running timng, firing order and cylinder sequence is the same as for battery ignition system.

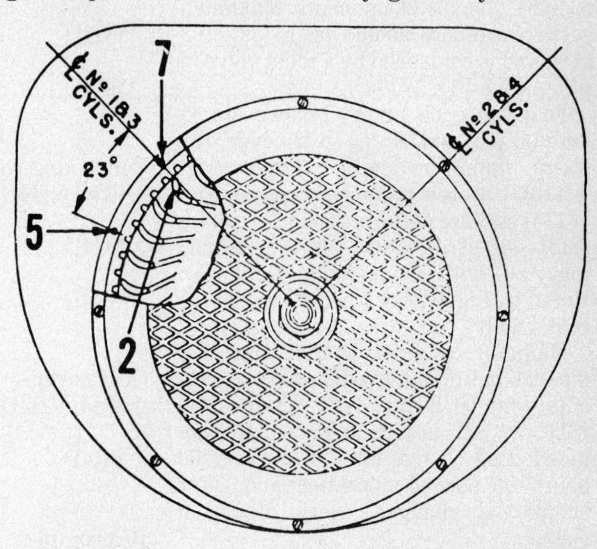

Fig. W106—Battery distributor timing marks for VH4, VH4D and VG4D is a depression (7) in air shroud.

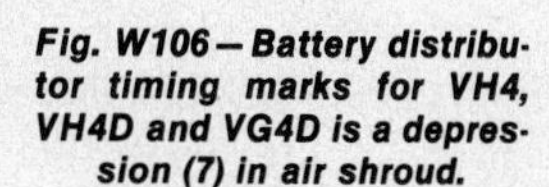

2. "X" marked flywheel vane
5. Running spark (full advance) timing index
7. Top dead center index (center line) for cylinders 1 and 3

7. Top center index (center line) for cylinders 1 and 3

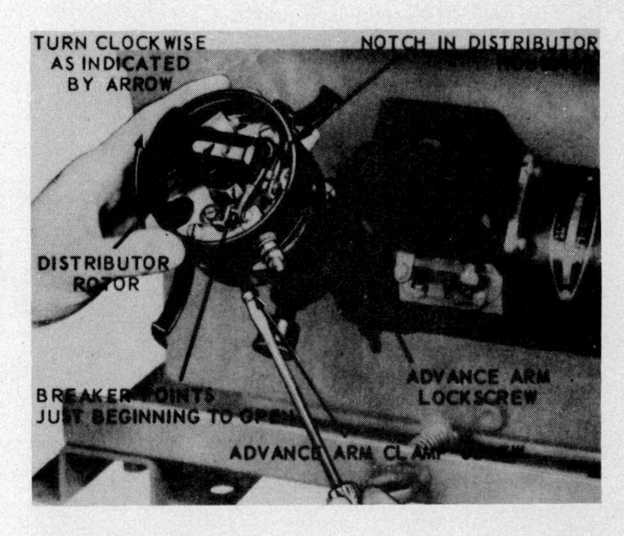

Fig. W107—When installing distributor on VH4, VH4D and VG4D, align rotor arm with center of notch in distributor body as shown. Notch also aligns with No. 1 spark plug wire terminal in distributor cap.

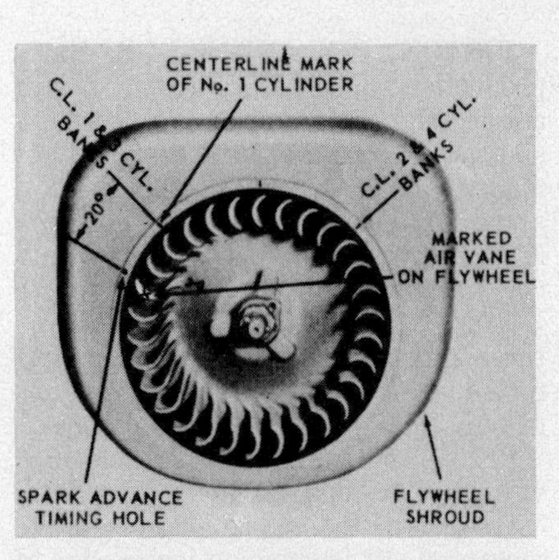

Fig. W104—Running spark timing for VR4D model is 20° BTDC for either battery or magneto ignition.

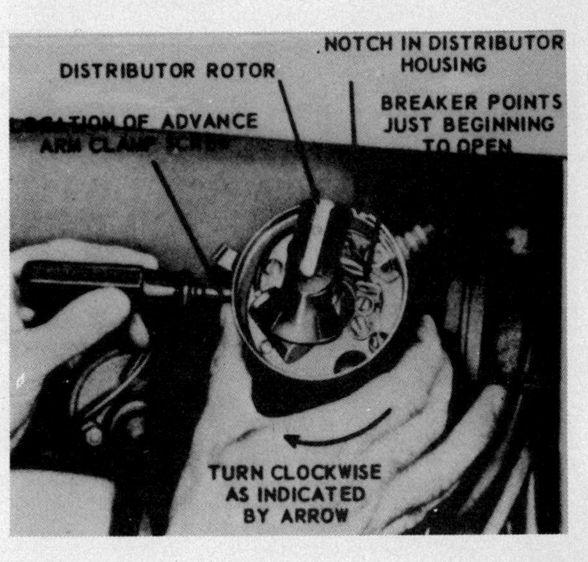

Fig. W108—On VR4D model, install distributor in position shown with rotor arm aligned with center of notch in distributor body. Notch also aligns with No. 1 spark plug wire terminal in distributor cap.

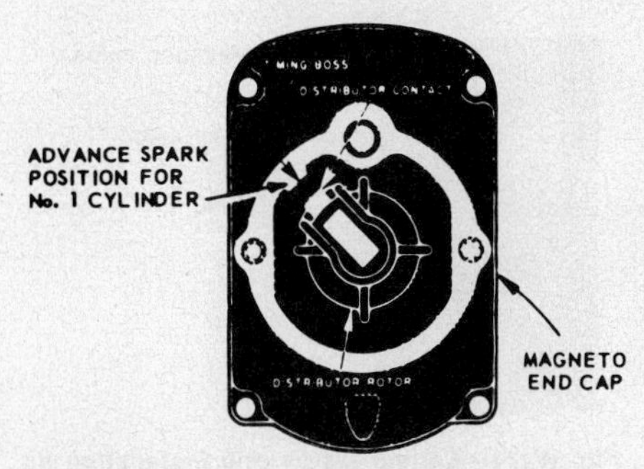

Fig. W105—Align rotor with timing boss when installing magneto on VR4D model.

LUBRICATION. Pressure spray lubrication is provided by a gear type oil pump which is driven by crankshaft gear via an idler gear. Oil is pumped through an external oil line at 4-5 psi (28-35 kPa) to lubricate governor and gear train and nozzles spray connecting rods and surrounding engine components. Maximum system pressure should be 15 psi (103 kPa) and is controlled by a relief valve in pump. See Fig. W114.

To remove oil pump, remove crankcase oil pan and timing gear cover. Remove pump retaining screw (Fig. W113). Unscrew pump gear nut (1 – Fig. W114) and use soft brass punch to drive shaft out of gear. Oil pump assembly may be withdrawn from crankcase. Refer to Fig. W114 for disassembly or reassembly.

Maintain oil level at full mark on dipstick but do not overfill. High quality detergent oil having API classification "SD" or "SE" is recommended. Recommended oil change interval is every 50 hours of normal operation and filter should be changed every other oil change.

Crankcase capacity is 4 quarts (3.8 L) for VH4 and VH4D models, 5 quarts (4.7 L) for VG4D model and 8 quarts (7.6 L) for VR4D model.

Use SAE 30 oil when operating temperature is between 120°F (49°C) and 40°F (4°C), SAE 20 oil between 40°F (4°C) and 15°F (−9°C), SAE 10 oil between 15°F (−9°C) and 0°F (−18°C) and SAE 5W-20 oil for below 0°F (−18°C) temperature.

REPAIRS

TIGHTENING TORQUES. Recommended torque specifications are as follows:

Cylinder head –
- VH4, VH4D .... 22-24 ft.-lbs. (30-33 N·m)
- VG4D .... 25-32 ft.-lbs. (35-43 N·m)
- VR4D .... 25-30 ft.-lbs. (34-40 N·m)

Cylinder block –
- VH4, VH4D .... 40-50 ft.-lbs. (54-68 N·m)
- VG4D .... 62-78 ft.-lbs. (84-106 N·m)
- VR4D .... 95-110 ft.-lbs. (129-149 N·m)

Connecting rods –
- VH4, VH4D .... 22-24 ft.-lbs. (30-33 N·m)
- VG4D .... 26-32 ft.-lbs. (35-43 N·m)
- VR4D .... 40-50 ft.-lbs. (54-88 N·m)

Gear cover –
- VH4, VH4D, VG4D ..... 14-18 ft.-lbs. (19-24 N·m)
- VR4D .... 15-20 ft.-lbs. (20-33 N·m)

Bearing plate –
- All models .... 25-30 ft.-lbs. (34-40N·m)

Manifold –
- VH4, VH4D .... 18-23 ft.-lbs. (24-31 N·m)
- VG4D .... 40-50 ft.-lbs. (54-68 N·m)
- VR4D .... 25-30 ft.lbs. (34-40 N·m)

Oil pan –
- VH4, VH4D, VG4D .... 6-9 ft.-lbs. (8-12·m)
- VR4D .... 18-20 ft.-lbs. (24-27 N·m)

Spark plug –
- All models .... 25-30 ft.-lbs. (34-40 N·m)

CONNECTING RODS. Connecting rods and piston assemblies can be removed from above after removing cylinder head and crankcase oil pan.

Early model engine connecting rods ride directly on crankpin journal and running clearance for VH4 and VH4D models should be 0.0007-0.002 inch (0.0178-0.0508 mm) and for VG4D and VR4D models clearance should be 0.0015-0.0028 inch (0.0381-0.0711 mm).

Late model engine connecting rods and most replacement rods, have renewable insert type bearings and running clearance should be 0.0012-0.0033 inch (0.0305-0.0838 mm) for VH4 and VH4D models and 0.0013-0.0035 inch (0.0330-0.0889 mm) for VG4D model.

Connecting rod side play for all models should be 0.009-0.016 inch (0.2286-0.4064 mm).

Number stamped on side of rod should correspond with number stamped on side of rod cap and oil hole in rod cap must be toward oil spray nozzles when rods are reinstalled (Fig. W109).

Crankpin standard journal diameter for VH4 and VH4D models should be 1.875-1.876 inches (47.625-47.650 mm), standard diameter for VG4D model should be 1.317-1.322 inches (43.4518-43.5788 mm) and standard diameter for VR4D model should be 2.750-2.751 inches (69.85-69.88 mm). Connecting rods or bearing inserts are available for a variety of undersize crankshafts.

PISTON, PIN AND RINGS. Early production engines were equipped with a four ring piston which was either a split skirt or cam ground design. Late model VH4D engines beginning with serial number 5538322, and some replacement pistons are a three ring, cam ground piston.

Four ring split skirt, cam ground or three ring pistons should not be intermixed in one engine. Always use pistons of the same type in individual engines.

Pistons and rings are available in a variety of oversizes as well as standard.

Piston to cylinder bore clearance when measured 90° from pin at lower edge of skirt should be 0.0035-0.004 inch

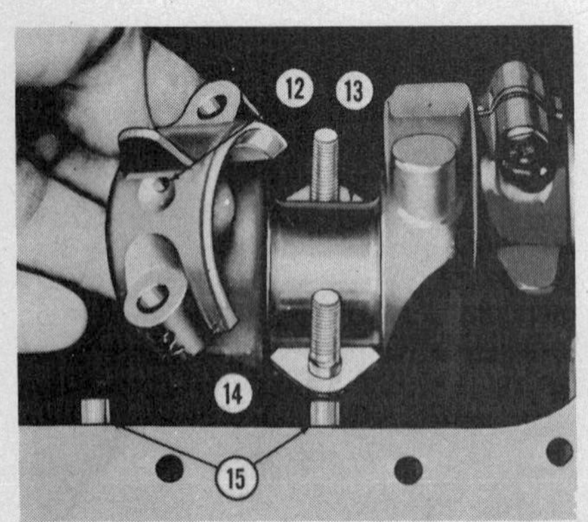

Fig. W109 – Installing connecting rod and cap on Wisconsin V4 engine. Numbers on connecting rod and rod cap must face oil spray nozzle side of engine.

12. Oil hole
13. Crankshaft
14. Number on rod cap
15. Oil spray nozzles

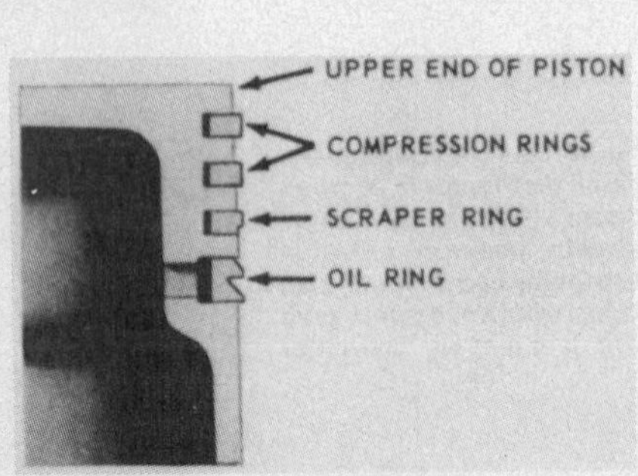

Fig. W110 – On VG4D model, piston rings for four ring piston must be installed on piston as shown.

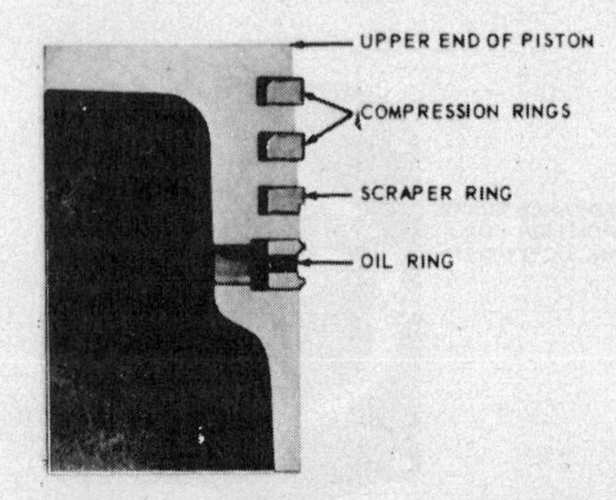

Fig. W111 – Correct piston ring installation on four ring piston used on VH4, VH4D and VR4D models.

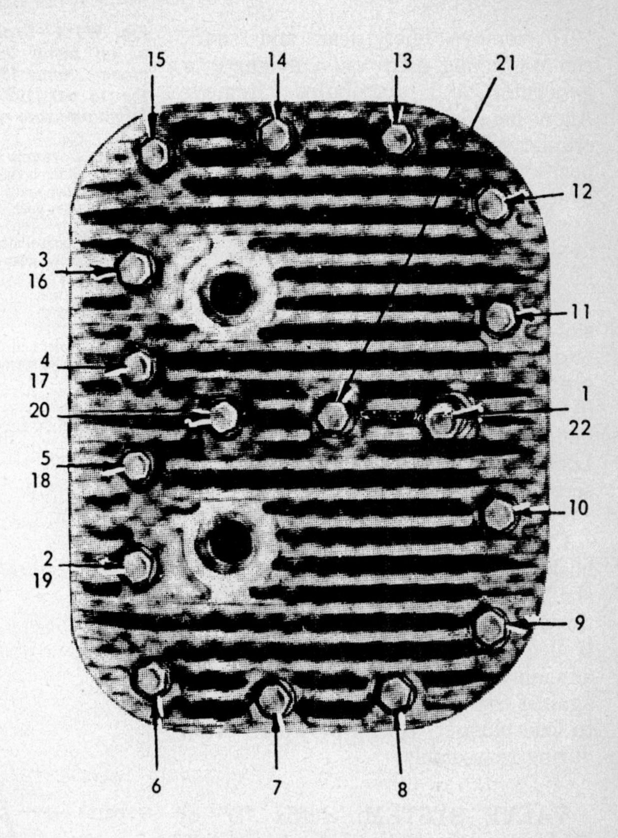

Fig. W112A—Torque cylinder head cap screws for VH4 and VH4D models in sequence shown.

(0.0889-0.1016 mm) for VH4 and VH4D models, 0.004-0.005 inch (0.1016-0.1270 mm) for VH4D model and 0.005-0.0055 inch (0.127-0.140 mm) for VR4D model.

Ring side clearance in piston grooves for VH4, VH4D and VG4D models four ring pistons should be 0.002-0.0035 inch (0.0508-0.0889 mm) for top ring, 0.001-0.0025 inch (0.0254-0.0635 mm) for second and third rings and 0.0025-0.004 inch (0.0635-0.1016 mm) for oil control ring.

Ring side clearance in piston grooves for three ring piston should be 0.002-0.004 inch (0.0508-0.1016 mm) for top ring and second ring and 0.001-0.003 inch (0.0254-0.0762 mm) for oil control ring.

Ring side clearance in piston grooves for VR4D model should be 0.0015-0.0035 inch (0.0381-0.0889 mm) for all rings except oil control ring. Oil control ring should have 0.002-0.004 inch (0.0508-0.1016 mm) side clearance in piston groove.

Wide thrust face skirt of cam ground piston must face thrust side of engine when installed, see Fig. W114A.

Split skirt piston should be assembled on rod so split is on opposite side of connecting rod from oil hole in connecting rod cap, see Fig. W114A.

Piston pin should be 0.000-0.0008 (0.000-0.0203 mm) tight fit in piston and piston pin should have 0.0004-0.0012 inch (0.0102-0.0305 mm) clearance in connecting rod bushing for VH4 and VH4D models, 0.0005-0.0011 inch (0.0127-0.0279 mm) clearance for VG4D model and 0.0003-0.001 inch (0.0076-0.0254 mm) clearance for VR4D model. Pins are available in a variety of oversizes as well as standard.

Fig. W112—Wisconsin V4 engine with gear cover removed showing gear train and location of timing marks.

2. Magneto, generator or accessory drive
3. Camshaft gear
4. Governor gear
5. Valve timing marks
6. Crankshaft gear
7. Idler gear
8. Oil pump gear

CYLINDER HEADS. Always use new head gaskets when installing cylinder heads and make certain head bolts of varying length are in correct locations. Tighten cap screws evenly to correct torque. See Fig. W112A.

CYLINDER. Two cylinders are cast as a unit and bolt to crankcase. Cylinders should be bored to nearest oversize for which pistons and rings are available if worn or out-of-round more than 0.005 inch (0.127 mm).

Standard cylinder bore diameter is 3.248-3.249 inches (82.4992-82.5246 mm) for VH4 and VH4D models, 3.498-3.499 inches (88.8492-88.8746 mm) for VG4D model and 4.248-4.249 inches (107.8992-107.9246 mm) for VR4D model.

CRANKSHAFT, MAIN BEARINGS AND OIL SEALS. Crankshaft is supported at each end by taper roller bearings. Crankshaft end play should be 0.002-0.004 inch (0.0508-0.1016 mm) for all models and is controlled by varying thickness of gaskets between main bearing plate and crankcase. Gaskets are available in a variety of thicknesses.

Main bearings are a press fit on crankshaft and rear bearng cup is pressed into bearing plate and front bearing cup is retained in crankcase by a retainer plate and cap screws.

Timing marks should be aligned as shown in Fig. W112 during crankshaft installation.

Oil seals should be installed in timing gear cover and rear bearing plate before they are bolted in place and care should be taken to avoid seal damage during installation.

If scored or out-of-round, crankpin journals can be reground and a variety of oversize connecting rods or bearing inserts are available. See **CONNECTING ROD** section for standard crankshaft rod journal diameters.

IDLER GEAR AND SHAFT. AN idler gear driven by crankshaft gear is used to drive magneto and oil pump on all models.

Fig. W113—Removing oil pump retaining screw on Wisconsin V4 engine. Allen head screw is accessible after removing slotted plug (94).

To remove idler gear and shaft, remove timing gear cover, magneto or generator and distributor. Remove Allen head set screws from side of crankcase which retain idler shaft in position and use a sleeve type puller over end of idler shaft and against idler gear, thread a 3/8-16 cap screw into end of shaft. Pull shaft from a crankcase and idler gear.

When reassembling, make certain oil slot in idler shaft is facing upward. Press or drive idler shaft into crankcase until a clearance of 0.003-0.004 inch (0.0762-0.1016 mm) exists between gear hub and shoulder on end of idler shaft. Lock shaft in position with Allen set screws.

CAMSHAFT. Camshaft runs in unbushed bores in crankcase casting. Camshaft gear is held on end of camshaft by three cap screws and end play is controlled by a spring loaded plunger assembly and wear button which bears against timing gear cover. Use care not to lose plunger, spring or wear button during reassembly.

VALVE SYSTEM. Suffix "D" of engine model number indicates stellite exhaust valve and seat. All models have renewable valve seats which should be ground at 45° angle. Seat width should not exceed 3/32-inch (2.381 mm).

Valve faces are ground at 45° angle and valves with oversize stems are available for some models.

Valve stem to guide clearance for VH4 and VH4D models should be 0.003-0.005 inch (0.0762-0.127 mm) and guides should be reconditioned, valves with oversize stems installed or guides renewed if clearance exceeds 0.007 inch (0.1778 mm).

Valve stem to guide clearance for VG4D model should be 0.004 inch (0.1016 mm) and guides should be reconditioned, valves with oversize stems installed or guides renewed if clearance exceeds 0.006 inch (0.1524 mm).

Valve stem to guide clearance for VR4D model should be 0.002-0.004 inch (0.0508-0.1016 mm) and guides should be reconditioned, valves with oversize stems installed or guides renewed if clearance exceeds 0.006 inch (0.1524 mm).

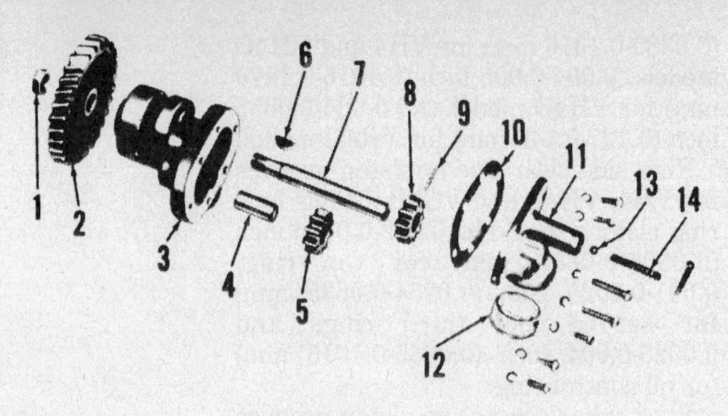

Fig. W114—Exploded view of oil pump used on all models. Items 13 and 14 are the 15 psi (103 kPa) maximum pressure relief valve.

1. Nut
2. Drive gear
3. Pump body
4. Idler shaft
5. Idler gear
6. Key
7. Pump shaft
8. Pump gear
9. Pin
10. Gasket
11. Cover
12. Screen
13. Relief ball
14. Relief spring

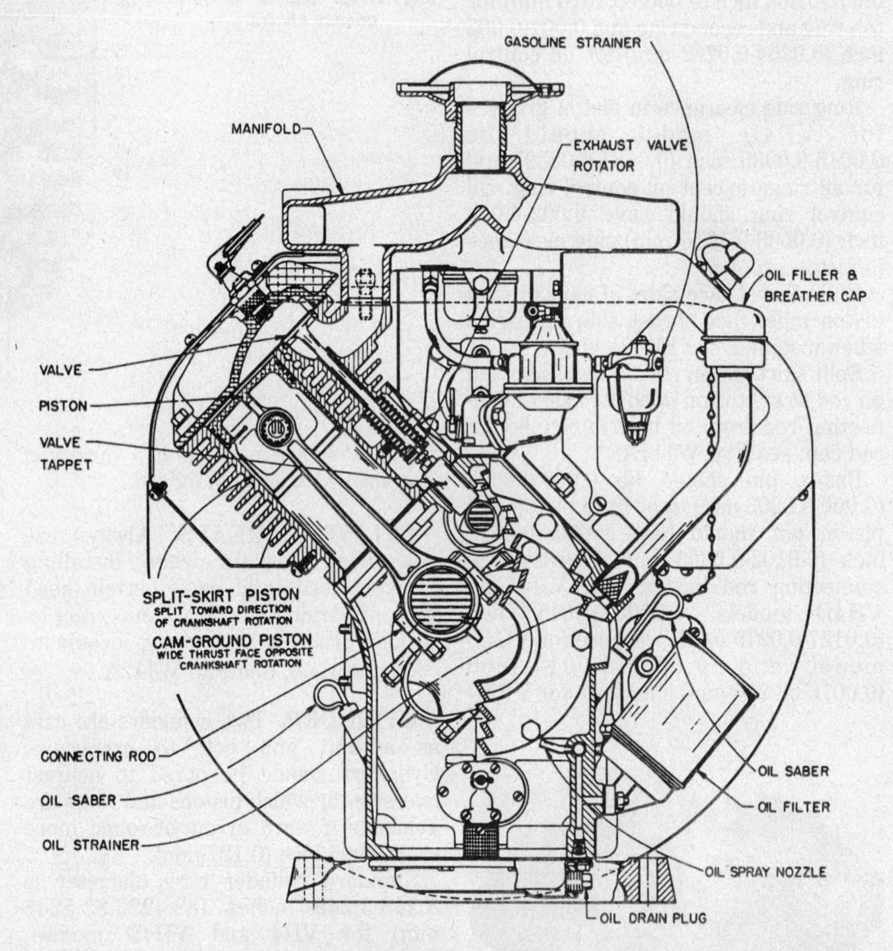

Fig. 114A—Sectional view of engine showing general location and relationship of component parts.

Valve tappet gap (cold) for VH4 and VH4D models should be 0.011-0.013 inch (0.2794-0.3302 mm) for intake and exhaust valves and valve gap (cold) for VG4D and VR4D models should be 0.008 inch (0.2032 mm) for intake valves and 0.016 inch (0.4064 mm) for exhaust valves.

Valve adjustment on all models is made by adjustable tappets.

WISCONSIN

TELEDYNE WISCONSIN MOTOR
Milwaukee, Wisconsin 53246

Model	No. Cyls.	Bore	Stroke	Displacement	Power Rating
V-460D, V-461D	4	3.5 in. (88.9 mm)	4 in. (101.6 mm)	154 cu. in. (2522.6 cc)	60.5 hp. (45.1 kW)
V-465D	4	3.75 in. (92.25 mm)	4 in. (101.6 mm)	177 cu. in. (2716.3 cc)	65.9 hp. (49.1 kW)

Engines in this section are four-cycle four cylinder "V" type with a horizontal crankshaft. Crankshaft for all models is supported at each end in taper roller bearings and at center journal by either a straight roller bearing with a two-piece outer race (early production) or a renewable insert type bearing (late production).

Connecting rods for all models have renewable insert type bearings which are lubricated by a gear type oil pump. Oil pump provides high pressure lubrication to crankshaft bearings and rod bearings and low pressure lubrication to camshaft bearings, tappets, valve train and governor gear train. Oil pump is driven by crankshaft gear via an idler gear.

Battery ignition with electric starting motor and generator is standard, however a magneto ignition is available for manual start models.

Zenith 87 model carburetor is used for all models and a mechanical type fuel pump is available.

Engine identification information is located on engine instruction plate and engine model, specification and serial number are required when ordering parts or service material. Special machine work on crankcase is identified by number stamped on crankcase above rear bearing plate. This number is required if crankcase is to be renewed.

MAINTENANCE

SPARK PLUG. Recommended spark plug is Champion N12Y or equivalent. Electrode gap should be 0.030 inch (0.762 mm).

CARBURETOR. Zenith 87 model carburetor equipped with an anti-dieseling solenoid is used on all models.

For initial adjustment of all models note main fuel jet, except in possible special applications, is fixed and non-adjustable.

For initial idle mixture screw adjustment, open idle mixture screw 1¼ turns

On all models make final adjustments with engine at normal operating temperature and running. Set engine at idle speed and adjust idle mixture screw for smoothest idle operation.

To check float level, invert carburetor throttle body and float assembly and measure from furthest face of float to machined gasket surface of throttle body. Measurement should be 31/32-inch (24.606 mm).

Adjust float level by bending float tang that contacts inlet valve.

Anti-dieseling solenoid shuts fuel supply off in carburetor when de-energized to help assure rapid shut down of engine. Engine should not run if ignition wire to solenoid is not supplied with electrical current.

GOVERNOR. Flyweight type governor driven by camshaft gear is used on all models. Refer to Fig. W115 for exploded view of typical governor component parts.

Speed changes on all models is obtained by varying spring (12-Fig. W115) tension by adjusting screw (14) or locating spring in alternate holes of governor control lever (11).

To correctly set governed speeds, use a tachometer to accurately record crankshaft rpm. Refer to Fig. W116 to determine correct hole in which to locate spring for desired rpm.

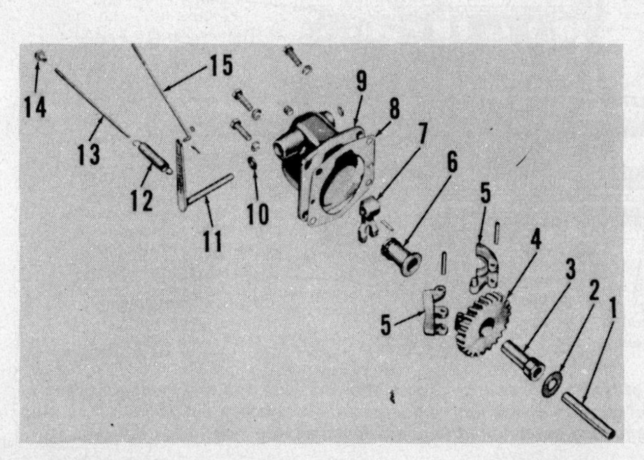

Fig. W115—Exploded view of typical governor assembly used on Wisconsin V4 engines.

1. Shaft
2. Thrust washer
3. Bushing
4. Governor gear
5. Flyweights
6. Thrust sleeve & bearing assy.
7. Yoke
8. Gasket
9. Housing
10. Oil seal
11. Governor lever & shaft
12. Governor spring
13. Control rod
14. Adjusting screw
15. Throttle rod

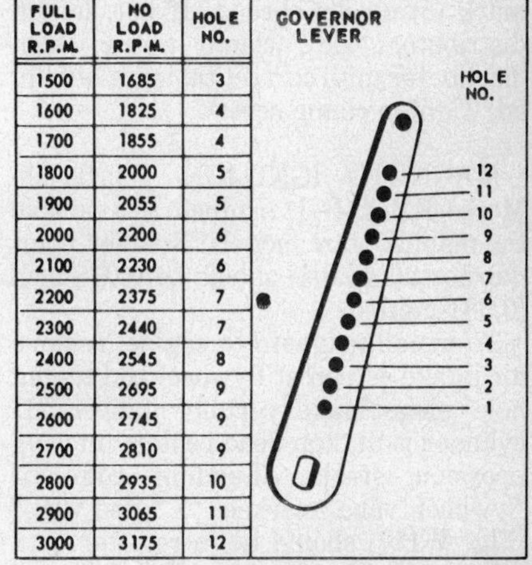

FULL LOAD R.P.M.	NO LOAD R.P.M.	HOLE NO.
1500	1685	3
1600	1825	4
1700	1855	4
1800	2000	5
1900	2055	5
2000	2200	6
2100	2230	6
2200	2375	7
2300	2440	7
2400	2545	8
2500	2695	9
2600	2745	9
2700	2810	9
2800	2935	10
2900	3065	11
3000	3175	12

Fig. W116—Governor speed table for V-460D, V-461D and V-465D models.

Governor-to-carburetor rod should be adjusted so it just will enter hole in governor lever when governor weights are "in" and carburetor throttle is wide open.

IGNITION SYSTEM. Battery type ignition system is standard, however a magneto type system is available for manual start models.

BATTERY IGNITION. Delco-Remy DR1112695 or Prestolite IAD-6004-2N distributor is used according to model and application.

Distributor is mounted on accessory drive shaft housing and turns at one half engine rpm in a counter-clockwise direction as viewed from above. Breaker point gap should be 0.020 inch (0.508 mm) and firing order is 1-3-4-2. Cylinders are numbered from flywheel side of engine and (facing flywheel) are as follows: Left bank, Nos. 1 and 3 and right bank, Nos. 2 and 4.

To install and time distributor, remove screen over flywheel air intake opening and make certain number "1" cylinder is at "top dead center" on compression stroke and leading edge of flywheel vane marked "X" is in line with centerline mark of number "1" cylinder (Fig. W118). Hold distributor so primary terminal wire (Delco-Remy) or terminal (Prestolite) is at "12 o'clock" position and rotate rotor arm to "2 o'clock" position (Delco-Remy) or "1 o'clock" position (Prestolite) and carefully install distributor. Rotate distributor counterclockwise until breaker points just begin to open and tighten clamp screw.

Running timing should be 23° BTDC at 2000 rpm or above for all models. Advance timing mark indicating this position is a ⅛-inch (3.165 mm) hole in air shroud (5-Fig. W118). With engine operating at 2000 rpm or above, timing light should flash when leading edge of "X" marked flywheel vane is in register with running (fully advanced) timing mark (5) on air shroud. If not, loosen distributor body clamp screw, turn distributor until correct timing is obtained. Tighten clamp screw.

MAGNETO IGNITION. Fairbanks-Morse FM-X4B7D magneto is available for manual start models. Breaker point gap for all models should be 0.015 inch (0.381 mm).

To install magneto to engine, remove air intake screen at flywheel and timing hole plug. Make certain number "1" cylinder is at "top dead center" on compression stroke. Leading edge of flywheel vane marked "X" and "DC" (Fig. W117) should be in register with static timing mark on air shroud. Install magneto so "X" marked tooth on

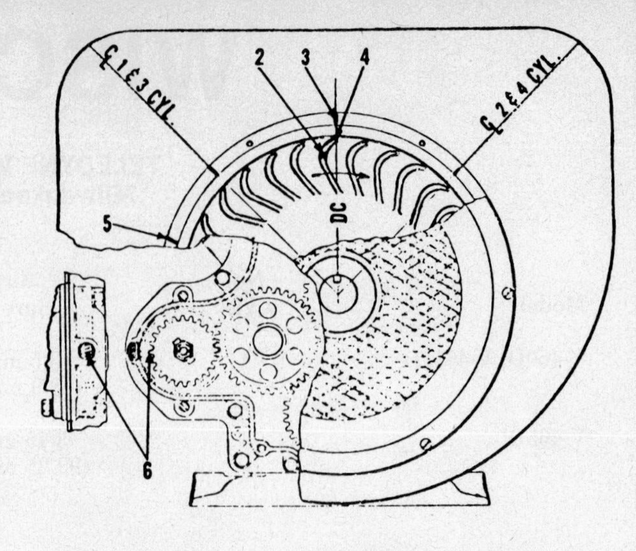

Fig. W117—Magneto ignition timing marks on manual start V-460D, V461D and V-465D engines.
2. "X" marked flywheel vane
3. Magneto installation timing mark
4. Leading edge of marked flywheel vane
5. Running timing mark
6. Marked tooth on magneto gear

Fig. W118—Battery distributor timing marks on V-460D, V-461D and V-465D models equipped with electric starting system.
2. "X" marked flywheel vane
5. Running spark (full advance) timing mark
7. Top center index (center line) for cylinders 1 and 3

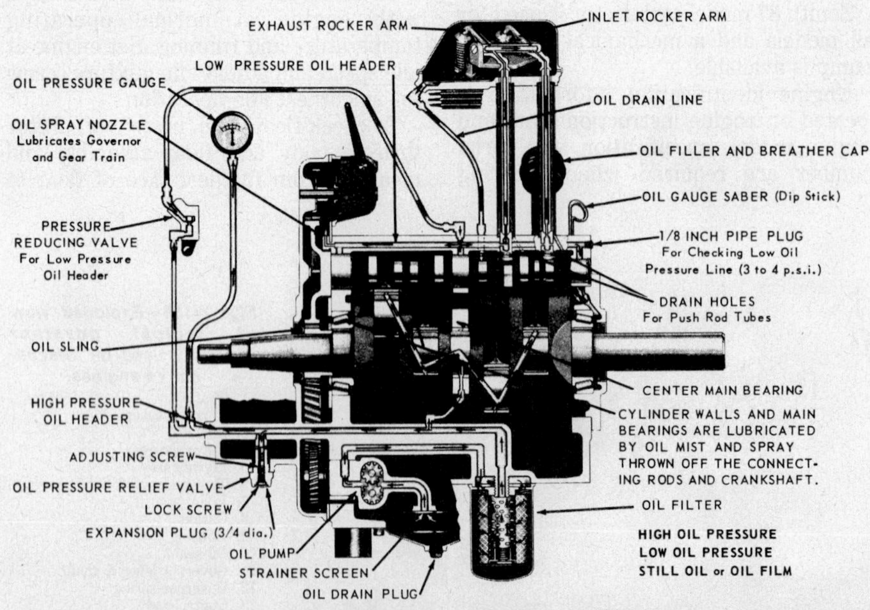

Fig. W118A—Sectional view of typical Wisconsin engine showing high and low pressure areas of lubrication. Note placement of gage to check high pressure side of system and 1/8-inch pipe plug where gage may be installed to check low pressure side.

magneto drive gear is visible through port in timing gear housing as shown. To check for correct installation, slowly crank engine. Impulse coupling should snap when leading edge of marked vane on flywheel is aligned with center line of cylinders.

Procedure for setting running timing, firing order and cylinder sequence is the same as for battery ignition system.

LUBRICATION. Lubrication is provided by a gear type oil pump driven by crankshaft gear via an idler gear. Oil pressure system consists of a high pressure system which provides oil at 40-45 psi (276-345 kPa) to center main and connecting rod bearings and low pressure system which provides oil at 3-4 psi (21-26 kPa) to camshaft bearings, tappets, valve train and governor gear train. See Fig. W118A).

High pressure in system is controlled by an adjustable pressure relief valve located beneath starter next to oil filter (Fig. W118A). To adjust high pressure relief valve remove expansion plug and lock screw from end of valve and with engine at normal operating temperature and running at 1800 rpm, turn adjusting screw in relief valve until 0-45 psi (276-345 kPa) pressure is reached. Reinstall lock screw and expansion plug.

Low pressure in system is controlled by a pressure reducing valve which is non-adjustable. Low pressure should be 3-4 psi (21-26 kPa) when engine is at normal operating temperature and running at 1800 rpm. Check pressure by installing gage in oil header (gallery) at rear of engine (Fig. W118A). If valve is faulty it must be renewed as a unit.

To remove oil pump, remove crankcase oil pan and timing gear cover. Remove pump retaining screw (Fig. W124). Unscrew pump drive gear nut (1-Fig. W124) and use a soft brass punch to drive shaft out of gear. Oil pump assembly may be withdrawn from crankcase. Refer to Fig. W124 for disassembly or reassembly.

Maintain oil level at full mark on dipstick but do not overfill. High quality detergent oil having API classification "SD" or "SE" is recommended. Recommended oil change interval is every 50 hours of normal operation and filter should be changed every other oil change.

Crankcase capacity is 6 quarts (5.7 L) without filter change and 7 quarts (6.6 L) with filter change for all models.

Use SAE 30 oil when operating temperature is between 120°F (49°C) and 40°F (4°C), SAE 20 oil between 40°F (4°C) and 15°F (−9°C), SAE 10 oil between 15°F (−9°C), and 0°F (−18°C) and SAE 5W-20 oil for below 0°F (−18°C) temperature.

REPAIRS

TIGHTENING TORQUES. Recommended torque specifications are as follows:

Item	Torque
Cylinder head	30 ft.-lbs. (40 N·m)
Connecting rod	32 ft.-lbs. (43 N·m)
Gear cover	18 ft.-lbs. (24 N·m)
Center main cap to hanger–	
Roller bearing	40 ft.-lbs. (54 N·m)
Insert bearing	32-35 ft.-lbs. (43-48 N·m)
Hanger to crankcase	60 ft.-lbs. (81 N·m)
Main bearing plate	32 ft.-lbs. (43 N·m)
Main bearing retainer	32 ft.-lbs. (43 N·m)
Manifold to head	25 ft.-lbs. (34 N·m)
Manifold to manifold	15 ft.-lbs. (20 N·m)
Oil pan	18 ft.-lbs. (24 N·m)
Spark plug	22 ft.-lbs. (30 N·m)

CONNECTING RODS. Connecting rods, piston assemblies and cylinders can be removed from above after removing cylinder head and crankcase oil pan.

Connecting rods have renewable insert type bearings and a renewable piston pin bushing.

Running clearance of connecting rod to crankpin should be 0.0005-0.0018 inch (0.0127-0.0457 mm).

Connecting rod side clearance should be 0.008-0.016 inch (0.2032-0.4064 mm).

Number stamped on side of rod should correspond with number stamped on side of rod cap when reinstalled.

Crankpin standard journal diameter should be 2.1233-2.1238 inches (53.932-53.945 mm) for all models.

Connecting rod bearings are available in a variety of sizes for undersize crankshafts.

PISTONS, PINS AND RINGS. Cam ground aluminum pistons are equipped with two compression rings, one scraper ring and one oil control ring on V-465D or three compression rings, one oil control ring and one scraper ring (on skirt) on V-460D and V-461D engines. Pistons and rings are available in a variety of oversizes as well as standard.

Piston to cylinder clearance should be measured at 90° to piston pin at lower edge of skirt. Clearance should be 0.0025-0.003 inch (0.0635-0.0762 mm) for all models.

Piston ring end gap should be 0.008-0.024 inch (0.2032-0.6096 mm) for all models. Rings must be installed on pistons as shown in Fig. W119. Stagger ring gaps 90° apart around piston.

Ring side clearance in piston grooves should be 0.002-0.004 inch (0.0508-0.1016 mm) for all rings.

Piston pin should be 0.000-0.0008 inch (0.000-0.0203 mm) tight fit in piston and is retained at each end by a snap ring in bore of piston. Pin should have 0.0005-0.0011 inch (0.0127-0.0279 mm) clearance in bushing in end of connecting rod. Pins are available in a variety of oversizes as well as standard.

CYLINDER HEADS. To remove one cylinder head, it is not necessary to remove manifold assembly from engine. Remove rocker arm cover and manifold to cylinder head stud nuts and carefully lift off cylinder head assembly, push rods and push rod tubes. Oil drain line will pull out of adapter in crankcase and

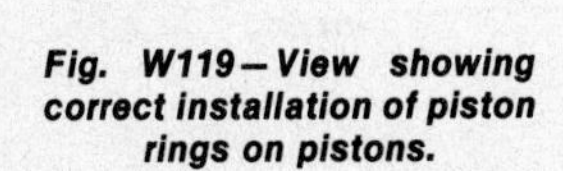

Fig. W119—View showing correct installation of piston rings on pistons.

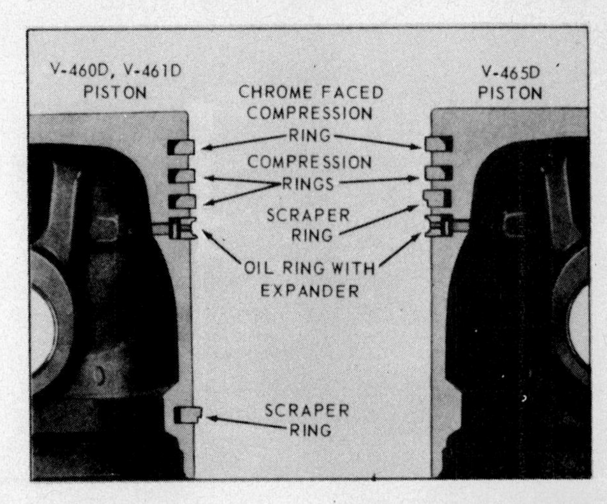

will be removed with cylinder head. Rocker arms, valves and springs can now be removed from cylinder head if necessary.

If all cylinder heads are to be removed, first unbolt and remove manifold and carburetor assembly, then remove each of the four cylinder heads as outlined previously.

When installing cylinder heads, always use new head gaskets, "O" rings on push rod tubes and "O" rings on oil drain lines. Align cylinder heads by placing a parallel steel bar across exhaust and inlet ports as shown in Fig. W120. Tap heads lightly with a rubber mallet to rotate heads until port flanges are aligned with steel bar. Tighten cylinder head retaining nuts alternately in three steps. First to 10 ft.-lbs (13 N•m) torque, to 20 ft.-lbs. (27 N•m) torque and finally to 30 ft.-lbs. (40 N•m) torque.

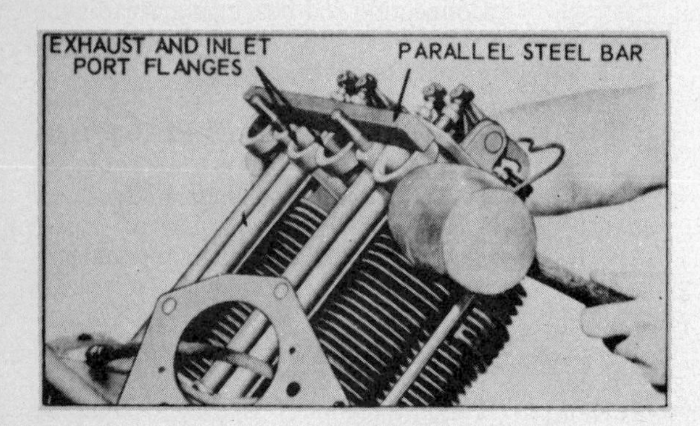

Fig. W120—View showing procedure for aligning exhaust and inlet ports when installing cylinder heads.

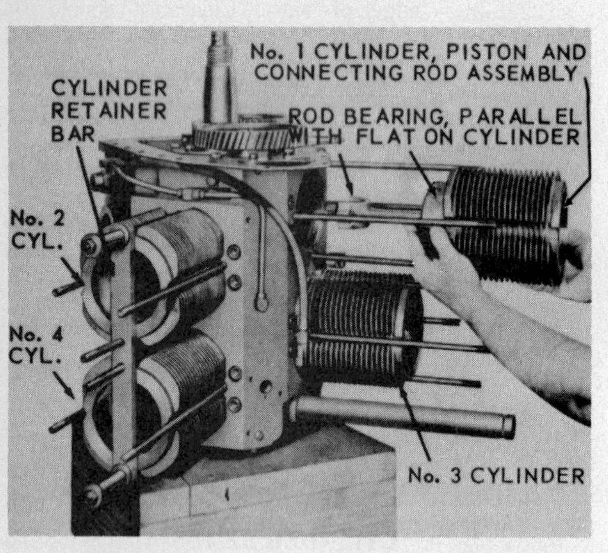

Fig. W121—View showing installation of cylinder, piston and rod assembly. Note cylinder retainer bar holding Nos. 2 and 4 cylinders in position.

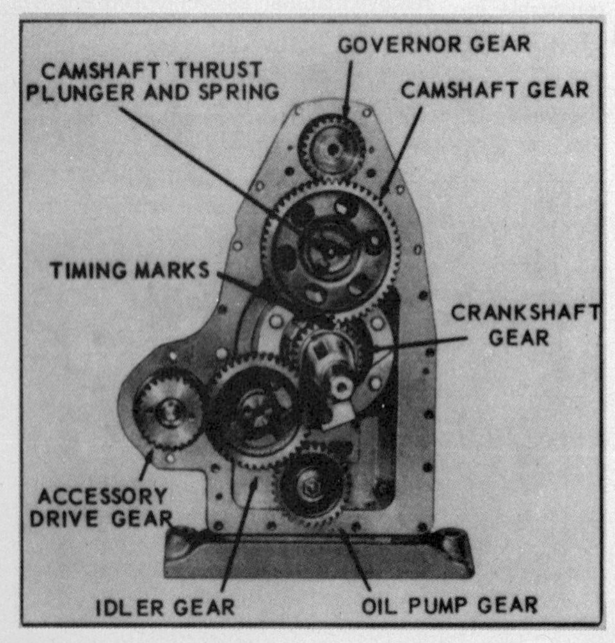

Fig. W122—Wisconsin V4 engine with timing gear cover removed showing gear train and location of timing marks.

When installing manifold assembly, align cast notches on numbers 2 and 3 inlet ports on manifold with similar cast notches on cylinder head ports. Tighten manifold to cylinder head nuts evenly to a torque of 25 ft.-lbs. (34 N•m).

To renew rocker arms, remove complete rocker arm bracket assembly from cylinder head. Remove set screw (beginning with engine serial number 5634850, set screw is installed with "Loctite 271" and heat may be required to remove) from bottom of bracket and tap shaft out toward set screw end of bracket using a brass rod.

To reassemble, flat surface on shaft must align with set screw hole in bracket. Apply "Loctite 271" to set screw threads and install. Lubricate shaft and rocker arms with engine oil before running engine.

Refer to **VALVE SYSTEM** section for valve tappet gap adjustment procedure.

CYLINDERS. Individually cast cylinders may be removed from crankcase after removing cylinder heads and manifold. Mark location of each cylinder so they may be reinstalled in their original positions.

Standard cylinder bore is 3.498-3.499 inches (88.849-88.875 mm) for V-460D and V-461D engines and 3.748-3.749 inches (95.199-95.255 mm) for V-465D engine.

When reassembling, install piston and rod assemblies in cylinders, then, using new cylinder base gaskets, install cylinders as shown in Fig. W121. A cylinder retainer bar should be attached to cylinder studs to hold cylinders in position on crankcase. Install cylinder heads as outlined in **CYLINDER HEADS** section.

CRANKSHAFT, MAIN BEARINGS AND OIL SEALS. Crankshaft is supported at each end in taper roller bearings and at center journal by either a straight roller bearing with a two-piece outer race (early production) or a renewable insert type bearing (late production).

Front and rear main bearings are a press fit on crankshaft and rear bearing cup is pressed into bearing plate and front bearing cup is retained in crankcase by a retainer plate and cap screws.

Timing marks should be aligned as

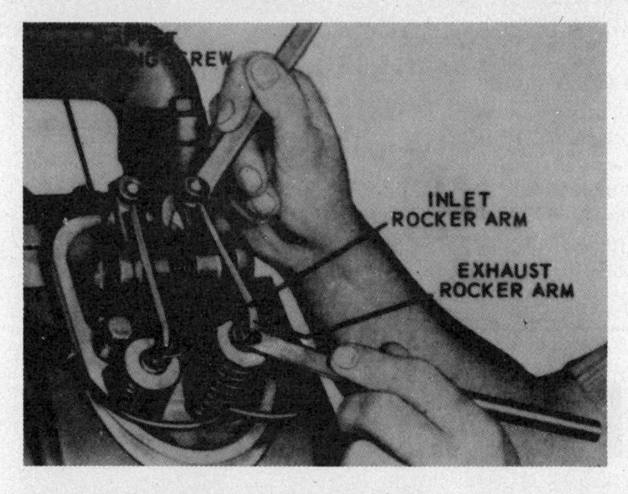

Fig. W123—Adjusting valve tappet gap in typical valve-in-head V4 Wisconsin engine.

Fig. W124—Remove slotted pipe plug and Allen head lock screw to remove oil pump.

shown in Fig. W122 during installation of crankshaft for all models.

SPLIT ROLLER CENTER MAIN BEARING. Center main bearing for early production engines (before serial number 4904657) was a split roller bearing type. To remove crankshaft, unbolt bearing hanger and remove oil coupling located behind crankshaft gear. Remove rear bearing plate and withdraw crankshaft with center main bearing assembly through bearing plate opening in crankcase.

To remove center bearing assembly, back bearing cap retaining screws out ¼-inch (12.7 mm) and tap cap screws lightly to separate cap from hanger. With cap and hanger removed, remove retaining ring, two-piece outer race and the 32 bearing rollers. Note crankshaft standard bearing journal diameter should be 2.300-2.3005 inches (58.42-58.43 mm) and a 0.030 inch (0.762 mm) oversize bearing is available for reground crankshafts.

For reinstallation, retain bearing rollers in split race with low melting point grease and install on crankshaft securing bearing halves with retaining ring. Install cap and hanger.

Reinstall crankshaft and set end play before securing center main bearing hanger to crankcase by varying thickness of gasket between bearing plate and crankcase until 0.002-0.004 inch (0.0508-0.1016 mm) end play is obtained.

Tighten center main bearing cap screws to 15 ft.-lbs. (20 N·m) torque, then 30 ft.-lbs. (40 N·m) torque and finally to 40 ft.-lbs. (54 N·m) torque, then tighten bearing hanger to crankcase cap screws to 60 ft.-lbs. (81 N·m) torque.

NOTE: Beginning with engine serial number 4052826, cap screws for mounting bearing hanger to crankcase were lengthened to 3 inches and a spacer added under screw heads to minimize possibility of improper assembly and there must be 0.004 inch (0.1016 mm) clearance between sides of bearing hanger and crankshaft cheeks.

Crankshaft designed for insert type bearing at center main will not interchange with crankshaft designed for split roller bearing center main without crankcase modifications.

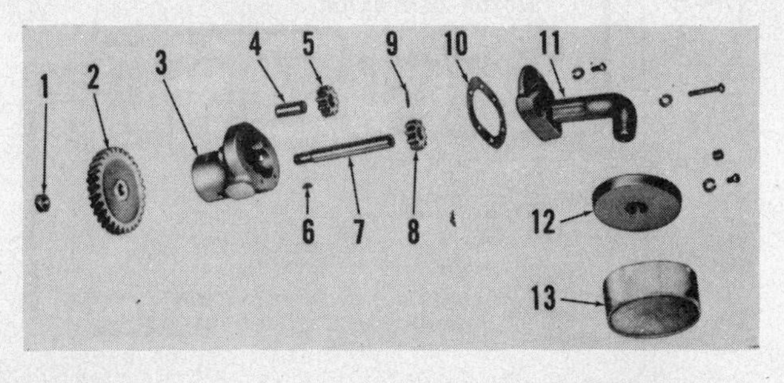

Fig. W125—Exploded view of oil pump used on V-460D, V-461D and V-465D models.

1. Nut
2. Drive gear
3. Pump body
4. Idler shaft
5. Idler gear
6. Key
7. Pump shaft
8. Pump gear
9. Pin
10. Gasket
11. Cover
12. Screen adapter
13. Screen

INSERT TYPE CENTER MAIN BEARING. Center main bearing for late model engines (serial number 4904657 or after) is an insert type bearing. To remove crankshaft, disconnect oil line at center bearing hanger. Remove bearing cap and hanger from engine. Remove rear bearing plate and withdraw crankshaft. Note crankshaft standard bearing journal diameter should be 2.3020-2.3025 inches (58.471-58.484 mm) and bearing inserts are available in a variety of sizes for reground crankshafts as well as standard.

For reinstallation, install crankshaft in crankcase and set end play before installing center main bearing assembly by varying thickness of gasket between bearing plate and crankcase until 0.002-0.005 inch (0.0508-0.1270 mm) is obtained.

Install bearing inserts in cap and hanger and install assembly on crankshaft. Tighten bearing cap screws to 32-35 ft.-lbs. (43-48 N·m), then tighten bearing hanger to crankcase cap screws to 60 ft.-lbs. (81 N·m) torque. Connect oil line.

See **CONNECTING ROD** section for standard crankshaft rod journal diameters.

On all models oil seals should be installed in timing gear cover and rear bearing plate before they are bolted in place and care should be taken to avoid seal damage during installation.

IDLER GEAR AND SHAFT. An idler gear driven by crankshaft gear is used to drive magneto or accessory drive and oil pump.

To remove idler gear and shaft, remove timing gear cover, magneto or accessory drive and distributor. Remove Allen head set screws from side of crankcase which retain idler shaft in position and use a sleeve type puller over end of idler shaft and against idler gear, thread a 3/8-16 cap screw into end of shaft. Pull shaft from crankcase and idler gear.

When reassembling, make certain oil slot in idler shaft is facing upward. Press or drive idler shaft into crankcase until a clearance of 0.003-0.004 inch (0.0762-0.1016 mm) exists between gear hub and shoulder on end of idler shaft. Lock shaft in position with Allen set screws.

CAMSHAFT. Camshaft runs in unbushed bores in crankcase casting. Camshaft gear is held on end of camshaft by three cap screws and end play is controlled by a spring loaded plunger assembly and wear button which bears against timing gear cover. Use care not to lose plunger, spring or wear button during reassembly.

VALVE SYSTEM. All models are equipped with Stellite exhaust valves and seats and rotocap on exhaust valves is standard.

Valve seats are not renewable but may be reground at 45° angle if seat width does not exceed 3/32-inch (2.381 mm).

Valve faces are ground at 45° angle and valve stem clearance in guide should be 0.002-0.004 inch (0.0508-0.1016 mm) and valve, guide or both should be renewed if clearance exceeds 0.006 inch (0.1524 mm). Press old guide out and allow 1/64 to 1/32-inch (0.397-0.794 mm) clearance between valve guide boss and valve guide shoulder. Standard inside diameter of valve guide should be 0.3440-0.3445 inch (8.7376-8.7503 mm).

Valve tappet gap (cold) for all models should be 0.008 inch (0.2032 mm) for intake valves and 0.014 inch (0.3556 mm) for exhaust valves.

Adjustment is made by turning tappet adjusting screw (Fig. W123) as required.

WISCONSIN

SERVICING WISCONSIN ACCESSORIES

12-VOLT MOTOR-GENERATOR

Combination motor-generator manufactured by Delco-Remy is used on some Wisconsin engines. Motor-generator functions as a cranking motor when starting switch is closed. When engine is operating and with starting switch open, unit operates as a generator. Generator output and circuit voltage for battery and various operating requirements are controlled by a current voltage regulator. See Figs. W130, W131 and W132.

Motor-generator belt tension should be adjusted until about 10 pounds (45 N) pressure applied midway between pulleys will deflect belt ½-inch (12.7 mm).

To determine cause of abnormal operation, motor-generator should be given a "no load" test or a "generator output" test. Generator output test can be performed with motor-generator on or off engine. No load test must be made with motor-generator removed from engine. Refer to Fig. W133 for exploded view of a typical motor-generator assembly. Parts are available from Wisconsin as well as authorized Delco-Remy service stations.

Motor-generator and regulator service test specifications are as follows:

Motor-Generator 1101696

Brush spring tension, oz. 22-26
Field draw,
Amperes 1.43-1.54
Volts 12
Cold output,
Amperes 10
Volts 14
Rpm 5750
No load test,
Volts 11
Amperes, max 17
Rpm, min 2350
Rpm, max 2850

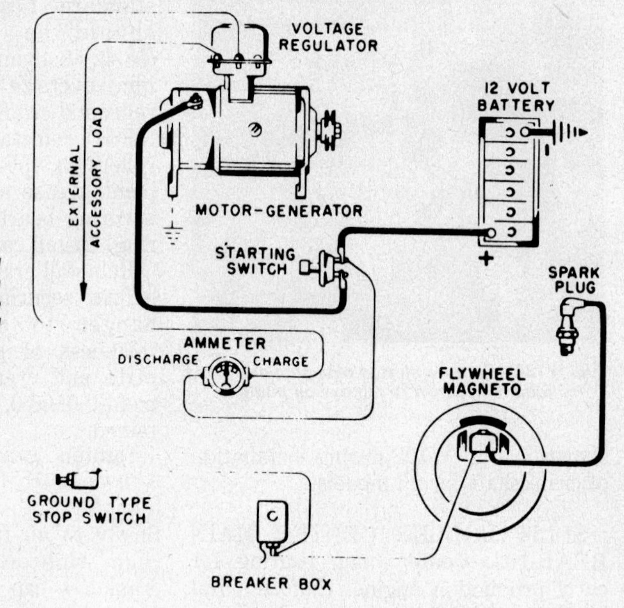

Fig. W130—Typical wiring layout for combination motor-generator on engine equipped with flywheel magneto ignition.

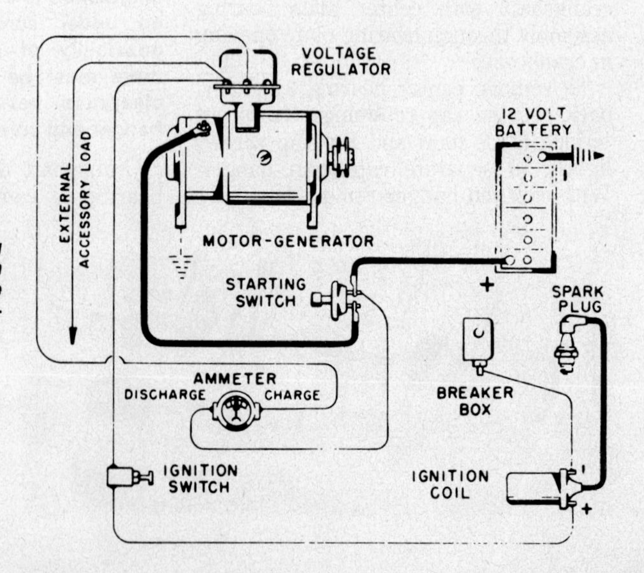

Fig. W131—Typical wiring layout for combination motor-generator on engine equipped with battery ignition.

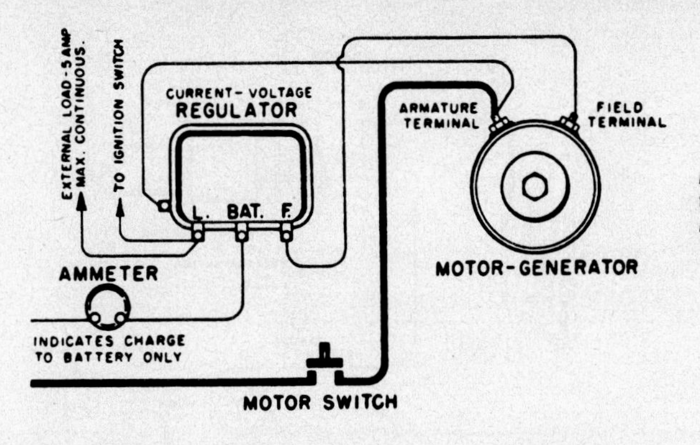

Fig. W132—Wiring connections for current voltage regulator used with motor-generator.

Motor-Generator 1101870

Brush spring tension, oz	22-26
Field draw,	
Amperes	1.52-1.62
Volts	12
Cold output,	
Amperes	12
Volts	14
Rpm	4950
No load test,	
Volts	11
Amperes, max	18
Rpm, min	2500
Rpm, max	2900

Motor-Generator 1101871 & 1101972

Brush spring tension, oz	24-32
Field draw,	
Amperes	1.43-1.54
Volts	12
Cold output,	
Amperes	10
Volts	14
Rpm	5450
No load test,	
Volts	11
Amperes, max	17
Rpm, min	2500
Rpm, max	3000

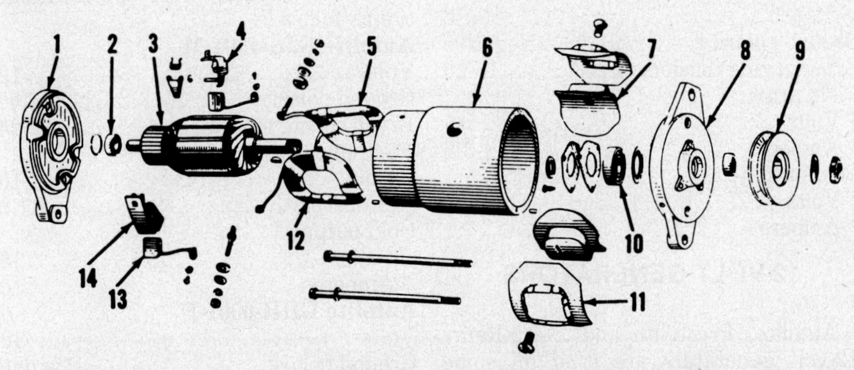

Fig. W133—Exploded view of typical Delco-Remy motor-generator.

1. Commutator end frame
2. Bearing
3. Armature
4. Ground brush holder
5. Field coil (L.H.)
6. Frame
7. Pole shoe
8. Drive end frame
9. Pulley
10. Bearing
11. Field coil insulator
12. Field coil (R.H.)
13. Brush
14. Insulated brush holder

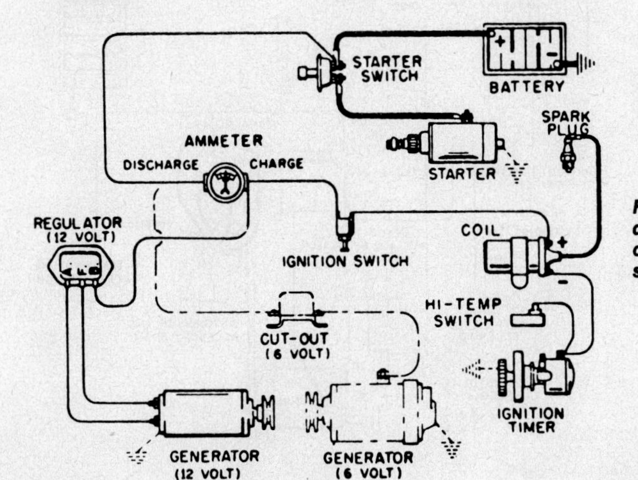

Fig. W134—Typical wiring diagram for starting and charging system used on some Wisconsin one cylinder engines.

Regulators 1118791 & 1118985

Ground Polarity	Positive
Cut-out relay,	
Air gap	0.020 (0.508 mm)
Point gap	0.020 (0.508 mm)
Closing voltage, range	11.8-14.0
Adjust to	12.8
Voltage regulator,	
Air gap	0.075 (1.905 mm)
Setting voltage, range	13.6-14.5
Adjust to	14.0

Regulators 1118983 & 1118984

Ground polarity	Negative
Cut-out relay,	
Air gap	0.020 (0.508 mm)
Point gap	0.020 (0.508 mm)
Closing voltage, range	11.8-14.0
Adjust to	12.8
Voltage regulator,	
Air gap	0.075 (1.905 mm)
Setting voltage, range	13.6-14.5
Adjust to	14.0

6-VOLT STARTER MOTORS

Autolite and Prestolite 6-volt starter motors are used on some Wisconsin engines. Test specifications are as follows:

Autolite MZ-4118, MZ-4175 & MZ-4184

Volts	6
Brush spring tension, oz	42-53
No load test,	
Volts	5
Amperes	68
Rpm	4000

Autolite MAK-4008

Volts	6
Brush spring tension, oz	38-61
No load test,	
Volts	4
Amperes	70
Rpm	4700

Autolite MDH-4001M

Volts	6
Brush spring, tension, oz.	42-66
No load test,	
Volts	5
Amperes	52
Rpm	8100

Prestolite MZ-4213

Volts	6
Brush spring tension, oz.	42-53
No load test,	
Volts	5.5
Amperes	60
Rpm	5000

12-VOLT STARTER MOTORS

Autolite, Prestolite and Delco-Remy 12-volt starters are used on some Wisconsin engines. Test specifications are as follows:

Autolite MDL-6001
Volts 12
Brush spring tension, oz. 52-65
No load test,
Volts 10
Amperes 100
Rpm 4700

Autolite MDH-4002M
Volts 12
Brush spring tension, oz 42-66
No load test,
Volts 10
Amperes 38
Rpm 10000

Autolite MDL-6011
Volts 12
Brush spring tension, oz. 31-47
No load test,
Volts 10
Amperes 60
Rpm 3200

Prestolite MBG-4140
Volts 12
Brush spring tension, oz. 42-53
No load test,
Volts 10
Amperes 55
Rpm 5200

Prestolite MDY-7006
Volts 12
Brush spring, tension, oz 42-53
No load test,
Volts 10
Amperes 60
Rpm 4200

Delco-Remy 1107246
Volts 12
Brush spring tension, oz. 35
No load test,
Volts 10.3
Amperes 75
Rpm 6900

6-VOLT GENERATORS

Autolite and Prestolite 6-volt generators are used on some Wisconsin engines. Test specifications are as follows:

Autolite GAS-4301, GAS-4302, GAS-4303, GAS-4305 & GAS-4306
Volts 6
Ground polarity .... Positive
Brush spring tension, oz. 15-20
Field draw,
Volts 5
Amperes 3.4-3.8
Cold output,
Volts 8
Amperes 12.5

Autolite GAS-4177
Volts 6
Ground polarity .... Negative
Brush spring tension, oz 15-20
Field draw,
Volts 5
Amperes 3.4-3.8
Cold output,
Volts 8
Amperes 12.5

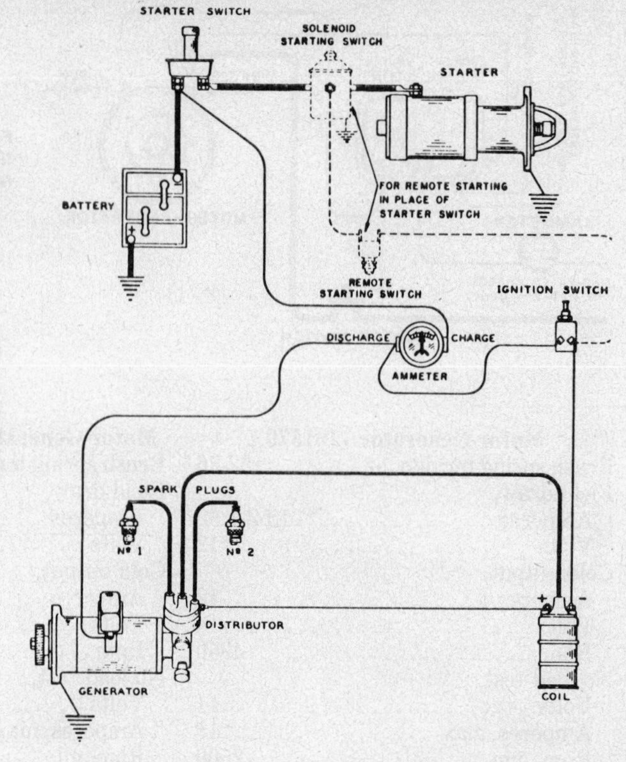

Fig. W135—Typical wiring diagram for starting and charging systems used on some Wisconsin two cylinder engines. Note distributor mounting on rear of gear driven generator.

Prestolite GAS-4301-1
Volts 6
Ground polarity .... Negative
Brush spring tension, oz 15-20
Field draw,
Volts 5
Amperes 3.4-3.8
Cold output,
Volts 8
Amperes 7.1

12-VOLT GENERATORS

Autolite, Prestolite and Delco-Remy 12-volt generators are used on some Wisconsin engines. Test specifications are as follows:

Autolite GJG-4001-M
Volts 12
Ground polarity .... Negative
Brush spring tension, oz. 12-24
Field draw,
Volts 10
Amperes 1.7-1.9
Cold output,
Volts 15
Amperes 10

Autolite GHH-6001-F
Volts 12
Ground polarity .... Positive

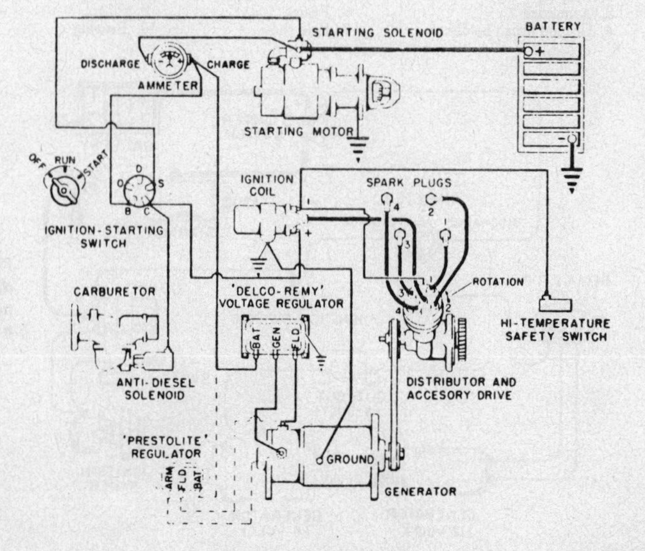

Fig. W136—Typical wiring diagram for starting and charging systems used on some Wisconsin V4 engines.

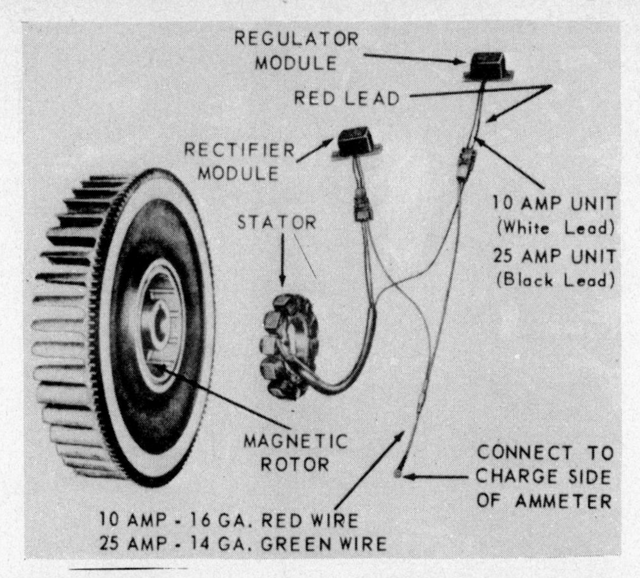

Fig. W137—View of 10 or 25 amp flywheel alternator charging system used on some engines.

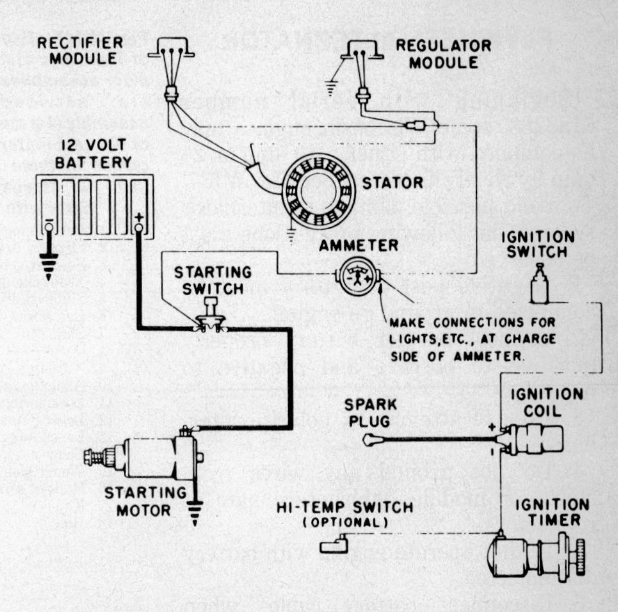

Fig. W138—Typical wiring diagram for ignition, starting and alternator charging systems used on some single cylinder engines.

Brush spring tension, oz 26-46
Field draw,
Volts 10
Amperes 1.6-1.7
Cold output,
Volts 15
Amperes 14.7

Prestolite GJG-4001-MP
Volts 12
Ground polarity ....Positive
Brush spring tension, oz 12-24
Field draw,
Volts 10
Amperes 1.7-1.9
Cold output,
Volts 15
Amperes 10

Prestolite GJG-4010-M
Volts 12
Ground polarity ....Negative
Brush spring tension, oz. 12-24
Field draw,
Volts 10
Amperes 1.7-1.9
Cold output,
Volts 15
Amperes 10

Prestolite GJY-7401-S
Volts 12
Ground polarity ....Positive
Brush spring tension, oz. 35-53
Field draw,
Volts 10
Amperes 1.53-1.70
Cold output,
Volts 15
Amperes 17

Prestolite GJY-7401-SN
Volts 12
Ground polarity ....Negative
Brush spring tension, oz 35-53
Field draw,
Volts 10
Amperes 1.53-1.70

Cold output,
Volts 15
Amperes 17

Delco-Remy 1102225
Volts 12
Ground polarity ....Positive
Brush spring tension, oz 28
Field draw,
Volts 12
Amperes 1.67-1.79
Cold output,
Volts 14
Amperes 17

Delco-Remy 1102343
Volts 12
Ground polarity ....Negative
Brush spring tension, oz 28
Field draw,
Volts 12
Amperes 1.67-1.79
Cold output
Volts 14
Amperes 17

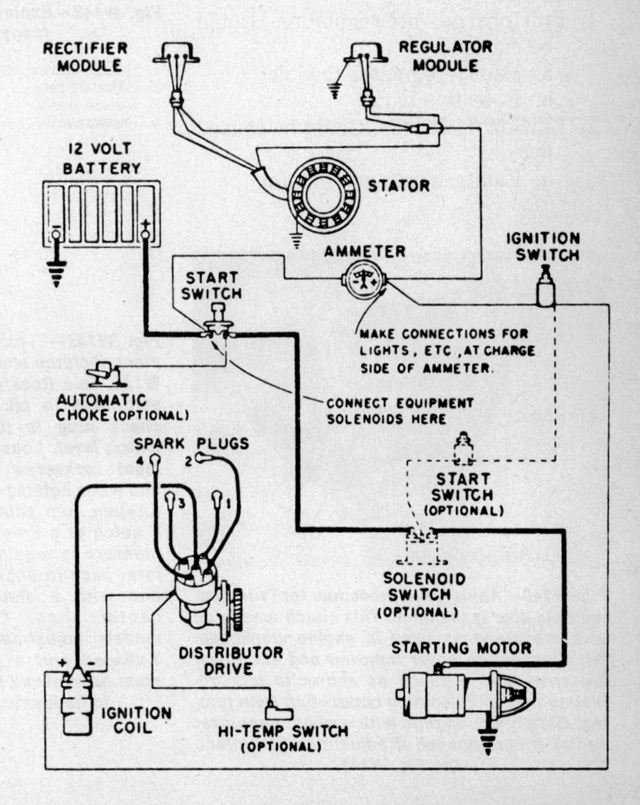

Fig. W139—Typical wiring diagram for ignition, starting and alternator charging systems used on some four cylinder engines. Wiring diagram for two cylinder engines is similar.

FLYWHEEL ALTERNATOR

Beginning with serial number 5188288, some Wisconsin engines may be equipped with either a 10 amp or 25 amp flywheel alternator. See Fig. W137. To avoid possible damage to alternator system, the following precautions must be observed:

1. Negative post of battery must be connected to ground on engine.
2. Connect booster battery properly (positive to positive and negative to negative.)
3. Do not attempt to polarize alternator.
4. Do not ground any wires from stator or modules which terminate at connectors.
5. Do not operate engine with battery disconnected.
6. Disconnect battery cables when charging battery with a battery charger.

OPERATION. Alternating current (AC) produced by alternator is changed to direct current (DC) in rectifier module. See Figs. W138 and W139. Current regulation is provided by regulator module which "sense" counter-voltage created by battery to control or limit charging rate. No adjustments are possible on alternator charging system. Faulty components must be renewed. Refer to following trouble shooting paragraph to help pin point faulty component.

TROUBLE SHOOTING. Trouble conditions and their possible causes are as follows:

1. Full charge–no regulation. Could be caused by:
   a. Faulty regulator module
   b. Defective battery
2. Low or no charge. Could be caused by:
   a. Faulty windings in stator

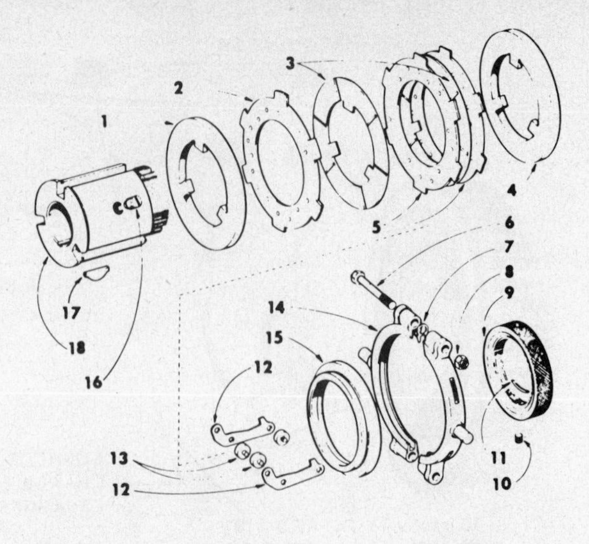

Fig. W141—Exploded view of Twin Disc clutch hub and disc assemblies. All parts are serviced. Collar assembly (14) may be bronze or die cast material. Key (17) comes in three styles. Furnish full details and order parts with care.

1. Back clamping plate
2. Sintered drive plate
3. Steel driven plates
4. Front clamp plate
5. Sintered drive plates
6. Cap screw (2)
7. Shim (2)
8. Locknut (2)
9. Adjusting ring
10. Adjuster set screw
11. Lock spring
12. Lever (6 used)
13. Lever rollers (9)
14. Collar assy.
15. Wedge sleeve
16. Hub set screw
17. Key
18. Hub

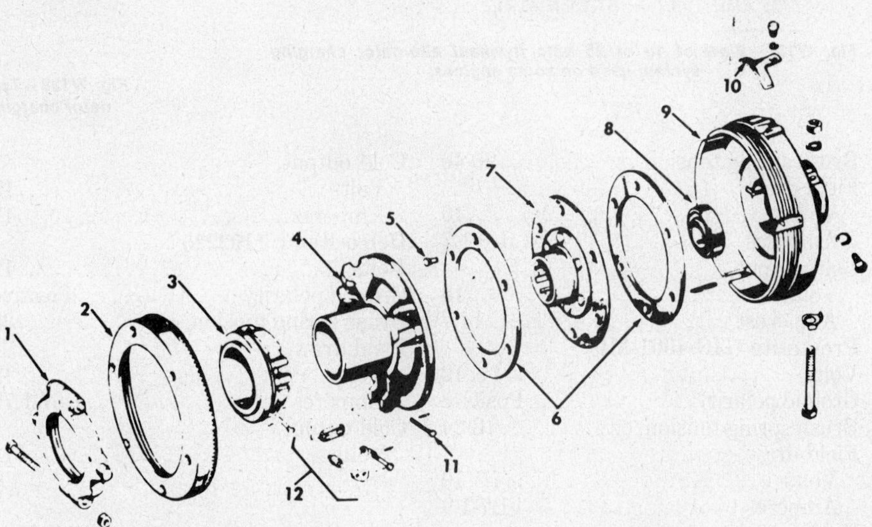

Fig. W142—Exploded view of an often used Rockford clutch design. This unit will appear under many different model numbers. See Fig. W143 for adjustment procedure.

1. Release bearing collar
2. Adjusting ring
3. Release sleeve
4. Pressure plate
5. Brass facing rivet (6)
6. Facings (2)
7. Driven plate assy.
8. Pilot bearing
9. Housing
10. Adjusting lock
11. Lever pin (3)
12. Link lever & roller assy. (3)

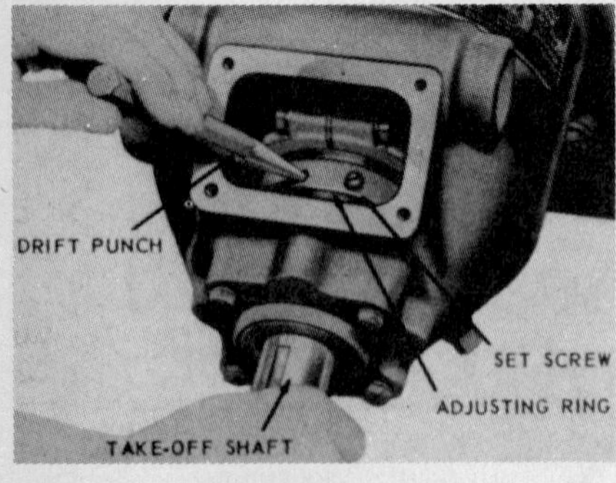

Fig. W140—Adjustment procedure for Twin Disc multiple-disc type clutch. This clutch runs in oil of same grade as used in engine crankcase. With inspection cover removed and set screw loosened, use a punch as shown to turn adjusting ring while holding ouput shaft from turning. Clutch will engage with a slight snap over center when engaged if adjustment is correct. See Fig. W141.

Fig. W143—Typical adjustment of clutch shown in Fig. W142. This Rockford clutch also runs in oil. Oil level check plug is just below clutch lever. Loosen adjustment lockscrew one turn and while holding to prevent rotation, turn adjusting ring a notch at a time until firm pressure is required to pull lever back to engaged position with a distinct over-center snap. On some models, adjustment lock is T-shaped and a pipe plug must be removed for access to its lockscrew.

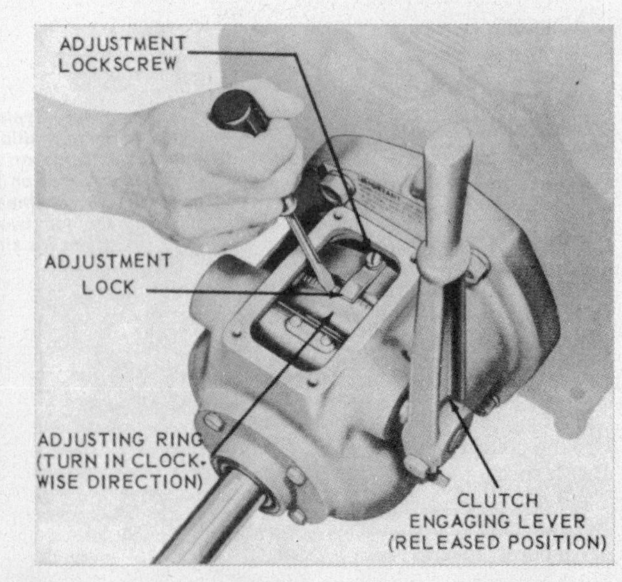

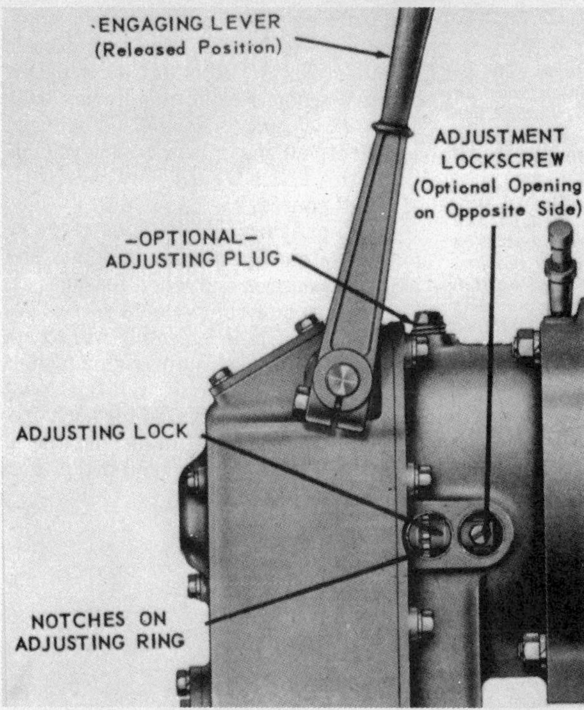

Fig. W144—View to show alternate adjustment access points when same clutch as shown in Figs. W142 and W143 is used with reduction gears. Technique for adjustment is the same.

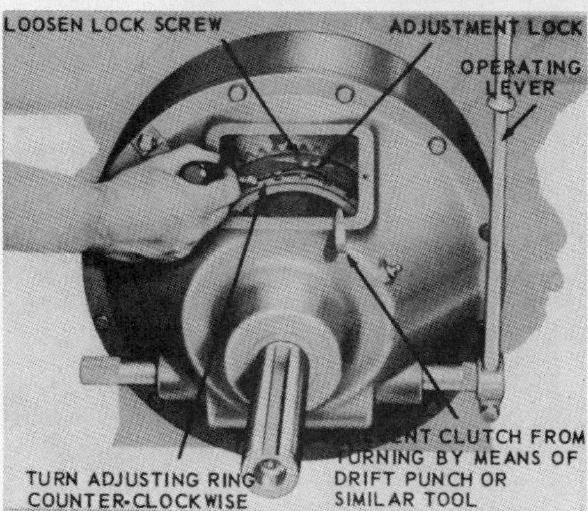

Fig. W145—One type of Rockford dry pto clutch used on heavy duty engines. Visible grease fitting is used to lubricate housing bearing every 50 hours. Other fitting (not shown) at left of inspection plate is for daily greasing of throwout bearing. Clutch adjusts similarly to other models: Release lock screw from adjustment lock and tighten adjusting ring until firm pressure is required to engage clutch over center.

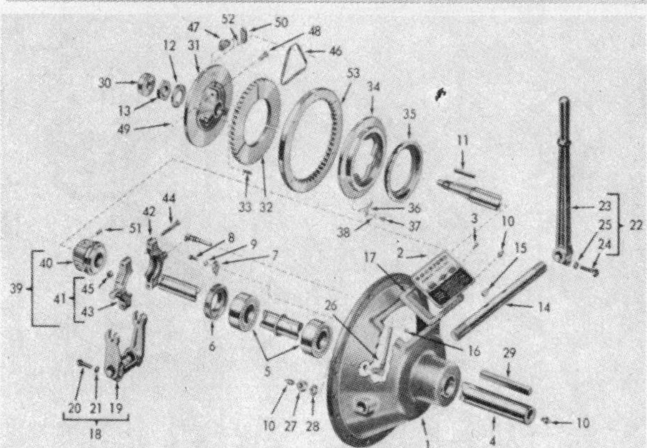

Fig. W146—Exploded view of clutch shown in Fig. W145.

1. Housing
2. Specification plate
4. Output shaft
5. Main bearings
6. Bearing retainer
7. Lock plate
11. Key
14. Yoke shaft
15. Woodruff key
17. Gasket
18. Yoke assy.
22. Lever assy.
26. Grease tube
29. Shaft key
30. Pilot bearing
31. Clutch body
32. Facing
33. Separator spring
34. Pressure plate
35. Adjusting ring
36. Adjustment lock
37. Lock screw
39. Release sleeve
41. Release bearing
46. Lever spring set
47. Lever (3)
48. Pin (3)
50. Lever link (6)
53. Driving ring

b. Faulty rectifier module
c. Regulator module not properly grounded or regulator module defective.

If "full charge–no regulation" is the trouble, use a DC voltmeter and check battery voltage with engine operating at full rpm. If battery voltage is over 15.0 volts, regulator module is not functioning properly. If battery voltage is under 15.0 volts and over 14.0 volts, alternator, rectifier and regulator are satisfactory and battery is probably defective (unable to hold charge).

If "low" or "no charge" is the trouble, check battery voltage with engine operating at full rpm. If battery voltage is more than 14.0 volts, place a load on battery to reduce voltage to below 14.0 volts. If charge rate increases, alternator charging system is functioning properly and battery was fully charged. If charge increases, permanently install new rectifier module. If charge rate does not increase, stop engine and unplug all connectors between modules and stator. Start engine and operate at 2400 rpm. Using an AC voltmeter, check voltage between each black stator lead and ground. If either of the two voltage readings is zero or there is more than 10% difference between readings, stator is faulty and should be renewed.

CLUTCH. A wide variety of clutches are used on WISCONSIN engines. By type, they may be wet (running in oil) or

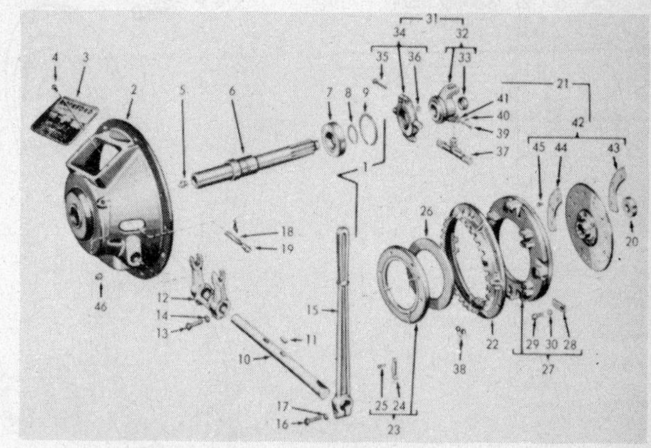

Fig. W147—Another Rockford dry pto clutch design used on some models. Note different style pressure plate (27), facings and disc (42) and sleeve and bearing (31). On this model, adjusting lock (24) must be disengaged from notches in back plate (22) so adjusting ring (23) can be turned.

2. Housing
3. Instruction plate
6. Shaft
7. Bearing
8. Snap ring
9. Snap ring
10. Yoke shaft
11. Key
12. Yoke
15. Lever
20. Pilot bearing
21. Clutch assy. (22-45)
22. Back plate
23. Adjusting ring
24. Adjusting lock
26. Adjusting ring plate
27. Pressure plate assy.
32. Sleeve assy.
34. Release bearing
37. Camshaft assy.
38. Return spring (4)
39. Lever pin (2)
42. Driven plate assy.

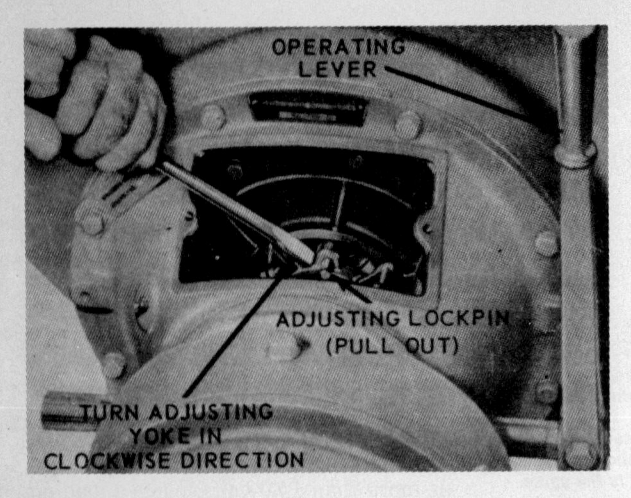

Fig. W148 — Heavy duty dry Twin Disc clutch used on some models. To adjust, pull lock pin out as shown and hold by inserting a piece of 1/16-inch (1.588 mm) wire through hole in pin. With lockpin so disengaged, turn yoke to increase pressure on operating lever as required for firm clutch action. Remove wire after adjusting and shift yoke slightly so lockpin can snap back into one of the holes in floating plate. Also see Fig. W149.

dry plate equipped and of either single or multiple disc design. Units in use are manufactured by Rockford Clutch Div. (Borg-Warner) Rockford, Illinois, or by Twin Disc, Inc., Racine, Wisconsin. However, all parts service is available through TELEDYNE WISCONSIN MOTOR sources.

Accompanying illustrations are intended to show adjustment points and work procedures and when feasible, exploded views of components so relative position of parts in assembly will be apparent. Because so many engine models are covered, specific clutch part numbers are not identified for each one, but rather, typical characteristics of each clutch style are pointed out. For

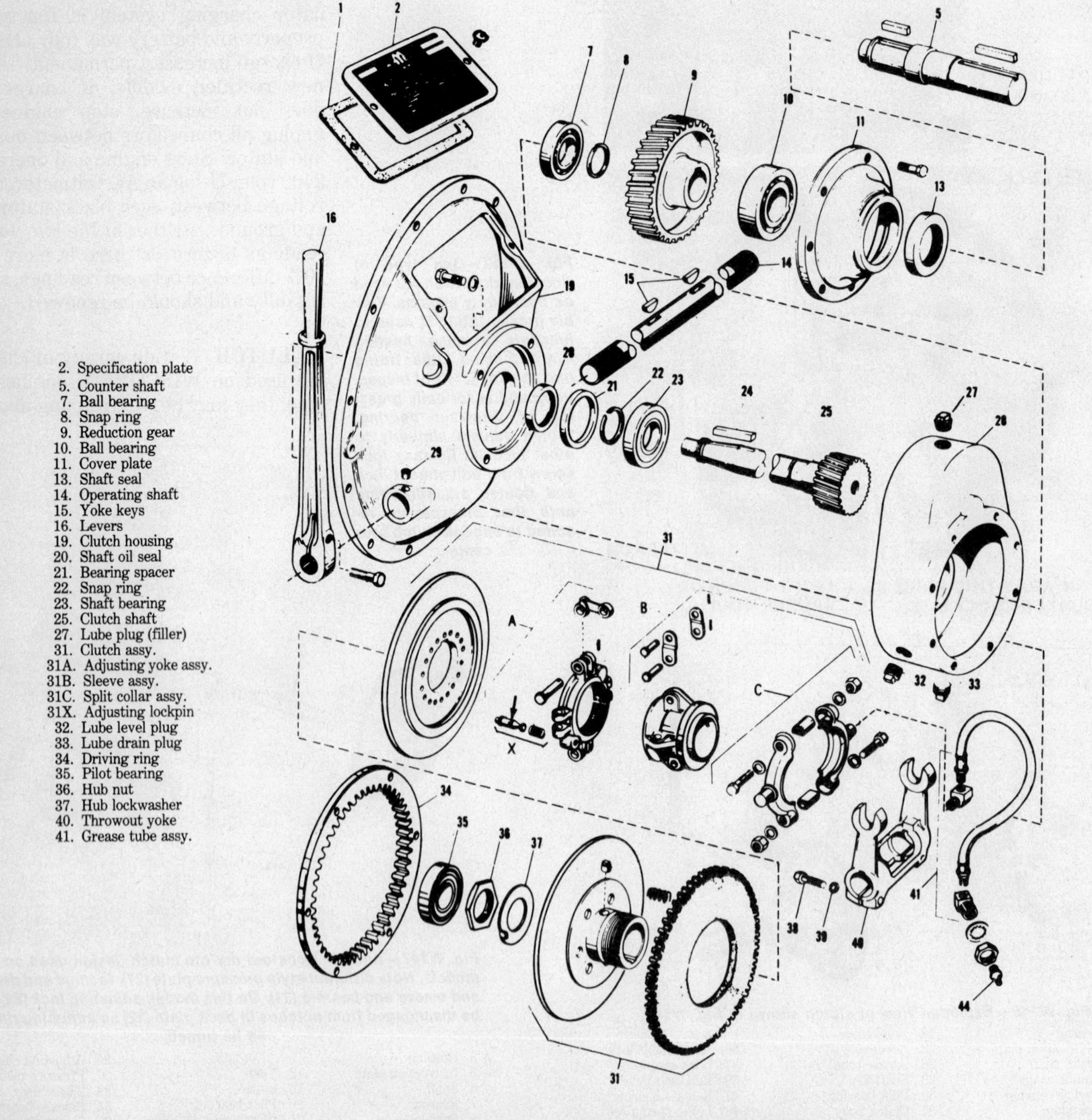

Fig. W149 — Exploded view of Twin Disc dry clutch of design shown in Fig. W148 fitted with reduction gears. Adjusting yoke is shown at 31A. Note hole (arrow) in lockpin (X) and circle of holes in floating plate for entry of lockpin.

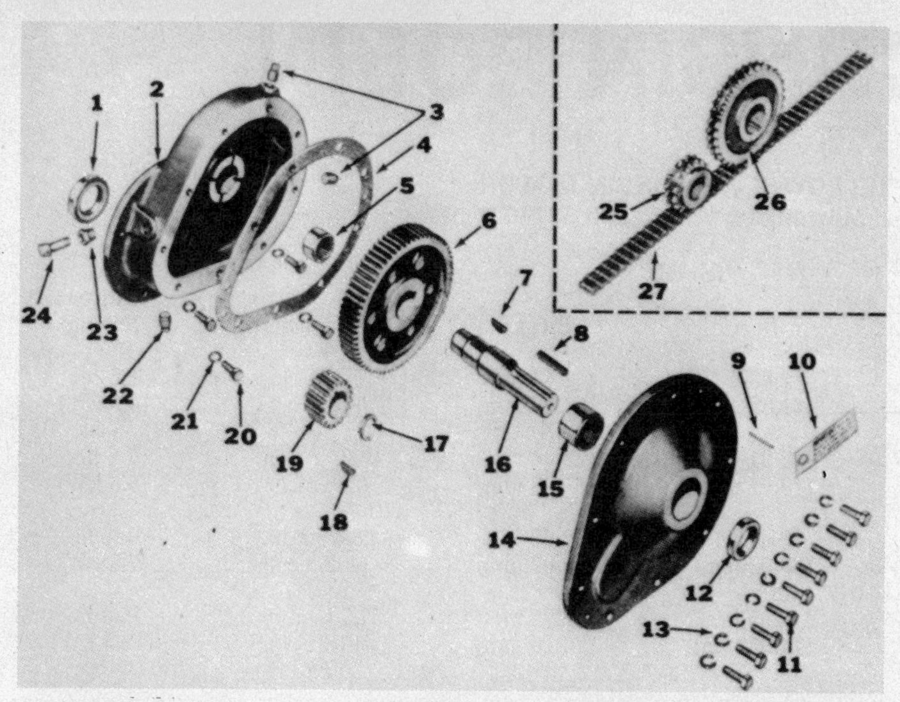

Fig. W150—Exploded view of typical reduction gear set used with some Wisconsin engine models. Inset shows chain-sprocket arrangement which may be used in some cases instead of spur gears shown.

1. Shaft oil seal
2. Main housing
3. Pipe plugs
4. Cover gasket
5. Inner bearing
6. Driven gear
7. Woodruff key
8. Square key
9. Cover pin
10. Specification tag
11. Cover screws
12. Output shaft seal
13. Lockwashers
14. Housing cover
15. Outer bearing
16. Output shaft
17. Drive gear retainer
18. Woodruff key
19. Drive gear (pinion)
20. Housing screw
21. Lockwasher
22. Drain plug (same as 3)
23. Reducer bushing
24. Breather (vent)
25. Drive sprocket
26. Driven sprocket
27. Drive chain

practical purposes, appearance and shape of parts used is identical, with dimensions (size measurements) being the only actual difference between clutch models used with separate engine models. It is extremely important **exact identification** be taken from specification plates and tags when it becomes necessary to order renewal parts such as bearings and shaft seals.

Following are accepted principles pertaining to service for manual-operated over-center clutches:

1. Do not allow a new clutch to slip when engaged under load. Several early adjustments may be necessary if clutch begins to slip during break-in period.

2. Wet-type clutches must have oil level maintained in housing. This oil is an essential cooling medium to prevent destruction of friction surfaces. Observe service instructions printed on specification plates.

3. Dry clutches require scheduled lubrication at grease fittings mounted in clutch bell housings. One fitting is for outer bearing (output shaft) in housing and should be serviced every fifty hours operation. Throwout (release) bearing should be lubricated daily. Twin Disc manufactured units have an external fitting for throwout bearing. Rockford clutches require an inspection plate (marked) be removed for access to throwout bearing grease fitting.

4. In event of poor performance, lack of power or undue vibration from power unit, do not overlook possibility of faulty clutch. These clutches are extremely rugged, but neglect or lack of routine maintenance may cause problems. Check for looseness or leakage at output shaft and remove inspection plate to examine clutch body and linkage for defects or damage.

REDUCTION GEARS

Reduction gear sets in a wide choice of ratios are offered for all WISCONSIN models except TR series. Some are furnished with manually operated clutches as shown in Fig. W149 while others, as in Fig. W150, are coupled to and driven directly from engine crankshafts. Some reduction gear sets are lubricated from an oil supply stored within their own housing while others share oil supply from engine crankcase. Units with separate oil supply call for use of high grade gear oil of SAE 90 to SAE 110 grade. Seasonal requirements for changes in engine oil viscosity determine weight of oil used for gear sets which share a common oil supply with engines. Gear oil should be changed when warm, observing 200 hour service intervals.

Many of the reduction gear combinations offered may be mounted with output shaft at a position left, right, above or below engine crankshaft or clutch pto centerline. For this reason, several pipe plugs are fitted into outer circumference of gearcase for convenience in draining and filling and for checking lube level in housing.

Printed specifications come with every reduction unit manufactured, usually on a metal tag or on inspection plate. Be sure to check for lubrication instructions and for identification by number when renewal parts are needed for necessary repairs.

WISCONSIN ROBIN

TELEDYNE WISCONSIN MOTOR
Milwaukee, Wisconsin 53246

Model	No. Cyls.	Bore	Stroke	Displacement	Power Rating
EY25W	1	2.83 in. (71.88 mm)	2.44 in. (61.98 mm)	15.4 cu. in. (251.5 cc)	6.5 hp. (4.9 kW)
EY27W	1	2.91 in. (73.91 mm)	2.44 in. (61.98 mm)	16.26 cu. in. (265.5 cc)	7.5 hp. (5.6 kW)
EY44W	1	3.54 in. (89.92 mm)	2.68 in. (68.07 mm)	26.4 cu. in. (432.3 cc)	10.5 hp. (7.8 kW)
EY21W	2	2.95 in. (74.93 mm)	2.76 in. (70.10 mm)	37.74 cu. in. (618.2 cc)	16.8 hp. (12.5 kW)

EY25W, EY27W and EY44W models are four cycle, horizontal crankshaft, one cylinder engines and EY21W models are four cycle, horizontal crankshaft, two cylinder opposed engines. Crankshaft of one cylinder engines is supported at each end in ball bearings and crankshaft for two cylinder engine is supported by a ball bearing at front of shaft and by a roller bearing at rear of shaft.

Connecting rod of EY25W and EY27W model one cylinder engines and EY21W model two cylinder engine ride directly on crankshaft. Connecting rod for EY44W model is equipped with renewable insert type connecting rod bearings. All models are splash lubricated by a dipper attached to connecting rod cap.

One cylinder models have a magneto ignition system with points and condenser located beneath flywheel. Two cylinder models have battery type ignition with points and condenser located in an ignition timer located on top of engine.

A side draft, float type carburetor is used on one cylinder models and a downdraft, float type carburetor is used on two cylinder models. A mechanical type fuel pump is available for some models.

Engine model and specification number are stamped on name plate located on flywheel shroud and engine serial number for one cylinder models is stamped on crankcase base left mounting flange. Serial number for two cylinder models is stamped on crankcase just below fuel pump.

Always give specification, model and serial number when ordering parts or service material.

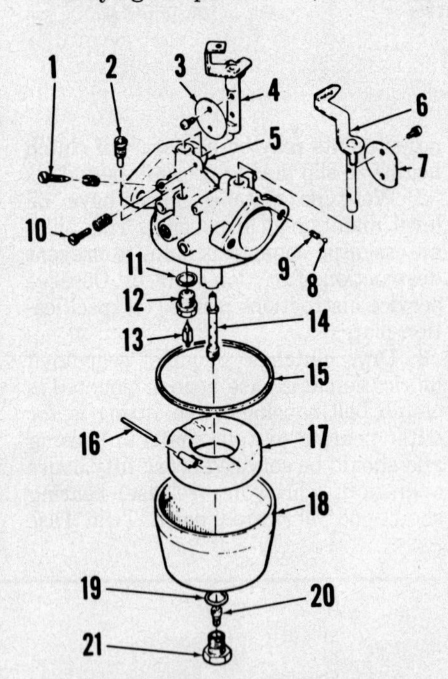

Fig. WR1 – Exploded view of typical Mikuni float type carburetor used on EY25W, EY27W and EY44W models.

1. Idle mixture needle
2. Idle jet
3. Throttle plate
4. Throttle shaft
5. Carburetor body
6. Choke shaft
7. Choke plate
8. Choke detent ball
9. Detent spring
10. Idle speed stop screw
11. Gasket
12. Fuel valve seat
13. Inlet fuel valve
14. Main nozzle
15. Bowl gasket
16. Float pin
17. Float
18. Fuel bowl
19. Gasket
20. Main fuel jet
21. Jet holder

MAINTENANCE

SPARK PLUG. Recommended spark plug is Champion L86, or equivalent, for one cylinder models and Champion J8, or equivalent, for two cylinder models. Electrode gap is 0.020-0.025 inch (0.508-0.635 mm) for all models.

CARBURETOR. Mikuni float type carburetors are used on all models. Side draft style carburetor (Fig. WR1) is used on one cylinder models and downdraft style carburetor (Fig. WR2) is used on two cylinder models.

For initial carburetor adjustment, open idle mixture screw 1-5/8 turns for EY25W model, 1-1/2 turns for EY27W model, 1-3/4 turns for EY44W model and 1-1/4 turns for EY21W model.

Main fuel mixture is controlled by jet size and is non-adjustable.

Make final adjustment with engine at normal operating temperature and running. Set engine at idle speed (1000 rpm for EY21W model, 1200 rpm for EY44W model and 1250 rpm for EY25W and EY27W models) no load and adjust idle mixture screw to obtain smoothest idle operation and acceleration.

To check float level on one cylinder models, position carburetor as shown in Fig. WR3.

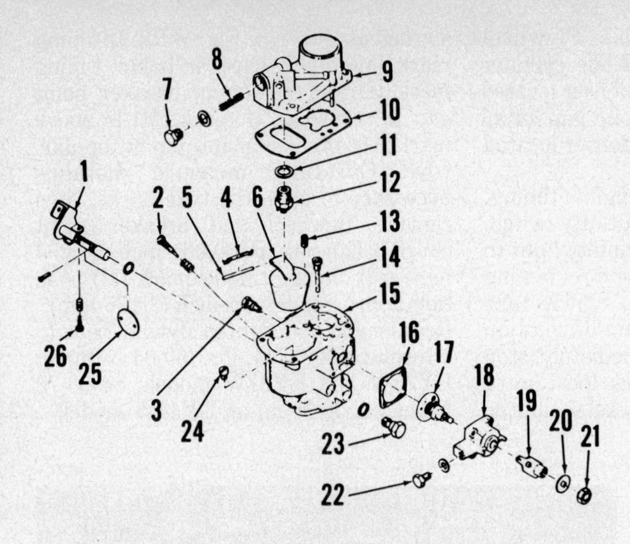

Fig. WR2—Exploded view of Mikuni float type carburetor used on EY21W engines.

1. Throttle shaft
2. Idle mixture needle
3. Idle jet
4. Float pin retainer
5. Float pin
6. Float
7. Plug
8. Fuel screen
9. Upper body
10. Gasket
11. Gasket
12. Fuel inlet valve assy.
13. Idle air jet
14. Main nozzle
15. Lower body
16. Gasket
17. Starting valve
18. Valve housing
19. Starting lever
20. Washer
21. Nut.
22. Plug
23. Starting jet
24. Main fuel jet
25. Throttle plate
26. Idle speed stop screw

NOTE: Needle valve is spring loaded and float tab should just contact needle valve pin, but should not compress spring.

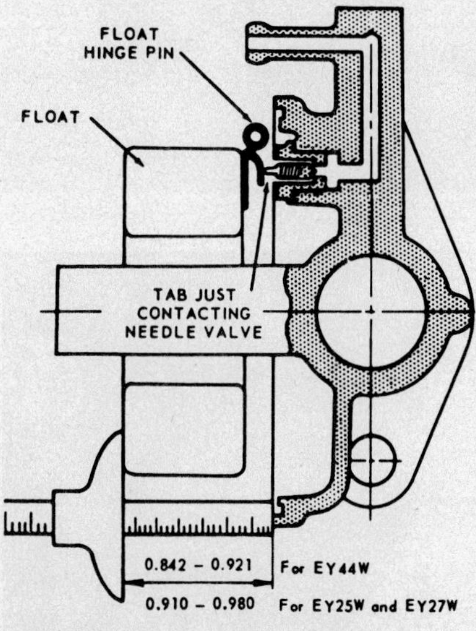

Fig. WR3—With fuel bowl removed, stand carburetor on manifold flange and measure float setting as shown for EY25W, EY27W and EY44W models.

Fig. WR4—Float setting on EY21W models will be correct if distance from top of float lever to bottom of float is 1-13/64 inches (30.56 mm). Refer to text.

Use a depth gage to measure distance between body flange and free end of float (Fig. WR3). Distance should be 0.842-0.921 (21.39-23.39 mm) for EY44W model or 0.910-0.980 inch (23.11-24.89 mm) for EY25W and EY27W models.

Adjust float by bending float lever tank that contacts inlet valve.

To check float level on two cylinder models, refer to Fig. WR4 and note float setting will be within specified range if distance from top of float lever to bottom of float is 1-13/64 inches (30.56 mm) as shown. If necessary, bend float lever to obtain correct setting. Float lever must be parallel with float in area where fuel inlet valve contacts float lever.

Mikuni downdraft carburetor used on two cylinder model has a special starting fuel valve instead of a choke. Fuel from fuel bowl in lower body (15–Fig. WR2) is metered through starting jet (23) and is mixed with air from inside vent. Mixture then flows into starting valve assembly where additional air from main air intake maintains a correct air/fuel ratio for easy starting.

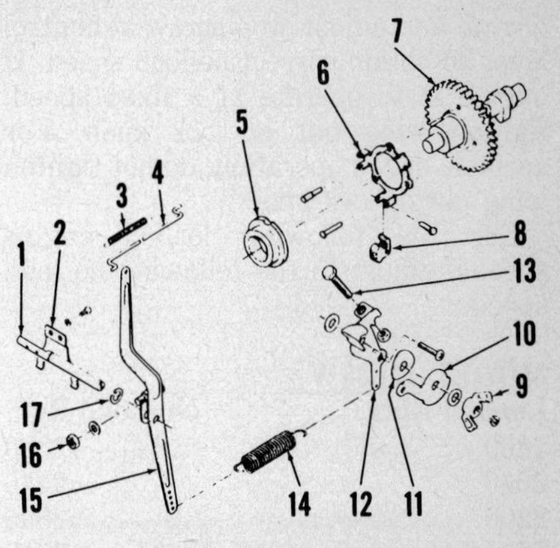

Fig. WR5—Exploded view of typical mechanical governor and linkage used on EY25W, EY27W and EY44W models.

1. Governor lever shaft
2. Yoke
3. Link spring
4. Carburetor link
5. Thrust sleeve
6. Governor plate
7. Camshaft assy.
8. Flyweights (3 used)
9. Wing nut
10. Stop plate
11. Wave washer
12. Control lever
13. Speed stop screw
14. Governor spring
15. Governor lever
16. Clamp nut
17. Retaining ring

GOVERNOR. Mechanical flyweight governor used on all models is mounted on and operated by camshaft gear (Fig. WR5 and WR6). Engine speed is controlled by tension on governor spring.

Before attempting to adjust governed speed on one cylinder models, synchronize governor linkage by loosening clamp nut (16-Fig. WR5) and turning governor lever (15) counter-clockwise until carburetor throttle plate is in wide open position. Insert screwdriver in slot in end of governor lever shaft (1) and rotate shaft counter-clockwise as far as possible. Tighten governor clamp nut.

No governor linkage adjustment is possible on two cylinder models as a slotted hole in governor lever link (Fig. WR6) fits on flat sides of governor shaft end.

To adjust all models for a particular loaded rpm, loosen wing nut (9-Fig. WR5) or lock knob (Fig. WR6). Start

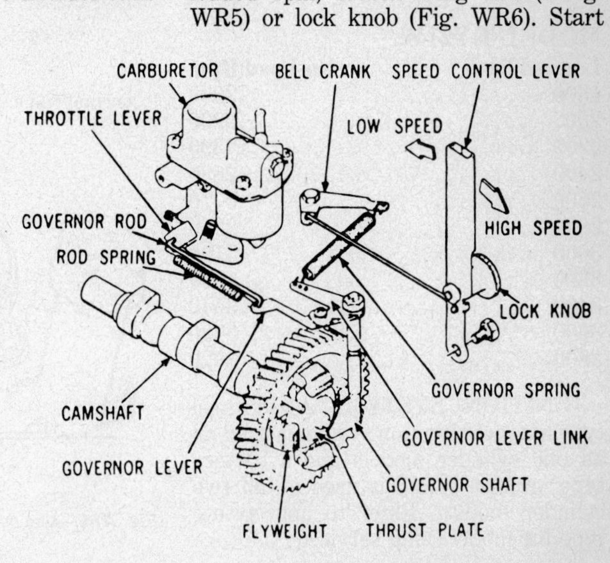

Fig. WR6—View showing mechanical governor and linkage used on EY21W engines.

engine and adjust stop screw at control lever to obtain correct no-load speed. If engine is to operate at a fixed speed, tighten wing nut or lock knob. For variable speed operation, do not tighten wing nut or lock knob.

For the following loaded engine speeds, adjust to the following no load speeds:

MODEL EY25W

Loaded Rpm	No-Load Rpm
1800	2330
2000	2445
2200	2595
2400	2745
2600	2900
2800	3065
3000	3230
3200	3400
3400	3580
3600	3765

MODEL EY27W

Loaded Rpm	No-Load Rpm
1800	2210
2000	2375
2200	2500
2400	2660
2600	2850
2800	3020
3000	3210
3200	3385
3400	3590
3600	3760

MODEL EY44W

Loaded Rpm	No-Load Rpm
1800	2200
2000	2365
2200	2510
2400	2675
2600	2850
2800	3025
3000	3250
3200	3420
3400	3605
3600	3770

MODEL EY21W

Loaded Rpm	No-Load Rpm
1800	2000
2000	2200
2200	2390
2400	2580
2600	2770
2800	2960
3000	3150
3200	3350
3400	3540
3600	3730
3800	3920

IGNITION SYSTEM. A flywheel type magneto ignition system is used on all one cylinder models and a battery type ignition sytem is used on all two cylinder models. Refer to appropriate type for model being serviced.

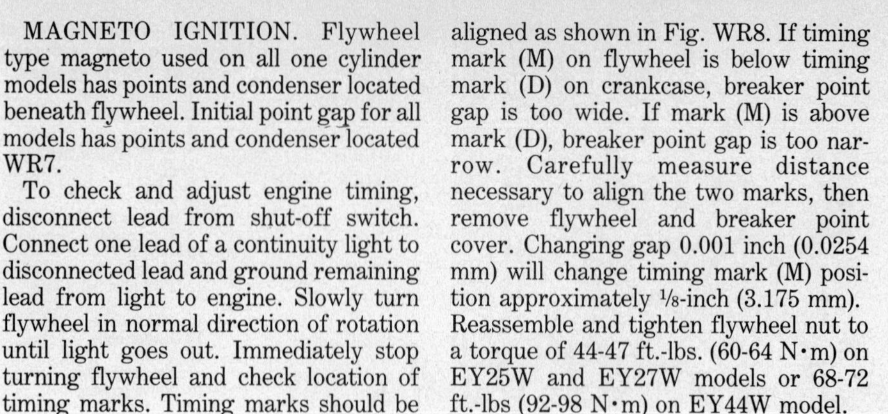

MAGNETO IGNITION. Flywheel type magneto used on all one cylinder models has points and condenser located beneath flywheel. Initial point gap for all models has points and condenser located WR7.

To check and adjust engine timing, disconnect lead from shut-off switch. Connect one lead of a continuity light to disconnected lead and ground remaining lead from light to engine. Slowly turn flywheel in normal direction of rotation until light goes out. Immediately stop turning flywheel and check location of timing marks. Timing marks should be aligned as shown in Fig. WR8. If timing mark (M) on flywheel is below timing mark (D) on crankcase, breaker point gap is too wide. If mark (M) is above mark (D), breaker point gap is too narrow. Carefully measure distance necessary to align the two marks, then remove flywheel and breaker point cover. Changing gap 0.001 inch (0.0254 mm) will change timing mark (M) position approximately 1/8-inch (3.175 mm). Reassemble and tighten flywheel nut to a torque of 44-47 ft.-lbs. (60-64 N·m) on EY25W and EY27W models or 68-72 ft.-lbs (92-98 N·m) on EY44W model.

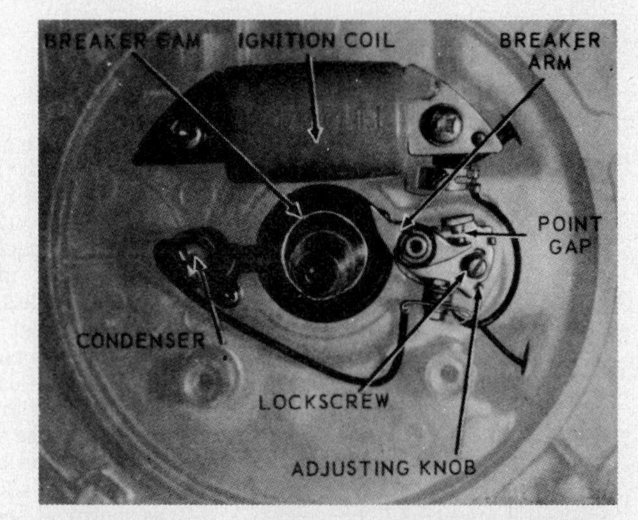

Fig. WR7—View showing flywheel magneto used on EY25W and EY27W models. Magneto on EY44W model is similar. Flywheel and breaker cover are removed.

Fig. WR8—View showing 23°BTDC timing mark (M) on flywheel aligned with timing mark (D) on crankcase. Refer to text for timing procedure.

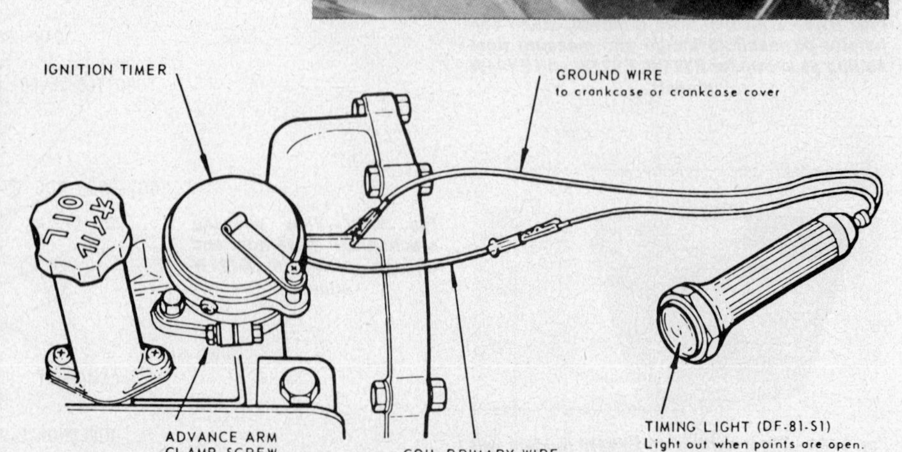

Fig. WR9—Use continuity light (Wisconsin No. DF-81-S1 or equivalent) to check and adjust ignition timing on EY21W engines.

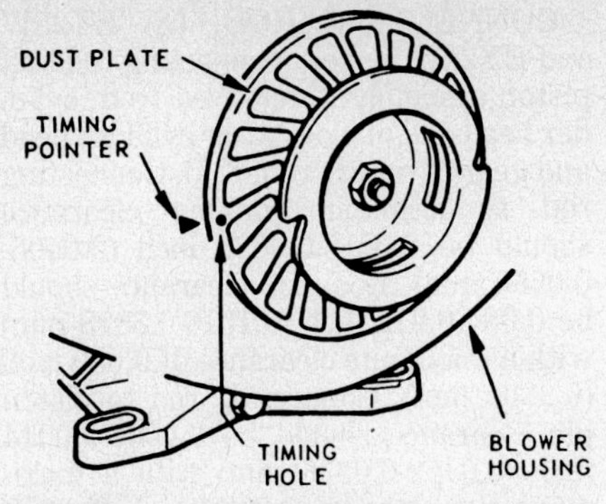

Fig. WR10 – View showing timing hole (8° BTDC static timing mark) in dust plate aligned with timing pointer on blower housing.

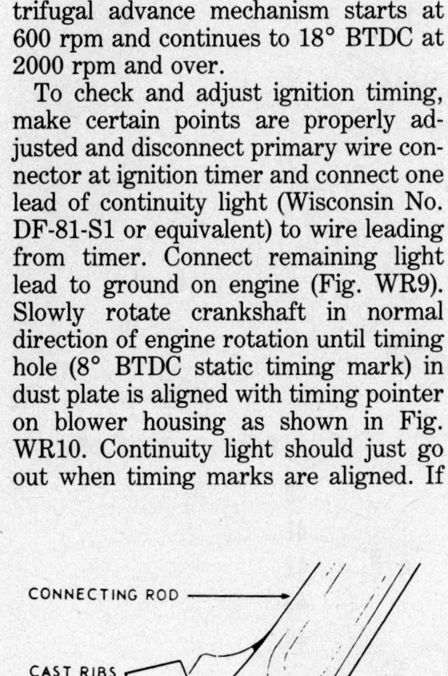

BATTERY IGNITION. A 12 volt battery ignition system is standard on EY21W engines. Breaker points and condenser are located in ignition timer on top of engine (Fig. WR9). Timer is driven by a pinion on engine camshaft. Two lobe breaker cam rotates at half crankshaft speed and is equipped with a centrifugal spark advance mechanism. Initial point gap is 0.014 inch (0.36 mm).

Static timing is 8° BTDC and centrifugal advance mechanism starts at 600 rpm and continues to 18° BTDC at 2000 rpm and over.

To check and adjust ignition timing, make certain points are properly adjusted and disconnect primary wire connector at ignition timer and connect one lead of continuity light (Wisconsin No. DF-81-S1 or equivalent) to wire leading from timer. Connect remaining light lead to ground on engine (Fig. WR9). Slowly rotate crankshaft in normal direction of engine rotation until timing hole (8° BTDC static timing mark) in dust plate is aligned with timing pointer on blower housing as shown in Fig. WR10. Continuity light should just go out when timing marks are aligned. If light goes out before or after timing marks line up, align timing marks and loosen timer advance arm screw (Fig. WR9). Rotate timer body slowly until breaker points are just beginning to open and light just goes out. Tighten clamp screw. Slowly rotate crankshaft in normal direction of rotation and recheck to make certain continuity light goes out just as timing marks are aligned.

LUBRICATION. Splash lubrication is provided by oil dippers attached to connecting rod caps (Figs. WR11, WR12 and WR13) for all models. Oil dippers on one cylinder models pick up oil directly from crankcase. Oil dipper on two cylinder models pick up oil from a trough located directly under crankpins. Trough oil level is maintained at a constant level by oil supplied by a small capacity oil pump located inside crankcase.

To remove oil pump, remove blower housing, blower and timing gear cover. Withdraw oil pump gear and shaft (2-Fig. WR14) with thrust washer (1). Unbolt and remove engine base, loosen pump adapter lockscrew (11) and remove oil pump assembly from inside crankcase. Remove the four screws and separate pump from adapter (3). Shaft (5), rotor (6), stator (7) and housing (8) are serviced as an assembly only.

When reassembling pump, use new gasket (4) and make certain shaft (5), rotor (6) and stator (7) turn free when screws are tightened securely.

Oils approved by manufacturer must meet requirements of API service classification MS or SD.

For one cylinder models use SAE 30 oil in temperatures above 40°F (4°C), SAE 20 oil in temperatures between 15° and 40°F (-9° and 4°C) and SAE 10W-30 oil in temperatures below 15° F (-9° C).

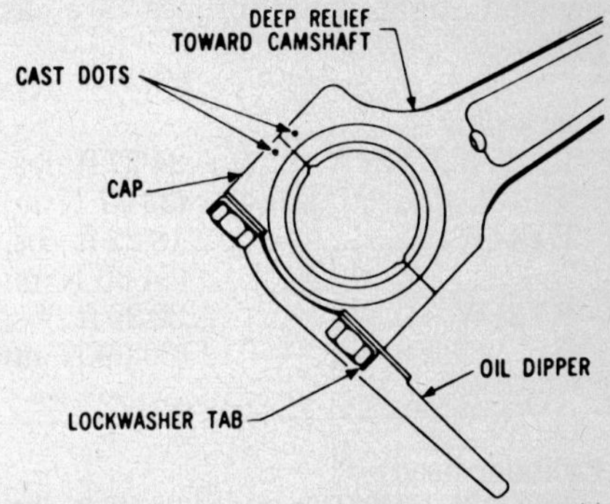

Fig. WR13 – Install connecting rods and rod caps with cast dots (match marks) together on EY21W engines.

For two cylinder models use SAE 30 oil in temperatures above 60°F (16°C), SAE 10W-30 oil in temperatures between 0° and 60°F (-18° and 16°C) and SAE 5W-20 oil in temperatures below 0°F (-18°C).

Maintain oil level at "FULL" mark on dipstick. **DO NOT** overfill.

Recommended oil change interval for all models is every 50 hours of normal operation.

Crankcase oil capacity is 1.6 pints (0.76 L) for EY25W and EY27W models, 2.4 pints (1.14 L) for EY44W model and 5.25 pints (2.48 L) for EY21W model.

CRANKCASE BREATHER. A floating poppet type breather valve is located in breather plate at valve cover. A breather tube connects breather into air cleaner. Restricted or faulty breather is indicated when oil seeps from gasket surfaces and oil seals.

REPAIRS

TIGHTENING TORQUES. Recom-

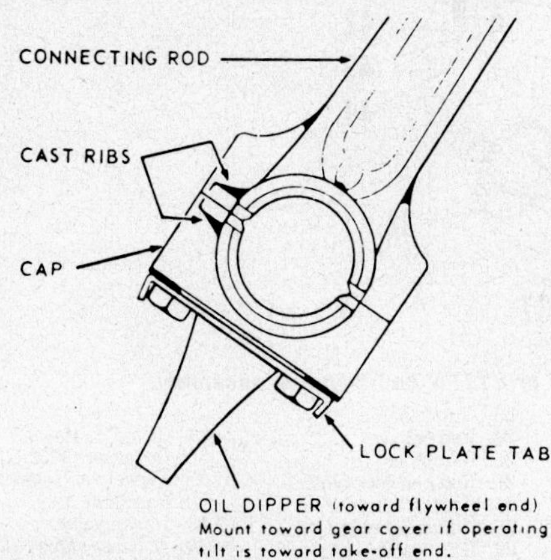

Fig. WR11 – Connecting rod and cap must be installed with cast ribs (match marks) together on EY25W and EY27W models.

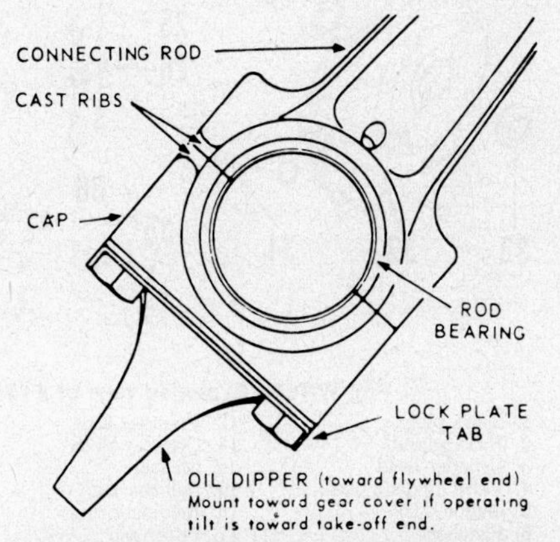

Fig. WR12 – Renewable insert type connecting rod bearings are used on EY44W models. Cast ribs (match marks) on connecting rod and cap must be installed together.

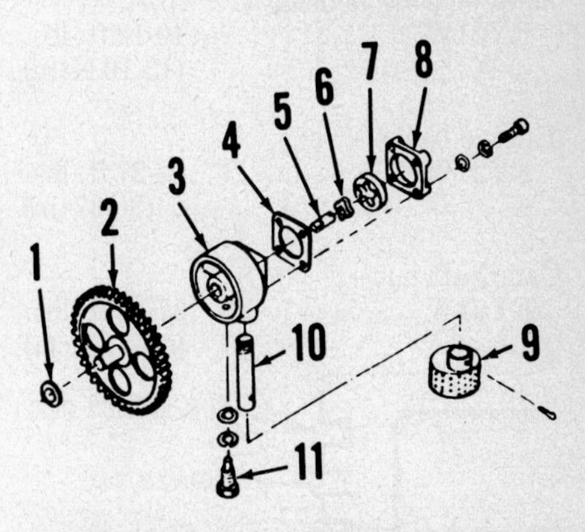

Fig. WR14 – Exploded view of oil pump used on EY21W engines. Shaft (5), rotor (6), stator (7) and housing (8) are serviced as an assembly only.

1. Thrust washer
2. Pump drive gear
3. Adapter
4. Gasket
5. Shaft
6. Rotor
7. Stator
8. Housing
9. Oil strainer
10. Oil intake pipe
11. Lockscrew

mended tightening torques are as follows:

Spark plug–
EY25W,EY27W .... 24-27 ft.-lbs. (33-37 N•m)
EY44W .... 18-22 ft.-lbs. (24-30 N•m)
EY21W .... 22-29 ft.-lbs. (30-39 N•m)

Cylinder head–
EY25W, EY27W .... 25-27 ft.-lbs. (34-37 N•m)
EY44W .... 36-37 ft.-lbs. (49-50 N•m)
EY21W .... 33-36 ft.-lbs (45-49 N•m)

Cylinder block–
EY44W .... 25-29 ft.-lbs (34-39 N•m)
EY21W .... 23-26 ft.-lbs. (31-35 N•m)

Connecting rod–
EY25W, EY27W .... 15-18 ft.-lbs. (20-24 N•m)
EY44W, EY21W .... 18-22 ft.-lbs. (24-30 N•m)

Flywheel nut–
EY25W, EY27W .... 44-47 ft.-lbs. (60-64 N•m)
EY44W .... 68-72 ft.-lbs. (92-98 N•m)
EY21W .... 60-70 ft.-lbs. (81-95 N•m)

Gear cover–
EY25W,
EY27W, EY44W .... 13-14 ft.-lbs. (18-19 N•m)
EY21W .... 10-12 ft.-lbs. (13-16 N•m)

Main bearing housing–
EY21W .... 10-12 ft.-lbs. (13-16 N•m)

Engine base–
EY21W .... 22-27 ft.-lbs. (30-37•m)

Camshaft nut–
EY21W .... 37-45 ft.-lbs. (50-61 N•m)

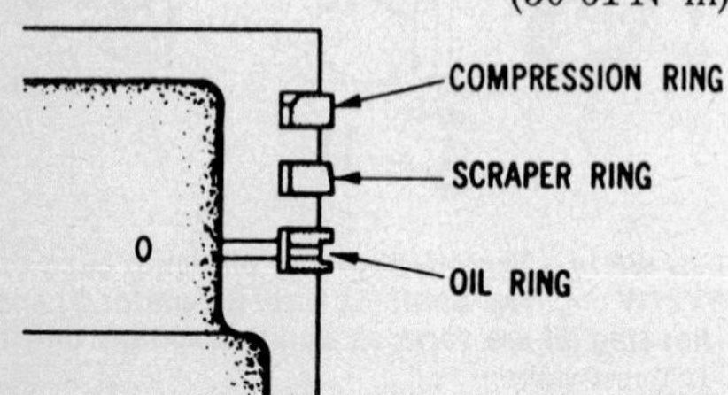

Fig. WR15—Install piston rings on piston as shown for EY25W, EY27W one cylinder engines and EY21W two cylinder engines.

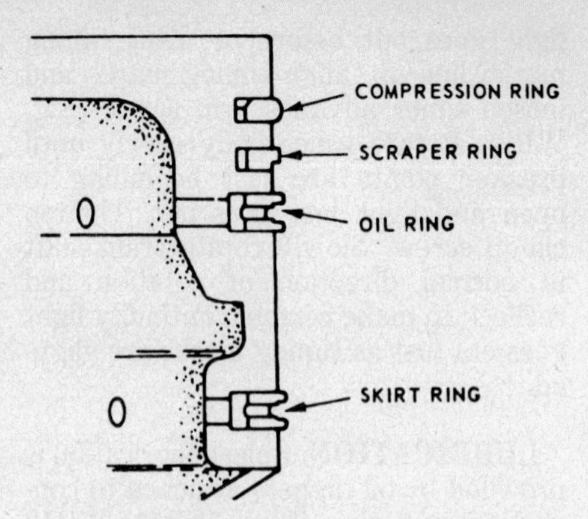

Fig. WR16—Install piston rings on piston as shown for EY44W one cylinder engines.

CYLINDER HEAD. When removing cylinder head bolts from two cylinder models, note locations of bolts as they must be reinstalled in their original positions.

On all models, check cylinder head for distortion. If warpage is 0.006 inch (0.1524 mm) or more, renew cylinder head. Always use new head gasket when installing cylinder head. Tighten cylinder head retaining nuts or head bolts evenly, to correct specified torque.

CONNECTING ROD. On EY25W and EY27W models, connecting rod and piston assembly is removed from cylinder head end of block after cylinder head and gear cover are removed. Connecting rod to crankpin running clearance should be 0.0016-0.0026 inch (0.0406-0.0660 mm). Rod side clearance should be 0.004-0.012 inch (0.1016-0.3018 mm) with a maximum clearance of 0.039 inch (0.9906 mm). Connecting rod to piston pin clearance should be 0.0006-0.0014 inch (0.0152-0.0356 mm) with a maximum clearance of 0.0032 inch (0.0813 mm). When installing connecting rod in engine, make certain match marks (cast ribs) on rod and cap are together. Refer to Fig. WR11 for correct installation information. Tighten rod cap screws to 15-18 ft.-lbs. (20-24 N•m) torque.

On EY44W model, connecting rod and piston assembly is removed from cylinder head end of block after cylinder head and gear cover are removed. Connecting rod is equipped with renewable type bearings. Connecting rod bearing to crankpin running clearance should be 0.0016-0.0042 inch (0.0406-0.1067 mm). Rod side clearance should be

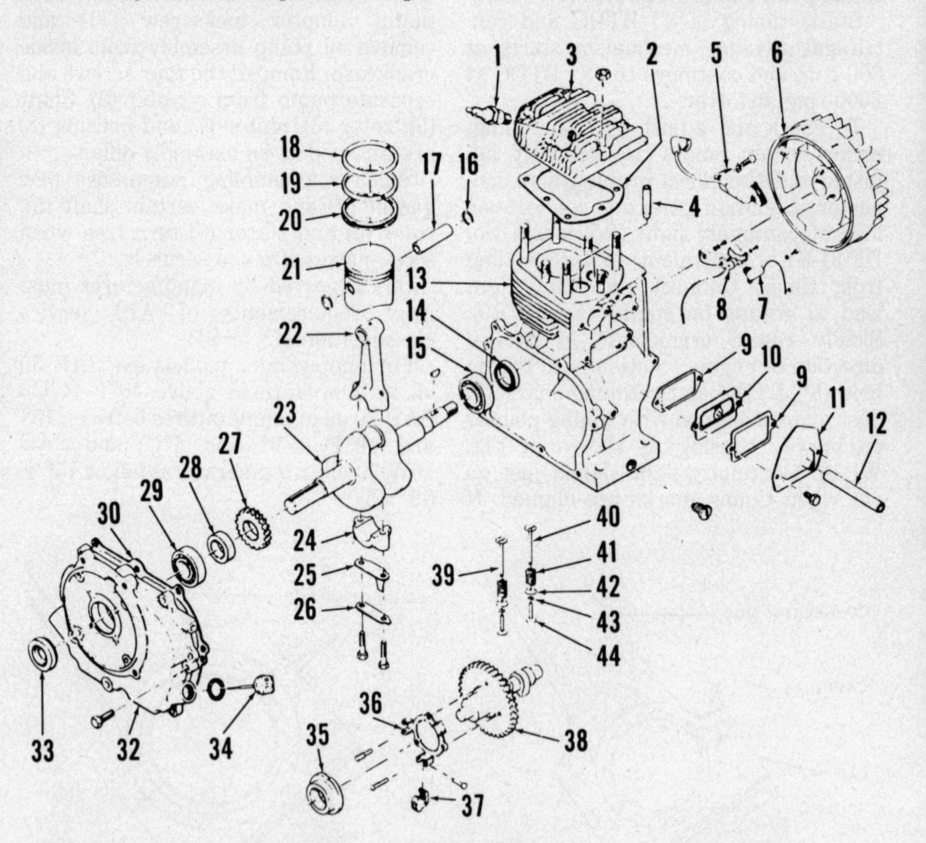

Fig. WR17—Exploded view of EY25W or EY27W basic engine assembly.

1. Spark plug
2. Head gasket
3. Cylinder head
4. Valve guide (2 used)
5. Magneto coil
6. Flywheel
7. Condenser
8. Breaker points
9. Gaskets
10. Breather plate
11. Valve cover
12. Breather tube
13. Cylinder block
14. Oil seal
15. Ball bearing
16. Retaining ring
17. Piston pin
18. Compression ring
19. Scraper ring
20. Oil control ring
21. Piston
22. Connecting rod
23. Crankshaft
24. Rod cap
25. Oil dipper
26. Rod bolt lock plate
27. Crankshaft gear
28. Adjusting collar
29. Ball bearing
30. Gasket
32. Gear cover
33. Oil seal
34. Dipstick
35. Governor sleeve
36. Governor plate
37. Flyweight (3 used)
38. Camshaft assy.
39. Intake valve
40. Exhaust valve
41. Valve spring
42. Spring retainer
43. Locks
44. Valve tappets

0.0079-0.0197 inch (0.2007-0.5004 mm) with a maximum clearance of 0.034 inch (0.8636 mm). Piston pin to connecting rod clearance should be 0.001-0.0019 inch (0.0254-0.0483 mm) with a maximum clearance of 0.0039 inch (0.0991 mm). Connecting rod bearing is available for a 0.010 inch (0.254 mm) undersize crankshaft crankpin journal. When installing connecting rod in engine, make certain match marks (cast ribs) on connecting rod and cap are together (Fig. WR12) and connecting rod is correctly installed. Use new lock plate and tighten connecting rod cap screws to 18-22 ft.-lbs. (24-30 N•m) torque.

On EY21W model, connecting rod and piston assemblies can be removed after removing engine base and cylinder heads. Identify each rod and piston unit so they can be reinstalled in their original position and do not intermix connecting rod caps. Connecting rods ride directly on crankpin journal and connecting rod to crankpin running clearance should be 0.0016-0.0032 inch (0.0406-0.0813 mm). Rod side clearance should be 0.004-0.020 inch (0.1016-0.508 mm) with a maximum clearance of 0.039 inch (0.9906 mm). Piston pin to connecting rod clearance should be 0.001-0.0018 inch (0.0254-0.0457 mm) with a maximum clearance of 0.0047 inch (0.1194 mm). Connecting rod is available for a 0.010 inch (0.254 mm) undersize crankpin journal. When reinstalling connecting rod in engine, make certain match marks (cast dots) on connecting rods and caps are together and that "V1" mark on top of pistons is toward blower end of engine. Refer to Fig. WR13 for correct connecting rod installation and tighten cap screws to 18-22 ft.-lbs (24-30 N•m) torque.

PISTON, PIN AND RINGS. On EY25W, EY27W and EY21W models, pistons are equipped with one compression ring, one scraper ring and one oil control ring (Fig. WR15). On EY44W models, a fourth ring is used on piston skirt (Fig. WR16).

Ring end gap should be 0.002-0.010 inch (0.0508-0.254 mm) for EY25W model, 0.008-0.016 inch (0.2032-0.4064 mm) for EY27W model, 0.002-0.012 inch (0.0508-0.3048 mm) for EY21W model and ring end gap for EY44W lower oil control ring should be 0.012-0.020 inch (0.305-0.508 mm).

On all models, if side clearance of new ring in piston ring groove is 0.006 inch (0.1524 mm) or more, renew piston. Stagger ring end gaps at 90° intervals around piston.

Recommended piston to cylinder bore clearance, measured at thrust face of piston, is 0.0024-0.0039 inch (0.0610-0.0991 mm) for EY25W model, 0.0028-0.0052 inch (0.0711-0.1321 mm) for EY27W model, 0.006-0.007 inch (0.1524-0.1778 mm) for EY44W model and 0.0047-0.0071 inch (0.1194-0.1803 mm) for EY21W model.

Standard piston diameter when measured at skirt thrust face is 2.8315-2.8323 inches (71.92-71.94 mm) for EY25W model, 2.9090-2.9105 inches (73.89-74.13 mm) for EY27W model, 3.5382-3.5390 inches (89.87-89.89 mm) for EY44W model and 2.9468-2.9480 inches (74.85-74.88 mm) for EY21W model.

Pistons and rings are available in a variety of oversizes as well as standard.

Standard piston pin diameter is 0.6297-0.6300 inch (15.95-16.00 mm) for EY25W and EY27W models, 0.7870-0.7874 inch (19.99-20.00 mm) for EY44W model and 0.7081-0.7084 inch (17.986-17.993 mm) for EY21W model.

Piston pin to piston pin bore clearance should be 0.00035 inch (0.0089 mm) tight to 0.00039 inch (0.0099 mm) loose with a maximum clearance of 0.0024 inch (0.061 mm) loose. Refer to **CONNECTING ROD** section for piston pin to connecting rod clearance.

To renew rings only on EY44W and EY21W models engine need not be totally disassembled. Unbolt and remove cylinder block unit from crankcase to service piston and rings while connecting rod remains attached to crankshaft. When installing either piston on EY21W model, make certain identifying mark "V1" on top piston is toward cooling blower end of engine.

CYLINDER. Aluminum crankcase and cylinder with cast-iron liner is a one piece unit on EY25W and EY27W models. Cast iron cylinder unit is removable from aluminum crankcase on EY44W and EY21W models.

On all models, if cylinder wall is scored or out-of-round more than 0.003 inch (0.0762 mm) or if cylinder bore taper ex-

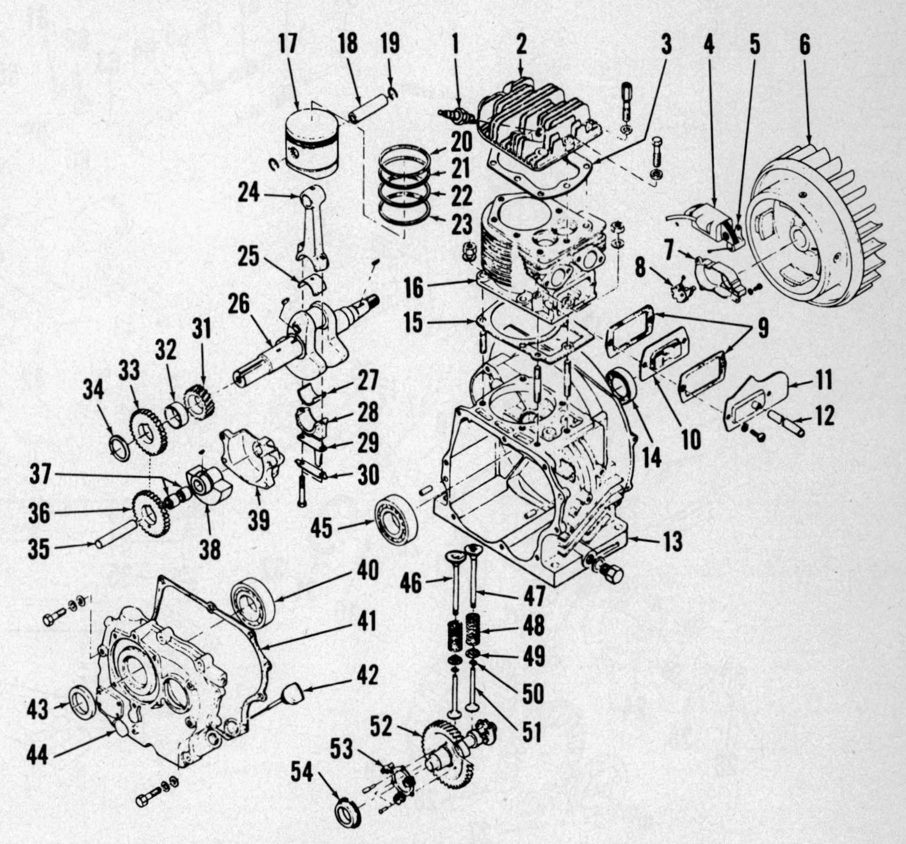

Fig. WR18 — Exploded view of EY44W model basic engine assembly.

1. Spark plug
2. Cylinder head
3. Head gasket
4. Magneto coil
5. Condenser
6. Flywheel
7. Cover
8. Breaker points
9. Gaskets
10. Breather plate
11. Valve cover
12. Breather tube
13. Crankcase
14. Oil seal
15. Cylinder gasket
16. Cylinder
17. Piston
18. Piston pin
19. Retaining ring
20. Compression ring
21. Scraper ring
22. Oil ring
23. Skirt ring
24. Connecting rod
25. Rod bearing (upper half)
26. Crankshaft
27. Rod. bearing (lower half
28. Rod cap
29. Oil dipper
30. Rod bolt lock plate
31. Crankshaft gear
32. Spacer
33. Balancer drive gear
34. Adjusting collar
35. Balancer shaft
36. Balancer gear
37. Needle bearings
38. Balancer weight
39. Balancer housing
40. Ball bearing
41. Gasket
42. Dipstick
43. Oil seal
44. Gear cover
45. Ball bearing
46. Intake valve
47. Exhaust valve
48. Valve spring
49. Spring retainers
50. Locks
51. Valve tappets
52. Camshaft assy.
53. Governor plate & flyweights
54. Governor sleeve

ceeds 0.006 inch (0.1524 mm), cylinder should be rebored to nearest oversize for which piston and rings are available.

Standard cylinder bore is 2.8346-2.8354 inches (71.9988-72.0192 mm) for EY25W model, 2.9134-2.9141 inches (74.0003-74.0181 mm) for EY27W model, 3.5447-3.5456 inches (90.0354-90.0582 mm) for EY44W model and 2.9528-2.9539 inches (75.001-75.029 mm) for EY21W model.

When installing a cylinder unit on EY44W or EY21W models, use a new cylinder base gasket and tighten cylinder retaining nuts to specified torque.

CRANKSHAFT AND MAIN BEARING. EY25W, EY27W and EY44W models crankshaft is supported at each end in ball bearings. See 15 and 29, Fig. WR17 or 40 and 45, Fig. WR18.

EY21W model crankshaft is supported by a ball bearing (15-Fig. WR19) at front of crankshaft and a roller bearing (58) at rear of crankshaft.

On all models, ball bearings are a press fit on crankshaft and should be renewed if any indication of roughness, noise or excessive wear is found.

Crankshaft end play for EY25W and EY27W models should be 0.001-0.009 inch (0.0254-0.2286 mm) and is controlled by adjusting collar (28-Fig. WR17) located between crankshaft gear (27) and gear cover main bearing (29). Three lengths of adjusting collars, 0.740-0.748, 0.748-0.756 and 0.756-0.764 inch, are available. To determine correct length of adjusting collar with gear cover removed, refer to Fig. WR20 and measure distance (A) between machined surface of crankcase face and end of crankshaft gear. Measure distance (B) between machined surface of gear cover and end of main bearing. Compressed thickness of gear cover gasket (C) is 0.007 inch (0.1778 mm). Select adjusting collar that is 0.001-0.009 inch (0.0254-0.2286 mm) less in length than total of A, B and C. Install adjusting collar on crankshaft with recessed side toward crankshaft gear. After reassembly, end play can be checked with a dial indicator.

Crankshaft end play for EY44W model should 0.001-0.008 inch (0.025-0.2032 mm) and is controlled by adjusting collar (34-Fig. WR18) located between balancer drive gear (33) and gear cover main bearing (40). Three lengths of adjusting collars, 0.0381-0.0443, 0.0443-0.0502 and 0.0502-0.0561 inch, are available. To

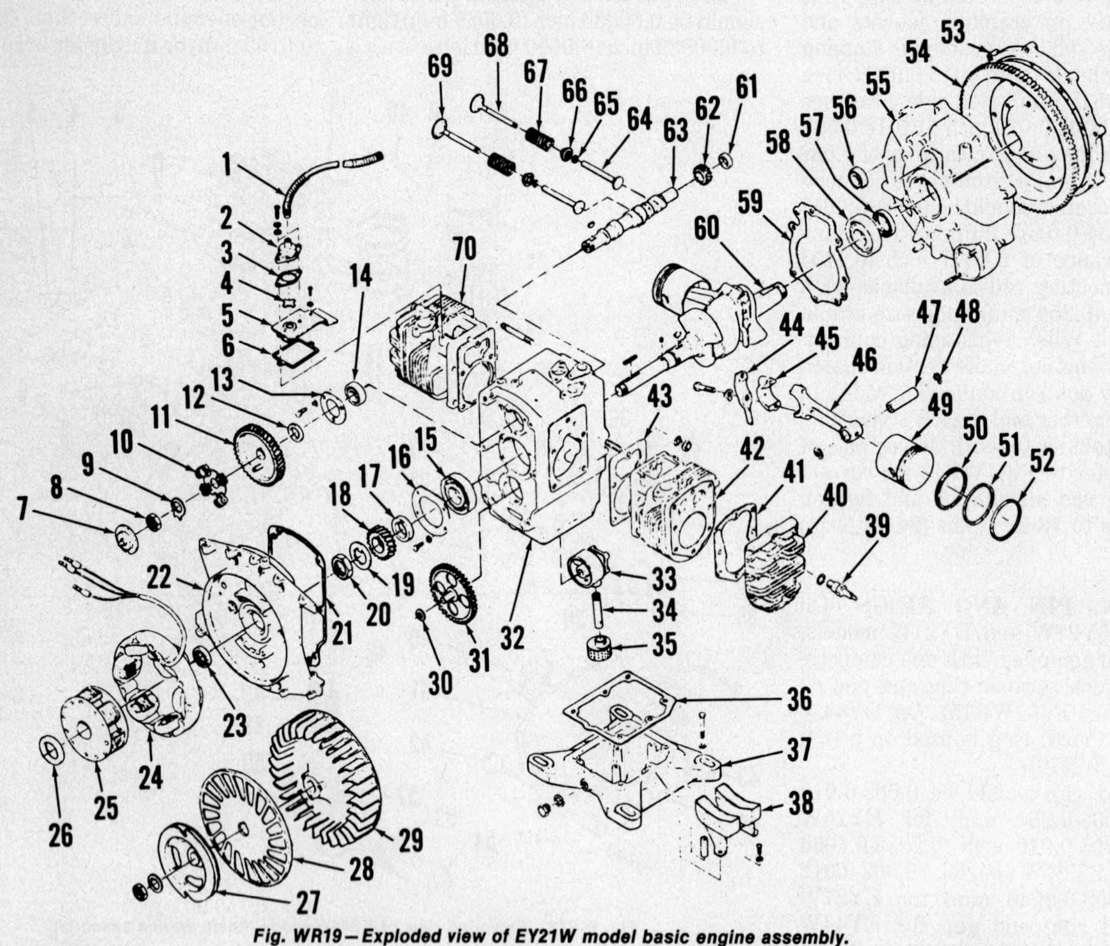

Fig. WR19—Exploded view of EY21W model basic engine assembly.

1. Breather tube
2. Adapter
3. Gasket
4. Breather valve
5. Valve cover
6. Gasket
7. Governor thrust plate
8. Nut
9. Lock
10. Governor plate & flyweights
11. Camshaft gear
12. Spacer
13. Bearing retainer
14. Camshaft bearing (front)
15. Crankshaft bearing (front)
16. Bearing retainer
17. Spacer
18. Crankshaft gear
19. Lock
20. Nut
21. Gasket
22. Gear cover
23. Oil seal
24. Alternator stator
25. Rotor
26. Spacer
27. Pulley
28. Blower dust plate
29. Cooling blower
30. Thrust washer
31. Oil pump drive gear
32. Crankcase
33. Oil pump assy.
34. Oil suction tube
35. Oil strainer
36. Gasket
37. Engine base
38. Oil trough assy.
39. Spark plug
40. Cylinder head
41. Head gasket
42. Cylinder
43. Cylinder gasket
44. Oil dipper
45. Rod cap
46. Connecting rod
47. Piston pin
48. Retaining ring
49. Piston
50. Oil control ring
51. Scraper ring
52. Compression ring
53. Flywheel cover
54. Flywheel assy.
55. Flywheel & bearing housing
56. Camshaft bearing (rear)
57. Oil seal
58. Crankshaft bearing (rear)
59. Gasket
60. Crankshaft, rod & piston assy.
61. Spacer
62. Timer drive gear
63. Camshaft
64. Valve tappets
65. Locks
66. Spring retainers
67. Valve springs
68. Exhaust valve (2 used)
69. Intake valve (2 used)
70. Cylinder & head assy.

determine correct length of adjusting collar with gear cover removed, refer to Fig. WR21 and measure distance (A) between machined surface of crankcase face and face of balancer drive gear. Measure distance (B) between machined face of gear cover and end of main bearing. Compressed thickness of gear cover gasket (C) is 0.008 inch (0.2032 mm). Select adjusting collar that is 0.001-0.008 inch (0.0254-0.2032 mm) less in length than total of B plus C, minus distance A. Install adjusting collar on crankshaft with recessed side toward balancer drive gear. After reassembly, end play can be checked with a dial indicator.

Crankshaft end play for EY21W model should be 0.002-0.010 inch (0.0508-0.254 mm) and is controlled by condition of front main (ball) bearing (15). If end play is excessive, renew front bearing.

Standard crankpin journal diameter is 1.1003-1.1008 inches (27.95-27.96 mm) for EY25W and EY27W models, 1.3754-1.3760 inches (34.94-34.95 mm) for EY44W model and 1.3750-1.3760 inches (34.93-34.95 mm) for EY21W model. If crankpin journal is worn or scored more than 0.0025 inch (0.0635 mm) on EY25W and EY27W models or 0.004 inch (0.1016 mm) for EY44W and EY21W models or if journal is tapered or out-of-round more than 0.0002 inch (0.0051 mm) on all models, crankshaft should be renewed or crankpin journal reground according to availability of oversize connecting rod or bearing inserts for model being serviced.

When reinstalling crankshaft in engine, make certain punch marked tooth on crankshaft gear is between two punch marked teeth on camshaft gear.

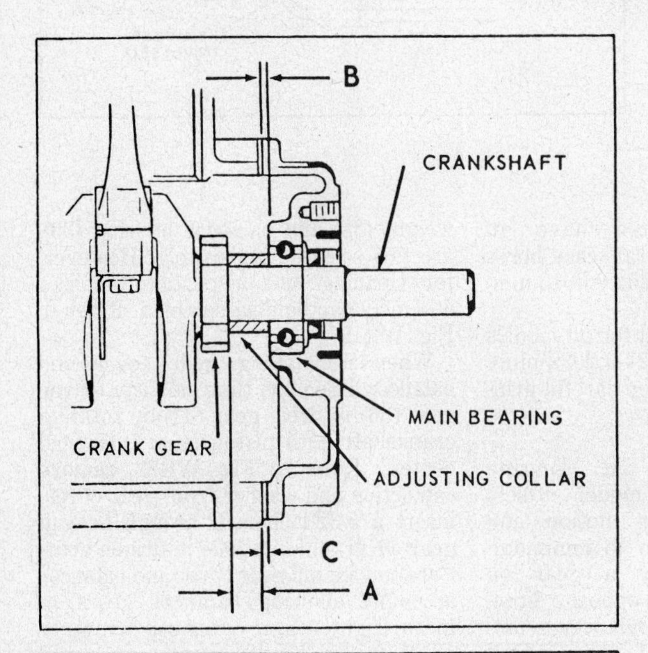

Fig.WR20 – On EY25W and EY27W models, crankshaft end play is controlled by adjusting collar. Refer to text for procedure to determine length of collar to be used.

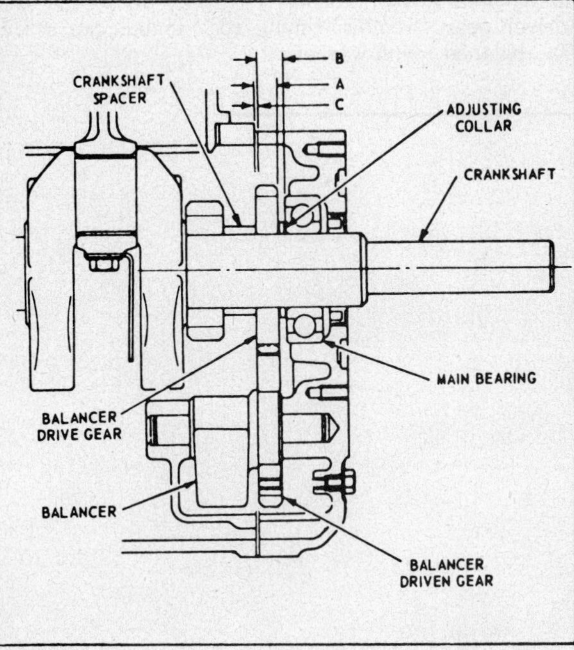

Fig. WR21 – On EY44W model, crankshaft end play is controlled by adjusting collar. Adjusting collars are available in three lengths. Refer to text for procedure to determine correct length collar to be installed.

CAMSHAFT. EY25W, EY27W and EY44W models camshaft rides directly in bores machined into crankcase and gear cover. When removing camshaft assembly, lay engine on side to prevent tappets (44 – Fig. WR17 or 51 – Fig. WR18) from falling out. If valve tappets are removed they must be reinstalled in their original positions. Camshaft journal diameter is 0.5889-0.5893 inch (14.96-14.97 mm) for each end for EY25W and EY27W models and 0.7467-0.7472 inch (18.97-18.98 mm) for journal at gear cover end and 0.6682-0.6687 inch (16.98-16.99 mm) for journal at flywheel end for EY44W model.

EY44W model camshaft is equipped with an automatic compression release which at cranking speed, holds exhaust valve slightly open during first part of compression stroke. This reduces compression pressure and allows easier cranking of engine. As engine rpm increases, centrifugal action overcomes compression release mechanism and engine operates with normal compression.

When reinstalling camshaft install governor thrust sleeve (35-Fig. WR17) or (54-Fig. WR18) on governor flyweights. Slide camshaft into position making certain punch marked tooth on crankshaft gear is between two punch marked teeth on camshaft gear.

EY21W model camshaft (63-Fig. WR19) rides in a ball bearing at each end (14 and 56). To remove camshaft, remove blower housing, pulley (27), dust plate (28), blower (29), spacer (26), rotor (25), stator (24) and gear cover (22). Release valve springs and move tappets away from camshaft. Unbolt and remove fuel pump and ignition timer. Remove governor sleeve (7), nut (8) and lock (9). Use suitable puller to remove camshaft gear (11). Woodruff key and spacer (12). Unbolt retainer (13) and withdraw camshaft assembly. Check camshaft bearings (14 and 56) for excessive wear or other damage and renew as necessary. If valve tappets are removed they must be reinstalled in their original positions.

Reinstall camshaft by reversing removal procedure, making certain punch marked tooth on crankshaft gear is between two punch marked teeth on camshaft gear. Tighten camshaft gear retaining nut (8) to 40 ft.-lbs. (54 N·m) torque and secure with lock tab.

VALVE SYSTEM. Valve seats are ground at 45° angle and seat width should be 3/64 to 1/16-inch (1.191 to 1.588 mm) for all models.

Valve faces are ground at 45° angle for all models and valve stem diameter is 0.273-0.274 inch (6.93-6.96 mm) for intake and exhaust valves for EY25W and EY27W models, 0.313-0.314 inch (7.95-7.98 mm) for intake valve and 0.310-0.311 inch (7.87-7.90 mm) for exhaust valves for EY44W model and 0.3118-0.3128 inch (7.93-7.95 mm) for intake valves and 0.3106-0.3114 inch (7.89-7.91 mm) for exhaust valve for EY21W model.

Valve stem to guide clearance should be 0.0015-0.004 inch (0.0381-0.1016 mm) for intake and exhaust valves for EY25W and EY27W models, 0.001-0.003 inch (0.0254-0.0762 mm) for intake valves and 0.004-0.006 inch (0.1016-0.1524 mm) for exhaust valves for EY44W model and 0.0022-0.0047 inch (0.0559-0.1194 mm) for intake valves and 0.0036-0.0059 inch (0.0914-0.1499 mm) for exhaust valves for EY21W model.

Valve guide bore diameters are 0.2755-0.2770 inch (7.00-7.04 mm) for EY25W and EY27W models, 0.315-0.316 inch (8.00-8.03 mm) for EY44W model and 0.315-0.3165 inch (8.00-8.04 mm) for EY21W model.

If valve stem to guide clearance is excessive for EY25W or EY27W models, renew valves and/or valve guides. If valve stem to guide clearance is excessive for EY44W or EY21W models, valve guides are not renewable and valves and/or cylinder unit must be renewed.

Valve spring free length should be 1.3582-1.4173 inches (34.50-36.00 mm) for EY25W and EY27W models and 1.750-1.811 inches (44.45-46.00 mm) for EY44W and EY21W models. Renew springs which are rusted, pitted or do not meet free length specifications.

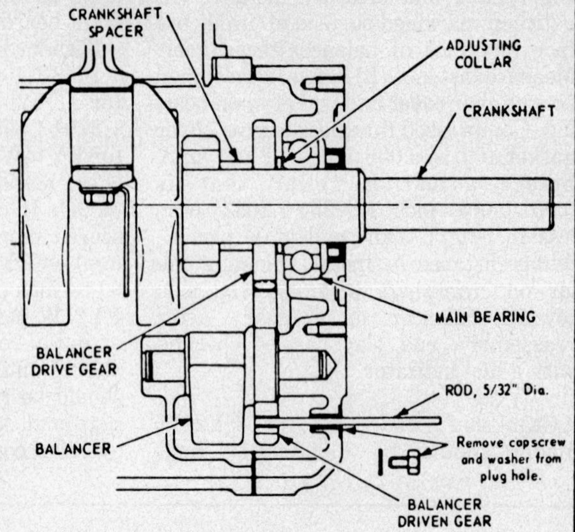

Fig. WR22—Sectional view showing method of timing engine balancer to crankshaft. Refer to text.

Valve tappets should have an operating clearance in crankcase bores of 0.001-0.0024 inch (0.0254-0.0610 mm) on all models.

Valve tappet gap (cold) for all models is 0.006-0.008 inch (0.1524-0.2032 mm). Adjustment is obtained by careful grinding of valve stem tips.

ENGINE BALANCER. Engine balancer on EY44W model offsets forces of reciprocating motion and reduces engine vibration to minimum. Balancer is driven by a gear on crankshaft and rotates in opposite direction of crankshaft. Balancer components, consisting of drive gear (33-Fig. WR18), shaft (35), driven gear (36), needle bearings (37), balance weight (38) and balancer housing (39), are not serviced separately. However, for cleaning and inspection balancer assembly can be disassembled. Refer to Fig. WR18.

When reinstalling gear cover and balancer assembly, time balancer driven gear (36) to drive gear (33) by rotating crankshaft until piston is at "top dead center". Refer to Fig. WR22, remove cap screw and washer from gear cover, insert a 5/32-inch rod through hole in gear cover and into hole in driven gear. Carefully install gear cover and balancer assembly allowing balancer gears to mesh. Tighten gear cover cap screws to 13-14 ft.-lbs. (18-19 N•m) torque and remove timing rod. Install cap screw and washer.

WISCONSIN ROBIN

SERVICING WISCONSIN ROBIN ACCESSORIES

REWIND STARTER

OVERHAUL. To disassemble rewind starter, refer to Fig. WR23. Release spring tension by pulling rope handle until about 18 inches of rope extends from unit. Use thumb pressure against ratchet retainer to prevent reel from rewinding and place rope in notch in outer rim of reel. Release thumb pressure slightly and allow spring mechanism to slowly unwind. Twist loop of return spring and slip loop through slot in ratchet retainer. Refer to Fig. WR24 and remove nut, lockwasher, plain washer and ratchet retainer. Reel will completely unwind as these parts are removed. Remove compression spring, three ratchets and spring retainer washer. Slip fingers into two of the cavity openings in reel hub (Fig. WR25) and carefully lift reel from support shaft in housing.

Fig. WR23—View showing method of releasing spring tension on rewind starter assembly.

CAUTION: Take extreme care that power spring remains in recess of housing. Do not remove spring unless new spring is to be installed.

If power spring escapes from housing, form a 4½ inch (114.3 mm) I.D. wire ring and twist ends together securely. Starting with the outside loop, wind spring inside ring in a counter-clockwise direction.

NOTE: New power springs are secured in a similar wire ring for ease in assembly.

Place spring assembly over recess in housing so hook in outer loop of spring is over tension tab in housing. Carefully press spring from wire ring and into recess of housing.

Using a new rope of same length and diameter as original, place rope in handle and tie a figure eight knot about 1-½ inches (38 mm) from end. Pull knot into top of handle. Install other end of rope through guide bushing of housing and through hole in reel groove. Pull rope out through cavity opening and tie a slip knot about 2-½ inches (64 mm) from end. Place slip knot around center bushing as shown in Fig. WR26 and pull knot tight. Stuff end of rope into reel cavity. Spread a film of light grease on power spring and support shaft. Wind rope ¼-turn clockwise in reel and place rope in notch on reel. Install reel on support shaft and rotate reel counter-clockwise until tang on reel engages hook on inner loop of power spring. Place outer flange of housing in a vise and use finger pressure to keep reel in housing. Then, by means of rope hooked in reel notch, preload power spring by turning reel 7 full turns counter-clockwise. Remove rope from notch and allow reel to slowly turn clockwise as rope winds on pulley and handle returns to guide bushing on housing.

Install spring retaining washer (Fig. WR24), cup side up and place compression spring into cupped washer. Install return spring with bent end hooked into hole of reel hub. Place the three ratchets in position so they fit contour of

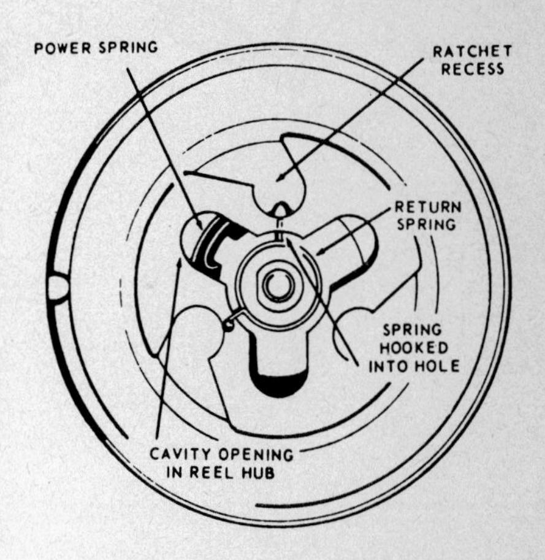

Fig. WR25—Use fingers in reel hub cavities to lift reel from support shaft.

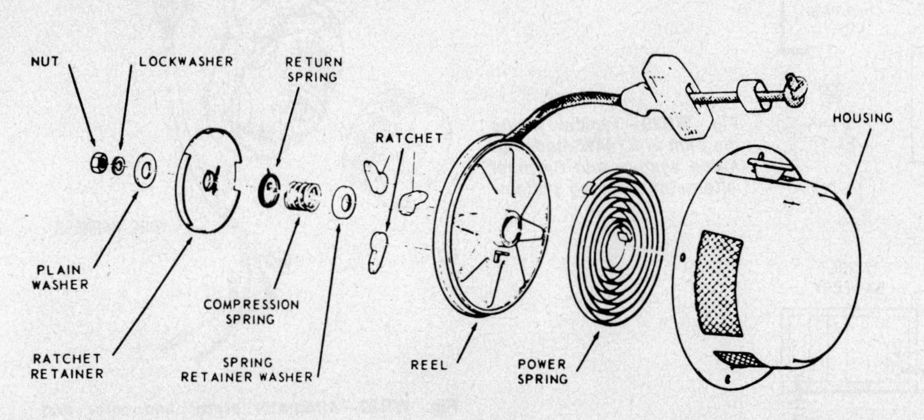

Fig. WR24—Exploded view of rewind starter assembly used on some Wisconsin Robin engines.

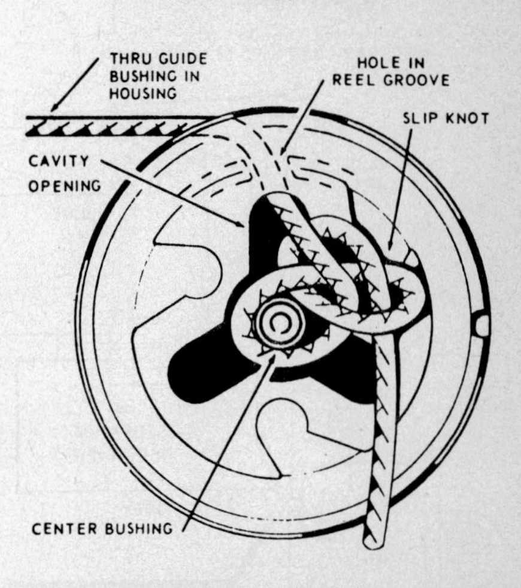

Fig. WR26—Install rope through guide bushing and hole in reel groove, then tie slip knot around center bushing.

recesses. Mount ratchet retainer so loop end of return spring extends through slot. Rotate retainer slightly clockwise until ends of slots just begin to engage the three ratchets. Press down on retainer, install flat washer, lockwasher and nut, then tighten nut securely.

12-VOLT STARTER MOTORS

The 12 volt starting motors used on Models EY44W and EY21W are magnetic shift type in which the solenoid provides positive drive pinion engagement with ring gear. Two brush type starter shown in Fig. WR27 is used on Model EY44W. Refer to Fig. WR28 for the four brush starter used on Model EY21W. Disassembly and reassembly is conventional and procedure is obvious after examination of units and reference to Fig. WR27 or WR28. Parts are available from Wisconsin distributors or service centers.

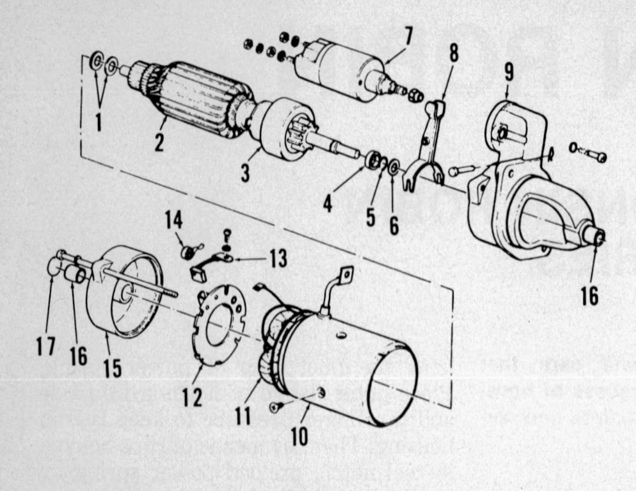

Fig. WR27—Exploded view of 12 volt starter motor used on EY44W model.

1. Thrust washers
2. Armature
3. Drive assy.
4. Pinion stop collar
5. Snap ring
6. Thrust washer
7. Solenoid
8. Shift lever
9. Drive end housing
10. Frame
11. Field coils
12. Brush holder plate
13. Brushes (2 used)
14. Brush spring
15. End cover
16. Bushings
17. Expansion plug

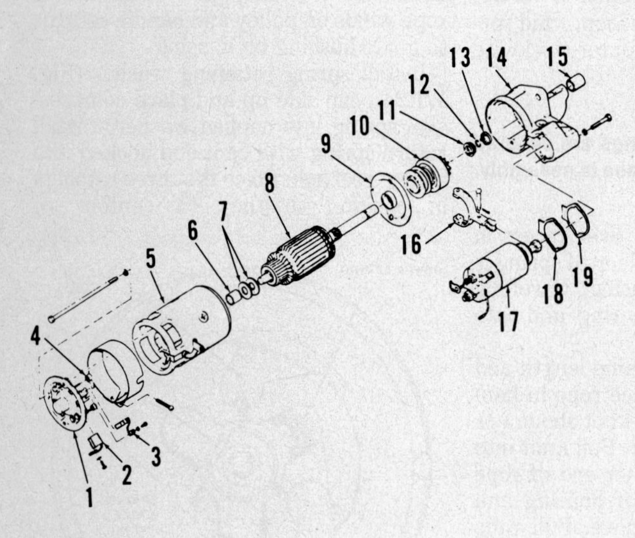

Fig. WR28—Exploded view of 12 volt starter motor used on EY21W model.

1. End cover & brush holder assy.
2. Negative brush (2 used)
3. Positive brush (2 used)
4. Brush spring
5. Frame & field coil assy.
6. Bushing
7. Thrust washer
8. Armature
9. Center bearing
10. Drive assy.
11. Pinion stop collar
12. Snap ring
13. Thrust washer
14. Drive end housing
15. Bushing
16. Shift lever
17. Solenoid
18. Gasket
19. Dust plate

ALTERNATORS

On models equipped with an alternator, the following precautions must be observed:

A. Do not reverse battery connections. System is for negative ground only.

B. Connect booster batteries properly (positive to positive and negative to negative).

C. Do not attempt to polarize alternator.

D. Do not ground any wires from stator or rectifier which terminate at connectors.

E. Do not operate engine with battery disconnected.

F. Disconnect battery cables when charging battery with a battery charger.

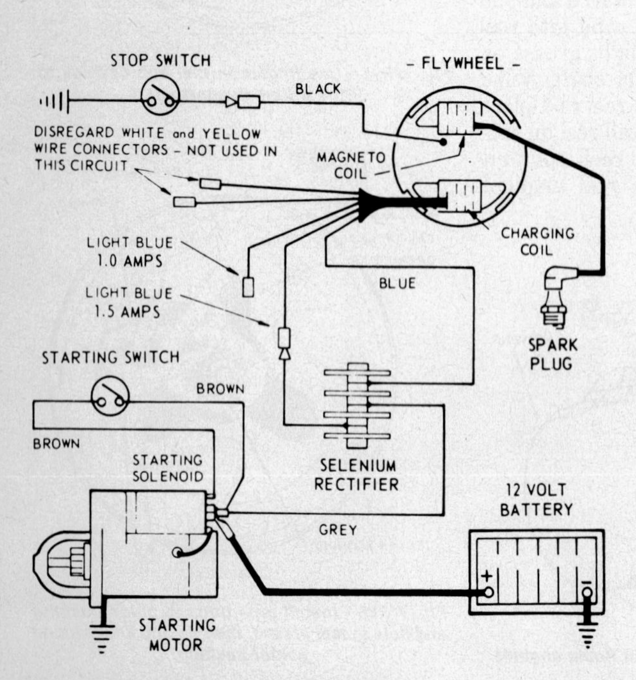

Fig. WR29—Typical wiring diagram of EY44W model ignition system and flywheel alternator charging system.

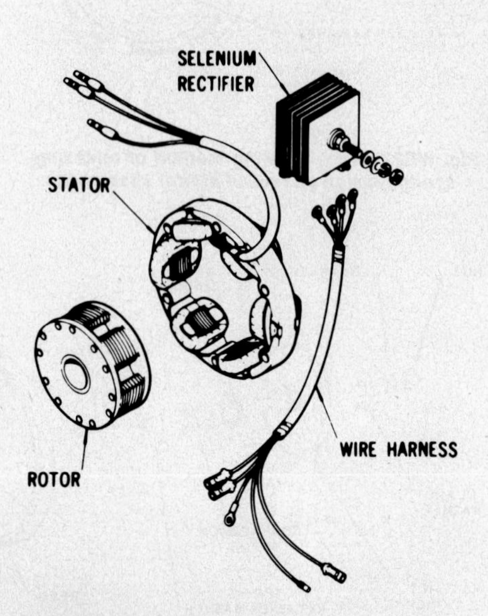

Fig. WR30—Alternator stator and rotor and selenium rectifier used on EY21W model.

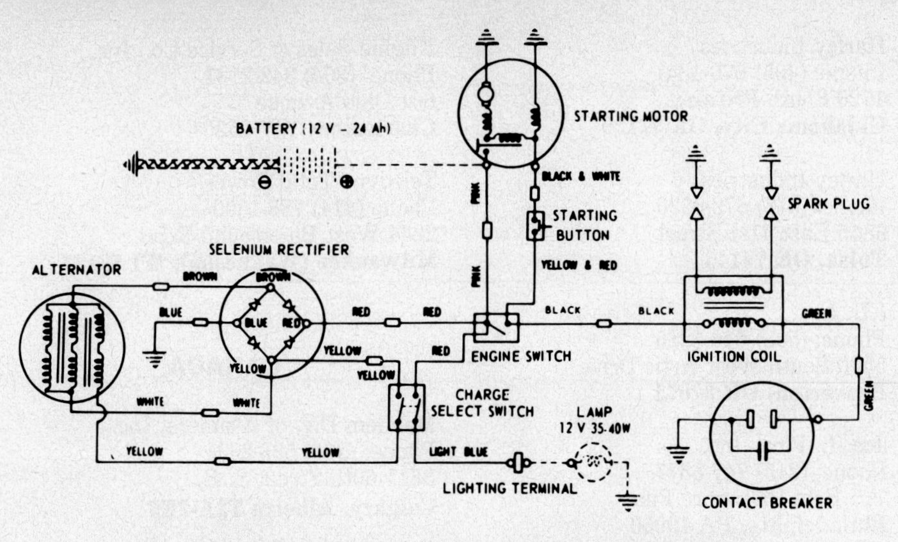

Fig. WR31 – Typical wiring diagram of EY21W model electrical system.

On Model EY44W, alternator charging coil (Fig. WR29) is located behind flywheel and rotating magnets are bolted to inner side of flywheel. A selenium rectifier is used to convert AC current to DC current to charge battery. If battery overcharging is indicated, disconnect light blue wire marked 1.5A from rectifier and connect light blue wire marked 1.0A. Alternator system is rated at 12 volts, 1.5 amps.

On model EY21W, alternator stator and rotor (Fig. WR30) are located behind cooling blower. AC current from alternator is converted to DC current by a selenium rectifier. A manual Hi-Lo charge select switch (Fig. WR31) is used to prevent overcharging battery. Alternator system is rated at 12 volts, 3 amps.

TELEDYNE WISCONSIN CENTRAL PARTS DISTRIBUTORS

(Arranged Alphabetically by States)

These franchised firms carry extensive stocks of repair parts. Contact them for name of dealer in their area who will have replacement parts.

Parts Service Company
Phone: (205) 262-4485
12 Randolph Street
Montgomery, AL 36101

Sahlberg Equipment, Inc.
Phone: (907) 276-5494
1702 Ship Avenue
Anchorage, AK 99501

Southwest Products Corporation
Phone: (602) 269-3581
2949 North 30th Avenue
Phoenix, AZ 85061

Keeling Supply Company
Phone: (501) 945-4511
4227 East 43rd Street
North Little Rock, AR 72115

Lanco Engine Services, Inc.
Phone: (213) 772-2471
12915 Weber Way
Los Angeles, CA 90250

E.E. Ricter & Son, Inc.
Phone: (415) 658-1100
6598 Hollis Street
San Francisco (Emeryville), CA 94608

Central Equipment Company
Phone: (303) 388-3696
4477 Garfield Street
Denver, CO 80216

Highway Equipment & Supply Co.
Phone: (904) 783-1630
5366 Highway Avenue
Jacksonville, FL 32206

P.H. Neff & Sons, Inc.
Phone: (305) 592-5240
5295 N.W. 79th Avenue
Miami FL 33144

Highway Equipment & Supply Co.
Phone: (305) 843-6310
1016 West Church Street
Orlando, FL 32805

American Outdoor Corporation
Phone: (813) 522-5502
4475 28th Street North
St. Petersburg, FL 33714

Georgia Engine Sales & Service
Phone: (404) 446-1100
5715 Oak Brook Parkway
Norcross, GA 30093

Lanco Engine Service, Inc.
Phone: (808) 841-5896
3140 Koapaka Street
Honolulu, HI 96819

Teledyne Total Power
Phone: (208) 522-3872
1230 North Skyline Drive
Idaho Falls, ID 83402

Port Huron Machinery Company
Phone: (515) 266-1136
555 North East 16th
Des Moines, IA 50308

Harley Industries
Phone: (316) 262-5156
1607 Wabash
Wichita, KS 67203

Wilder Motor & Equipment Company
Phone: (502) 966-5141
4022 Produce Road
Louisville, KY 40218

Wm. F. Surgi Equipment Corporation
Phone: (504) 293-0556
5707 Siegen Lane
Baton Rouge, LA 70814

Wm. F. Surgi Equipment Corporation
Phone: (504) 733-0101
221 Laitram Lane
Harahan, LA 70183-0715

Wm. F. Surgi Corporation
Phone: (318) 233-6322
220 Industrial Parkway
Lafayette, LA 70502

Diesel Engine Sales Division
Phone: (617) 341-1760
199 Turnpike Street
Staughton, MA 02072

Engine Supply
Phone: (313) 349-9330
44455 Grand River Avenue
Detroit (Novi), MI 48050

Teledyne Total Power
Phone: (612) 425-7200
8575 County Road, 18
Minneapolis (Osseo), MN 55429

Allied Construction Equipment Co.
Phone: (314) 371-1818
4015 Forest Park Avenue
St. Louis, MO 63108

Central Motive Power, Inc.
Phone: (505) 884-2525
3740 Princeton Drive, North East
Albuquerque, NM 87107

John Reiner & Company
Phone: (201) 460-9444
145 Commerce Road
Carlstadt, NJ 07072

John Reiner & Company
Phone: (315) 474-5741
946 Spencer Street
Syracuse, NY 13208

King-McIver Sales Inc.
Phone: (919) 294-4600
6375 New Burnt Poplar Road
Greensboro, NC 27420

Northern Engine & Supply Company
Phone: (701) 232-3284
2710 3rd Avenue North
Fargo, ND 58102

Cincinnati Engine & Parts Co., Inc.
Phone: (513) 221-3525
2863 Stanton Avenue
Cincinnati, OH 45206

Allied Farm Equipment, Inc.
Phone: (614) 486-5283
1066 Kinnear Road
Columbus, OH 43212

Harley Industries
Phone: (405) 670-1341
1720 South Prospect
Oklahoma City, OK 72129

Harley Industries
Phone: (918) 672-9220
6845 East 41st Street
Tulsa, OK 74145

I.D. Inc.
Phone: (503) 646-8285
5500 South West Artic Drive
Beaverton, OR 97075

Jos. L. Pinto Inc.
Phone: (215) 747-3877
719 East Baltimore Pike
Philadelphia, PA 19050

Wilder Motor & Equipment Company
Phone: (803) 799-1220
1219 Rosewood Drive
Columbia, SC 29201

RCH Distributors, Inc.
Phone: (901) 345-2200
3150 Carrier Street
Memphis, TN 38101

Wilder Motor & Equipment Co., Inc.
Phone: (615) 329-2365
301 15th Avenue
Nashville, TN 37203

Harley Industries
Phone: (512) 851-1991
1918 Holley Road
Corpus Christi, TX 78408

Harley Industries
Phone: (214) 638-4504
8005 Sovereign Row
Dallas, TX 75247

Harley Industries
Phone: (713) 492-6445
17150 Park Row
Houston, TX 77084

Harley Industries
Phone: (915) 337-8676
3220 Kermit Highway
Odessa, TX 79760

Harley Industries
Phone: (512) 342-4255
8406 Speedway Drive
San Antonio, TX 78320

Teledyne Total Power
Phone: (703) 752-9395
1127 Ind. Pkwy.
Fredricksburg, VA 22150

Engine Sales & Service Co., Inc.
Phone: (304) 342-2131
601 Ohio Avenue
Charleston, WV 25301

Teledyne Total Power
Phone (414) 786-1600
2244 West Bluemound Road
Milwaukee (Waukesha), WI 53187

CANADA

Mandem Div. of Asamera, Inc.
Phone: (306) 523-2631
3611 60th Street S. E.
Calgary, Alberta T2A-2E5

Mandem Div. of Asamera, Inc.
Phone: (403) 465-0244
5925 83rd Street
Edmonton, Alberta T6E-4Y3

Pacific Engines
Phone: (604) 254-0804
1391 William Street
Vancouver, British Columbia V5L-2P6

Mandem Div. of Asamera, Inc.
Phone: (204) 885-4440
21 Murray Park road
Winnipeg, Manitoba R3J-3S2

Mandem Div. of Asamera, Inc.
Phone: (506) 854-0982
146 Albert Street
Moncton, New Brunswick E1C-1B2

Mandem Div. of Asamera, Inc.
Phone: (416) 255-8158
3 Bestobell Road
Toronto, Ontario L4W-1A4

Mandem Div. of Asamera, Inc.
Phone: (514) 342-9233
8550 Delmeade Road
Montreal, Quebec H4T-1L7

Mandem Div. of Asamera, Inc.
Phone: (306)523-2631
1250 St. John Street
Regina, Saskatchewan S4R-1R9

Mandem Div. of Asamera, Inc.
Phone: (306) 244-1505
1729 Ontario Avenue
Saskatoon, Saskatchewan S4R-1R9

Mandem Div. of Asamera, Inc.
Phone: (403) 667-6939
114 Calcite Road
Whitehorse, Yukon Y1A-4S2

NOTES

NOTES

NOTES

NOTES